ADVANCES IN
X-RAY ANALYSIS

Volume 13

ADVANCES IN X-RAY ANALYSIS

Volume 13

Edited by

Burton L. Henke

Department of Physics and Astronomy
University of Hawaii
Honolulu, Hawaii

and

John B. Newkirk and Gavin R. Mallett

Denver Research Institute
The University of Denver
Denver, Colorado

Proceedings of the Eighteenth Annual Conference on
Applications of X-Ray Analysis
Held August 6-8, 1969

Sponsored by
University of Denver
Denver Research Institute
Metallurgy Division

℗

PLENUM PRESS
NEW YORK
1970

Plenum Press, New York
A Division of Plenum Publishing Corporation
227 West 17th Street, New York, N. Y. 10011

United Kingdom edition published by Plenum Press, London
A Division of Plenum Publishing Corporation, Ltd.
Donington House, 30 Norfolk Street, London W. C. 2, England

Library of Congress Catalog Card Number 58-35928
SBN 306-38113-3

ISBN-13: 978-1-4613-9965-0   e-ISBN-13: 978-1-4613-9963-6
DOI: 10.1007/978-1-4613-9963-6

FOREWORD

This conference has attempted to achieve a balance in the presentation of papers on the application of current methods to established problem areas and on the introduction of new methods and applications. It has recognized the relevance of papers on basic physics and chemistry and on the total interaction of x-rays with matter. In order to achieve sufficient depth, a topic is chosen each year for special emphasis. This conference had as its central theme, "The Interactions and Applications of Low Energy X-Rays."

Those who were invited as speakers and as contributors to this volume are among the outstanding workers in the application of low energy x-ray and the associated photo-Auger electron interactions. These include A. K. Baird and W. L. Baun on Light Element Analysis and Long Wavelength Instrumentation; J. E. Holliday, D. W. Fischer, R. J. Liefeld and D. J. Nagel on Bonding and Valence State; H. Friedman and W. P. Reidy on X-Ray Astronomy; and R. Nordberg on Photo-Auger Electron Spectroscopy.

Upon reading over the papers as presented here, one cannot help but be impressed by the steady, dynamic growth and expansion of the field of applied x-ray analysis, beginning about thirty years ago with quantitative elementary analysis and extending to the present time with dramatic and exciting applications to x-ray astronomy.

It has been most appropriate and indeed a privilege to have Dr. Herbert Friedman as a speaker and contributor to this volume. He was among the few who pioneered the application of x-ray spectroscopy as an analytical technique and now he is noted as one of the first to open the new field of x-ray astronomy.

Burton L. Henke
Invited Co-Chairman

v

PREFACE

Consistent with our policy for the past few years, a subject within the broad field of x-ray analysis has been chosen for emphasis at the annual Denver x-ray conference. This year the uses of <u>low energy x-rays</u>, sometimes referred to as soft or long wavelength x-rays, appeared to be a timely subject for emphasis. We were most fortunate in being able to enlist the cooperation of Professor Burton L. Henke, a long time and widely recognized leader in the field of long-wavelength x-ray phenomena. His pre-conference efforts resulted in the contribution of a large number of high quality papers on that subject. The papers give ample evidence of the current activity in this field and they document important progress that has been made in a variety of subjects involving soft x-rays. The local committee wishes to thank Dr. Henke for his wholehearted cooperation during the planning and execution of the conference and for his help in editing of parts of this volume.

The three day conference consisted of the following technical sessions. The assistance of the following persons, who served as session chairmen, is acknowledged with thanks:

LOW ENERGY X-RAYS AND X-RAY ASTRONOMY. B. L. Henke, University of Hawaii, Honolulu, Hawaii and J. B. Newkirk, University of Denver, Denver, Colorado.

X-RAY DIFFRACTION APPLICATIONS. H. Yakowitz, National Bureau of Standards, Washington, D.C. and S. B. Miller, General Electric Company, West Lynn, Massachusetts.

BONDING AND VALENCE EFFECTS ON X-RAY SPECTRA. G. R. Mallett, Ker-McGee Nuclear Division, Oklahoma City, Oklahoma and A. E. Hutchings, Jarrell-Ash Company, Waltham, Massachusetts.

X-RAY FLUORESCENCE APPLICATIONS. M. L. Salmon, Fluo-X-Spec Analytical Laboratory, Denver, Colorado and J. F. Croke, Philips Electronic Instruments, Mount Vernon, New York.

STUDIES FOR EMISSION SPECTRA. E. N. Sickafus, Ford Motor Company, Dearborn, Michigan and T. A. Whatley, Applied Research Laboratories, Inc., Sunland, California.

ABSORPTION AND SPECIAL DIFFRACTION APPLICATIONS. E. C. de Wys, University of Denver, Denver, Colorado and E. M. Proctor, Picker X-Ray Corporation, Cleveland, Ohio.

It is my pleasure again to acknowledge the help of Mrs. Carolyn Anderson, Conference Secretary, and Mr. Frank Rivera, both of whom served efficiently during and before the conference, to coordinate activities and events, and who performed ably during the editing of the conference proceedings. The cover designs for the Advanced Program and for the Abstract Book were produced by Mr. Eldon Bright of the Denver Research Institute. His talent and cooperation are much appreciated.

It is with great regret that I must note the departure from Denver of a close personal friend and valued colleague, Mr. Gavin R. Mallett. Gavin was an active cochairman of these conferences since 1963 and he contributed heavily to this one. After several productive years with the University of Denver Research Institute he was hired away to a very good position in another state. His contributions to this conference were outstanding and he will be missed by all of us.

I must also note in sadness the passing of Mrs. Marie Fay, known to all conferees during the nineteen sixties. Her influence as conference coordinator and coeditor of the proceedings was great and lasting. Probably more than any other single person she was responsible for the Denver X-Ray Conference's becoming known as "the friendly conference". She will be greatly missed and long remembered.

> J. B. Newkirk
> Conference Cochairman

CONTENTS

PREFACE   v

AN INTRODUCTION TO LOW ENERGY X-RAY AND ELECTRON ANALYSIS . .   1
    B. L. Henke

LIGHT ELEMENT ANALYSIS   26
    A. K. Baird

DETECTION AND SPECTROSCOPY OF LONG WAVELENGTH X-RAYS   49
    W. L. Baun

QUANTITATIVE X-RAY FLUORESCENCE ANALYSIS WITH VARIABLE TAKE-
    OFF ANGLE   68
    H. Ebel

EVALUATION OF SOFT AND HARD SCATTERED X-RAYS AS AN INTERNAL
    STANDARD FOR LIGHT ELEMENT ANALYSIS   80
    D. Taylor and G. Andermann

AN IMPROVED X-RAY FLUORESCENCE METHOD FOR THE ANALYSIS OF
    MUSEUM OBJECTS   94
    M. E. Salmon

ON-STREAM ANALYSIS OF LIQUID SAMPLES CONTAINING ELEMENTS OF
    HIGH AND LOW ATOMIC NUMBER BY X-RAY FLUORESCENCE WITH AIR
    AND VACUUM SPECTROGRAPH 105
    F. L. Chan

SIMULTANEOUS X-RAY EMISSION ANALYSIS OF P, Si, Ca, Fe, Al,
    AND Mg IN PHOSPHATE ROCK USING A SMALL COMPUTER TO CORRECT
    FOR MATRIX VARIATIONS 125
    C. N. McKinney and A. S. Rosenberg

SOFT X-RAY VALENCE STATE EFFECTS IN CONDUCTORS 136
    J. E. Holliday

CHEMICAL BONDING AND VALENCE STATE--NONMETALS 159
    D. W. Fischer

ix

INTERPRETATION OF VALENCE BAND X-RAY SPECTRA 182
    D. J. Nagel

A VACUUM SPECTROMETER FOR STUDYING THE CHEMICAL EFFECT ON
    SOFT X-RAY SPECTRA 237
    W. L. Baun and E. W. White

POINT SCATTERING THEORY OF X-RAY K-ABSORPTION FINE STRUC-
    TURE 248
    D. E. Sayers, F. W. Lytle, and E. A. Stern

A VERSATILE VACUUM SCANNING DOUBLE CRYSTAL SPECTROMETER FOR
    SOFT X-RAY ABSORPTION EDGE STUDIES 272
    A. S. Bhalla and E. W. White

X-RAY ASTRONOMY 289
    H. Friedman

X-RAY INSTRUMENTATION FOR SPACE EXPERIMENTS 313
    W. P. Reidy

SYSTEM FOR NON-DISPERSIVE ANALYSIS OF LUNAR X-RAYS FROM
    APOLLO 330
    P. Gorenstein, H. Gursky, I Adler, and J. Trombka

DEVELOPMENT OF A SLITLESS SPECTROGRAPH FOR X-RAY ASTRONOMY. . 342
    T. Zehnpfennig

X-RAY INTERACTION COEFFICIENTS: EFFECT ON INTERPRETATION OF
    SOLAR X-RAY DATA 352
    R. L. Blake

X-RAY SPECTROMETRIC PROPERTIES OF POTASSIUM ACID PHTHALATE
    CRYSTALS 373
    R. J. Liefeld, S. Hanzely, T. B. Kirby, and D. Mott

GRATING STUDIES AT X-RAY WAVELENGTHS 382
    R. J. Speer

ELECTRON SPECTROSCOPY FOR STUDYING CHEMICAL BONDING 390
    R. Nordberg

IEE - A NEW TYPE OF X-RAY PHOTOELECTRON SPECTROMETER 406
    N. H. Weichert and J. C. Helmer

α-EXCITED AUGER SPECTRA 418
    Lo I Yin, I. Adler, and R. E. Lamothe

Contents

THE APPLICATION OF X-RAY DATA TO THE DETERMINATION OF
ATOMIC ENERGY LEVELS 426
   A. F. Burr

ON THE SYMMETRY OF ORIENTATION DISTRIBUTION IN CRYSTAL
AGGREGATES 435
   D. W. Baker

AUTOMATED LATTICE PARAMETER DETERMINATION ON SINGLE CRYSTALS. 455
   A. Segmuller

CORRELATION OF RESIDUAL STRESS LEVEL AND FATIGUE DAMAGE IN
B.C.C. METALS 468
   G. Koves

APPLICATION OF THE X-RAY TWO-EXPOSURE STRESS MEASURING TECH-
NIQUE TO A CARBURIZED STEEL 487
   B. A. MacDonald

X-RAY DIFFRACTION FROM VIBRATING QUARTZ PLATES 507
   W. J. Spencer and G. T. Pearman

X-RAY TOPOGRAPHIC STUDY OF VIBRATING DISLOCATIONS IN ICE
UNDER AN AC ELECTRIC FIELD 526
   K. Itagaki

AN APPROACH TO THE SOLID SOLUTION PROBLEM USING A COMPUTERIZED
IDENTIFICATION TECHNIQUE 539
   G. G. Johnson, Jr. and F. L. Chan

A VERSATILE BRAGG-BRENTANO/SEEMAN-BOHLIN POWDER DIF-
FRACTOMETER 550
   H. W. King, C. J. Gillham, and F. G. Huggins

MEASUREMENT OF LONG RANGE ORDER IN γ' PHASE OF NICKEL-BASE
SUPERALLOYS 598
   J. R. Mihalisin

MEASUREMENT OF THE MOLECULAR SIZE OF A SODIUM HUMATE
FRACTION 609
   R. L. Wershaw, S. J. Heller, and D. J. Pinckney

A NEW ABSOLUTE-SCALE SMALL-ANGLE X-RAY SCATTERING INSTRU-
MENT 618
   H. Pessen, T. F. Kumosinski, S. N. Timasheff, R. R.
   Calhoun, Jr., and J. A. Connelly

MASS ABSORPTION COEFFICIENT MEASUREMENTS USING THIN FILMS . . 632
   P. Lublin, P. Cukor, and R. J. Jaworowski

X-RAY ABSORPTION TABLES FOR THE 2-to-200 A REGION   639
     B. L. Henke and R. L. Elgin

AUTHOR INDEX   667

SUBJECT INDEX   677

AN INTRODUCTION TO LOW ENERGY X-RAY AND ELECTRON ANALYSIS

Burton L. Henke

University of Hawaii

Honolulu, Hawaii 96822

ABSTRACT

This is an introductory review of the physics and applications
of low energy x-rays and electrons in the 10-1000 ev region. The
basic interactions of these radiations within matter are discussed
and typical de-excitation spectra, fluorescent x-ray and photo-
Auger electron, are presented. Specific examples of spectrographic
methods and instruments for the low energy region are described as
based upon the use of long-spaced, Langmuir-Blodgett type of multilayers for ultrasoft x-ray analysis and the use of the hemispherical electrostatic analyzer for photo-Auger electron spectroscopy.
Some examples of spectrographic signal, signal/background, and
resolution are presented for applications to light element fluorescence, valence emission band, and photo-Auger electron analysis.
The special aspects of the low energy x-ray analysis of high temperature plasmas and of x-ray astronomical sources in general are
described. Finally, the requirement for an adequate model for the
sample system on which to base a quantitative analysis of the spectral data is introduced, the associated strong need for obtaining
more precise interaction coefficient tables is described, and a new
method for the measurement of low energy electron attenuation
coefficients is proposed and illustrated.

ORIGINS AND INTERACTIONS OF LOW ENERGY X-RAYS

X-radiations of photon energies less than one or two thousand
electron volts (such as in the 10-to-100 A region) may originate as
the relatively strong K-characteristic radiations of the lightest
elements, Be-through-Mg, or as the L radiations of the next elements,

1

Si-through-Ge. For all elements, low energy x-ray emissions follow
electronic transitions from the valence levels to the first inner
levels. Most of the radiant energy that is characteristic of high
temperature plasma in the one-to-ten million degree range lies in
this low energy x-ray region.

   The most probable interaction of a low energy x-ray photon
with matter is the ejection of the innermost electron for which
it is energetically capable (photoelectric absorption). Incoherent
scattering is negligible. Coherent scattering is measurable only
when concentrated into narrow angular regions as caused by low
angle diffraction, by low angle specular reflection, or by diffraction by strongly scattering, long-spaced planes of a crystal.

   Following low energy excitation and the filling of an inner
shell vacancy by an electron transition from an outer level,
de-excitation energy is emitted most probably as an Auger electron
(from an outer level). The competing process is relatively improb-
able for low energy de-excitation, viz., the fluorescence emission
of a photon. The magnitude of the fluorescence yield, w_k, for
K-de-excitation as shown in Figure 1 is of the order of one percent at Z for Mg or Al, decreasing exponentially with Z for the
lighter elements. In Figure 2 the interaction probabilities are
summarized for an Al-K_α photon (1487 ev) with an oxygen atom.

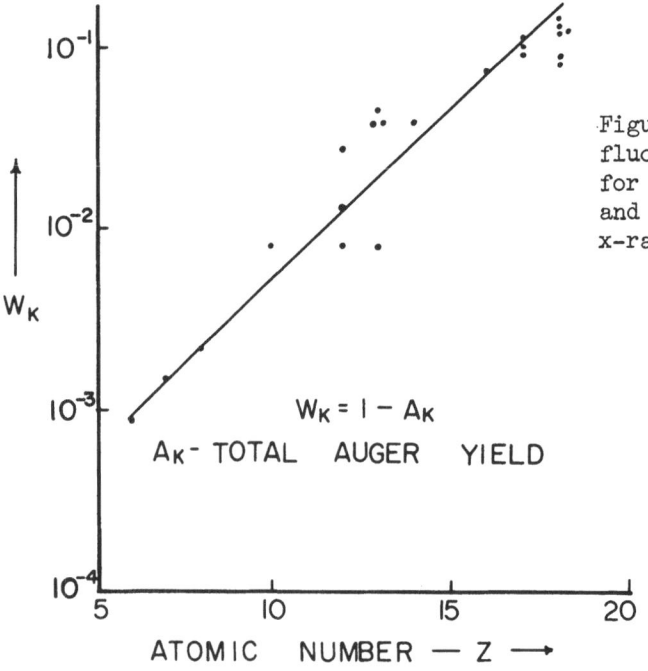

Figure 1. Typical
fluorescent yield data
for the lighter elements
and for the low energy
x-ray region.

ULTRASOFT X-RAY
INTERACTIONS

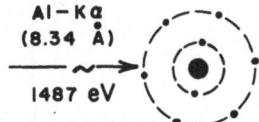

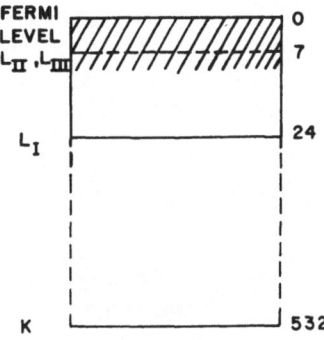

RE-EMISSION	PROBABILITY	ENERGY ev
INCOHERENT SCATTERING	0 %	
COHERENT SCATTERING	0.1%	1487 (hν)
(PHOTOIONIZATION)		
L-ELECTRON EJECTION	5%	1463,1480 (e)
K-ELECTRON EJECTION	95%	955 (e)
(DE-EXCITATION)		
FLUORESCENT PHOTON	1%	525 (hν)
AUGER ELECTRON	99 %	~505 (e)

Figure 2. Interaction and re-emission probabilities for
the Al-K$_\alpha$ (1487 ev) photon and the oxygen atom.

In order to compare the photo-Auger electron and the fluor-
escent photon spectra for a given sample, Al-K$_\alpha$ radiation was used
to excite samples of stearic acid and the spectra shown in
Figure 3 were recorded (using measurement methods described below).
The photo-Auger spectra were taken from thick multilayer crystals
(Langmuir-Blodgett type) of stearic acid and the fluorescent
K-spectrum was from a thick, pressed sample of stearic acid.

The strong photoelectron peak measures the difference between
the energy of the exciting photon and the binding energy of an
atomic energy level. A precise knowledge of the photon energy can
thus be used to determine equally precise values of atomic energy
levels and their absolute shifts as resulting from changes in the
physical or chemical state of the atom. (Conversely, with a known
sample as converter, the photo peak can be used to measure the
photon energy in "photo-conversion" spectroscopy.)

When a monoenergetic electron, such as the photoelectron,
leaves the sample surface, there is a certain probability that it
will give up a discrete amount of energy, typically 10-20 ev, that
is used to raise a valence electron to a probable higher state or
to excite collective plasma electron oscillations. The first characteristic energy loss peak appears here about 20 ev below the
photoelectron peak.

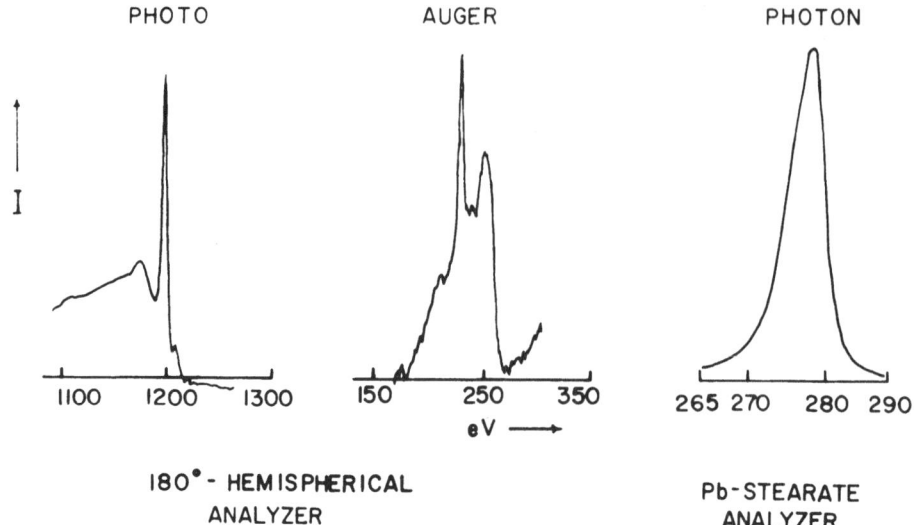

Figure 3. De-excitation spectra for carbon following photo-
ionization by Al-K_α (1487 ev). Electron spectrograph
resolution - 0.3%. X-ray spectrograph resolution - 1.0%.

        The characteristic K (photon) spectrum, due to a transition
from an outer level into the K-vacancy, measures differences in
the atomic energy levels. These characterize the elements (fluorescence analysis) and normally are insensitive to the chemical
state which usually induces approximately equal, absolute changes
in the inner levels. However, since the inner levels are relatively very sharp as compared to the valence band, spectral bands
resulting from transitions from the valence levels into the
nearby, sharp inner levels do reflect rather sensitively the states
of the valence electrons. After unfolding the effect of instrumental resolution, the measured width for the carbon-K emission
band for this stearic acid sample was about 3.5 ev. (A spectrum
was taken under identical measurement conditions from carbon in
a graphite sample, which yielded a band width of about 6 ev.)

        The origin of the Auger spectrum is somewhat more complicated
than either the photoelectron or the photon spectrum. An inner
vacancy is filled by a transition from an outer level and then the
available de-excitation energy appears in the emission of another
(Auger) electron. What is measured is a difference in certain
atomic level energies minus a binding energy corresponding to
another outer electron. Because there are usually multiple possibilities for the original and for the final vacancy sites in the

outer levels for Auger transitions, these spectra are usually more
complex (see below). Nevertheless, the Auger spectral structure
is often distinctive and sharp. When the outermost electronic
levels are involved, there is clearly a strong sensitivity to the
chemical and physical state.

The Auger spectra, as with the photon spectra, can also be
excited by electron or heavy particle bombardment (protons or alpha
particles, for example). Two very important advantages of photon
over particle excitation for generating Auger spectra are that
considerably lower background is observed and the complementary
photoelectron spectra are gained at the same time. By appropriate
selection of the exciting radiation, the photoelectron lines may
be "placed" in energy away from possible interference lines and
where the measurement may be optimized.

EXAMPLES OF MEASUREMENT METHODS

Relatively high intensities of monoenergetic, low energy
x-rays can be generated by excitation sources such as shown in
Figure 4. The anode structure shown in this photograph is inserted
into a cylindrical housing as illustrated, for example, for a
vacuum x-ray spectrograph in Figure 5. These 2-4 kilowatt tubes
have been used for many years and have been described in detail
previously.   The anode structure shown in Figure 4 is a new modification which permits the use of quickly interchangeable anode
cylinders. With this tube, full advantage is taken of the fact

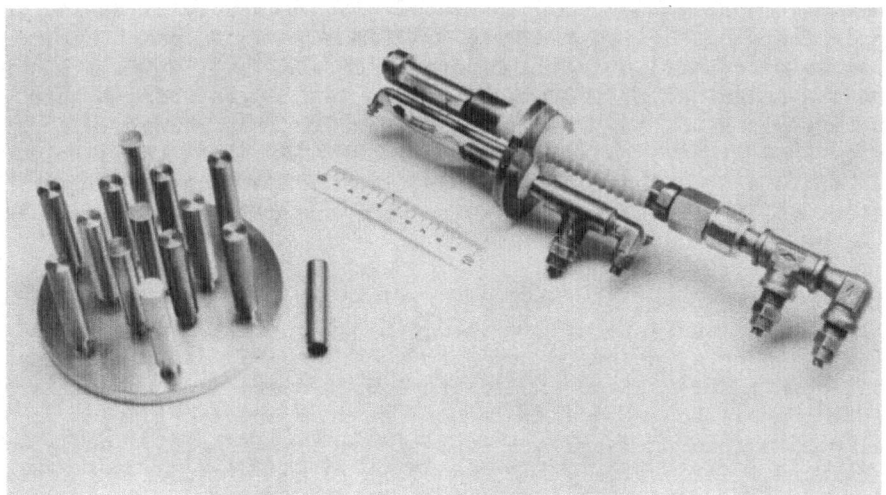

Figure 4.   Anode structure for demountable ultrasoft x-ray source.

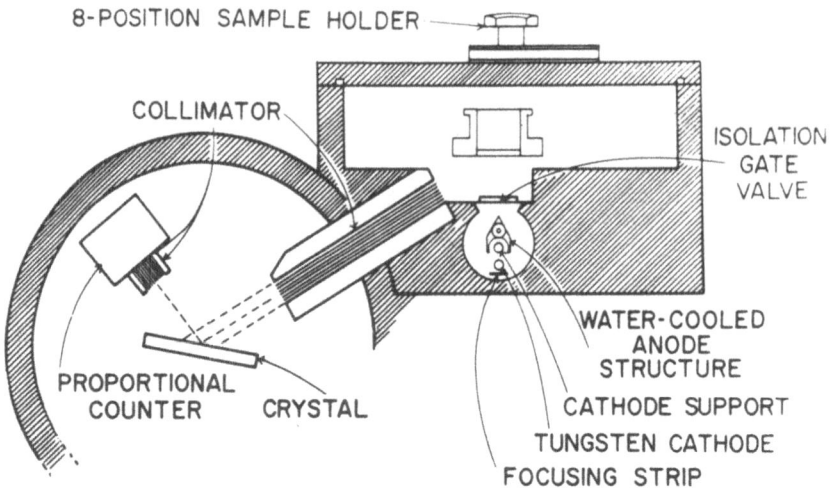

Figure 5. Adaptation of vacuum x-ray spectrograph,
employing demountable excitation source, long-spaced
Bragg analyzers and reduced pressure, flow proportional
counter.

that low energy x-ray "lines" are very strong relative to the associated continuum background and, for a given anode, are few in
number.

Strong sources of monoenergetic radiations can usually be isolated for wavelengths from 113 A (Be-K) through 8.34 A (Al-K$_\alpha$) by
the use of specific anode cylinders, such as beryllium, magnesium,
aluminum, iron, copper, copper plated with a variety of surfaces,
copper impregnated with graphite or diamond powder, and other non-
platable materials, etc. In fluorescence analysis, optimum excitation can be gained by choosing a source wavelength shorter than,
but very close to, the line being measured. In this way, higher
energy (and higher order) spectra are not excited in the sample,
thus minimizing background and possible interferences. Maximum
excitation efficiency is also gained for the desired line radiation
by using such excitation.

For many years, diffraction gratings have been used successfully for the analysis of low energy x-ray spectra.[1] The method
illustrated here, however, has also been very successful and is
based upon a straightforward adaptation of a standard vacuum x-ray
spectrograph. Long-spaced crystals, natural and fabricated, in the
20-160 A 2d spacing range are employed as the dispersive analyzers.
Background is effectively suppressed by pulse height discrimination
with the use of flow proportional counters having very thin, high
transmission windows (supported on metal grid structures). The

counter gas path absorption is "tuned" for the low energy photons
by using methane, for example, at atmospheric pressure, or various
heavier counter gases at reduced pressures.[2] When possible, the
x-ray tube windows (on a vacuum isolation gate) and the counter
windows are made of materials that can provide an enhancement of
signal/background through appropriate filtering.

     In order to fabricate "crystals" of 2d-spacing of 70 to 160 A,
special techniques have been developed for making Langmuir-Blodgett
multilayer systems of lead laurate, lead stearate, lead lignocerate and lead melissate (2d-spacings of 70, 100, 130 and 160 A,
respectively). A tank which has been developed in recent years in
this laboratory for dipping such multilayers is described schematically in Figure 6. Typical energy resolution of these analyzers

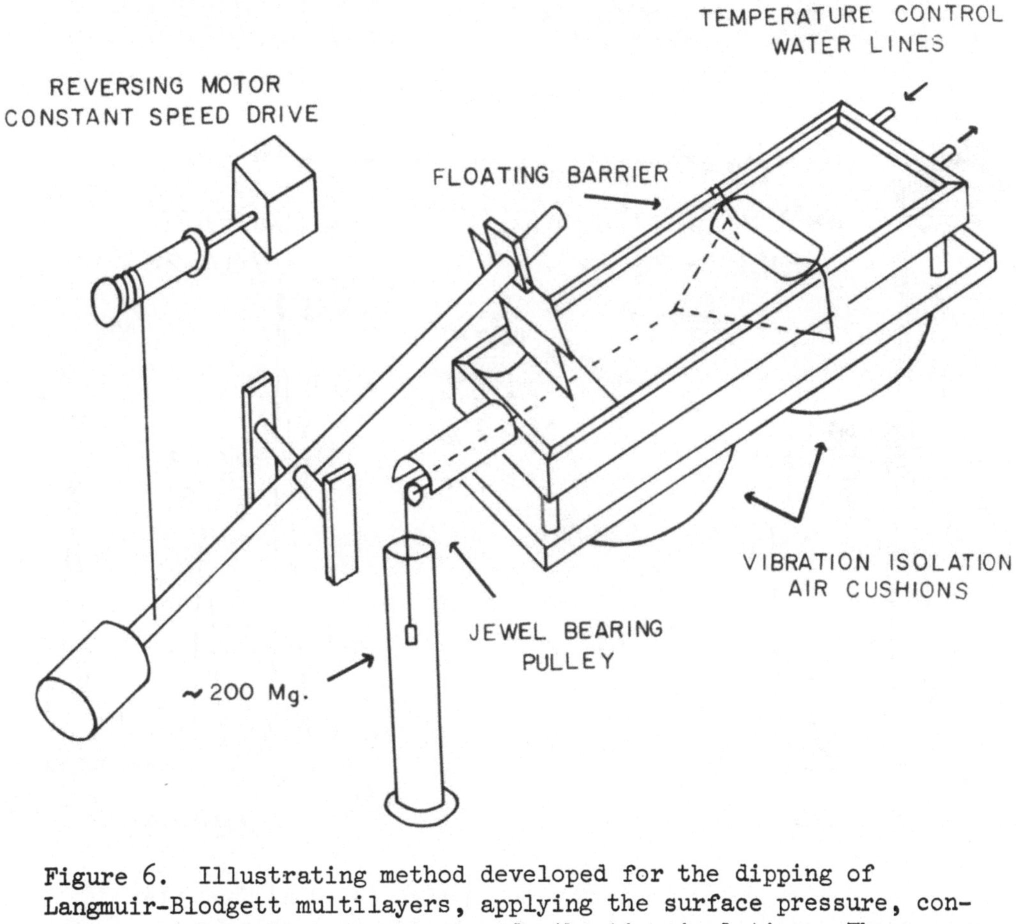

Figure 6. Illustrating method developed for the dipping of
Langmuir-Blodgett multilayers, applying the surface pressure, controlling substrate temperature and vibration isolation. The apparatus is mounted on a clean bench at positive pressure of filtered
air.

is about two percent in first order and one percent in second order.
The carbon-K spectrum shown in Figure 3 was taken in second order
diffraction with a lead stearate multilayer of 75 d-spacings in
thickness.

In the use of diffraction gratings for low energy x-rays, near-
grazing incidence must be employed and advantage is taken of the
consequently strong specular reflection. It is important to recognize that specular reflection can compete with the desired Bragg
reflection intensity in such a way as to place an inherent limitation on the effectiveness of Bragg analyzers for the low energy
x-ray region. In order to illustrate this fact, a C-N-O fluorescence analysis of a urea sample is shown in Figure 7, as taken with
a lead stearate and a lead laurate analyzer of 100 A and 70 A 2d-
spacings.

The major residual background, which increases rapidly with
decreasing 2θ-angle, is the "tail" of the specularly reflected long

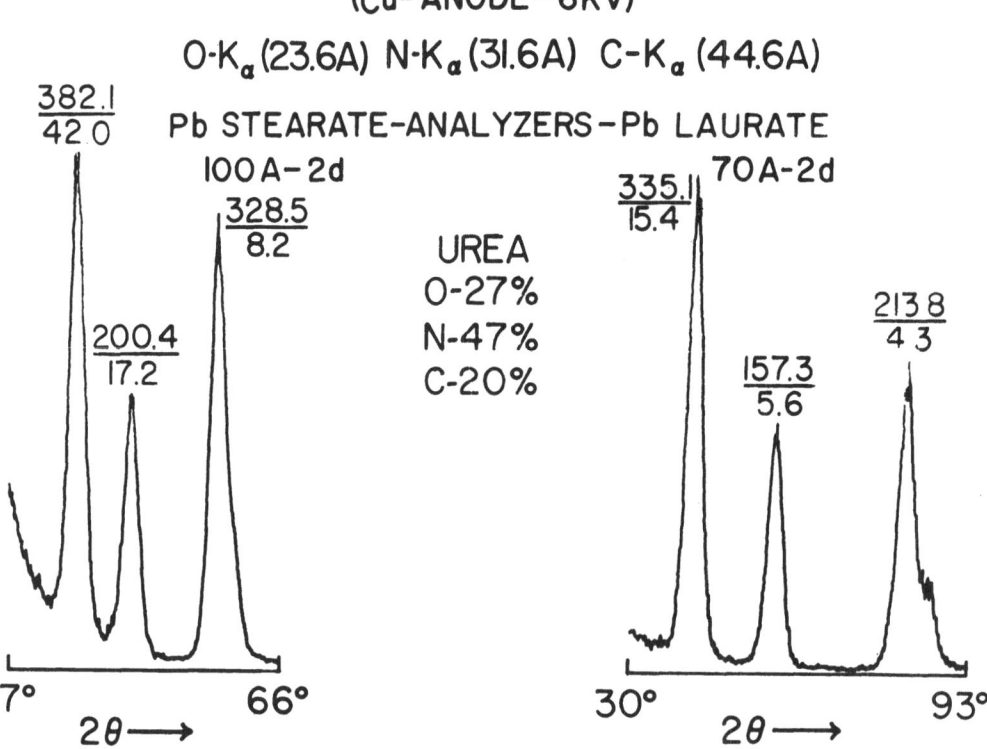

Figure 7. Presence of long wavelength [C-K (44.6 A)] specular
reflection "tail" background in Bragg reflection analysis of
ultrasoft x-rays.

wavelength components from the sample [C-K (44.6 A) radiation in
this instance]. This background is not present when the sample
does not contain the lighter element carbon, as is shown in the
oxygen fluorescence measurement on a quartz sample presented in
Figure 8. A method for avoiding the competitive specular background, suggested in Figure 7, is to use crystals of 2d-spacing
not much larger than the wavelengths being measured so as to dif-
fract at the largest possible angles.

For the analysis of beta spectra in the energy region above
a few kilovolts, it has been conventional to use 180° or sector
magnetic field spectrographs. However, for low energy photo-Auger
electron spectroscopy, electrostatic energy analyzers have been
found to be very efficient and usually simpler and less expensive
to construct. In Figure 9 are shown schematically the electrostatic analyzer geometries which are being used to obtain focussed
electron spectra.

The author has been using for about ten years the hemispherical
type of electrostatic spectrograph.[3] A recent version of this
instrument has been previously described and is shown schematically

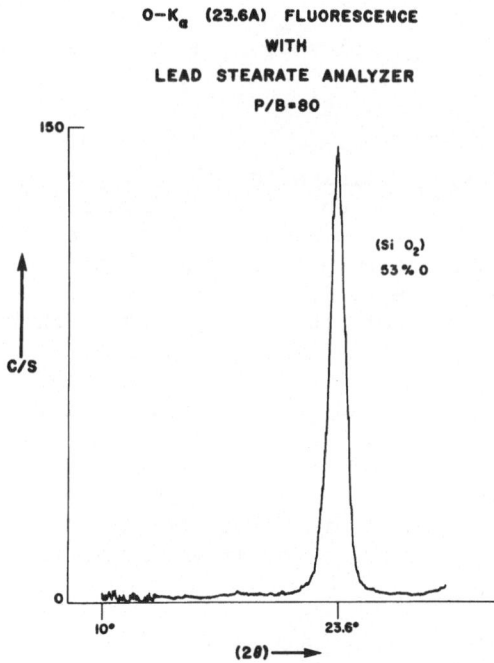

Figure 8. Absence of specular background in Bragg reflection
analysis for oxygen-K radiation in a sample not containing the
lighter element carbon.

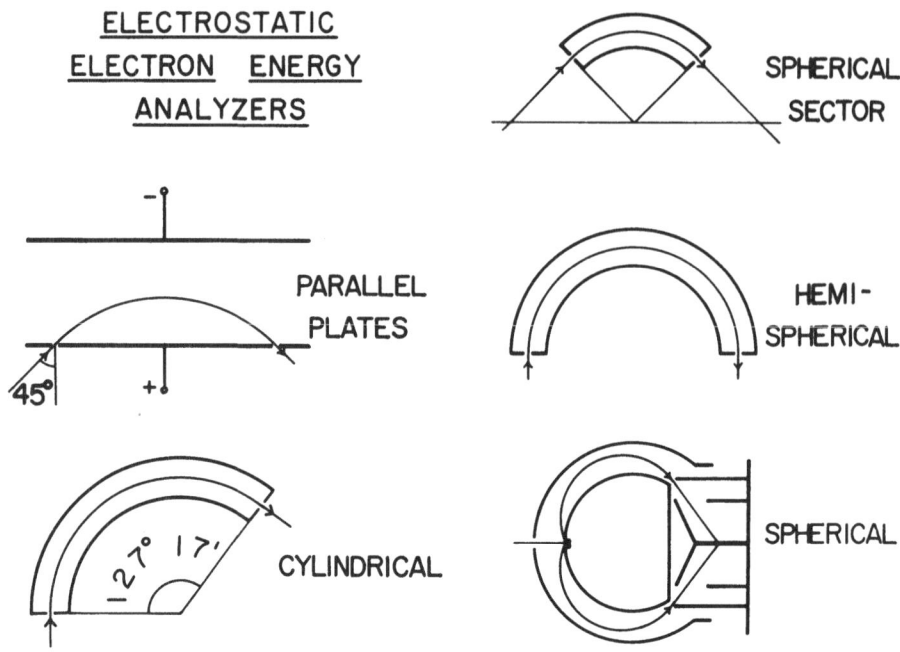

Figure 9. Commonly employed, focussing electrostatic
energy analyzers.

in Figure 10. The spun stainless steel hemispheres are of 18.5"
and 21.5" diameters. These establish an inverse square field that
point-focusses charged particle radiation to the 180° position.
In the design of this instrument, annular slits are used with
their width and length set by the desired resolution. For the
spectra shown in this paper, the energy resolution was set at
about 0.3 percent.

Four samples are mounted on a rotating holder. These may be
heated from behind by radiation or electron bombardment and/or
ion-cleaned by a gun in the top of the sample chamber. Typically,
an aluminum or magnesium target demountable x-ray source (as
described above) with an aluminum foil filter are utilized to
provide a monoenergetic excitation of energy width equal to about
one electron volt.

The 80 liter/sec Vacion pump is used to achieve fast pump-
down. Its magnet is removed when the spectrograph is being
operated. Any significant effect of the residual field of the
lower pump magnet and of the earth's field is easily eliminated
by a pair of Helmholtz coils.

LOW   ENERGY   ELECTRON   SPECTROMETER

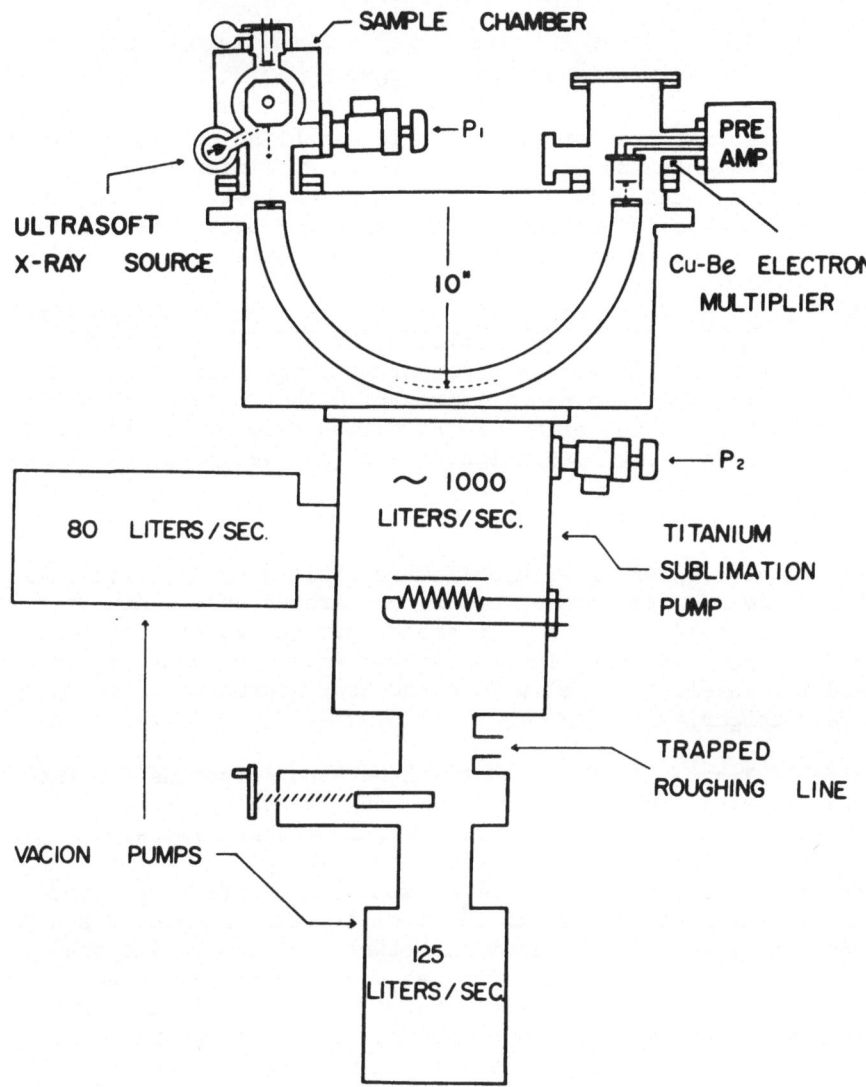

P₁, P₂ — MOLECULAR SIEVE PUMPS

Figure 10. Electron spectrographic system, as used in this laboratory, which employs hemispherical plates to establish an inverse
square electrostatic field that point focusses at the 180° position.

An analysis for the determination of the optimum slit and sample geometries for maximum luminosity per unit resolution for this particular instrument has been carried out.

SOME APPLICATIONS

As a first requirement, in order to establish the feasibility for low energy x-ray and electron analysis, sufficient signal, signal/background and resolution must be demonstrated. Using the specific methods described above, typical fluorescent spectra have been measured for both elementary and valence band analysis.[4,5]

Usually, a maximum sensitivity for light element analysis is gained by using relatively open collimation and a low resolution analyzer as the multilayers in first-order Bragg reflection. In this way, an instrument window is presented that matches the relatively broad band widths of the K-radiations from oxygen through Be. Also, for this low resolution, there is essentially no sensitivity to changes in the shape or position of these emission bands as due to chemical differences.

The spectra shown in Figure 11 were excited by Cu-L (13.3 A) radiation from a copper anode for oxygen through carbon analysis, and by C-K (44.6 A) from a copper anode impregnated with diamond powder for boron and beryllium analysis. A lead stearate analyzer was used for wavelengths below 90 A and lead lignocerate for those which were longer.

Also in Figure 11 are shown the $L_{2,3}$ band emissions for the elements chlorine, sulphur and phosphorous, which appear in this long wavelength region and possibly as interference lines.

In Figure 12 the chlorine $L_{2,3}$ band was recorded again under the same measurement conditions and from the same sample of sodium chloride, but with 0.3° collimation between the sample and the crystal instead of 3.7° collimation as used for the measurements of Figure 11. In this instance, the energy resolution is determined essentially by the lead stearate analyzer at about two percent. Here the characteristically more complex band shape associated with the $L_{2,3}$ radiations is expressed in the appearance of a second peak at the longer wavelength, 70 A. A remarkable sensitivity of the valence emission band structure to the chemical state is also illustrated here in the $L_{2,3}$ spectra of chlorine from a polymer (Saran), sodium chlorite, sodium chlorate and sodium perchlorate.

Historically, such ultrasoft x-ray valence band emission spectra have been obtained from samples imbedded in x-ray tube targets and excited under electron bombardment. Here the feasibility is demonstrated for obtaining such spectra in a more

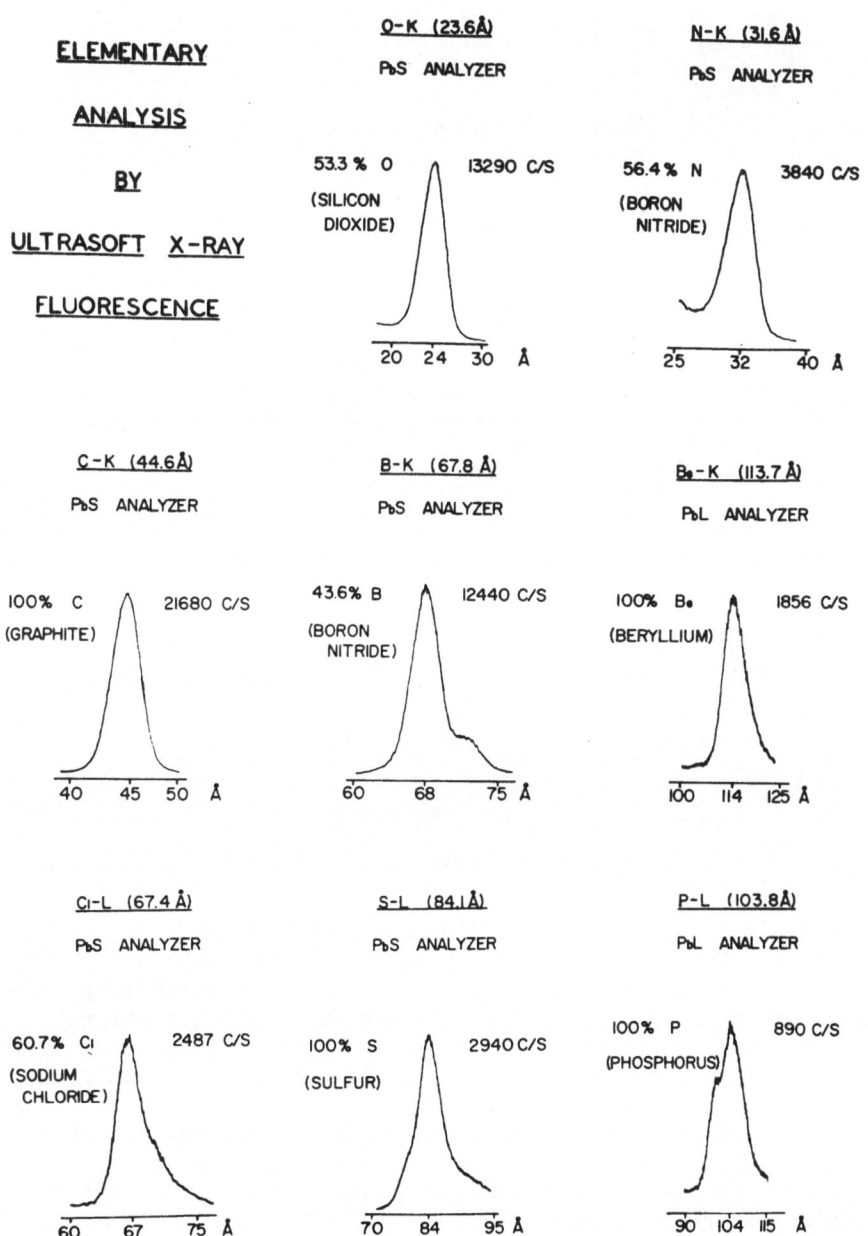

Figure 11. Indicating signal, signal/background and resolution typically obtained in light element fluorescence analysis using open collimation and multilayer analyzers as shown in Figure 5.

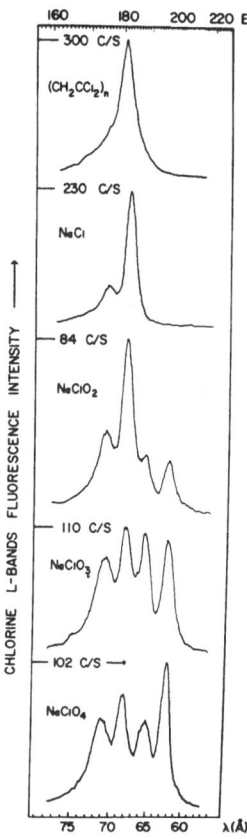

Figure 12.   Chlorine $L_{2,3}$ valence
emission band spectra as excited in
fluorescence by C-K (44.6 A) radiation
and from samples of a polymer (Saran),
sodium chloride, sodium chlorite,
sodium chlorate and sodium perchlorate.

convenient and controllable method using photon excitation [C-K
(44.6 A) in this example].   In fact, it would be very difficult, if
not impossible, to gain such spectra of the last three compounds
because of their instability under electron beam excitation.

Finally, it should be noted that low energy x-ray analysis
involves a considerably increased mass sensitivity over that for
conventional x-ray analysis because the effective sample depth is
greatly reduced as a result of the very high photoelectric cross
sections.   Typical effective sample sizes for the fluorescence
analysis measurements described here are of the order of one milligram spread over an area of one or two square centimeters.

Low energy photo-Auger electron spectra characterize samples of
effective depths one hundred or more times smaller than for the
associated low energy x-ray spectra.   Sample thicknesses are in
micrograms per cm^2 instead of milligrams per cm^2.   In fact, as shown

below, measurable signals for photoelectron peaks from one-atom-thick samples can be obtained in low energy electron analysis.

Total spectral plots, taken directly from the X-Y recorder and with the electron spectrograph described above, are presented in Figures 13 and 14 for samples of BeO and MgO on beryllium and magnesium and excited by Al-K$_\alpha$ (1487 ev) x-radiation. (There is no background suppression in these measurements.) These samples were cleaned by ion bombardment. A sensitive indication of the elimination of carbonaceous contamination is the absence of the carbon Auger line at the 250 ev position. In both of these spectra, the oxygen Auger and photo peaks are very strong, appearing at 500 ev and 955 ev, respectively. Also prominent in the MgO spectra are the magnesium photo and Auger peaks at 180 ev and 1170 ev, respectively.

The percentage energy resolution of the spectrograph was set by the slits chosen for these measurements at 0.3 percent. This corresponds to one electron volt analyzer width at the 300 ev position, where the energy width of the exciting Al-K$_\alpha$ radiation is just matched. As described in a previous report,[2] such spectra can be obtained by setting the hemisphere potentials so as to fix the energy acceptance at, say, 300 ev (where instrument window and the exciting line width are matched) and scanning by varying the potential of the sample relative to the entrance slit. In the measurements illustrated here, the hemisphere potentials are varied to accept electrons in the 0-1500 ev range. A constant relative energy resolution, rather than absolute energy resolution, was thus obtained.

The magnesium KLL Auger spectrum was re-run on an expanded scale, as shown in Figure 15, in order to illustrate more fully the structure that can be resolved in these relatively high sensitivity measurements. Here we have categorized the KLL structure, following Asaad,[6] by the numerals I, II and III, corresponding respectively to the particular pair of ionization sites left in the L levels after the Auger transition,

$$(2s)^0 \, (2p)^6, \quad (2s)^1 \, (2p)^5 \quad \text{or} \quad (2s)^2 \, (2p)^4.$$

The sodium and magnesium KLL spectra have been analyzed in detail by an Uppsala group.[7] The characteristic energy loss peaks mentioned earlier, which appear on the low energy side of any sharp photo or Auger line, are in evidence here as the two small humps at each side of the 1165 ev position.

A relatively new application of low energy x-ray analysis which is rapidly becoming of very considerable interest should also be introduced here, viz., the analysis of high temperature plasmas (10^7 degree region, for example) as encountered in controlled nuclear

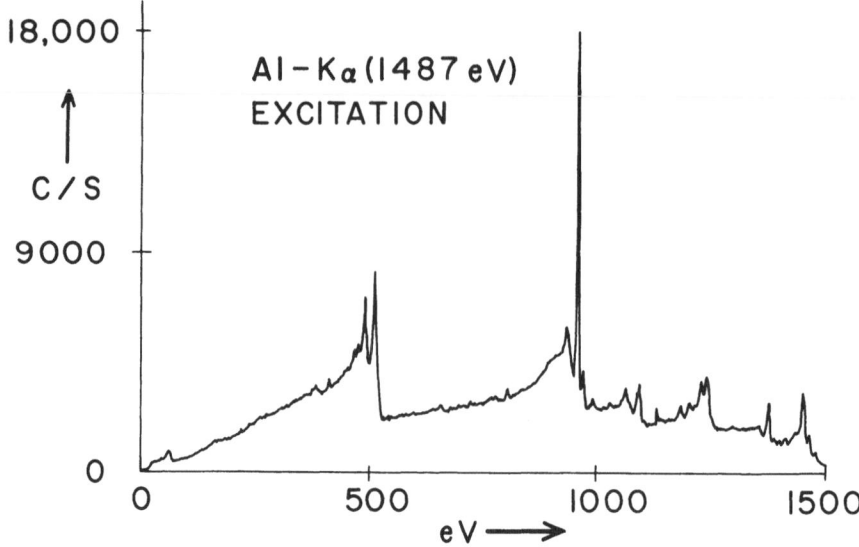

Figure 13. Total photo-Auger spectrum from Be-O on Be as excited
by Al-Kα radiation. Hemisphere potential scanned. Energy resolution set at 0.3%. Background was not suppressed.

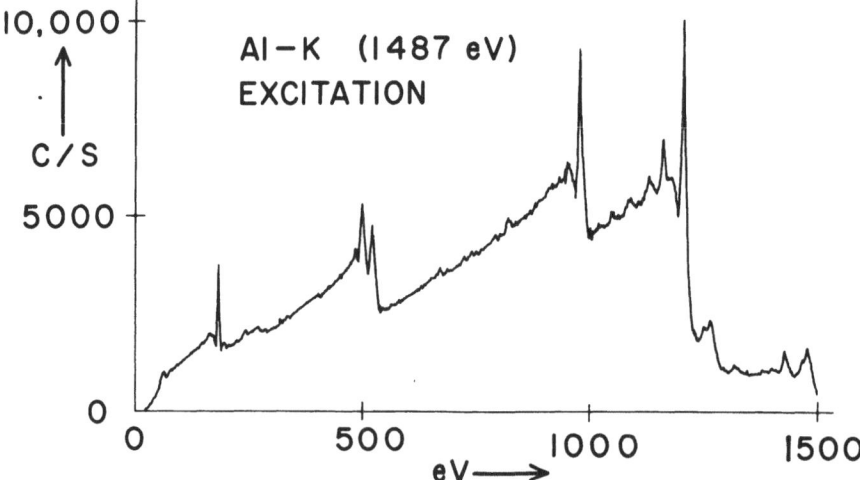

Figure 14. Total photo-Auger spectrum from MgO on Mg, as for the
spectrum of Figure 13. In addition to the Auger and photo lines
for oxygen (at about 500 ev and 955 ev and appearing also in the
BeO spectrum), the photo and Auger lines for Mg are shown here at
180 ev and 1170 ev.

fusion research and in x-ray astronomy. The types of low energy
x-radiations involved would include characteristic line radiations
from highly ionized atomic species, and continuum radiations resulting from recombinations with external energetic electrons, collisions of thermal and nonthermal electrons with heavy charge centers
(bremstrahlung), accelerations of high energy electrons in magnetic
fields (synchrotron radiation or magnetic bremstrahlung), and from
electrons imparting energy to photons of relatively low initial
energy (inverse Compton effect).

   Astronomical sources of low energy x-rays include solar,
galactic, extra-galactic discrete sources, and a nearly isotropic
background. The atmospheric "blanket" about the earth completely
absorbs the x-radiations from such sourses. Their analysis can
only be made with instruments carried above the atmosphere by
either rockets or satellites. The atmospheric absorption for low
energy x-rays and typical flight parameters involved in rocket
astronomy are depicted in Figure 16. Tables for the transmission
through the upper atmosphere of extra-terrestrial x-radiations in
the 2-200 A region, as a function of altitude, are presented later
in this volume.

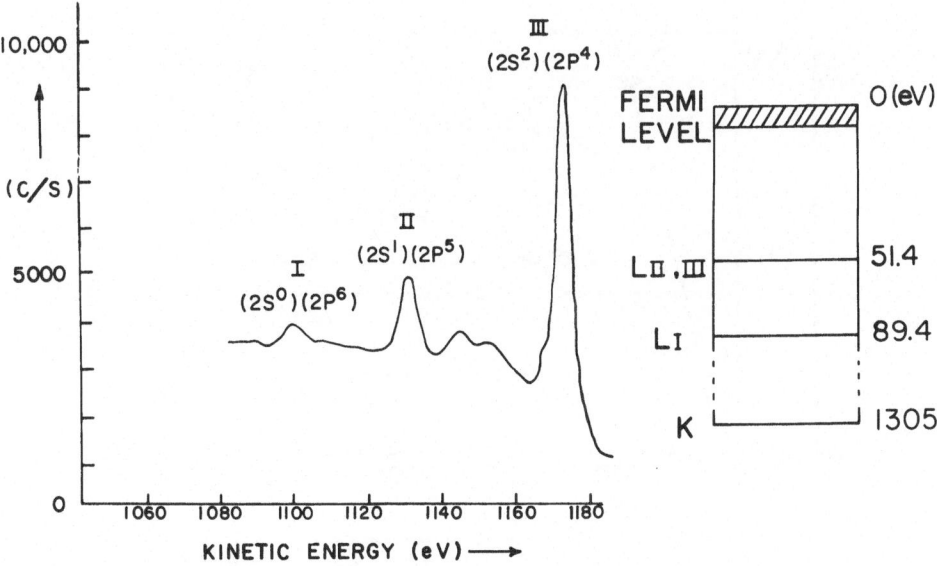

Figure 15. Re-run of the Mg Auger line of Figure 14 on expanded
scale illustrating signal and resolution; characteristic structure
corresponding to particular pair of vacancy sites in the L levels
following the KLL Auger process.

Rocket measurement of solar corona line spectra in the ultra-soft x-ray region, using spectroscopic methods similar to those reviewed here, have been made, for example, by groups at the Naval Research Laboratory[8] and at the Los Alamos Scientific Laboratory.[9]

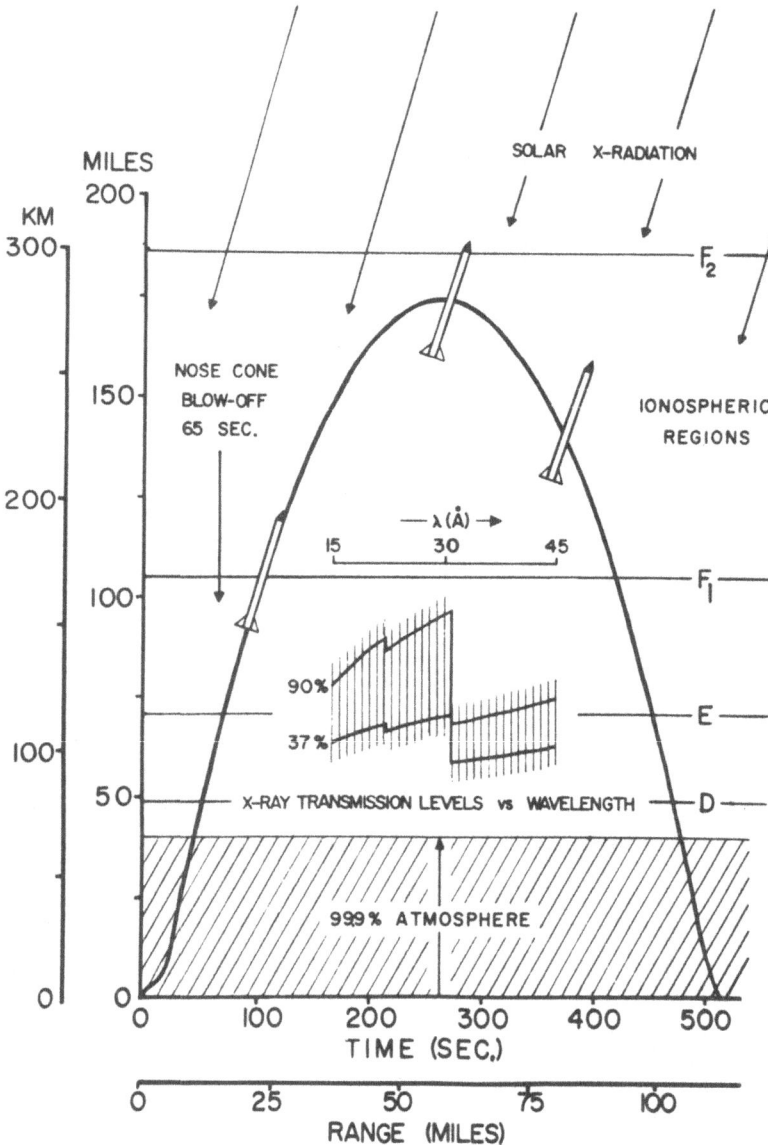

Figure 16. Typical parameters for rocket flights involved in solar x-ray measurements, illustrating the absorption of low energy x-rays within the atmosphere.

NEED FOR INTERACTION COEFFICIENTS

As noted above, a first requirement for analysis is instrumentally gaining sufficient signal, signal/background and resolution. The next requirement is establishing an adequate model for
the sample and its interaction mechanisms which will permit an
accurate interpretation of the spectral data and yield the desired
information. Often the models are actual sample analogs (known,
similar standards), comparison with which allows a simple extrapolation or interpolation (calibration curve) for the analysis.

The theory of x-ray and electron interactions has developed
to such a point in recent years, along with the rapid increase in
the availability of high-speed computers, that it is now becoming
feasible to use mathematical models for given sample systems which
will permit accurate and efficient analysis. Entering the spectral
data into a computer which is programmed to carry out the analysis
would seem to be the ideal, ultimate approach. Not only does this
approach require good theory, but it also introduces a strong need
for accurate and detailed interaction parameters or coefficients.

As described in the first section of this paper, the fundamental interaction coefficients for the low energy region are the
photoelectric cross sections. In a paper presented later in this
volume, the photoelectric absorption coefficient data reported to
this time have been compiled. From these, best-fit functions have
been determined and "state of the art" tables for cross sections
and for the 2-200 A region are obtained.[10,11]

In order to present some indication of the precision of currently available measurements, their deviations from the values as
predicted by the averaging, best-fit functions are also presented.
Another indication of precision is the comparison of these averaged
cross sections to those predicted by another independent compilation and fitting by the Lawrence Radiation Laboratory group[12] in
the overlapping region of one and six kev absorption. (Their
tables are for the conventional x-ray region and for energies down
to one kev.) This comparison is shown in Figure 17.

Finally, to demonstrate the capability of present theory to
permit the calculation of photoelectric cross sections, the completely theoretical values for the photoelectric cross sections by
McGuire[13] (for the 20, 40 and 80 A regions) are compared in
Figure 18 to those derived from the tables of Henke et al.[10,11]
as based upon the averaging of their data along with other available experimental data. In this impressive work of McGuire, he
has fit the potential function, $rV(r)$, of Herman and Skillman
(using the Hartree-Fock model) with straight line segments, which
permitted an exact solution of the radial Schrodinger equation.
It seems very likely that even better agreement with the experi-

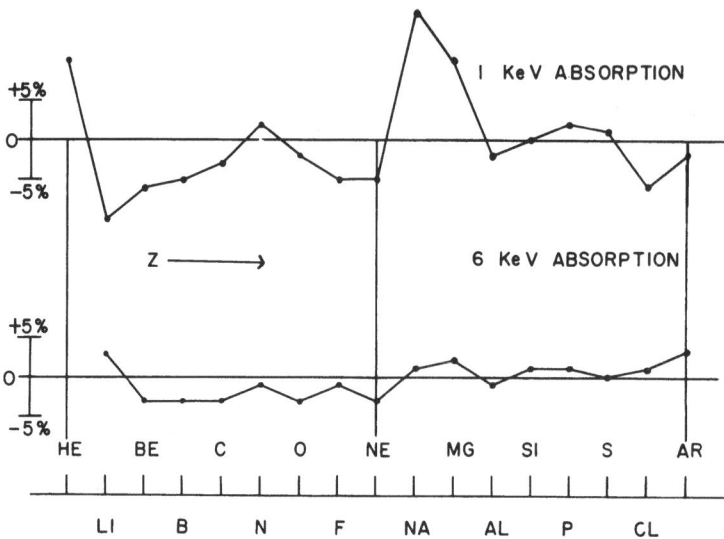

Figure 17. Comparing the values of predicted photoelectric cross sections in the overlap region of the Lawrence Radiation Laboratory best-fit tables to those of Henke et al. (both based upon compilations of experimental data).

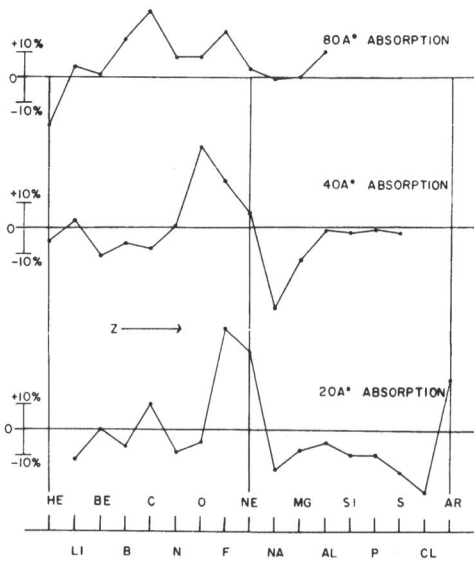

Figure 18. Comparison of the theoretical values of McGuire to the experimentally based values of Henke et al. for the ultrasoft x-ray region.

mentally based values would have been obtained if more detailed
straight line fitting had been undertaken. In fact, in certain
ultrasoft wavelength regions for which very few measurements have
been made, McGuire's theoretical values are probably better than
those predicted by the extrapolations of measurements as given by
Henke et al.

Electron absorption within matter is generally by a much more
complicated combination of processes than those for photons in the
low energy region. As noted earlier, the most likely interactions
for the photon are complete absorption (photoelectric) or coherent
scattering with no change in the photon energy. In contrast, the
electron in the 100-1000 ev region will give up its energy, a small
amount at a time, in many atomic collisions and along a circuitous
path. The great difficulty involved with attempts to define this
process precisely, both by theory and by experiment, is attested
to by the large disagreements among the reported range-energy
relationships for electrons in this energy region.

It would seem important to concentrate upon obtaining a precise description of the differential process, viz., determining the
probability (per unit path length) for an electron of given energy
to suffer its first energy loss in a given sample. This parameter
is a constant; it may be expressed alternatively as a mean free
path; and it is analogous to the linear photoelectric cross section
for the low energy photon. The fraction of a monoenergetic beam of
electrons that survives without energy loss is given by the relation,

$$I/I_o = \exp\,(-x/\lambda),$$

in which x is the sample thickness and λ is the mean free path,
or reciprocal linear absorption cross section. This relation is
similar to that for the transmission of x-ray photons.

The author would like to propose here a method for the determination of the cross sections for this primary process, based upon
the use of the electron spectrograph and multilayer samples constructed by such as the Langmuir-Blodgett technique. The method is
illustrated by some preliminary measurements which are described in
Figures 19, 20 and 21.

Four samples consisting of thick barium stearate multilayers
are placed on the four-position holder of the electron spectrograph. Three of these have deposited upon them one, two and three
double-layers of stearic acid by the Langmuir-Blodgett dipping
process. Al-K_α (1487 ev) radiation is used to excite the Ba-M_V
(705 ev) photoelectron emission from each of the substrates.
The initial photoelectron intensity is equal for each sample
because the photon absorption within the stearic acid overlayers
is negligible. However, the 705 ev electron emission which survives

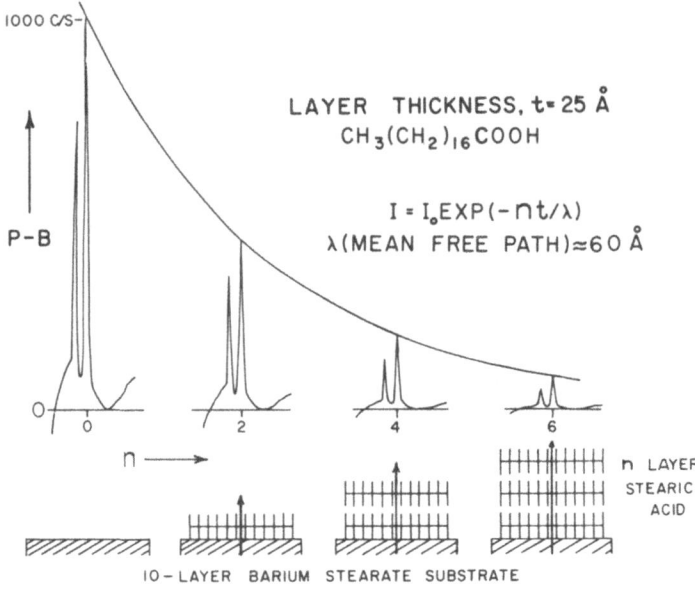

Figure 19. Determination of mean free path in stearic acid layers by the measurement of the attenuation of the 705 ev photo line from substrate crystal of barium stearate.

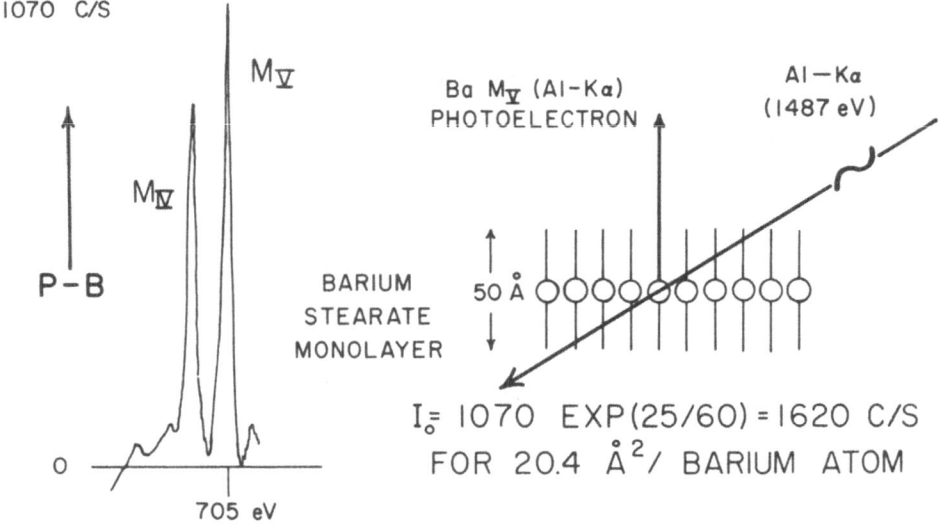

Figure 20. Measured signal of Ba-M_V (705 ev) photo line from a one-atom-thick layer of barium.

depends upon the number, n, of stearic acid monolayers and upon the
mean free path for these electrons as summarized in Figure 19. The
thickness, density and chemical composition are well known for the
stearic acid monolayers and a sensitive measurement for the mean
free path can be obtained. (Because of the similarity of the four
sample systems, their preparation and their exposure to the spectrograph vacuum environment, any contamination layers present
should be identical for each. Consequently, the effect of their
attenuation is cancelled in this comparison measurement technique.)

In Figure 20 is shown the configuration of a double layer of
barium stearate (in-and-out dipping) as deposited on a thick lead
stearate multilayer substrate. The Ba-M_V (705 ev) photoelectron
signal now measured is that due to a monatomic layer of barium
(of known density) as attenuated by essentially one monolayer of
stearic acid. Using the mean free path value for electrons of this
energy and through stearic acid monolayers as described in Figure 19, the signal that is measured from a single layer of barium

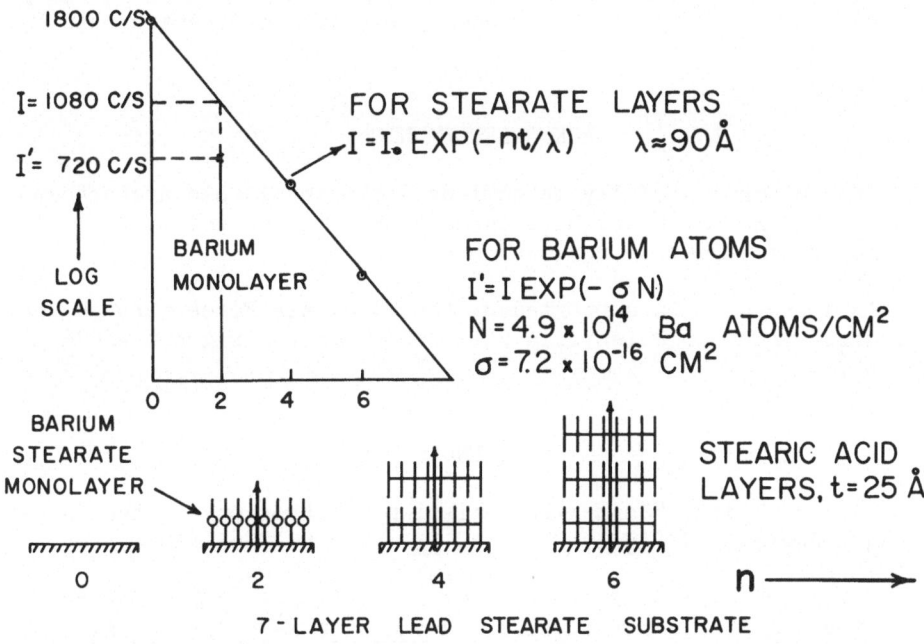

Figure 21. Measurement of the atomic cross section for barium
atoms for the primary absorption process and for the Pb-$N_{VI,VII}$
(1350 ev) photo line.

atoms in this spectrographic system is noted to be 1620 counts/second.

If now, as shown in Figure 21, Pb-$N_{VI,VII}$ (1350 ev) photoelectron emission, excited by Al-K_α (1487 ev) photons in the thick lead stearate substrate crystal, is measured as attenuated by one barium stearate double-layer, two stearic acid double-layers and three such double-layers, the attenuation for the monolayer of barium can be deduced. Since the density of the barium layer is known in these monolayer systems, the atomic cross section for barium and for 1350 ev electron interaction can be determined. Because such monolayer systems can be constructed using any bivalent metal cation, many such cross sections over a large atomic number range can be obtained. The measurements depicted here are intended to be illustrative only, but such measurements are now being undertaken in this laboratory with the objective of obtaining electron interaction coefficient tables.

There is a definite and important need for many laboratories to contribute to the completion of precise and detailed cross section tables for the low energy x-rays and electrons. Having such information would considerably advance the field of quantitative measurements by fluorescence x-ray, and photo-Auger electron analysis, as well as analysis by low energy electron diffraction and by microprobe analyzers.

ACKNOWLEDGEMENTS

The author gratefully acknowledges the invaluable assistance of Mr. Jerel A. Smith and Miss Therese Chen in the recent work of this laboratory that is reviewed here.

This research is supported by the U. S. Air Force through the Air Force Office of Scientific Research under their grant AFOSR 1262-67.

REFERENCES

1. J. E. Holliday, "Soft X-Ray Emission Spectroscopy in the 13- to 44 A Region," J. Appl. Physics, 33, 11:3259-3265 (1962).

2. B. L. Henke and R. E. Lent, "Some Recent Work in Low Energy X-Ray and Electron Analysis," Advances in X-Ray Analysis, Vol. 12, Plenum Press, New York, 1969, pp. 480-495.

3. B. L. Henke, "Microanalysis with Ultrasoft X-Radiations," Advances in X-Ray Analysis, Vol. 5, Plenum Press, New York, 1962, pp. 285-305.

4. B. L. Henke, "Some Notes on Ultrasoft X-Ray Fluorescence Analysis--10 to 100 A Region," Advances in X-Ray Analysis, Vol. 8, Plenum Press, New York, 1965, pp. 269-284.

5. B. L. Henke, "Application of Multilayer Analyzers to 15-150 A Fluorescence Spectroscopy for Chemical and Valence Band Analysis," Advances in X-Ray Analysis, Vol. 9, Plenum Press, New York, 1966, pp. 430-440.

6. W. N. Asaad, "A New Treatment of the K-LL Auger Spectrum," Nuclear Physics, 66, 494-512 (1965).

7. A. Fahlman, R. Nordberg, C. Nordling, and K. Siegbahn, "Auger Spectra for Elements of Low Atomic Number," Zeitschrift fur Physik, 192, 476-483 (1966).

    Also see: K. Siegbahn, et al., Atomic, Molecular and Solid State Structure Studied by Means of Electron Spectroscopy, Almqvist & Wiksells Boktryckeri AB, Uppsala, 1967.

8. R. L. Blake, T. A. Chubb, H. Friedman, and A. E. Unzicker, "Spectral and Photometric Measurements of Solar X-Ray Emission Below 60 A," The Astrophysical Journal, 142, 1:1-12 (1965).

9. H. V. Argo, J. A. Bergey, W. D. Evans, and S. Singer, "16-40 A Coronal X-Ray Emission during the 12 November 1966 Eclipse," Solar Physics, 5, 551-563 (1968).

10. B. L. Henke, R. L. Elgin, R. E. Lent and R. B. Ledingham, "X-Ray Absorption in the 2 to 200 A Region," Technical Report, AFOSR 67-1254, June 1967.

11. B. L. Henke, R. L. Elgin, R. E. Lent and R. B. Ledingham, "X-Ray Absorption in the 2 to 200 A Region," Norelco Reporter, XIV, No. 3-4, 1967, p. 112.

    NOTE: The compilations of absorption data first presented in references 10 and 11 have been updated and appear later in this Volume 13 of Advances in X-Ray Analysis.

12. W. H. McMaster, N. K. Del Grande, J. H. Mallett, and J. H. Hubbell, "Compilation of X-Ray Cross Sections," Lawrence Radiation Laboratory Report UCRL-50174 Section II, University of California, Livermore (1969).

13. E. J. McGuire, "Photo-Ionization Cross Sections of the Elements Helium to Xenon," Phys. Rev., 175:20-30 (1968).

LIGHT ELEMENT ANALYSIS

A. K. Baird

Geology Department, Pomona College

Claremont, California

ABSTRACT

Qualitative and quantitative analyses of elements below atomic number 20, and extending to atomic number 4, have been made practical and reasonably routine only in the past five to ten years by advances in: 1) excitation sources; 2) dispersive spectrometers; 3) detection devices; and 4) reductions of optic path absorption. At present agreement is lacking on the best combination of parameters for light element analysis. The principal contrasts in opinion concern excitation.

Direct electron excitation, particularly as employed in microprobe analysis (but not limited to such instruments), provides relatively high emission intensities of all soft X-rays, but also generates a high continuum, requires the sample to be at essentially electron gun vacuum, and introduces practical calibration problems ("matrix effects"). X-ray excitation of soft X-rays overcomes some of the latter three disadvantages, and has its own limitations. Sealed X-ray sources of conventional or semi-conventional design can provide useful (if not optimum) light element emission intensities down to atomic number 9, but with serious loss of efficiency in many applications below atomic number 15 largely because of window-thinness limitations under electron bombardment. Such devices, however, probably provide the most stable sources available. Demountable, continuously pumped X-ray sources and open-window X-ray tubes can largely or completely overcome the window absorption problem. Demountable sources permit selection of target materials optimally suited to excite emission lines of interest and to *not* excite unwanted and possibly interfering lines of heavier elements. Required vacuum techniques, and emission instability due

26

to vacuum variations, are disadvantages of these sources.

Within the spectrometer, dispersive geometries (crystals and diffraction gratings) have been used almost exclusively. In particular, the development of "constructed crystals" has permitted high efficiency of dispersive reflection in the very soft X-region (stearate type), and the recently developed graphite monochromator has substantially improved reflectivity, with little loss of resolution, in the harder, soft X-ray region (atomic number 15-20). The gas-flow proportional counter is the only detector in wide use. The development of techniques for forming thin organic (formvar, collodion, polypropylene) detector windows permits reasonable counting efficiency. Until very recently, non-dispersive detection systems were limited to analyses of elements heavier than atomic number 20; extension of this method into the light element region offers promise of much more efficient detection.

Sample preparation, and calibration for quantitative analysis, require specialized (*and careful*) design depending upon the application and upon instrumental conditions (e.g., electron excitation vs. X-ray excitation). In general, the present poor knowledge of basic theoretical correction factors has required close referencing to standards or has required empirically derived correction procedures for best results.

Today, instruments and techniques are available which permit routine analyses of light elements heavier than neon (atomic number 10) with nearly the "push-button" ease long associated with heavy element X-ray analysis. Below atomic number 10, very soft X-ray analysis is now definitely possible and practical, but requires more specialized and sophisticated techniques which may remove the method from a "routine" classification in some laboratories.

INTRODUCTION

In 1968 Campbell and Brown[1] wrote "... that the future of low energy X-rays lies in the interpretation of spectra rather than the determination of concentration." A glance through the literature of the past two or three years, especially the proceedings of this Conference series, seems to support their conclusion. Thus this review paper *might* be an obituary for X-ray methods of analytical chemistry, at least as far as light elements are concerned! However, in the belief that soft X-ray analysis is a powerful quantitative tool, this review will attempt to outline what is presently possible and, perhaps, to encourage further applications. The intent here will be to emphasize X-ray spectrometry as such--that is, bulk sample analysis measuring fluorescent X-rays generated

[1]References are at the end of the paper.

either by primary X-ray photons or electrons, or both. Electron microprobe analysis, though certainly an important method employing X-ray measurements in the light element region, has additional problems (and advantages) related to the sample volume being considered, and this method will not be dealt with.

PROBLEMS OF LIGHT AND VERY LIGHT ELEMENT ANALYSIS

For the purposes of this review, I will use the term "light elements" to mean those elements of atomic number 10 (neon) through atomic number 20 (calcium); "very light elements" to mean those elements below atomic number 10. The K-spectra for the former group of elements are soft X-rays and for the latter group of elements are very soft X-rays. These definitions differ somewhat from the usages of other workers, but have practical distinctions in instrumental conditions of analyses: at about atomic number 20 the air-path absorption of fluorescent K-spectra X-rays becomes a serious matter and marks a convenient boundary between "hard" and "soft"; commercially available sealed X-ray sources (as will be discussed below) can provide reasonably efficient excitation for light elements, but not for very light elements, thus marking a convenient boundary between "soft" and "very soft."

The problems encountered in the X-ray analysis of calcium and lighter elements, which increase with decreasing atomic number (though not regularly), can be summarized as follows:

1) a reduction in excitation efficiency;

2) an increase in absorption of fluorescent X-rays throughout the optic path of the spectrometer from sample to detector;

3) a resulting decrease in signal and signal to noise ratio and thus sensitivity;

4) a change from line spectra to band spectra, characteristics of which may be (or are) concentration independent;

5) an increase in the criticalness of sample preparation;

6) a decrease in the knowledge of X-ray interactions (matrix effects).

Problems 1) and 2) are instrumental and much has been accomplished to solve them, thus alleviating problem 3). Problems 4), 5), and 6) are highly dependent upon the analyses attempted and generalization is difficult. At one extreme, sample character and complexity may preclude quantification; in many situations, however, useful quantitative information in the soft and very soft X-ray region can be obtained--especially with limited compositional ranges, samples having similar chemical combination, and with adequate standards for comparison.

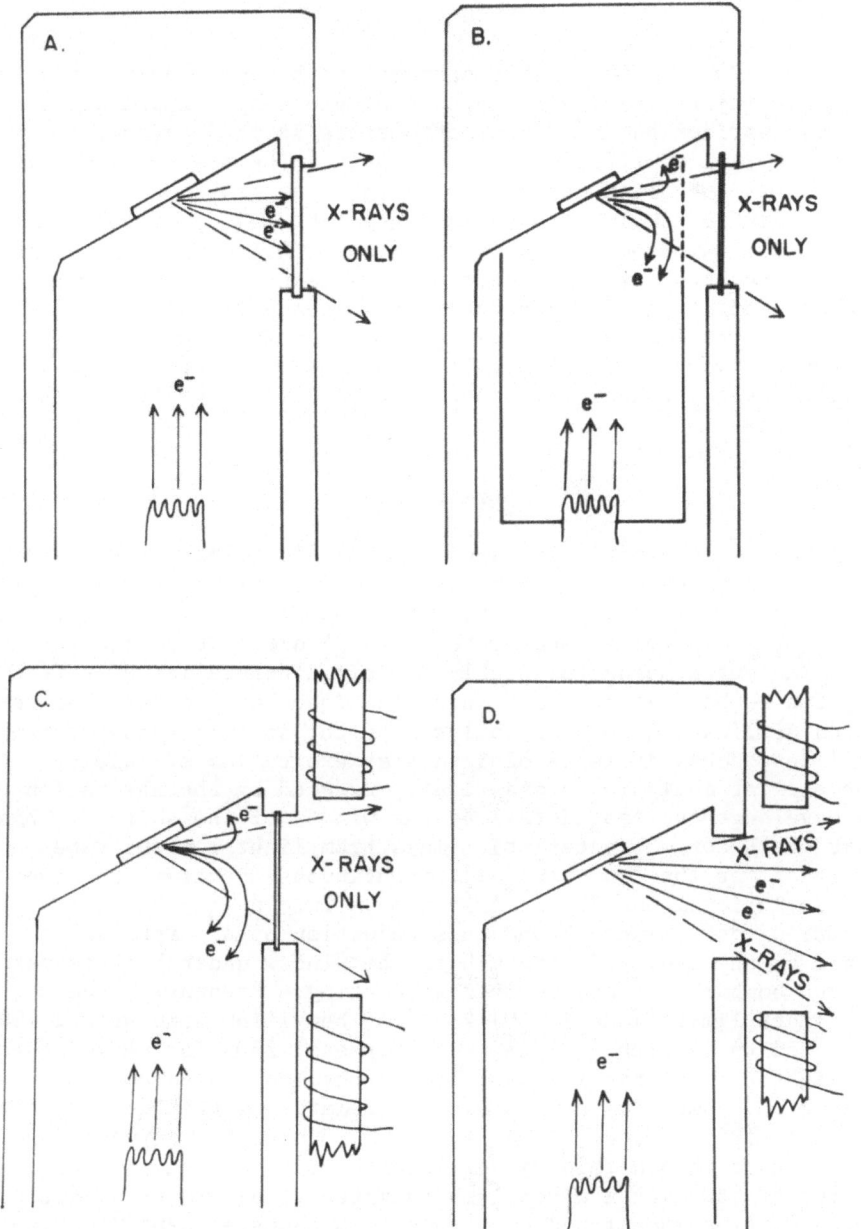

Figure 1. X-ray tube configurations. A) conventional sealed tube
with heavy element target and relatively thick window; B) sealed
tube with cathode grid to deflect electrons from thin window; C)
sealed tube with magnetic deflection of electrons from thin window;
and D) windowless tube with or without magnetic deflection for X-
ray or X-ray/electron excitation. (From R. Jenkins.)

INSTRUMENTATION

After 1959, vacuum-path spectrometers became widely available, offering a major improvement in convenience, expense and perform- ance over earlier hard X-ray spectrometers adapted to soft X-ray analysis by helium-filled tunnels or bags. The sealed tungsten target X-ray tube of about 2.5 kW rating, the gas-flow proportional counter with a 6μ Mylar window, and the LiF, EDDT, and ADP analyz- ing crystals composed the basic equipment. With such a spectrometer practical results could be obtained on aluminum and heavier elements, and under conditions of moderate to high concentration, on magnesium.[2] Sodium and all very light elements could not be detected. Attempts to extend the method have centered primarily upon excitation, thin-window construction, and large-spacing analyz- ing crystals.

Excitation

Advances in excitation have, in general, taken three divergent paths: 1) improvements in conventional sealed X-ray sources, pri- marily through reduction in window thickness; 2) demountable, thin- window (or windowless) X-ray tubes; and 3) direct electron excita- tion. (A fourth means for highly efficient generation of soft X-rays is proton bombardment[3,4], but this technique has, so far, had very limited application to bulk analysis of the sort discussed here.) In all approaches it is recognized that excitation efficiency in light element analysis is drastically reduced by the absorption of long wavelength primary X-rays by the window of the source. This is true both for characteristic lines from light-element tube targets or for the long wavelength continuous spectrum from the more conventional tungsten target tube. In conventional X-ray sources, window thinness, and thus reduction of absorption, is limited by the required strength of the window under backscattered electron bombardment and consequent heating. Presently, the approximate limitations in thickness of beryllium windows are 1500μ for sealed tubes with heavy element targets, 250μ for sealed tubes with light element targets, and 15-25μ for non-sealed sources. Thatcher and Campbell[5] have shown the relative excitation efficien- cies (for AlKα) increase by a factor of 10 over the conventional tungsten tube when a thin-window chromium target tube is used; by a factor of 150 with a windowless tungsten tube, and by a factor over 1000 if backscattered electrons plus tungsten continuum is employed. (Designs of several forms of X-ray tubes using grids or magnets for electron/X-ray excitation are shown in figure 1.)

These results (and results of other workers with windowless sources)[6-9] seem to suggest that electron excitation is *the* answer to light element analysis problems. However, there are distinct disadvantages to this approach:

1) In direct electron excitation the sample becomes the target of a form of X-ray tube and must be in a "hard" vacuum, must be electrically conductive, and is subject to heating and possible damage.

2) In demountable windowless X-ray tubes, employing both primary X-rays and electrons, similar vacuum and sample heating problems exist, and emission stability (crucial for quantitative analytical work) appears to be much worse than in sealed tubes.

3) In any form of direct electron excitation, the facility to discriminate against excitation of fluorescent spectra from possibly interfering elements in the sample is limited. The gain in excitation efficiency is at the expense of a relatively poor signal to noise ratio.

The demountable, reverse-potential Henke tube design[10] (figure 2) overcomes window heating problems because the window and housing are kept at ground potential and deflect electrons. Windows of beryllium as thin as 25μ and windows of aluminum as thin as 6μ may be employed which will support atmospheric pressure while the tube is in full operation. In effect this permits analytical techniques similar to those with sealed sources:

SECTION ⬑

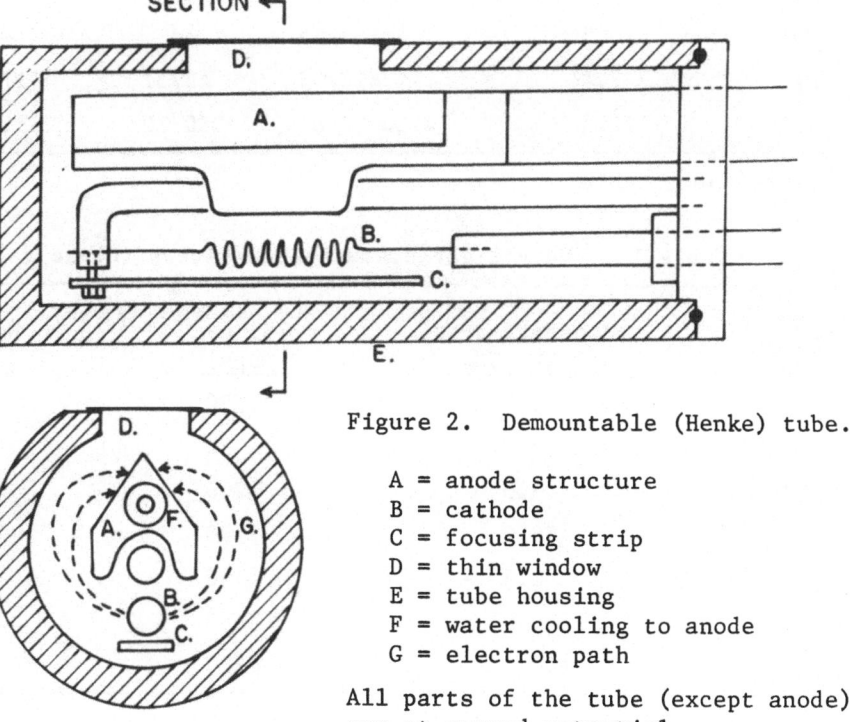

Figure 2. Demountable (Henke) tube.

A = anode structure
B = cathode
C = focusing strip
D = thin window
E = tube housing
F = water cooling to anode
G = electron path

All parts of the tube (except anode) are at ground potential.

samples at roughing vacuum only, sample changing without turning
off tube power, and relatively high emission stability because of
continuous tube operation. In this design targets may be selected
which emit characteristic spectra close to the absorption edge of
the element of analytical interest permitting high excitation
efficiency. Further, by using light-element targets (e.g., aluminum emitting AlKα for Na and Mg analyses) very efficient discrimination against unwanted X-ray fluorescence from the sample, or
from other parts of the optic path of the spectrometer, can be
achieved. For example, a gypsum ($CaSO_4 \cdot 2H_2O$) analyzing crystal can
be used for the efficient dispersion of NaKα without calcium and
sulfur fluorescence from the crystal. Measured intensities in the
light element region are compared for a 1kW sealed chrome tube, a
2.7kW sealed chrome tube, and a Henke tube with fixed window in
Table I.

Table I. Light element intensities (silicate rock).

Element	Concentration Wt. %	Signal/Noise (cps)		
		Henke (1.5kW)	Chrome (1kW)	Chrome (2.7kW)
Na	3.96	297/16	nd	47/23
Mg	0.50	182/15	nd	49/18
Mg	3.44	820/32	48/37	178/27
Al	16.8	10025/8	950/70	8542/42

nd = not detectable

Table II. Very light element intensities (Henke).

Element	Concentration Wt. %	Signal (cps)	Noise (cps)
F	5.7	150	10
O	2.2	30	8
O	48.0	7200	600
N	56.0	2500	170
C	0.5	40	4
B	44.0	4140	140
Be	100	580	30

In the very light element region absorption by the fixed tube window described above is too great for transmission of optimum primary CuLα or CKα X-rays and Henke[11] has devised a sliding window gate which supports ultra-thin parlodion or Formvar films capable of isolating the hard vacuum (10^{-6}mm) of the X-ray tube from the roughing vacuum (10^{-2}mm) of the spectrometer. A blanking plate is slid in place before air is admitted to the spectrometer for sample changing. This requires power shut-down to the tube and is a source for increased instability in quantitative analysis where numerous specimens must be run. Intensities in the very light element region are shown in Table II.

Dispersive Devices

In the light element region, EDDT, PET, ADP, KAP, and gypsum have been the standard crystals employed in dispersive spectrometers. Table III summarizes the useful ranges of these crystals.

A recent development of major importance for light element analysis is the highly oriented polycrystalline graphite[12] ($2d = 6.7$Å), suitable for dispersion of the K-spectra from phosphorus and heavier elements. Gould and others[13] have shown that integrated intensities 21 times greater than for EDDT are

Table III. Useful Ranges and 2θ Values for Several Analyzing Crystals.

Element	K-line Å	LiF 4.03	PET 8.74	EDDT 8.81	ADP 10.65	Gypsum 15.16	KAP 26.6	Stearate 100
Ca	3.35	113.1	45.1					
K	3.73	136.6	50.5					
Ar	4.19		57.3	56.8	46.4			
Cl	4.73		65.5	54.9	52.7			
S	5.37		75.8	75.1	60.6			
P	6.15		89.4	88.6	70.6			
Si	7.13		109.3	108.0	84.0			
Al	8.34		145.1	142.0	103.1			
Mg	9.89				136.5	81.1	43.7	
Na	11.9					103.0	53.2	
Ne	14.6					148.0	66.6	
F	18.3						86.9	
O	23.6						125.0	27.3
N	31.6							36.7
C	44.6							52.8
B	67.8							85.1

possible for PKα. Their crystal (obtained in 1967) has a rocking
curve half-maximum breadth of 0.60°. We obtained a graphite
crystal in 1969 which has an improved angle of 0.35° (nearly as
good as EDDT) and a signal 30 times that of EDDT for PKα. Fortunately, in most analytical work in the light element region
resolution is not a critical parameter and coarse collimation
(4 in. x 0.035 in.) can be employed with this crystal. The
graphite monochromator is very stable in the vacuum path and is not
temperature sensitive.

In dispersive light element analysis today the greatest need
is for a stable, efficient crystal for the diffraction of NaKα
through SiKα spectra. PET, with high reflectivity for AlKα and
SiKα, is particularly 2θ- sensitive to temperature variations; ADP,
KAP, and gypsum all contain elements which fluoresce strongly if a
chromium target X-ray tube is used. This radiation can only be
partially removed by pulse height discrimination and results in
relatively high background levels. Several laboratories have
reported that gypsum crystals dehydrate in the spectrograph vacuum,
though we have used one such crystal for hundreds of hours over six
years at a vacuum of about 10^{-2}mm without deterioration.

In the very light element region, constructed crystals of metal
stearates have found wide application[9,14,15,16] as have grazing
incidence diffraction gratings[17,18]. The former are produced by
building up mono-layers using the Langmuir-Blodgett dipping technique. Crystals of lead myristate ($2d$ = 80Å), lead stearate
($2d$ = 100Å), lead lignocerate ($2d$ = 130Å), and lead melissate
($2d$ = 160Å) have been made. Henke[15] concludes that these crystals
are to be preferred over grazing incidence diffraction gratings
for very light element spectroscopy because they are directly interchangeable with conventional crystals in conventional spectrometers,
they have resolutions consistent with most soft X-ray analytical
requirements (though poorer than gratings), they are more efficient
with the large X-ray sources than gratings, and they are easier to
align than gratings.

Stearate crystals can be used for sodium analyses, but I find
no significant improvement in signal over a gypsum crystal and a
poorer signal to noise ratio, probably due to the low angle of 2θ
required. With a chromium target tube fluorescence of PbM would
be an additional problem.

Detection

The only detector in wide use today for light element analysis
is the gas-flow proportional counter[19] of conventional design, but
greatly improved in efficiency by newly developed types of thin
windows capable of supporting atmospheric pressure in the detector

against the roughing vacuum of the spectrometer. Relatively low
absorption of fluorescent X-rays from the heavier light elements
permits the use of high strength 6μ- thick aluminized Mylar. In
analyses for aluminum and magnesium, however, the use of 6μ
aluminum foil, or polypropylene stretched to near interference
color thickness greatly reduces window absorption. An added
advantage of aluminum foil is the marked discrimination against
SiKα in all silicate analyses for light elements. Both the foil
and stretched polypropylene may be used directly against the
collimator blades of the detector which provide adequate support.
P-10 gas (90% argon, 10% methane) is used in the light element
region. In the very light element range, two double layers of
Formvar, 30μg/cm each, are supported by a 70% transmission 200-mesh
nickel grid. This material has good transmission characteristics
throughout the very light element range. Procedures for casting
thin films are described in detail by Henke.[11] Methane, or P-75
gas (75% methane, 25% argon) is used in this region.

Sterk[3] has described a windowless photoelectric detector with
electron multiplier which has detection efficiencies of better than
50% for very soft X-rays, but its output is not energy proportional.
In many applications non-proportionality would be a serious handicap.

Very marked improvements have been made in lithium-drifted
silicon non-dispersive detection systems in the past year, which

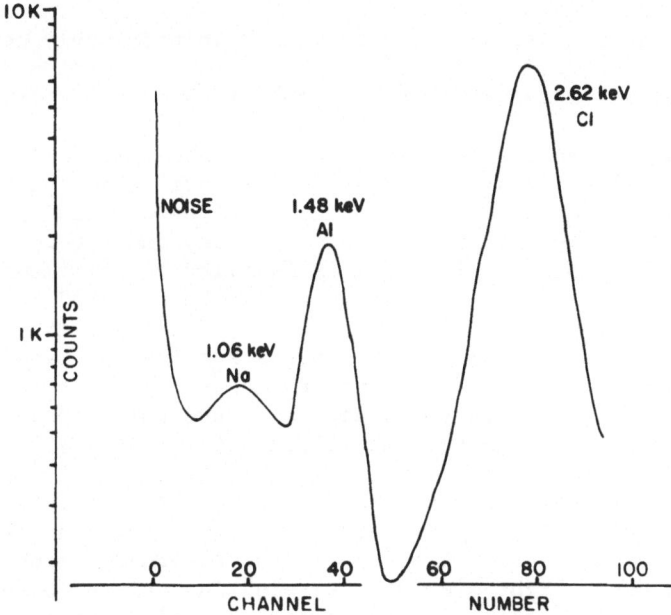

Figure 3. Non-dispersive detection of X-ray spectra, lithium-
drifted silicon detector.

have extended the light element detection from about atomic number
20 to atomic number 11. In addition, improvements have been made
in resolution (better than 250 eV at 6.7 keV). Though I know of no
detailed report on light-element capabilities at the time of this
writing, figure 3 shows a spectrum[20] which indicates the promise
of this type of non-dispersive detection system.

SOME COMPARISONS OF EQUIPMENT

An attempt will be made here to compare and contrast the
capabilities of presently available types of X-ray equipment from
the standpoint of practical analytical work in the light element
region. This attempt is fraught with difficulty for at least
three reasons:

1) Analyses performed in different laboratories are rarely, if
ever, done under the same instrumental conditions, and no single
laboratory has available all the possible combinations of equipment
for a controlled test.

2) The majority of light elemental analyses that have been
performed to date have been done on specialized, "home-made"
spectrographs, often not by analysts, but by researchers in the
soft X-ray field who are particularly adept at squeezing-out high
instrumental performance, sometimes at the expense of practicality.

3) My own biases developed by experiences in using only certain
types of equipment (and by the lack of experience in not using other
equipment!) on a limited variety of problems.

Instrumental factors of prime importance in practical X-ray
analysis are: 1) a signal and signal/noise ratio which permits an
analysis to be performed at an adequate level of counting precision
within an adequate period of time; 2) a primary and fluorescent
emission stability of high level; 3) a facility to accommodate and
analyze a variety of sample types, which can be prepared easily and
can be rapidly changed; and 4) a high level of reliability. The
soft X-ray literature abounds with reports showing adequate results
for 1) by various means of X-ray and/or electron excitation and
proton bombardment. But very few papers in the soft X-ray
literature even mention factors 2), 3), and 4).

Table IV summarizes my opinions of the present state of light
and very light element bulk analysis. It is apparent that a rather
marked distinction exists between what I have termed "light element
analysis" and "very light element analysis" in terms of ease of
analytical operation, stability and sample handling aspects. In
support of this opinion a few specific examples can be cited:

Table IV. Some instrumental comparisons; Light and very light element analyses.

Parameter	Electron Excitation	Demountable Windowless Electron/X-ray	Demountable Window X-ray	Sealed Chrome X-ray	Sealed Tungsten X-ray
Applicable Elemental Range	Wide	Wide	Light/very	Light elem.	Heavy light elem.
Stability	Low-moderate	Low-moderate	Moderate-high	High	High
Counts/sec/watt	High	High	Moderate-high	Low-moderate	Very low-to zero
Signal	High	High	High	Low-moderate	Very low
Noise	High	High	Low	Low	Low
Sensitivity	Moderate	Moderate(?)	High	Low	Very low
Reliability	Moderate(?)	Moderate(?)	Moderate	Very high	Very high
Sample Character	Limited	Fairly broad	Wide range	Wide range	Wide range
Changing Facility	Difficult	Difficult	Easy	Easy	Easy
Possible Damage?	Yes	Yes	No	No	No

1) Low concentrations of sodium and magnesium in silicates.
Results from a demountable (Henke) X-ray source operated at 1.5 kW
with an aluminum target and fixed aluminum foil window are compared
with results from a sealed chromium target X-ray source operated at
2.7 kW in Table V. An order of magnitude improvement in the
theoretical minimum detection limits is evident for the demountable
source operated at about half the total loading of the sealed
source. To achieve this improvement, however, requires dealing
with high vacuum isolated by a relatively fragile window. The
results indicate, moreover, that practical analytical results can
be obtained in the light element region using highly convenient
and stable sealed sources.

Table V. Minimum Detection Limits (Silicate Rock).

Oxide	Wt. %	1kW Chrome	2.7kW Chrome	1.5kW Henke
MgO	3.44	0.167	0.035	0.007
Na_2O	3.96	–	0.154	0.029

$$M.D.L.(wt.\%) = \frac{(0.3)(wt.\%)(Background)^{\frac{1}{2}}}{(Peak) - (Background)} \text{ for 100 sec. counting.}$$

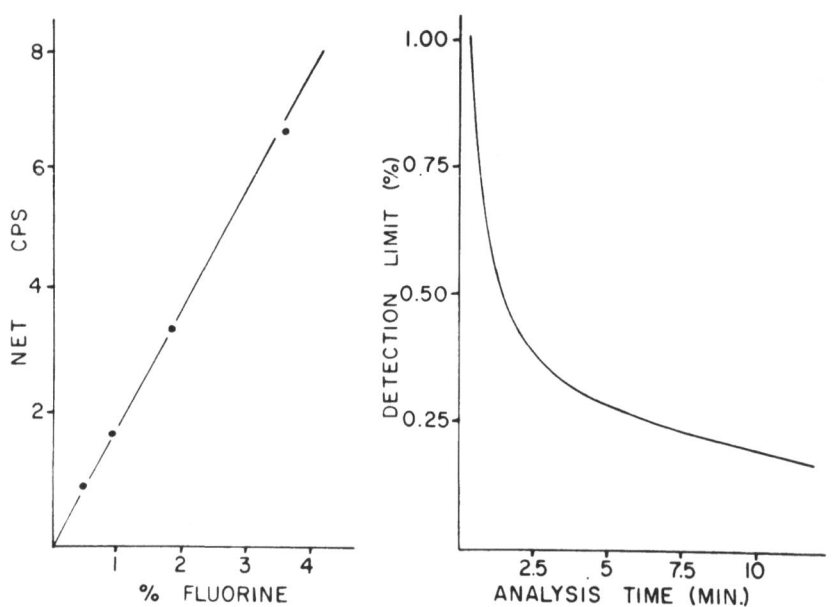

Figure 4. Analyses of fluorine in calcium fluoro-aluminate.
Calibration line (left) and limits of detection as a function of
time devoted to analysis (right).

2) Low concentrations of fluorine in a calcium fluoro-aluminate. An interesting study by Jenkins and van Gelder[21] shows that a sealed chromium source of 2 kW rating can be effectively used in the upper very light element region by very careful consideration of window and detector absorption characteristics. Figure 4 shows the derived calibration line and limits of fluorine detection as a function of required analysis time. It seems obvious that this application is pressing the use of sealed chromium sources to the practical analytical limit.

3) Oxygen in silicates. We have shown[22] that the major element oxygen can be determined in finely ground silicate rock powders with an analytical precision better than 1% relative standard deviation with a demountable source employing CuLα excitation. The minimum detection limit indicated is very similar to that for the sodium analysis cited in 1) above, but the oxygen analyses require ultra-thin X-ray tube and detector windows with the former on a sliding vacuum isolation gate. Convenience of operation and stability is, therefore, distinctly poorer than for light element operation with a fixed window.

4) Carbon in steel (figure 5). Carbon in amounts below 0.5% has been determined quantitatively by Henke[23] in NBS standard steels. The limits of detection for a two minute count time were estimated at 0.006%C, again using ultra-thin X-ray tube and detector windows.

It is probable that combined X-ray and electron excitation or direct electron excitation, of the samples in the examples cited above would provide much higher total count rates, but probably at the expense of poorer signal to noise ratios. None of the instruments designed for electron excitation (to my knowledge) provides the convenience of bulk sample changing found with a conventional vacuum-path X-ray spectrometer, nor do they permit the direct use of non-conductive samples. Unlike X-ray excitation, samples can suffer electron beam damage and/or contamination.

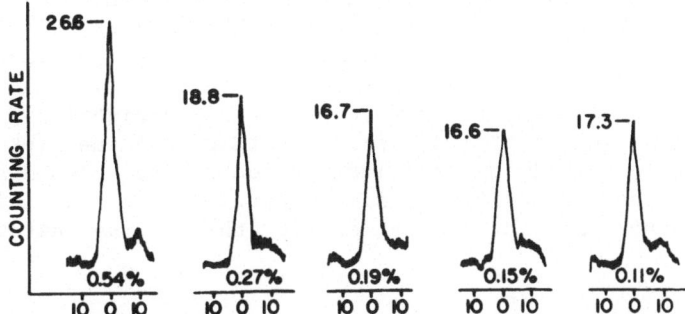

Figure 5. Carbon in NBS standard steels. Peak values are counts per second.

SAMPLE PREPARATION

In most light, and all very light element analyses, samples
must be in the solid state because of the vacuum environment and/or
the inverted or inclined mode of presentation of the sample to the
excitation source in most commercial spectrographs. Liquids containing heavier light elements can be analyzed directly through
supporting Mylar films or in spectrographs having upright optics
combined with helium tunnels. In many applications liquids
(e.g., blood serums[10]) can be conveniently treated as dried pre-
cipitates on a neutral substrate.

Problems presented by solids are highly diverse depending
upon the application. In heavier element analysis, effective
infinite sample thickness from which fluorescent X-rays are
generated and detected can range up to hundreds of microns.
Depending upon excitation, element of interest and sample matrix,
effective thickness can be as little as a micron or less in very
light element analysis. Absorption of long wavelength X-rays is
much higher than for shorter wavelengths, but since effective
sample thickness is so small, matrix interference problems and
total absorption are similar to heavier element analyses. Much
more critical in comparison, however, is the physical state of the
surface and near surface of the sample. Building upon the results
of Bernstein[24], Claisse and Samson[25], and Gunn[26] with heavier
elements, Madlem[27] investigated particle size and matrix effects
in a variety of light and very light element sample mixtures. It
was found that some intensity variations commonly ascribed to
inadequately established absorption corrections were actually
caused by varying particle sizes. Critical for maximum intensity
and reproducibility is a uniformly fine (1-10μ) grain size which
is difficult to achieve in impact mills with samples containing
phases of differing physical properties (hardness, cleavage, etc.).

Various forms of fluxing techniques[28-30] have been devised to
overcome variations of physical characteristics of heterogeneous
samples and to provide a uniform matrix for the element(s) of
analytical interest. These range from methods using only light
element fluxes (e.g., lithium tetraborate) in minimum quantity, to
methods in which a strongly absorbent compound (e.g., lanthinum
oxide) is also added to further reduce matrix variability. As
pointed out by Volborth[31], all fusion methods run the risk of
differentially volatizing (and losing) one or more elements in
the sample, require additional sample handling with possibilities
for gross error, and, of course, destroy the original nature of
the sample.

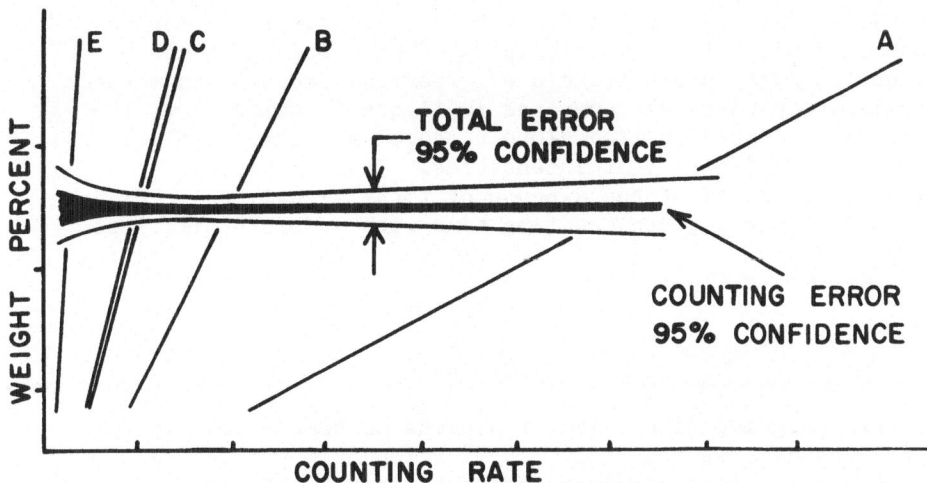

Figure 6. Relationships between counting statistics error, total error, and sample preparation techniques for silicates[29]. A) ground powders; B) moderate dilution fusions; C) high dilution fusions with heavy absorber; D) high dilution fusions; and E) very high dilution fusions.

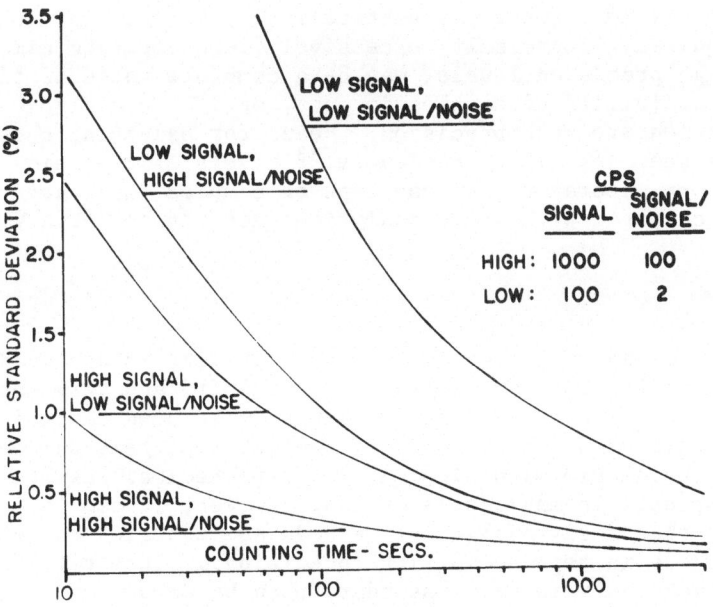

Figure 7. Examples of counting precision versus counting time for several combinations of signal and signal to noise ratios.

If, however, destruction of the sample can be tolerated significant improvements in the reproducibility of samples and in the reduction of the criticalness of grinding often can be achieved through fusion techniques. Several preparation methods for silicate powders are compared in figure 6, which shows the changes in calibration line slope produced by various dilutions, and the reproducibility of intensities. It is obvious that higher precision is gained at the expense of total signal, and thus sensitivity, for a given counting time with moderate dilutions.

QUANTIFICATION

Precision

The foregoing discussion of aspects of sample preparation strongly emphasizes that every application of X-ray analysis to light and very light elements should be thoroughly tested for reproducibility. A level of precision should be sought (primarily by the effort expended in sample preparation) consistent with the counting statistics[32] employed and, obviously, with the requirements of the analytical job. Reference to figure 7 indicates the time limitations imposed by spectrometers providing low signal or high signal/noise ratio or both for a given level of counting precision. An analysis within the theoretical sensitivity limits of a given X-ray instrument may be totally impractical in view of the time required. Conversely, excessively long counting times, achieving high precision levels, may be a complete waste of time if the reproducibility of samples is poor, or if the analytical job does not require such precision. Means for balancing these efforts have been described previously.[33] Levels of precision obtained in our laboratory, on one type of samples (silicate rocks), are compared with other analytical methods in Table VI.

Accuracy

Assuming an adequate level of reproducibility is achieved, quantification of all X-ray analyses, hard, soft, and very soft, ultimately depends upon standards of known or assumed composition; and usually this means that the X-ray analyst must rely on the chemist to provide him with high quality wet-chemical results on standard samples. In many applications, however, it can be demonstrated that X-ray analytical methods are superior to wet-chemical methods in reproducibility (precision). Thus reliance on a single wet analysis of a "standard" can be dangerous. Further, it can be demonstrated that a series of standards, wet-analyzed by two laboratories, gives two calibration lines (figure 8), each internally consistent, but offset from each other.[33]

Table VI. Precisions of Silicate Analyses by Several Methods[22].

| Element | Wt. % | Intralaboratory Precision, std. dev., Wt. % | | | | |
		X-ray	Wet Chemistry	Emission	Flame	Neutron Activation
O	48.1	0.4	0.3	–	–	0.9
Na	2.6	0.02	0.15	0.11	0.07	–
Mg	0.5	0.01	0.15	0.12	–	–
Al	9.0	0.04	0.19	0.53	–	–
Si	30.0	0.10	0.14	1.1	–	–
K	3.7	0.03	0.21	0.15	0.07	–
Ca	2.5	0.01	0.07	0.14	–	–
Ti	0.2	0.003	–	–	–	–
Fe	3.1	0.02	0.21	0.14	–	–

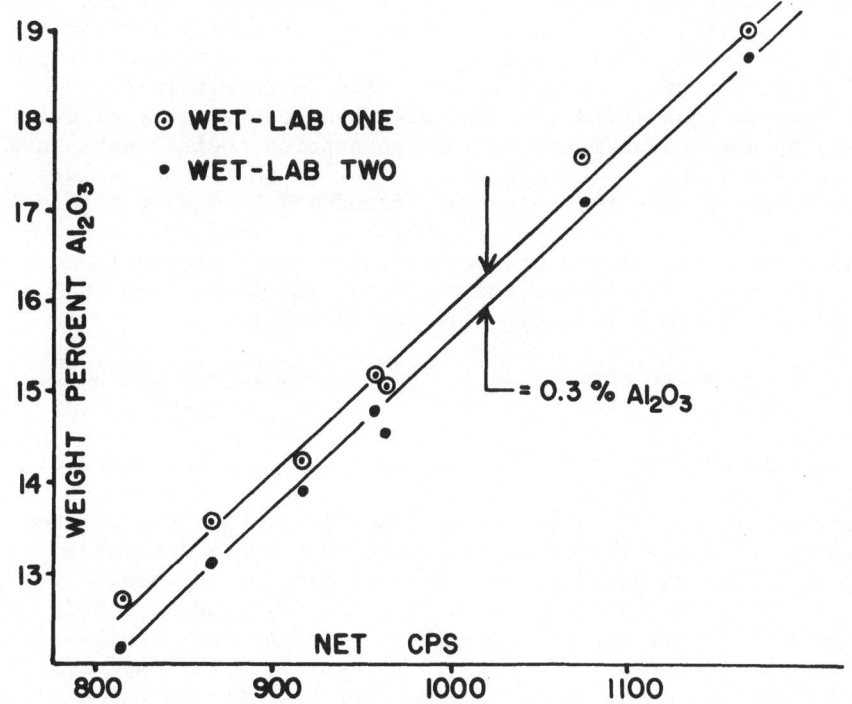

Figure 8. Calibration lines derived for six standards wet-analyzed by two different laboratories.

Standards are used today in two ways: 1) as the basis for ratioing of X-ray intensities, unknown sample to standard sample, either directly or through calibration lines or curves; and 2) as the basis for judging whether or not experimentally or theoretically derived correction factors for various matrix effects are actually yielding "correct" answers. In either usage the caution cited in the prior paragraph holds.

In light and very light element quantitative analyses all the "matrix" problems exist that exist with heavier element analysis, plus some added problems: 1) separating biases introduced by inadequate sample preparation from biases introduced by inter-element effects[27]; 2) band spectra instead of line spectra[34,35] and 3) a poorer knowledge of interelement effects. The first added problem has its best solution in the use of standards as physically and chemically similar to the unknown samples as possible. The effect of the second problem is somewhat analogous to pulse-shift and counter dead-time in heavy element analysis at high count rates. Here, the relative low count rates which usually must be tolerated in light element analyses reduce the latter corrections to negligibility; but a change from a single emission line to several lines, because of concentration-independent chemical combination, can also apparently reduce count rates. Again, the best solution lies with using standards as chemically similar as possible to the unknowns.   Further, the spectrometer resolution should be as coarse as possible, consistent with resolving any interferences. The third added problem has been greatly alleviated recently by new measurements of mass absorption coefficients in the very soft X-ray region[36], though agreement on the "best" values is not complete. Errors that might be introduced by making absorption corrections using inadequately established coefficients are obviously minimized if the primary intensity data are from standards closely similar to the unknowns, merely because the magnitude of the required correction(s) is small.

If interelement correction procedures are required (and it should be emphasized that they are not *always* required[33]) they commonly take one of two forms: 1) primary intensity ratios are theoretically adjusted for secondary matrix-element(s) absorption and enhancement, for primary absorption (assuming an effective excitation wavelength), and for the geometry of the spectrometer[37]; 2) primary intensity ratios are adjusted by factors determined empirically for the particular system and samples in question.[38] The second form involves no assumptions nor any possibly inadequately established coefficients, but is, of course, limited in scope to the concentration ranges which went into the empirically derived factors. It is clearly a superior technique for the quantitative analyses of large numbers of similar samples where interelement effects preclude use of simple calibration lines directly.

ACKNOWLEDGMENTS

I am indebted to the Co-chairman of this Conference, Burton L. Henke, for nearly ten years of patient tutelage in the application of soft X-ray instrumentation and methods of his design to my problems of light element analysis. This paper reflects that tutelage, at least in the biases it demonstrates, if not in its clarity and completeness! To those whose work and contributions I have slighted out of ignorance, my apologies. I also thank Ron Jenkins of the Eindhoven Laboratories of N. V. Philips for providing several comparative light element analyses using sealed sources and for information on the present status of light-element work in Europe, the latter from an unpublished paper presented at the 50th Conference of X-ray Analysis for Industrial Use, Society of Japan Chemists, February, 1969.

REFERENCES

1. W. J. Campbell and J. D. Brown, "X-ray Absorption and Emission," Anal. Chem. 40:346R-375R, (1968).

2. A. A. Chodos and C. G. Engel, "Fluorescent X-ray Spectrographic Analyses of Amphibolite Rocks and Constituent Hornblendes," Advances in X-ray Analysis 4:401-413, (1961).

3. A. A. Sterk, "X-ray Techniques in the 1 to 400Å Range," Advances in X-ray Analysis 9:410-419, (1965).

4. A. A. Sterk, C. L. Marks, and W. P. Saylor, "Production Efficiencies of X-ray Emission Spectra by Proton Bombardment," Advances in X-ray Analysis 10:399-409, (1966).

5. J. W. Thatcher and W. J. Campbell, "Instrumentation for Primary and Secondary Excitation of Low-Energy X-ray Spectral Lines," U. S. Bureau of Mines, Report of Investigations 6689:1-29, (1965).

6. R. A. Mattson, "Some Measurements of Carbon K Excitation in a New Ultrasoft X-ray Spectrometer," Advances in X-ray Analysis 8:333-340, (1964).

7. F. Bernstein and R. A. Mattson, "Quantitative X-ray Emission Analysis of Magnesium through Fluorine with X-ray and Electron Excitation," Advances in X-ray Analysis 10:494-505, (1966).

8. R. W. G. Wyckoff and F. D. Davidson, "Windowless Tubes for X-ray Spectroscopy," Rev. Sci. Instr. 34:572, (1963).

9. D. W. Fischer and W. L. Baun, "Experimental Dispersing Systems
    for soft X-rays," Advances in X-ray Analysis 7:489-496, (1963).

10. B. L. Henke, "Sodium and Magnesium Fluorescence Analysis--
    Part I: Method," Advances in X-ray Analysis 6:361-376, (1962).

11. B. L. Henke, "Some Notes on Ultrasoft X-ray Fluorescence
    Analysis--10 to 100Å Region," Advances in X-ray Analysis
    8:269-284, (1964).

12. M. Canon, "A Polycrystalline X-ray Analyzing Crystal,"
    Advances in X-ray Analysis 8:285-300, (1964).

13. R. W. Gould, S. R. Bates and C. J. Sparks, "Application of
    the Graphite Monochromator to Light Element X-ray Spectroscopy,"
    Applied Spectroscopy 22:549-551, (1968).

14. R. C. Ehlert, "The Diffraction of X-rays by Multilayer
    Stearate Soap Films," Advances in X-ray Analysis 8:325-332,
    (1964).

15. B. L. Henke, "Application of Multilayer Analyzers to 15-150Å
    Fluorescence Spectroscopy for Chemical and Valence Band
    Analysis," Advances in X-ray Analysis 9:430-440, (1965).

16. F. D. Davidson and R. W. G. Wyckoff, "L and M X-ray Spectra
    in the Region 2-85Å," Advances in X-ray Analysis 9:344-353,
    (1965).

17. J. B. Nicholson and D. B. Wittry, "A Comparison of Gratings
    and Crystals in the 20-115Å Region," Advances in X-ray
    Analysis 7:497-511, (1963).

18. J. E. Holliday, "Determination of Electron Distribution and
    Bonding from Soft X-ray Emission Spectroscopy," Advances in
    X-ray Analysis 9:365-375, (1965).

19. H. Spielberg, "Characteristics of Flow Proportional Counters
    for X-rays," Advances in X-ray Analysis 10:534-545, (1966).

20. Personal communication from Nuclear Diodes, Inc. (1969).

21. R. Jenkins and S. van Gelder, "Determination of Low Concen-
    trations of Fluorine in Cement by X-ray Fluorescence Spectros-
    copy," Scientific and Analytical Equipment Bulletin 79.177/
    FS19, H. V. Philips, Eindhoven, The Netherlands (1969).

22. A. K. Baird and B. L. Henke, "Oxygen Determinations in
    Silicates and Total Major Elemental Analysis of Rocks by
    Soft X-ray Spectrometry," Anal. Chem. 37:727-729, (1965).

23. B. L. Henke, "X-ray Fluorescence Analysis for Sodium,
     Fluorine, Oxygen, Nitrogen, Carbon, and Boron," Advances in
     X-ray Analysis 7:460-488 (1963).

24. F. Bernstein, "Application of X-ray Fluorescence Analysis to
     Process Control," Advances in X-ray Analysis 5:486-499, (1962).

25. F. Claisse and C. Samson, "Heterogeneity Effects in X-ray
     Analysis," Advances in X-ray Analysis 5:335-354, (1961).

26. E. L. Gunn, "The Effect of Particles on Surface Irregularities
     on the X-ray Fluorescent Intensity of Selected Substances,"
     Advances in X-ray Analysis 4:382-400, (1960).

27. K. W. Madlem, "Matrix and Particle Size Effects in Analyses
     of Light Elements, Zinc through Oxygen by Soft X-ray
     Spectrometry," Advances in X-ray Analysis 9:441-455, (1965).

28. F. Claisse, "Accurate X-ray Fluorescence Analysis Without
     Internal Standard," Quebec Dept. of Mines P.R. 327, (1956).

29. E. E. Welday, A. K. Baird, and K. W. Madlem, "Silicate Sample
     Preparation for Light-Element Analyses by X-ray Spectrography,"
     Am. Mineral. 49:889-903, (1964).

30. H. J. Rose, I. Adler, F. J. Flanagan, "Use of La_2O as a Heavy
     Absorber in the X-ray Fluorescence Analysis of Silica Rocks,"
     U. S. Geol. Surv. Prof. Paper 450-B:80-82, (1962).

31. A. Volborth, "X-ray Spectrographic Determination of All Major
     Oxides in Igneous Rocks and Precision and Accuracy of a
     Direct Pelletizing Method," Nevada Bureau of Mines Rept.
     6:1-72, (1963).

32. R. C. Stanley, "Counting Statistics in X-ray Spectroscopy,"
     British Jour. Appl. Physics 12:503, (1961).

33. A. K. Baird and E. E. Welday, "Precision and Accuracy of
     Silicate Analyses by X-ray Fluorescence," Advances in X-ray
     Analysis 11:114-128, (1967).

34. W. L. Baun and D. W. Fischer, "The Effect of Valence and
     Coordination on K series Diagram and Nondiagram Lines of
     Magnesium, Aluminum, and Silicon," Advances in X-ray Analysis
     8:371-383, (1964).

35. G. R. Mallett, M. J. Fay, and W. M. Mueller (editors), "The
    Effect of Chemical Combination on X-ray Spectra: Open Dis-
    cussion," Advances in X-ray Analysis 9:393-397, (1965). See
    also papers preceding this discussion, pages 323-392.

36. B. L. Henke, R. L. Elgin, R. E. Lent, and R. B. Ledingham,
    "X-ray Absorption in the 2- to -200Å Region," Norelco
    Reporter XIV:112-134, (1967).

37. B. M. Gunn, "Matrix Corrections for X-ray Fluorescence
    Spectrometry by Digital Computer," Canadian Spectroscopy
    12:1-7, (1967).

38. R. J. Traill and G. R. Lachance, "A New Approach to X-ray
    Spectrochemical Analysis," Canadian Dept. Mines Technical
    Survey Paper, p. 57-64, (1965).

ADDED NOTE

In July, 1969, the Isomet Corporation announced two new analyzing crystals for the dispersion of long wavelength X-rays. One
offers special promise as a substitute for gypsum in analyses for
sodium and magnesium: sorbitol hexaacetate (SHA) with a 2d spacing
of 13.98Å. The other, rubidium acid phthalate (RAP) has a 2d
spacing of 26.121Å. It is claimed to have a resolution similar to
that of KAP, but approximately twice the reflectivity.

DETECTION AND SPECTROSCOPY OF LONG WAVELENGTH X-RAYS

William L. Baun

Air Force Materials Laboratory (MAYA)

Wright-Patterson Air Force Base, Ohio   45433

ABSTRACT

Instrumentation for spectrometry in the long wavelength X-ray region has been improved significantly in the last several years. In the area of excitation of soft X-rays, many changes have taken place. Secondary excitation source windows have been made much thinner and high power windowless and ultrathin windowed continuously pumped X-ray tubes have been developed. One source has been described which allows a choice of either primary or secondary excitation. Cold cathode techniques have been developed in which pressure control is achieved using an automatic pressure controller. Primary electron excitation has been revived and has become very important mainly due to the popularity of electron microbeam probe analysis. In the area of spectral dispersion, many improvements and innovations have been seen recently in both grating and crystal spectrometers. Organic single crystals have been grown having 2d spacings up to about $100\text{\AA}$ and soap films have been fabricated with 2d spacing as large as $165\text{\AA}$, allowing crystal spectrometers to be used to about $160\text{\AA}$. High resolution two-crystal spectrometers have been developed which allow soft X-ray emission and absorption measurements to be obtained on low atomic number materials. In the area of soft X-ray detection, designs and methods have been improved for the use of thin windowed flow proportional detectors. New detectors based on the magnetic electron multiplier (MEM) and channel electron multiplier (CEM) show promise for use in the $100\text{\AA}$ region and beyond. MEMs with replaceable and rotatable cathode structures have been constructed to allow optimatization for X-rays. Multiple CEM arrays and curved CEMs with funnel inputs intercept a large solid angle of the X-ray beam. Other detectors, such as one measuring the

charge distribution in a photoemissive surface, show promise for
future use in soft X-ray spectroscopy.

INTRODUCTION

In the period thirty to forty years ago, soft X-ray spectroscopy was of keen interest to a large group of research workers.
Then the technique lay fallow for twenty years or more with few
technological improvements appearing during that period. In recent
years, however, we have seen a resurgence of the technique and many
improvements have been made in instrumentation for soft X-ray spectroscopy. It is the purpose of this paper to briefly review some
recent advances in the areas of soft X-ray excitation, dispersion
and detection.

EXCITATION

Photon (X-Ray) Excitation

Only several years ago, the common secondary X-ray excitation
source was a side or end window X-ray tube having a tungsten target
and a window of 0.030 or 0.040 inch beryllium. These tubes are
quite efficient in the hard X-ray region but are extremely inefficient for exciting soft X-rays. The absorption of radiation by the
beryllium window make these tubes useless at wavelengths beyond about
3Å as can be seen from figure 1 which shows the X-ray transmission

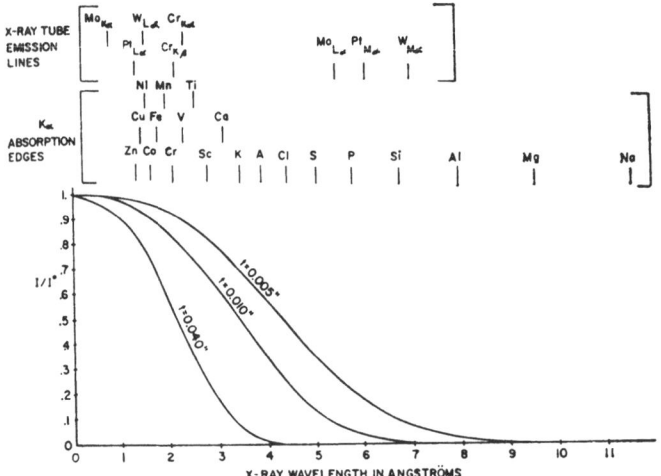

Figure 1. Absorption of Various Thicknesses of Beryllium[1]

characteristics of three different beryllium windows[1]. From this
work by Kirkendall and Varadi[1] which is summarized in this slide,
it was decided that a platinum target tube with a 0.005 inch beryllium window would be useful for excitation of a broad range of
elements by the use of PtL for excitation of hard X-rays and PtM
for excitation of low energy X-rays. In more recent work, Kirkendall, Varadi and Naill showed that a sealed X-ray tube having a
0.002 inch beryllium window could be used to excite radiation as
soft as carbon K in a non-dispersive system where high power is not
needed[1a]. Dual anode tubes and special anode tubes are useful for
improvement of sensitivity of specific elements. In systems using
a spectrometer (where high power is required) when it is desired to
work beyond 10Å or so, it is necessary to use a windowless or very
thin windowed demountable tube of the type described by Henke[2]. In
this Henke design, the tube is continuously ion pumped and is isolated from the low vacuum of the spectrometer by a sliding gate
valve as seen in figure 2[2]. When the spectrometer is pumped to
rough vacuum, an opening which is covered by a thin window is slid
into place over the port. Since the window only separates the high
vacuum side from the low vacuum side, it experiences little mechanical stress and can be extremely thin. Windows prepared by Henke
consisting of stretched polypropylene and a thin film of formvar
transmit about 60% of the radiation on the long wavelength side of
the carbon K absorption edge. The Henke tube is normally operated
at high power levels, with typical settings of 6KV and 330 ma[3]. A
somewhat similar high vacuum, high power tube was described by

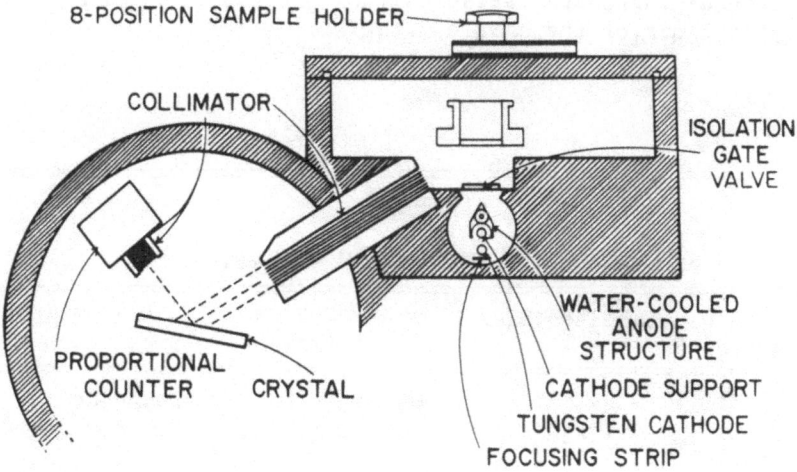

Figure 2. Ultrathin Window X-Ray Tube and Spectrograph[2]

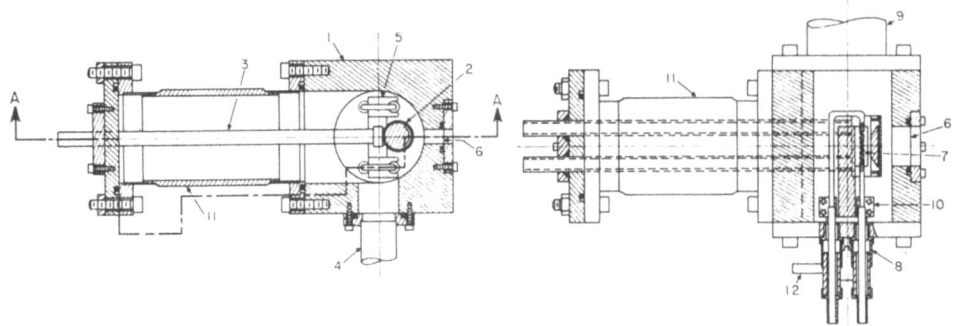

Figure 3. High Power Demountable X-Ray Tube[4]

Deslattes and Simson[4]. The geometry of their tube is such that both
the anode and window are shielded from evaporated filament material
as seen in the diagram in figure 3. The source utilizes a dual filament cathode assembly and the anode is a cylindrical surface cooled
very efficiently by a laminar flow of water resulting in operation of
this tube at a power level of over 3KW without anode distortion.
This large power input gives a very strong beam of X-rays which efficiently excite the characteristic X-rays in the sample without the
heating effects that are obtained with electron excitation. This is
a definite advantage in that unstable materials may be investigated
using secondary excitation. The target of these high power tubes is
chosen to most efficiently excite the sample material. Table I gives
tube targets, the wavelength of the exciting line and the elements
that are most efficiently excited by each target. Target materials
are chosen having characteristic lines just shorter than the characteristic lines of the sample material.

TABLE I

Demountable X-Ray Tube Targets

Target and Series	Wavelength, Å	K - Series Excitation of	L - Series Excitation of
Al(K)	8.34	Mg,Na,F	Se,As,Ge
Cu(L)	13.33	F,O,N,C	Ni,Co,Fe, etc.
Cr(L)	21.71	O,N,C	V,Ti,Sc, etc.
C(K)	44.8	B,Be	Cl,S,P, etc.

Direct Excitation

The increasing popularity of the electron microbeam probe and the introduction of commercial instruments using electron excitation has rapidly made direct excitation methods important. In the early history of X-ray spectroscopy, nearly all measurements were made using direct excitation. The resurgence of the technique has come about due to increased interest in analysis of low atomic number elements by X-ray techniques. A disadvantage of the technique is that direct excitation requires the use of high vacuum. In addition, decomposition of unstable compounds is a possibility under electron bombardment.

A convenient source in which one can switch back and forth between electron and X-ray excitation was described by Mattson and Ehlert[5]. Their source, which may be pumped separately with an ion pump, has been used for X-ray excitation, electron excitation, electron excitation of a heated wire, and electron excitation of a gas.

When using direct excitation, it is imperative that high vacuum techniques be used. It is well known that carbon contamination is a serious problem in soft X-ray spectroscopy. This carbon contamination is produced by a breakdown in the electron beam of the hydrocarbon vapors present from mechanical and diffusion pump oils, cleaning solvents, and rubber "O" rings. A source in which utmost care has been taken to lessen contamination is shown in figure 4[6].

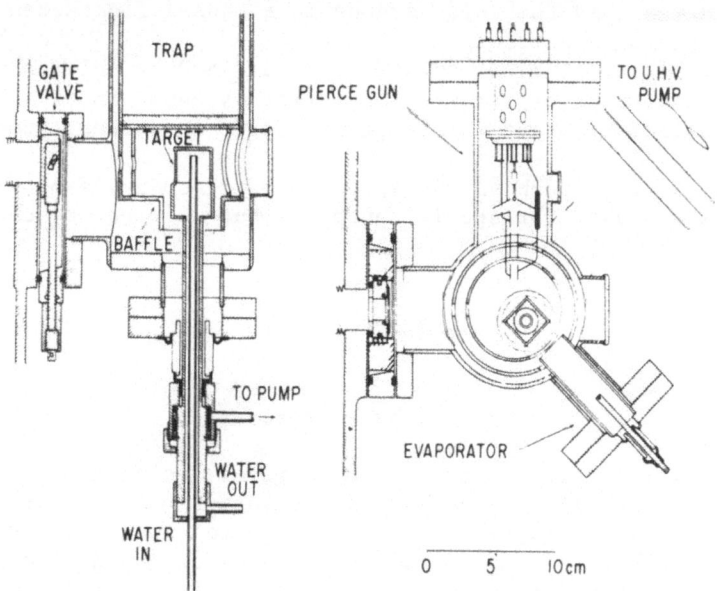

Figure 4. High Vacuum Direct Excitation Source

Flanges are sealed with gold "O" rings, the target is water-cooled and surrounded by a copper baffle cooled by a liquid nitrogen trap. Just opposite the Pierce-type electron gun is a small furnace for evaporating target material. After baking, the tube attains a vacuum of 3 x 10^{-9} torr, and operates at about 10^{-8} torr.

Other workers expend considerable effort to assure that the sample does not change and that carbon contamination is kept to a minimum. For instance, Liefeld maintains his samples at high temperatures when possible in a working vacuum of about 10^{-9} torr obtained by titanium-ion and sublimation pumping[7]. Watson, Dimond and Fabian use a source in which the sample may be rotated to a position where a bellows mounted scraper is used to clean the sample under vacuum[8]. Holliday pumps both spectrometer and X-ray tube with mercury vapor diffusion pumps and the tube is operated in 10^{-8} torr range. An arrangement is provided for cleaning the sample by ion bombardment after the sample has been outgassed in vacuum[9].

Protons and other heavy ions have been used to produce X-rays, but these techniques are not widely used, probably because of the complexity of generating, accelerating and controlling ion beams. Proton excitation is more efficient than electron excitation in low atomic number elements and the line radiation is virtually mono-chromatic since little white radiation is produced.

The use of cold cathode tubes removes the necessity for high vacuum but has seldom been used in recent work except by Wyckoff and Davidson[10] who used the cold cathode in a "quasi fluorescent" (X-rays + scattered electrons) mode, Hayasi and Hayasi[11] who obtained spectra from beryllium, boron and aluminum, and by Solomon and Baun[12] who used an automatic pressure controller as shown in figure 5 to obtain aluminum K X-rays with approximately one percent stability.

Other sources such as the spark, sliding spark, exploding wire, and plasma are used for special purposes but will not be considered here.

DISPERSION

Crystal Spectrometers

The crystal spectrometer is an extremely versatile instrument that can best be used in the soft X-ray region but is also usable in a portion of the ultrasoft region. Crystal spectrometers may take many forms depending on source geometry or the degree of sen-sitivity or resolution desired. Basically, crystal spectrometers are classified in figure 6 according to the configuration of the

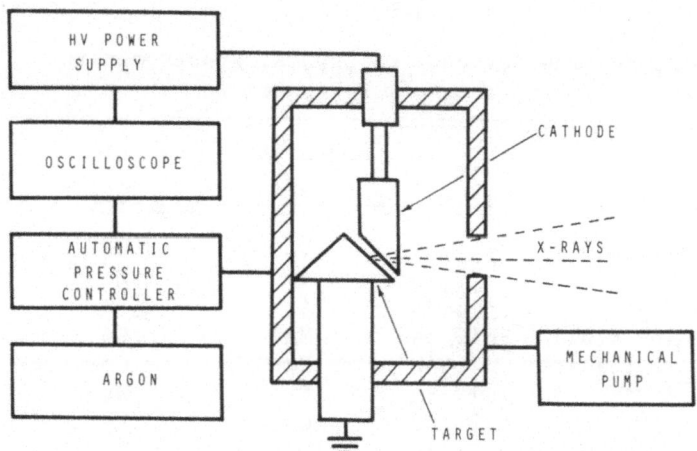

Figure 5.   Cold Cathode Soft X-Ray Source

crystal, i.e. "flat" for use with Soller slits, "cylindrical" for
use with line focus sources, and "toroidal" for use with point
focus sources. The two-crystal spectrometer generally uses plane
crystals to achieve very high resolution.

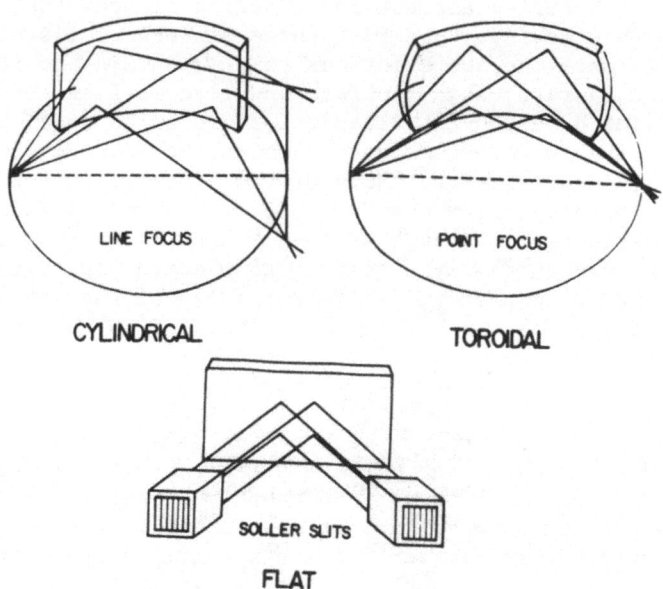

Figure 6.   Types of Crystal Spectrometers

TABLE II

Dispersing Devices for Crystal Spectrometers
In The Soft X-Ray Region

Crystal	2d, A	Long-Wavelength Limit K Series	Long-Wavelength Limit L Series
ADP (200)	7.50	Si	Rb
PET	8.75	Al	Br
EDDT	8.80	Al	Br
ADP (101)	10.64	Mg	Ge
BiTitanate	16.40	Na	Co
MICA	19.91	F	Mn
KAP	26.63	O	V
Clinochlore	28.39	O	Ti
Pb-C_{12}	70	B	Cl
Pb-C_{18}	100	B	S
Pb-C_{30}	165	Be	Si

Table II lists useful dispersing devices for crystal spectrometers in the region 5-165Å, along with the long wavelength limits
in terms of the K series element and L series element that can be
dispersed by each device. The last three dispersers listed in table
II are soap films. The abbreviations Pb-C_{12}, Pb-C_{18} and Pb-C_{30}
represent lead laurate, lead stearate and lead mellisate, respectively.
These soap films have proven valuable for use up to 165Å, but this
appears to be the longest (largest 2d) soap that can be prepared at
the current state-of-the-art. Many of these soap films may be rather
easily built up, especially near the C_{18} or stearate spacing. The
much shorter spacing films such as Pb-C_{12} and the much longer films
such as Pb-C_{30} are extremely difficult to prepare. Whatley recently performed a comprehensive investigation of the chemistry
of lead soap films which resulted in the ability to make films of
very high quality[13].

Commercial flat crystal spectrometers are being used on the
fringe of the soft X-ray region with sealed X-ray tubes as previously
discussed and in the soft and ultrasoft regions with ultrathin windowed tubes such as Henke uses[4]. However, when direct excitation is
used and high vacuum is required, it has generally been necessary to
design and build a spectrometer for the specific purpose that is to
be made of the equipment. Such a spectrometer system pumped by a
turbomolecular vacuum pump and utilizing semiautomatic step scanning is described in this volume[14].

Spectrometers of the two-crystal type have been used extensively in X-ray spectroscopy for studying high resolution spectra
and the perfection of crystals. Much of the early work using two-
crystal spectrometers is summarized by Compton and Allison[15]. A
recent design for a two-crystal vacuum monochromator was described
by Deslattes[16]. This spectrometer was designed for use at long
wavelengths and provides for coordinated rotations of both crystals,
source and detector, and allows continuous scans of up to 8°, 2θ
without beam displacement. The design goals were to use present
day bearing technology to obtain sensitive stable axes and to mechanically decouple the drive system from distortion normally associated
with evacuation of the spectrometer vacuum enclosure. Another very
successful design for a two-crystal spectrometer is described by
Liefeld[7]. Liefeld has equipped his spectrometer with high quality
potassium acid phthalate (KAP) crystals to study satellite emission
and self absorption effects in L emission spectra from first transition elements.

Grating Spectrometers

Grating spectrometer design technology is highly advanced,
since, before the discovery of long spacing crystals and the advent of soap films, they were the only devices that were used for
measurement of soft and ultrasoft X-rays. In this technique a
spectrum is obtained by reflecting the X-rays from a ruled grating
at glancing angles within the range of total reflection. Early
work and theory is covered by Compton and Allison[15]. Holliday[9]
has pioneered the modern use of the grating instrument in the
United States. Since about 1960, Holliday has used various types
of gratings to obtain high quality spectra over a very wide wavelength region. Recently, he has made improvements in his spectrometer to allow measurements (in the 10-20Å region) that are
comparable to some crystal spectrometer measurements. The basic
design of Holliday's instrument is seen in figure 7 which shows the
recent addition of a precision lead screw (B) which is driven from
a point directly beneath the diffraction grating. Holliday obtains high resolution by the use of a 3600 groove/mm grating with
a platinum surface and a one degree blaze angle and 25μ slits.
A traveling microscope is used to align the slits and the grating
in the horizontal plane, providing better precision than obtained
earlier with a telescope arrangement. Watson, Dimond and Fabian[8]
have developed the instrument shown in the spectrometer layout in
figure 8, where a moiré-fringe system for angular measurement has
been mounted outside the vacuum chamber, with the reading head
attached to the detector carriage arm through a one inch diameter
shaft that also forms the pivot for adjustments of the grating
and source slit. These workers use a 600 line/mm platinum coated
2°4' blazed grating with a radius of curvature of 99.88 cm. A

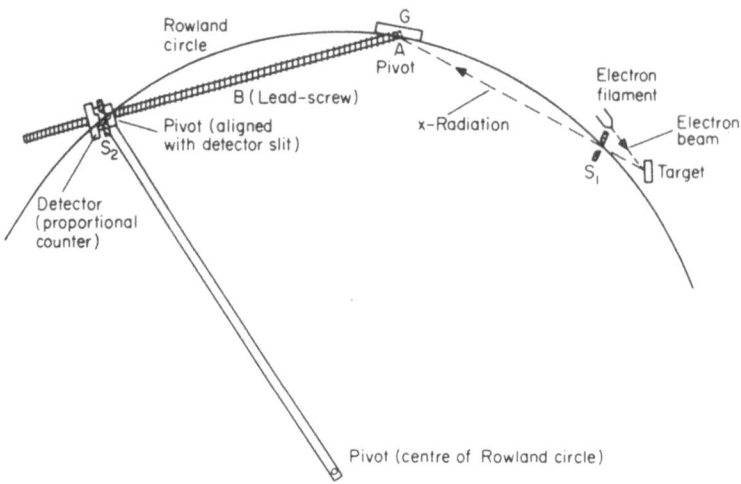

Figure 7.   Grating Spectrometer by Holliday[9]

novel design of a grating spectrometer was recently described by
Sawada[17] in which a limited scanning range can be achieved without
moving the detector or grating.   Rather, this design uses a fixed
counter and a moving X-ray source.   Weich[18] uses a grating spectro-

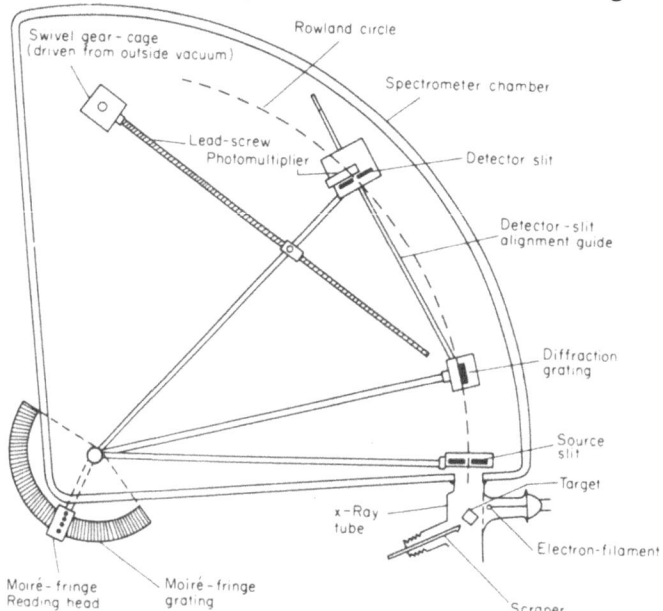

Figure 8.   Grating Spectrometer Utilizing a
Moiré Fringe Reading Head[8]

meter in which the source is mounted in a separate source inside
the main chamber, rather than flanged onto the outside as is commonly done. As a result, vacuum as good as 7×10^{-9} torr may be
achieved in the source while the pressure in the outer (spectrometer)
chamber is in the 10^{-6} torr range. From Weich's work on carbon
contamination and decomposition of oxides, he concludes that most
of the emission bands previously recorded with photographic film
will have to be remeasured.

DETECTION

When X-ray photons are absorbed in matter, a number of interactions can occur which leave the absorbing material changed in some
way. The nature of the interaction is dependent on the chemical and
physical state of the absorbing material and the energy of the photon. Phenomena which have been used for soft X-ray detectors include ionization (and emission of electrons), photo chemical and
photoluminescent effects.

Photographic Plate

The photochemical effect on silver halide salts occurs throughout the soft X-ray region, which accounts for the widespread use of
the photographic plate in early work. Due to the high absorption of
soft and ultrasoft X-rays, the emulsion must necessarily be very
thin, with virtually no gelatin in the emulsion. The photosensitive
material may be vacuum evaporated, so that no gelatin is necessary,
but this is not a routine operation. The photographic plate suffers from the disadvantage that the information must be extracted
from the plate by a microdensitometer. Since the densitometer
measures only photographic blackening which is not always linearly
related to intensity, the operator is forced into making complicated
corrections and calibration curves for quantitative work.

Windowless Electron Multipliers

Windowless electron multipliers have been successfully used
in the ultrasoft X-ray region with a grating instrument. Work by
Crisp[19] in Australia using an open photomultiplier is especially
notable. Some open photomultipliers suffer the disadvantage that
their gain changes with age and sometimes exposure to air poisons
these devices. Another possible disadvantage to the open electron
multiplier is that it requires high vacuum ($<10^{-6}$ mm Hg) for
operation.

Accelerated Electron-Scintillation Counter

The accelerated electron-scintillation counter is based on the principle of electron emission from a metal surface irradiated with low energy photons and the subsequent acceleration and scintillation counting of these secondary electrons. This detector system has been described by Grodski[20] who points out that the advantage of this detector is that the electrons are given sufficient energy to be counted by a convential closed scintillation crystal-photomultiplier assembly.

Magnetic Electron Multiplier (MEM)

The magnetic electron multiplier is a device using a continuous surface of high resistance semiconductor material under the action of a crossed electric and magnetic field to obtain high gain electron multiplication. The device, as shown in figure 9, was originally developed for ion detection in the time-of-flight mass spectrometer. More recently, MEM's have been built which are much smaller and more convenient for use with soft X-rays. Also, the MEM has been built with a demountable cathode assembly to allow the use of an optimum photocathode material for the region of interest. To obtain best results and highest efficiency, it is also desirable to be

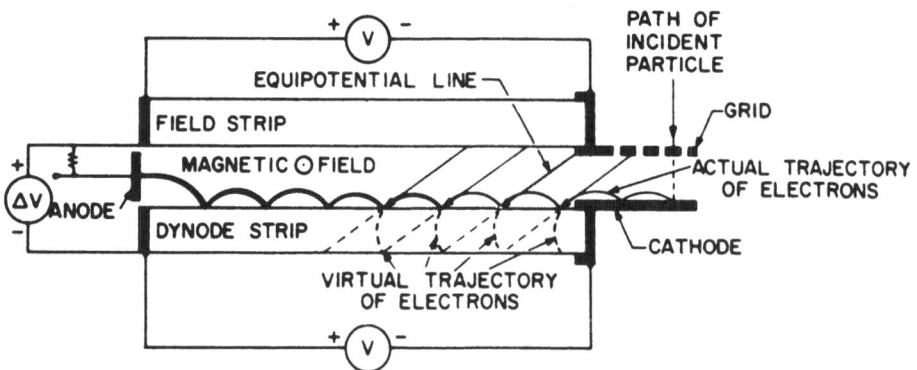

Figure 9. Magnetic Electron Multiplier (MEM)

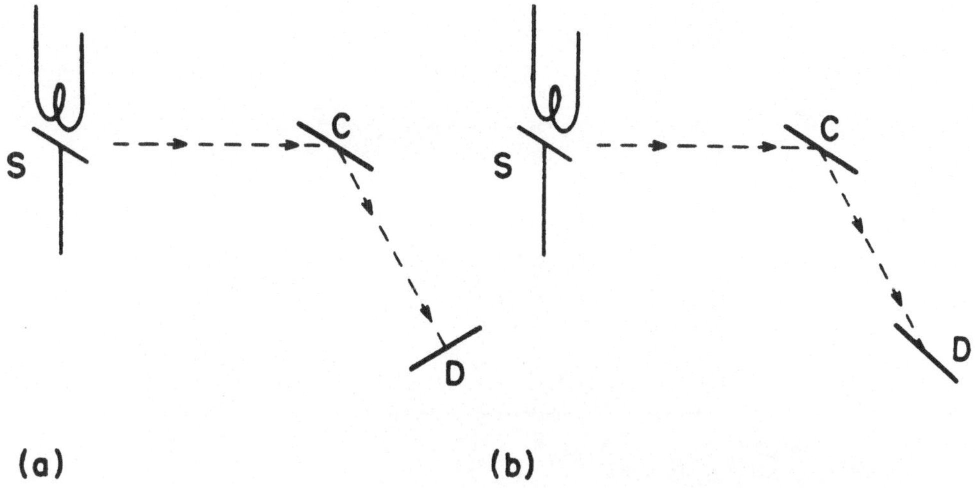

(a)   **(b)**

Figure 10. Use of Photocathode at Normal
(a) and Grazing Incidence (b)

able to vary the angle of incidence of the beam to the photocathode.
In the original MEM, it was necessary to use it in the configuration shown in (a), figure 10. Used in normal incidence as in (a)
with a tungsten photocathode the device, in much of the soft X-ray
region, was very inefficient. Used at grazing incidence as shown
in (b) with the optimum photocathode efficiences can approach
80-90%. The commercial MEM manufactured by Bendix (Model M-306) is
used by Watson, Dimond, and Fabian[8] for detection of soft X-rays
with a grating spectrometer.

Channel Electron Multiplier (CEM)

The continuous channel photomultiplier, developed by Bendix
and others, is in the form of a hollow glass tube with a highly
resistive inner surface. It has an internal bore diameter, d,
(see figure 11) of typically a few tenths of a millimeter, and has
a length to bore diameter ratio of about 50. A potential difference of about 2000V is maintained between the ends, causing a
current on the inner surface. The initial electrons are generated
either from the coating at the input end or from an external photocathode material. The emitted electrons cascade down the tube as
shown in figure 11 producing gains of 10^5 or more. The multiplier
can be curved to allow higher gain and minimize feedback. Curved
channels may be fitted with a cone shaped input to increase the
efficency and acceptance angle. Multiple arrays of straight chan-

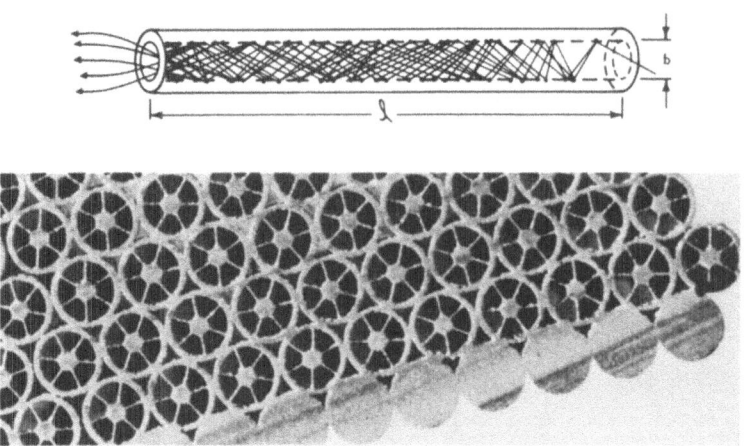

Figure 11.   Channel Electron Multiplier (CEM), Top
             and Multiple Array, Bottom

nels may be fabricated for special purposes such as making the
active detection element just the same size as the receiving slit
on a spectrometer. A cross section of a multiple array (Spiraltron)
is shown at the bottom of figure 11[21]. The complete detector
including anode, spiraltron matrix, focusing electrode and demountable photocathode is shown in figure 12[21]. Typical operating voltages are shown on each of the elements.

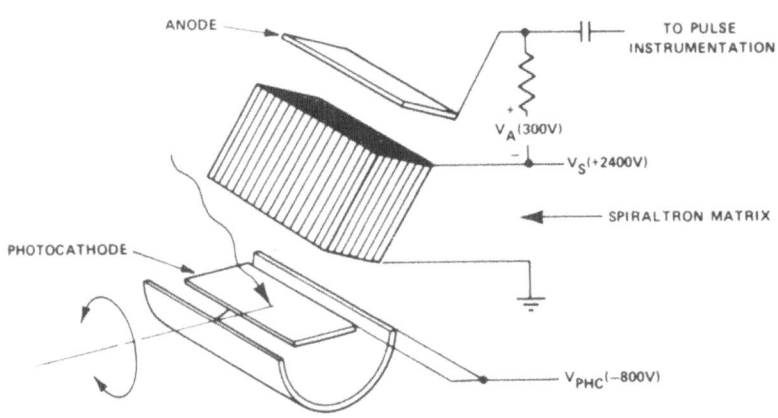

Figure 12.   Complete Multiple CEM[21] Array
             for Soft X-Ray Detection

Flow Proportional Detector

The most widely used detector for soft X-ray spectroscopy at this time is the flow proportional counter. An example of such a detector is shown in another paper in this volume[14]. This detector is supplied with commercial X-ray spectrometers and electron microbeam probes as well as being used in most custom equipment installations in the United States. The major difficulty with the flow proportional detector is found in installing and maintaining a leak free window between the counter and the spectrometer vacuum, especially when the counter is operated at or near a pressure difference of one atmosphere. The best solution appears to be the use of thin double formvar films supported on copper or nickel electroform grids having 60-70% transmission. The preparation of such films is described by Henke[4]. It is also useful to evacuate the counter as the spectrometer chamber is evacuated and then operate the counter at a pressure of only 100-200 mm Hg above the chamber pressure. Caruso and Kim[22] have shown a unique method of stretching very thin polypropylene windows which need no supporting grid. Deslattes, Simson and LaVilla[23] have designed a gas density controller to gain stabilization in flow proportional counters. Such a controller is a necessity when operating the detector at reduced pressure in a vacuum chamber in order to minimize gas gain changes.

A wide choice of flow gases are available for use with proportional counters. Mixtures such as argon-methane, helium-butane, and helium-isobutane are used as well as pure gases such as methane and carbon dioxide. The choice of gas and the pressure at which it is used will depend on the energy of the radiation to be detected, among other things. A recent publication by Henke and Lent[24] shows curves of pressure vs wavelength for 95% absorption in several gases in the 10-50$\overset{\circ}{A}$ region.

A Detector of the Future

A new detector for recording spectral information in the form of a spatial distribution of charge has been described by Boksenberg[25]. The charge image is first recorded on the surface of an insulating layer by direct photoemission as shown in figure 13. The photon beam strikes the active surface where electrons are liberated and collected on the mesh which is just above the photoemissive surface. The positive charges left behind on the surface constitute the recorded image, corresponding to the photographic image in a conventional spectrograph.

To extract the information from the charge distribution on the plate, one uses a process similar to the vibrating reed (dynamic

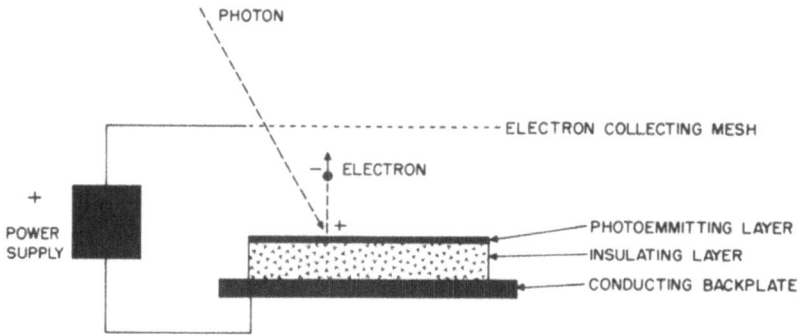

Figure 13. Recording the Image[25]

capacitor) electrometer except that here Boksenberg uses a fine
closely scanning, vibrating, conducting probe connected to a current
amplifier as shown in the schematic of figure 14.   During a reading
the probe never touches the surface and may remain over any spot on
the plate sufficiently long to integrate however long it is necessary
to obtain the desired measuring statistics.   When the operator is

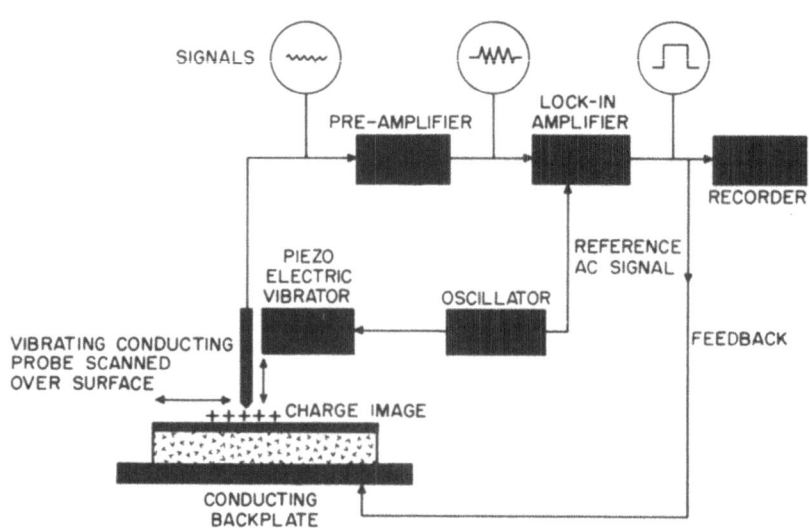

Figure 14. Reading the Image[25]

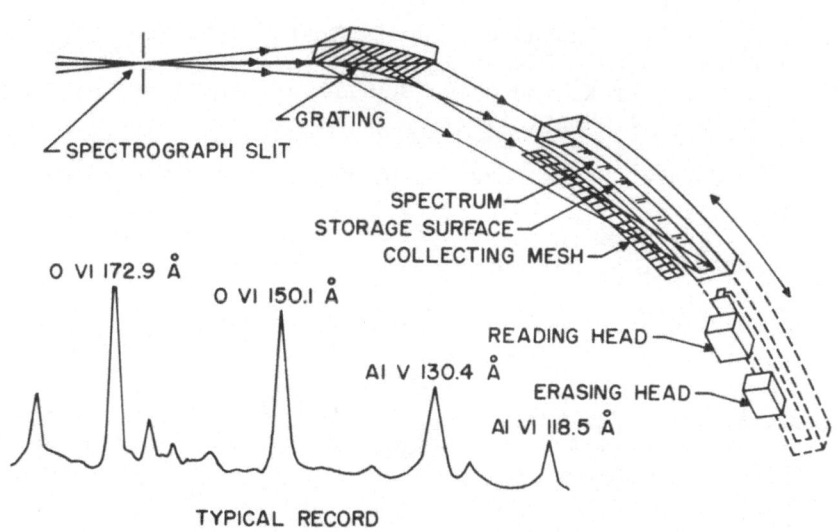

Figure 15. The Use of Charge Distribution
Detector in a Grating Spectrograph[25]

finished with the image, he may very simply erase it by spraying
the surface with electrons, which replaces the electrons lost during
initial exposure. An example of the way in which such a detector
would be used is seen in figure 15 where the photographic plate of
a grating spectrometer has been replaced by the charge distribution
device. The sample spectrum is from an air filled flash tube source.
The many advantages of this detector are obvious. When used with a
grating spectrograph, all of the spectrum is simultaneously recorded
and no moving parts are necessary. The probe then is scanned over
the charge. It should be possible to scan very accurately and precisely around the Rowland circle.

REFERENCES

1. T. D. Kirkendall and P. F. Varadi, "A Sealed Thin Window X-Ray
    Tube for Light and Heavy Element Excitation," Anal. Chem. 39,
    1342 (1967).

1a. T. D. Kirkendall, P. F. Varadi, and R. F. Naill, paper #5,
    "Energy Dispersion Analysis Using Sealed X-Ray Excitation
    Sources," Abstracts, Pittsburgh Conference on Analytical Chem-
    istry and Applied Spectroscopy, Cleveland, Ohio, March, 1969.

2. B. L. Henke, "Techniques of Low Energy X-Ray and Electron
    Physics: 50 to 1000 eV Region," Proceedings of 2nd Symposium
    on Low Energy X- and Gamma Sources and Applications (ORNL-II
    C-10, p. 523, Sept. 1967).

3. B. L. Henke, "Some Notes on Ultrasoft X-Ray Fluorescence
    Analysis - 10 to 100Å Region," in Mueller, Mallett and Fay,
    Eds., Advances in X-Ray Analysis, Vol. 8, Plenum Press, New
    York, 1965, p. 269-284.

4. R. D. Deslattes and B. Simson, "Demountable High Power Source
    for Soft X-Ray Region," Rev. Sci. Inst. 37, 753 (1966).

5. R. A. Mattson and R. C. Ehlert, "The Application of a Soft
    X-Ray Spectrometer to Study the Oxygen and Fluorine Emission
    Lines from Oxides and Fluorides," in Mallett, Fay and Mueller,
    Eds., Advances in X-Ray Analysis, Vol. 9, Plenum Press, New
    York, 1966, p. 471-486.

6. T. Sagawa, "Soft X-Ray Emission and Absorption Spectra of
    Light Metals, Alloys, and Alkali Halides," in D. J. Fabian,
    Ed., Soft X-Ray Band Spectra, Academic Press, London, 1968,
    p. 29-43.

7. R. J. Liefeld, "Soft X-Ray Emission Spectra at Threshold
    Excitation," in D. J. Fabian, Ed., Soft X-Ray Band Spectra,
    Academic Press, London, 1968, p. 133-149.

8. L. M. Watson, R. K. Dimond, and D. J. Fabian, "Soft X-Ray
    Emission Spectra of Magnesium and Beryllium," in D. J. Fabian,
    Ed., Soft X-Ray Band Spectra, Academic Press, London, 1968,
    p. 45-58.

9. J. E. Holliday, "Soft X-Ray Emission Bands and Bonding for
    Transition Metals, Solutions and Compounds," in D. J. Fabian,
    Ed., Soft X-Ray Band Spectra, Academic Press, London, 1968,
    p. 101-131.

10. F. O. Davidson and R. W. G. Wyckoff, "L and M Spectra in the
    Region 2-85Å," in Mallett, Fay, and Mueller, Eds., Advances
    in X-Ray Analysis, Vol. 9, Plenum Press, New York, 1966,
    p. 344-354.

11. T. Hayasi and Y. Hayasi, "Long Wavelength X-Ray Spectra of
    Beryllium, Boron, and Aluminum Emitted from Gas-Ion X-Ray
    Tube," Sci. Rep., Tohoku Univ., 50, no. 4, 228 (1967).

12. J. S. Solomon and W. L. Baun, "Performance and Applications
    of a Cold Cathode Soft X-Ray Source," Technical Report
    AFML-TR-69-41, June, 1969.

13. T. A. Whatley, "Improved Diffraction Elements for Analysis in the Soft X-Ray Region," paper #6, Abstracts, Pittsburgh Conference on Analytical Chemistry and Applied Spectroscopy, Cleveland, Ohio, March, 1969.

14. W. L. Baun and E. W. White, "A Vacuum Spectrometer for Studying the Chemical Effect on Soft X-Ray Spectra," in Advances in X-Ray Analysis, Vol. 13, Plenum Press, New York, 1970.

15. A. H. Compton and S. K. Allison, X-Rays in Theory and Experiment, 2nd Ed., D. Van Nostrand Co., Inc., New York, 1935.

16. R. D. Deslattes, "Two Crystal, Vacuum Monochromator," Rev. Sci. Instr. $\underline{38}$, 616 (1967).

17. M. Sawada and K. Tsutsumi, "Source Scanning Type X-Ray Grating Spectrometer," J. Appl. Phys. $\underline{40}$, 1950 (1969).

18. G. Weich, "Investigations of the $L_{II,III}$ X-Ray Emission Band of Al with a New Concave Grating Spectrograph," Z. fur Physik $\underline{193}$, 490 (1966).

19. R. S. Crisp, "The Soft X-Ray L_{23} Emission Spectrum of Mg From Solid and Evaporated Targets," Aust. J. Phys. $\underline{11}$, 449 (1958).

20. J. J. Grodski, "Photoelectric Detectors for 10-400 Å Photons," Norelco Reporter XIV, 107 (1967).

21. T. A. Somer and P. W. Graves, "Spiraltron Matrices as Windowless Photon Detectors for Soft X-Ray and Extreme UV," 15th Nuclear Science Symposium, Oct. 1968.

22. A. J. Caruso and H. H. Kim, "Method of Stretching Polypropylene for Use as Soft X-Ray Detector Windows," Rev. Sci. Instr. $\underline{39}$, 1059 (1968).

23. R. D. Deslattes, B. G. Simson, and R. E. LaVilla, "Gas Density Stabilizer for Flow Proportional Counters," Rev. Sci. Instr. $\underline{37}$, 596 (1966).

24. B. L. Henke and R. E. Lent, "Some Recent Work in Low Energy X-Ray and Electron Analysis," in Advances in X-Ray Analysis, Vol. 12, Plenum Press, New York, 1969.

25. A. Boksenberg, "A New Detector for Spectrometry," Anal. Chem. $\underline{41}$, #7, p. 87A (1969).

QUANTITATIVE X-RAY FLUORESCENCE ANALYSIS WITH VARIABLE TAKE-OFF ANGLE

H. Ebel

Institut für Angewandte Physik, University of Technology and Ludwig Boltzmann-Institut für Festkörperphysik, Vienna, Austria

ABSTRACT

A solution of the intensity equation of X-ray fluorescence analysis is possible either by the assumption of an effective wavelength or by a variation of the take-off angle. When the intensity of the fluorescence radiation is measured as a function of the take-off angle, the extrapolation of this curve against a take-off parallel to the flat sample surface offers a simple evaluation of the intensity equation. Since the influence of the secondary excitation is also taken into account the composition and the mass per unit area of samples can be investigated as well as mass absorption coefficients and mean wavelengths of the polychromatic primary radiation.

INTRODUCTION

The fluorescence intensity does not only depend on the constitution of the sample, the intensity distribution of the primary radiation, the mass absorption coefficients and the mass per unit area but as well upon the angles between the normal on the sample surface and the primary beam and take-off direction, respectively. In this paper a discussion of the informations obtained by a variation of the take-off angle is given.

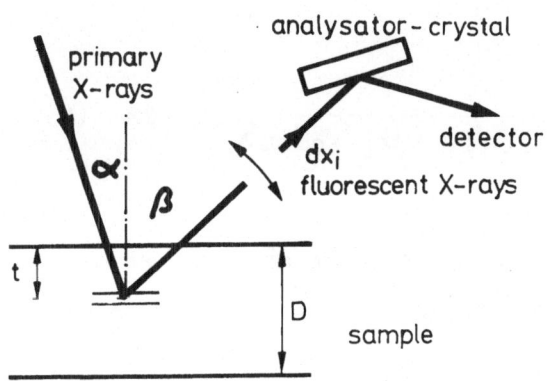

primary X-rays

analysator-crystal

detector

dx_i
fluorescent X-rays

α

β

t

D

sample

Figure 1. Schematic representation of the experiments with variable take-off angle.

THEORY

According to figure 1 the fluorescence intensity of the element i due to a thin layer of thickness dt is

$$dx_i = x_\lambda \varepsilon_{\lambda i}(z_i/n_i)d\lambda dt \, \exp\left[-(\mu_{\lambda c}/\cos\alpha + \mu_{ic}/\cos\beta)t\right] \qquad (1)$$

The ratio z_i/n_i can be replaced by

$$z_i/n_i = c_i \, \rho_c/\rho_i \qquad (2)$$

The conversion factor $\varepsilon_{\lambda i}$ in common notation[1,2] is given by

$$\varepsilon_{\lambda i} = qW_i p_i \mu_{\lambda i}(S_{Ki} - 1)/S_{Ki} \qquad (3)$$

By integration and application of the first mean value theorem of integral calculus follows:

$$x_i = c_i(\rho_c/\rho_i)\frac{1 - \exp\left[-(\mu_{\overline{\lambda}c}/\cos\alpha + \mu_{ic}/\cos\beta)D\right]}{\mu_{\overline{\lambda}c}/\cos\alpha + \mu_{ic}/\cos\beta}$$

$$\times \int_{\lambda_o}^{\lambda_{Ki}} x_\lambda \varepsilon_{\lambda i}d\lambda \qquad (4)$$

Bulk materials are characterized by $D \to \infty$. So the intensity caused by a pure element is

$$x_i = \frac{1}{\mu_{\overline{\lambda}i}/\cos\alpha + \mu_{ii}/\cos\beta} \int_{\lambda_o}^{\lambda_{Ki}} x_\lambda \varepsilon_{\lambda i}d\lambda \qquad (5)$$

and that of an alloy

$$x_i = c_i(\rho_c/\rho_i) \frac{1}{\mu_{\bar{\lambda}c}/\cos\alpha + \mu_{ic}/\cos\beta} \int_{\lambda_0}^{\lambda Ki} x_\lambda \varepsilon_{\lambda i} d\lambda \quad (6)$$

For pure element films

$$x_i = \frac{1 - \exp\left[-(\mu_{\bar{\lambda}i}/\cos\alpha + \mu_{ii}/\cos\beta)(m/F\rho_i)\right]}{\mu_{\bar{\lambda}i}/\cos\alpha + \mu_{ii}/\cos\beta}$$
$$\times \int_{\lambda_0}^{\lambda Ki} x_\lambda \varepsilon_{\lambda i} d\lambda \quad (7)$$

is obtained and for multicomponent films we have

$$x_i = c_i(\rho_c/\rho_i) \frac{1 - \exp\left[-(\mu_{\bar{\lambda}c}/\cos\alpha + \mu_{ic}/\cos\beta)(m/F\rho_c)\right]}{\mu_{\bar{\lambda}c}/\cos\alpha + \mu_{ic}/\cos\beta}$$
$$\times \int_{\lambda_0}^{\lambda Ki} x_\lambda \varepsilon_{\lambda i} d\lambda \quad (8)$$

Since the value of film density is smaller in comparison to bulk material of the same composition[3], D has been replaced in equations (7) and (8) by

$$D = m/F\rho_{if} \quad (9)$$

and the linear absorption coefficients of the film material by

$$\mu_{\bar{\lambda}if} = (\mu_{\bar{\lambda}i}/\rho_i)\rho_{if} \qquad \mu_{iif} = (\mu_{ii}/\rho_i)\rho_{if}$$
$$\mu_{\bar{\lambda}cf} = (\mu_{\bar{\lambda}c}/\rho_c)\rho_{cf} \qquad \mu_{icf} = (\mu_{ic}/\rho_c)\rho_{cf} \quad (10)$$

The unknown mean wavelengths $\bar{\lambda}$ in the equations (5) to (8) as well as the excitation integral can either be determined or eliminated by a variation of the take-off angle and the use of intensity ratios.

DETERMINATION OF MEAN WAVELENGTH BY
VARIATION OF THE TAKE-OFF ANGLE

The mean wavelengths in equations (5) to (8) are defined differently and therefore not identical[4]. These mean wavelengths also differ from the effective wavelength $\tilde{\lambda}$ defined by

$$x_i = c_i(\rho_c/\rho_i) \frac{x_{\tilde{\lambda}} \varepsilon_{\tilde{\lambda}i} (\lambda_{Ki} - \lambda_o)}{\mu_{\tilde{\lambda}c}/\cos\alpha + \mu_{ic}/\cos\beta} \qquad (11)$$

As an example we will show the determination of $\bar{\lambda}$ for a pure element sample. It is necessary to know x_i as a function of ß. Since only x_i, $dx_i/d\beta$ and $\mu_{\bar{\lambda}i}$ depend on ß, differentiation of equation (5) gives the following differential equation for $\mu_{\bar{\lambda}i}/\rho_i$

$$d(\mu_{\bar{\lambda}i}/\rho_i)/d\beta + [d(\ln x_i)/d\beta](\mu_{\bar{\lambda}i}/\rho_i)$$

$$+ (\mu_{ii}/\rho_i)(\cos\alpha/\cos\beta)d[\ln(x_i/\cos\beta)]/d\beta = 0 \quad (12)$$

The solution is

$$\mu_{\bar{\lambda}i}/\rho_i = -\cos\alpha(\mu_{ii}/\rho_i)\left[(dx_i/d\beta)_{\beta=90°}/x_i + 1/\cos\beta\right] (13)$$

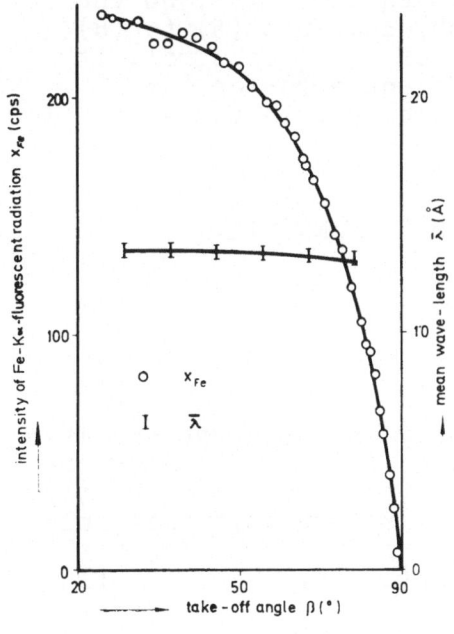

Figure 2. FeKα-fluorescence intensity x_{Fe} as a function of the take-off angle ß for pure iron. $\bar{\lambda}$ is the according mean wavelength of the primary unfiltered Cu-radiation (36 kV, 20mA).

Figure 2 shows a measured $x_i(\beta)$-curve for a bulk sample of pure iron, excited by unfiltered Cu-radiation (36 kV, 20 mA). The $\bar{\lambda}$-curve has been calculated from equation (13).

DETERMINATION OF MEAN WAVELENGTH
BY THE USE OF THIN FILMS

In the case of thin films having a thickness $D < 2000$ Å, equation (7) can be approximated by

$$x_i = (m/F\rho_i) \int_{\lambda_o}^{\lambda_{Ki}} x_\lambda \varepsilon_{\lambda i} \, d\lambda \tag{14}$$

The intensity ratio

$$r_i = x_i(14)/x_i(5) \tag{15}$$

can be used to calculate $\bar{\lambda}$ according to

$$\mu_{\bar{\lambda}i}/\rho_i = \left(r_i F/m - \mu_{ii}/\rho_i \cos\beta\right)\cos\alpha \tag{16}$$

DETERMINATION OF THE EXCITATION INTEGRAL

The knowledge of $\mu_{\bar{\lambda}i}/\rho_i$ enables us to find values of the excitation integral from equation (5). On the contrary to the mean wavelength, equations (5) to (8) give the same values for the excitation integral. It depends neither on the take-off angle nor on the composition and mass per unit area but on the excitation conditions.

MASS PER UNIT AREA FOR AN ELEMENT FILM
(FILM RADIATION)

In the intensity ratio

$$r_i = x_i(7)/x_i(5) \tag{17}$$

besides the unknown mass per unit area the unknown mean wavelengths according to equations (5) and (7) are contained. Therefore m/F can be derived from equation (17) by introduction of an apparent mass per unit area only

$$(m/F)_a = -\left[\cos\beta/(\mu_{ii}/\rho_i)\right] \ln(1 - r_i) \tag{18}$$

m/F is than described as limes as follows

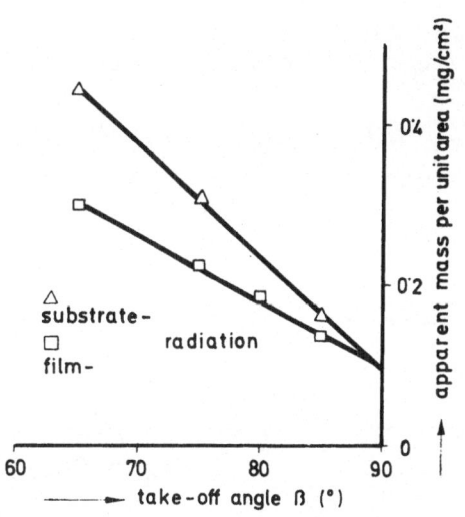

Figure 3. Extrapolation
of apparent mass per
unit area against ß = 90°
for a thin Au-film on a
Cu-substrate.

$$\lim_{\substack{ß \to 90^{\circ}}} (m/F)_a = m/F \qquad\qquad (19)$$

Theoretical and experimental investigations have shown
that the dependence $(m/F)_a = f(ß)$ is linear in a good
approximation and can also be extrapolated linearly
against ß=90°. Equation (18) contains measurable or
known quantities only. Therefore it is necessary to
measure r_i for different take-off angles. The value of
extrapolation is - as can be gathered from equation (19)-
identically with the unknown mass per unit area of the
film. In this way we are able to determine m/F of an
element film by comparison with bulk material without
the use of reference films. The accuracy of the film
radiation method is better than 5%. Figure 3 shows the
result for an Au-film deposited on copper[5].

MASS PER UNIT AREA OF AN ELEMENT FILM
(SUBSTRATE RADIATION)

The fluorescence intensity of a bulk alloy sample
coated with a film of the element j can be obtained
from equation (6) as follows

$$x_i = \frac{(c_i \rho_c / \rho_i)\, \exp\left[-(\mu_{\bar{\lambda}j}/\cos\alpha + \mu_{ij}/\cos ß)m/F\,\rho_j\right]}{\mu_{\bar{\lambda}c}/\cos\alpha + \mu_{ic}/\cos ß} \int_{\lambda_0}^{\lambda_{Ki}} x_\lambda \varepsilon_{\lambda i}\, d\lambda \quad (20)$$

By comparison with the uncoated sample of the same
composition an intensity ratio can be defined again

$$r_i = x_i(20)/x_i(6) \tag{21}$$

The solution of the problem is comparable to that of
the film radiation method. An apparent mass per unit
area is introduced

$$(m/F)_a = \frac{-\cos\beta}{\mu_{ij}/\rho_j}\ln r_i \tag{22}$$

The dependence $(m/F)_a = f(\beta)$ is linear again and can
also be extrapolated linearely to $\beta=90$. $(m/F)_a = f(\beta)$
follows from a measurement of $r_i(\beta)$. Since

$$\lim_{\beta\to 90^{\circ}} (m/F)_a = m/F \tag{23}$$

is valid, the extrapolated value is the unknown m/F.
Figure 3 shows the results of the substrate radiation
method applied to the same Au-film deposited on a
copper substrate. The accuracies of the two methods
can be compared. According to our experiences the
methods are applicable in a thickness range from 50 to
20000 Å.

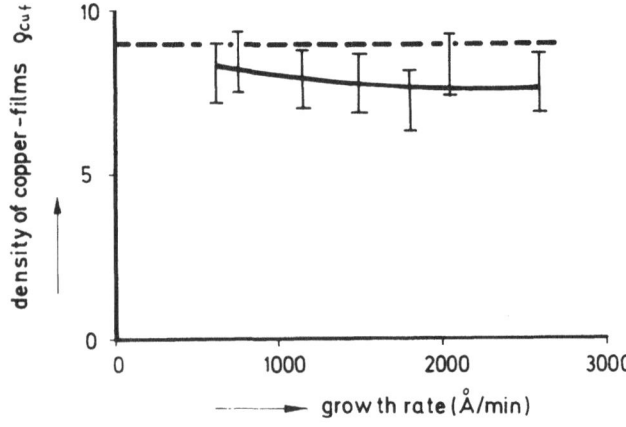

DENSITY OF FILMS

Figure 4. Density
of thin Cu-films
(1500 Å) as
 function of
growth rate. The
broken line
represents the
density of bulk
copper for
comparison.

Together with m/F determined by the methods above and the thickness D determined by optical interferometry, for instance, the density of films can be calculated from equation (9). Figure 4 shows the density of Cu-films with thickness of about 1500 Å as a function of the growth-rate. It is an apparent effect that the density of the film is smaller than that of the bulk material, as already mentioned.

MASS PER UNIT AREA AND CONCENTRATION
OF THIN MULTICOMPONENT FILMS

For this method thin element films with known mass per unit area $(m/F)_i$ are necessary. The fluorescence intensity of a thin multicomponent film is similar to that of a thin element film given by

$$x_i = (c_i/\rho_i)(m/F)_c \int_{\lambda_0}^{\lambda_{Ki}} x_\lambda \varepsilon_{\lambda i} d\lambda \qquad (24)$$

From that the intensity ratio r_i is

$$r_i = x_i(24)/x_i(14) = c_i(m/F)_c/(m/F)_i \qquad (25)$$

If the multicomponent film consists of n different elements, n different intensity ratios are to be measured. Together with

$$\sum_{i=1}^{n} c_i = 1 \qquad (26)$$

n+1 linear equations for the n unknown concentrations c_i and the unknown mass per unit area $(m/F)_c$ are obtained. This method needs of course no variation of the take-off angle but the $(m/F)_i$-values of the reference films can be found either by the film radiation method or by the substrate method as described above.

Investigations on binary Cu-Ag-films with different concentrations and thickness up to 10000 Å showed deviations if compared with the results of the chemical analysis within 3 wt%[6].

MASS ABSORPTION COEFFICIENTS

The film methods allow the determination of mass absorption coefficients as soon as the mass per unit area values of the film are known[7]. Mass absorption

coefficients can be obtained again with an accuracy
of 5%. Since thin absorber films are within the range
of the film methods, large values of mass absorption
coefficients can be easily investigated.

CONCENTRATION OF BULK ALLOY SAMPLES

A comparison with bulk element samples is
necessary. The intensity ratio r_i is given by

$$r_i = x_i(6)/x_i(5)$$

$$= c_i \frac{\mu_{ii}/\rho_i}{\mu_{ic}/\rho_c} \times \frac{1 + (\mu_{\bar{\lambda}i}/\mu_{ii})(\cos\beta/\cos\alpha)}{1 + (\mu_{\bar{\lambda}c}/\mu_{ic})(\cos\beta/\cos\alpha)} \qquad (27)$$

The intensity ratio according to $\beta=90^\circ$ is
independent of $\bar{\lambda}$ once more

$$\lim_{\beta \to 90^\circ} r_i = c_i(\mu_{ii}/\rho_i)/(\mu_{ic}/\rho_c) \qquad (28)$$

and enables us to calculate the unknown concentrations c_i
together with

$$\mu_{ic}/\rho_c = \sum_{j=1}^{n} c_j(\mu_{ij}/\rho_j) \qquad (29)$$

If the sample consists of n different elements, n- 1
different values of $\lim r_i$ have to be determined and
together with equation (26) a system of n linear
equations for the n unknown concentrations c_i is
obtained. Again the measurement of r_i-values for a
number of different take-off angles is sufficient.
Here $r_i(\beta)$ is not linear, as can be seen for example
from figure 5, showing the intensity ratios of different
Cu-Ni-alloys.
A very useful approximation of equation (27) for
a numerical extrapolation is

$$r_i = (1 + A\cos\beta) \lim_{\beta \to 90^\circ} r_i \qquad (30)$$

This approximation was used for computer evaluations.
The concentrations of Cu-Ni-alloys found by the
measurement of $CuK\alpha$ - and $CuK\beta$-radiation agree
within 1 wt%.

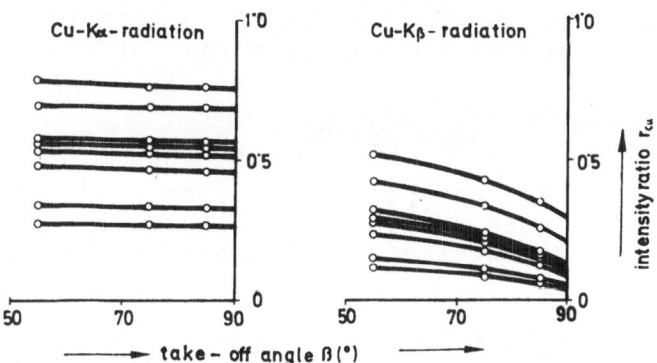

Figure 5. Intensity ratio r_{Cu} as a function
of ß. Since the m.a.c. μ_{CuCu}/ρ_{Cu} and
μ_{CuNi}/ρ_{Ni} are comparable for CuKα-radiation,
in this case only, a linear dependence $r_{Cu}(\beta)$
results.

SECONDARY EXCITATION

Intensity formulas taking into account this effect
have already been described by several authors[1], [8], [9].
Using these the limes of the intensity ratio - quotient
of the intensities of the alloy and the pure element -
at ß=90° can be approximated to

$$\lim_{\beta \to 90^\circ} r_i = \lim_{\beta \to 90^\circ} r_i(27) + \sum_{j=1}^{n} c_i c_j K_j \qquad (31)$$

The constants K_j vanish either for j=i or if the
element j can not excite the i-fluorescence radiation.
They are determined by the aid of reference samples
of known composition. The composition of binary reference
samples can be found by the method described above,
applied to the radiation of the heavier element. For
elements far distant in the periodic system, the influence of secondary excitation is comparable small.

Equation (31) is an approximation since the
quantities K_j are assumed to be independent of concentration. This is not the case but measurements[4]
of K_{Fe} in the binary system Fe-Cr gave variations
less than 10%. Since this afflicts a correction term
only, the approximation is still good. The equations

(28) and (31) are the fundaments of quantitative analysis
of alloys by means of the variable take-off technique.

SOME REMARKS

A quantitative analysis without reference to
chemical analysis is possible and also values of mass
per unit area are obtained without the use of reference
films.

The samples have to be homogeneous and with a
smooth surface. The detection limit compared with
the usual methods is one order of magnitude worse.
The accuracy depends on the knowledge of mass absorption
coefficients and independence of mass absorption coefficients and conversion factors of the bond is assumed.
This is at least not the case for light elements.

For our experiments we used a standard diffraction
unit in the way outlined by figure 1 and also a standard
XRFA-unit together with special sample holders. The time
necessary for a whole quantitative analysis lies in the
range of 10 to 60 minutes according to the problem. Now
we develop a goniometer especially suited for the variable
take-off technique in order to improve the accuracy and
to simplify the experiments.

NOTATION

x_i intensity of i-fluorescence radiation

x_λ intensity distribution of the primary radiation

μ_{ij} linear absorption coefficient
         (i denotes the radiation, j the absorbing
         element or c an obsorbing alloy)

μ_{ij}/ρ_j mass absorption coefficient

ρ_i, ρ_c density of the element i or an alloy

c_i i-element concentration (wt%)

ACKNOWLEDGEMENTS

This research has been supported by the "Ludwig Boltzmann-Gesellschaft". I wish to thank Prof. Dr. Malissa, Prof. Dr. Wagner and Dr. Pell for the chemical analysis of our samples and Prof. Dr. Lihl, Prof. Dr. Tomiser, Dipl.Ing. Ebel, Dr. Skalicky and Dr.Wagendristel for discussions and their assistance.

REFERENCES

1   T. Shiraiwa and N. Fujino, "Application of theoretical calculations to X-ray fluorescent analysis", The Pittsburgh Conference on Analytical Chemistry and Applied Spectroscopy (1967).

2   R. Jenkins and J.L. de Vries, "Practical X-ray Spectrometry", Philips Technical Library, 19 - 25 (1967).

3   H. Ebel, A.Wagendristel and H. Judtmann, "Messungen der Dichte dünner Aufdampfschichten", Zeitschr. f.Naturforschg 23a: 1863 - 1864 (1968).

4   W. Meizer, "Die effektive Wellenlänge in der quantitativen Röntgenfluoreszenzanalyse", Dissertation, Technische Hochschule Wien (1969).

5   P.R. Perez, "Zur röntgenfluoreszenzanalytischen Absolutbestimmung der Dicke ebener dünner Elementschichten", Diplomarbeit, Technische Hochschule Wien (1966).

6   W.C. Ho, "Quantitative Röntgenfluoreszenzanalyse ebener dünner Legierungsschichten", Dissertation, Technische Hochschule Wien (1969).

7   H. Ebel, "Röntgenfluoreszenzanalytische Bestimmung von Massenschwächungskoeffizienten", Spektrometertagung, Bern (1968).

8   E. Gillam and H.T. Heal, Some problems in the analysis of steels by X-ray fluorescence", Brit. J.Appl.Phys. 3, 353 (1952).

9   J. Sherman, "The theoretical derivation of fluorescent X-ray intensities from mixtures" Spectrochim.Acta 7, 284 (1955).

EVALUATION OF SOFT AND HARD SCATTERED X-RAYS AS

AN INTERNAL STANDARD FOR LIGHT ELEMENT ANALYSIS

David L. Taylor and George Andermann

University of Hawaii

Honolulu, Hawaii 96822

ABSTRACT

In the research described, the use of scattered x-rays has been successfully applied as an internal standard for the analysis of calcium in aqueous specimens containing a wide range of matrix components. In addition to the demonstration of the utility of scattered x-rays for light element analysis, some comments are offered on the fundamental aspects of this technique, since to date the method has not been explained thoroughly. The present research represents a continued effort to determine the fundamental importance of various parameters intrinsic to any collection of atoms undergoing scattering, such as the Rayleigh-Compton ratio, the scattering angle, the wavelength utilized, and the presence or absence of discontinuities in the matrix absorption coefficient. It has been concluded that large values of the scattering angle coupled with short wavelength tend to yield improved internal compensation. The results also indicate that for light matrices the Compton component of the scattered continuum is of particular importance in achieving good internal standardization for matrix effects.

INTRODUCTION

The purpose of this paper is to demonstrate that the technique of using scattered radiation as an internal standard in x-ray fluorescence analysis is a method which has its basis in well understood physical principles.

As such, this technique, which Hasler, Kemp, and Andermann[1] first reported at the 1955 Denver X-Ray Conference, and Andermann and Kemp[2] later described in detail, cannot properly be termed as being identical to the peak-to-background ratio method which is well known in optical emission spectroscopy[3]. Experimental evidence indicates that the compensation achieved by use of Andermann's method cannot be understood if one adopts the very restrictive assumption that only the scattered background at or near the analytical emission peak should be used as an internal standard. The results also indicate that the internal standardization obtained by this method is dependent upon the wavelength as well as the Rayleigh-Compton ratio of the scattered radiation. Theoretical treatments of the technique to date[2,4,5] have not clearly indicated that such dependences should be important.

In the original report by Hasler, et al., the fluorescent intensity I_F of the analyte was compared to that of the radiation scattered by the specimen I_{SR} at a wavelength of 0.6 A. By means of this scheme of analysis good results were obtained for the determination of lead and nickel in complicated ore matrices containing variable quantities of iron, cobalt, copper, and zinc. Later[2,6], it was also reported that variations in physical properties of the specimen such as particle size and surface roughness, as well as variations in spectroscopic factors such as tube-to-sample distance and tube kV and mA, could be compensated for by the I_F/I_{SR} method.

The above results indicate that the ratio I_F/I_{SR} is less sensitive to atomic number variations in specimen composition than is I_F alone. This fact has been successfully exploited by numerous investigators in determining a large number of elements ranging from sulfur to plutonium and encompassing almost every type of specimen. However, as was explicitly pointed out by Campbell and Brown in the 1964 Fundamental Analytical Reviews[7], the utilization of the scattered radiation method has been mainly empirical. The selection of the scattered wavelength has usually been arrived at taking into account practical features such as intensity, resolution, spectral interference, etc. In general, the spectroscopist could select a wavelength occupied by a coherently scattered primary tube line, a Compton band on the long wavelength side of a tube line, or the bremsstrahlung "white radiation" continuum (figure 1.) Examples of all these approaches have been reported. However, these efforts have not had the benefit of a complete unifying theoretical justification based upon physical principles.

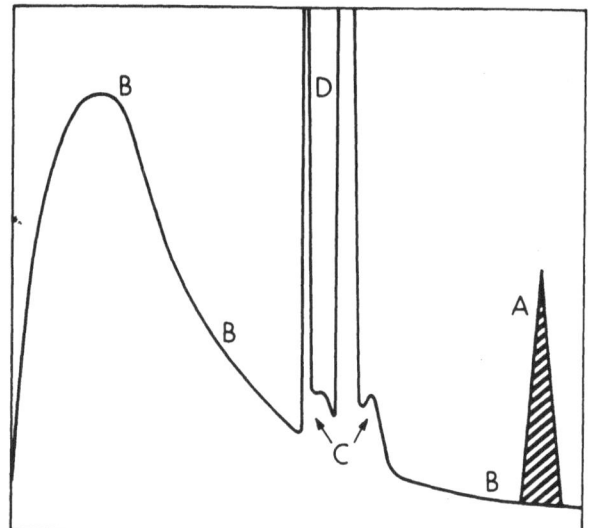

Figure 1

Radiation emitted

(A) and scattered

(B,C,D) by a specimen

Fundamental studies of x-ray scattering, particularly as
the phenomenon relates to spectroscopic analysis, have
been called for as a necessary addition to our present
understanding.[7]

A brief, purely qualitative explanation of the scattered radiation method was first attempted by Andermann
and Kemp[2], whereby I_F and I_{SR} were shown to have atomic
number dependences to the (-4) and (-3 to -2) powers,
respectively. The atomic number dependence of the ratio
I_F/I_{SR} is thus (-1 to -2). There was nothing in the explanation that called for the inclusion or exclusion of
any particular wavelength of scattered radiation as the
prospective internal standard. It is of interest, however, that a scattered wavelength relatively far removed
from the analyte emission peaks was used successfully for
internal standardization in this case. Kalman and
Heller[4], in describing the analysis of traces of transition metals in light rock matrices, derived complete
quantitative expressions for the relevent portions of the
fluorescent and scattered spectrum. Specifically, they
dealt with the Compton and Rayleigh scatter at a given
wavelength, the analyte fluorescent radiation, and the
reflection and scattering from the analyzing crystal.
The foregoing were combined into an overall expression
for the intensity observed on a given fluorescent peak
and at a given wavelength of scattered background, such
that a nomograph comparing I_F to I_{SR} yielded a family of
intersecting straight lines whose slopes were a function
of analyte concentration. This was an approximation of

the restrictive peak-to-background approach, in that the scattered wavelengths were chosen relatively near to the analyte emission peaks. Champion, et al.[5], dealt with the analysis of traces of strontium in light matrices, in which the intensity of the $SrK\alpha$ emission and that of the underlying continuum represented I_F and I_{SR}. A quantitative expression for I_F/I_{SR} was derived in terms of geometric and instrumental constants, sample composition factors, matrix absorption coefficients, and Compton and Rayleigh scattering factors. This approach, moreso than in the preceding case, characterizes the viewpoint that the use of scattered radiation as an internal standard is essentially a peak-to-background ratio method.

There are two very significant theoretical considerations basic to this argument. One is that the expression for I_F/I_{SR} has no wavelength uniqueness. Although the statement was derived specifically for the case of $SrK\alpha$ and the underlying background at 0.88 A, there is nothing inherent in the theoretical derivation that excludes the use of the expression for other analyte peaks and their corresponding backgrounds. Internal standardization should be possible, then, for the analysis of other elements in variable light matrices by measuring the intensity of the characteristic emission as well as that of the underlying scattered background.

Another point is that the intensity of scattering at any given wavelength was assumed to be strictly additive. This assumption states that the intensity of the Rayleigh as well as the Compton scatter from the entire assemblage of atoms composing the specimen is expected to be the sum of all the individual elemental effects. This can be surmised from the simplified[5] equation (1):

$$I_{SR} = K \left\{ \sum \frac{\dfrac{C_i F_i^2}{A_i}}{\mu_C} + \sum \frac{\dfrac{C_i (Z_i - f_{ni}^2) R}{A_i}}{\mu_I} \right\} \quad (1)$$

where

K   = instrumental, etc., constant

R   = Breit-Dirac recoil factor

A_i, Z_i, C_i = Atomic weight, atomic number, weight fraction of element i

F_i, f_{ni} = coherent and incoherent scattering
             factors of element i

μ_c, μ_I = coherent and incoherent matrix mass
             absorption coefficients, including geo-
             metric factors for incoming and outgoing
             radiation

Σ = summation over all individual elemental
             effects

The scattering factors in the above equation determine the character and intensity of the observed scatter;
they are functions of the angle of scattering 2θ and the
wavelength λ. The numerical value of the parameter (sin
θ)/λ is of primary importance in scattering phenomena.
Compton and Allison[8] point out that for values of (sin
θ)/λ greater than approximately 0.5 the total observed
scatter from a substance can generally be considered to
be simply the sum of the coherent and incoherent components. At relatively large values of (sin θ)/λ the observed coherently scattered x-rays from n atoms have the
same total intensity as that produced by the same number
of atoms that scatter x-rays independently of each other[9].

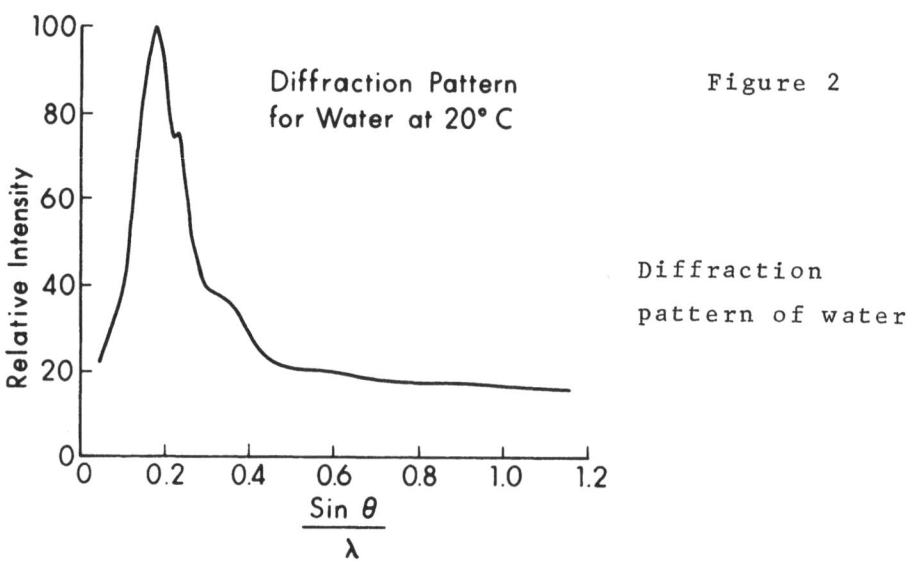

Figure 2

Diffraction
pattern of water

In the case of incoherent scattering, the total intensity can usually be taken to be the sum of the intensities
scattered by each separate atom, independent of $(\sin \theta)/\lambda$,
the intensities tending toward a constant proportional
to atomic number when $(\sin \theta)/\lambda$ is large[10].

The above assumption regarding additivity is inherently limited, however, since the observed scatter from
a medium is not strictly additive in all spectral regions.
Namely, the diffraction properties of matter become apparent when the combination of scattering angle and wavelength produce a relatively small numerical value of
$(\sin \theta)/\lambda$; for instance, when 2θ is 90 degrees and λ is
greater than 2 A. This is illustrated in figure 2, which
shows the diffraction pattern for water.

In evaluating the fundamental aspects of scattering,
therefore, it is important to realize that the intensity
and character of the x-ray scatter are predictable functions of the scattering angle, wavelength, and the atomic
number constituency of the medium undergoing scattering.
In the case of x-ray fluorescence spectrometry with conventional instruments, the scattering angle is generally
fixed in the region of 90 degrees. The spectroscopic
variable, therefore, is wavelength. As the selected
wavelength becomes shorter, the observed scatter from any
given element becomes more incoherent (figure 3.) As the
atomic number of the specimen becomes small, the percentage of Compton scatter at any given wavelength increases
(figure 4.)

The true analytical significance of the above physical factors governing x-ray scattering generally has
been overlooked in practical spectrometry, although in
addition to the original work by Andermann and his coworkers there have been a number of examples of the use
of a short scattered continuum wavelength as an internal
standard. The spectral conditions (emission peak and
scattered wavelength) utilized in those instances were:
$SK\alpha$ and $CaK\alpha$ with 0.45 A [11]; $CaK\alpha$ with 0.5 A [12]; $TiK\alpha$
with 0.53 A [13]; and the $K\alpha$ of the elements managanese
through gallium with 0.5 A [14]. Also, there are some examples of the selection of predominately Compton scatter
as an internal standard. The conditions in these cases
were: $SK\alpha$ and $CaK\alpha$ with $WL\alpha 1$ Compton at 1.51 A [15], and
$PuL\alpha$ with $MoK\alpha$ Compton at 0.75 A [16]. Heavy element analysis by means of gamma-ray excitation has been carried
out using Compton scatter from the primary radiation
beam [17, 18]. The common feature of the above examples is
that in each case the scattered wavelength was selected

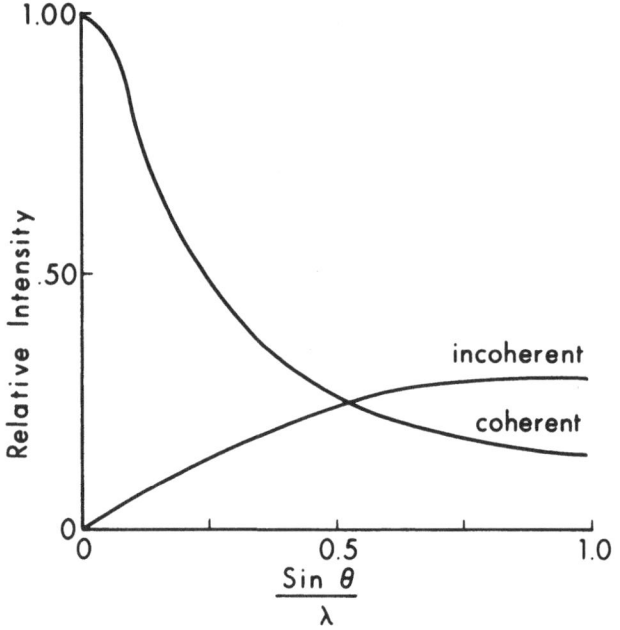

Figure 3

Distribution of
coherent and
incoherent radiation
scattered by a
given element.

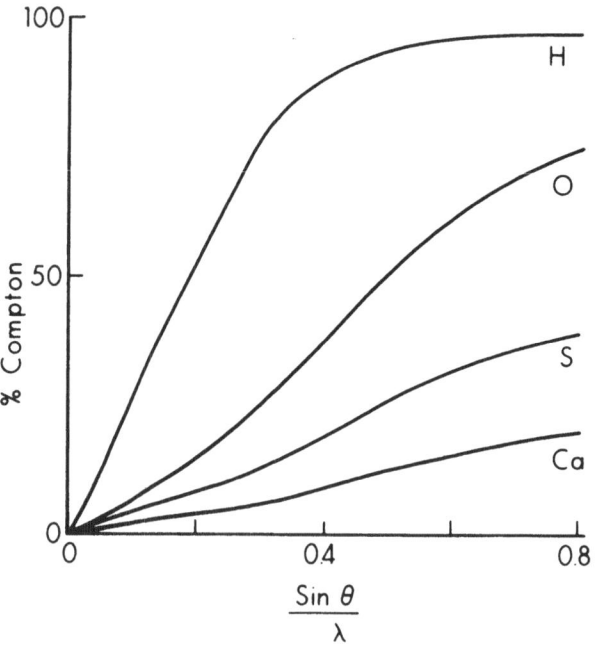

Figure 4

Compton contribution
to the total scatter
as a function of
atomic number.

pragmatically, with no argument put forth to justify that selection on clearly theoretical grounds.

EXPERIMENTAL

Experiments were devised and carried out which were intended to answer questions of fundamental importance in establishing the physical basis of the scattered radiation technique. It was decided to test the peak-to-background approach at a relatively long wavelength for a light matrix; namely, to use CaKα and the scattered background near 3.36 A for internal standardization in the analysis of calcium in variable aqueous matrices. In the case that these results showed very poor compensation at 3.36 A, a number of longer and shorter wavelengths would be tried to determine whether a consistent relation obtains between the wavelength used and the quality of internal standardization achieved.

Aqueous solutions of calcium ion were prepared to provide concentrations ranging from zero to 100 parts per million. To test the simple absorption effect, potassium chloride was added to the solutions to produce concentrations ranging from zero to 16.1 weight percent. Another series of solutions containing calcium, cupric and strontium ions was prepared to test effects due to absorption edges. The final calcium concentrations ranged from zero to 125 ppm, the strontium from zero to 4.1 weight percent, and the cupric from zero to 3.3 weight percent.

A North American Electric Corp. Universal Vacuum Spectrometer was fitted with an OEG-50 chromium target x-ray tube which was operated at 45 kV and 20 mA. A helium atmosphere was used in the presence of the liquid specimens contained in standard aluminum sample holders with 6-micron bottom windows. The fluorescent and scattered background intensities at various wavelengths were measured, with at least ten thousand counts accumulated in virtually every case; accumulation of greater numbers of counts was not considered necessary for reasons of overall instrument stability.[19]

RESULTS

The experimental results show that the scattered background at or near the analyte emission peak does not act as an effective internal standard in all situations.

In the case of calcium solutions containing variable quantities of potassium chloride, the absorption effect is very pronounced, since these ions produce a significant alteration in the sample mass absorption coefficient at 3.36 A. Figure 5 is a plot of calcium concentration versus CaK_α intensity and versus I_F/I_{SR} at several scattered wavelengths. The calcium versus CaK_α plot shows the extent to which the CaK_α intensity is depressed by the potassium chloride. When the ratio I_F/I_{SR} is plotted on the abscissa, various degrees of standardization result. When a scattered wavelength (3.2 A) near the CaK_α peak is used, very little improvement in the array of data is seen. If 1.54 A is used, the results appear somewhat better, while at 0.75 A they are very good, especially considering the range of matrix variation. Similar experiments performed over a range of longer scattered wavelengths extending up to 5.1 A showed no useful internal standardization.

In the preceding set of experiments the major matrix components--hydrogen, oxygen, chlorine, and potassium-- each have absorption edges at wavelengths greater than CaK_α. It may be expected that the presence of discontinuities in the sample mass absorption coefficient between the emission peak and the scattered wavelength used for internal standardization would have a negative effect on the analytical results; e.g., the presence of copper and strontium as minor matrix constituents would have a serious effect upon the intensity of the scattered continuum, as well as upon the intensity of the CaK_α. Figure 6 shows the results of experiments with solutions containing calcium, strontium, and cupric ions. The plot of calcium concentration versus CaK_α shows the resulting matrix effect. Internal standardization afforded by using the ratio I_F/I_{SR} at 3.2 A appears very poor. Similarly, the results at 0.75 A are very poor, showing the serious depression of the scatter at that wavelength. However, very good internal standardization is provided by the use of the Compton scatter at 2.32 A attending the very intense coherently scattered CrK_α primary tube line.

CONCLUSIONS AND DISCUSSION

The experimental indications are that the degree of internal standardization obtained is related in a very fundamental way to the wavelength of scattered radiation used. The results imply that, in the absence of sample mass absorption discontinuities, the use of a comparatively short scattered wavelength yields relatively

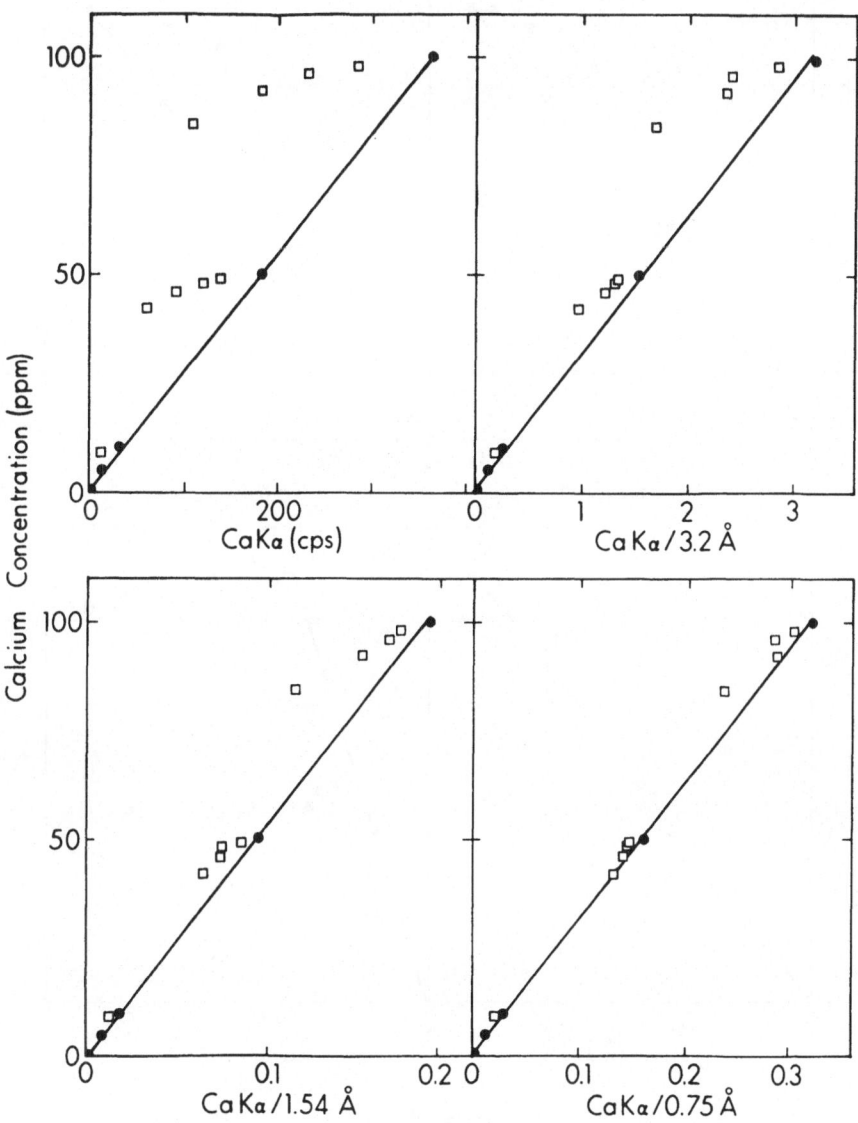

Figure 5:   Internal standardization achieved
at various scattered wavelengths.

LEGEND: Ca (ppm)   KCl (wt%)

●   0-100   - - - -

□   0-100   0 - 16.1

D. Taylor and G. Andermann

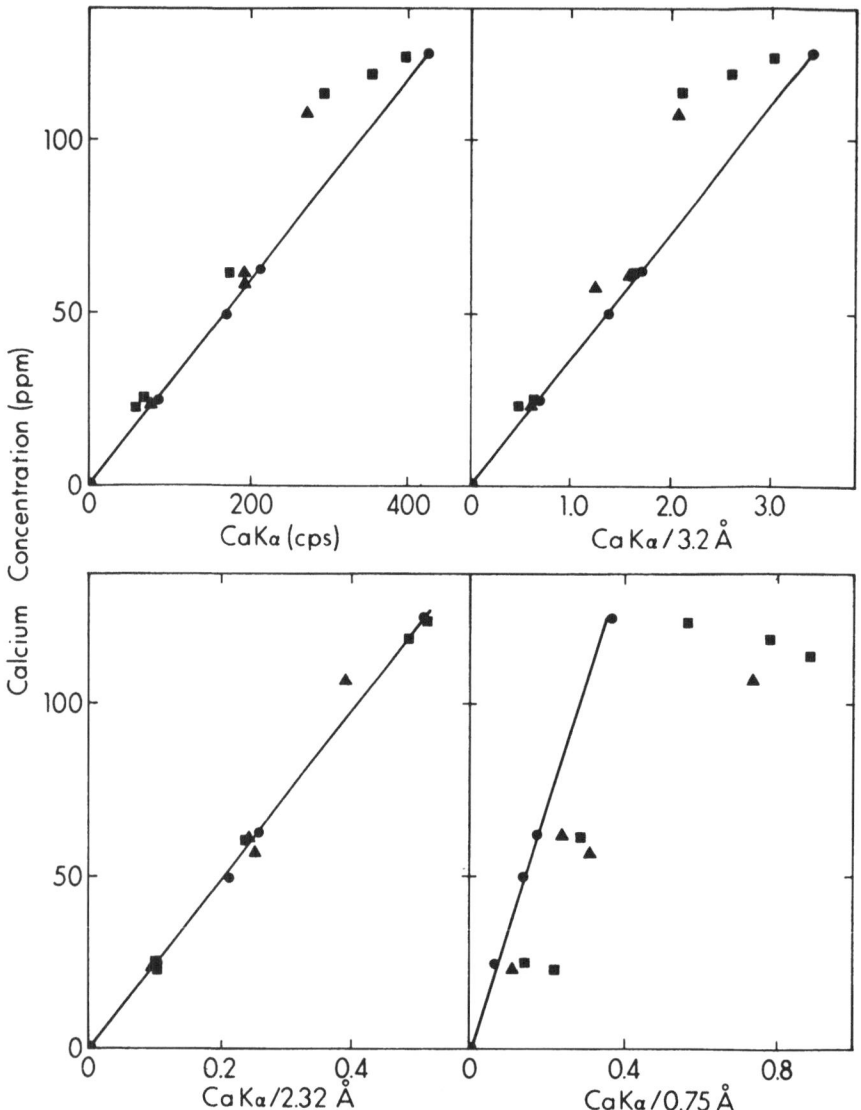

Figure 6:   Internal standardization achieved
at various scattered wavelengths.

LEGEND: Ca (ppm)   Cu (wt%)   Sr (wt%)

   ●   0-125   ---   ---

   ■   0-125   ---   0 - 4.1

   ▲   0-125   0 - 3.3   ---

superior results. Another indication is that when the observed intensity has a large Compton component a relatively long scattered wavelength (2.32 A) serves as a very effective internal standard. Keeping these results in mind, it is possible to explain the internal standard efficacy of a short wavelength or a Compton band in terms of fundamental principles.

An internal standard works most perfectly when it accurately relates the atomic number constituency of the specimen to the spectroscopic measurements. In this sense, scattered radiation containing a large Compton component is a better internal standard than is coherent scatter, since Compton scatter is a more linear function of the elemental composition of the scattering medium[20]. Hence 0.75 A x-rays scattered from an aqueous medium would be expected to act as an effective internal standard, as would the Compton band associated with an intense tube line.

It has been pointed out previously that the diffractive properties of the scattering medium become pronounced at values of $(\sin \theta)/\lambda$ less than approximately 0.5. There is a possibility, therefore, that interference effects can exist under certain conditions, and also that these effects may be variable as the elemental composition changes from one sample to the next. This should certainly be taken into account in the analysis of relatively low atomic number elements when the scattered radiation technique of internal standardization is used. Since interference effects do not accompany incoherent scatter, this can be taken as another reason dictating the choice of a Compton-rich wavelength as an internal standard.

ACKNOWLEDGMENTS

This research was made possible by research contracts awarded by the Pacific Biomedical Research Center (NIH FR 0702603) and the National Science Foundation (NSF GP 9596). The use of the support facilities of the Hawaiian Institute of Geophysics is gratefully acknowledged. Hawaii Institute of Geophysics Contribution No. 271.

REFERENCES

1. M. F. Hasler, J. W. Kemp and G. Andermann, "New Instruments and Techniques Applicable to X-Ray Spectrochemical Analysis," Proceedings, Conference on Industrial Applications of X-Ray Analysis, Fourth Annual, University of Denver Press, Denver, (1955).

2. G. Andermann, and J. W. Kemp, "Scattered X-Rays as Internal Standards in X-Ray Emission Spectroscopy," Anal. Chem., 30, 1306-1309, (1958).

3. C. E. Harvey, A Method of Semi-Quantitative Spectrochemical Analysis, Applied Research Laboratories, Glendale, California, 1947.

4. Z. H. Kalman and L. Heller, "Theoretical Study of X-Ray Fluorescent Determination of Traces of Heavy Elements in a Light Matrix," Anal. Chem, 34, 946-951, (1962).

5. K. P. Champion, J. C. Taylor and R. N. Whittem, "Rapid X-Ray Fluorescence Determination of Traces of Strontium in Samples of Biological and Geological Origin," Anal. Chem., 38, 109-112, (1966).

6. J. W. Kemp and G. Andermann, "Refinements in X-Ray Emission Techniques," Spectrochim. Acta, 8, 114A, (1956), (abstract.)

7. W. J. Campbell and J. D. Brown, "X-Ray Absorption and Emission," Anal. Chem. Analytical Reviews, 36, 312R-328R, (1964).

8. A. H. Compton and S. K. Allison, X-Rays in Theory and Experiment, D. Van Nostrand, Princeton, New Jersey, 1935, p. 192-193.

9. N. S. Gingrich, "The Diffraction of X-Rays by Liquid Elements," Rev. Mod. Phys., 15, 90-110, (1943).

10. D. P. Riley, "Background Scattering in Powder Photographs," in H. S. Peiser, H. P. Rooksby and A. J. C. Wilson, Eds., X-Ray Diffraction by Polycrystalline Materials, Reinhold, New York, 1960, 0. 433-435.

11. A. H. Smallbone, "Briquetting, X-Ray Techniques Refine On-Stream Analysis," Rock Products, 68, 60-63, (1965).

12. K. R. Stever, J. L. Johnson and H. H. Heady, "X-Ray
     Analysis of Tungsten-Molybdenum Metals and Electro-
     lytes," in W. M. Mueller, Ed., Advances in X-Ray
     Analysis, Vol. 4, Plenum Press, New York, 1962,
     p. 474-487.

13. T. J. Cullen, "X-Ray Spectrographic Analysis of
     Alloyed Copper," in J. E. Forrette and E. Lanterman,
     Eds., Developments in Applied Spectroscopy, Vol. 3,
     Plenum Press, New York, 1964, p. 97-103.

14. F. W. Lytle, W. B. Dye and H. J. Seim, "Determination
     of Trace Elements in Plant Materials by Fluorescent
     X-Ray Analysis," in W. M. Mueller, Ed., Advances in
     X-Ray Analysis, Vol. 5, Plenum Press, New York, 1962,
     p. 433-446.

15. B. R. Boyd and H. T. Dryer, "Analysis of Nonmetallics
     by X-Ray Fluorescence Techniques," in J. R. Ferraro
     and J. S. Ziomek, Eds., Developments in Applied
     Spectroscopy, Vol. 2, Plenum Press, New York, 1963,
     p. 335-349.

16. M. Ganivet and T. Arnal, "Dosage du Plutonium dans
     les Alliages Plutonium-Aluminium par Fluorescence-X,"
     Anal. Chim. Acta, 39, 73-79, (1967).

17. A. Lubecki, M. Wasilewska and L. Gorski, "On the
     Application of Compton Scattering to the Elimination
     of Matrix Effects in Non-Dispersive X-Ray Fluore-
     scence Analysis," Spectrochim. Acta, 23A, 831-840,
     (1967).

18. P. Martinelli and P. Blanquet, "Application des
     Sources de Rayons Gamma au Dosage des Elements
     Lourds par Fluorescence-X," Symp. Radiochem. Methods
     Anal, Salsburg, 1964, International Atomic Energy
     Agency, Vienna, 1965, p. 451-467.

19. R. Jenkins and J. L. DeVries, Practical X-Ray
     Spectrometry, Springer-Verlag, New York, Inc.,
     New York, 1967, p. 95.

20. A. H. Compton and S. K. Allison, X-Rays in Theory
     and Experiment, D. Van Hostrand, Princeton, New
     Jersey, 1935, p. 139-140.

AN IMPROVED X-RAY FLUORESCENCE METHOD FOR THE ANALYSIS OF MUSEUM OBJECTS

Maurice E. Salmon

Conservation-Analytical Laboratory

Smithsonian-Institution, Washington, D. C.   20560

ABSTRACT

A method is described for the x-ray fluorescence analysis of small samples, taken from museum objects, to determine alloy composition. The samples are dissolved in an appropriate reagent and absorbed on cellulose powder. The resulting powder is formed into a film of less than critical thickness and the effective absorption of the sample for the characteristic wavelength of the element being measured is determined. The effective absorption coefficient is used to correct the observed intensities in order to obtain quantitative results.

INTRODUCTION

X-ray Fluorescence Spectrography is being used in museum laboratories as a method of analysing artifacts. In the method commonly used, an unprepared or minimally prepared surface is exposed to the x-radiation. The method is non-destructive. However, major errors can be present in the results if the surface of the area exposed to the beam of x-rays is not flat and smooth. The macro-probe approach, where the area of irradiation is limited to compensate for lack of flatness, has been taken by some workers. This method is also subject to errors because of inhomogeneities and variations in composition between interior and surface layers, caused by the manufacturing process and by preferential dissolution and corrosion prior to recovery. The lack of adequate, well characterized, solid standards, with appropriate composition, is another serous problem of this method. Ideally, the standards should have a physical condition like that found in the artifact

94

and should have the same matrix in order to establish calibration curves capable of yielding accurate non-destructive analyses. Most of these difficulties can be circumvented by the method to be described.

The effects of surface enrichment can be circumvented by careful sampling. The drilling-out of a small sample provides material which is more nearly representative of the inconstant composition of the usual object. Furthermore, the sample can then be manipulated and prepared in a standardized manner in order to yield a sample which has a constant physical configuration, is homogeneous, and is representative of the bulk of the artifact. Finally, the drilled-out hole may be filled so that the restored sample-site is virtually undetectable.

A successful analysis of museum artifacts should satisfy the following requirements:

1. The technique must be non-destructive or minimally destructive. If a sample is taken it must be small and the sample site must be easily restorable.

2. The precision and relative accuracy of the method must be known.

3. The technique should be relatively free of interferences.

4. The technique should be relatively independent of standards, i.e., the calibration should not be dependent upon standards which approximate the composition and physical state of the artifact to be analysed.

In view of the fact that there are inhomogeneous layers present on the surface of most artifacts, the only practical method of obtaining an analytical result which is representative of the average composition of the bulk material is to remove a sample by drilling into the interior of the artifact.

BACKGROUND

Rose, Cuttitta, and Larson[1] have demonstrated that it is feasible to dissolve milligram amounts of silicate rocks in an appropriate reagent and then absorb the resulting solution in an accurately weighed portion of chromatographic grade cellulose powder. The resulting pulp is then dried, formed into a disc-shaped pellet under pressure between the polished parallel faces of a die, and exposed to the x-ray beam in that form. This method of preparation results in a homogeneous distribution of the sample in a

uniform, light element matrix. In consequence the matrix effects
of the sample are minimized because of dilution. Furthermore, the
standards required for calibration are easily made from solutions
which are readily available commercially.

    Rose and co-workers found detection limits of their method
to be in the micro-gram and sub-micro-gram range, thus permitting
detection of 0.1 percent of an element or less in a one milligram
sample. Furthermore, the calibration curves were found to be
linear, so long as the pellet thickness remained constant. This
characteristic permits calibration for a wide range of composition
by means of a relatively small number of standards. Samples prepared in this manner are of less than critical thickness. The
intensity, therefore, is a function of the concentration of the
element, the thickness of the pellet, and the absorption of the
pellet for the characteristic wavelength of the element being
analysed.

    Direct adaptation of this method to the estimation of
metallic elements in ancient materials resulted in an unexpected
difficulty. The first alloy system to which the method was
applied experimentally and satisfactorily contained only silver
and copper, but samples taken from actual antiquities were found
to contain significant quantities of gold. In the quantities
found (about 2 percent) the gold failed to dissolve in the re-
agents chosen and potassium cyanide had to be added to achieve
complete dissolution. This addition of non-volatile material
changed both the x-ray absorption characteristics and the thickness
of the pellets. The desirable linear relation between intensity
and elemental content no longer prevailed.

    Carr-Brion[2,3] has published a method of analysing mineral
powders for zinc and tin which involves the measurement of the
effective absorption coefficient of the sample, in the form of a
thick film, for the wavelength of the element being analysed.
This technique involves the mixing of the mineral powder to be
analysed in a wax base with a dilution 1:3. The resulting wax
mixture is then formed into a thick film, of less than critical
thickness, and the effective absorption coefficient is calculated
from intensity measurements made on the thick film alone, the thick
film backed with a disc of the pure element being measured and from
an infinitely thick sample of the pure element being measured.
From the measurements he also calculated theoretically, the intensity of an infinitely thick sample of the same composition. The
ratio of the calculated intensity of an infinitely thick sample
and the intensity of an infinitely thick sample containing one percent of the element sought with the same characteristic absorption
was compared by means of a calibration curve to obtain the
analytical result in weight percentage. This method also relies
upon the maintenance of a constant film thickness.

The theory by which Carr-Brion measured the effective absorption of the sample is fundamentally sound. However, a method of overcoming the dependence on sample thickness was needed. This has now been sought and found.

THEORETICAL BACKGROUND

The intensity I_∞ of the characteristic (secondary) x-rays emitted by an "infinitely" thick sample containing a given concentration of an element, when excited by an incident (primary) beam of monochromatic x-rays of wavelength λ_p is given by:

$$I_\infty = K / (\mu_p \csc\theta_p + \mu_s \csc\theta_s)\rho \qquad (1)$$

where   K   is a proportionality constant that measures the absorption and conversion of incident to characteristic radiant energy

μ_p and μ_s are mass absorption coefficients for the incident (λ_p) and the characteristic (λ_s) wavelengths

ρ is the density

θ_p and θ_s are the angles made by the incident and emergent beams with the (plane) sample surface.

It has been shown[4] that under constant conditions of excitation, the characteristic radiation excited in a sample d cm thick will have an intensity I_d whose value can be calculated from the consideration that the contribution dI to I_d of an element of constant area and thickness dx, located at a depth x is:

$$dI = K \exp[-(\mu_p \csc\theta_p + \mu_s \csc\theta_s)\rho x]dx \qquad (2)$$

The integrated equation may be written:

$$I_d = K \int_0^d \exp(-ax)\,dx = K[1- \exp(-ad)]/a \qquad (3)$$

where

$$a = (\mu_p \csc\theta_p + \mu_s \csc\theta_s)\rho \qquad (3a)$$

At infinite thickness (d=∞) this reduces to eq. (1),

$$I_\infty = K/a$$

The ratio of the intensity from a thin pellet I_d to the intensity from an infinitely thick pellet I_∞ is:

$$I_d/I_\infty = 1-\exp(-ad) \qquad (4)$$

It has been further shown[5] that the intensity I of the characteristic x-radiation from a substrate covered by a thin film <u>not</u> containing elements present in the substrate obeys the exponential absorption law in the form

$$\ln[I_0/I] = ad \tag{5}$$

where I_0 is the intensity from the bare substrate,
    I is the intensity from the substrate covered by the thin film
    a is the same as in equation (3)
    d is the thickness of the film in cm.

Kaelble and McEwan[6,7] have demonstrated that when the substrate and the film contain the same elements, the intensity, I, of the substrate with the film in place consists of two components, namely, I_1 the intensity from the substrate modified by the absorption of the film and I_2 the intensity from the film alone

$$I = I_1 + I_2 \tag{6}$$

From eq. (5)

$$I_1 = I_0 \exp(-ad) \tag{7}$$

and from eq. (4)

$$I_2 = I_\infty - I_\infty \exp(-ad) \tag{8}$$

where I_2 is identical to I_d of eq. (3)

$$I = I_0 \exp(-ad) - I_\infty \exp(-ad) + I_\infty \tag{9}$$

and analogous to eq. (5) the exponential absorption takes the form

$$ad = \ln[(I_0-I_\infty)/(I-I_\infty)] \tag{10}$$

From eq. (4) it is apparent that the intensity that would be observed from an infinitely thick sample can be calculated from the intensity observed on a thin film of the same material, provided that the absorption characteristics are known from eq. (8)

$$I_\infty = I_2 / [1-\exp(-ad)]$$

From eq. (7)

$$\exp(-ad) = I_1/I_0$$

therefore

$$I_\infty = I_2/[1-(I_1/I_0)] = I_2 I_0/(I_0-I_1) \tag{11}$$

and substituting for I_∞ in eq. (10) gives us

$$ad = \ln[(I_0^2 - I_0 I_1 - I_2 I_0)/(II_0 - II_1 - I_2 I_0)] \tag{12}$$

Dividing both sides of eq. (12) by d, the thickness of the sample film, gives us the effective absorption coefficient of the sample with the effects of sample thickness accounted for.

If we then let

$$A^s = \ln[(I_0^2 - I_1 I_0 - I_2 I_0)/(II_0 - II_1 - I_2 I_0)]/d \tag{13}$$

then

$$A^s = a = (\mu_p \csc\theta_p + \mu_s \csc\theta_s)\rho \tag{14}$$

Jenkins and De Vries[8] have demonstrated that it is possible to linearize a calibration curve which exhibits strong deviations from linearity caused by differences in the mass absorption coefficients of the elements making up the matrix. The linearization is accomplished by multiplying the observed intensity of the element of interest by the ratio of the absorption coefficient of the sample to the absorption coefficient of the pure element being analysed. A^s is the effective absorption coefficient of the sample and A^e, the absorption coefficient of the pure element can be calculated from eq. (3a) by assuming that the effective wavelength of excitation lies just on the short wavelength side of the absorption edge of the element of interest and choosing μ_p accordingly, μ_s is chosen at the characteristic wavelength, θ_p and θ_s are the incident and take-off angles respectively and ρ is the density of the pure element. I_∞, calculated according to eq. (11), is multiplied by the ratio, $A^s:A^e$ to give us a corrected intensity, I_c.

$$I_c = I_\infty(A^s/A^e) \tag{15}$$

EXPERIMENTAL

Calibration

A series of standard solutions were prepared with known weights of copper and silver per unit volume. Aliquots of these solutions were pipetted into teflon beakers and weighed portions of chromatographic-grade cellulose powder were added and thoroughly mixed. The resulting moist pulp was dried overnight at

80° C. The dried powders were thoroughly mulled in an agate mortar and pressed into discs in a hydraulic press at 87,500psi.

The intensities of the appropriate K_α lines were measured on the sample disc alone, I_2; on the sample disc backed by a solid bulk sample of the pure element being determined, I; and on the bulk sample of the pure element, I_0. The observed intensities were corrected for instrumental dead time but no correction was made for background. I_∞ was calculated for each sample according to eq. (11) and A^s according to eq. (13). A^e, for the element of interest, was calculated according to eq. (3a). I_∞ was multiplied by the ratio A^s/A^e to obtain the corrected intensity I_c. A simple linear regression was done to obtain the slope and Y intercept of the calibration curve in the equation

$$W_S = MI_c + B$$

where W_S is the weight fraction of the element in the prepared sample disc. Table I presents a summary of these results for copper and silver.

Sample Preparation

In practice, the solid sample is accurately weighed and dissolved in a reagent appropriate to the alloy being analysed. A weighed portion of chromatographic grade cellulose powder is added and mixed into the solution. The resulting pulp is dried overnight at 80° C and the resulting dried powder is thoroughly mulled in an agate mortar and finally pressed into a pellet at 87,500psi.

For the purpose of the preliminary evaluation of this technique, standards were prepared from solutions with a known weight of silver and copper per unit volume. Aliquot portions of these standards were transferred to teflon beakers and evaporated to dryness. The samples were then dissolved in a solution of potassium cyanide, with the addition of four drops of 30 percent hydrogen peroxide, to approximate the conditions which would be met in the dissolution of silver-copper-gold alloys. The total weight of potassium cyanide added was intentionally varied in order to extend the effective absorption over a wider range. The final stages of sample preparation were identical with the preparation of the standards for calibration.

The samples were weighed and the thickness was measured immediately prior to the measurement of the appropriate K_α intensities. I_c was calculated from these measurements and W_S was determined by means of the regression coefficients determined during calibration. From W_S and the weight, W, of the original sample, before dissolution, the concentration, in weight percentage, of the element being analysed is easily calculated.

TABLE I

CALIBRATION CURVES

Fit of the points I_c, W_s to the straight line $W_s = M\ I_c + B$ by a simple linear regression.*

Silver

Sample Number	I_c	W_s in standard	W_s calculated
152	40.04	1.282×10^{-2}	1.268×10^{-2}
153	34.72	1.147×10^{-2}	1.084×10^{-2}
155	33.14	1.051×10^{-2}	1.029×10^{-2}
156	27.39	8.309×10^{-3}	8.302×10^{-3}
157	35.88	1.069×10^{-2}	1.124×10^{-2}
158	32.47	9.505×10^{-3}	1.006×10^{-2}
159	24.69	7.727×10^{-3}	7.366×10^{-3}
160	19.33	5.530×10^{-3}	5.506×10^{-3}
161	23.26	6.602×10^{-3}	6.869×10^{-3}

*Regression coefficients: $M=3.4680 \times 10^{-4}$ and $B=-1.1988 \times 10^{-3}$. The standard deviations of the coefficients are: M 2.2316×10^{-5} and B 6.8683×10^{-4}.

Copper

Sample Number	I_c	W_s in standard	W_s calculated
152	4.517	1.389×10^{-3}	1.414×10^{-3}
153	3.890	1.242×10^{-3}	1.199×10^{-3}
155	0.554	5.842×10^{-5}	5.543×10^{-5}
156	1.076	2.374×10^{-4}	2.343×10^{-4}
157	0.794	1.336×10^{-4}	1.375×10^{-4}
158	0.736	1.188×10^{-4}	1.177×10^{-4}
159	0.995	2.207×10^{-4}	2.066×10^{-4}
160	1.764	4.645×10^{-4}	$4.703\ \ 10^{-4}$
161	2.098	5.546×10^{-4}	5.848×10^{-4}

*Regression coefficients: $M=3.4285 \times 10^{-4}$ and $B=-1.3470 \times 10^{-4}$. The standard deviations of the coefficients are: M 5.6058×10^{-6} and B 1.2769×10^{-5}.

Tables II and III present a compilation of some typical results obtained by repeated measurements made on the same sample pellets and repeated preparation of sample pellets from the same standard solution. From these results it appears that quantitative measurements can be made on 5mg samples with an accuracy on the order of ± four percent of the amount present.

The probable source of a portion of this error lies in sample

TABLE II

Results of Copper Determinations

Reproducibility of measurements (repeated analysis of sample number 152 containing 9.46% copper)

I_e	A^s	% found	difference
4.517	29.247	9.64	0.18
4.461	29.330	9.51	0.05
4.477	29.439	9.55	0.09
4.519	28.981	9.64	0.18
4.561	29.576	9.74	0.28
	average	9.62	0.16

The average percentage found is 9.62 with a standard deviation of 0.092 and a coefficient of variation of 0.01. The average difference from the standard value is 0.16 or 1.69% of the amount present.

Reproducibility of sample preparation (multiple samples prepared from the same standard solution containing 17.0% copper)

Sample number	Weight of KCN in mg	I_c	A^s	% found	difference
124	135	4.340	76.087	17.83	0.83
125	135	4.472	78.249	18.43	1.43
126	185	4.373	81.678	18.60	1.60
127	185	4.201	84.833	17.89	0.89
128	85	4.646	55.995	17.11	0.11
			average	17.97	0.97

The average percentage found is 17.97 with a standard deviation of 0.588 and a coefficient of variation of 0.033. The average difference from the standard value is 0.97 or 5.70% of the amount present.

preparation. The liquid standards utilized in this exploratory study were subject to the usual errors attendant to volumetric transfer operations. No effort was made to minimize these errors by dilution in order that transfers could be made in larger volumes. Most transfers were made in 1ml portions with a probable error of ± five percent. Improved sample preparation will improve the accuracy of the technique.

TABLE III

Results of Silver Determinations

Reproducibility of measurements (repeated analysis of sample number 152 containing 87.41% silver)

I_c	A^s	% found	difference
40.04	3.3224	86.49	-0.92
39.12	3.0915	84.32	-3.09
38.99	2.7632	84.00	-3.47
39.75	3.1676	80.36	-7.05
39.56	2.9849	85.35	-2.06
	average	85.19	-2.22

The average percentage found is 85.19 with a standard deviation of 1.031 and a coefficient of variation of 0.012. The average difference from the standard value is -2.22 or 2.53% of the amount present.

Reproducibility of sample preparation (multiple samples prepared from the same standard solution containing 79.6% silver)

Sample number	Weight of KCN in mg	I_c	A^s	% found	difference
124	135	21.89	5.5481	84.22	4.62
125	135	21.64	5.5148	83.11	3.51
126	185	21.17	5.7823	83.77	4.17
127	185	21.50	6.1017	85.78	6.18
128	85	23.29	4.3164	80.71	1.11
			average	83.52	3.92

The average percentage found is 83.52 with a standard deviation of 1.852 and a coefficient of variation of 0.022. The average difference from the standard value is 3.92 or 4.92% of the amount present.

CONCLUSIONS

The proposed method of analysis offers a technique of obtaining quantitative information on the alloy composition of museum artifacts. The technique is economical in terms of the amount of sample required, its precision and relative accuracy are predictable, it is relatively free of interferences and the calibration is not dependent upon standards that are difficult to obtain.

REFERENCES

1. J. J. Rose, F. Cuttitta, and R. R. Larson, Use of X-ray Fluorescence in Determination of Selected Major Constituents in Silicates, U. S. Geological Survey Professional Paper 525B, 1965, p. B155

2. K. G. Carr-Brion, The X-ray Fluorescence Determination of Zinc in Samples of Unknown Composition, Analyst 89: 346, 1964

3. K. G. Carr-Brion, The Determination of Tin in Powder Samples by X-ray Fluorescence Analysis, Analyst 90: 9, 1965

4. H. A. Liebhafsky, H. G. Pfeiffer, E. H. Winslow and P. D. Zemany, X-ray Absorption and Emission in Analytical Chemistry, John Wiley & Sons, Inc., 1960, p. 154

5. Ibid, p. 150

6. E. F. Kaelble and G. J. McEwan, "X-ray Spectrographic Measurment of Wax Film Thickness", presented at the 11th Annual Symposium on Spectroscopy, Society for Applied Spectroscopy, Chicago, 1960

7. W. D. Johns, Measurement of Film Thickness, in: E. F. Kaelble, Handbook of X-rays, McGraw-Hill, New York, 1967, p. 44-1

8. R. Jenkins and J. L. De Vries, Practical X-ray Spectrometry, Springer-Verlag New York Inc., New York, 1967, (Philips Technical Library), p. 136

ON-STREAM ANALYSIS OF LIQUID SAMPLES CONTAINING ELEMENTS OF HIGH AND LOW ATOMIC NUMBER BY X-RAY FLUORESCENCE WITH AIR AND VACUUM SPECTROGRAPH

Frank L. Chan

Aerospace Research Laboratories

Wright-Patterson Air Force Base, Ohio 45433

ABSTRACT

Recently on-stream analysis by various methods and instruments has been greatly and rightfully emphasized. This is particularly true in such research as air and stream pollutions, desalting of ocean water and ore processing.

Among the various methods, instruments, and procedures adopted, the X-ray fluorescence method is perhaps the best among them. This is because the X-ray fluorescence method neither consumes nor destroys the samples taken for analysis. It also has the merits of simplicity and specificity. However, on-stream analysis of liquid samples containing elements having an atomic number below 20 by the X-ray fluorescence method is not an easy task particularly when vacuum spectrograph is being used.

In this study liquid cells of different dimensions for the on-stream analysis operated in air, helium and in vacuum have been successfully constructed. The cells have been tested for several months.

These cells have been used for the on-stream analysis of liquid samples containing elements of high atomic numbers, as well as elements of low atomic numbers such as calcium, potassium, chlorine, sulfur and others. X-ray target material, analyzing crystal, and the detector were selected so as to give the highest count rate using the existing instrument available. Because of the necessary materials required for the construction of the cell located inside the vacuum chamber,

the count rate for these elements monitored is lower than
expected. However, the performance of the cells constructed
has been good from the standpoint of leakage of liquid inside
the air-path, helium-path and vacuum spectrograph is concerned.

During the on-stream analysis of the elements cited above,
the intensity of the element monitored is recorded on a strip
chart as well as the numerical print-out. The numerical
print-out can be in interval or cumulative counts.

Results of this study with respect to optimum concentration, rate of flow, limit of detectivity and others will
be presented in this paper.

INTRODUCTION

On-stream analysis is indeed important to the process
and mineral industries in todays plant operation. It is no
wonder that the XXII International Congress of Pure and
Applied Chemistry and XII International Conference on Coordination Chemistry planned to have a scientific program
especially for on-stream analysis in the mineral industry to
be held in Sidney, Australia, in the latter part of August
1969. On-stream analysis is usually made continuous for
control purposes. In reality the instrument for analysis in
the control laboratory is moved to the plant in a process
stream. It analyzes continuously the sampling line of certain
interested components in the flowing stream. By such analysis
it is possible (a) to have a close control, (b) to increase
material utilization or (c) to detect tolerance limits on-
the-spot.

In general, the instrument outlay for on-stream analysis
is more extensive. It provides accurate records such as strip
charts or computer print-outs for future references pertain-
ing to stream conditions, concentration ranges and accuracy
of analysis.

The set-up described in this paper utilizing the liquid
cells constructed for on-stream analysis has met the requirements mentioned. It is based on the measurement of the
intensity of X-ray fluorescence produced by the monitoring
element incident with a primary X-ray beam. It is capable
of monitoring elemental analysis of those elements with
atomic number 16 or greater. Elements with atomic number
greater than 20 for the on-stream analysis requires no vacuum.
Using the commercial available equipment with liquid cells
so designed they meet the required criteria of reliability,
simplicity of operation, and ease of maintenance. Aqueous

solution of the elements distributed widely in the periodic table have been used. Such elements as molybdenum, zinc, calcium, potassium, chlorine, sulfur and others have been monitored.

The importance of on-stream analysis has been realized by analytical chemists and design engineers. Books by various chemists have begun to point out the different facets dealing with on-stream analysis.[1]

EXPERIMENTAL

Instrumentation

For the on-stream analysis study the instrument used was the spectrometer (SPG-3) made by the General Electric Company. It has the provision to operate the spectrometer in vacuum, helium or air depending on the element monitored. It is the conventional dispersive system using a flat crystal for reflecting the characteristic fluorescence X-ray. Although there are two detectors in the system the SPG 9 flow counter tube was used for all runs. In general the spectrometer is part of the XRD-6 installation that includes an XRD-6 high-voltage assembly, an EA-75 X-ray dual target tube and an SPG-4 detector.

The instrument is provided with a special cable and transformer installed to avoid a sudden change of voltage. The normal operating condition is 5-50 mA and 60 kVP. Although the goniometer in the vacuum chamber can be used to scan the two-theta angle accurately, for monitoring the element, the two-theta angle remains the same during the on-stream analysis.

The original spectrometer has been improved to shorten the distance between the target and the surface of the liquid cell. This cell is held in a position 30° to the X-ray tube for optimum intensity. In order to leave enough space for the tubing connecting the liquid cell inside the spectrometer, three of the four circular frames for the sample holders were removed, (see figure 2). In this manner ample space in the vacuum spectrometer is provided for the liquid inlet and outlet tubings when the liquid cell revolves from the loading position to the testing position. The revolving mechanism for the liquid cell and the bottom plate can be removed when necessary. The EA-75 X-ray tube has two targets, one made of tungsten and the other of chromium. The crystal changers inside the vacuum chamber can accommodate four analyzing crystals.

For evacuation of the spectrometer a Welch Duo-Seal
pump was used. A vacuum of 10µ or less can be attained in
3 to 4 minutes. The whole assembly can be moved to the plant
where the process stream is situated. The spectrometer provides
also a constant-potential unit. The SPG No. 4 detector has a
scaler-timer combination, an amplifier, pulse height selector,
rate meter, strip chart recorder and digital printer. For
on-stream analysis the strip chart and digital printer were
both running at the same time checking one against the other.
The SPG No. 9 flow proportional counter was operated at 1530 V,
with the pulse height selector adjusted to E = 2 V and ΔE = 4 V.

The Liquid Cell

The cell with its general appearance is shown in figure 1.

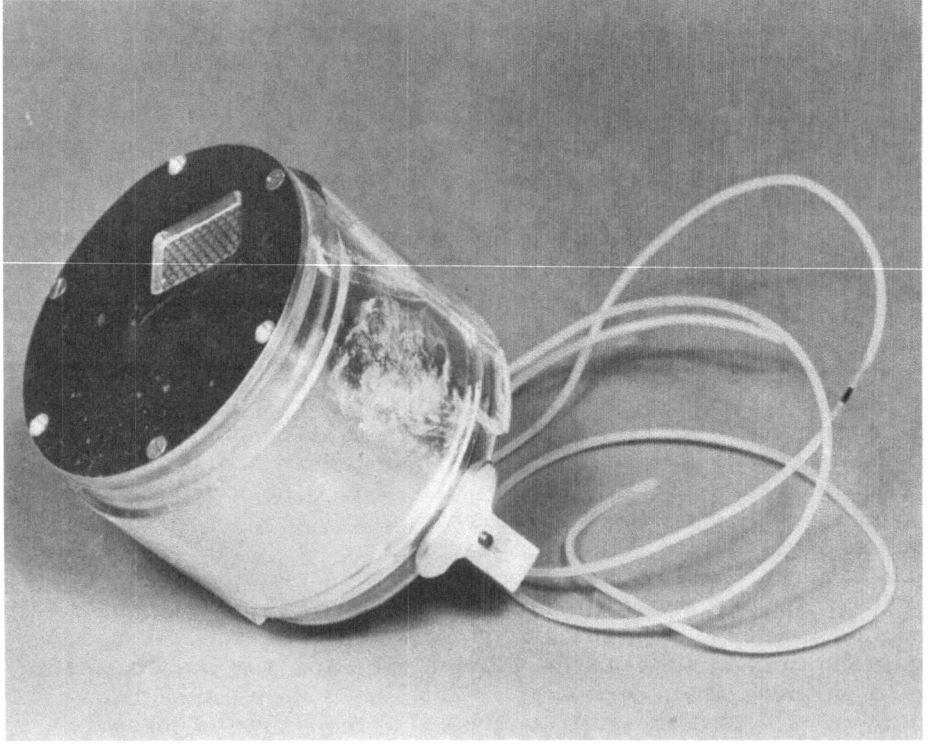

Figure 1. Liquid cell for on-stream analysis by X-ray fluorescence

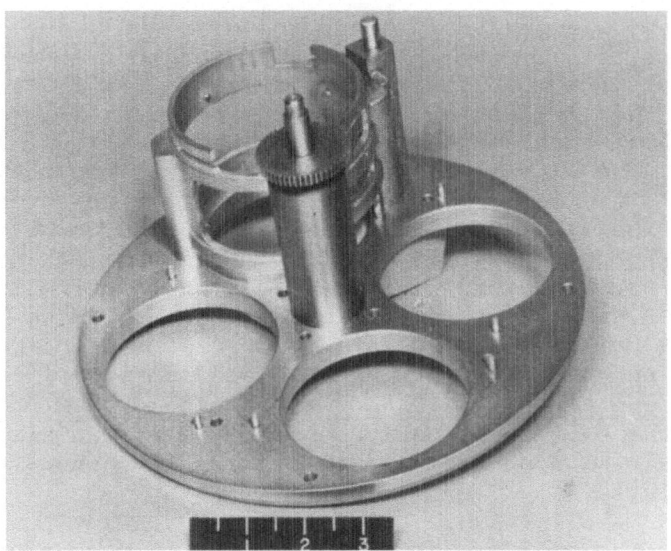

Figure 2.   Holder for liquid cell.

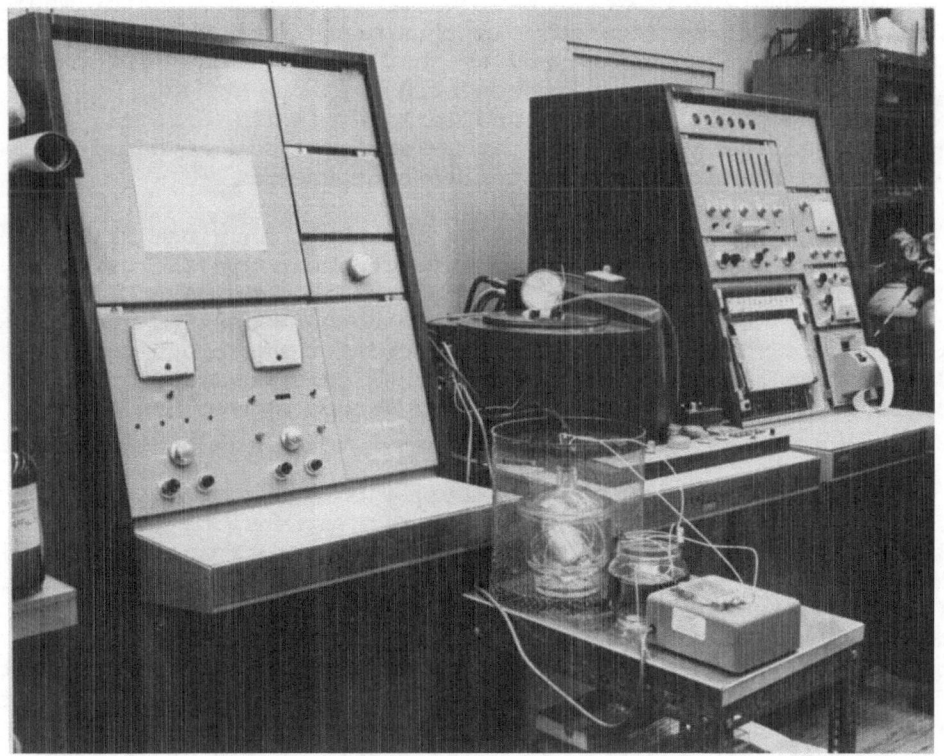

Figure 3. X-ray spectrometer used with the liquid cell.

There are two openings for liquid to enter and leave the cell.
Liquid flows into the cell from the bottom through a flexible
polyethylene tube of 1.5 mm inside diameter. Since the
surface of this cell is exposed to the incident X-ray at an
angle 30° horizontal to the X-ray target, the liquid leaves
the cell at the extreme tip of the tilted cell so that the
liquid fills the cell at all times during the on-stream
analysis. The value of using this cell for a vacuum system
is its leak proof capability so that no liquid leaks to
the vacuum spectrometer. This cell has an outer collar so
that two to three sheets of Mylar film can be placed tightly
between the cell and the collar. Two screens of different
meshes, as shown in figure 1, were placed on top of the Mylar,
the finer mesh screen on the lower, and the coarse one above
it. When the cell is properly constructed it can also be
used with air-path or helium-path besides in vacuum for on-
stream analysis.

Before the cell is actually used for in vacuum spectrometer it is tested for leakage by the set-up shown in figure 3.
To facilitate the checking of leakage, one of the two inlets
to the vacuum spectrometer was connected to a desiccator where
the liquid cell is located. In this manner the vacuum existing
in the desiccator is the same as that in the vacuum spectrometer.
If the cell is in operating condition without defect, liquid
pumped through it will not show leakage. When this cell is
used for air-path or helium-path, leakage could be carried
out in the manner described but without vacuum.

The pumping of liquid varies from less than one milli-
liter to as much as 5 ml per minute depending on the setting
of the rheostat attached to the pump. The pumping of liquid
made use of two durable and flexible hoses squeezing the
liquid forward as they revolved around a three prone-turning
disc. Each hose delivers a separate liquid to different
cells. On-stream analysis with the vacuum system has been
operating ten to twelve hours at a run without failure.

The sample holder for the liquid cell described in
figure 1 has a rectangular opening with dimensions 19 mm x
25 mm as normally used with this type of spectrometer. The liquid
cell described above is two thirds this area below the screen.
The X-ray fluorescence intensity of an element such as
molybdenum from the liquid monitored gave sufficient counts
per second in concentration as low as one-tenth of one percent.
However, a larger diameter liquid cell has also been constructed
and used with helium atmosphere. The diameter of this cell
is 30 mm and its depth 10 mm. The screen on top of this cell
is approximately 12 mesh. In this manner the liquid in the
cell covers the entire area of the rectangular opening cited.[2]

Procedure for On-stream Analysis

For simulated continuous on-stream analysis, stock
solutions containing the monitoring element having the highest
concentration were prepared. The stock solutions were put
in a smaller container with 200 ml capacity and used for the
on-stream analysis. Solutions having a concentration less
than the stock solutions were prepared by dilution using
volumetric flasks and pipettes. For a complete run of a
monitoring element, solutions of different concentrations
were first prepared. Each solution was in turn pumped into
the liquid cell placed in the spectrometer with the incident
X-ray on. The strip chart recorder as well as the digital
print-out were in operation simultaneously in conjunction
with the pumping of the liquid into the cell. The speed
of the pump usually ran 30 rev/min. delivering three ml/min.
By setting the two theta angle at the K_α peak of the monitoring element or elements the strip chart and the digital
print-out recorded the intensity of the peak as shown in some of
the examples cited in this paper. The strip chart ran 2.5
minutes per inch and the print-out, either interval or cumulative,
printed once in every ten seconds. When the instrument was in
working order and with the complete changing over of the liquid
from one concentration to the next the strip chart and the
print-out recorded a constant intensity as shown in figure 4
to figure 7 inclusive. To change from one solution to the
next the pumping of liqud was momentarily stopped and the
inlet tube cleaned and inserted into the new solution to be
pumped. To obtain net counts per second, distilled water
was subsequently pumped into the liquid and its intensity
likewise recorded. Net counts per second of K_α of a given
solution is the difference between total counts and the
background counts. For points in an intensity-concentration
curve as shown in figure 4 to figure 7 inclusive, 30 digital
print-out figures were averaged.

RESULTS AND DISCUSSION

It has been shown that the liquid cells and the instrument
described in this paper can be used for the on-stream analysis
of liquid. The only modification to the instrument besides
the liquid cell is the fabrication of a cover at the bottom
of the sample holders for the liquid to enter and leave the
spectrometer, and the removal of three sample holders to
provide enough space for the polyethylene tubes. Additional
equipment needed are the pumping set-up and the liquid
reservoir where the liquid is being stored for the on-stream
analysis. It is believed that for actual plant operation
the set-up can run continuously when vacuum, helium-path or

F. L. Chan

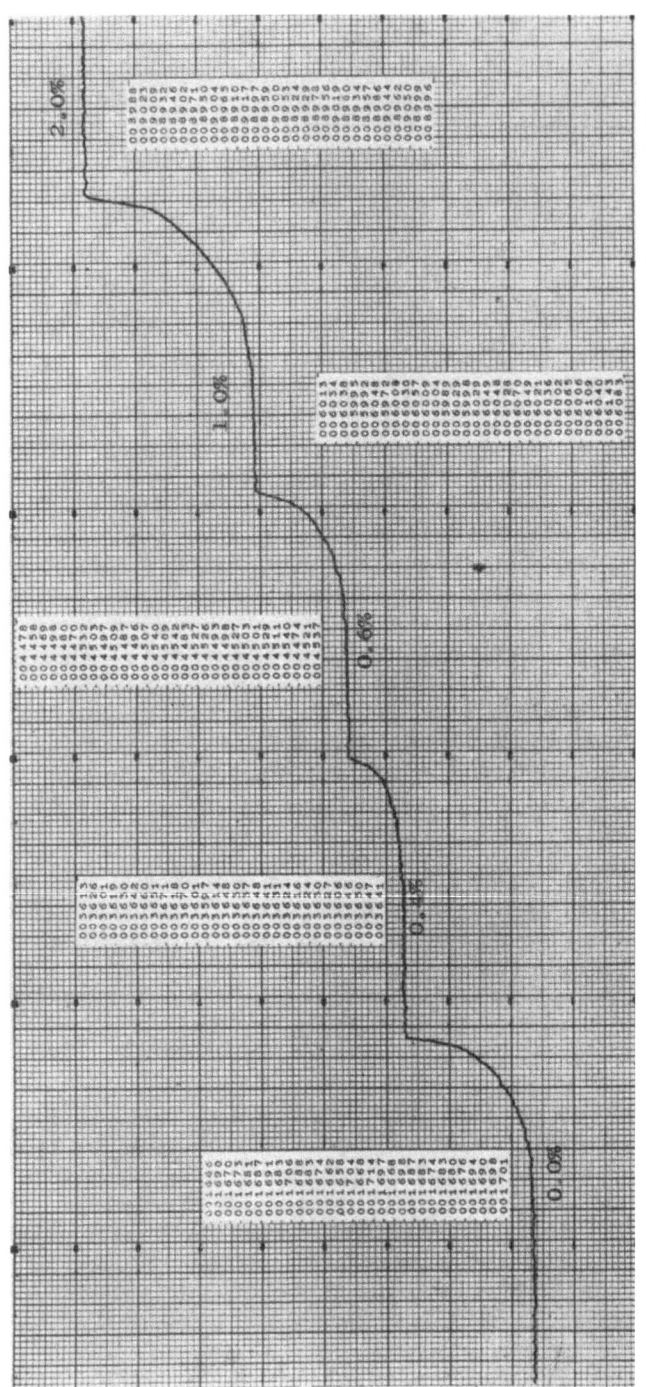

Figure 4. Recorder chart showing actual tracing of continuous on-stream analysis of molybdenum K_α in aqueous solution with concentration indicated on chart using air-path spectrometer. Conditions: 60 kVP; 50 mA; tungsten target; no PHS; analyzing crystal, LiF; No. 9 SPG flow proportional counter; full scale, 10,000 counts/sec.

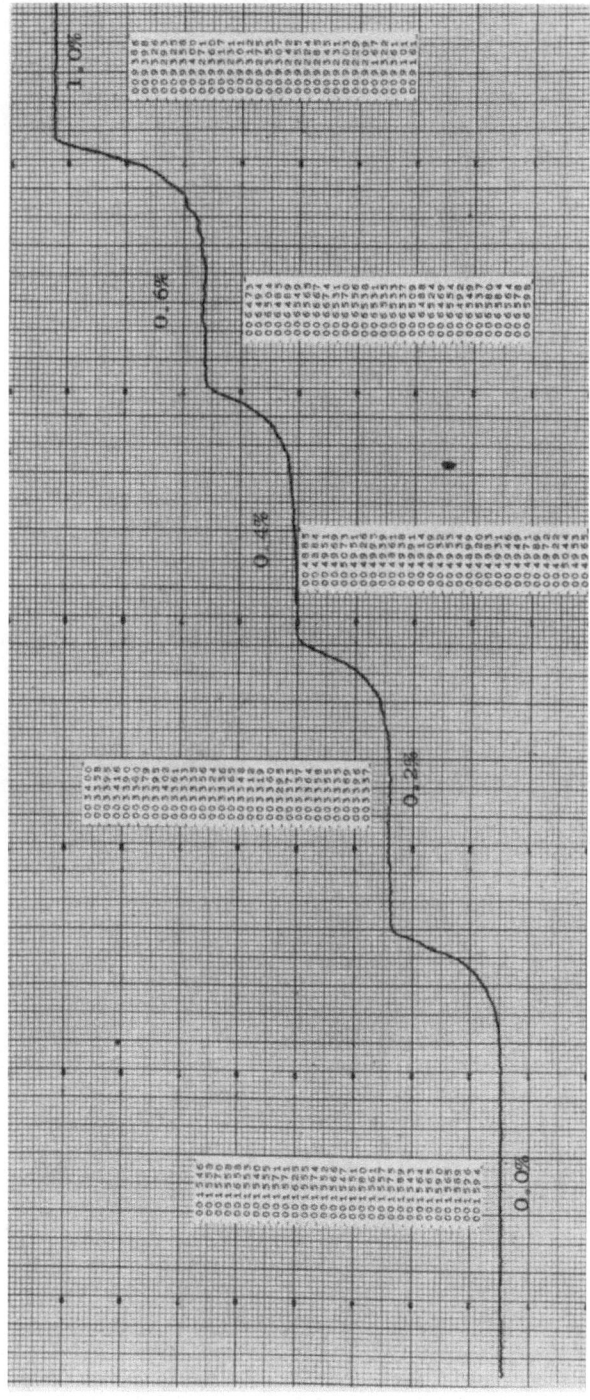

Figure 5.   Recorder chart showing actual tracing of continuous on-stream analysis of calcium K$_\alpha$ in aqueous solution with concentration indicated on chart using vacuum spectrometer. Conditions: 60kVP; 50mA; chromium target; no PHS; analyzing crystal, PET; No. 9 SPG flow proportional counter; full scale 10,000 counts/sec.

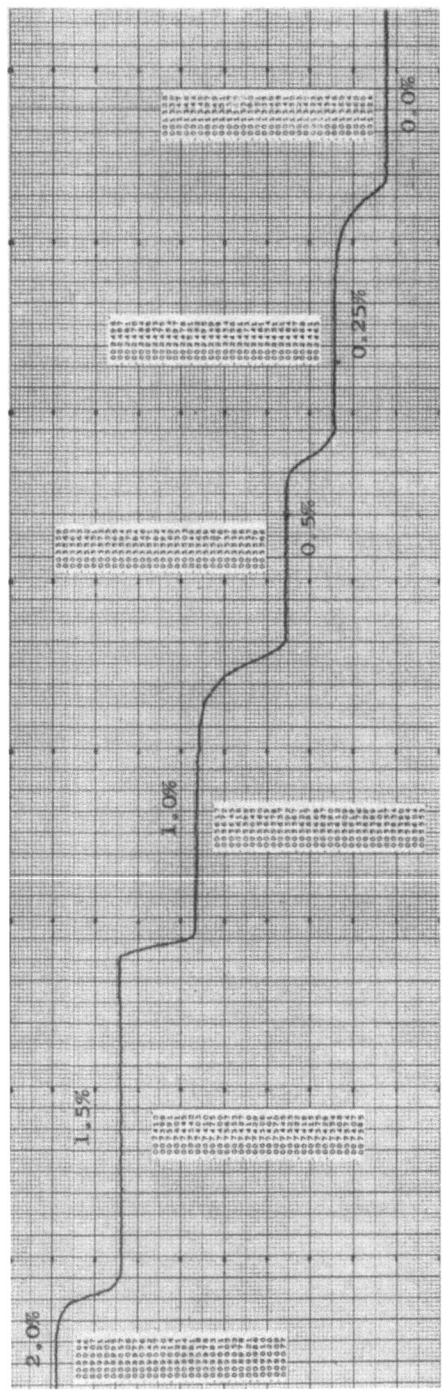

Figure 6. Recorder chart showing actual tracing of continuous on-stream analysis of potassium K_α in aqueous solution with concentration indicated on chart using vacuum spectrometer. Conditions: 60 kVP; 50 mA; chromium target; no PHS; analyzing crystal, PET; No. 9 SPG flow proportional counter; full scale 10,000 counts/sec.

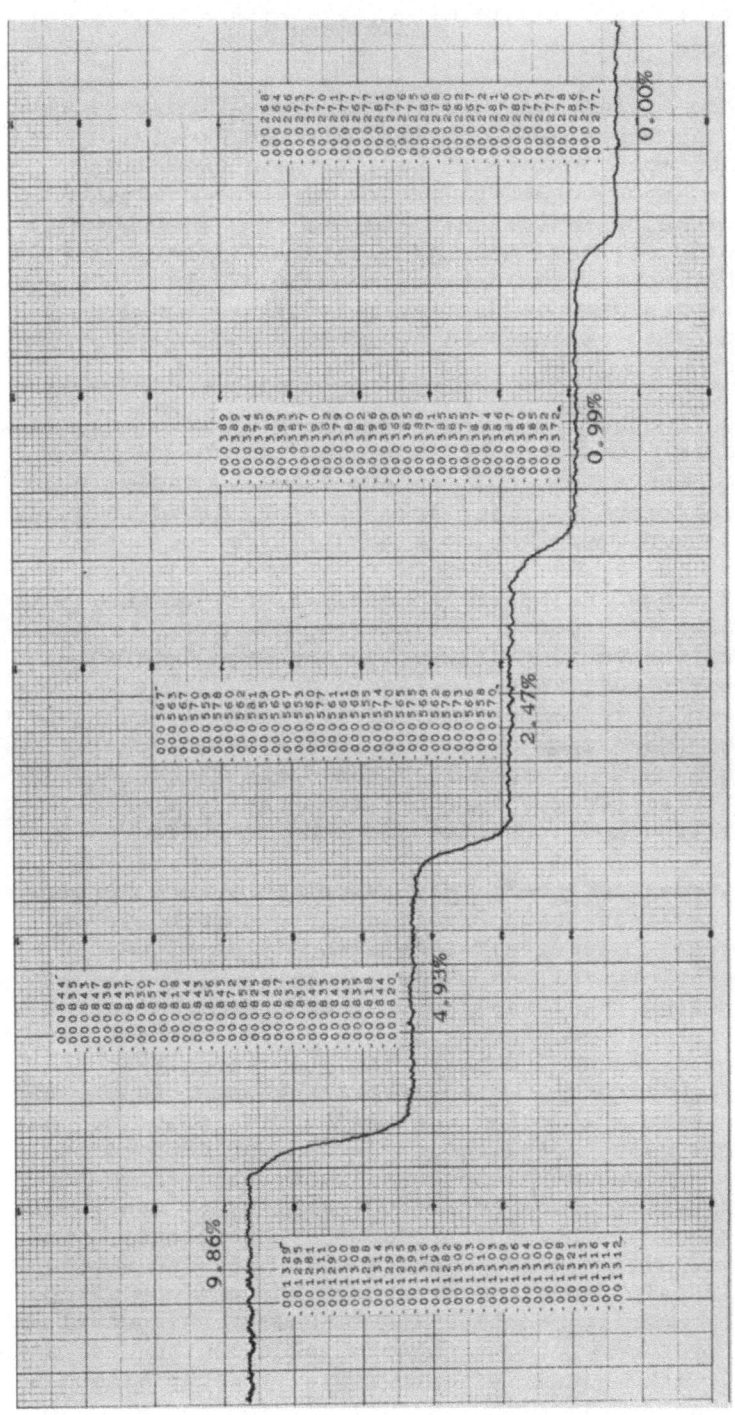

Figure 7. Recorder chart showing actual tracing of continuous on-stream analysis of chlorine K_α in aqueous solution with concentration indicated on chart using helium-path spectrometer. Conditions: 60 KVP; 50 mA; chromium target; no PHS; analyzing crystal, PET; No. 9 SPG flow proportional counter; full scale 2,000 counts/sec.

air-path is being used. If leakage occurs, the few ml. liquid
that is liable to spill could be quickly cleaned and a spare
cell introduced.

For the on-stream analysis by the X-ray fluorescence
spectrometer, a number of elements scattered widely in the
periodic table were monitored. For elements producing
fluorescent X-ray of comparative long wavelengths such as
sulfur, chlorine, potassium and calcium, the spectrometer
with the liquid cell was operated with helium atmosphere or
in vacuum. Elements with an atomic number greater than 20
(calcium) air-path were used. In this latter category
molybdenum and zinc with atomic numbers 42 and 30 respectively
were monitored.

Chlorine in solution cannot be monitored with the small
liquid cell described above because the fine and coarse
screens placed above the cell contain elements close to the
first order chlorine K_α. The large size of liquid cell was
used with a screen over the top made of different material.

Figure 8 shows an on-stream analysis of aqueous solutions
of molybdenum having concentrations ranging from 0.0 to 2.0%.
The procedure for running this series has been described.
Part of the print-out, when the liquid cell attained a constant
state with respect to concentration, is shown with the strip
chart tracing, (see figure 4). The spectrometer for the on-
stream analysis used no vacuum. Other operating conditions
are shown in this figure. For on-stream analysis of zinc
solution similar conditions were adopted, (see figure 10).

Some interesting preliminary experiments were performed
with calcium analysis using Brazilian gypsum macro-crystal
(23.28% calcium) to ascertain the intensity of K_α using PET
as the analyzing crystal and chromium or tungsten as the
target. Operating the X-ray fluorescence spectrometer at
50 mA and 60 kVP and no vacuum in the system chromium target
gave 6,695 counts/second compared to 1,636 counts/second
with a tungsten target. On the other hand when vacuum was
used in the system with other conditions remaining the same
chromium target gave 64,851 counts/second as compared to
28,145 counts/second with tungsten target. Therefore for
on-stream analysis PET, as described earlier,[3,4] was used
as the analyzing crystal and chromium as the target. Under
this optimum condition counts per second, net, for calcium K_α with
vacuum and air-path spectrometer under the same operating
conditions for on-stream analysis are shown in figure 11 and
figure 12. For a one percent calcium solution the net intensity
of calcium K_α with vacuum is approximately twenty times as
compared to that of air-path. Actual tracing by the recorder

of continuous on-stream analysis of calcium K_α with concentration and print-out are shown in figure 5. Likewise,
similar analysis for aqueous solution of potassium is shown
in figure 6 starting from 0.0% potassium on the right to 2.0%
on the extreme left.

Helium atmosphere was used with the on-stream analysis
for chloride using a bigger size liquid cell as described.
Actual tracing of continuous on-stream analysis of K_α in
aqueous solution is shown in figure 7.

Without the use of vacuum, sufficient intensity in counts
per second of K_α for molybdenum and zinc were recorded in the
continuous on-stream analysis as shown in figure 8 and figure
9. Likewise, with vacuum and two layers of Mylar (0.00025"
thick) sufficient intensities of K_α of calcium and of potassium
have also been noted as shown in figure 12 and figure 13.
However as the atomic number becomes smaller for elements such
as sulfur and phosphorus, the counts per second become less
and less. With the present set-up it is not practical to
run on-stream analysis on solution of aluminum unless the
dispersive system is modified. One of the main obstacles is
the opacity of the Mylar toward soft X-ray when used for
construction of the cell. Figure 15 shows the decrease of
silicon K_α intensity on a sample quartz as a result from
increasing the number of layers of Mylar placed over the
surface exposed to the incident X-ray. For example with the
vacuum spectrometer running at 50 mA and 60 kVP 38,184 (average
of five readings) counts/second were recorded as compared to
only 152 (average of five readings) counts/second when six
layers of Mylar were placed over the surface, a reduction of
99.6% of K_α intensity.

One of the criterions for a successful on-stream continuous
analysis is the electronic stability so that the count rate
recorded in the strip chart and the print-out give true
intensity of the element monitored. In carrying out the on-
stream analysis, the two theta angle once set at the peak
will remain in that position during the operation. To check
the electronic stability a method is hereby described to
detect the change of intensity of the incident X-ray beam.
It is based on the fact that when an element is permanently
present in the liquid cell such as chromium is present on the
two screens placed over the surface of the liquid cell for
reinforcing the Mylar film, an analyzing crystal is selected
so that the two theta angle for K_α of chromium is the same
as the two theta angle for K_α of the interested element
monitored during the on-stream continuous analysis. In this
manner a mere switching of one analyzing crystal to the other
will enable one to detect the stability of the incident X-ray

used for the on-stream analysis. Figure 10 shows the on-stream analysis of zinc using a quartz crystal (2d = 6.682$\overset{o}{A}$). The K_α for zinc with the crystal is 24.81^o two theta. Using a second analyzing crystal such as ADP (2d = 10.642 $\overset{o}{A}$) K_α for chromium is 24.86^o two theta. Therefore any doubt as to the stability of incident X-ray can be checked quickly. At the present time the number of crystals used in X-ray fluorescence is limited and therefore to find suitable crystal for a given on-stream analysis is somewhat difficult. Efforts to search as many crystals for this purpose as possible is one of the objectives for the X-ray fluorescence study. Two crystal systems can also be extended to on-stream analysis involving two or more elements.

Normally one would select an analyzing crystal that has the high count rate of an interested element. It is commonly recognized that LiF, 2d = 4.026$\overset{o}{A}$, is the best analyzing crystal for elements having atomic numbers between 20 and 50. Figure 8 and figure 9 show a marked difference between quarts crystal and LiF for on-stream analysis of molybdenum under the same operating conditions.

Examples cited thus far may be summarized as a case of variation of one element in aqueous solution where water is the matrix. In such instances the net intensity versus concentration, as shown, in the calibration curves are sufficient for monitoring the interested element. However in the present study zinc was monitored not only as a one element variation as shown in figure 10, but also monitored in the presence of copper having concentrations varying from one to four percent. With a constant zinc concentration the chart tracing and the print-out indicated different zinc count rate when the copper content varies. Therefore when two or more elements vary in concentration the element monitored should take into consideration the variation of the other element in solution. Various authors[5,6,7] have suggested compensation for the interelement effect correction factor derived experimentally or by using published mass absorption coefficient. The various methods pointed out by these authors could possibly be adopted if proper modifications were made. The requirement for a successful on-stream analysis as with other analysis of liquid, bubbles should not form on the Mylar used for the cell.[8] Freezing of liquid as described in earlier publication is not applicable with the on-stream analysis dealt with in this paper.[9] X-ray fluorescence analysis as exemplified in the present study has high background intensity because of excessive scattering from the light elements.[5,8]

Advantages by X-ray spectrometric analysis involving liquid have been pointed out. [5] Some of these advantages

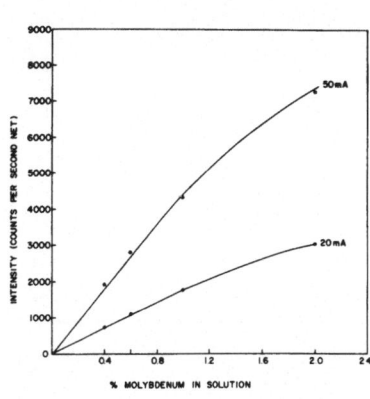

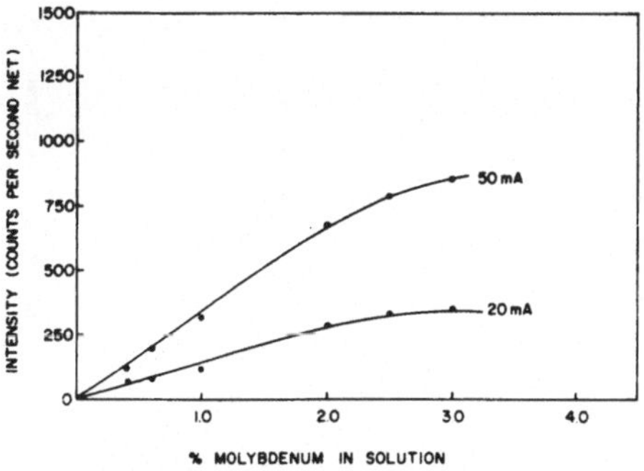

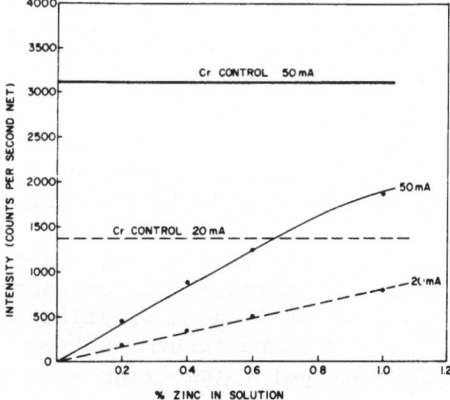

Figure 8 (upper left). Curves
      showing intensity of
      molybdenum K$_\alpha$ taken
      during the on-stream
      analysis versus con-
      centration of molyb-
      denum in aqueous solu-
      tion using air-path
      spectrometer. (See
      figure 4 for conditions.)

Figure 9 (upper right). Curves showing intensity of molybdenum K$_\alpha$
      taken during the on-stream analysis versus concentration
      of molybdenum aqueous solution using air path spectrometer.
      Conditions the same as figure 4, except quartz (2d=6.682Å)
      as analyzing crystal in lieu of LiF.

Figure 10 (lower right). Curve showing intensity of zinc K$_\alpha$
      taken during the on-stream analysis versus concentration
      of zinc in aqueous solution using air-path spectrometer.
      The constancy of the incident beam was monitored by the
      presence of chromium peak from chromium plated screen
      placed over the top of the liquid cell. 2θ angle in set-
      up remains the same for both peaks. Conditions: 60 kVP;
      tungsten target; no PHS; analyzing crystal, quartz
      (2d=6.682Å) for zinc, ADP (2d=10.648Å) for chromium; No. 9
      SPG flow proportional counter.

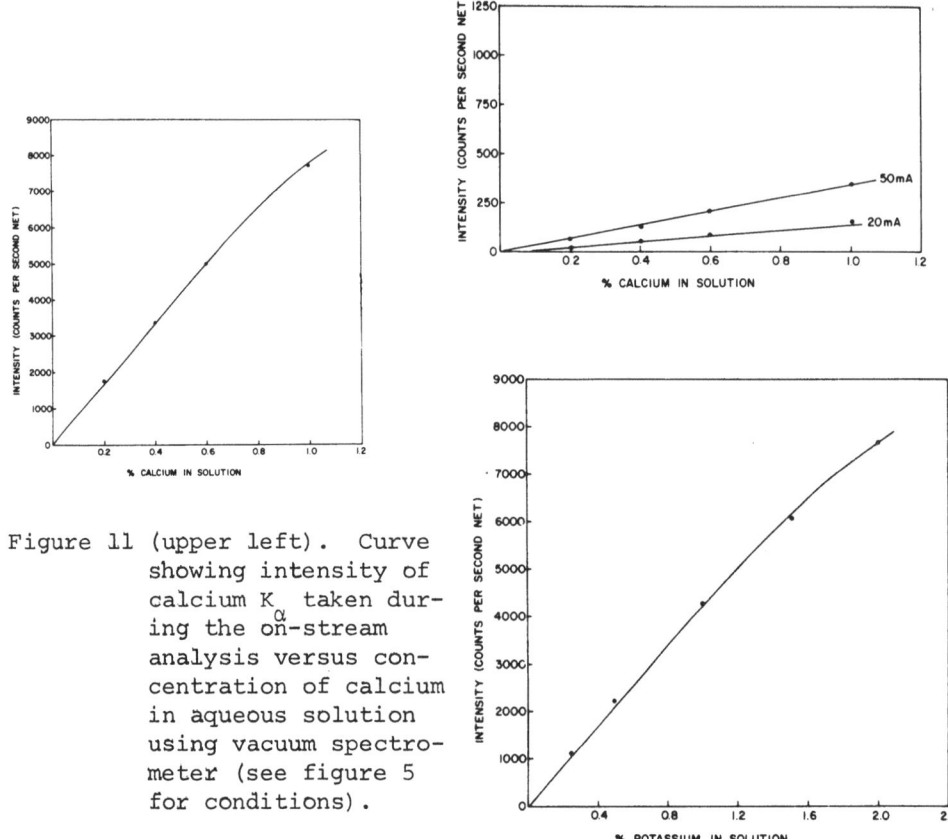

Figure 11 (upper left). Curve
     showing intensity of
     calcium K$_\alpha$ taken dur-
     ing the on-stream
     analysis versus con-
     centration of calcium
     in aqueous solution
     using vacuum spectro-
     meter (see figure 5
     for conditions).

Figure 12 (upper right). Curves showing intensity of calcium K$_\alpha$
     taken during the on-stream analysis versus concentration
     of calcium using air-path x-ray spectrometer in a typical
     on-stream analysis. Conditions: 60 kVP; chromium tar-
     get; no PHS; analyzing crystal, PET; No. 9 SPG flow
     proportional counter.

Figure 13 (lower right). Curve showing intensity of potassium K$_\alpha$
     taken during the on-stream analysis versus concentra-
     tion of potassium in aqueous solution using a vacuum
     spectrometer. (See figure 6 for conditions.)

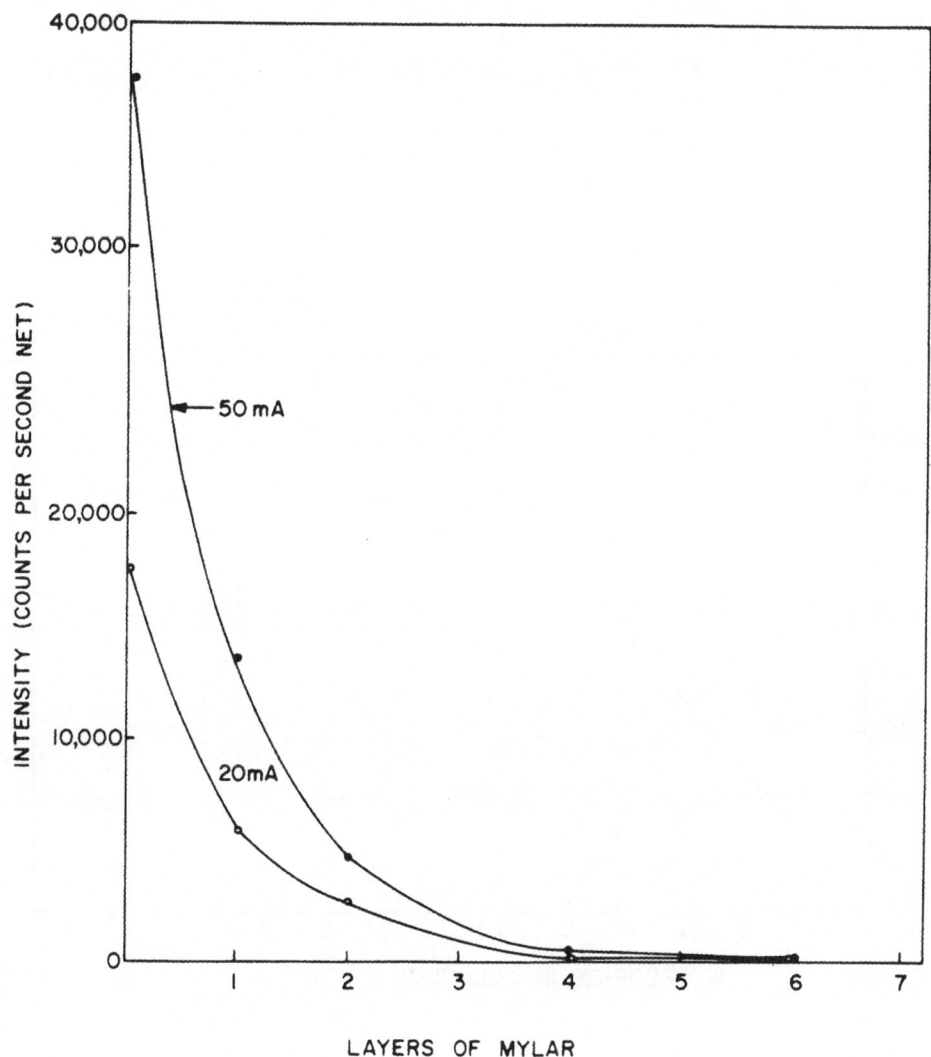

Figure 14.   Curve showing the relationship between layers of Mylar,
             0.00025" thick, placed over the quartz sample and the
             intensity of silicon K_α. Conditions: 60 kVP; chro-
             mium target; no PHS; analyzing crystal PET; No. 9 flow
             proportional counter; vacuum spectrometer.

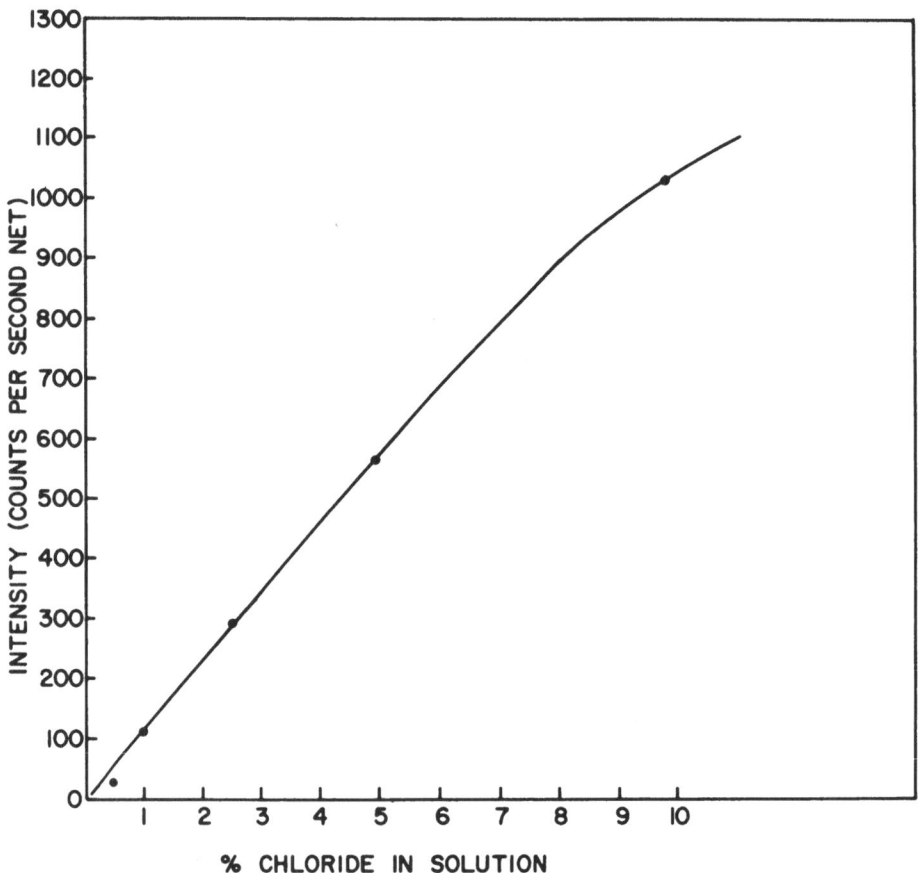

Figure 15.   Curve showing intensity of chlorine K_α taken during
             the on-stream analysis versus concentration of chloride
             in aqueous solution using a helium-path spectrometer.
             (See figure 7 for conditions.)

apply also to on-stream analysis with the liquid cell described.

SUMMARY

1.   Special emphasis has been made in this paper to
employ the X-ray fluorescence method for on-stream analysis
of liquid containing elements of high or low atomic number.

2.   Liquid cells have been designed for this purpose and
could be used with air or helium atmosphere and in vacuum
depending on the element monitored.

3.   These cells are designed to be used with existing
X-ray spectrometer which could be operated continuously for
the on-stream analysis.

4.   Mylar film was used as overlays for these cells.
Even with two to three layers of these film, elements such as
molybdenum or zinc could be monitored using air-path with
sufficient K_α intensity. With calcium. potassium or chlorine
sufficient intensity could be had for monitoring purposes if
vacuum or helium were to be used.

5.   Strip charts and digital print-outs are presented
which show the operating characteristics of the on-stream
continuous analysis.

6.   In the presence of other elements which affect the
intensity of the spectra peak of the element monitored, it
is necessary to determine the concentration of the interfering
element and corrections made. Presence of copper which affect
the monitoring of zinc has been cited as an example.

7.   Because of the presence of two to three layers of
Mylar on the liquid cell, the count-rate for silicon and
aluminum is low with the set-up using flat crystal. Undoubtedly the count-rate for these elements could be increased
by the use of curved crystal for the set-up.

REFERENCES

1.   H. H. Willard, L. L. Merritt, Jr., and J. A. Dean, "In-
     strumental Methods of Analysis," 4th Edition, D. Van
     Nostrand Co., Inc., New York, 1965, p. 746-760.

2.   General Electric X-ray Department, "Vacuum Liquid Sample
     Cells," Report No. 1, Publication No. 7A-4022, 1965.

3. F. L. Chan, "Some Observations on the Use of Certain
    Analyzing Crystals for the Determination of Silicon and
    Aluminum," in G. R. Mallet, Marie J. Fay and W. M.
    Mueller, Editors, Advances in X-ray Analysis, Vol. 9,
    Plenum Press, New York, 1965, p. 515-527.

4. F. L. Chan, "A Study of Silicon Determination in Organo-
    silicon Compounds by X-ray Fluorescence with Vacuum
    Spectrograph," in Proceedings of the SAC Conference,
    Nottingham, 1965, Published by W. Heffer and Sons Ltd.,
    Cambridge, England, p. 89-101.

5. T. J. Cullen, "X-ray Spectrometric Analysis of Solutions,
    A Comparison of Techniques," Paper presented at the 20th
    Annual Mid-America Symposium on Spectroscopy, May 12-15,
    1969, Chicago, Ill.

6. H. V. Carter, "Simplified Mathematical Matrices for
    Solution Analyses," Norelco Reporter, Vol. XIII, No. 2,
    1966, p. 45-47.

7. G. Andermann and J. W. Kemp, "Scattered X-rays as
    Internal Standards in X-ray Emission Spectroscopy,"
    Analytical Chemistry, Vol. 30, 1958, p. 1306-1309.

8. L. S. Birks, "X-ray Spectrochemical Analysis," Inter-
    science Publishers, Inc., New York, 1959, p. 79.

9. F. L. Chan, "An Apparatus for the Analysis of Liquid
    Samples by the X-ray Fluorescence Method with a Vacuum
    Spectrograph," in L. R. Pearson and E. L. Grove, Editors,
    Developments in Applied Spectroscopy, Vol. 5, Plenum
    Press, New York, 1966, p. 59-75.

SIMULTANEOUS X-RAY EMISSION ANALYSIS OF P, Si, Ca, Fe, Al, AND Mg

IN PHOSPHATE ROCK USING A SMALL COMPUTER TO CORRECT FOR MATRIX

VARIATIONS

Charles N. McKinney and Arthur S. Rosenberg

Continental Oil Company

Ponca City, Oklahoma 74601

ABSTRACT

Phosphate rock from different sources varies widely in composition. This results in a large interelement effect. A mathematical model was developed to correct for matrix variations in
each sample. This model utilizes the X-ray intensity from each of
the six elements considered and the mass absorption coefficients
of the elements to calculate concentrations. Accurate analyses for
P, Si, Ca, Fe, Al, and Mg are obtained using this model. The X-ray
instrument is interfaced with a small computer that performs the
calculations on each sample and prints the results.

Sample preparation, standardization and the computer programs
used for sample calculation and standardization will be discussed.

INTRODUCTION

Chemical analysis of P, Si, Ca, Fe, Al, and Mg in phosphate
rock requires the use of several separate procedures and is very
time consuming. The same six elements can be determined by X-ray
emission analysis in less than ten minutes. The difficulty encountered in X-ray analysis is in correcting for the large interelement effects. The X-ray analysis of phosphate rock is especially
difficult in that four of the elements normally determined are adjacent to each other in the periodic table; thus large mass absorption coefficients are involved. Calculations using mass absorption
coefficients with an empirical correction, as described in this
paper gave excellent results despite these problems.

EXPERIMENTAL

Sample Preparation

The phosphate samples were dried and ground to approximately −100 mesh. Ten grams of sample, 0.3 g of polyethylene glycol and 0.2 g of aspirin were ground for two minutes in a Bleuler rotary mill. The polyethylene glycol served as a binder and the aspirin made cleaning of the grinding dishes easier. The ground samples were then pressed into 1.25 in. diameter pellets at 32,000 psig. The prepared samples were held at a temperature of about 50°C to keep the absorption of moisture to a minimum until the X-ray analysis was performed.

X-ray Analyzer Conditions

An Applied Research Laboratories Vacuum X-ray Quantometer was used to make the X-ray measurements. A chromium target X-ray tube was operated at 50 kV and 25 mA. The sample chamber was evacuated and a one minute counting time was used. A PDP-8S computer (Digital Equipment Corporation) was interfaced to the X-ray instrument to perform the calculations. Spectrometer conditions are listed in table I.

Table I.   Spectrometer Conditions for the X-ray
Analysis of Phosphate Rock

Element	Line	Wavelength, Å	Crystal	Detector
Phosphorus	Kα	6.155	Ge	Proportional
Silicon	Kα	7.126	EDT	Proportional
Calcium	Kα	3.359	LiF	Proportional
Iron	Kα	1.937	LiF	Proportional
Aluminum	Kα	8.338	EDT	Geiger
Magnesium	Kα	9.889	ADP	Geiger

Calibration

Samples of phosphate rock and of Ottawa sand, each of which had been analyzed chemically in triplicate, were used to set up the initial instrumental conditions. The sensitivity and integrator bias controls on the X-ray instrument were set such that the observed intensities agreed with the chemical concentrations of the oxides for these two samples with the following exceptions.

The Si intensity was set to one half of the chemical concentration as it was not possible to instrumentally adjust the slope to unity when the SiO_2 concentration was close to 100%. This was compensated for in computer program with the following:

$$\%SiO_2 = (2)(Si\ Intensity)\ . \qquad (1)$$

Interference of the third order Ca Kα line on the Mg Kα line could be corrected empirically by

$$\%MgO = (Mg\ Intensity) - (0.25)(Ca\ Intensity)\ . \quad (2)$$

Mathematical Model

The interelement effect was compensated for by designing a more comprehensive algorithm. The non-linear relationship between the observed emission intensity for an element and its actual concentration is primarily caused by absorption of the emitted radiation by the sample as a whole. Our model therefore involves the assumption that component concentration is a function not only of observed intensity of the element in question, but of the absorption characteristics of the sample. A correction term is added to the observed intensity which compensates for the attenuation of emitted radiation by the sample. This term involves the mass absorption coefficients and intensities of all six elements. For simplicity the word "intensity" is used with the understanding that equations 1 and 2 are used to correct the Si and Mg intensities respectively. The equation used was

$$\%\ Component_i = Intensity_i + (Intensity_i)(R_i)(K_i), \quad (3)$$

where

$$R_i = \frac{\sum\limits_{j=1}^{6} I_j \mu_j(\lambda_i) \quad Sample}{\sum\limits_{j=1}^{6} I_j \mu_j(\lambda_i) \quad Standard}, \qquad (4)$$

I_j is the intensity of the j-th element and $\mu_j(\lambda_i)$ is the mass absorption coefficient of the j-th component for the radiation of element i. R_i is an approximation to the ratio of the total mass absorption coefficient of the sample to that of the standard phosphate sample used to set up the instrumental conditions. Since attenuation is a function of the amount of the elements present, it is assumed that, for the interfering elements, intensity is proportional to concentration. This undoubtedly introduces some inaccuracy, however, since R_i appears only in the correction term, its effects are minimal.

The mass absorption coefficients used were those given by Heinrich.[1] K_i can be evaluated by rearranging equation 3.

$$K_i = \frac{\%Component_i - Intensity_i}{(R_i)(Intensity_i)} . \qquad (5)$$

A plot of K_P vs. R_P is shown in figure 1 for P_2O_5 concentration ranging from 2 to 36%. The solution of equation

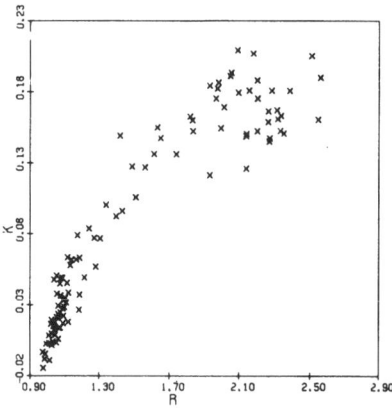

Figure 1. Plot of K vs. R for P_2O_5

5 for K_i as a function of R_i is accomplished readily with the aid of a digital computer. This makes it practical to use a large number of samples in the calibration procedure. In our model, K_i is expressed as a polynomial in R_i. The program calculated the coefficients of the polynomial by the method of least squares. Polynomials of degree 1, 2, 3 or 4 may be used. The calibration coefficients are determined and listed. Equation 6 indicates the method by which these calibration coefficients, listed as a_{ij}, are used to calculate component concentrations:

$$\text{\% Component}_i = \text{Intensity}_i + \text{Intensity}_i (R_i) \left(\sum_{j=0}^{4} a_{ij} R_i^j \right) \qquad (6)$$

This equation is simply the substitution of the least square solution for K_i into equation 3. These calculated values are compared to the known chemical values and the standard deviation is calculated. Polynomials larger than second order did not improve results, therefore, all results reported are for the quadratic equation.

RESULTS

Calibration Using Production Samples

Calibration coefficients for production samples were determined by using 109 samples that had been analyzed chemically at least twice. The equation determined for P_2O_5 was:

$$\%P_2O_5 = \text{Intensity}_p + (\text{Intensity}_p)(R_p)(-0.0970678 +$$
$$0.123790R_p - 0.0225284R_p^2). \qquad (7)$$

When the sample being analyzed has the same total mass absorption coefficient as the sample used to calibrate the X-ray instrument; that is, when $R_P = 1$, the sum of the coefficients is approximately zero and equation 7 becomes:

$$\%P_2O_5 \stackrel{\sim}{=} \text{Intensity}_p. \qquad (8)$$

This condition must be true or the method would not give satisfactory results. The standard deviation of calibration is shown in table II.

Table II. Standard Deviation of Calibration

Component	Concentration Range %	Standard Deviation
P_2O_5	22.6 – 33.8	0.19
SiO_2	2.4 – 33.3	0.45
CaO	32.8 – 49.2	0.42
Fe_2O_3	0.8 – 2.5	0.08
Al_2O_3	0.1 – 2.0	0.19
MgO	0.2 – 2.0	0.08

In order to show that the analytical method would be useful for samples not used in the calibration program, 27 new samples were analyzed using the calibration coefficients derived from the 109 calibration samples. These results were compared to the chemical results and the standard deviations are listed in table III. Also listed in table III are the average standard deviations between different chemical laboratories.

Table III. Comparison of Errors in X-ray Analysis With Those Between Different Chemical Laboratories

| Component | X-ray Analysis | | Chemical Analysis* | |
	Range	Standard Deviation	Range	Average Standard Deviation
P_2O_5	27.4–33.6	0.18	30.2–36.7	0.18
SiO_2	2.6–19.5	0.38	2.3–9.5	0.27
CaO	40.7–49.6	0.26	44.8–52.1	0.56
Fe_2O_3	0.8– 2.3	0.07	0.78–3.6	0.08
Al_2O_3	0.3–1.3	0.17	0.9–3.0	0.16
MgO	0.3–1.1	0.05	---	---

*The average standard deviation obtained for check samples numbered 168, 268, 368, 468, 568, 668, 768, 868, 968, 1068, 1168 and 1268 submitted by the International Minerals and Chemical Corporation to over 50 different laboratories.

Calibration of Exploration Samples

Exploration samples are separated into several fractions by physical tests such as screening and floating. The various fractions have larger concentration ranges for the six elements than do the production samples. The interelement effect thus becomes much more serious. The calibration program was used for 112 samples that had been analyzed one to four times chemically. Figure 2 compares X-ray intensity and calculated X-ray results vs. chemically determined P_2O_5 concentrations. Figures 3, 4, 5, 6, and 7 give the same comparisons for SiO_2, CaO, Fe_2O_3, Al_2O_3, and MgO respectively.

Figure 2 indicates that a more linear relationship is obtained between calculated X-ray results and chemical concentration (A) than is obtained between X-ray intensity and chemical concentration (B) for P_2O_5. Comparisons of standard deviations of calibration for these two models are listed, respectively, under columns (A) and (B) of table IV. Table IV also shows the improvement obtained using our model (A) over that obtained when chemical concentration is expressed as a quadratic function of the X-ray intensity (C).

Comparison of calculated X-ray results (A) and X-ray intensity (B) vs. chemical concentrations for SiO_2 and CaO are shown in figures 3 and 4 respectively. There is little, if any, improvement in calculated X-ray results over a linear plot of X-ray intensity vs. chemical concentration. This is also indicated by the standard deviations of calibration listed under columns (A) and (B) of figure IV.

A marked improvement is shown in figures 5 and 6 for Fe_2O_3 and Al_2O_3, respectively, using our model (A) over a plot of X-ray intensity vs. chemical concentration (B). In the case of Al_2O_3, however, a large scatter in the data points remains. This could be due to the fact that chemical methods are used to determine acid soluble Al_2O_3 while the X-ray method measures the total aluminum X-ray radiation. Also, the relative error in the chemical determination of Al_2O_3 is larger than for any of the other components determined as can be seen from table III. A comparison of standard deviations of calibration in table IV shows that our model (A) gives a significant improvement over both the straight line relationship (B) and the quadratic relationship (C).

Table IV. Comparisons of Standard Deviations
Obtained with Models A,B,C and D

Component	Range	Standard Deviation			
		A	B	C	D
P_2O_5	2.4 - 36.8	0.40	0.94	0.47	0.29
SiO_2	1.1 - 90.6	1.94	1.97	1.95	1.00
CaO	3.0 - 51.0	0.57	0.59	0.58	0.60
Fe_2O_3	0.1 - 4.8	0.26	0.60	0.57	0.25
Al_2O_3	0.0 - 2.9	0.32	0.47	0.48	0.17
MgO	0.0 - 1.9	0.07	0.15	0.08	0.03

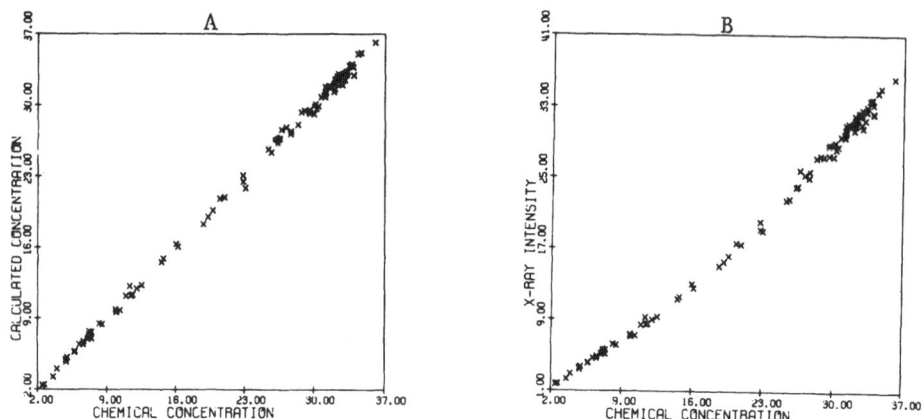

Figure 2. Comparison of calculated X-ray results (A) and X-ray intensity (B) vs. chemical concentration for P_2O_5.

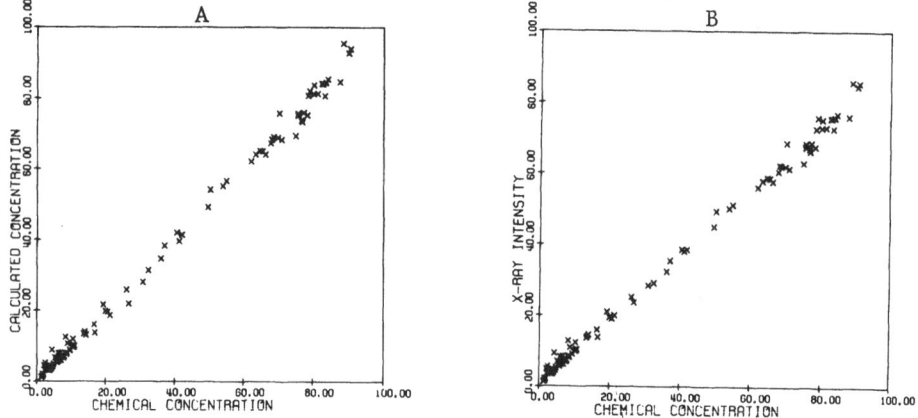

Figure 3. Comparison of calculated X-ray results (A) and X-ray intensity (B) vs. chemical concentration for SiO_2.

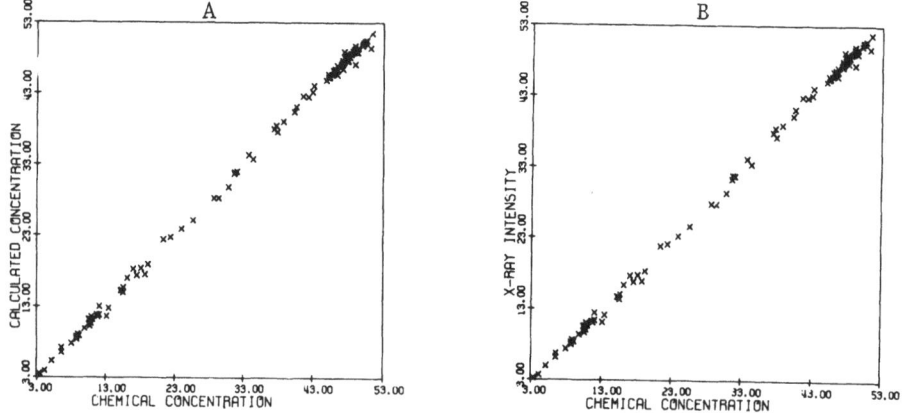

Figure 4. Comparison of calculated X-ray results (A) and X-ray intensity (B) vs. chemical concentration for CaO.

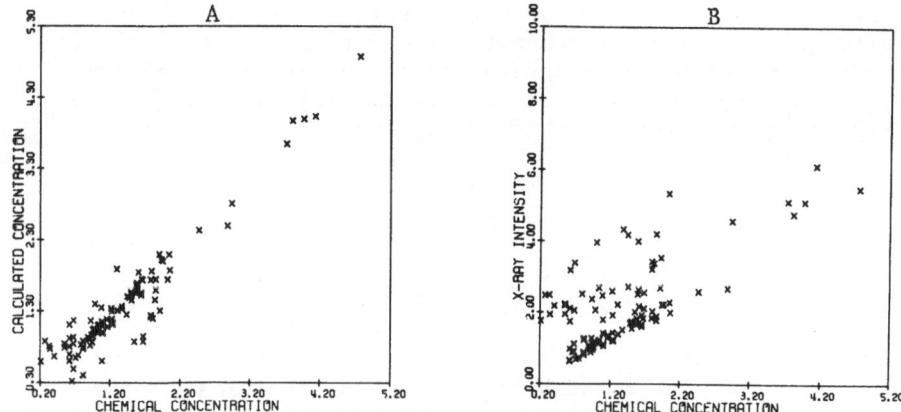

Figure 5. Comparison of calculated X-ray results (A) and X-ray intensity (B) vs. chemical concentration for Fe_2O_3.

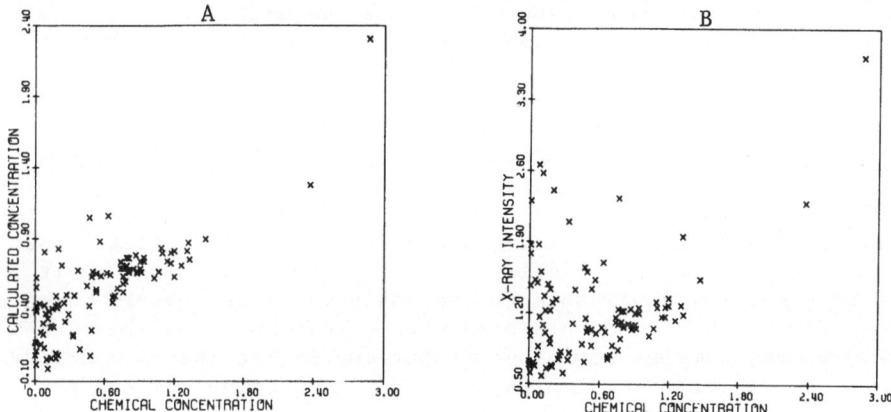

Figure 6. Comparison of calculated X-ray results (A) and X-ray intensity (B) vs. chemical concentration for Al_2O_3.

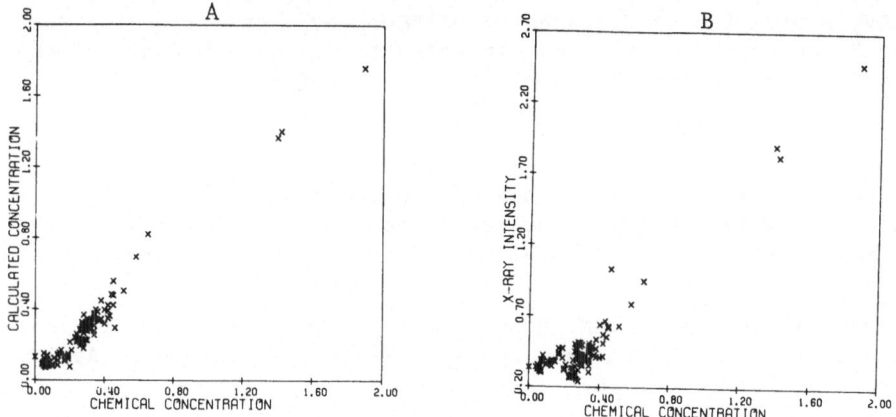

Figure 7. Comparison of calculated X-ray results (A) and X-ray intensity (B) vs. chemical concentration for MgO.

Figure 7 compares our calculated results (A) and X-ray intensity (B) vs. chemical concentration for MgO. A two fold improvement is shown in the standard deviations of model (A) over model (B) in table IV. The results from models (A) and (C) are essentially equivalent.

A 28 term polynomial model, as described by Mitchell[2,3], was also evaluated in this study. The equation used was

$$C_i = a_i + \sum_{j=1}^{6} (a_{ji} I_j + I_j \sum_{k=j}^{6} b_{jki} I_k), \qquad (9)$$

where the coefficients a and b were determined by least squares analysis. Standard deviations of calibration for this method are shown in column (D) of table IV. Although it would appear that this is the method of choice, the range of component concentrations in our standards was not broad enough to make the method reliable. Small perturbations in the X-ray intensity of one or more elements often introduced extremely large errors. For this reason, we rejected the 28 term model.

CONCLUSIONS

A mathematical model has been developed that allows an accurate and rapid X-ray emission analysis of phosphate rock. The method was developed specifically for phosphate rock, however, it should be generally applicable to the analysis of any system where the majority of the components are known. This model has been used to analyze many samples having elemental concentrations far outside the range of calibration samples, without introducing serious errors.

ACKNOWLEDGMENT

We wish to thank Mr. John F. Erving of the AGRICO Chemical Company, a division of Continental Oil Company, for collecting most of the X-ray data. We are also indebted to Mr. Joseph Padar of the AGRICO Chemical Company for supplying the chemical analyses.

REFERENCES

1. K. F. J. Heinrich, "X-ray Absorption Uncertainty" in
    E. D. McKinley, K. F. J. Heinrich, and D. B. Wittry (eds.), The
    Electron Microprobe, John Wiley and Sons, New York, 1966, pp. 296-
    377.

2. B. J. Mitchell and F. N. Hopper, "Digital Computer Calculation
    and Correction of Matrix Effects in X-ray Spectroscopy," Appl.
    Spectry. 20: 172, 1966.

3. B. J. Mitchell, "Applications of Computerized Statistical Techniques in Quantitative X-Ray Analysis," in: J. B. Newkirk, G. R. Mallett and H. G. Pfeiffer (eds.), Advances in X-ray Analysis, Vol. II, Plenum Press, New York, 1968, pp. 129-149.

SOFT X-RAY VALANCE STATE EFFECTS IN CONDUCTORS

J. E. Holliday

United States Steel Corp. Research Center

Monroeville, Pennsylvania 15146

ABSTRACT

Investigation of the NiL_{III} band at various excitation voltages and the use of Ni and Ti thin films in the path of the $NiL_{II,III}$ and $TiL_{II,III}$ radiation, has shown that with proper electrode geometry and using target voltages of about 4 kv the effect of self absorption on emission edge shift and band shape is negligible for these elements. However, the L_{II}/L_{III} intensity ratio of titanium is strongly effected by self absorption, but it is not effected for nickel and iron by self absorption.

A survey of several investigators' soft x-ray band spectra from alloys and conducting compounds showed that the observed band changes were related to the electronegativity difference of the combining elements or the ionic character of the band. Peaks appearing on the low energy side of transition metal bands when combined with 2nd period elements appear to be a cross transition between the 2s or 2p band of the non-metal and inner level of the metal. Although peak shift is the band change that is easiest to measure quantitatively and is related to electron transfer or valance, the whole band change must be considered and compared with electronic structure theory and other measurements before the amount and direction of electron transfer is determined. It is shown that the ionic character of the transition metal carbide bond, as measured by soft x-ray band spectra and electron spectroscopy (ESCA), is a strong factor in the bond strength of these carbides.

INTRODUCTION

Alterations in soft x-ray emission spectra due to chemical combination have been observed for a number of years. It is only recently that systematic studies have been made as to the basic types and reasons for these alterations. There are three basic types of changes: (1) shift in peak position, (2) change in intensity distribution, (3) appearance of additional peaks on the high and low energy side of the band

It is generally found that these changes, relative to the pure element, are more pronounced in insulators than in conducting compounds and alloys. However, significant changes are observed in the soft x-ray emission spectra from elements forming conducting compounds and alloys. In this paper only soft x-ray emission spectra from conducting materials will be considered.

Self Absorption

There has been some controversy over the extent to which self absorption alters the shape of the emission bands. For this reason it is necessary to consider the effects of self absorption on the shape of the emission band before discussing the changes in soft x-ray emission bands due to chemical combination. Liefeld[1] has argued, from his experiments on the NiL_{III} band, that the corresponding emission edge becomes sharper with increasing target voltage above 2 kv. Below 2 kv the edge becomes sharper with decreasing target voltage due to a reduction in the contribution from satellites on the high energy side of the band. Holliday[2] however, has shown that alterations in the shape of the emission band can also occur for increasing target voltages because of surface contamination. With proper baking of the target and ion bombardment cleaning it is possible to keep the target clean below the first few monolayers for a vacuum in the 10^{-8} region. Also, self absorption can be kept to a minimum by using a large incident electron angle (about 60° in this experiment) and a 90° x-ray takeoff angle.

Using the above precautions, Liefeld's measurements on the NiL_{III} band were repeated to observe if there were still changes in the emission band with increasing excitation voltage. The NiL_{III} band edge was measured with a 3600 groove/mm, blazed grating. The NiL_{III} edge from Ni was measured at three different target voltages and compared to the NiL_{III} emission edges measured by Liefeld in figure 1. The NiL_{III} emission edge measured by Liefeld at 1.5 kv agrees best with the present data. In contrast to Liefeld's work there is no significant effect of voltage on the NiL_{III} edge indicating that self absorption is negligible at the target voltage (4 kv) used in the present measurements. The reduction of satellite

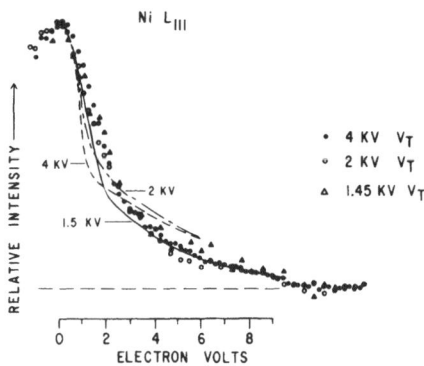

figure 1. Comparison of normalized NiL$_{III}$ edges from Ni. Plotted
points obtained by Holliday and the smooth curves obtained by
Liefeld[1].

intensity below 2 kv reported by Liefeld was also not observed in
the present measurements,because the NiL$_{III}$ edge measured at 1.5 kv
is not below that measured at 2 kv but is slightly above it. No
shift of the NiL$_{III}$ peak in wavelength was observed with increasing
target voltage as reported by Liefeld.

The lack of any noticeable self absorption in the present
measurements could be due in part to the differences in depth of
the effective x-ray source. Differing surface conditions might also
contribute to the disagreement since Liefeld operated his targets
at 700°C to eliminate carbon contamination, which could have
resulted in oxidation of the Ni surface.[2] Cuthill et al[3] state
that self absorption is not a problem in measuring the NiM$_{II,III}$
band from Ni.

A more direct method of determining the effect of self absorption on emission bands is to place a film of the material under
investigation between the grating and the detector as reported by
Holliday[2] at the Glasgow Conference. The NiL$_{II,III}$ bands from Ni
with and without a 500 Å Ni filter are compared in figure 2. The
position of the NiL$_{III}$ edge of the filter (arrow) with respect to
the NiL$_{III}$ emission edge shows there is overlap between the
emission and absorption edges. From the results it will be seen
that the 500 Å Ni filter has had no noticeable effect on the NiL$_{III}$
edge and the NiL$_{II}$/L$_{III}$ intensity ratio. In a similar experiment
no effect was observed on the FeL$_{II}$/L$_{III}$ intensity ratio. According
to Anderson's[4] experimental studies for the present electrode
geometry and accelerating voltage (4 kv) 50% of the x-rays in Ni
will originate at a depth of approximately 500 Å. Thus, the failure
of a 500 Å Ni filter to show any appreciable absorption is consistent
with the absence of self absorption with a 4 kv primary beam.

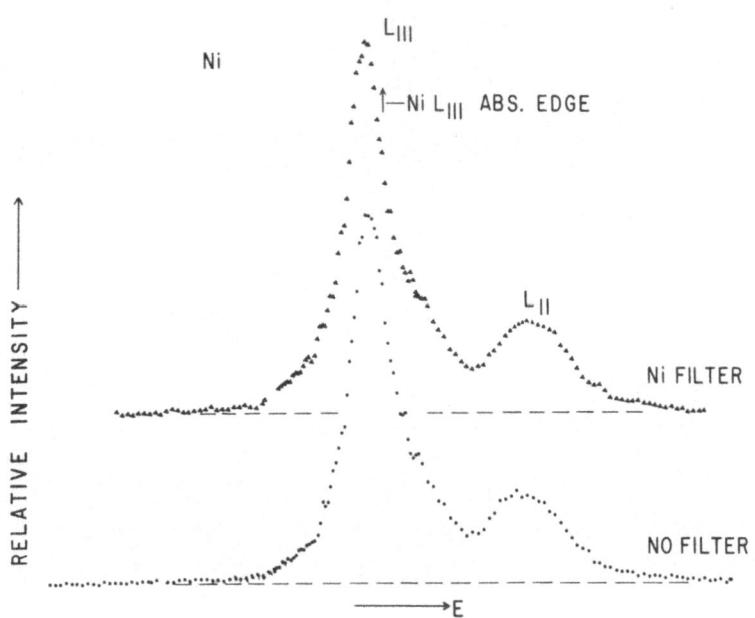

figure 2. The effect of a 500 Å Ni film on the NiL$_{II,III}$ band from Ni (normalized). The arrow indicates the wavelength position of the NiL$_{III}$ absorption edge for the 500 Å Ni film.

Greater thicknesses of Ni were found to reduce the NiL$_{II,III}$ intensity to too low a level for good statistics (a 500 Å Ni filter reduced the peak intensity of the NiL$_{III}$ band by a factor of 10).

To check for self absorption in the case of thicker films a 1000 Å film of Ti was placed in the path of the TiL$_{II,III}$ x-ray emission. Since Ti is a good getter for oxygen, it is probable that the evaporated Ti film would be partially oxidized and the TiL$_{III}$ absorption edge of the filter would be at higher energy relative to that of pure Ti.* To insure sufficient overlap of the TiL$_{III}$ absorption and emission edges a TiO$_{1.02}$ specimen target was used. It will be seen from the wavelength position of the TiL$_{III}$ absorption and emission edges in figure 3 that there is a sufficient overlap between them. The effect of the Ti filter is to

---

* The Ti L$_{III}$ absorption edge of titanium oxide is at a higher energy than the Ti L$_{III}$ absorption edge of titanium.

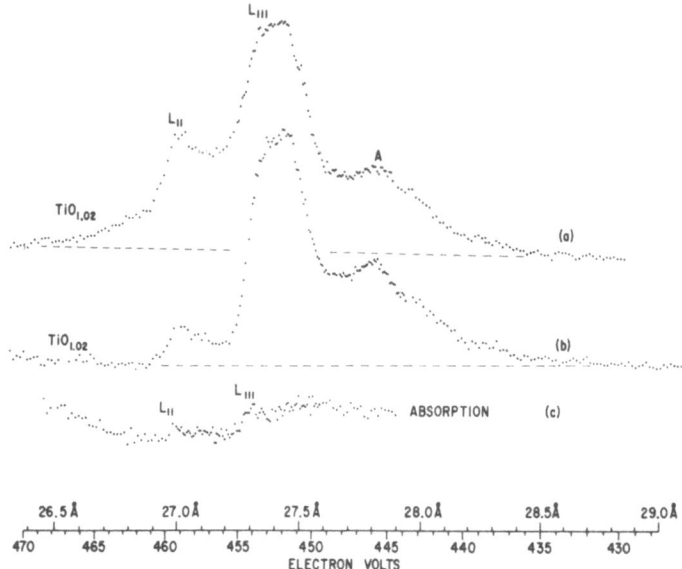

figure 3. The effects of a 1000 Å Ti film on the TiL$_{II,III}$ emission band from TiO$_{1.02}$ (normalized). Curve a is without the Ti filter and curve b is with the Ti filter. Curve c is the absorption spectra of the 1000 Å Ti film. [Reprinted by permission of J. E. Holliday in reference 2.]

cause a reduction in the TiL$_{II/III}$ intensity ratio, and reported changes in the TiL$_{II}$/L$_{III}$ intensity with chemical combination would be due in part to self absorption.

Fischer and Baun[5] have also demonstrated the effects of self absorption on the TiL$_{II,III}$ emission band which are caused by a 600 Å Ti filter. Their results are presented in figure 4 and also show that the TiL$_{II}$/L$_{III}$ intensity ratio is reduced by self absorption.

From figures 3 and 4 it would appear that self absorption has caused the TiL$_{III}$ edge to have a greater slope. Liefeld has argued that self absorption was the reason Skinner's[6] emission band showed sharp edges. However, normalizing the base and the peak of the TiL$_{III}$ emission edges obtained with and without the Ti filter shows that self absorption has not changed the slope of the emission edge. This is seen in both the present work in figure 5 and in Fischer and Baun's results in figure 6. The TiL$_{III}$ edge measured with the filter lies directly over that measured without the filter; thus, self absorption reduces the background on the high energy side of the absorption edge but does not change the wavelength position of the emission edge.

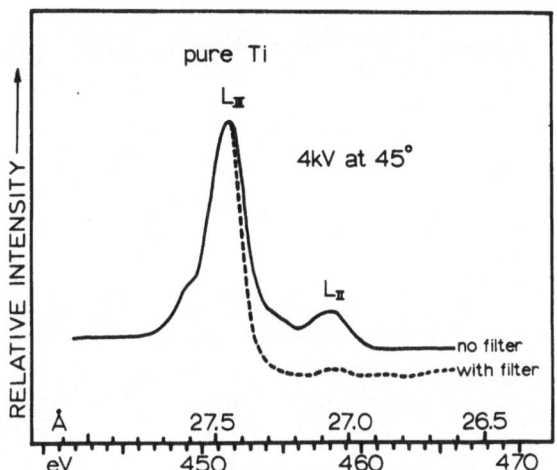

figure 4. Titanium L$_{II,III}$ emission spectrum from pure Ti metal.
[Reprinted by permission of David W. Fischer and William L. Baun,
J. Appl. Phys. 39, 4757 (1968)].

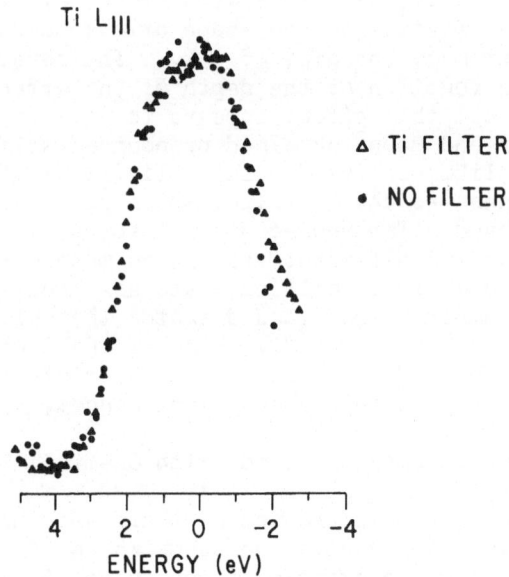

figure 5. Normalization of the base and the peak of the TiL$_{III}$
edges for curves a and b in figure 3.

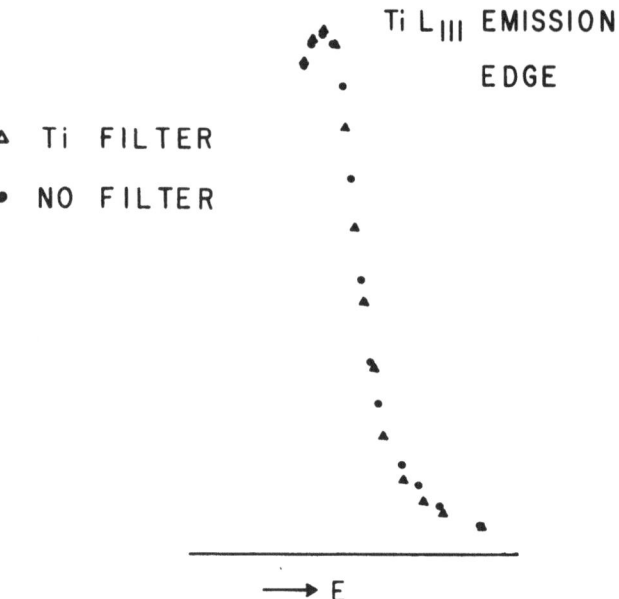

figure 6. Normalization of the base and the peak of the TiL$_{III}$ emission edges from Fischer and Baun's data in figure 4.

The above results show that with proper electrode geometry, self absorption effects (on the shape of the band) can be negligible even at primary energies of 4 kv. The severity of self absorption is a function of the depth of the effective x-ray source in the target, but this effect appears to vary for different elements. Fischer and Baun[5] obtained pronounced self absorption changes on the TiL$_{II,III}$ band using a Ti filter of approximately the same thickness as the Ni filter which produced no changes on the NiL$_{II,III}$ band. The degree of self absorption effects on the shape of the band by different transition metals appears to be related to the number of unfilled d states. Thus, titanium, which has a greater number of unfilled d states than Ni, shows a greater self absorption effect[1] than Ni. Fischer and Baun[7] have shown that the emission bands from rare earth elements (which have empty 4f states) show strong self absorption changes.

Alterations in Emission Bands with Chemical Changes

As outlined in the introduction there are three basic types of changes in soft x-ray emission bands which can occur when elements are combined. An example of the change in intensity distribution is shown by the work of Appleton and Curry[8] on the MgL$_{II,III}$ band from Mg$_2$Ni in figure 7 where it is evident that a radical change occurred when Mg was combined with Ni. The occurrence of additional peaks in the emission bands, as a result of

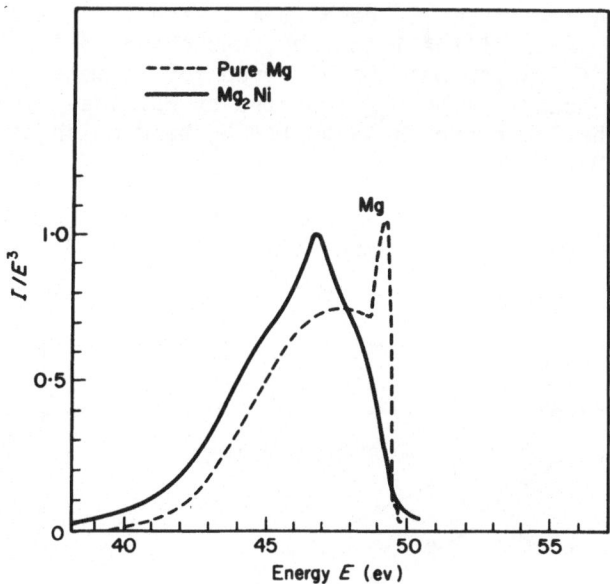

figure 7.   Magnesium L$_{II, III}$ emission from Mg$_2$Ni and from pure Mg.
[Reprinted by permission from A. Appleton and C. Curry, Phil. Mag.
<u>16</u>, 1031 (1967)].

chemical combination is illustrated in figure 8 from previous work

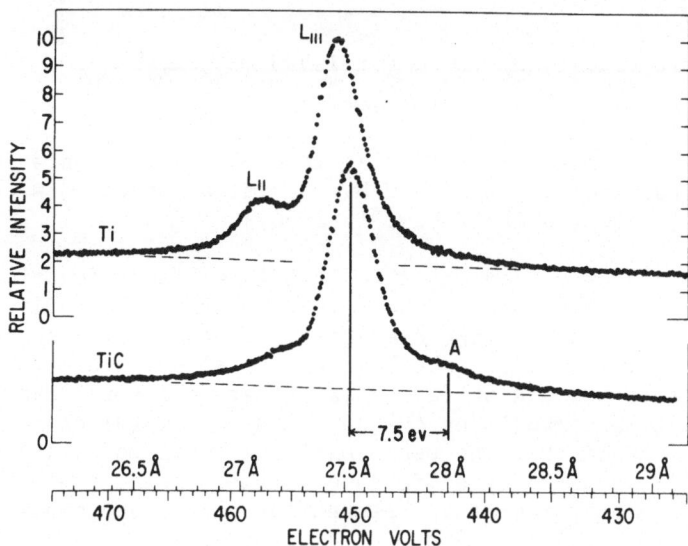

figure 8.   Titanium L$_{II, III}$ emission bands (4s+3d→2p transition) for
Ti, TiC$_{.95}$ (peaks normalized). Peak A is a cross transition.
[Reprinted by permission of J. E. Holliday in reference 9a.]

(Holliday[9-9a]) on the TiL$_{II,III}$ band from Ti and TiC. A peak A,
on the low energy side of the main peak, was observed for the
TiL$_{III}$ band from TiC which was not observed for Ti metal. In
addition, the M$_V$ band from NbC$_{.85}$ reported by Holliday[10] has a
peak 9.6 ev on the low energy side of the M$_V$ band which is not
present in pure Nb.

    The third type of change is shown by the work of Fischer and
Baun[11] on the AlK shift for a number of Al alloys in figure 9. In

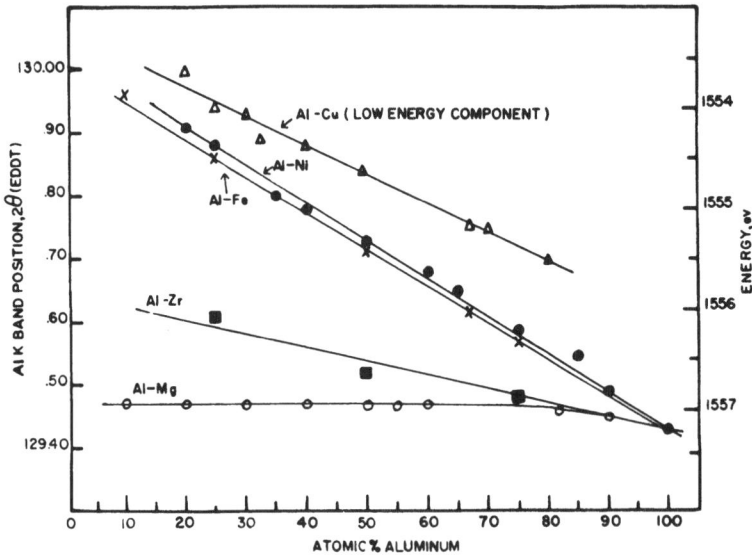

figure 9. Relationship of shift in AlK band position as a function
of alloy composition. ⌊Reprinted by permission of David W. Fischer
and William L. Baun, AFML-TR-66-191 (1966)⌋.

these alloys the amount of the AlK shift is a linear function of
at % Al even though several phases exist between 0-100% Al. For
a given percentage of Al, the AlK shift increases with the increase
in electronegativity difference between the alloying elements.
These results on the AlK shift indicate that soft x-ray bands are
more sensitive to composition changes than structural changes.
Fassler[12] has shown that the peak shift for insulators is determined by electronegativity difference of the combining elements,
but it has been only recently observed for alloys and conducting
compounds.

    In some cases, the soft x-ray bands from an alloy have not
changed relative to the pure element. This was shown by Cleft,

Curry and Thompson[13] for the Cu and Ni $M_{II,III}$ band from Cu Ni alloys in figure 10. The shape and peak wavelength of the emission band

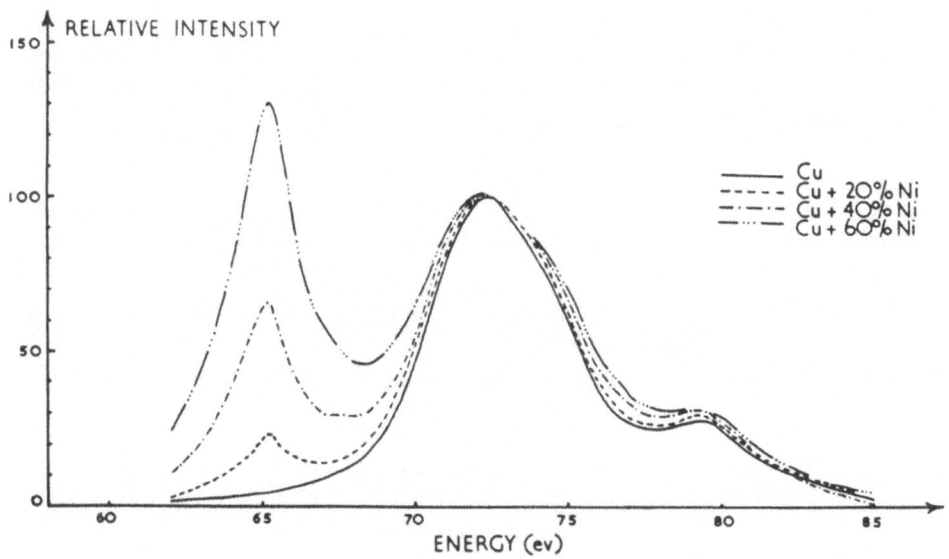

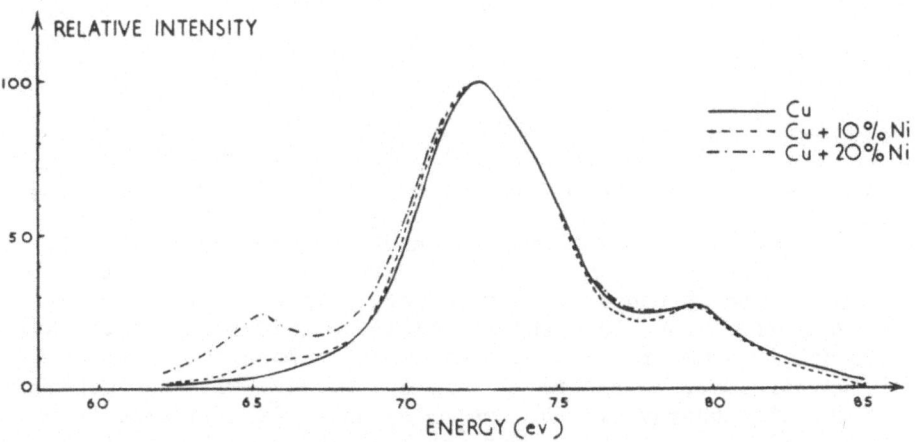

figure 10. The $CuM_{II,III}$ and $NiM_{II,III}$ emission band from 0,10,20, 40,60% nickel in copper. [Reprinted by permission of J. Clift, C. Curry and B. J. Thompson, Phil. Mag. 8, 593 (1963)].

from the element were retained on alloying. Here the lack of any
change is probably due to the fact that there is only a small electronegativity difference between the alloying elements. This result was
a surprise to the band theorists who held that the density of states
did not vary when close to an individual atom.

Further evidence for the effect of electronegativity difference
on soft x-ray emission bands of alloys and conducting compounds was
observed when an element is combined with elements in a given group
of the periodic table. The work of Fischer and Baun[11], figure 11,

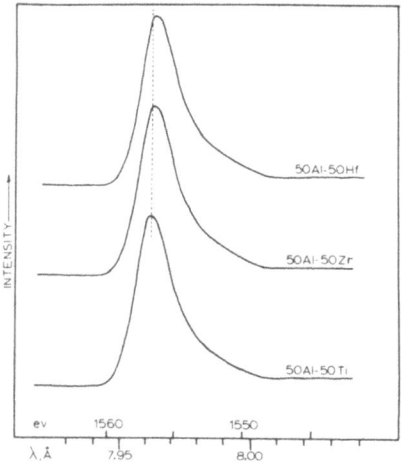

figure 11. Aluminum K emission bands for Al combined with Group IV
transition metals. [Reprinted by permission of David W. Fischer
and William L. Baun, AFML-TR-66-191, p. 77 (1966)].

shows that when Al is combined with Group IV transition metals the
AlK band has the same intensity distribution for all three alloys.
This effect was also observed by Holliday[9], in figures 12 and 13,
for the CK band when carbon is combined with Group IV and V transition metals. The CK bands for a given group have the same shape.
The CK bands from Group IV transition metal carbides appear to have
a single nearly symmetrical peak, while the CK band for Group V
transition carbides, in figure 13, are all similar with an additional
peak on the high energy side of the main peak. It will also be observed from figures 12 and 13 that the CK band intensity distribution from transition metal carbides has changed radically compred
to the CK band from graphite. In figure 13 Cr_3C_2 is included with
Group V carbides because its CK band shape and bonding properties
are more similar to Group V than Group VI carbides.

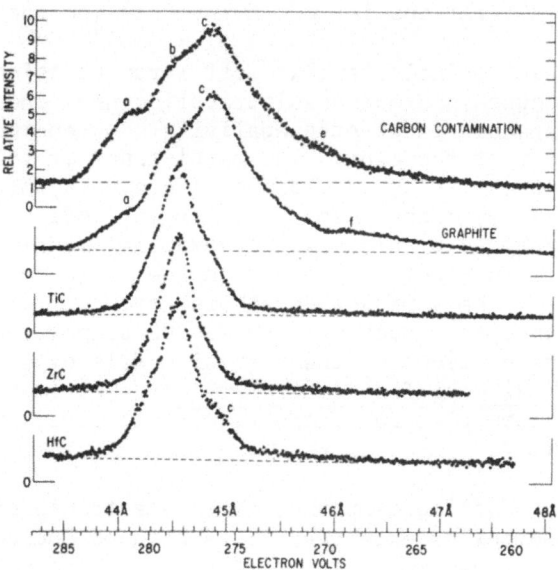

figure 12. Carbon K-emission bands (normalized): for carbon contamination for graphite; and for Group IV carbides, TiC, ZrC, and HfC. [Reprinted by permission of J. E. Holliday in reference 2.]

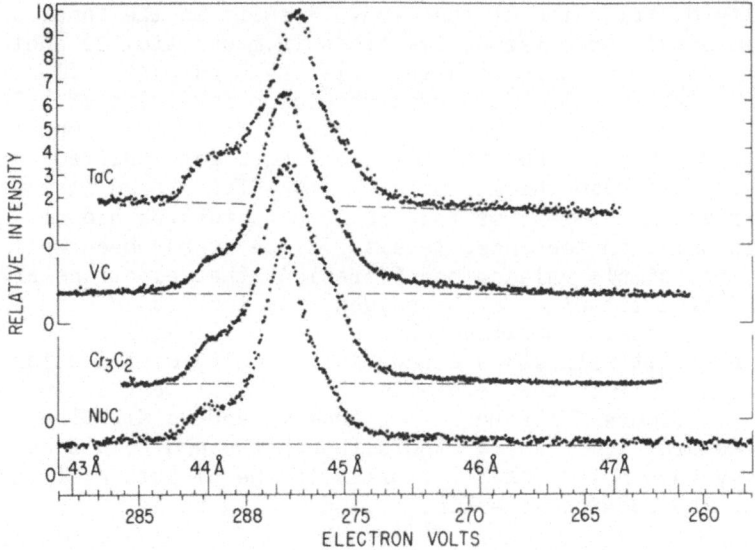

figure 13. The carbon K-emission bands (normalized) for the Group V transition metal carbides TaC, VC, and NbC and Group VI Cr_3C_2. [Reprinted by permission of J. E. Holliday, J. Appl. Phys. 38, 4720 (1968).]

Discussion of Alterations in Soft X-ray Emission Bands

The above results indicate that soft x-ray emission bands are
influenced by changes in electron distribution near individual atoms
produced by differences in electronegativity between the elements or
the ionic character of the bond. The persistence of the electron
distribution characteristics of individual elements, which has been
revealed by soft x-ray measurements, has already modified theories
of electronic structure of alloys. For example, Nagel[14a] has proposed that changes in electron density in the vicinity of solute
atoms might explain the apparent departures from the classical density of states concepts. Much more extensive comparison of soft
x-ray data with predictions of theoretical models of electronic
structure as well as data from other types of experiments is needed
before a satisfactory picture of the electronic structure of alloys
can be developed.

One of the conducting compounds whose electronic structure has
been studied by number of experimental techniques including soft
x-rays, and one that a number of theoretical studies has been made
on is TiC. Figure 8 shows that the TiL_{III} band from TiC is more
symmetrical than from Ti, the peak has shifted towards lower energy,
and as indicated above, a peak A appears on the low energy side.
These alterations indicate a change in the electron density around
the Ti atom, and some ionic character to the Ti-C bond. A shift in
the peak is either due to a shift in the inner levels or a change
in intensity distribution of the band. A shift in the inner level
reflects an actual transfer of electrons from one atom to another.
In order to determine whether a shift in the band peak reflects a
shift in the levels, the whole band should be investigated. In
figure 14 is shown the $TiL_{II,III}$ band for Ti from TiO_x compounds
obtained by Holliday.[2] Not only is the TiL_{III} peak shifted towards
higher energy but also the end points of the TiL_{II} band indicating
that the shift in the peak of the TiL_{III} band towards higher energy
is due to a shift in the inner levels. Since it has been well established from simple valance considerations that electrons are
transferred from titanium to the oxygen, the increasing band shift
reflects an increasing positive charge on the Ti metal atom. A case
where a peak shift reflects a change in intensity distribution
rather than a shift in the levels is shown by the work of Appleton
and Curry[8] in figure 7 for $MgL_{II,III}$ from Mg and Mg_2Ni. The peak
is 2.5 ev towards lower energy and although the end points have
shifted they have not shifted uniformly in the direction of the
peak as was the case for the TiL_{III} band in figure 14.

However, the $TiL_{II,III}$ band end points from TiC in figure 8
appear to have shifted uniformly with the peak towards lower energy,
indicating a change in the energy separation between the 3d, 4s
bands and the $2pj_{1/2}$, $2pj_{3/2}$ levels due to an electron transfer.
The TiL_{III} peak shift of TiC_x compared to that of TiO_x, as x

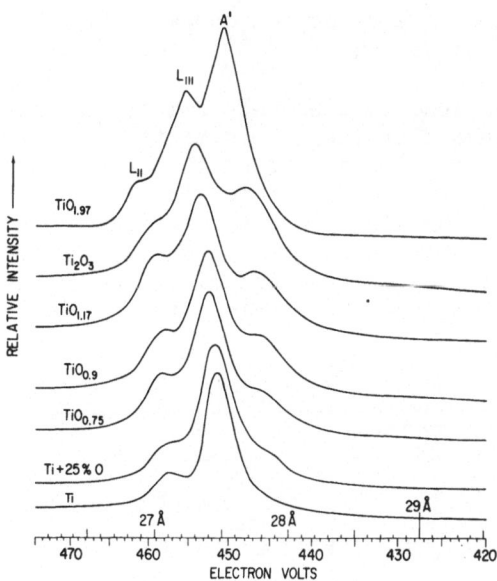

figure 14. TiL$_{II,III}$ emission bands (normalized) for Ti, Ti + 25% O in TiO$_x$ oxides. [Reprinted by permission of J. E. Holliday in reference 2].

is increased is shown by Holliday[9] in Table I. The TiL$_{III}$ band from

Table I

Shifts in peak of Ti and V L$_{III}$ bands.[*]

Material	Shift in L_{III} peak	
	ΔE (eV)	Δλ (Å)
Ti L_{III}[a]		
TiC$_{0.95}$	−1.2	+0.07
TiC$_{0.93}$	−1.0	+0.065
TiC$_{0.83}$	−0.6	+0.04
Ti	...	...
TiO$_{0.9}$	+0.9	−0.06
TiO$_{1.17}$	+1.8	−0.11
Ti$_2$O$_3$	+2.3	−0.14
TiO$_{1.97}$	+3.5	−0.20
V L_{III}[b]		
VC	−1.0	+0.05
V	...	...
V$_2$O$_5$	+1.9	−0.09

[a] Ti L_{III} peak shift relative to Ti L_{III} band from Ti.
[b] V L_{III} peak shift relative to V L_{III} band from V.
[*]Reprinted by permission of J.E.Holliday in Ref. 9

TiC_x shifts uniformly with a direction opposite to that of TiO_x.
This has also been found for VC and V_2O_5.

This result could lead to the speculation that the transfer of
electrons in TiC is in a direction opposite to that in TiO. The
direction of electron transfer according to Lye and Logothetis[14]
LCAO band calculations is from carbon to titanium (which is opposite to that expected from electronegativity considerations), because the 2p band of carbon was higher in energy than the 3d band
of titanium. However, the APW band calculated on TiC by Ern and
Switendick,[15] shown in figure 15, predicts the 2p band of carbon

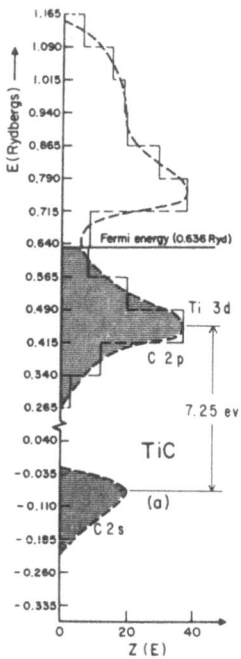

figure 15. Density of states histogram for TiC. Calculated by Ern
and Switendick using A.P.W. method. ⌊Reprinted by permission of
V. Ern and A. C. Switendick, Phys. Rev. 137A, 1927 (1965)⌋.

to be at the bottom of 3d band of titanium, and any electron transfer would be from the 3d band of Ti to the 2p band of carbon. Although they did not assume any ionic character in there self consistent field for TiC, their final calculations indicated that they
should have assumed some ionic character to the bond.

Since the direction of electron transfer depends on the position of the carbon 2p band relative to the 3d band, the direction
of transfer could be established if the energy position of the carbon 2p band was determined experimentally. Blochin and Shuvaev[16]

have shown that some of the peaks that appear on the low energy side of the K emission bands of TiX compounds, (where X is a 2nd period element) is a cross transition between the 2p band of the non metal and the 2p level of titanium. Fischer and Baun[5] have presented arguments that peak A' for TiO_x oxides in figure 14 is a cross transition between the 2p band of oxygen and the 2p level of titanium. It does not seem likely that peak A for TiC in figure 8 is a cross transition between the 2p band of carbon and the 2p level of titanium. Peak A' in figure 14 is 6.5 ev from the main peak while peak A in figure 8 is approximately 7.5 ev. If peak A was the 2p band of carbon then TiC would be more ionic than TiO (because the 2p level of carbon for TiC would be lower than the 2p band of oxygen in TiO) which is contrary to theory and experiments on these compounds. Since 7.5 ev is close to the 7.25 ev calculated by Ern and Switendick[15] for the separation of the 3d and 2s bands in TiC, it would appear that peak A is a cross transition from the 2s band of carbon to the 2p level of Ti.

Further support for electrons being transferred from carbon to titanium is obtained from the electron spectroscopy (ESCA) measurements by Lars Ramquist.[17] The advantage of electron spectroscopy measurements for determining level shifts is that the absolute energy value of the level is measured and not a difference between levels as in the case for soft x-ray spectroscopy. In Table II is

TABLE II

Shifts of the metal $1s$, $2p_{3/2}$, $3d_{5/2}$, $4f_{7/2}$ and carbon $1s$ binding energies and the metal $K\alpha_1$, $L\alpha_{1,2}$ and $L\beta_1$ lines in cubic carbides with varying carbon content relative to pure elements *

Carbide	Shift, eV					
	Ti $1s$	Ti $2p_{3/2}$	C $1s$	Ti $K\alpha_1$		
$TiC_{0.97}$	1.0 ± 0.4	1.3 ± 0.1	-3.3 ± 0.1	-0.33 ± 0.02		
$TiC_{0.86}$		$1.3_5 \pm 0.1$	-3.3 ± 0.1			
$TiC_{0.78}$		0.9 ± 0.1	-3.3 ± 0.1			
$TiC_{0.68}$		$0.9_5 \pm 0.1$	-3.3 ± 0.1	-0.09 ± 0.04		
$TiC_{0.59}$		0.6 ± 0.1	-3.3 ± 0.1			
	V $1s$	V $2p_{3/2}$		V $K\alpha_1$		
$VC_{0.85}$	1.5 ± 0.4	1.8 ± 0.1	-2.6 ± 0.2	-0.21 ± 0.03		
$VC_{0.75}$		1.5 ± 0.1	-2.6 ± 0.2	-0.09 ± 0.03		
		Nb $3d_{5/2}$			Nb $L\alpha_{1,2}$	Nb $L\beta_1$
$NbC_{0.94}$		$1.2_4 \pm 0.1$	-2.9 ± 0.2		0 ± 0.04	0 ± 0.06
$NbC_{0.90}$		$1.2_8 \pm 0.1$	-2.9 ± 0.2		0 ± 0.04	0 ± 0.06
$NbC_{0.77}$		$1.4_0 \pm 0.1$	-3.0 ± 0.2		0 ± 0.04	0 ± 0.06
		Ta $4f_{7/2}$			Ta $L\alpha_{1,2}$	
$TaC_{0.99}$		1.6 ± 0.2	-2.9 ± 0.2		0 ± 0.04	
$TaC_{0.96}$		1.7 ± 0.2	-2.9 ± 0.2		0 ± 0.04	
$TaC_{0.90}$		1.6 ± 0.2	-2.9 ± 0.2			
$TaC_{0.74}$		1.8 ± 0.2	-3.1 ± 0.2			

*Reprinted by permission Lars Ramquist, Jernkont. Ann. 153, 159 (1969)

shown the shift of the inner levels of the metal atom from TiC_x,
VC_x NbC_x and TaC_x by Lars Ramquist. The shifts of the inner level
of the metal atom is toward the nucleus while the 1s level of carbon
is moved away from the nucleus relative to the levels in the uncom-
bined elements. The direction of the level shifts indicate that
carbon is transferring electrons to titanium.

The change in the carbon 1s level is not readily observed in
the CK band peakshift from the transition metal carbides because
there are large changes in intensity distribution for the CK bands.
The peak shift was found to be due to a change in weighting of the
subpeaks a, b', b, c, d, e, anf f in figure 16 rather than a change

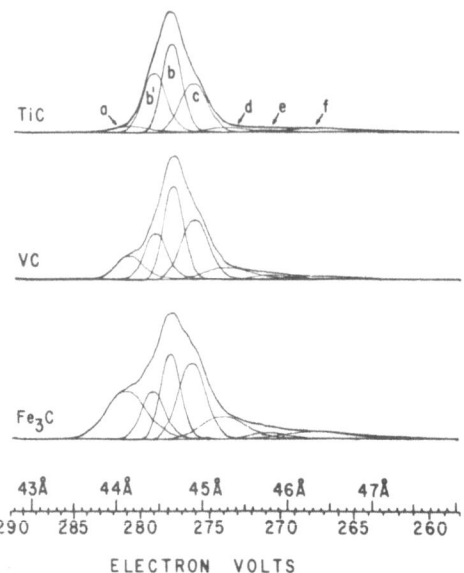

figure 16. The CK-emission bands for TiC, VC and Fe_3C resolved into
Gaussian curves a, b', b, c, d, e and f. [Reprinted by permission
of J. E. Holliday in reference 2.]

in the energy difference of the levels. The method of resolving
the CK band into these subpeaks has been described elsewhere by
Holliday.[2] The intensity distribution of the CK bands of graphite,
TiC, VC and Fe_3C could be reproduced by changing the relative heights
of the sub-bands without changing their energy positions. The electron transfer is, however, reflected in the intensity distribution
of the CK band. From the electronegativity difference of the metal
and the carbon atoms, the ionic character of the bond, for transition metal carbides, increases with decreasing group number carbide.
Thus, from figure 16, the more ionic the carbide the narrower is

the CK band and the more a single subpeak tends to predominate.

Bonding

From the above discussion of the alterations in the soft x-ray emission bands it is clear that electronegativity difference between the combining elements, or ionic character of the bond, is a determining factor in the observed changes in emission bands. It has been shown by Holliday[9] that when the ionic character of the bond is a factor in bond strength, changes in soft x-ray emission bands can be related to the strength of the bond. In Table III

Table III

CK Band Parameters for Transition Metal Carbides

Material	Peak Shift[a] Δ (ev)	Peak Shift[a] Δλ Å	Peak Wavelength	Half-width[b] $W_{1/2}$ (ev)	Asymmetry[b]	$\Delta H^{\circ}_{298}/C$ atom Enthalpy	Melting[c] Point °C
Graphite	—		44.85 Å	6.	0.83		
Diamond	+2.1	—0.33	44.52	8.1	1.25		
Group A							
TiC	+1.95	—0.31		3.0	1.1	—43	3200
VC	+1.8	—0.29		3.3	1.45	—28	2850
Cr_3C_2	+1.9	—0.30		3.3	1.6	—10.5	1870
ZrC	+2.05	—0.325		2.4	0.85	—44	3530
NbC	+1.9	—0.30		2.4	1.05	—33	3500
HfC	+2.0	—0.32		3.0	0.80	—50	3800
TaC	+1.2	—0.19		3.0	0.80	—33	3880
$TaC_{.8}$	+1.7	—0.27		2.7	1.7		3400
Group B							
$(MnCo)_4C$	+1.2	—0.19		5.2	0.9		
Fe_3C[d]	+1.8	—0.29		4.4	1.25	+ 5.98	1650
80%[e] Martensite } 20% Austenite }	+1.0	—0.105		5.0	0.7	+ 4.2	(1200)
Mo_2C	+1.2	—0.19		4.4	1.2	— 6	2600
WC	+1.8	—0.29		6.7		— 5	2850

(a) peak shift relative to graphite
(b) not corrected for instrumental error
(c) taken mostly from Peter T. B. Shaffer, High Temperature Materials, *Plenum Press, New York, 1964, and Lawrence S. Darken and Robert W. Curry,* Physical Chemistry of Metals. *McGraw-Hill Book Co., N.Y. (1953), p. 364, and Ref. 19*
(d) as second phase in Fe-1.83 C alloy
(e) Fe-1.83 C alloy

the transition metal carbides have been divided into Groups A and B. Group A consists of carbides with high negative heats of formation and high melting temperature, while Group B is composed of transition metal carbides with lower or positive heats of formation and melting temperatures close to that of the pure metals. From the $W_{1/2}$ in Table III and in figures 12, 13 and 16, it will be seen that Group A consists of Group IV and V transition metal carbides whose CK bands are narrow with the predominance of a single subpeak while the Group VI and higher carbide CK bands in Group B have broader $W_{1/2}$'s with complex structure close to that of the

CK band from graphite. Since the narrower the $W_{1/2}$ of the CK band the more ionic is the transition metal carbide, then it can be seen from Table III that the carbides with large ionic character to the bond have high negative heats of formation and those with less ionic character have lower or positive heats of formation.

A plot of the carbon ls level shift, which is a measure of the ionic character of the bond, by Lars Ramquist[17] as a function of the heat of formation in figure 17 supports the above statement on the relation between ionic character of the bond and heat of formation. In addition, the relation between the CK band intensity distribution and ionic character is also confirmed because from figure 17 the carbides whose CK band is the narrowest have the highest ls carbon level shift. The relation between the carbon ls

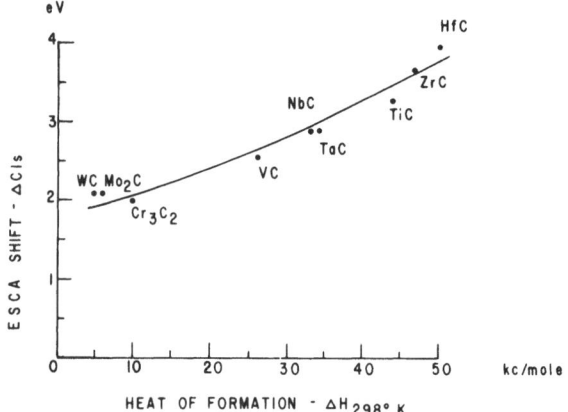

figure 17. Relationship between carbon ls level shift obtained from ESCA measurements by Lars Ramquist and heat of formation of transition metal carbides.

level shift and heat of formation in figure 17 is the same as that given in Table III, i.e., the greater the ionic character of the bond the higher the bond strength. This is also shown by the work of Lars Ramquist et al[18] in figure 18 where the shift in the inner level of the metal atom (measured by electron spectroscopy) for TiC, VC, NbC and TaC carbides follows the change in the heat of formation.

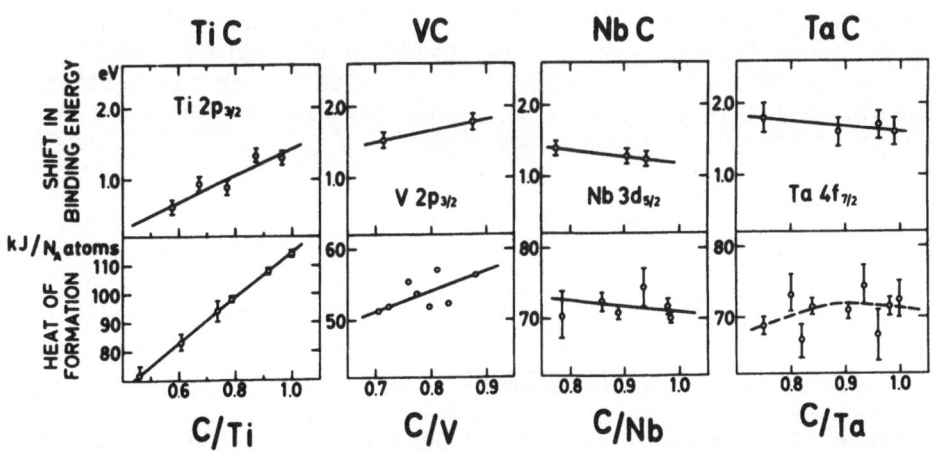

figure 18. Relationship between metal inner level shift of transition metal carbides and heat of formation as a function of C/metal atom ratio. [Reprinted by permission of Lars Ramquist et al.[18]]

Conclusion

The analysis of soft x-ray band spectra has shown that there are large changes in the band from elements in alloys or conducting compounds relative to the pure element. For a few elements such as the rare earths and titanium, self absorption can account for some of the observed changes but in most cases the present results show that self absorption effects are much less than investigators had inferred from previous results. If the proper x-ray target geometry is used and the excitation voltage is about 4 kv, the contribution of self absorption to the changes in emission bands appears to be negligible.

The band changes have been shown to be due to alterations in the electronic distribution of the atom or the ionic character of the bond. The degree of this change relative to the pure element appears to be related to the electronegativity difference of the combining elements. The retention of the individual character of the elemental bands for alloys and conducting compounds was a surprise result and has resulted in revision of the electronic structure theory for alloys. This shows that a study of soft x-ray band changes will lead to a better understanding of electronic structure of alloys. Observations on secondary peaks due to cross transitions and band shift measurements of band shifts by electron spectroscopy will be very important in determining the amount and direction of electron transfer in alloys and conducting compounds.

The relation between the ionic character of the bond and bond strength found from both soft x-ray band spectra and electron spectroscopy measurements indicates that, with proper calibration, it should be possible to determine bond energies from soft x-ray band spectra changes for some materials.

REFERENCES

1. Robert J. Liefeld in Soft X-ray Band Spectra and The Electronic Structure of Metals and Materials, D. J. Fabian, ed., Academic Press, London and New York (1968), pp. 133-149.

2. J. E. Holliday in Soft X-ray Band Spectra and the Electronic Structure of Metals and Materials, D. J. Fabian, ed., Academic Press, London and New York (1968) pp. 101-32.

3. J. R. Cuthill, A. J. McAlister, M. L. Williams, and R. E. Watson, Phys. Rev. 167, 1006 (1967).

4. C. A. Anderson in The Electron Microprobe, T. D. McKinley, K. F. J. Heinrich and D. B. Wittry, Eds., Wiley, New York, 1966, pp. 58-74.

5. David W. Fischer and William L. Baun, J. Appl. Phys. 39, 4757 (1968).

6. H. W. B. Skinner, Phil. Trans. Roy. Soc. (London), A239, 95 (1940).

7. David W. Fischer and William L. Baun, J. Appl. Phys. 38, 4830 (1967).

8. A. Appleton and C. Curry, Phil. Mag. 16, 1031 (1967).

9. J. E. Holliday, J. Appl. Phys. 38, 4720 (1967).

9a. J. E. Holliday, in the Proceedings of the Electronic Density of States Conference, National Bureau of Standards, Gaithersburg, Maryland, Nov. 3-6, 1969.

10. J. E. Holliday, in The Electron Microprobe, T. D. McKinley, K. F. J. Heinrich, and D. B. Wittry, Eds., Wiley, New York, 1966, pp. 3-22.

11. David W. Fischer and William L. Baun, "The Effects of Elec-
     tronic Structure and Interatomic Bonding on the Soft X-ray
     Emission Spectra from Aluminum Binary System", Tech. Report
     AFML-TR-66-191 (1966).

12. A. Faessler, in Colloq. Spectr. Int'l 19th Univ. Maryland,
     1962, E. R. Lippincot and M. Margoshes, Edgs., Spartan Books,
     Washington, D. C., p. 307.

13. J. Clift, C. Curry and B. J. Thompson, Phil. Mag. $\underline{8}$, 593 (1963).

14. R. G. Lye and E. M. Logothetis, Phys. Rev. 147, 622-635 (1966).

14a. D. J. Nagel, in Advances in X-ray Analysis, B. L. Henke,
     G. R. Mallett and J. B. Newkirk, eds., Plenum Publishing Co.,
     New York (this volume).

15. V. Ern and A. C. Switendick, Phys. Rev. 137A, 1927 (1965).

16. M. A. Blochin and A. T. Shuvaev, Bull. Acad. Sci. U.S.S.R.
     Phys. Ser. $\underline{2q}$, 429 (1962).

17. Lars Ramquist, Jernkontoretc Annaler $\underline{153}$, 159 (1969).

18. Lars Ramquist, Kjell Hamrin, Gunilla Johansson, Anders Fahlman
     and Carl Nordling, Uppsala University Institute of Physics
     Rep. No., UVIP 609 (1968).

19. Edmund K. Storms, The Refractory Carbides, Academic Press,
     New York and London (1967).

CHEMICAL BONDING AND VALENCE STATE--NONMETALS

David W. Fischer

Air Force Materials Laboratory (MAYA)

Wright-Patterson Air Force Base, Ohio   45433

ABSTRACT

The L_{III} x-ray emission bands from some 3d transition metal oxides and the K emission bands from some second and third period metal oxides and nitrides are shown and discussed in relation to the energy band structure of these compounds. It is demonstrated that the L_{III} emission bands can be grossly distorted by excitation conditions. Many of the compound spectra show strong evidence of a crossover transition which originates in the anion and terminates in the metal ion. For the 3d compounds this crossover transition appears to be related to the electrical and magnetic properties. In oxides of some second and third period elements it appears that the crossover transitions are responsible for the entire K band spectrum.

INTRODUCTION

Soft x-ray valence band spectroscopy has long been recognized as a potentially valuable tool for probing the band structure of solids. Significant pioneering work was done in this respect by Skinner and his associates about thirty years ago[1,2,3,4]. In more recent years, however, soft x-ray applications to band structure analysis were in somewhat of a state of disrepute. This condition apparently stemmed both from uncertainties in the experimental data abounding in the literature and in serious questions about correct interpretation of the experimental curves.

The situation was often marred by strong disagreement on band shapes obtained by various investigators from what were supposed

to be the same materials. The reasons for this are not always clear
but some of the more recent results, especially for the transition
metal spectra, have shown the problems to be largely experimental.
Different excitation conditions can produce different shapes for
emission bands. It is now known that the L_{III} emission bands of
the 3d elements, for instance, can be very seriously distorted by
satellite emission, exciton emission, and self-absorption effects[5,6,7,8,9,10,11]. Chemical changes in the target surface during
the course of electron beam bombardment can also have a significant
effect on the spectral shape. By taking these effects into account,
along with the other more well-known distortions which occur in
spectra[7], one may now have a reasonable chance of obtaining x-ray
emission band spectra which are related to the valence-conduction
band of the material.

At one time it was thought that the intensity distribution $I(E)$
of the measured emission band was a faithful reproduction of the
density-of-states. This is now known to be incorrect. The transition probabilities between the x-ray levels involved are at least
as important as the density-of-states distribution[12]. The relation
is often written as $I(E) \sim P(E) \cdot N(E)$ where $P(E)$ is the transition
probability and $N(E)$ the density-of-states. There is still considerable argument extant about the exact relationship between the
measured experimental curve and the calculated energy band structure. Nevertheless, it is believed that the relative energy positions of different characteristics of the emission and absorption
spectra are equal to the corresponding distances on the density-of-
states curve. This assumption will be applied to some of the experimental results discussed in this report.

Another question concerning the interpretation of soft x-ray
data has to do with the effect of the electron-hole interaction[13].
There is considerable debate on just how significantly the hole
affects the measured emission band.

The experimental soft x-ray data potentially contains considerable information from both the theoretical and practical standpoints.
For the theoretician, the most important items which could be obtained from the data are the bandwidth and the Fermi-energy of the
material. This information is extremely useful in calculating band
structures. From the practical standpoint, the major interest is
in the changes which occur in the spectra as a function of chemical
bonding and valence state. Since soft x-ray bands originate in the
outermost levels of an atom they are often quite sensitive indicators of any changes which occur in the immediate vicinity of an
atom. There can be significant shifts in the energy position of
the intensity maxima, gross changes in the band shape, increases or
decreases in relative intensities, and the appearance of new bands
in compounds which do not appear in the pure element. Often these
changes can relay important information which is not normally asso-

ciated with soft x-ray spectroscopy. Many of the spectra from compounds shown in this report exhibit a crossover transition which
can be used to infer some significant points concerning cation-cation and anion-cation orbital overlap[10,11]. Certain physical and
chemical properties of the compounds can also be correlated with
the observed spectral changes.

Good examples of the strong influence that bonding and valence
state can have on x-ray spectra are provided by the K bands of the
second and third period elements and the L bands of the first row
transition elements. These elements also happen to be the ones for
which the bulk of the energy band calculations in the literature
have been done. Most of the theoretical and experimental work has
been done on the pure elements, however, and not on the compounds.
It is the compounds such as oxides and nitrides which are of primary interest in this report. The spectra from the compounds will
be compared with those of the pure elements and also with each other.
From these comparisons certain assumptions will be made concerning
the energy band structure of the compounds and some correlations
will be made with the wide variations in physical and chemical properties. Some new ideas on interpretation of the data will be discussed. Most of the data referred to is quite recent. Much of the
older data in the literature, especially for 3d element L spectra,
is not entirely reliable because of the unknown experimental conditions and the therefore unknown effects of chemical state, satellite
emission and self-absorption on the band shapes. Also, the bulk of
the work reported in the literature is on the pure elements and good
data on transition metal compounds is quite scarce. Most of the
spectra shown in this report are therefore ones which were obtained
in the author's own laboratory during the last three years. Emphasis
will be placed on the K and L emission bands from oxides, nitrides,
and carbides which fall in the 7 to 30Å wavelength region. Some of
these compounds are quite good electrical conductors while others are
insulators. They are usually thought of as "non-metals", however,
hence the title of this paper.

From the practical application standpoint, the chemical analyst
may find many of these spectra of good use. For most of the simple
compounds, each chemical state yields a spectrum which is immediately
distinguishable from spectra of other chemical states. This holds,
even under conditions of considerable self-absorption and instrumental broadening.

EXPERIMENTAL

Since many of the experimental curves discussed in this report
were obtained by the author, a brief description of the apparatus used
seems in order. A description of the apparatus used by other workers
whose data are presented here will be found in the cited literature

references. In the 7 to 30Å wavelength region a good crystal spectrometer is capable of providing better resolution than the grating
spectrometer. especially at the lower end of the region. At longer
wavelengths the grating spectrometer is superior, mainly because of
the lack of good dispersing crystals. The crystals used to obtain
many of the spectra shown in this report are listed in table I.

table I

Window width of author's spectrometer at various wavelengths.

Emission band	wavelength(Å)	dispersing crystal	window width(eV)
SiK	6.7	ADP(200)	0.62
AlK	8.0	EDT	0.52
MgK	9.5	Mica(2nd order)	0.28
NiL	14.6	Sucrose	0.34
CoL	16.0	BiTitanate	0.33
FeL	17.6	Mica	0.39
MnL	19.4	Mica	0.24
CrL	21.7	RAP	0.38
VL	24.3	RAP	0.28
TiL	27.4	Clinochlore	0.28

A detailed description of the author's plane crystal vacuum
spectrometer will be found elsewhere[10]. Characteristic emission
spectra are produced by electron beam bombardment of the target
material. The brass, copper, or aluminum anode assembly can be rotated perpendicular to the entrance slit, making the takeoff angle
continuously variable between 0° and 90°.

The detector is a flow-proportional counter with an ultra-thin
formvar window and using an argon-methane (P-10) flow gas at a reduced pressure of 120 Torr. All recording electronics are commercial
Picker items except for the low-noise Tennelec preamplifier. The
accumulated data has a statistical deviation of no more than $\pm 2\%$.

Under normal operating conditions the spectrometer vacuum is
about 1×10^{-6} Torr.

For optimum resolution two Soller collimators are usually employed. The entrance collimator is 4.5 inches long with a 20-mil
vane spacing and the receiving collimator is 4.5 inches long with a
5-mil vane spacing. The width of the spectrometer window at half
maximum intensity for various wavelength positions is indicated in
table I.

The wavelengths of the spectral features measured have a probable error of ± 0.02Å (± 0.3eV) but wavelength differences could be
measured to ± 0.005Å (± 0.1eV).

For the compounds, fine powder was mixed into a slurry with ethanol and spread in a thin layer on the anode surface.

RESULTS AND DISCUSSION

Compounds of 3d Transition Metals

From the band structure standpoint, the 3d transition metals and their compounds have attracted considerable interest due to their wide variety of magnetic, electrical, thermal, and mechanical properties. These materials are characterized by an incomplete 3d shell which, in the compounds, is usually assumed to be responsible for electrical conduction as well as the magnetic and optical properties. Energy-band calculations have been made for most of the 3d elements but very little has been done for their compounds. Much of the band theory is based on measurements of transport, optical absorption, magnetic susceptibility, Hall effect, thermoelectric power, etc. The use of soft x-rays for band structure analysis has been largely overlooked primarily because of the problems mentioned in the introduction.

It is well known that the valence-conduction band of the 3d metals consists primarily of admixed 3d and 4s states. Ideally, to obtain information about the distribution of d and s states, one must study spectra arising from transitions to an inner level of p symmetry. For the transition metals, therefore, the $L_{II,III}$ emission bands $(3d4s \rightarrow 2p)$ would appear to be well suited for band structure studies. In many of the compounds, however, such as oxides and nitrides, the anion 2p levels also assume considerable importance. To obtain full information about the total range of the occupied band structure it therefore becomes necessary to measure both the K and L emission bands. The emission bands contain information regarding the filled portion of the energy band structure while the absorption spectra provide information concerning the vacant portion. Due to space limitations only certain L emission spectra will be presented here, however.

Over the last 35 years or so many different investigators have studied the $L_{II,III}$ spectra of the 3d elements but very little has been done with the compounds. Gwinner[14] and Skinner et al.[4,15] have shown spectra from a few selected oxides but apparently they were obtained under conditions of considerable self-absorption. Bonnelle[8] has also reported a few oxides and has shown that the band shape is significantly affected by excitation conditions. More recently, Fischer[10,11] has made a systematic study of the oxides, nitrides, and carbides of titanium and vanadium. Holliday[16,17] has also studied some oxides and carbides.

Much work needs to be done yet on the spectra from transition

metal compounds but the data accumulated so far appears to warrant
a few interesting observations. Experimentally, it is known that
some oxides such as TiO, Ti_2O_3, VO, V_2O_3 and VO_2 are good conductors
above their Neel temperatures while most of the oxides of the elements manganese through nickel are insulators both above and below
their Neel temperatures. Various explanations of this phenomenon
have been offered by Morin[18], Mott[19], Goodenough[20] and Adler and
Brooks[21]. It appears that the semiconductor-to-metal transition in
the lower oxides of titanium and vanadium is due to the collapse of
a narrow forbidden gap at a certain temperature (Neel temperature)
caused by a change in crystal symmetry[21]. This striking difference
in behavior of the titanium and vanadium oxides on the one hand and
the heavier 3d oxides on the other might therefore be expected to
have some effect on the appearance of the soft x-ray L spectra.
The magnitude of orbital overlap (both cation-cation and cation-
anion overlap) and the degree of occupancy of the 3d levels could
both be expected to have significant influence on the x-ray valence
emission bands and absorption edges. Fischer[10,11] has already
shown that the spectra of the titanium and vanadium oxides are indeed considerably different from the heavier 3d oxides. This section will be devoted to exploring the reasons for this difference
and suggesting an interpretation of the spectra from the compounds.

First of all, if meaningful results are expected, one must
have reliable x-ray emission bands and must know all the factors
which can distort his spectra. Most spectroscopists are aware of
the normal instrumental distortions which are usually present[7] but
unfortunately they overlook effects such as satellite emission and
self-absorption. Liefeld[6,7], Chopra[5], Bonnelle[8], Hanzeley[9], and
Fischer[10,11] have shown that these effects can be quite gross for
the L_{III} bands of the 3d metals.

The effect is just as significant in the spectra of the compounds as it is for the pure elements as can be seen in figure 1
for TiO. This is the $TiL_{II,III}$ emission spectrum obtained under
two widely different excitation conditions[10]. The solid line spectrum was obtained using a bombarding electron beam energy of 2kV
and a takeoff angle of 80°. The dashed line spectrum was obtained
at 6kV and 30° takeoff. Notice the severe intensity loss beginning
at the high energy side of the L_{III} band (peak B) as the beam voltage is increased and/or the takeoff angle decreased. The main cause
of this intensity loss is self-absorption in the target material.

The very fact that self-absorption cuts off part of the emission structure as seen in figure 1, indicates that the L_{III} absorption edge overlaps part of the L_{III} emission band. As pointed out
previously by several authors[5,6,7,8,9,10,11], the emission-absorption overlap occurs because much of the structure at the high energy
side of the L_{III} emission band edge is satellite emission from
multiply ionized atoms. These satellites appear to reach maximum

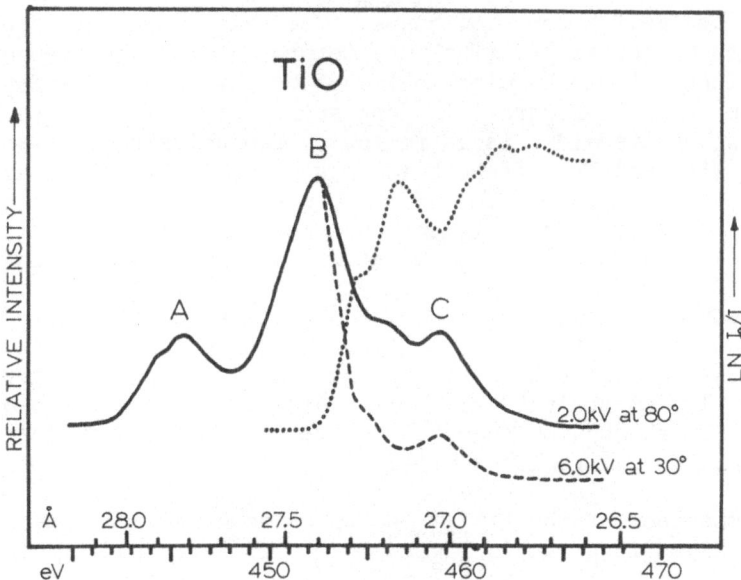

Figure 1 - Titanium $L_{II,III}$ x-ray emission and absorption spectra from TiO. Numbers on the high energy tails indicate the beam voltage and takeoff angle. Peak heights are normalized.

intensity at beam voltages of two to three times threshold[6,10]. We therefore have a rather curious phenomenon occurring as the beam voltage is increased. Satellite emission is growing and at the same time it is being suppressed by the self-absorption effect.

Obviously then, if one is attempting to obtain an L_{III} emission band which is related to the energy band structure, both satellite emission and self-absorption must be accounted for in addition to the usual corrections. Satellite emission is going to be present at any beam voltage higher than L_{II} threshold because of the $L_{II} \rightarrow L_{II}$ Auger transition. Liefeld[7] and Hanzeley[9] have shown that at beam voltages less than L_{II} threshold but greater than L_{III} threshold the satellite emission is no longer present for the L_{III} bands of copper, nickel, cobalt, and iron. Obtaining reliable emission bands at threshold voltage is quite difficult, however, and it has not been done for the compounds discussed in this report. Unless otherwise stated, the L_{III} bands shown here were obtained under conditions of negligible self-absorption but full satellite saturation. The full interpretation of points A, B, and C of the emission spectrum in figure 1 will be discussed a little later. The primary point to notice for now is that self-absorption can grossly distort the L_{III} band shape, not only for TiO but for other transition metals and their compounds as well.

Although self-absorption causes a certain amount of problems,
it can also be put to practical use to provide us with absorption
edge replicas which otherwise could not be obtained. The dotted
absorption curve in figure 1 is just such a replica. The technique,
first shown by Liefeld[6], is as follows: Two emission band spectra
afflicted with widely different amounts of self-absorption are obtained (as in figure 1). The spectrum with the lesser amount of
self-absorption is used as I_0 and the spectrum with the larger
amount of self-absorption is used as I. A point-by-point intensity
ratio plotted as $\ln\frac{I_0}{I}$ or as $\frac{I_0}{I}$ is constructed. In cases where comparisons could be made, the absorption replica was virtually iden-
tical in appearance to the normal photon absorption curve[9,10,11].
It should be pointed out, however, that the two types of curves
are not really equivalent[7,9]. Nevertheless, in the absence of
normal photon absorption spectra, the self-absorption curves can
provide very useful replicas[10,11].

A comparison of the $L_{II,III}$ emission spectra from a series of
titanium oxides is given in figure 2[10]. The left hand side of the
figure is an energy level diagram which will be used shortly to
aid in interpreting the spectra. The spectra on the right hand
side of the figure have all been normalized for comparison purposes
and were obtained at an 80° takeoff angle. For the various materials
the bombarding electron beam voltage was between 2.0 and 2.5kV.

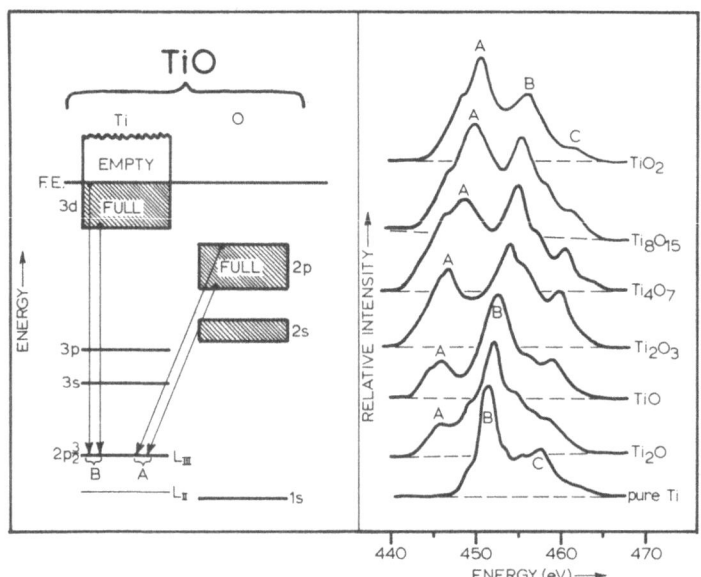

Figure 2 – Titanium $L_{II,III}$ x-ray emission spectra from several
titanium oxides and simplified energy level diagram showing
origin of emission peaks in TiO. Peak heights are normalized.

These spectra are therefore relatively free of self-absorption
distortion but L_{III} satellite emission is at saturation.

The most noticeable change in the spectra is the appearance of
a new emission band in the oxides (labeled A) which is not present
in the pure metal spectrum. This band grows in intensity and shifts
to a higher energy position as the oxidation state increases. Bands
B and C also shift to higher energy with an increase in oxidation
state. In the spectrum from pure Ti, band B is the normal L_{III} band
and C is the normal L_{II} band. This is not the case for all the
oxides, however[10].

Peak A has been identified as due to a transition from the oxygen
2p band to a vacancy in the titanium L_{III} shell. Although such a
p→p transition is normally forbidden by the dipole selection rules,
it probably occurs here because of strong cation-anion orbital overlap which admixes other symmetries into the anion 2p states. Such
a transition also occurs in vanadium compounds[11] and in certain
other transition metal compounds as will be shown presently. A
list of reasons for identifying peak A as a crossover transition
has been discussed in a previous report[10].

The major oxidation states of titanium are TiO, Ti_2O_3, and
TiO_2 which correspond to titanium electron configurations of $3d^2$,
$3d^1$, and $3d^0$, respectively[20]. Since energy band calculations have
been accomplished for TiO[22] it makes a good example to use in interpreting the spectra. The energy level diagram on the left side of
figure 2 is based on the calculations of Ern and Switendick[22] but
it is not necessarily drawn to scale.

The interpretation of the TiO $L_{II,III}$ spectrum is as follows:
Band A represents the O2p→TiL_{III} crossover transition; band B is
predominantly the normal L_{III} emission (Ti3d→TiL_{III}) but there is
also some contribution to it from the O2p→TiL_{II} transition; band C
is the normal L_{II} emission (Ti3d→TiL_{II}); the structure between the
B and C intensity maxima is mainly satellite emission arising from
multiply ionized atoms.

As the oxidation state is increased, band B contains more of a
contribution from O2p→TiL_{II} and less of a contribution from
Ti3d→TiL_{III}. At TiO_2 the d band is empty and the whole spectrum
is due to the crossover transitions since there are no d electrons
available[10]. This same type of process is also observed in the
vanadium oxides[11]. From such a series of oxides, therefore, one
can see just how important the anion 2p level is in contributing
to the metal ion emission spectrum. Similar effects can be observed
also in nitrides and carbides[10,11].

With certain reservations, as mentioned in the introduction,
the experimental curves can be compared with features on the calcu-

lated density-of-states. This is done in figure 3 for TiO. In the
upper part of the figure is Ern and Switendick's calculated structure using the APW method[22]. Fischer's experimental TiL_{III} emission
and absorption curves are in the lower half. The zero of energy is
placed at the Fermi energy. Since transition probabilities have not
been taken into account, one can only compare relative energy positions of certain features on the curves. Most importantly, it
appears that the width of the filled portion of the energy band is
in very reasonable agreement in the calculated and experimental
curves. The separation of the intensity maxima of the 2p band from
the Fermi edge is also in good agreement. The position of the first
maxima in the unfilled portion of the band agrees quite well also.
The major point of disagreement is in the filled portion of the 3d4s
band. Experimentally, there is only one intensity maximum here
while the calculated curve shows two. All in all, the results of
the comparison are quite encouraging. Perhaps even better results
could be obtained by combining the K and L emission bands as was
done by Wiech for silicon[23]. Incidentally, the window width of the
spectrometer in this wavelength region (28Å) is such that the instrumental broadening of the experimentally measured curves is virtually negligible[10](table I).

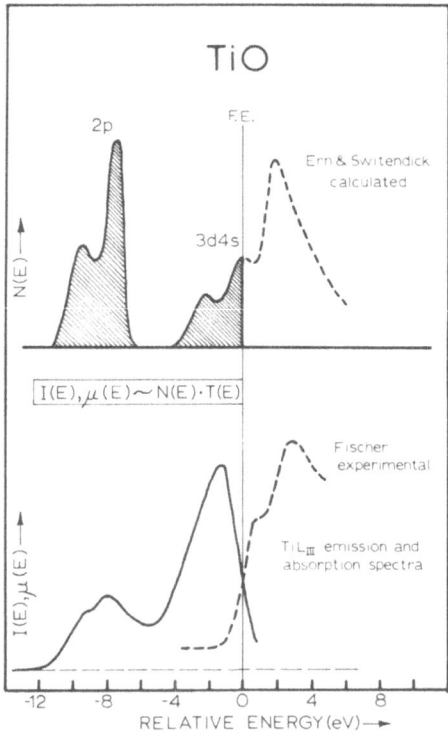

Figure 3 - Comparison of experimental L_{III} x-ray spectrum with
calculated density-of-states histogram for TiO.

As mentioned earlier, the appearance of the crossover transition in the $L_{II,III}$ spectra of transition metal oxides may be somewhat inter-dependent on the electrical and magnetic properties. The lower oxides of titanium and vanadium, for instance, exhibit metallic conduction while the oxides of heavier 3d metals are good insulators at all temperatures. Significant differences are also observed in the soft x-ray spectra of these compounds as can be seen in figure 4. These are all metal ion L_{III} spectra from sesquioxides. On the far right side of the figure are given the bombarding electron beam voltage and the takeoff angle at which the spectra were obtained. The dispersing crystals used and the instrumental resolution for each of the spectra were listed earlier in table I. All of the spectra were obtained in the author's laboratory since he could find none of them reliably reported in the literature. The assumed electronic configuration of each oxide is noted at the left edge of the spectra. Peak A is the $O2p \rightarrow$ metal L_{III} crossover transition. It is strongest in Ti_2O_3 and as the atomic number of the metal ion increases, peak A decreases. Notice that the crossover transition is seen in Ti_2O_3, V_2O_3, Cr_2O_3 and Mn_2O_3 but it is not seen in Fe_2O_3, Co_2O_3 or Ni_2O_3. Its appearance

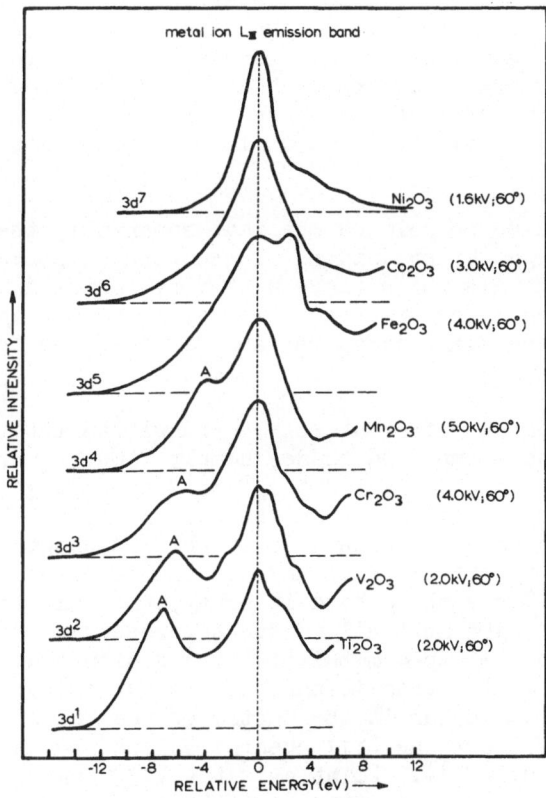

Figure 4 - Metal-ion L_{III} emission band from series of 3d metal sesquioxides. Peak heights are normalized.

therefore seems to depend on the degree of occupancy of the 3d shell. If the 3d shell is less than half filled, the crossover transition occurs; it does not occur if the 3d is more than half filled. Although not all oxidation states of all the transition metals have been studied as yet, this rule apparently applies to all the monoxides, sesquioxides and dioxides which the author has investigated.

As mentioned earlier, the O2p→ metal L_{III} crossover transition probably occurs because of strong anion-cation orbital overlap which mixes d and/or s symmetry into the 2p states of the anion. The more vacancies there are in the d band, the stronger is the overlap. Once the d band becomes half filled or more, the screening does not permit sufficient overlap and the crossover transition cannot occur. The crossover transition could then be a direct measure of the overlap integral. Morin, in fact, considers the anion-cation $(d\epsilon, p\pi)$ overlap to be just as important as cation-cation $(d\epsilon, d\epsilon)$ overlap in determining whether a compound is an insulator or a metal[24]. He succeeded in showing that metals and insulators fall into two distinct groups based on their overlap integrals. There appears to be much significance, therefore, in the fact that crossover transitions are found in the spectra from compounds such as TiO, Ti_2O_3, VO, V_2O_3 and VO_2 which are conductors, but are not found in the spectra of oxides such as Fe_2O_3, CoO and NiO which are insulators. Such correlation between spectral detail and physical properties of materials could be an important use for soft x-ray emission bands.

It should also be pointed out that apparently these materials can also be grouped by bandwidth. Those with relatively wide d bands are good metals while those with very narrow d bands are insulators. From the x-ray spectra, however, it is not apparent that the bandwidths are changing appreciably as one goes from Ti_2O_3 to Ni_2O_3.

There are three different stable transition metal oxides which are assumed to have empty 3d bands, namely TiO_2, V_2O_5, and CrO_3. These oxides are insulators and the forbidden band gap is believed to represent the energy difference between the empty 3d band and the filled oxygen 2p band. A schematic of this is shown in figure 5 along with the $L_{II,III}$ emission and absorption spectra from TiO_2 and V_2O_5[10,11]. Since the d band is empty, the emission band spectra of both TiO_2 and V_2O_5 must all be due to crossover transitions. Peak A is due to the O2p→metal L_{III} transition and peak B is due to the O2p→ metal L_{II} transition. The lowest empty state into which L_{III} absorption can occur is the bottom of the 3d band. The dotted curves in figure 5 are the L_{III} absorption spectra which have been reported previously[10,11]. Point a is the L_{III} edge position. A comparison of the energy positions of the L_{III} emission and absorption edges should give a measure of the size of the forbidden gap.

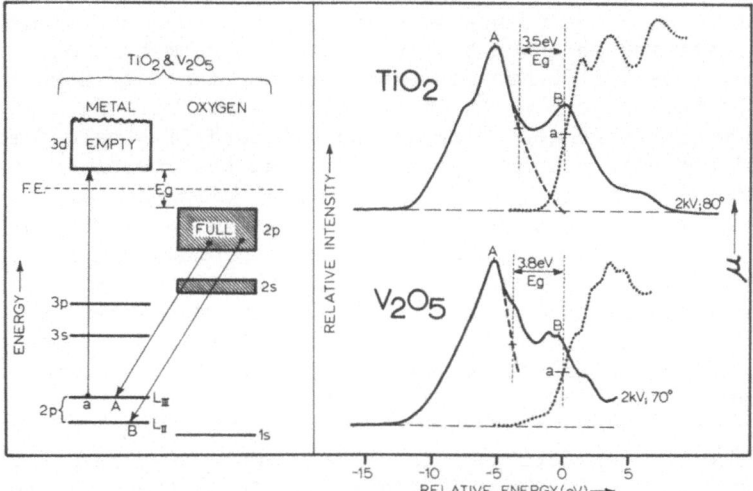

Figure 5 - Measurement of forbidden band gap in TiO$_2$ and V$_2$O$_5$ by means of L$_{III}$ emission and absorption spectra.

The results indicate a gap of 3.5eV for TiO$_2$ and 3.8eV for V$_2$O$_5$. These values are in good agreement with optical and thermal measurements[10,11]. The significance of these measurements is that they increase our confidence both in the assumed energy band structure of the oxides and in the interpretation of the soft x-ray spectra. Apparently the ionic bonding model is a pretty good one for the 3d oxides. There is some question as to whether the same model can be applied to the compounds of the elements beryllium, boron, magnesium, aluminum and silicon. The author thinks it can and the next section is devoted to discussing this point.

Compounds of Second and Third Period Elements

Unlike the compounds of the 3d transition elements, for which very little soft x-ray data was available in the literature, compounds of the second and third period elements have had their K emission band spectra investigated by many different workers. This especially applies to the elements magnesium, aluminum and silicon. One of the reasons for this is that the K band spectra from these elements is in a wavelength region (7 to 10Å) which is relatively easy to work with. Also, chemical bonding effects can have a quite large influence on these spectra and the changes are not difficult to observe.

Despite this abundance of data, however, no real agreement has been reached on interpretation of the energy shifts and shape changes which are observed in the K band spectra. The author feels that to explain the spectra from the compounds it is necessary to identify

most of the emission structure as being due to crossover transitions
which originate in the anion and terminate in the metal ion.   This
is not a new idea but it has not even been mentioned by most investigators who have attempted to interpret the spectra.   In a way
this seems rather strange.   O'Bryan and Skinner[4], in their famous
paper published nearly thirty years ago, used the crossover transition idea to explain the spectra from several compounds of the
second and third period elements.   Many later investigators have paid
lip service to O'Bryan and Skinner's work by pronouncing it part of
"the bible" of soft x-ray spectroscopy but they then overlooked some
of the important points made in it.   So let us now examine some of
these spectra and also keep in mind the interpretation suggested in
the previous section for the $L_{II,III}$ spectra of the 3d transition
metal oxides.

The effect of oxidation on the K band spectra of magnesium,
aluminum and silicon is shown in figure 6.   These spectra were obtained in the author's laboratory using the dispersing crystals
listed in table I.   Self-absorption effects are not nearly as significant for these spectra as for the L_{II} emission bands shown in
the previous section.   There is no strong satellite emission at the
K edge so that self-absorption affects principally the background
to the high energy side of the edge and not the overall shape of
the band.   The spectra in figure 6 were obtained with a beam potential of 4kV and a 60° takeoff angle.   Peak heights are all normalized
for comparison purposes.

Notice, in figure 6, that there is a considerable change in the
K band when going from metal to oxides for all three elements.
There is a gross change in the band shape, a large shift in energy

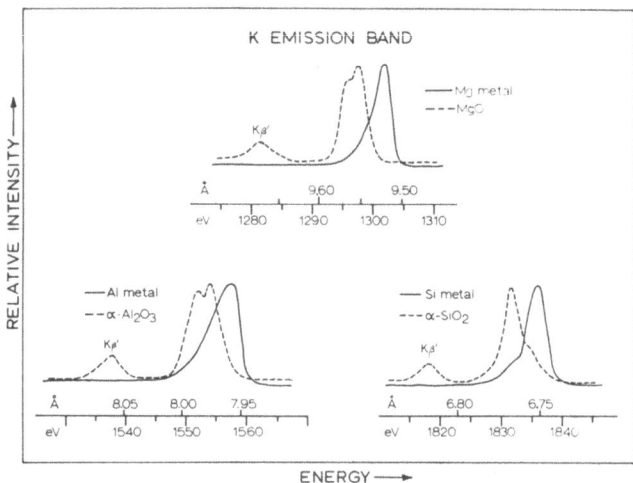

Figure 6 - K emission band spectra from metal and oxide for magnesium, aluminum, and silicon.

position and the appearance of a new band Kβ' in the oxides which is not present in the pure element. Recently, Dodd and Glen[25] have attempted to interpret the main band in the oxides, based on molecular orbital theory, as being due to 3p→1s transitions. They consider Kβ' to be a satellite line much the same as the Kα_3 or Kα_4 lines. This author disagrees with Dodd and Glen. While molecular orbital theory has been successfully applied to organic compounds and transition metal complexes, it appears more difficult to apply to simple inorganic compounds such as oxides. Specifically, the points that this author finds difficult to accept are the interpretation of Kβ' and the conclusion that the bonding in both MgO and Al_2O_3 has a high degree of covalent character[25].

An alternative interpretation proposed by this author, by Mendel[26], and by Cauchois and Bonnelle[27] is based on a highly ionic bonding model. With this model, both the main band and Kβ' in the compounds are assumed to be due to crossover transitions which originate in the anion and terminate in the metal ion. This is shown schematically in figure 7 for Al_2O_3. The upper part of the figure is a simplified energy level diagram which, while not quantitatively accurate, serves as an aid for the interpretation. If one assumes ionic bonding, then in Al_2O_3 the three valence electrons of aluminum are lost to oxygen, leaving the aluminum 3s3p band empty. The oxygen 2p band then becomes the valence band of the compound

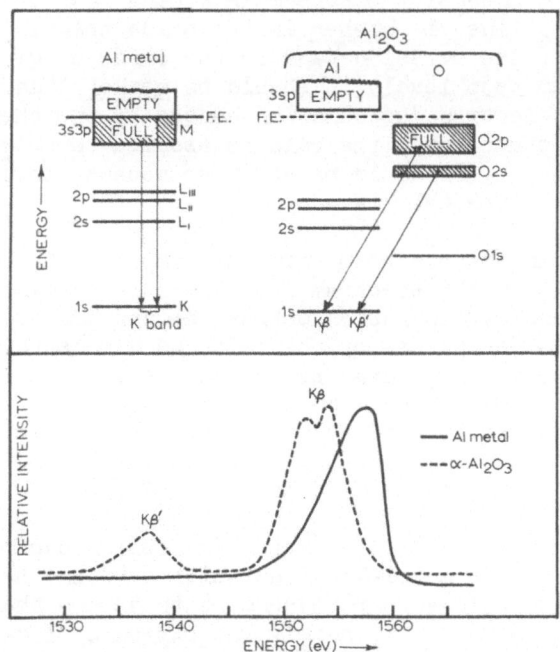

Figure 7 - Aluminum K emission bands from metal and oxide and simplified energy level diagram showing origin of peaks. Peak heights are normalized.

with the forbidden band gap representing the energy difference between the top of the 2p band and the bottom of the vacant aluminum
3s3p band. This is similar to the situation described in the previous section for TiO_2, V_2O_5 and CrO_3. The K band spectrum from
Al_2O_3 would then have the following interpretation: the main emission band (Kβ) is due to transitions from the oxygen 2p band to a
vacancy in the aluminum K shell. This transition is fully permitted
by the dipole selection rules. Kβ' would then be due to transitions
from the oxygen 2s level to a vacancy in the aluminum K shell.
Such an s→s transition is normally forbidden by the dipole selection rules but it probably occurs here because of strong anion-cation
orbital overlap which admixes p symmetry into the anion 2s levels.

Several factors appear to support this interpretation. For
instance, the following points concerning the x-ray K emission bands
from magnesium, aluminum, and silicon are noted:

1. When going from metal to oxide, the main K band shifts to a
   lower energy position. If the band represented a 3p→1s transition in both materials, the shift would be to higher energy
   because the charge flow is from aluminum to oxygen. An O2p→
   metal 1s crossover transition can, on the other hand, readily
   explain the shift observed.

2. The relative intensity of the main K band (measured relative
   to the K$\alpha_{1,2}$ line) is higher in the oxide than in the pure
   element. If the emission band in the oxide originated in
   the aluminum 3s3p level, it should be weaker than in the metal
   because of electron depopulation caused by the chemical bond.
   If, on the other hand, the band is assumed to originate in the
   oxygen 2p level, it would be stronger because more electrons
   are available for the transition.

3. Kβ' occurs in the spectrum from the compounds but does not
   occur at all in the spectrum from the pure elements. Furthermore, the energy separation between Kβ and Kβ' is directly
   dependent on the nature of the anion as demonstrated in figures 8 and 9. This energy separation is about 12eV for nitrides,
   about 14.5eV for oxides, and about 20eV for fluorides[26].
   These values are in good agreement with optical absorption
   data for the energy separation between the 2s and 2p levels
   of nitrogen, oxygen and fluorine[28].

4. As shown in figure 10, the oxygen K emission band has the
   same width as the metal-ion K emission band in the oxides.
   From this it would appear reasonable to assume that both originate in the same level, namely the oxygen 2p level.

When considered together, the above four points appear to give
more merit to the author's band model than to the molecular orbital

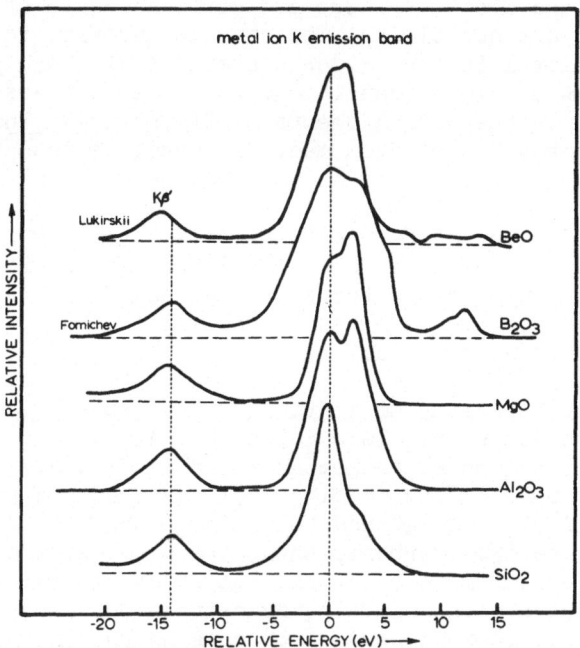

Figure 8 – Metal-ion K emission band spectra from some oxides of second and third period elements. Peak heights are normalized.

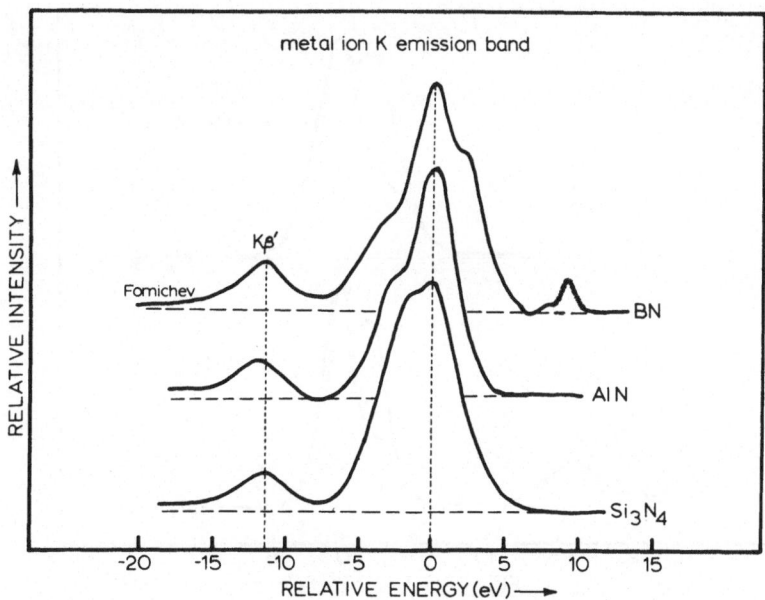

Figure 9 – Metal-ion K emission band spectra from some nitrides of second and third period elements. Peak heights are normalized.

model of Dodd and Glen[25]. There are still a couple of points, however, which are not clear. One obvious question is this: if the oxygen 2s level in the oxides contains sufficient p symmetry to allow the O2s→ metal 1s transition, why do we not observe the O2s→ O1s transition in the oxygen spectrum (figure 10)? The author has studied the oxygen K band from many different oxides[10,11] and in no case is there any evidence of an emission peak in the vicinity of 15eV to the low energy side of the main band. Perhaps the answer lies in what O'Bryan and Skinner[4] called a "symmetry-interchange of wave-functions", wherein those levels of the lattice which have one type of symmetry when referred to the oxygen atoms, have a different symmetry when referred to the metal atoms. Possibly too, our model is over-simplified.

Another point worthy of mention is that some oxides, such as SiO_2, are known to be much less ionic than Al_2O_3 and MgO. In such instances the ionic model as demonstrated in figure 7 may not be entirely applicable. Yet, in SiO_2, $K\beta'$ has the same relative intensity as in Al_2O_3 and MgO and the main $K\beta$ shift is comparable in magnitude (figure 6). Perhaps, then, the K emission band alone does not provide sufficient information about the energy band structure. Just such a point has already been demonstrated by Wiech[23]. He concludes that it is necessary to combine the information present in both K and L emission-bands to obtain a complete picture of the band structure. Wiech's $L_{II,III}$ spectrum from SiO_2[23] shows evidence of the crossover transitions from the oxygen 2p and 2s

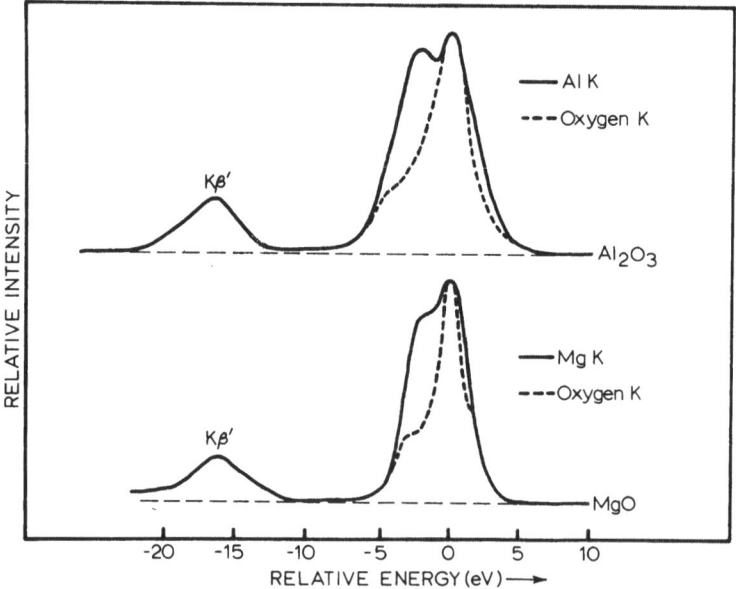

Figure 10 - Comparison of metal-ion K emission band with oxygen K emission band for Al_2O_3 and MgO. Peak heights are normalized.

levels but, in addition, it shows an extra high energy emission
peak. This extra peak may arise from occupied silicon states of
mostly 3s character which would therefore not be observed in the
K spectrum. Perhaps these occupied silicon states are responsible
for the hump on the high energy side of the main K band from SiO_2
(figures 6 and 8).

It will be noticed in figures 8 and 9 that the main K bands
from compounds of second and third period elements each have two or
more intensity maxima. The K bands from the corresponding pure
elements have only one intensity maximum. In oxides such as MgO
and Al_2O_3 this doubled maximum may be a result of the crystal field
splitting of the oxygen 2p levels. The oxygen K band, as shown in
figure 10, shows the same two maxima although the intensity distribution is not the same. In support of this view, it is worth noting
that the splitting is seen in compounds where aluminum is octahedral-
ly coordinated but not in compounds where it is tetrahedrally coordinated[25].

The spectra in figures 8,9, and 10 were obtained in the author's
laboratory with the exception of the K bands for BeO, B_2O_3 and BN.
The BeO curve was taken from a report by Lukirskii and Brytov[29],
and the boron compound curves from Fomichev[30].

Very few energy band calculations have been made for compounds
of second and third period elements which makes it difficult to
estimate how well our experimental data agrees with theory. Wiech
has shown that it is probably necessary to combine the K and L spectra
in order to better observe the distribution of both s and p symmetries which are the most important parts of the valence and/or conduction bands[23]. Very encouraging results have been obtained in
this way for elemental silicon but so far the theoretical data has
not become available for oxides or nitrides.

Since the K bands of the light elements are extremely sensitive
to any changes in the chemical bond, they are quite useful from the
analytical standpoint. Specific examples of practical applications
have been discussed in several published reports[31,32,33].

SUMMARY AND CONCLUSIONS

If there is an over-riding theme to this report it surely must
be the importance attributed to the crossover transition in interpreting the soft x-ray bands from simple compounds. Although the
crossover transition idea is not a new one it has very seldom been
applied to specific cases. In this paper it was applied to some
simple compounds of the second and third period elements and of the
3d transition series. The transitions of prime interest were
those which originate in the anion and terminate in the metal ion

such as O2p→ metal L_{III}, O2p→ metal K, and O2s→ metal K. At
first glance some of these transitions appear to violate the normal dipole selection rules but it is suggested that strong anion-
cation orbital overlap sufficiently admixes other symmetries into
the anion 2p and 2s levels to permit the transitions. As a result
of these interpretations it appears that compounds such as BeO,
B_2O_3, MgO, Al_2O_3 and all the 3d transition metal oxides are highly
ionic.

The appearance of the O2p→ metal L_{III} crossover transition in
the 3d oxides appears to be correlated with certain electrical
properties. This transition occurs in the oxides of titanium and
vanadium, for instance, which exhibit metallic conductivity but
not in oxides such as Fe_2O_3, CoO and NiO which are insulators.
The intensity of this transition appears to be dependent on the
strength of the anion-cation (dϵ, pπ) orbital overlap.

In TiO_2 and V_2O_5 the crossover transition makes possible the
measurement of the forbidden band gap. This gap represents the energy difference between the top of the filled 2p band and the
bottom of the empty 3d band which is equal to the energy difference
between the soft x-ray L_{III} emission and absorption edges. For
oxides of second and third period elements such a measurement is
not easily made because of the symmetry characters of the levels
involved.

Self-absorption and satellite emission can grossly distort the
true shape of the L_{III} emission band from the 3d metals and their
compounds. This distortion occurs in the immediate vicinity of the
L_{III} edge and must be corrected for if the results are to be meaningful in relation to the band structure of the material. The K
emission bands of the second and third period elements and their
compounds are not as significantly affected by self-absorption due
to the lack of strong satellite emission at the edge.

For the L_{III} bands of the 3d compounds, the self-absorption
distortion can be used to construct absorption edge replicas. Due
to the difficulty in making specimens for photon absorption measurements in the soft x-ray region, the self-absorption curves can
provide information not easily obtainable otherwise. The combination
of emission bands and absorption edges reflects a lot of vital information about both the filled and vacant portions of the valence/
conduction band.

Unfortunately, for most of the compounds discussed in this
paper there are no theoretical band calculations available for comparison with the experimental results. It would certainly be helpful
if our brother theoreticians would try their hand on a few of these
materials. The few comparisons made here and by Wiech[23], for
instance, look encouraging even though the experimental curves do

not, by their nature, faithfully reflect the density-of-states.

One of the important uses of soft x-ray spectroscopy is the extraction from the data of information about the band structure which would otherwise not be available. Such information includes, for instance, bandwidths and Fermi energies. By working much closer together the experimentalist and the theoretician should be able to produce many interesting results in the near future where band structure of simple compounds is concerned. So get to work fellows!

REFERENCES

1. H. W. B. Skinner and H. M. O'Bryan, "Soft X-Rays and Energy States of the Conduction Electrons", Phys. Rev. 45, 293(1934)

2. H. W. B. Skinner and H. M. O'Bryan, "Soft X-Ray Emission Spectra of Valence Electrons", Phys. Rev. 55, 604(1939)

3. H. W. B. Skinner, "The Soft X-Ray Spectroscopy of Solids. I. K and L Emission Spectra from Elements of the First Two Groups", Phil. Trans. Roy. Soc. (London) A239, 95(1940)

4. H. M. O'Bryan and H. W. B. Skinner, "The Soft X-Ray Spectroscopy of Solids. II. Emission Spectra from Simple Chemical Compounds"., Proc. Roy. Soc. (London) A176, 229(1940)

5. D. R. Chopra, "The NiLα X-Ray Emission Line", Ph.D. Dissertation, New Mexico State University (1964)

6. R. J. Liefeld, "Lα X-Ray Lines of Nickel, Copper, and Zinc", Bull. Amer. Phys. Soc. 10(4), 549(1965)

7. R. J. Liefeld, "Soft X-Ray Emission Spectra at Threshold Excitation", in D. J. Fabian, Editor, Soft X-Ray Band Spectra and the Electronic Structure of Metals and Materials, Academic Press, New York, 1968, p. 133-149.

8. C. Bonnelle, "Contribution to the Study of Transition Metals of the First Group, of Copper and Their Oxides by X-Ray Spectroscopy in the Range of 13 to 22Å", Ann. Phys. (Paris)1, 439(1966)

9. S. Hanzeley, "L Series X-Ray Spectra of Metallic Iron, Cobalt, Nickel, Copper, and Zinc", Ph.D. Dissertation, New Mexico State University (1968)

10. D. W. Fischer and W. L. Baun, "Band Structure and the Titanium $L_{II,III}$ X-Ray Emission and Absorption Spectra from Pure Metal, Oxides, Nitride, Carbide, and Boride", J. Appl. Phys. 39, 4757 (1968)

11. D. W. Fischer, "Vanadium $L_{II,III}$ X-Ray Emission and Absorption Spectra from Metal, Oxides, Nitride, Carbide and Boride", to be published in J. Appl. Phys., vol 40, 1969.

12. W. A. Harrison, "Electronic Structure and Soft X-Ray Spectra", in D. J. Fabian, Editor, Soft X-Ray Band Spectra and the Electronic Structure of Metals and Materials, Academic Press, New York, 1968.

13. N. H. March, "The Theory of Soft X-Ray Emission", in D. J. Fabian, Editor, Soft X-Ray Band Spectra and the Electronic Structure of Metals and Materials, Academic Press, New York, 1968, p. 224-226.

14. E. Gwinner, "Die Lα und Lβ Linien der Elemente 32 Ge bis 26 Fe und ihrer Verbindung und Legierungen", Zeit f. Phys. 108, 523 (1938)

15. H. W. B. Skinner, T. G. Bullen, and J. E. Johnston, "Notes on Soft X-Ray Spectra, Particularly of the Fe Group Elements," Phil. Mag. 45, 1070(1954)

16. J. E. Holliday, "Soft X-Ray Emission Bands and Bonding for Transition Metals, Solutions, and Compounds" in D. J. Fabian, Editor, Soft X-Ray Band Spectra and the Electronic Structure of Metals and Materials, Academic Press, New York, 1968, p. 101-132.

17. J. E. Holliday, "Investigation of the Carbon K and Metal Emission Bands for Stoichiometric and Nonstoichiometric Carbides", J. Appl. Phys. 38, 4720(1967)

18. F. J. Morin, "Oxides Which Show a Metal-to-Insulator Transition at the Neel Temperature", Phys. Rev. Letters 3, 34(1959)

19. N. F. Mott, "The Transition to the Metallic State", Phil. Mag. 6, 287(1961)

20. J. B. Goodenough, "Direct Cation--Cation Interactions in Several Oxides", Phys. Rev. 117, 1442(1960)

21. D. Adler and H. Brooks, "Theory of Semiconductor-to-Metal Transitions", Phys. Rev. 155, 826(1967)

22. V. Ern and A. C. Switendick, "Electronic Band Structure of TiC, TiN, and TiO", Phys. Rev. 137, A1927(1965)

23. G. Wiech, "Soft X-Ray Emission Spectra and the Valence-band Structure of Beryllium, Aluminum, Silicon and Some Silicon Compounds", in D. J. Fabian, Editor, Soft X-Ray Band Spectra and the Electronic Structure of Metals and Materials, Academic Press, New York, 1968, p. 59-70.

24. F. J. Morin, "Halides, Oxides, and Sulfides of the Transition
     Metals", J. Appl. Phys. 32, 2195(1961)

25. C. G. Dodd and G. L. Glen, "Chemical Bonding Studies of Silicates
     and Oxides by X-Ray K-Emission Spectroscopy", J. Appl. Phys.
     39, 5377(1968)

26. H. Mendel, "A Theoretical Interpretation of Satellite Lines in
     the X-Ray K-Emission Spectra of Compounds", Koninkl. Nederl.
     Akademie van Wetenschappen (Amsterdam) 70B, 276(1967)

27. Y. Cauchois, "Sur Les Spectres X des Metaux--Quelques Commen-
     taires et Exemples" in D. J. Fabian, Editor, Soft X-Ray Band
     Spectra and the Electronic Structure of Metals and Materials,
     Academic Press, New York, 1968, p. 71-79.

28. Landolt-Börnstein Tables, Volume I, Atomic and Molecular Physics,
     Part 1, Atoms and Ions, Springer-Verlag, Berlin, 1950.

29. A. P. Lukirskii and I. A. Brytov, "Investigation of the Energy
     Structure of Be and BeO by Ultralongwave X-Ray Spectroscopy",
     Fiz. Tverd. Tela (U.S.S.R.) 6, 43(1964)

30. V. A. Fomichev, "Study of Energy Structures of Boron and Boron
     Nitride by the Ultralongwave X-Ray Spectroscopic Method", Izv.
     Akad. Nauk (U.S.S.R.), Ser Fiz. 31, 957(1967)

31. D. W. Fischer and W. L. Baun, "The Influences of Chemical Com-
     bination and Sample Self-Absorption on Some Long Wavelength
     X-Ray Emission Spectra", Norelco Reporter XIV(3-4), 92(1967)

32. D. W. Fischer and W. L. Baun, "The Effects of Electronic Struc-
     ture and Interatomic Bonding on the Soft X-Ray Al K Emission
     Spectrum From Aluminum Binary Systems", in J. B. Newkirk and
     G. R. Mallett, Editors, Advances in X-Ray Analysis, Vol. 10,
     Plenum Press, New York, 1966, p. 374-388.

33. E. W. White and G. V. Gibbs, "Structural and Chemical Effects
     on the SiKβ X-Ray Line for Silicates", Amer. Mineralogist 52,
     958(1967)

INTERPRETATION OF VALENCE BAND X-RAY SPECTRA

David J. Nagel

Naval Research Laboratory

Washington, D. C. 20390

ABSTRACT

X-ray spectra arising from the valence bands of solids are useful for basic studies of the electronic structure of most materials and for practical measurements of unknowns to obtain information on local atomic structure and material properties as well as chemical composition. Understanding the characteristics of valence band spectra is prerequisite to their fullest use. One-electron and many-body aspects of the x-ray emission process, and the effects of experimental conditions, must be understood and are reviewed. Interpretation of spectral features and determination of electronic structure are complementary parts of one procedure which is based on the use of bonding theory. The various band and bond theories which are finding use for spectral interpretation are briefly reviewed. Calculation of the electronic structure of aluminum metal and quartz, and interpretation of their x-ray spectra, are examples which illustrate basic work with valence band spectra. Most practical studies of local atomic structure, material properties and composition have been based on experimental correlations between material characteristics and spectral features such as peak position. But these practical aspects of a material are related to its electronic structure. Hence, bonding theory is now being used to understand the nature and limitations of observed correlations and as a guide to finding new relationships. An extensive listing of the practical information available from valence band spectra is given and discussed in terms of bonding theory. Examples include bond lengths and coordination in minerals, electrical, magnetic and thermodynamic properties and chemical composition of thin and inhomogeneous samples. The need for a bonding correction in low atomic number x-ray chemical analysis and the characteristics of such a correction are examined. It is concluded that such a correction, if required, would be impractical to calculate. In general, interpretation of valence band spectra requires attention to the fact that x-rays arise mainly from those regions where the inner level electron is concentrated; that is, x-ray spectroscopy probes only local regions in a material. Because of this, local bond theories should be especially appropriate for interpretation of x-ray valence band spectra of metals and alloys as well as insulators.

INTRODUCTION

Measurement and interpretation of valence band x-ray spectra are powerful means to obtain information on the electronic and atomic structure, properties, and composition of materials. Valence band spectra arise from transitions of valence electrons to holes in an inner level of an excited atom. Figure 1 is a conventional energy level diagram which shows schematically the transitions which lead to valence band spectra in the elements aluminum and silicon. Valence band x-ray energies range from about 10 keV to below 60 eV. Hence we are concerned in this paper with spectra in the wavelength region from 1 to beyond 200 A.

The present paper treats interpretation of valence band spectra in the broadest sense of the word. That is, we consider both understanding of the origin of valence band spectral features and the use of spectra to study materials. It is emphasized that the understanding of valence band spectra is fundametal to their use to obtain both basic and practical information on specific materials. Most aspects of soft x-ray spectra

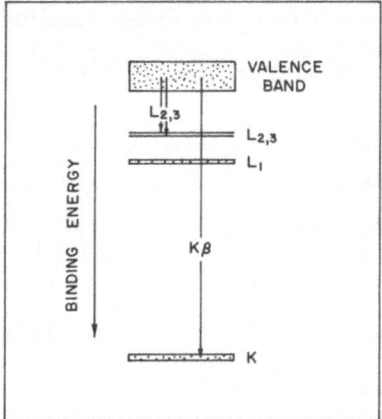

Figure 1. Conventional energy level diagram for the elements aluminum and silicon (not to scale). The transitions which give rise to valence band spectra are indicated.

interpretation are surveyed in this paper. The discussion is limited to emission spectra which are used for both basic and practical studies of materials. Absorption spectra are very useful for basic electronic structure studies but are seldom used to obtain practical information. There is no restriction on the types of materials discussed, although examples have been drawn primarily from the third row of the periodic table because of their relative clarity and also because of recent advances in understanding of materials containing these elements. The examples chosen are meant to be illustrative. No attempt at a comprehensive discussion of the literature on valence band spectra could be made here.

The first section of this paper contains general considerations which are applicable to all materials. The goals of valence band spectroscopy and the methods used to attain these goals are discussed in detail. Also listed are the factors which determine the spectra and which are basic to interpretation of all spectra. Since electronic structure is fundamental to both understanding and use of valence band spectra, the second section is concerned with the calculation of electronic structure using bonding theories, and the verification or refinement of the theory using x-ray data. This section lays the groundwork for understanding why it is possible to extract practical information from valence band spectra. Discussion of practical determination of local atomic structure, material properties, and chemical composition of unknowns constitutes the third section. X-ray spectra from disordered alloys are discussed briefly in the fourth section. Details of factors which determine valence band spectra, and of bonding theories are given in appendices following the summary and discussion.

The aims of this paper are (1) to provide some appreciation for the fundamental factors which underlie the understanding and the basic and applied use of x-ray valence band emission spectra and (2) to summarize and give examples of the kinds of information which can be obtained from such spectra.

GENERAL CONSIDERATIONS

Goals and Methods of Valence Band X-Ray Spectroscopy

The basic goal of x-ray valence band work is determination of the electronic structure of a material. Practical goals are determination of local atomic structure, material properties and chemical composition. For each of these goals there are three important questions: (1) what information is desired, (2) why is it wanted, and (3) how is it obtained? There are two general answers to the third question; information on material characteristics may be obtained (1) by a combination of theory and experiment or (2) by solely empirical means. This section contains a discussion of the what, why, and how for each of the goals of valence band x-ray measurements.

Electronic structure consists of the energies and spatial distribution of valence electrons in a material. More specifically, the electron wave functions $\psi(\bar{r})$ and electron energies E are the basic quantities constituting electronic structure. E is a function of the electron wave number $\overline{k}$, which can be thought of as the electron momentum or velocity. $E(\overline{k})$ is the band structure of a material, so called because for given directions in the lattice $\overline{k}/|\overline{k}|$, there are allowed and forbidden ranges (bands) of energies. Electrons occupy the energy bands up to the Fermi level (energy). The Fermi surface is the boundary surface in $\overline{k}$ space everywhere on which electrons have the Fermi energy and inside of which all states are filled. The electron spatial distribution (equal to ψ^2) and the density of states $\rho(E)$ [derived from $E(\overline{k})$] are also part of the electronic structure. $\rho(E)$ is the number of valence electron states per unit energy interval at energy E. The electronic structure of a material is determined by the types of atoms bonded together and their locations relative to each other (local atomic structure), the number of electrons (valence) which they each contribute in bond formation and the type of bonding (metallic, covalent, ionic) which gives the extent of electron sharing. Valence and bond type can be considered as aspects of electronic structure. The number of valence electrons available and the extent of their sharing determine the effective charge on an atom in a solid.

Ground state electronic structure is basic to the study of materials. It determines all perfect crystal properties and also the defects which can exist in a material given its composition and thermal and mechanical history. Hence, electronic structure ultimately influences even those properties which depend on defect behavior. The understanding and calculation of many material properties are predicted on knowledge of electronic structure.

The full electronic structure of a material consists of electron distributions and velocities in three dimensional (real) space. Since x-ray spectra are one dimensional (intensity as a function of the one variable energy), it is impossible in principle to calculate the complete electronic structure from x-ray spectral data.[1] To obtain the full electronic structure, it is necessary first to calculate the electronic structure using bonding theory and then to compare calculated spectra based on the theoretical electronic structure with (corrected) experimental spectra. This comparison of theoretical and experimental spectra is more complete than comparison of density of states with the x-ray spectra and serves to more thoroughly verify the calculated electronic structure, or to indicate needed refinements. Nikiforov and Blokhin[2] have noted that x-ray spectroscopy provides one means (of several) to test the correctness of bonding theories, and that it is best to compare calculated and experimental spectra even though much useful information comes from less complete comparisons.

In some cases partial electronic structure information can be obtained directly (empirically) from x-ray spectra without the use of bonding theory. For example, the valence band width equals the spectrum width (less tails and satellites). Also, valence, bonding type, and effective charge can often be correlated with peak position or intensity of valence band spectra. It should be noted, however, that bonding theory is needed to fully understand and use valence band spectra to obtain electronic structure information.

Local atomic structure consists of the types and arrangements (coordination, bond distances, bond angles) of atoms in a material. Hence, while atomic structure is a three dimensional entity, it is characterized by scalar quantities. Material properties (e. g. electrical resistivity, magnetic susceptibility, etc.) are also given by scalars. Chemical composition is measured in weight percent with no attention to how atoms are combined. The more limited information required to obtain the practical goals means that direct relations between the desired information and measured spectra can be obtained.

Atomic structure, properties and composition are termed "practical" because they characterize or identify a material (structure and composition) and tell how it will behave (properties). The practical problems which arise in the manufacturing, testing, and use of materials often require knowledge of these factors. Samples measured in basic valence band x-ray work have to be well known to insure that the experiment corresponds to the theoretical situation. In contrast to this, practical measurements are usually made on unknowns (except when constructing a calibration curve). If enough material is available, techniques other than x-ray spectroscopy may be more informative or efficient. For instance, if the electrical conductivity of a material is desired, a direct measurement is best. But if the unknown is a small inclusion which must remain in its matrix, a valence band spectra measurement with an electron microprobe may be the only way to obtain even an estimate of the conductivity. Batyrev and Shatunova[3] have advocated the combined use of the electron microprobe and valence band spectroscopy for practical work.

Sometimes a sample may be large enough but conventional techniques cannot be used. For example, local atomic structure can be determined from valence band x-ray spectra in materials (e. g., gels) which are so finely divided or poorly crystalline that the structure cannot be determined by x-ray diffraction. Thin and inhomogeneous samples are difficult to analyze accurately by the usual x-ray spectroscopy methods. The composition and thickness of thin oxidation and corrosion films can sometimes be obtained from valence band peak shifts and intensities. Quantitative analysis of thin films and phases in inhomogeneous samples can be done in some cases by measuring valence band spectra. Clearly, x-ray valence band spectroscopy is the best technique to obtain the desired information on any unknown when it is the most efficient of available means.

The relations which are used to determine material characteristics (electronic and atomic structure, properties, and composition) from x-ray data (spectral energy, shape, intensity) must be quantitative to be of practical use, in which case they are expressed as "calibration curves". These relations are obtained by either of two means: (1) a theoretical bonding calculation plus measurement of a single standard or (2) empirical correlations based on chemical shifts from a series of known standards. The chemical shifts used to construct empirical

calibration curves are differences in the valence band spectra of a
particular element (especially in peak position) which are observed
in going from one compound or alloy composition to another.   The
strictly empirical approach to valence band work requires more
standards and is less broadly applicable than a combination of theory
and experiment.   The use of theory is more complex but besides being
more general it also allows rationalization of observed material-
spectral correlations.   Since chemical shifts can be caused by changes
in any of several aspects of the electronic and atomic structure, it is
important to understand from theory the conditions and limits of applicability of empirical correlations.

What Determines Valence Band X-Ray Spectra?

The goals of valence band spectroscopy and the two major methods
to obtain practical information from spectra have been discussed.   It
is next necessary to examine the factors which determine and influence
valence band spectra.   This study is basic to an understanding of x-ray
spectra and also indicates what theoretical calculations, experimental
precautions, and corrections are required to interpret x-ray data.

The factors which determine measured valence band spectra are
naturally related to the characteristics of the emitting material and
the experimental conditions.   It is convenient to examine the factors
in relation to the sequence of steps in the generation and measurement
of an x-ray photon, namely (a) excitation from the ground state to an
x-ray (ionized) state, (b) decay of the x-ray state, with photon emission,
(c) photon exit from the emitting material, and (d) measurement with
some kind of spectrometer.   Table I lists the factors which enter at
each step in an x-ray experiment.   Characteristic times for each step
are also given.   See Appendix 1 (page 217).

TABLE I

Factors Which Determine X-Ray Spectra

Step	Material Factors	Experimental Factors	Time (sec)
Excitation	Ground and x-ray state electronic structure	Type and energy of exciting radiation	10^{-17}
Emission	Electronic Structure of x-ray states	Depth of x-ray production and takeoff angle	10^{-15}
Exit	Overlap of emission and absorption spectra	- - - - - - - - - - -	10^{-15} per micron
Measurement	- - - - - - - - - - -	Spectrometer	10^{-11}/cm.

The particular x-ray state which is formed upon ionization depends on the ground and excited state electronic structure and on the
manner of excitation. The normal x-ray state is most probable, but
excitation and multiple ionization x-ray states also form. Each type
of x-ray state has a main mode of decay, with the normal state of a
singly ionized atom leading to the most prominent and useful main valence band and the others giving less intense satellite structure. We
will be primarily concerned with the most intense main band. A more
detailed discussion of x-ray states and the various less intense spectral features is contained in Appendix 1 (page 217).

   Experimental factors include, in addition to sample considerations,
the type and energy of the ionizing radiation used to excite the sample,
the exit of the emitted x-ray photons from the material, and the spectrometer. All of these can have a strong influence on the experimental
results. Appendix 2 gives more detail on the experimental factors
(including self-absorption) which are important in valence band spectroscopy. A list of corrections to measured spectra is also given in
Appendix 2 (page 222).

   It is important to determine the effect of experimental conditions
on the spectra emitted by the material under study since the goal is
to obtain information on the material and not the means of measurement. Differences in type and energy of the exciting radiation lead to
x-ray production at different depths in the material. Spectra produced
by high energy electron or x-ray excitation are more subject to self-
absorption.[4,5] Observed differences in direct and x-ray-excited aluminum $K\beta$ spectra were ascribed to greater relative excitation of
surface oxide by electrons compared to the fluorescent-excited spectrum which is produced deeper in the sample.[6]

   Aside from differences in emission due to self-absorption or surface effects, the question remains: does the spectrum of a clean sample
emitted from near the surface (approximately free of self-absorption)
depend on the excitation conditions? That is, does the x-ray state
formed depend on the type and energy of ionizing radiation used to excite
it? As discussed in Appendix 1, there are some conditions which favor
formation of other than the predominant normal x-ray state. These involve electrons with energy near the excitation threshold or heavy particles. However, valence band spectra excited by the usual means
(electrons well above threshold energy and x-ray fluorescence) reflect
material properties rather than the excitation conditions.

   The central importance of electronic structure in determining x-ray
spectra follows from the equation for x-ray intensity. The spectrum
(photons per unit energy) emitted by singly ionized atoms is calculated
as a function of photon energy $h\nu$ according to

$$I(h\nu) = 2 \int_{E = constant} (d^2 k / \nabla_{\overline{k}} E) \, T[E(\overline{k})]$$

The integration for $I(h\nu)$ is over a surface of constant energy $E(\overline{k})$ in $\overline{k}$ space. E is measured from the bottom of the valence band and given by $(h\nu - E_i)$ where E_i is the energy separation of the valence band and the x-ray state with an inner electron missing. The differential density of states is $(d^2 k / \nabla_{\overline{k}} E)$ with the transition probability given by[7]

$$T[E(\overline{k})] = \frac{4 e^2}{3h^4 c^3} (h\nu)^3 \int_{\text{all } \overline{r}} \psi_f^*(\overline{r}) \, \overline{r} \, \psi_i(\overline{k}, \overline{r}) \, d^3 r$$

ψ_i and ψ_f are the wave functions for the initial (inner hole + valence electron) and final (valence electron hole) excited states separated by energy $(h\nu)$. Electrons of two spins occupy each state, giving the factors of 2 in $I(h\nu)$. The overall density of states at a particular

energy is $\rho(E) = \int_{E=\text{constant}} (d^2 k / \nabla_{\overline{k}} E)$. This has a definite physical

meaning, namely the number of electron states per unit energy with E. The integral

$$T(h\nu) = \int_{E=\text{constant}} T[E(\overline{k})] \, d^2 k \quad \text{can be viewed as a transition prob-}$$

ability in some sense but constancy of this integral does not insure that the spectra will reflect the density of states. This is true because mathematically, the integral $\int f(x) \, g(x) \, dx$ is not in general equal to the product of integrals $\int f(x) \, dx \int g(x) \, dx$.

It is worth emphasizing the importance of the transition probability in the study and use of valence band spectra. When the spatial variables $\overline{r}$ in the integral for $T[E(\overline{k})]$ are given in spherical coordinates, integrations over the angular coordinates θ and φ give the selection rules. These reflect the angular symmetry of the wave functions about the center of the emitting atom. This symmetry is often destroyed in solids so that crossover transitions from an orbital primarily on one atom to an inner level hole in another need not follow the usual atomic selection rules.

When a transition is allowed, $T[E(\overline{k})]$ is important in determining the emitted intensity at each energy and hence the shape of a spectrum. While it was once thought that $T[E(\overline{k})]$ could be taken as independent of $\overline{k}$ and constant over a valence band, the calculation of the $FeK\beta_5$ spectrum by Nikiforov and Blokhin in 1963 clearly demonstrated the major effect $T[E(\overline{k})]$ has in determining valence band spectra.[2] The importance of $T[E(\overline{k})]$ was emphasized at the 1967 Strathclyde conference.[8]

The integral for $T[E(\overline{k})]$ is zero for those regions in the material where either the initial or final electron orbital (electron density) is zero. X-ray spectra arise only from those regions of the valence electron distribution where the valence and inner electron wave functions overlap. Hence, the transition probability causes x-ray spectroscopy to be spatially selective. It is not possible to probe the entire electronic structure of a material with x-ray spectroscopy. K spectra, for example, sample only the valence electrons in the small spherical region around the nucleus where the 1s electrons have appreciable density. The extent of inner level wave functions can be obtained from atomic calculations.[9] Fomichev has mentioned the need for spatial considerations in valence band spectroscopy.[10] The ability to make measurements in particular localities within a material, and the ability to distinguish electrons of different symmetry, give x-ray spectroscopy particular advantages over other methods. These factors offset difficulties in extracting information from spectra and the inability of x-ray spectra to probe all of the electronic structure.

All the quantities in the equation for $I(h\nu)$ concern the ionized states which exist before and after emission. It is assumed that the electron level relaxation which occurs between ionization and emission due to the change in screening accompanying ionization does not alter the density of states. The calculated ground state density of states is used in the intensity integral even though electrons in orbitals of different spatial extent (e.g. s and d) may shift somewhat relative to each other following ionization. Similarly, it is assumed that all electronic wave functions other than those involved in the transition are unchanged during emission. Hence, only the wave functions of the electron which "switches orbitals" are needed. Further, for calculation of the intensity of the main valence band spectrum these wave functions are commonly replaced by the corresponding wave functions in the ground state atom.

The assumptions used in the calculation of $I(h\nu)$ amount to ignoring the effect of the missing electron (hole) on the other electron orbitals. The effect of this approximation can be assessed by comparing results based on it with intensity calculations using ionized state wave functions. Some of the required ionized state wave functions (for isolated atoms) have been calculated by Bagus.[11] He gave both energy levels and transition probabilities calculated using wave functions for ground state and for ionized atoms. The $K\alpha$ transition probabilities in atoms and ions with neon and argon configurations differed by up to a factor of two between the two cases. No similar intensity calculations using ionized state wave functions in solids are known to the author.

The discussion of x-ray emission to this point has been based on the one-electron picture. The hole in an inner atomic level of a conductor and the subsequent x-ray emission involve many-body effects which are receiving increased attention now. Two of the many-body effects, the intensity change near the Fermi edge and the low energy Auger lifetime tail, alter the shape of the main spectral band. See Appendix 1 (page 221).

BASIC STUDY OF ELECTRONIC STRUCTURE

Calculation of Electronic Structure in Ordered Materials

Starting with the goals of valence band spectroscopy it was argued that a combination of theory and experiment is most useful for both practical and basic work. Further, it was seen that electronic structure is basic to the calculation and understanding of spectra. Since the full electronic structure cannot be derived from experimental results, we now consider the ways in which it is calculated. Theories of atomic bonding (cohesion) are used to calculate the valence electron energies and spatial distribution in a material. Such information is fundamental to practical valence band x-ray spectra work, as well as being the starting point for basic electronic structure studies.

An understanding of bonding theories depends on an appreciation of the types of bonding which hold atoms in a solid together. Table II summarizes the bonding types and some of their features.

TABLE II
Bonding Types and Their Characteristics

Category	Bond Strength	Bond Types	Valence Electrons	Example
Within molecules	Strong	Metallic	Delocalized	Al
		Covalent	Localized between atoms	SiO_2
		Ionic	Localized on negative ions	KCl
Within and between molecules	Weak	Hydrogen	Limited sharing	Organic crystals
Between molecules	Weak	van der Waal's	No sharing	Solid Argon

For this table, "molecule" is taken to mean a collection of atoms which share or exchange electrons. In this sense, a piece of metal or ionic crystal, no matter how large, constitutes only one molecule.

It is the sharing of electrons in strong bonds among atoms within a molecule which primarily determines the ground state electronic structure of interest in x-ray work. Hence, only metallic, covalent, and ionic types of bonding will be considered further. Figure 2 indicates the increasing valence electron localization in going from the metallic through covalent to ionic cases. It should be noted that

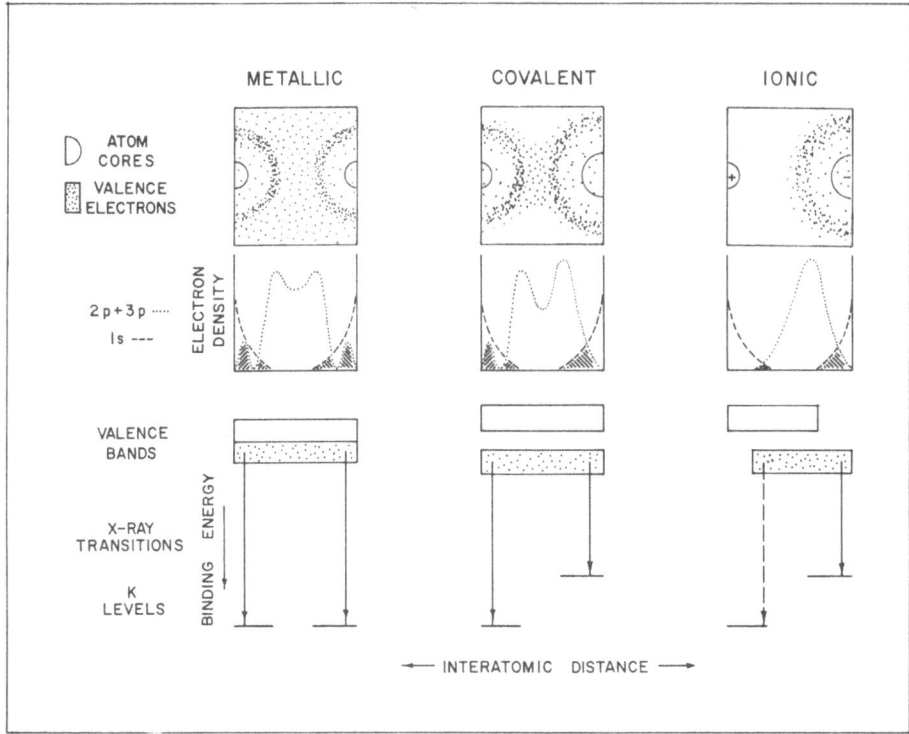

Figure 2.   Valence electron distributions between atoms for the
three major bonding types.   Electron density distributions for third
row cations (e. g. Al, Si) and second row anions (e. g. C, N, O) are
indicated.   The hatched regions show where the valence and 1s
electron densities overlap.   The K spectra shown below arise from
these regions.   The crossover transition is dashed.

the three bonding types are extremes in the sense that few materials
in nature exhibit only one type of cohesion.   Most materials have a
mixture of ionic and covalent, or of covalent and metallic forces.

Bonding theories   for all types of bonding can be grouped into two
major categories, namely those applicable to ordered or disordered
materials.   Calculations of electronic structure using bonding theories
begin with the types and arrangement of atoms in a volume over which
electron sharing occurs.   An ordered material is, for present purposes, one having a known chemistry and structure in the region of
electron sharing.   In insulating covalent and ionic materials the range
of electron sharing (mobility) is less than in metals due to greater
electron localization.

It is useful to further distinguish "slightly" and "grossly" disordered materials. Very dilute alloys have relatively little disorder despite the large range of electron sharing. Theories of dilute alloys are based on the theory for pure ordered metals.[12] Dilute alloy effects are not easily measured with x-ray spectroscopy and will not be considered further. Glass lacks long range order but since the region of electron sharing is small in most glasses, the disorder is only slight within that region. For example, there is x-ray evidence that the electronic structure around silicon atoms in glass is unchanged from that in quartz.[13] The usual bonding theories can be applied with small alterations in these cases if the extent of the small, local disorder is known. In the case of disordered concentrated alloys, the structure is unknown, non-periodic and grossly different from the pure metal end members. Hence the usual bonding theories which are based on the symmetry of periodic, adequately-known structures are doubly inapplicable to disordered alloys. New theoretical approaches are required. This is why disordered alloys are considered separately in a later section.

The bonding theories which apply to ordered materials are further divided into two major groups, band and bond theories. Band theories tend to focus on the distribution of valence electron energies. Bond theories are somewhat more concerned with the spatial electron distribution as well as the electron energies. Both kinds of theories start with the type and arrangement of atoms involved in the bonding. The differences between band and bond theories when applied to solids lie in the areas of completeness and calculational ease. Table III shows that only band calculations give complete information on solids and, therefore, allow complete interpretation of valence band spectra. In many cases, however, the partial comparison of an incomplete (bond) theory with x-ray spectra is sufficient to understand what causes the major features of the spectra, or to use the spectra to obtain practical information. In these instances, the results of bond theory in covalent materials are not only adequate, but have the advantage of being obtained with less calculational effort. The reason for the difference can be traced to the fact that bond calculations do not require solving Schrodinger's differential equation to obtain wave functions, while such a solution is usually necessary in band calculations.

The various specific band and bond theories are listed and discussed briefly in Appendix 3 (page 225). In the remainder of this section we will be concerned with how electronic structure is calculated. Inquiry into how calculations are made, as well as what results are obtained, is necessary for basic interpretation of spectra, to obtain an understanding of why there are relations between material characteristics and spectral features, and to discuss the characteristics of these relations. Two examples will be used to illustrate the technique and results of bonding calculations; aluminum metal for band theory and quartz for bond theory.

Band theory calculations often follow a pattern which is shown in the flow diagram in figure 3. The electron potential is the potential energy of an electron at various points within the solid. It arises

TABLE III

Comparison of Band and Bond Theories of Electronic Structure

	Band Theory	Bond Theory	
	Solids	Separate Molecules	Solids
Resulting electronic structure:	Complete	Complete	Partial
Band or level energies:	Complete	Complete	Partial
Energy Band Widths:	Yes	- - -	Estimated
Energy Band Structure:	Yes	- - -	No
Density of States	Yes	- - -	No
Wave Functions	Yes	Yes	Partial
Mainly Applied to:	Metallic and ionic materials	Covalent Materials	

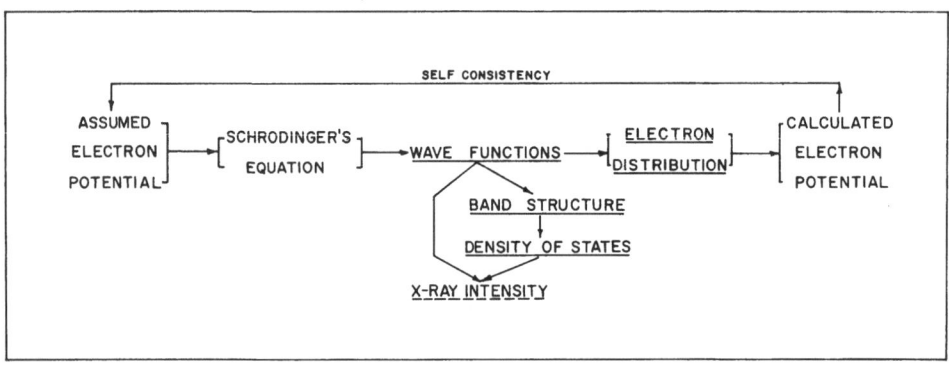

Figure 3. Flow diagram for the calculation of electronic structure (underlined) by use of band theories. The results of the band theory calculation can then be used to calculate x-ray spectra.

from the nuclear charge and the effects of all other electrons, and
determines, through the Schrodinger equation, the electron wave
functions and energies. The wave functions contain essentially all
the information about the solid (exclusive of defects). They can be
used to calculate the electron distribution which, as just mentioned,
determines the potential. If the potential derived from the wave
functions is the same as the starting potential, self-consistency has
been attained and the wave functions are used with the density of
states to calculate various properties and quantities for comparison
with experiment, such as x-ray spectra. Band calculations are often
terminated after calculation of the band structure $E(\bar{k})$, with no attempt being made to determine the density of states, accurate wave
functions, or even to attain self-consistency. Frequently, experimental data are used along with or instead of the self-consistency
requirement as a means to refine the theory. The complexity of
band theory essentially limits the calculation of electronic structure to specialists.

As an example of a band calculation, figure 4 gives some of the
results of Segall's band calculation for aluminum metal. [14] The FCC
structure in real space $\bar{r}$ contrasts with the Brillouin zone structure
in wave number (momentum) $\bar{k}$ space. While atom positions are
points in real space, the points in momentum space essentially give
the speed and direction of electron travel. The band structure $E(\bar{k})$
is shown for the (100) and (111) directions of electron motion. The
relation of E and $\bar{k}$ is parabolic near k = 0, that is, $E \sim (velocity)^2$
as in ordinary mechanics. Departures from this relation near the
Brillouin zone boundary are due to mobile valence electron interaction with the lattice. At the zone boundary for a particular direc-
tion, the wavelength $\lambda = 2\pi/|\bar{k}|$ is such that Bragg reflection of the
electrons occurs. Standing waves are set up, no electron motion is
possible at this point, and energy gaps occur. The difference in
lattice spacing in different directions determines through the usual
Bragg law the position of the zone boundary. The density of states
obtained by Rooke[15] from Segall's band structure is also shown in
figure 4. The use of these results for interpretation of the x-ray
spectra of aluminum will be discussed in the next section after we
examine the features of bond calculations.

As with band theory, bond theory calculations tend to follow a
general pattern with a wide variety of variations possible. Atomic
orbitals provide the starting point for most bond calculations. These
are formed into molecular or composite (hybrid) orbitals which form
the bonds. The energies of such orbitals are calculated from simple
equations, with physical effects other than bond formation included
to refine the energy levels.

The stepwise calculations of localized-bond theory are more
transparent than band calculations. Reilly[16] has obtained the energy
bands in SiO_2 and Al_2O_3 using a localized-bond theory. We will be

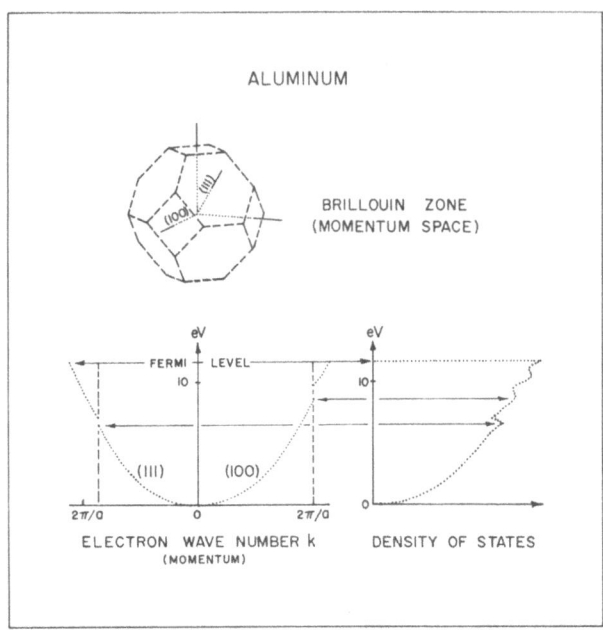

Figure 4. The first Brillouin zone, partial band structure, and total density of states for aluminum. Electrons occupy states up to the Fermi level. Vertical dashed lines show the Brillouin zone boundaries for the (100) and (111) directions.

concerned with use of the SiO_2 calculation for interpretation of x-ray spectra. The tetrahedral structure of quartz (αSiO_2) is shown in figure 5. Bonds between the central silicon atoms and the oxygen atoms at the corners of the tetrahedra are shown as solid lines. The two bonds linking the pair of silicon atoms and single oxygen atom which constitute the basic unit chosen for the calculation are shown by heavy lines. The choice of a basic unit is largely

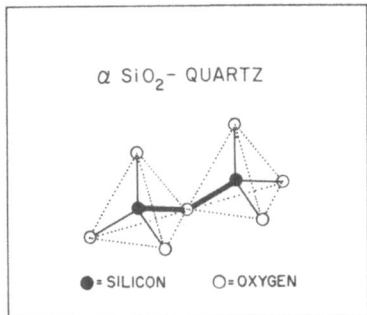

Figure 5. The tetrahedral structure of SiO_2 is indicated by dotted lines. Bonds are shown as solid lines. Heavy lines connect the atoms in the basic Si-O-Si unit used in calculating the electronic structure.

a matter of convenience and clarity. The energy levels at each
stage of the calculation are shown in figure 6. Each step will be
discussed in detail to demonstrate the simplicity of local-bond
theory and to prepare for consideration of local atomic structure
in a later section.

Step 1. Hybrid orbital formation: The formation of directional
covalent bonds requires the overlap of electron concentrations or
lobes from the bonded atoms. The tetrahedral structure of SiO_2
dictates that the silicon valence electrons must be concentrated towards the corners of a tetrahedron. It is seen from figure 6 that
neither the silicon 3s nor 3p valence electrons have the required
symmetry so that it is necessary to form a combination or hybrid
of the available electrons. Since the combination of one of the s
and three p electrons (sp^3) has tetrahedral symmetry such a hybrid

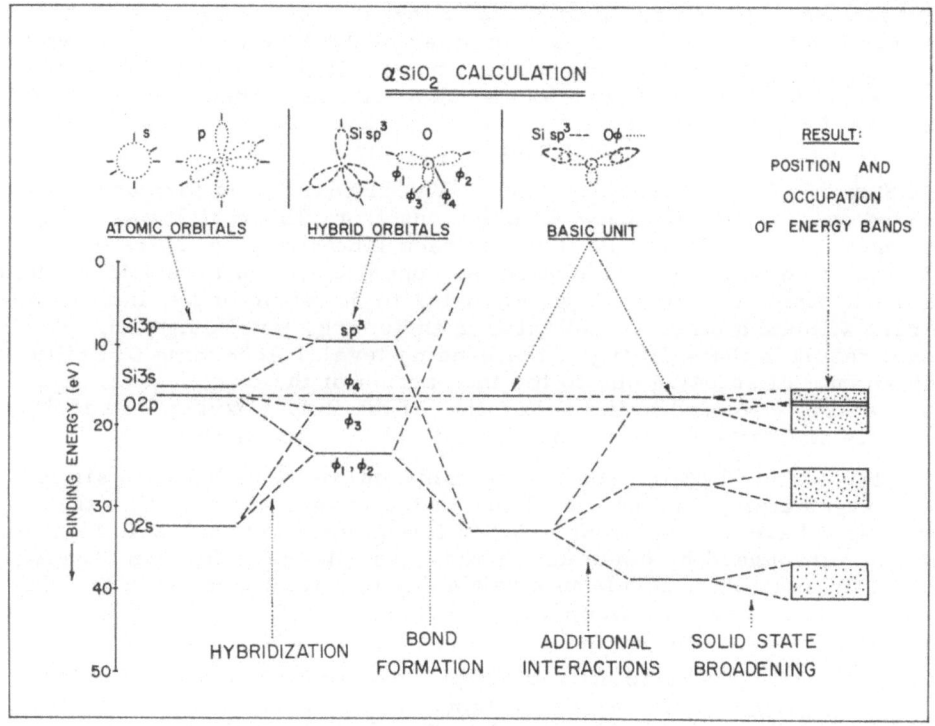

Figure 6. The orbitals, steps, and results of Reilly's local-bond
theory calculation of the energy bands of SiO_2.

is formed mathematically. Its energy is three-quarters of the way from the starting 3s atomic orbital toward the original atomic 3p orbital. This energy is higher than the average energy of the initial atomic orbitals (midway between the atomic levels) but it is recognized that the loss of binding energy due to sp^3 hybridization will be more than recouped on bond formation. In a similar fashion, the oxygen atomic orbitals do not have the needed spatial distribution, but it is possible to form ϕ hybrid orbitals as shown. These have two equivalent lobes ϕ_1 and ϕ_2 which will overlap with the lobes from silicon atoms. Their energy is lowest of the ϕ orbitals. Two "lone pair" orbitals, one in the plane of the diagram (ϕ_3) and the other perpendicular to it (ϕ_4) have the energies indicated. Calculation of the energies of the oxygen hybrid orbitals is not as trivial as for the silicon hybrid but is nonetheless straightforward.

Step 2. Bond formation: The overlap of the silicon (sp^3) and oxygen ϕ_1 and ϕ_2 orbitals to form covalent bonds within the basic unit is also illustrated. This leads to low energy bonding and higher energy non-bonding orbitals, with neither of the lone pair oxygen orbitals entering into the bonding at this stage. High energy anti-bonding orbitals, not shown in figure 6, are also formed. These are unfilled and are the lowest energy states of the ground state electronic structure probed in an x-ray absorption experiment.

Step 3. Additional interactions: The primary bond formation involved only interaction between electrons from different atoms. We are now concerned with the fact that each lobe shown in figure 6 is accompanied with a small lobe on the opposite side of the nucleus which is not shown. The three oxygen hybrids in the plane of the diagram interact with each other as indicated in the energy level diagram. The main result is the splitting of the bonding level. According to Reilly, the amount of splitting due to the interaction of the oxygen lobes with each other is related to the separation of the initial oxygen 2s and 2p orbitals and reflects the partial ionicity of bonding in SiO_2.

Step 4. Solid state broadening: Only interactions between silicon and oxygen atoms and between lobes of the oxygen atoms within the basic unit have been accounted for to this point. But the basic Si-O-Si unit is surrounded by other such units. This leads to further broadening of the calculated levels into bands due to interactions with the electrons in other nearby silicon and oxygen atoms. The band widths shown in figure 6 are estimated rather than being calculated rigorously.

The result of this relatively simple local-bond calculation is the relative energies of the bands, without any information on the distribution of electrons within the bands. The SiO_2 local-bond calculation is not much different, in complexity and labor, than complete correction of x-ray intensity data to obtain chemical composition. In contrast to the calculation of composition, however, such a calculation is not routine and requires a working knowledge of Quantum Mechanics.

All of the band and bond theories discussed in Appendix 3 are more or less useful for understanding spectra. In particular, there is a rapidly-growing use of Molecular Orbital theory for interpretation of both x-ray absorption[17, 18] and emission spectra.[19-25] Examples of the use of both band and bond theories for interpretation of x-ray spectra to obtain electronic structure information are discussed next.

X-Ray Study of Electronic Structure

Electronic structure calculations in ordered materials, such as those just discussed for aluminum and SiO_2, can be viewed as un-tested theory. There are many kinds of experiments which can verify such results.[26, 27] Most experimental methods sample only the mobile electrons near the top of a valence band. Some techniques probe the full range of valence electron energies. Among these methods, x-ray spectroscopy has the advantage of involving one simple (inner) level in contrast to optical and near-ultraviolet spectroscopy which often involve transitions between two bands.[28] Electron spectroscopy[26, 29] lacks the useful x-ray selection rules but allows measurement of valence electrons of any energy and distribution. We noted that x-ray spectra probe electrons of select symmetry in specific regions of a material, allowing local though incomplete study of electronic structure.

The examples in this section will illustrate the varying completeness of electronic structure studies using x-rays. Effects due to the missing inner electron are not important in the examples cited, leaving the density of states and transition probabilities as the major factors to consider. Rooke's comparison of aluminum valence band spectra calculated from band theory and his measurements exemplifies a relatively thorough test of the calculated electronic structure. Work as complete as this is rare. Most comparisons are made between x-ray spectra and some aspect of electronic structure, especially the density of states or band structure. We will compare silicon x-ray spectra from SiO_2 with Reilly's theoretical energy bands obtained using bond theory. Still less complete electronic structure studies have been based on chemical shifts and relative intensities, yielding information on valence and bonding type. The results of a few studies of this type will be briefly mentioned to demonstrate the kinds of partial electronic structure information which can be obtained from x-ray spectra with little or no use of bonding theory.

The result of Segall's band theory calculation[14] was used by Rooke[15] to obtain the density of states shown earlier in figure 4. Rooke used this and the (ground state) wave functions given by Segall to obtain the "effective density of states", that is, the density of states as altered by the varying transition probabilities. This was folded with the spectrometer function, the lifetime broadening of the inner level and the valence band Auger lifetime tail to obtain the expected x-ray spectra. Figure 7 shows that the calculated and experimental $L_{2,3}$ spectra obtained by Rooke are in good agreement. Neither of these spectra is the same as the density of states because of the transition probability. However, it is seen that the spectral fine

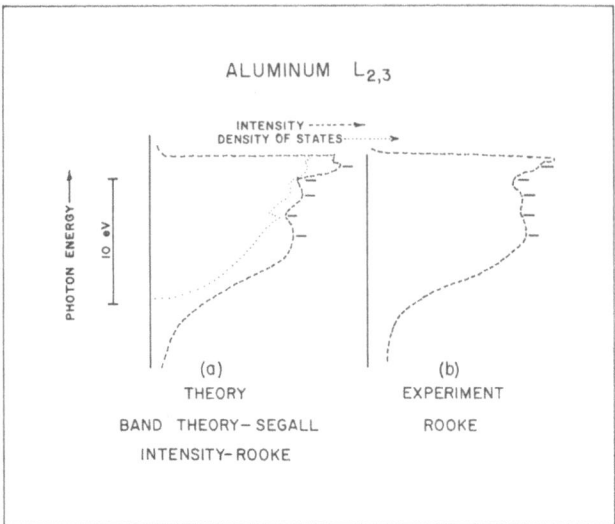

Figure 7.   The calculated and experimental Al L$_{2,3}$ spectra and Al
density of states obtained by Rooke.   The short bars mark the same
energy on both spectra for comparison of the fine structure.

structure comes from the details in the density of states curve which
in turn are related to the energies at band discontinuities indicated in
figure 4.   These energies are useful for further electronic structure
model calculations. [30]   The good agreement between calculated and experimental L$_{2,3}$ spectra lends confidence to the electronic structure
calculations of Segall and to Rooke's method for predicting the spectrum
from the band calculation results.   Rooke found that even better agreement between theory and experiment could be obtained by increasing the
density of states near the Fermi energy.   This work on aluminum demonstrates how comparison of x-ray theory and experiment can both verify
and indicate potential refinement of the calculated electronic structure.
Other thorough studies such as this are needed.

Turning now to SiO$_2$, figure 8 compares Reilly's theory[16] with the
spectra recently obtained by Wiech. [31]   The current comparison for SiO$_2$
demonstrates the advantages of comparing bonding theory and x-ray data,
even if unlike quantities (band energies and x-ray spectra) are being con-
trasted.

The first concern is understanding the spectra.   It is possible (and
the step-wise diagram of figure 6 makes it convenient) to calculate from
Reilly's local bond theory which of the starting atomic levels contributes,
and how much, to each of the bands which give rise to the spectra. [33]
That is, it is possible to understand the origin of the spectral features
in terms of the energy levels of the bonded atoms.   The central band is
largely of 2p, 3s, and 3p character, while the bottom band arises mainly
from the oxygen 2s level with a small admixture of p character.   The
Kβ_1 and Kβ' bands and their L$_{2,3}$ counterparts are therefore associated

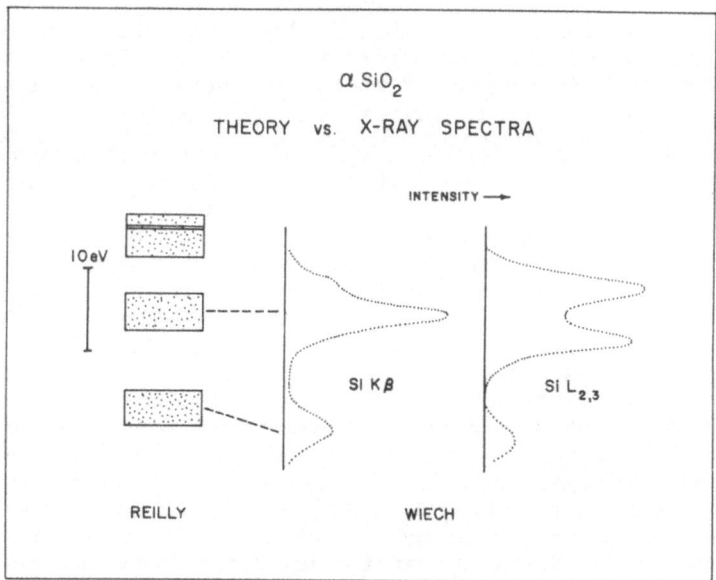

Figure 8. Comparison of the SiO$_2$ energy bands calculated by Reilly (figure 6) with the spectra of Wiech. The Kβ spectrum consists of the higher energy, more intense Kβ_1 peak and the less intense Kβ' peak. The main spectral peaks are aligned with the band from which they originate so the energy scale is relative.

with the central and lower calculated energy bands. Since the separation of the lower bands is related to the oxygen (anion) 2s-2p separation, the observed variation of Kβ_1-Kβ' separation with anion (ligand) type[22, 32] is accounted for by Reilly's model. Hence, Reilly's model, free of any need for invoking ionic-model crossover transitions,[32] explains the appearance and variation of the low energy bands in the spectra of third row elements such as silicon. Urch reached the same conclusion using a somewhat different theory[22] and showed that the Kβ' intensity can only be explained by a model including covalent bonding.[23] The ionic model gives more electrons available for transition[32] but, as indicated in figure 2, the small wave function overlap leads to a very small transition probability. The upper band in figures 6 and 8 is associated with the lone-pair orbitals which make little contribution to the silicon spectrum because these orbitals are concentrated near the oxygen atoms. The oxygen K spectrum arises from the lone-pair orbitals.[22, 33, 35]

The discussion so far provides an understanding of the spectra by going from theory to experiment. Going next from experiment to theory, the matching of the bands in the K and L spectra immediately verifies the mixing of s and p symmetry into each of these bands. Further, it is seen that the calculated spacing and widths of the lower two bands are in only rough agreement with the x-ray data. Hence the comparison of theory and experiment shows that the local bond theory must be extended to treat the additional interactions (figure 6) in more detail.[33] The shoulder in the SiKβ spectrum and the double SiL$_{2,3}$ main peak seem to be due to further level splittings.

Reilly's local bond theory is a Molecular Orbital approach[16] which involves the hybridization of atomic levels and yields a stepwise diagram as in figure 6. This contrasts with the more conventional SiO$_2$ Molecular Orbital theories of Dodd and Glen[21] and Urch[22] in which atomic orbitals are used without hybridization and only the original atomic and final molecular ground state energy levels are given. Ruffa has done a Valence Bond calculation[34] for SiO$_2$ to explain ultraviolet spectra. He recently extended this work to obtain the relative energies of the final valence band x-ray (ionized) states.[35] The number and spacing of levels which Ruffa calculates agree with x-ray data in a manner similar to the results of Reilly and Urch. It is noteworthy that Ruffa's calculation of final (ionized) state energies is basically different from the calculations of Reilly and Urch which give ground state energies. The situation which Ruffa treats is closer to what actually occurs in the x-ray measurement. Ruffa makes the point that it is important in insulators to consider the disruption of the local valence electron structure due to the hole in the valence band created by the radiative electron transition. Such final state effects would be observable only if the valence electron relaxation times are comparable to or less than the time for electron transition from the valence band to the inner level. The x-ray state lifetime is much shorter than minimum lattice vibration periods so that changes in atomic positions in response to the missing valence electron cannot be reflected in the x-ray spectra. See Appendix 1 (page 218). The SiO$_2$ theories of Reilly,[16,33] Urch,[22] and Ruffa[34,35] enjoy comparable success in predicting the presence and relative spacing of the major features in both silicon x-ray spectra. Calculation of the relative valence band spectral intensities based on each theory would be useful for judging the merits of these approaches. The lack of detailed density of states information and the approximate nature of the wave functions in these bond theories may make accurate intensity calculations difficult, however.

The comparison of theoretical energy bands (or calculated spectra) with experimental spectra would be more valuable if the theoretical quantities could be put on the same absolute energy scale as the measured x-ray spectra. This is difficult, however, because levels such as the silicon 2p and oxygen 1s are themselves subject to chemical shifts. For a particular material, if the energies of the inner levels were known relative to the vacuum level against which the theoretical bands are plotted, then subtraction of the x-ray energies from the inner level energy would allow a direct comparison of x-ray spectra with the calculated energy bands. Electron spectroscopy is capable of yielding the needed inner level energies on an absolute energy scale.[26] The values obtained are not those in the ground state atom but rather the shifted energies appropriate to an atom ionized in an inner level. This is precisely what is needed to place the x-ray data on an absolute scale. Hence it is possible to put bonding theories and x-ray data on the same energy scale.[26]

So far in this section we have been concerned with the detailed use of bonding theory to obtain electronic structure information. It is possible to obtain less complete electronic structure information with little use of bonding theories. For example, the widths of a filled band in a solid equals the corrected x-ray band width. The valence and type of bonding can also be obtained by use of x-ray peak shifts and relative intensities, as indicated in the remainder of this section.

Valence gives the number of electrons per atom which are available for bonding. Shuvaev[36] has discussed the relation between differences in valence and x-ray peak shifts in going from one compound to another. Changes in the valence of an atom imply that more (or less) electrons are removed from or added to the atom during bond formation. This alters the screening of inner electrons by the tails of the valence electron distribution and results in changes in their binding energies. Hence, the valence band x-ray energies (being the difference between electron binding energies) changes with valence. Blokhin and Shuvaev showed that shifts due to changes in valence can amount to several electron volts in titanium compounds.[37]

The type of bonding which exists in materials ranges from metallic through covalent to ionic (figure 2). In general, valence electron binding energies increase and x-ray energies decrease in going from metallic to covalent to ionic bonding. Hence peaks at different energies in a spectrum can sometimes be ascribed to electrons involved in different types of bonds. The relative amounts of different bonding types is important practical information. For instance, to understand the properties of transition metals and their compounds and alloys, it is necessary to know the extent of collective (metallic) bonding compared to the local (covalent) bonding. As an example of studies of this type, Korsunskii and Genkin found 1.2 metallic electrons and 3.8 local valence electrons in niobium metal from the intensity of valence band spectra.[38] The high energy peak in the Nb $L\beta_2$ spectrum is associated with the collective electrons while the low energy peak intensity reflects the covalent electron density. In a similar manner, Men'shikov and Nemnonov found that two $K\beta_5$ peaks reflected the collective and covalent electrons in chromium compounds, while the $K\beta''$ peak provided an indication of the ionic bonding in these materials.[39] The relative intensity of the three peaks thus allows study of the basic changes in bonding in going from one compound to another. Variations in transition probabilities complicate quantative studies of bonding type, however.

PRACTICAL MEASUREMENTS OF UNKNOWNS

The discussion just completed dealt with more or less thorough determination of electronic structure by theoretical and x-ray methods.
The theoretical aspects of that discussion provide a foundation for understanding why practical information on atomic structure, properties
and composition can be obtained from valence band x-ray spectra.
Bonding theory can also be used to determine the conditions and limitations of x-ray measurements to obtain practical information.

Local Atomic Structure from X-Ray Spectra

Local atomic structure consists of the type and geometry (coordination, bond distance and angle) of atoms around the emitting atom.
Diffraction techniques are commonly used to measure atomic structure. However, if an unknown is too small or imperfect to obtain
diffraction data, valence band spectroscopy offers an alternate means
of obtaining some information on atomic structure.

Valence band spectra from materials with open atomic structures
tend to be more sensitive to structure than spectra from materials consisting of close-packed atoms of nearly equal size. The structure a
material will assume is naturally that with the minimum free energy.
Enthalpy and entropy terms contribute to the free energy, with the enthalpy being largely determined by the valence electron energies probed
in a valence band spectral measurement. Vibrational entropy depends
on the openness of the structure while configurational (arrangement)
entropy depends on atomic disorder in a material. Crystalline (insulating) materials with open structures often have covalent or mixed bonding which allows little possibility for disorder in the atomic arrangement. Alternate structures have similar vibrational entropies. Hence,
the electron energies are dominant in determining the structure and the
valence band spectra are quite sensitive to structure. In contrast to
this, close packed metallic alloys can have large configurational entropy
contributions. Also, different structures have widely different amounts
of free space and large vibrational entropy differences. Hence the enthalpy is less important in determining the structure of ordinary alloys.
The valence electron energy bands and x-ray spectra of alloys are accordingly less sensitive to structure and do not change greatly across a phase
boundary.[40] In this section we will be concerned with the more open-
structured insulators whose spectra are sensitive to atomic structure.

We noted earlier that valence changes can cause peak shifts of several electron volts. The tabulation by Urch[22] of peak separations in spectra from third row elements ($K\beta_1$-$K\beta'$ as in figure 8) shows that the $K\beta'$
band can shift ten electron volts relative to $K\beta_1$ as the anion (ligand)
is varied. Blokhin and Shuvaev[37] showed in their study of titanium compounds that differing neighboring atoms lead to shifts of about 1 eV while
changes in structure give shifts of about 0.2 eV. Holliday[41] has also observed that emission bands sometimes are not most sensitive to crystal
structure. The magnitude of chemical shifts due to different causes is
important information for practical x-ray work. A summary of shift
data is given by Faessler.[42]

Despite the relatively small peak shifts with changes in local atomic structure, the shifts can be measured accurately. It is easy to see from the local bond SiO_2 calculation discussed earlier how local atomic structure influences the energy levels at each step, and hence the energy of the final bands and spectra. Table IV lists the factors which enter at each step in the calculation. This list establishes from theory the possibility of relationships between structural and spectral features. It does not indicate the direction or magnitude of differences in spectra which accompany changes in structure.

TABLE IV

Material Characteristics which Determine Valence Band Spectra

Step in Local Bond Calculation	Factors which Influence Result
Starting energy levels	Bonded atoms
Hybridization	Valence electron availability Coordination
Bond formation	Bond distances and angles
Additional interactions	Electron spatial distribution

White and Gibbs have observed correlations of silicon-oxygen[43] and aluminum-oxygen[44] bond distances with peak positions in the $SiK\beta$ and $AlK\beta$ valence band spectra from minerals. We will use the silicon case to illustrate the straightforward use of bonding theory to rationalize the observed correlations. Figure 9a shows the experimental relation between silicon-oxygen bond distance and the peak position of the $SiK\beta$ band. This correlation is a completely empirical calibration curve, and it allows measurement of bond distances to an accuracy comparable to that of crystal structure analysis using x-ray diffraction. Figure 9b illustrates the basis for the observed correlation. For a covalent model, a greater atomic separation means less overlap of orbitals and less binding energy. According to an ionic model, greater bond distances imply less coulombic attraction and again less binding. Referring to the diagram of figure 6, an increased silicon-oxygen distance results in a less-tightly-bound level after the bond formation step. Hence the position of the valence band at the end of the calculation is at a higher energy (less negative). This clearly leads to greater x-ray energies, if the position of the inner level is unchanged. The SiK level should be constant for the minerals measured by White and Gibbs because the nearest neighbors configuration around the silicon atoms does not vary. This correlation is being examined to see if the observed slope in figure 9a can be obtained from theory.[33] If this can be done, it may be possible to calculate calibration curves for other cases.

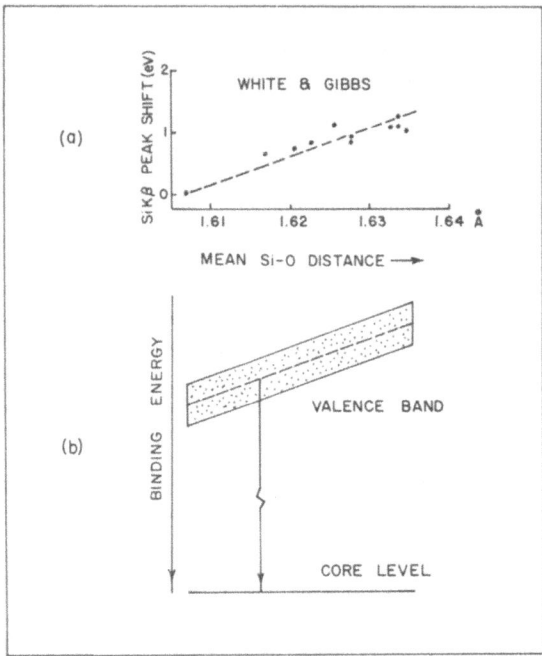

Figure 9. (a) Correlation of SiKβ peak energy with mean silicon-oxygen bond distance in a number of minerals. (b) Schematic illustration of the decrease in valence electron binding energy and the corresponding increase in x-ray energy with increasing bond distance.

Relations between coordination of aluminum in minerals and the AlKβ peak position have been studied by several authors. White and Gibbs[44] determined that the AlKβ peak energy tends to increase with increasing coordination. This trend is consistent with the observation that higher coordination requires greater bond distances and the fact discussed above, that longer bond distances lead to greater x-ray energies. Arrhenius[45] referenced early work in this area and interpreted the stabilization of aluminum in octahedral coordination in terms of d-orbital hybridization.

Material Properties and X-Ray Valence Band Spectra

A property of a material is a characteristic which gives its response to some applied force or field. For instance, the property of electrical conductivity tells how a material responds to an applied electrical field (voltage). If a property and some feature of the x-ray spectra are determined by (sensitive to) the same aspects of the electronic structure, relations may exist between the property and x-ray spectral features.

This section contains examples which show that (1) property-spectral correlations exist and can be explained by use of bonding theory, (2) it is possible to calculate the value of some properties using x-ray data, (3) x-ray spectra can lead to an understanding of some properties, and (4) information from x-ray spectra may be used to search for materials with desired characteristics.

The correlation between electrical conductivity and AlKβ peak position in a large number of aluminum compounds and alloys[46] which was observed by Baun and Fischer exemplifies a qualitative property-spectra relationship. The peak position was in one range of lower energies for insulators and in another higher energy range for semi-conductors and metals. No conductivities, and hence, no quantitative relations were given. The observed grouping is simply understood in terms of our earlier discussion of the greater binding energies of valence electrons in insulators compared to metals. In a manner similar to figure 9b, the greater binding of electrons in insulators leaves less energy available for x-ray emission while the more lightly bound metallic valence electrons yield higher x-ray energies.

The correlation of heats of formation with x-ray chemical shifts is an example of a quantitative property-spectra relationship. Das Gupta[47] calculated heats of formation for compounds $A_m B_n$ according to the equation $Q_{XR} = Q_A + Q_B = mV_A \Delta E_A + nV_B \Delta E_B$ where Q = heat of formation, V = valence, ΔE = the x-ray valence band peak shift which occurs on compound formation. The heats of formation calculated from x-ray data and obtained from thermodynamic measurements are given in the last two columns of Table V.

TABLE V
Heats of Formation Obtained from X-Ray Spectra[47]

Material	ΔE_A (eV)	Q_A (kcal)	ΔE_B (eV)	Q_B (kcal)	Q_{XR}	Q_{thermo}
MgO	4.5	207	0.52	72	135	145
Al_2O_3	4.5	618	0.52	215	403	400
SiO_2	3.9	359	0.52	144	215	200
SiC	0	0	0	0	0	~ 1
$AlFe_3$	2.0	138	0.24	132	6	~10

The agreement shown in Table V is surprisingly good since valence is only an approximate guide to the actual electron sharing and the x-ray spectra sample only part of the electronic structure. The qualitative relation between heat of formation and peak position is easy to understand. If the position of the inner level involved in an x-ray transition is constant, the shift in the x-ray energy of a valence band spectra

reflects the change in the energy of the valence electrons. Again
the situation is similar to that in figure 9b except that the valence
band shift is caused by compound formation instead of bond distance
variation. Since this change determines the heat (enthalpy) of formation of the alloy, there is a direct relation between x-ray energies
and the heat of formation. Shifts to lower x-ray energies go with
greater valence electron binding and exothermic compound formation.

The recognition that some properties and spectral features are
mainly dependent on the same part of the electronic structure provides a method for seeking correlations which allow calculation of the
values of properties from x-ray data. For example, in simple metals
the Fermi energy relative to the bottom of the band, that is, the
band width, is approximately related to the electron concentration
n_e by the nearly-free-electron equation

$$E_F = (\hbar^2/2m)(3\pi^2 n_e)^{2/3} \quad \text{where } m \text{ is the electron mass.}$$

The plasma frequency ω_p is given by $\omega_p^2 = \dfrac{4\pi n_e e^2}{m}$. Since the
plasma frequency also depends on n_e, we might expect a relation
between x-ray spectra band widths and ω_p, namely

$$\hbar\omega_p = 2.54 \, E_F^{3/4}$$

Figure 10 shows a plot of $E_F^{3/4}$ vs $\hbar\omega_p$. The slope of 2.68 is larger
than the free electron value. This correlation may prove useful in
the study of alloy plasmon energies.

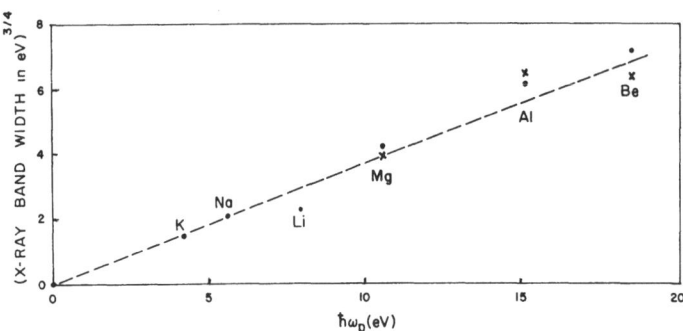

Figure 10. Correlation between x-ray band width and plasmon energy
$\hbar\omega_p$. Plasmon energies are from Rooke's tabulation[1] while the bandwidths are from Rooke[1] (·) and Tombulian[48] (x).

One further example of the use of x-ray data to calculate the value of properties will be mentioned. The ability to extract information on the number of collective and covalent electrons in transition metals was noted earlier. Korsunskii and Genkin[49] used their value of 1.2 collective electrons per niobium atom, and the nearly-free-electron approximation to calculate the paramagnetic susceptibility of niobium. Their result is in excellent agreement with the value measured by conventional means.

X-ray spectra sometimes provide the basis for understanding the origin of observed properties. For example, x-ray determination of the number of covalent electrons in Heusler alloys (CuMnAl) explains the appearance of ferromagnetism when aluminum is added to CuMn alloys. Shuvaev[36] showed that shifts in valence band spectra as composition is changed (or the temperature lowered through the Curie point) indicate that about one copper d electron goes into a p state while 1.5 of the manganese s and p electrons enter d states.

The final example concerns a case in which x-ray spectra may be useful in the search for materials with desired properties. Superconductivity is largely concerned with the mobile electrons near the top of a valence band. However, the effective ionic charge and effective number of electrons are also important in understanding superconductivity. These quantities depend on the entire valence electron structure. Ovsyannikova and Borovskii, following a study of cadmium compounds,[50] discussed how the density of states in a superconducting compound should be similar to that of the closest superconducting element in the periodic table. Since x-ray valence band spectra depend in part on the density of states, spectra from compounds may be useful in the search for superconducting materials.

Chemical Analysis with X-Ray Valence Band Spectra

In x-ray spectrochemical analysis, the weight percent of an element is usually sought without regard to how it is combined with the other elements present. Empirical calibration curves or calculated absorption, fluorescence and atomic number matrix corrections are used to obtain the composition with either fluorescence or microprobe techniques. The use of corrections assumes that the unknown sample is thick and homogeneous over the region of x-ray generation. Measurements with x-rays arising from transitions between inner levels are conveniently free of significant bonding effects. There is a one-to-one relation between the pair of levels involved in x-ray emission and the atoms in the sample. The spectral energy and shape are sufficiently constant from sample to sample for one element that peak intensity may be used as a measure of the amount of an element present.

For the elements boron through fluorine, only valence band x-ray spectra are emitted and available for analysis. X-ray measurements

of thick, homogeneous samples containing these elements have
three disadvantages relative to analysis with harder x-rays from
higher atomic number elements: (1) the matrix corrections, which
were designed for lines and definite absorption coefficients, not
bands over which the absorption coefficient may vary rapidly, are
larger and more uncertain, (2) the number, position, and shape of
the bands are not constant, requiring measurements of the entire
(integrated) band intensity and complicating the problem of background subtraction, and (3) even the integrated net intensity may
fail to give the correct composition if bonding effects change the
fluorescent yield. In short, the features which make possible the
study of electronic and atomic structure and material properties
enter to complicate x-ray chemical analysis of low atomic number
elements. The sensitivity of valence band spectra to local atomic
environment in insulators is a drawback when overall (average)
elemental composition is desired. In contrast to this, valence band
spectra from both low and high atomic number elements often contain features which allow analysis for compounds rather than just
elements. This is related to the ability to determine local atomic
structure, as discussed earlier. Some valence band spectra also
make possible convenient analysis of thin or inhomogeneous samples
which are usually difficult to analyze. This section is concerned with
the possible sensitivity of low atomic number fluorescent yields to
chemical bonding and the new analytical opportunities which valence
band spectra provide.

The electrons which give rise to valence band spectra are not
uniquely associated with the emitting atom due to bond formation.
Some valence electrons leave the locale of one atom in favor of nearby atoms. This alters the overlap of the inner and valence electron
states and should lead to a decrease in the x-ray intensity from atoms
which lose electrons because of bonding to other atoms. If the probability of non-radiative decay is not changed by the same percentage,
differences in bonding between an unknown and standard would result
in a different fluorescent yield for each. Hence, when using the correction approach to chemical analysis, the fluorescent yield would not
cancel out in expressions for relative x-ray intensity, leading to errors
in the analysis. There are two approaches to avoid errors due to possible differences in fluorescent yield. First, if the type of bonding of
the element of interest in the sample being analyzed is known, the use
of a standard in which that element is similarly bonded might avoid
differences in fluorescent yield between unknown and standard. On
the other hand, if a standard with different bonding must be used (e.g.
a metal standard for a covalent or ionic unknown compound), a bonding
correction might have to be made to x-ray data in addition to the usual
matrix corrections to obtain the correct composition.

It is necessary to examine the need for a bonding correction.
Direct measurement of small changes in the fluorescent yield is not
now possible in the low atomic number region due to large errors in
such experiments. A test for variation in the fluorescent yield with
bonding might be made theoretically. Fluorescent yield calculations

are difficult to make accurately but it might be possible to obtain changes which, along with the best experimental fluorescent yield values, would indicate the percent variation which bonding effects might introduce. An experimental demonstration of the need for a bonding correction could be made by comparing known compositions with the results of x-ray analysis only if uncertainties in the measurement and matrix correction of the data were small. But this is not likely to be the case. Hence it is concluded that, because of the other present experimental and theoretical uncertainties in soft x-ray analysis, a bonding correction is probably not important now.

Future advances in low atomic number x-ray analysis may reduce the measurement and data reduction errors below those errors which might be expected from changes in the fluorescent yield. In that case it would be desirable to have a bonding correction. What would be required to make a bonding correction? We have seen that the electronic structure and bonding depend on composition and atomic structure. Hence, it would be necessary to know the answer to the composition problem and much more (the atomic arrangement) in order to make the correction! The correction could not even be made in an iterative fashion due to the large number of variables which would be required but unknown. Also, calculations of electronic structure and then transition probabilities are too complex and costly for the slight improvement in the answer which might be realized. Clearly, if high accuracy low atomic number analysis is required, the indication is that some technique other than x-ray analysis should be used.

Although low atomic number quantitative analysis by conventional x-ray methods is complicated by bonding effects, the additional detail in valence band spectra do offer new analytical opportunities. Some spectra are essentially characteristic signatures of the emitting matrial. These allow unambigious identification of particular compounds when the number of possible unknowns is limited for some reason. For instance, oxidation or corrosion products can potentially contain only a few compounds which are distinguishable by the energy and shape of their valence band spectra. White and his collaborators have exploited this situation to analyze oxide films on silicon,[51, 52] using the SiKβ bands, and also the oxide, oxyhydroxide, and hydroxide corrosion products on aluminum[53] using both oxygen and aluminum bands.

Quantitative analysis of thin or inhomogeneous samples by standard hard x-ray methods is difficult because the usual matrix corrections alone are inadequate for accurate composition determination. Criss and Birks[54] have discussed the additional geometry corrections which can be applied if the details of the sample geometry are known. Measurement of the sample geometry, and calculation of thin or inhomogeneous sample corrections introduce extra steps in the analysis. Additional measurements for quantitative analysis can be avoided when

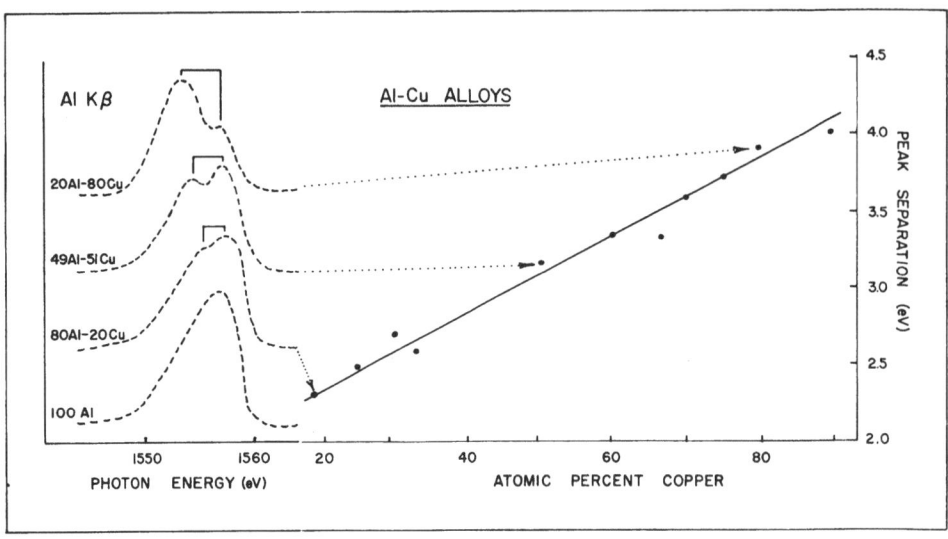

Figure 11. The AlKβ peak separation from aluminum-copper alloys measured by Baun and Fischer varies linearly with composition and provides a new method for x-ray chemical analysis.

valence band spectra contain features related to overall composition. This is likely to be the case only in alloys where the valence band spectra are insensitive to atomic structure. As an example, Baun and Fischer found that the peak separation in Al Kβ spectra[55] of aluminum-copper alloys varies directly with composition, yielding a calibration curve. Figure 11 shows their results. A single measurement on the unknown gives the composition through use of the published curve. There is no sensitivity to the spectrometer used as long as its resolution is adequate to clearly separate the AlKβ peaks. To use the Kβ peak separation method, the spacing between the two peaks has to be determined to about 0.1 electron volt (at 1.5 kev) to obtain five percent accuracy. Baun[56] has demonstrated the use of this peak separation method to analyze aluminum-copper thin films for which it would be difficult to prepare standards or use conventional corrections. We also note that, if aluminum and copper were present in an alloy in only one phase of an inhomogeneous system, the composition of that phase could be obtained from the Al Kβ spectrum. Understanding the occurrence and limitations of features useful for practical measurements on alloys, such as peak separations, will require the use of alloy bonding theory.

DISORDERED ALLOY SPECTRA

It is important to understand bonding in alloys in order to design materials with desired practical properties and to understand observed alloy properties. We saw earlier that the understanding of bonding in ordered materials was based on theory tested by a variety of means including prominently x-ray spectroscopy. But the unknown and nonperiodic

structures of concentrated disordered alloys make the usual bonding
theories inapplicable. Calculation of the electronic structure of alloys
is an area in which the basic theoretical tools are still under development. Many-body as well as one-electron theories of electronic structure and x-ray emission are required to understand alloying behavior
and interpret x-ray spectra. Despite complications of many-body effects (Appendix 1, page 221), valence band x-ray spectroscopy has particular potential as a test of and guide for development of alloy theories.
Little of this potential has been realized, possibly due in part to the extra
experimental problems in obtaining good spectra from alloys. These include more common overlap of emission[40] and absorption[57] spectra, the
difficulties of sample characterization (composition and homogeneity),
and changes in the sample during excitation and measurement. But
there is a rapidly increasing number of high quality x-ray alloy studies.
We will briefly mention a few aspects of basic x-ray studies of alloys.

Alloy x-ray spectra are more akin in some respects to the spectra
of compounds (even insulators) than to pure metal spectra. It is true
that alloy spectra have only a single band, while insulators (e.g. SiO_2)
may have a number of separate bands. However, the point is that, in
alloys as in compounds, the electronic structure in the neighborhood of
different atomic constituents is usually different. Curry,[58] in review-
ing many alloy x-ray measurements, concluded that the spectra generally indicated strong tendency for preferential grouping of valence electrons around the different species in an alloy. The interpretation of
alloy spectra requires consideration of (a) the spatially non-uniform
distribution of valence electrons and (b) the spatial and symmetry selectivity of x-ray spectroscopy. The second factor, namely the ability
of x-ray spectra to yield information on electrons in orbitals of different
symmetries in extended but small regions around the ions in an alloy
or compound is of prime importance.

The interpretation of alloy spectra ideally would involve a comparison of measured and calculated spectra similar to Rooke's study of aluminum which we reviewed earlier. But such thorough work is not possible now due to the lack of detailed electronic structure calculations
for disordered alloys. Hence at present, interpretation of disordered
alloy spectra is limited to comparison of spectra with whatever electronic structure information is available, either for the pure elements
in the alloy or the alloy itself.

Marshall et al.[59] based a valuable interpretation of Al L_{23} spectra
from aluminum-silver alloys on the pure aluminum and silver band
structures. The position of the silver d band relative to the Fermi
level indicates that the silver d orbitals are the origin of electrons
giving rise to a low energy peak in the spectrum. The second peak
which appears in the Al $K\beta$ spectrum from aluminum-copper alloys
(figure 11) may similarly arise from a band mainly associated with
copper. Use of pure metal band structure to interpret alloy spectra is
possible when there are highly distinctive features, such as the d bands,
in the electronic structure of the pure metals.

In most cases, consideration of the electronic structure of alloy itself is necessary to understand the x-ray spectra. Detailed electronic
structures calculated for ordered alloys (intermetallic compounds) by
conventional methods are increasingly available. Such information is
a valuable tool for interpreting disordered alloy spectra when few spectral changes occur due to disordering. The two general modern appoach-
es to the study of disordered alloy electronic structure are discussed in
Appendix 3 (page 225). There are, first,   multiple scattering models
and second, charge distribution models. The theories of electron pileup
(charging) are essentially local bond models for metallic alloys and seem
to have considerable potential for the interpretation of x-ray spectra.

The Electron Cell model of Bolsaitis and Skolnick[60] is a physically
thorough yet tractable charge model for simple metals. The redistribution of metallic electrons on alloying is easily calculated in this model
from the relative electronegativities of the constituent atoms. The result is the change in the electron density within a Wigner-Seitz cell due
to alloying. Stern's alloy charge model[61] requires this change in electron density on alloying to calculate the new potential in the alloy. This
gives, through the Virial theorem, the changes in electron energies due
to alloying. Such energy changes should be related to changes in x-ray
spectra. In addition, the piling up or depletion of electrons around constituent nuclei implies a change in x-ray intensity due to greater or lesser
electron density in the region of the core orbital. Study of intensity variation with alloying is complicated by absorption and fluorescence effects
and lack of transition probabilities. The intensity corrections commonly
used in chemical analysis[62] can be used to remove matrix effects. It is

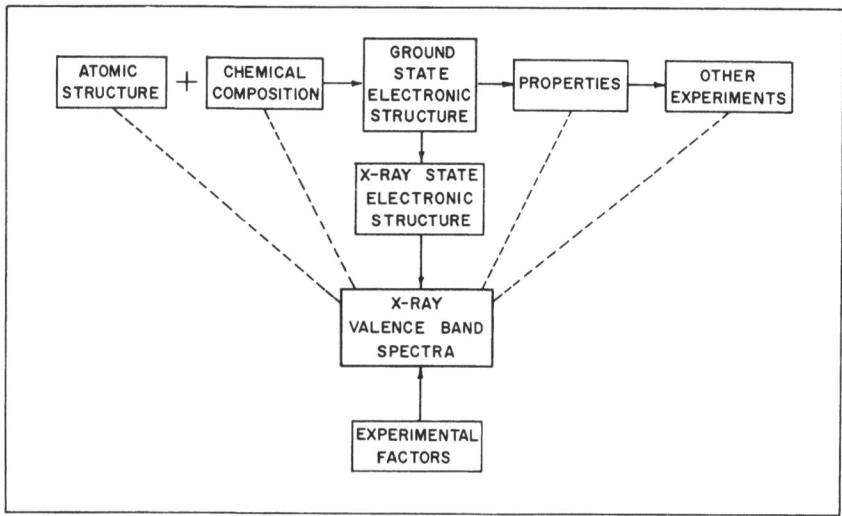

Figure 12. Relationship of material characteristics (top row) and
factors important in generation and measurement of valence band
x-ray spectra (center column).

anticipated that the more complex and thorough alloy bonding theories based on scattering will ultimately provide the most thorough interpretation of x-ray spectra. However, charging models should be very useful in the near future. The physical clarity and mathematical simplicity of charging models may compensate for their limited applicability and lack of detail compared to scattering models.

SUMMARY AND DISCUSSION

Figure 12 shows the relationships on which this paper is based. The factors along the top apply to the measured material while those in the center are related to the x-ray experiment. Direct relations are shown by solid lines while indirect relations, or correlations, are indicated by dashed lines. We saw that atomic structure and chemical composition determine electronic structure which provides the link between these quantities and spectra. Further, the properties of a material, and hence the results of experiments which measure its properties, depend on the electronic structure directly or indirectly. Since the x-ray spectra are also determined by the electronic structure, relations between properties and spectra are observed.

The following types of information available from valence band spectra were considered in this paper. The examples used are given in parentheses.

Electronic Structure: Thorough test of calculated electronic structure (aluminum), positions of energy bands (SiO_2), valence (titanium compounds) and bonding type (niobium and chromium compounds).

Local Atomic Structure: Type of neighboring atoms (titanium compounds), bond distance (silicate and aluminum minerals), and coordination (aluminum minerals).

Material Properties: Electrical conductivity (aluminum compounds), heats of formation (several compounds - Table V), plasmon energies (simple metals), magentic susceptibility (niobium), ferromagnetism (Heusler alloys), and superconductivity (cadmium compounds).

Chemical Analysis: Bulk samples (boron through fluorine), oxide films (on silicon), corrosion products (aluminum compounds) and thin alloy films (aluminum-copper).

Baun's review[63] mentioned some of these types of applications. His paper included a review of instrumentation which complements the present emphasis on interpretation of spectra. The expedient of concentrating on third row materials in this paper should be remembered. Valence band spectra from higher atomic number elements are more numerous.

The present review was limited to emission spectra because they arise from the filled electronic valence bands that determine bonding and most properties, and because only emission x-ray measurements are usually made to obtain practical information. Other techniques of x-ray spectroscopy, namely x-ray absorption, isochromat, and short-wavelength-limit spectroscopy probe the empty (anti-bonding) levels in a material directly from the ground state. These methods, especially x-ray absorption, are very useful in giving information which complements that from emission work.

The fact that x-rays arising from transitions between ionized states are used to seek information on the ground state electron structure is emphasized. It is a fortunate situation that the most intense spectral bands are not totally dominated by effects due to the missing electrons in core and valence orbitals, and also that spectra as emitted (before or without self absorption) are relatively insensitive to the means of excitation. These two factors simplify the theoretical interpretation and use of valence band spectra.

We noted that there are three aspects of transition probabilities which are important in valence band spectroscopy: (1) selection rules, (2) intensity as a function of energy, and (3) spatial limits on regions of x-ray origin. The second factor includes the overall intensity of a valence band spectrum, as well as its shape. Absolute emitted intensity measurements are difficult to make due to absorption and fluorescence in the sample and unknown spectrometer efficiencies. For chemical analysis, intensities from an unknown are commonly measured relative to the intensity from a standard in order to remove instrumental effects. However, intensities from more than one sample are seldom compared for basic valence band spectra interpretation. It seems that future electronic structure work should take advantage of information available from comparisons between different materials, as normally done in chemical analysis. The third factor, the spatial selectivity of x-ray spectroscopy, seems to be seldom considered. Since the source of x-rays in a material is localized, it is especially appropriate to use local bonding theories to interpret x-ray spectra.

We have discussed two local bond models; Reilly's calculation for the mixed covalent-ionic case and the Electron Cell model for metals and alloys. The step-wise energy level diagram resulting from Reilly's calculation is especially useful for the interpretation and use of x-ray spectra. Conventional Molecular Orbital local bond calculations are being more commonly used to understand spectra. Valence Bond theory is also flexible and broadly applicable, and should find increasing use for interpretation of spectra.

The field of x-ray valence band spectroscopy is well defined, with spectra being used for both basic and applied work. The information which results completes or competes with that from other techniques. Recent advances in both theory and experiment have clarified some problems, such as many-body effects and self-absorption. Other remaining problems, for example overlapping excitation structure, may not be avoidable. However, short-comings such as the scarcity of calculated spectra and lack of data on less intense spectra features will be overcome. Among the more active areas of basic work now are (a) the study of ionized-state effects, (b) the use of bond theory and (c) the interpretation of alloy spectra. Valence band x-ray spectroscopy should become a standard technique for the practical study of new materials with unusual properties which are due to their peculiar electronic structure.

Appendix 1

X-RAY STATES AND THEIR DECAY

The discussion in the text centered on the main valence band spectra which is most intense and useful for studying electronic structure and obtaining practical information. The non-uniqueness of both the x-ray state (with a hole in an inner shell) and its decay lead to additional spectral features, including excitation structure, many-body effects, and various satellites. The shape and intensity of these features is determined by the probabilities for excitation of a particular x-ray state and its decay by one of the possible modes. Relatively little information is available on these probabilities under various excitation conditions. We will discuss the types of excited states which occur, how they can decay and the less-intense spectral features which result. The less-intense structure is useful for studying the interaction of the ionizing radiation and matter, and the various decay modes. The study of ionized states and their dynamics complements study of the ground state electronic structure and can be viewed as a second basic goal of valence band spectroscopy.

Consideration of the formation, relaxation and decay times for x-ray states is a useful preliminary. Figure 13 gives estimates of the ionization, orbital relaxation and lifetimes for x-ray states. The time for ionization is the calculated time for an electron with a given kinetic energy to travel 2 A, that is, to the "exterior" of the atom, at its full escape velocity. This "sudden approximation" gives a minimum ionization time for the chosen 2 A. The relaxation time for electrons in the various filled shells is measured by and set equal to the orbital period of the electrons since slower moving electrons can be expected to take longer to relax. The orbital period is the time for an electron with kinetic energy equal to one-half its binding energy[64] to travel around a circle with radius equal to the distance from the nucleus to the maximum in the electron probability distribution.[65] The orbital periods are long compared to the 10^{-8} cm/$(3 \times 10^{10}$ cm/sec.$)$ $\sim 3 \times 10^{-19}$ sec. which is required for a change in an electromagnetic field to propagate 1 A. That is, the electrons "feel" the changing field due to inner electron ejection much faster than they can respond to it. The Uncertainty Principle $\Delta E \ \Delta t \sim \hbar =$ Planck's constant, and experimental inner electron level widths ΔE,[66,67] gives the lifetime Δt of the hole in the inner level. Lifetimes and relaxation times are plotted against the ground state kinetic energy of electrons in the various atomic orbital in figure 13. The horizontal scale merely provides a way to display the data.

The estimated ionization and relaxation times may well be wrong by a factor of two or so, particularly for valence electrons in solids. However the general conclusions to be drawn from the times are not particularly sensitive to such an uncertainty. It is seen that the relaxation times for electrons in outer shells (e.g. valence electrons) are indeed longer than those for inner shell electrons. Relaxation times are also longer than ionization times for outer shell electrons when the exiting electron leaves rapidly (i.e. with high kinetic energy). For high atomic number elements, however, the decay of the inner-shell-ionized x-ray state appears to occur even before the completion of outer electron relaxation.

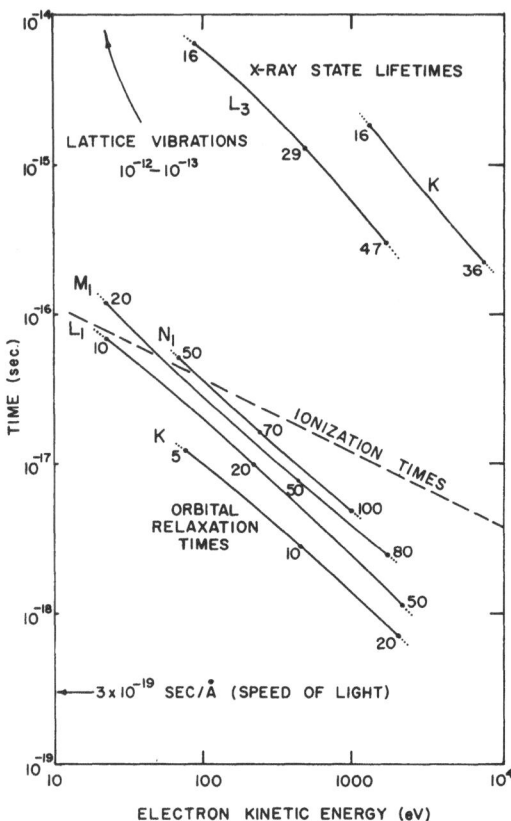

Figure 13. Ionization, relaxation, and x-ray excited-state times for the atomic numbers and shells indicated. The ionization time is taken to be the time required for an electron with the indicated kinetic energy to travel two angstroms. Relaxation times and liftimes are plotted against the kinetic energy of electrons in the atomic orbitals indicated.

In fact for threshold excitation the exciting electron may not be clear of the atom prior to radiative or Auger decay. Such rapid decay of the ionized state applies to hard x-ray emission. The x-ray states associated with the outer electron shells and low atomic number elements involved in valence band x-ray spectroscopy have decay times longer than ionization and relaxation times. This insures that the filling of the hole in an inner shell is independent of its formation, and allows distinguishing excitation and emission processes (as in Table I). The decay times for holes in outer orbitals approach the 10^{-12} to 10^{-13} sec. periods of lattice vibrations (ion motion).

With this background of times important in x-ray processes, we turn to the various x-ray states which can be formed. In his 1959 review paper, Parratt[66] argued that energy level diagrams based on a one-electron picture, such as those in figure 1, do not give a consistent or adequate picture of the energies of excited (inner level ionized) states. He suggested that the zero of energy should be taken as the normal atomic ground state. States with holes in any level would have positive energies because some ionization energy had to be expended in the ionization which created them. Since creation of a K shell vacancy requires more energy than ionization of an L shell level, the K state is plotted at a higher energy than L states. This is shown in figure 14. The natural tendency of K states to become L or M states, etc., is represented by downward transitions toward the ground state.

Lifetime broadening of the excited states is included in the diagram. The band of valence states is much wider than the other states due to solid state effects. There is continuum of ejected electron states associated with each major (K, L_1, ...) level. The continuum is important in absorption studies but, since we are concentrating on x-ray emission, the continuum is not shown in figure 14.

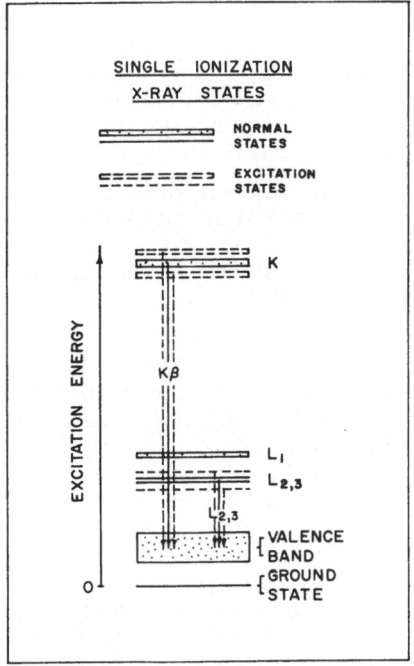

Figure 14. Parratt diagram for singly-ionized x-ray states in aluminum or silicon (not to scale).

Two of the three types of x-ray states are shown in figure 14. The singly ionized states which are most probable (occur most commonly) are termed normal x-ray states. Excitation states also occur in singly ionized atoms. According to Parratt[66] excitation states are of two types, Valence Electron Configuration (VEC) states and Bound Ejected Electron (BEE) states. VEC states are due to incomplete relaxation of the valence electrons as they adjust to the change in screening due to ionization. We saw from figure 13 that low atomic numbers and relatively energetic exciting radiation provide conditions favorable for completion of ionization prior to valence electron relaxation. It would seem that, since only partial relaxation would occur during VEC state formation, such states should have higher energies than the normal states.

BEE (or exciton) states are formed when the exciting electron does not have sufficient energy to leave the atom completely and is retained in a bound state. Since some binding energy is retained, BEE states occur at lower energies than normal states. Shielding of the exiting electron by mobile conduction electrons prevents the formation of "stable" BEE states in conductors (although the coulomb interaction between conduction electrons and the inner level hole is termed an "excitonic" effect[68]). BEE states are important in x-ray absorption spectra of insulators where they occur in the forbidden gap between valence and conduction bands. They are not thought to be important under the usual conditions of electron or high energy x-ray excitation but Blokhin and Demkin[69] discuss BEE state formation following normal excitation. BEE states are not important in most emission work although they have been observed using threshold excitation.[70]

The third type of x-ray state has one hole in an inner shell and a second or additional holes in either inner or outer shells. An $L_3 L_3$ state, if plotted on figure 14, would be at an energy slightly greater than twice that of L_3 state. The additional binding which follows creation of the first L_3 hole requires that more energy be expended in the second ionization.

The decay of x-ray states is no more unique than their formation. An excited state may decay by radiative, nonradiative (Auger) or a combination[71] of both processes. One-electron transitions occur in all materials while in conductors many-body effects[72] also enter. Table VI lists the spectral features which arise from radiative decay of normal, excitation and multiple-ionization x-ray states by electron transitions, with or without many-body effects. The origin of the normal band and excitation structure is indicated in figure 14. Holes in addition to the inner shell vacancy lead to higher energy multiple-ionization satellites, especially if there is a second hole in an inner level. Such satellites are sensitive to local atomic structure and sometimes composition, and may be used to measure these quantities. Production of additional holes in outer orbitals, e.g. under heavy atom bombardment,[73] yields peak shifts similar to but larger than chemical shifts. Fermi edge peaking ($L_{2,3}$ spectra) or rounding (K spectra) is due to scattering of conduction electrons by the inner electron vacancy in conductors.[74] Auger spectral tails arise from valence band energy level broadening due to the short lifetime of a hole in the valence band before it is filled in an Auger transition. Fermi edge effects and Auger tails are the main influence of many-body interactions on the shape of a valence band spectra. There are theoretical indications that the absolute intensity of the main band in conductors may be affected by many-body effects.[75] Comparison of measured intensities from a metal (e. g. sodium) and one of its compounds (not subject to such many-body effects) would, after correction for composition and matrix effects, provide a test for many-body influences on absolute intensity. Low energy satellites are actually part of the valence band spectrum being due to low-lying bonding levels. The first four features listed in Table VI can create problems in extracting electronic structure information from valence band spectra.

TABLE VI

The Origin, Energy, and Intensity of X-Ray Spectral Features
Compared to Energy and Intensity of the Main Band
(* = Conductors Only)

Spectral Feature	Origin	Relative Energy	Relative Intensity
Multiple ionization satellites	Multiple ionization	Higher	Sometimes over 30 %
Fermi edge effects*	Coulomb electron-hole interaction	High energy edge	Several %
Excitation structure	Non-unique excitation	Higher or Lower	Several %
Auger tails	Many-body effect	Lower	Several %
Low energy satellites	Bonding	Lower	Several %
Plasmon satellites*	Many-body effect	Lower	1 - 2 %
Radiative Auger structure	Simultaneous photon and electron emission	Lower	< 0. 1 %

Appendix 2

EXPERIMENTAL FACTORS AND CORRECTIONS TO SPECTRA

There are three major experimental considerations in valence band x-ray spectroscopy, those concerning (1) the sample, (2) how it is excited, and (3) the x-ray measurement. Each of these will be discussed briefly before a listing of spectral corrections is given.

Sample preparation, characterization, homogeneity, cleanliness and stability during excitation are all important in soft x-ray work. Since soft x-rays escape a sample from only a shallow depth, the condition of the surface is important. Vapor deposition, heating, sputtering, and scraping in the spectrometer vacuum have all been used to insure surface cleanliness. Intense or prolonged electron irradiation sometimes will cause surface films to grow, decompose the sample, or cause phase transitions, depending on the sample, vacuum conditions and the intensity and extent of electron irradiation. A clean vacuum and rapid measurement with low specific loading (electrons/mm^2/sec) are needed, if measurements can be made at all under these conditions.

In general, the basic questions concerning x-ray excitation are: (1) what radiation, (2) what energy (i.e., what spectrum), and (3) what intensity? The type, energy, and intensity of exciting radiation determine (a) which x-ray states are formed, as discussed in Appendix 1, (b) where in the material the x-rays are generated and (c) the associated background radiation.

Electron energy is easily controlled but, due to electron scattering in the sample, x-rays are not produced by monoenergetic electrons except very near threshold. Excitation at threshold also minimizes self absorption due to shallow depth x-ray production[4] and avoids generation of high energy satellites.[5] Fluorescent x-ray excitation usually yields lower intensities compared to electron excitation and is done with a broad spectrum which can scatter into the detector. But x-ray excitation does not damage the sample as much as incident electrons and yields a spectrum free of a bremssthralung background. Protons also excite little background but high energies (about 10^2 times electron energies) have to be used to yield comparable intensities.[76,77]

The effective depth of x-ray production, which determines the self absorption, depends on both the excitation conditions and x-ray take-off angle. If the absorption coefficient at the x-ray energies being measured is smoothly varying (away from absorption edges), the self absorption has negligible effect on the emission spectral shape (but does of course influence the intensity). If, however, the absorption coefficient varies rapidly with energy across the emission spectrum, self absorption alters the shape of the emitted spectrum.

This emission-absorption spectra overlap allows determination of
the absorption spectrum by ratioing the emission spectra obtained
at low and high incident electron energies which correspond to surface and bulk x-ray production. [4, 5] This is a convenient way to obtain emission and absorption data on the same energy scale. However, it is usually the emitted spectrum, free of absorption which
is desired for comparison with theory. Then low electron energies
and high take-off angles are used.

The experimental geometry is chosen to maximize intensity in
order to get good statistics in a short time while attaining resolution
adequate for the aim of the measurement. High resolution work can
be done with double-crystal and curved-single-crystal spectrometers
in the wavelength region below about 28 A, but it is necessary to go
to grating spectrometers to attain high resolution at longer wavelengths. Such high resolution is usually needed for studies of electronic structure and atomic processes and requires long measurement times. Measurements of local atomic structure, properties,
and composition frequently do not require such refined work since
often only peak positions, heights, or integrated intensities are
needed. Then it is possible to use flat-crystal spectrometers with
Soller slits or curved-single-crystal spectrometers for both wavelength regions. Most commercial equipment is of this type because
lower resolution generally goes with higher intensity. Reproducibility of an x-ray measurement depends on essentially all experimental factors, in particular on sample and excitation stability,
mechanical quality of the spectrometer and stability of the counting
electronics.

Corrections must be applied to a measured x-ray spectrum in order
to compare the spectrum as emitted with a calculated spectrum or to
extract information from the position, shape or intensity of spectral
features. Corrections are mainly concerned with the experimental
factors just considered since it is desireable to separate the influences
of making the measurement from the spectral features which are due
to the material under study. In general, corrections are necessary
for basic valence band spectra work but are often not required for
practical measurements. Also, not all corrections are necessary
in any one situation. For instance, relative intensity measurements
within a band do not require correction for spectrometer efficiency
if the efficiency is constant across the band. It should be remembered
that corrections which ostensibly concern only intensity can also affect
spectral shape. For instance, a rapidly varying spectrometer efficiency could skew a spectrum, changing the apparent shape and the
energies of peaks, or even introduce peaks, as shown by Liefeld et al.
for KAP crystals. [78]

Lists of corrections have frequently been given before. [5, 66] The
following list of corrections to valence band spectra is given roughly
in order application:

1. Backgrounds - Bremsstrahlung and scattered x-rays.

2. Spectrometer

    a. Nonlinearity with energy, i.e., conversion from distance on a strip chart, multichannel analyzer printout, or film to energy or wavelength,

    b. Nonlinearity with intensity due to counter-electronics dead time or film response,

    c. Efficiency of collection, dispersion and detection, and

    d. Window, or response function.

3. Broadening of both core level and valence band due to ionized state lifetimes.

Measurement of backgrounds is discussed by Liefeld. [5] Spectrometer linearity (dispersion and intensity) corrections are easily obtained and applied, in contrast to the spectrometer efficiency and window. The efficiency of the spectrometer depends on the geometry, and measured efficiencies of the crystal or grating and detector. Such information is usually unknown. The window function can be calculated from the geometry of the spectrometer and characteristics of the dispersing element, for instance the rocking curve width of a crystal. Core level widths are available from several sources. [66,67] Auger lifetime broadening of the valence band can be calculated by the method of Blokhin and Sachenko. [79]

The unfolding of the spectrometer function and core level broadening from experimental data can be done by a variety of mathematical techniques. It is essential to unfold spectra in order to compare predicted and measured x-ray spectra. Unfolding is preferable to folding the broadening functions into the theoretical spectrum because differences between unfolded and calculated spectra more accurately reflect the theoretical shortcomings which are sought. However, no technique for removing the Auger tail from a measured spectra is known to the author.

Multiple ionization satellites and self-absorption are not listed as corrections. Threshold excitation should be used to avoid them in the first place since it would be required anyway to determine a correction. [5] Excitation structure is not listed either. If it occurs, excitation structure is at least more difficult and maybe impossible to eliminate although varying the incident energy of the exciting radiation may provide a means to distinguish excitation structure.

The transition probability is not listed as requiring a correction. As discussed in the text, the transition probability and density of states are almost inextricably tied together in the spectrum. A correction for varying transition probability could be manufactured by ratioing, at each energy, calculated x-ray spectra (including the transition probability) and the calculated density of states. But, it is better to compare the calculated spectra directly with (corrected) experimental data so such an effective transition probability is of little use.

Appendix 3

BONDING THEORIES FOR ORDERED AND DISORDERED MATERIALS

Because the electronic structure of a material primarily determines its valence band x-ray spectra, the bonding theories which are used to calculate electronic structure are fundamental to the interpretation of spectra. Some characteristics of bonding theories were discussed in the text and are summarized in Table VII. In this

TABLE VII

Types of Bonding Theories

Material	Comment	Type of Theory
Ordered	Atomic structure and composition known over range of electronic sharing (e.g., ordered crystals).	
	More complete and complex, less transparent, mainly used metallic and ionic materials.	Band
	Simpler and clearer but less complete. Covalent materials mainly.	Bond
Slightly Disordered	Atomic structure not fully known over range of electron sharing (e.g., dilute alloys and glasses).	Band or bond
Grossly Disordered	Atomic structure not known over range of electron sharing (e.g., concentrated disordered alloys).	
	More complete and complex. Highly mathematical.	Scattering
	Simpler, clearer, and less complete. Applied mainly to alloys.	Charge distribution

appendix, the major specific theories of each type listed in the last column are discussed. Each bonding theory is based on a particular model, requires certain kinds of calculations, and is capable of yielding certain specific details of electronic structure. We begin with theories applicable to ordered materials, for which Slater's "Quantum Theory of Matter" is a good overall reference. [80]

Band theories[81, 82] are concerned with the electronic structure of materials with all types of bonding, no matter whether the electrons are localized or delocalized. There are two extreme types of band theories, those derived from atomic orbitals and those which are based on the nearly-free-electron approximation. In the first, the electrons are bound to the nuclei. There is only slight overlap between electrons on adjacent atoms. The Tight Binding method is the major example of this approach. In the second case, the valence electrons are highly mobile and little affected by the ions. Both extreme theories are inadequate in some respects. Since physical reality is a mixture of the two pictures, present-day band theories fall between the extremes.

Figure 15 contains a schematic illustration of the model for each of the band theories which are intermediate between tight binding and free electron approaches.[82]

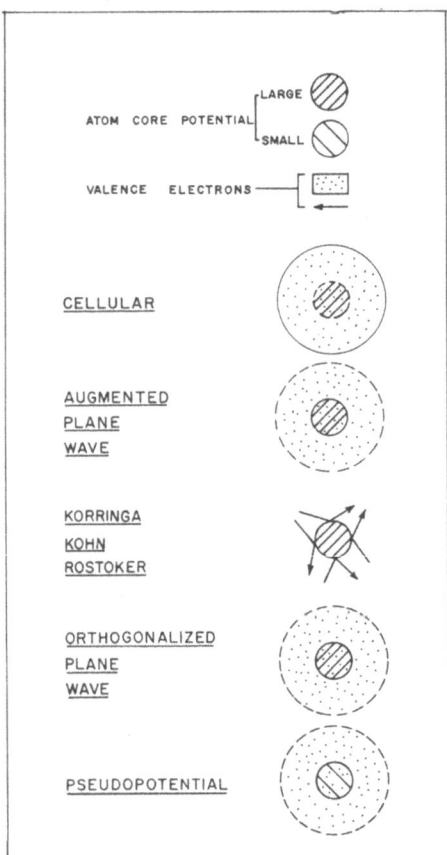

Figure 15. Schematic illustration of the major band theory models. Solid circles indicate where boundary conditions on the valence electron wave functions are specified.

Figure 15 is meant to summarize the major characteristics and differences of the theories. Cellular methods were the first developed but these involve a difficult boundary condition problem where the cells meet. Hence, cellular methods are not generally used for calculations of electronic structure now but they do provide the basis for cell models which are finding increasing use in disordered alloy work. The atom core picture used in the rest of the approaches provides a more tractable boundary condition problem. The differences in the electron potential used and the form of the wave functions distinguishes the models and determines their rapidity and accuracy of calculation, as well as their applicability to different materials. The Augmented Plane Wave (APW), Korringa Kohn Rostoker (KKR) and Orthogonalized Plane Wave (OPW) methods are the basic models, the Pseudopotential (PP) approach being an outgrowth of the OPW method. In PP theory, an effective potential which arises mathematically from the orthogonalization of core and plane wave valence electron states is found to cancel most of the potential due to the nucleus. The sum of the effective and nuclear potentials (the pseudopotential) is small. This both allows the use of more convenient theoretical (perturbation) techniques and provides an understanding of why the nearly free electron approach works as well as it does. Transition metals have been treated most successfully by the APW and KKR methods,[83] while the OPW and PP techniques have been used for a wide variety of simple metals, and III-V and II-VI compounds.[84]

Bond theories[85] are limited to localized electron situations. The electrons are localized around and between atoms as in the covalent picture. We will discuss the two general bond theories which are widely applicable and two other approaches which are more limited in their scope.

The general approaches to the calculation of bond characteristics are the Molecular Orbital (MO) or Hund-Mulliken, and Valence Bond (VB) or Heitler-London Theories. Both of these date back to the early days of Quantum Mechanics. The two types of theories are capable of treating any localized electron situation from the local electrons in a metal through the pure covalent situation to and including ionic bonding which is a limiting case of covalent bonding.

Figure 16 contrasts the MO and VB approaches to bonding in a simple diatomic molecule in order to compare the two models. The electron distributions and wave function are shown in each case for the two atoms A and B and the two electrons 1 and 2. The wave functions in both cases are made up of atomic orbitals such as that associated with atom A and occupied by electron 1, $\psi_A(1)$. The MO wave functions (molecular orbitals) encompass the entire molecule while VB wave functions are more atomic in character, at least initially. Both reflect the inability to distinguish electrons. The MO approach automatically includes both covalent and ionic character (equally weighted in the example). In the VB method a number of so-called resonant structures, one covalent and two ionic in this case,

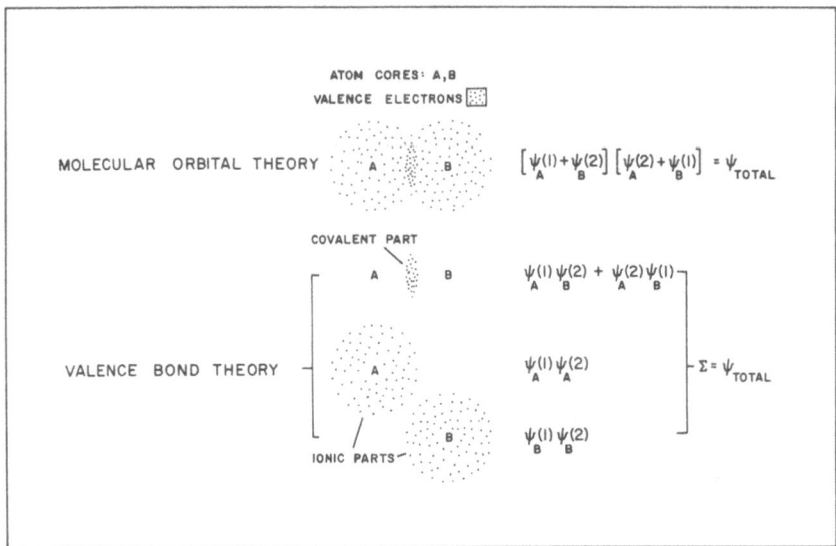

Figure 16. Schematic electron distribution and wave function for
like atoms A and B with electrons 1 and 2 in a diatomic molecule
as given by the Molecular Orbital and Valence Bond Theories.

are added together to obtain the total bonding picture, as illustrated
in figure 16. The wave functions obtained from both approaches are
identical in this simplest example.

It is most important that, because MO and VB theories use atomic
orbitals as starting wave functions, both approaches to bonding avoid
having to solve Schrodinger's equation in the course of the bonding calculation. The atomic orbitals which are used are already the result
of solving of this equation. These are readily available.[9] Hence, a
great deal of physical and mathematical complexity is avoided in bond
calculations compared to band theory, with the result that bond calculations tend to be easier and clearer than band calculations. The local
bond calculation for SiO_2 discussed in the text illustrates the relative
simplicity of bond calculations (page 195).

Some comments should be made concerning an alternative approach
to ionic bonding. Most solid state textbooks[86] contain a treatment of
the common approach to cohesion in ionic materials which consists of
a combination of electrostatic attraction and ion core repulsion. The
cohesive or bonding energy results from this method but no electron
wave functions are obtained. The cohesive energy is related to the
valence electron (and ion core contact) energies. Despite the broad
usefulness of this approach, the method is qualitatively different than
the MO and VB treatments; no allowance is made for partial covalent
bonding.

The two more specialized theories of bonding mentioned earlier are the Ligand Field (LF) and Crystal Field (CF) Theories.[87] While the MO and VB theories are capable of yielding the overall energy level structures, LF and CF theories are designed to give the splitting of the energy levels associated with inner (usually d but also f) electrons due to the presence of neighboring atoms or ions (ligands). LFT is usually applied to transition metal complex compounds. CFT was developed mainly to treat transition metal ions in a crystal lattice. The basic distinction between the two theories is not, however, the materials to which they are applicable. Either theory can be used for crystals or molecules. In the LF picture covalent interactions between the central atoms and perturbing neighbor atoms are taken into account. The CFT is distinguished as being the extreme ionic picture in which the perturbing neighboring ligands are treated as point charges with no overlap of wave functions. Since the splitting of a few electron volts due to the interactions treated by LF and CF theories is larger than the energy resolution attainable in x-ray measurements, LF and CF theories are of use in the interpretation of x-ray spectra.

This completes our discussion of theories applicable to ordered materials. Disordered alloys will now be considered. In pure metals, the electronic structure depends on the atomic structure, in particular on the interatomic spacing. The variety of atomic spacings present in alloys leads to a smearing and broadening of the band structure, e.g., the density of states, relative to the pure metal. The classical approach to alloy band structure, the Rigid Band Model, ignores such changes and demands that the bands remain unchanged upon alloying, the only effect being the change in the electron to atom ratio, i.e., the filling of the band. Modern alloy theories fall into two categories, those more basic theories based on the multiple scattering theory and those which are primarily concerned with the charges on atoms in an alloy.

Scattering theories of alloy electronic structure are similar to the KKR method in pure metals. The alloy is characterized by the "coherent potential" of Soven,[88] so chosen as to represent the average (coherent) properties of the lattice. A recent and thorough treatment of binary alloy electronic structure from the scattering viewpoint was given by Velicky et al.[89] The single site approximation was employed; this ignores the effects of clustering on the electron scattering potentials. The density of states, the contribution to the density of states by each component, and the Bloch wave spectral density are some of the results of this theory. It is shown to have the correct limiting behavior, for example, in the dilute alloy case. The scattering approach to alloy electronic structure is relatively new, with much work on developing the method as well as applying it still to be done.

Charge distribution theories are less complete but more tractable than scattering theories. Varley's early approach to concentrated alloy theory[90] is based on a two band model for which the charges on the atoms are easily calculated from the condition of Fermi Level equalization. The aim of Varley's work was the calculation of heats of alloy formation rather than quantities such as the density of states. Fair agreement between the predictions of the model and experimental heats was obtained. A number of physical effects due to Brillouin zone boundaries, ion core repulsive interactions, and van der Waal's forces were also discussed. It was noted by Varley "that the tendency to transfer charge expresses the magnitude of the electrochemical factor in an alloy system".

The Electron Cell model of alloys due to Bolsaitis and Skolnick[60] was based on precisely the factor just mentioned; the redistribution of charge upon alloying is taken to depend on the relative electronegativity of the constituent atoms. Heats and volumes of mixing and other alloy properties have been calculated using the Electron Cell model.[91] To date the model has been applied mainly to noble metals, but it is currently being examined for applicability to transition metals.[92]

Stern[61] has discussed several aspects of dilute and concentrated alloy charge models. In particular, he indicated how the density of states depends on the amount of charging (i.e., the unevenness of the electron distribution). Large electron redistribution goes with the development of two bands, similar to Varley's model in some respects. Stern asserts that the use of average potentials must be limited to cases of small charging. As pointed out in the text, the amount of charging for use in Stern's theory can be obtained from the Electron Cell model.

Finally, mention should also be made of an electronic structure theory due to Kohn and coworkers which is based on explicit use of the inhomogeneous electron density.[93] This theory has yielded good values for the lattice parameter and bulk modulus of sodium without the use of any parameters. This method seems to have potential for application to alloys.

ACKNOWLEDGMENTS

The author appreciates receiving preprints from P. Bolsaitis, D. W. Fischer, M. H. Reilly, D. S. Urch, and E. W. White, and the opportunity to use the revised Soft X-Ray Bibliography[94] prior to publication which was afforded by J. R. Cuthill and R. C. Dobbyn (National Bureau of Standards Alloy Data Section). Helpful discussions with D. W. Fischer, C. M. Gilmore, A. R. Ruffa, and especially M. H. Reilly are gratefully acknowledged. C. M. Dozier read the manuscript and made valuable comments. Special appreciation is due L. S. Birks for his support and guidance.

REFERENCES

Note: See reference 8 below, p. 369, for a recent bibliography on X-Ray Valence Band Spectra.

1. G. A. Rooke, "The Interpretation of X-Ray Band Spectra", in D. J. Fabian, Editor, Soft X-Ray Band Spectra, Academic Press, New York, 1968, p. 3.

2. I. Ya. Nikiforov and M. A. Blokhin, "Concerning the Shape of the $K\beta_5$ Emission Band of Iron", Bull. Ac. Sci. USSR Translation, 27, 323 (1963).

3. V. A. Batyrev and A. V. Shatunova, "Procedure for Investigating the Influence of Chemical Bonds on the Fine Sturcture of X-Ray Emission Spectra in Microvolumes", Bull. Ac. Sci. USSR Translation, 31, 896 (1967).

4. D. W. Fischer and W. L. Baun, "Self Absorption Effects in the Soft X-Ray $M\alpha$ and $M\beta$ Emission Spectra of the Rare Earth Elements", J. Appl. Phys., 38, 4830 (1967).

5. R. J. Liefeld, "Soft X-Ray Emission Spectra at Threshold Excitation", in D. J. Fabian, Editor, Soft X-Ray Band Spectra, Academic Press, New York, 1968, p. 133.

6. V. F. Demekhin and I. Ya. Kudraytsev, "Shape of the $K\beta_x$ Band of of Al Metal", Phys. Metals and Metallography Translation 26, 174(1968).

7. H. A. Bethe, "Intermediate Quantum Mechanics, Benjamin, New York, 1964, p. 143.

8. D. F. Fabian, Editor, "Soft X-Ray Band Spectra", Academic Press, New York, 1968.

9. F. Hermann and S. Skillman, "Atomic Structure Calculations", Prentice-Hall, Englewood Cliffs, 1963.

10. V. A. Formichev. "Investigation of the Energy Structure of B and BN and Ultrasoft X-Ray Spectroscopy", Bull. Ac. Sci. USSR Translation, 31, 972 (1967).

11. P. S. Bagus, "SCF Excited States and Transition Probabilities of Neon-and Argon-Like Ions", Argonne Nat. Lab. Rpt. 6959 (1964).

12. J. Friedel and A. Guinier, Editors, Metallic Solid Solutions, Benjamin, New York, 1963.

13. O. A. Ershov, D. A. Goganov, and A. P. Lukirskii, "X-Ray Spectra of Si in Crystalline and Glassy Quartz and in Li Silicate Glasses", Sov. Phys.-Solid State, Translation 7, 1903 (1966).

14. B. Segall, "Energy Bands of Aluminum", Phys. Rev., 124, 1797 (1961).

15.   G. A. Rooke, "Interpretation of Al X-Ray Band Spectra. I.
      Density Distribution", J. Phys. C (Proc. Phys. Soc.),
      Ser. 2, $\underline{1}$, 767 (1968).

16.   M. H. Reilly, "Temperature Dependence of Short Wavelength
      Transmittance Limits of Vacuum Ultraviolet Windows", J.
      Phys. Chem. Solids, to be published.

17.   W. Seka, Ph. D. Dissertation, Univ. of Texas (1965).

18.   G. L. Glen and C. G. Dodd, "Use of Molecular Orbital Theory
      to Interpret X-Ray Absorption Spectral Data", J. Appl. Phys. $\underline{39}$,
      5372 (1968).

19.   P. E. Best, "Electronic Structure of MnO_4^-, CrO_4^{2-} and VO_4^{3-} Ions
      From Metal K X-Ray Spectra", J. Chem. Phys., $\underline{44}$, 3248 (1966).

20.   P. E. Best, "Electronic Structures from X-Ray Spectra. II.
      Mostly ClO_3^- and ClO_4^-", J. Chem. Phys., $\underline{49}$, 2797 (1968).

21.   C. G. Dodd and G. L. Glen, "Chemical Bonding Studies of
      Silicates and Oxides by X-Ray Emission Spectroscopy", J.
      Appl. Phys., $\underline{39}$, 5377 (1968) and references therein.

22.   D. S. Urch, "The Origin of Low Energy Satellite Lines in X-Ray
      Emission Spectra; A Molecular Orbital Interpretation", to be
      published.

23.   D. S. Urch, "The Intensities of Low Energy Satellite Lines in
      X-Ray Emission Spectra", to be published.

24.   D. S. Urch, "Direct Evidence for $3d-2p\pi$ Bonding in Oxy-Anions",
      to be published.

25.   D. F. Lawrence and D. S. Urch, "Low Energy Satellites in
      X-Ray Fluorescence Spectrum of Fluoro-Anions", Spectrochem.
      Acta, to be published.

26.   K. Siegbahn, C. Nordling, A. Fahlman, et al., "Electron
      Spectroscopy for Chemical Analysis", Tech. Rpt. AFML-
      TR-68-189, 1968, p. 1 and 66.

27.   A. J. Freeman, "Band Structure in Metals and Alloys II: Sympo-
      sium on Experimental Methods", J. Appl. Phys. $\underline{40}$, 1386 (1969).

28.   F. Abeles, "Review of Optical Properties of Metals and Alloys
      Due to Interband Transitions", in ref. 8, p. 191.

29.   W. E. Spicer, "The Band Structure of Noble and Transition
      Metals; Photoemission Studies" in F. Abeles, Editor, Optical
      Properties and Electronic Structure of Metals and Alloys,
      Wiley and Sons, New York, 1966, p. 296.

30. G. A. Rooke, "Interpretation of Aluminum X-Ray Band Spectra II. Determination of Effective Potentials from $L_{2,3}$ Emission Spectra", J. Phys. C. (Proc. Phys. Soc.), Ser. 2, 1, p. 776 (1968).

31. G. Wiech, "Soft X-Ray Emission Spectra and Valence Band Structure of Be, Al, Si, and some Si Compounds", in ref. 8, p. 59.

32. W. Fischer, "Chemical Bonding and Valence State--Nonmetals", this volume.

33. M. H. Reilly and D. J. Nagel, in preparation.

34. A. R. Ruffa, "The Valence Bond Approximation in Crystals-Application to Analysis of Ultraviolet Spectrum of Quartz", Phys. Stat. Sol. 29, 605 (1968).

35. A. R. Ruffa, private communication.

36. A. T. Shuvaev, "Concerning the Interpretation of X-Ray Spectra", Bull. Ac. Sci. USSR Translation, 24, 434 (1960).

37. M. A. Blokhin and A. T. Shuvaev, "Concerning the Influence of Chemical Bonds on the X-Ray Emission Spectra of Titanium Compounds", Bull. Ac. Sci. USSR Translation, 26, 429 (1962).

38. M. I. Korsunskii and Ya. E. Genkin, "The Interpretation of the $L\beta_2$ Emission Band of Niobium", Sov. Phys. Doklady Translation, 7, 141 (1962).

39. A. Z. Men'shikov and S. A. Nemnonov, "Influence of the Chemical Bonds on the Valence State of Cr in Different Compounds", Bull. Ac. Sci. USSR Translation, 27, 402 (1963).

40. J. R. Cuthill, A. J. McAlister, and M. L. Williams, "Soft X-Ray Spectroscopy of Alloys: Ti-Ni and the Ni-Al System", J. Appl. Phys., 39, 2204 (1968).

41. J. E. Holliday, "The Use of Soft X-Ray Fine Structure in Bonding Determination and Light Element Analysis", Norelco Reporter, 14, 84 (1967).

42. A. Faessler, "Roentgenspektrum and Bindung Zustand", in Landolt-Bornstein Tables, Springer-Verlag, Berlin, Vol. 1/4, p. 769.

43. E. W. White and G. V. Gibbs, "Structure and Chemical Effects of the $SiK\beta$ X-Ray Line for Silicates", Am. Mineralogist, 52, 985 (1967).

44. E. W. White and G. V. Gibbs, "Structural and Chemical Effects on the $AlK\beta$ X-Ray Emission Band Among Al-containing Silicates and Al Oxides", Am. Mineralogist, 54, 931 (1969).

45. G. Arrhenius, "Chemical Bond Effects on Electron Transitions Between Inner Levels", in R. Castaing et al., Editors, X-Ray Optics and Microanalysis, Hermann, Paris, 1966, p. 328.

46. W. L. Baun and D. W. Fischer, "The Effect of Valence and Coordination on K Series Diagram and Nondiagram Lines of Mg, Al, and Si", in W. M. Meuller et al., Editors, Advances in X-Ray Analysis, 8, Plenum Press, New York, 1965, p. 371.

47. K. Das Gupta, "The Soft X-Ray Valence Band Spectra and the Heat of Formation of Chemical Compounds and Alloys", Phys. Rev. 80, 281 (1950).

48. D. H. Tomboulian, "Soft X-Ray Spectrometry", Handbuch der Physik, 30, Springer-Verlag, Berlin, 1957, p. 246.

49. M. I. Korsunskii and Ya. E. Genkin, "X-Ray Emission Bands and the Magnetic Properties of Nb", Bull. Ac. Sci. USSR Translation, 27, 740 (1964).

50. I. A. Ovsyannikova and I. B. Borovskii, "Investigation of the Fine Structure of the K Spectra of Some Sulphides", Bull. Ac. Sci. USSR Translation, 24, 444 (1960).

51. E. W. White and R. Roy, "Silicon Valence in SiO Films Studied by X-Ray Emission", Sol. St. Comm., 2, 151 (1964).

52. W. H. Knausenberger, K. Vedam, E. W. White, and W. Ziegler, "Thin Film Characterization by Electron Microprobe and Ellipsometry: SiO_2 Films on Silicon", Appl. Phys. Ltrs. 14, 43, (1969).

53. P. D. Gigl, G. A. Savanick, and E. W. White, "Characterization of Corrosion Layers on Al by Shifts in Al and O Emission Bands", submitted to J. Electrochemical Society.

54. J. W. Criss and L. S. Birks, "Formulas for Specimens Containing Spherical Particles", presented at 15th Colloq. Spect. Int., Madrid, Spain, 26-30 May 1969.

55. W. L. Baun and D. W. Fischer, "Effect of Alloying on AlK and CuL Emission Spectra in the Al-Cu System", J. Appl. Phys., 38, 2092 (1967).

56. W. L. Baun, "AlK X-Ray Emission Fine Features for Characterizing Al-Cu Films", J. Appl. Phys., 40, 4210 (1969).

57. R. K. Dimond, "Self Absorption of Soft X-Ray Spectra of Alloys", Phil. Mag., 15, 631 (1967).

58. C. Curry, "Soft X-Ray Emission Spectra of Alloys and Problems in Their Interpretation", in D. J. Fabian, Editor, Soft X-Ray Band Spectra, Academic Press, New York, 1968, p. 173.

59. C. A. W. Marshall, L. M. Watson, G. M. Lindsay, G. A. Rooke, and D. J. Fabian, "Interpretation of Soft X-Ray Emission Spectra of Al-Ag Alloys", Phys. Ltrs. 28A, 579 (1969).

60. P. Bolsaitis and L. P. Skolnick, "Electron Cell Model of Alloys", Trans. AIME, 242, 215 (1968).

61. E. A. Stern, "Requirements for a Theory of Disordered Alloys", in L. H. Bennett and J. T. Waber, Editors, Energy Bands in Metals and Alloys, Gordon Breach, New York, 1968, p. 151.

62. L. S. Birks, Electron Probe Microanalysis, Interscience, New York, 1963, p. 107ff. (Second Edition in preparation)

63. W. L. Baun, "Instrumentation, Spectral Characteristics, and Applications of Soft X-Ray Spectroscopy", Appl. Spect. Rev., 1, 397 (1968).

64.   J. A. Bearden and A. F. Burr, "Reevaluation of X-Ray
       Atomic Energy Levels", Rev. Mod. Phys., 39, 125 (1967).

65.   J. T. Waber and D. T. Cromer, "Orbital Radii of Atoms and
       Ions", J. Chem. Phys., 42, 4116 (1961).

66.   L. G. Parratt, "Electronic Band Structure of Solids by X-Ray
       Spectroscopy", Rev. Mod. Phys., 31, 616 (1959).

67.   Reference 26, p. 32.

68.   G. D. Mahan, "Excitons in Metals: Infinite Hole Mass", Phys.
       Rev., 163, 612 (1967).

69.   M. A. Blokhin and V. F. Demekhin, "Emission Spectrum of Sc
       in Sc_2O_3", Bull. Ac. Sci. USSR Translation, 27, 738 (1964).

70.   S. Hanzely, private communication.

71.   T. Aberg and J. Ultriainen, "Evidence for a Radiative Auger
       Effect in X-Ray Photon Emission", Phys. Rev. Letters, 22,
       1346 (1969).

72.   Reference 9, Part 4.

73.   P. Richard, I. L. Morgan, T. Furuta, and P. Burch, "Observed
       $K\beta$ Shift in Cu and Ni", Phys. Rev. Ltrs. 23, 1009 (1969).

74.   P. Nozieres and C. T. De Dominicos, "Singularities in X-Ray
       Absorption and Emission of Metals", Phys. Rev. 178, 1097 (1969).

75.   A. J. Glick, P. Longe, and S. M. Bose, "Effect of Electron In-
       teractions on Soft X-Ray Emission Spectra in Metals", in ref. 8,
       p. 319.

76.   L. S. Birks, R. E. Seebold, A. P. Batt, and J. S. Grosso, "Ex-
       citation of Characteristic X-Rays by Protons, Electrons, and
       Primary X-Rays", J. Appl. Phys. 35, 2578 (1964).

77.   A. A. Sterk, "X-Ray Generation by Proton Bombardment" in
       W. M. Meuller et al., Editors, Advances in X-Ray Analysis, 8,
       Plenum Press, New York, 1965, p. 189 and references therein.

78.   R. J. Liefeld, S. Hanzely, T. B. Kerby, and D. Mott, "Soft
       X-Ray Spectrometric Properties of Potassium Acid Phthalate
       Crystals", this volume.

79.   M. A. Blokhin and V. P. Sachenko, "Concerning the Shape of Energy
       Bands in Solids", Bull. Ac. Sci. USSR Translation, 24, 410 (1960).

80.   J. C. Slater, "Quantum Theory of Matter", 2nd Edition, McGraw
       Hill Book Co., New York, 1968.

81.   J. Callaway, "Energy Band Theory", Academic Press, New York
       1964.

82.   L. H. Bennett and J. T. Waber, Editors, "Energy Bands in
       Metals and Alloys", Gordon and Breach, New York, 1968.

83.   J. C. Slater, "Review of the Energy Band Problem with Recent Results", in ref. 82, p. 1.

84.   L. M. Falicov, "Orthogonalized Plane Waves and Pseudo-potentials; A Short Review", in ref. 82, p. 73.

85.   E. Cartmell and G. W. A. Fowles, "Valency and Molecular Structure", 3rd Edition, Van Nostrand, Princeton, 1966.

86.   C. Kittel, "Introduction to Solid State Physics", 3rd. Edition, Wiley and Sons, New York, 1966, p. 89.

87.   B. N. Figgis, "Introduction to Ligand Fields", Interscience, New York, 1966.

88.   P. Soven, "Coherent Potential Model of Disordered Substitutional Alloys", in ref. 82, p. 139.

89.   B. Velicky, S. Kirkpatrick, and H. Ehrenreich, "Single Site Approximation in the Electronic Theory of Simple Binary Alloys", Phys. Rev., 175, 747 (1968) and references therein.

90.   J. H. O. Varley, "The Calculation of Heats of Formation of Binary Alloys", Phil, Mag. 45, 887 (1954).

91.   K. A. Hsieh, "The Cohesive and Elastic Properties of Noble Metals in Terms of the Electron Cell Model", M. S. Thesis, University of Maryland, 1969.

92.   D. J. Nagel, "Theoretical Extension and Application of the Electron Cell Model to Correlation and Alloy Problems", unpublished.

93.   W. Kohn, "Electronic Structure from the Standpoint of the Inhomogeneous Electron Gas", in ref. 82, p. 65.

94.   H. Yakowitz and J. R. Cuthill, Nat. Bur. Stds. Mongraph 52 (1962).

A VACUUM SPECTROMETER FOR STUDYING THE

CHEMICAL EFFECT ON SOFT X-RAY SPECTRA

W. L. Baun

AFML (MAYA)

WPAFB, O. 45433

E. W. White

Materials Research Lab.

The Pennsylvania State Univ.

University Pk., Pa.   16802

ABSTRACT

Soft X-ray spectroscopy promises to be one of the most powerful techniques for solving a wide variety of materials
characterization problems. Although chemical analysis of the
light elements is commonly done by the electron microprobe and
X-ray fluorescence, the "chemical effect on X-ray spectra" has
been utilized in relatively few cases. This appears to result
from a general lack of appreciation for the applicability of chemical effect measurements and perhaps to a greater extent it is due
to the unavailability of appropriate soft X-ray spectrometers.
Most studies have been carried out using instruments that were not
particularly well suited to chemical effect studies. The purpose
of this paper is to describe a versatile new soft X-ray spectrometer and auxiliary equipment for the study of the chemical effect.
The spectrometer is housed in a versatile, high vacuum chamber 24
inches in diameter by 4 5/16 inch vertical clearance. Ten
standard flanges (eight on the circumference of the chamber and
one each on the top cover and base plate) facilitate a variety of
experimental setups. Each flange covers a 2 3/8 inch diameter
clearance hole in the chamber. The spectrometer is mounted in an
inverted position inside the top cover plate. A hoist lifts the
top cover assembly for 360° access of the spectrometer. Spectrometer motions are achieved through a high vacuum rotary feedthrough.
The motion is imparted to the main shaft by a 11.46 inch sine arm
that is driven by a precision step motor drive. The micrometer
drum reads angular displacements to one second of arc with
significant interpolation to better than ½ second. A graduated

circle with vernier is used to make settings to within 15" of arc throughout the entire 0-172° 2θ range. Most shift and fine structure studies are made using this precision sine arm but for certain wide range scans typically required for such broad peaks as B Kα and C Kα, a wide range worm gear drive is used and is also controlled by the HS-25 Slo-Syn motor and step driver. The 2:1 motion is provided by a unique gearless differential that operates unlubricated in vacuum. The main drive shaft may be rigidly clamped either to the detector arm plate or alternatively to the crystal plate as desired for a given set of experiments. For example, when the shaft is clamped to the detector plate the detector slit and sine arm become a rigid member, thus eliminating backlash inherent in gear drive assemblies. A Welch turbomolecular pump is used to provide a very rapid relatively clean vacuum on the order of 2×10^{-7} torr. The spectrometer sits on a special vibrationally isolated table. Isolation bellows and shock mountings are used to minimize vibrations generated by the turbomolecular and mechanical pumps.

A "NIM Standard" data acquisition system with printing scaler and Teletype and paper tape output is used with the spectrometer. Performance of the spectrometer with this data acquisition system is described and an example is shown of a computer plotted soft X-ray spectrum.

INTRODUCTION

Soft X-ray spectroscopy promises to be one of the most powerful techniques for solving a wide variety of materials characterization problems. Although chemical analysis of the light elements is commonly done by the electron microprobe and X-ray fluorescence, the "chemical effect on X-ray spectra" has been utilized in relatively few cases. This appears to result from a general lack of appreciation for the applicability of chemical effect measurements and perhaps to a greater extent it is due to the unavailability of appropriate soft X-ray spectrometers. Most studies have been carried out using instruments that were not particularly well suited to chemical effect studies. The purpose of this paper is to describe two slightly different versions of a versatile new spectrometer now in use in the authors' laboratories.

DESIGN CONCEPTS

A general view of the system is shown in figure 1. The spectrometer is housed in a versatile, high vacuum chamber 24 inches in diameter by 4 5/16 inch vertical clearance. Ten standard flanges (eight on the circumference of the chamber and one each on the top

Figure 1.   Soft X-Ray Spectrometer and Auxiliary Equipment

cover and base plate) facilitate a variety of experimental setups.
Each flange covers a 2 3/8 inch diameter clearance hole in the
chamber. The spectrometer drive is mounted in an inverted position
inside the top cover plate. A hoist lifts the top cover assembly
for 360° access of the spectrometer. The spectrometer assembly
and top cover may be turned upside down and supported by the hoist,
thus making modifications of the experimental setup quite easy for
the operator.

This inverted configuration simplifies manipulation of the
main shaft into the chamber. Motion is imparted to this shaft by
a 11.46 inch sine arm that is driven by a precision step motor
drive. The micrometer drum reads angular displacements to one
second of arc with significant interpolation to better than 1/2
second. A graduated circle with vernier is used to make settings
to within 15" of arc throughout the entire 0-172° 2θ range. Most
shift studies are made using this precision sine arm but for
certain wide range scans typically required for such broad peaks
as B Kα and C Kα a wide range worm gear drive (not shown in the
figure) is used and is also controlled by the HS-25 Slo-Syn motor
and step driver.

Up to 18 specimens (three on each demountable slide) may be
mounted simultaneously on the watercooled sample holder. With the

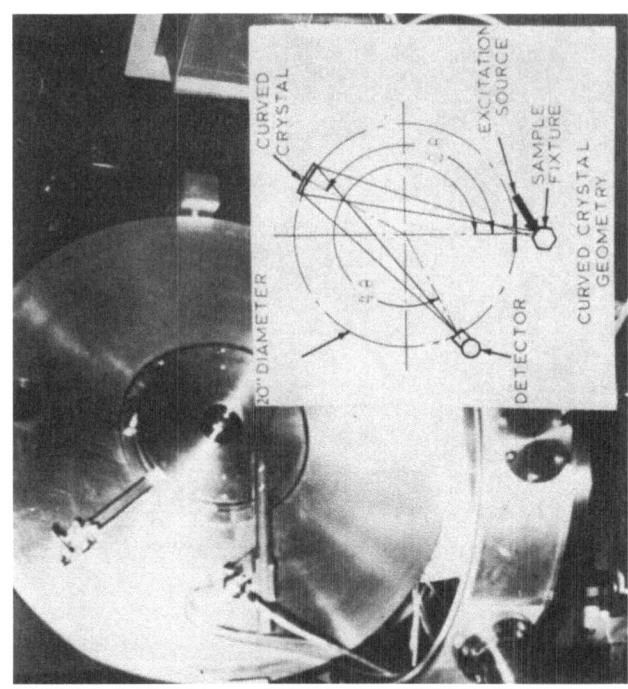

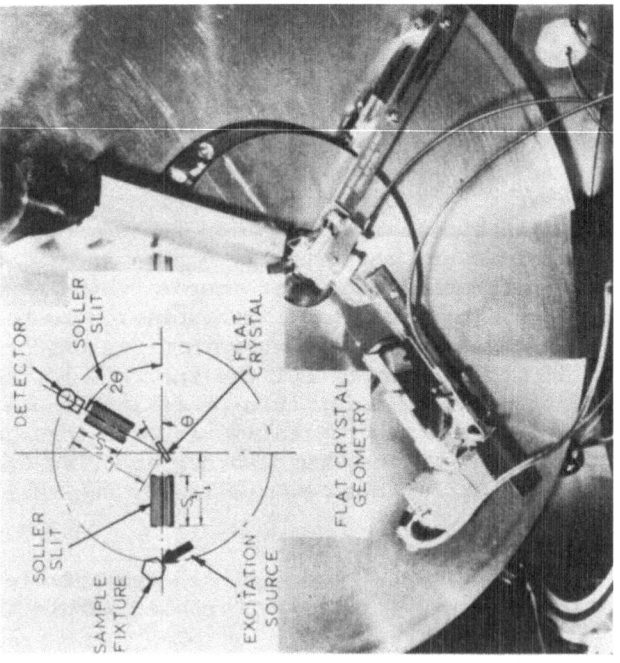

Figure 2.   Soft X-Ray Spectrometer Gearless Differential Used
With Flat And Curved Crystals

variety of crystals inside the vacuum chamber, together with this many specimens, one need not break vacuum very frequently.

The 2:1 motion is provided by a unique gearless differential that operates unlubricated in vacuum. The differential is shown in figure 2 applied to a flat crystal spectrometer on the left and a curved crystal spectrometer on the right. The main drive shaft may be rigidly clamped either to the detector arm plate or alternatively to the crystal plate as desired for a given set of experiments. For example, when the shaft is clamped to the detector plate the detector slit and sine arm become a rigid member, thus eliminating backlash inherent in gear drive assemblies. In this figure the top cover has been lifted with the hoist and is sitting at right angles to the chamber.

When operated as a flat crystal spectrometer, a six-position crystal changer makes available a variety of crystals which are selected and tuned (ω adjust) from outside the vacuum. Crystal selection and ω adjust are read from a ten-turn dial mounted on one of the standard flanges. External adjustment of the detector soller slit is also affected by an externally rotated shaft whose angular setting is read from a ten-turn dial.

The curved crystal spectrometer mode of operation provides for operation with fixed radius curved crystals having radii anywhere in a range of from five to eleven inches (10 to 22-inch diameter Rowland circle). Either curved and ground or simple curved crystals may be used.

Specimens may be excited by electron, X-ray, or proton beams. Figure 3 is a drawing of a top view of the chamber showing

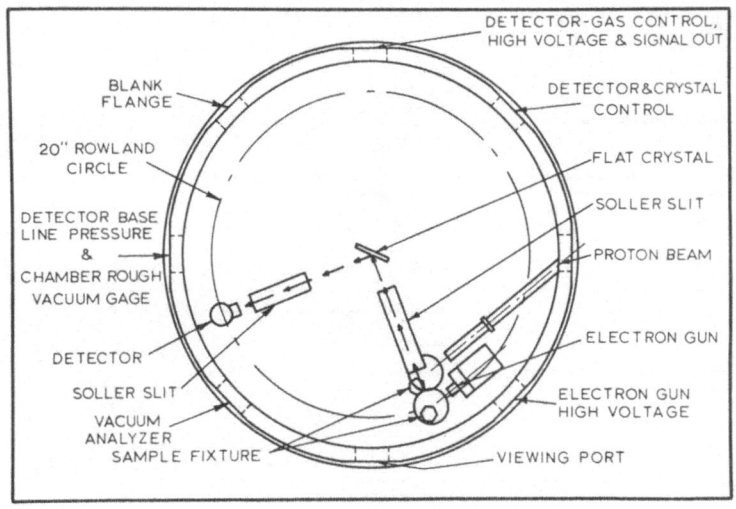

Figure 3.   Top View of Spectrometer Vacuum Chamber

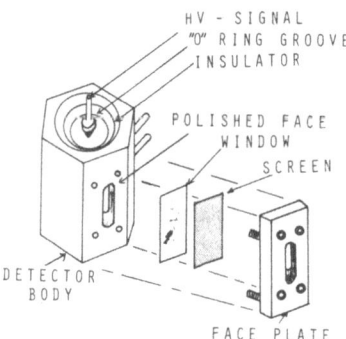

Figure 4.   Flow Proportional Detector

the position of critical parts of the spectrometer for electron and
proton excitation. In addition, a Siemens sealed X-ray tube may be
used for secondary excitation. The Siemens tube fits into any of
the ten chamber openings and is readily sealed with one "O" ring.

Several kinds of X-ray detectors have been used including flow
and sealed proportional counters, channel electron multipliers and
magnetic electron multipliers. The most useful detector has been
the thin windowed flow proportional counter shown in figure 4. The
construction and performance of this detector has been described
earlier[1]. Flow gas connections are made to one flange and provisions
for operation at reduced pressure have been made by using a manometer
in the system as shown in the plumbing diagram of the spectrometer
(figure 5).

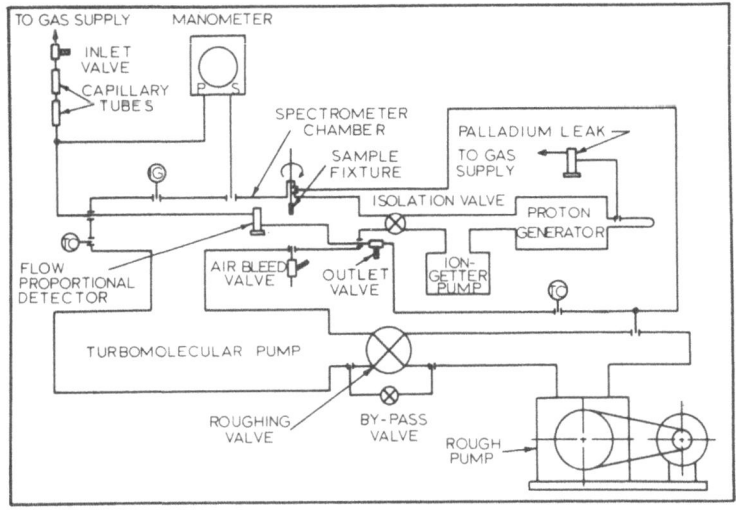

Figure 5.   Soft X-Ray Spectrometer Plumbing System

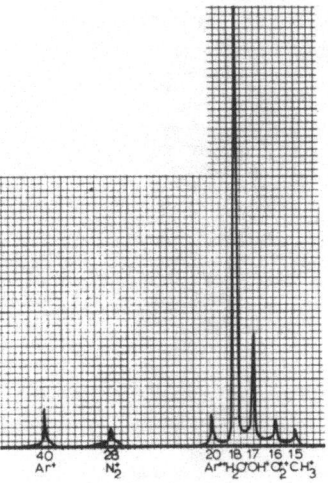

Figure 6.   Residual Gas Analyzer Trace

The Welch turbomolecular pump is used to provide relatively clean vacuums on the order of 2×10^{-7} torr. Minor carbon contamination buildup has been observed on a few samples when primary excitation was being used. Leak detection and measurement of partial pressure of various gaseous species (H_2O, N_2, O_2, etc.) are accomplished by a residual gas analyzer that mounts directly on one of the standard flanges. An example of a typical residual gas analyzer trace is seen in figure 6. Here we see that the major gases are due to water vapor. The turbomolecular pump has a large surface area on walls and vanes on which water vapor collects. The chamber is vented with dry nitrogen to minimize water vapor condensation. It would be beneficial to have a gate valve between the turbo pump and the chamber so that the turbo pump need not be vented each time the chamber is opened.

DATA ACQUISITION SYSTEM

One data acquisition system that may be used for X-ray spectroscopy is based on "NIM Standard" electronics. "NIM Standard" is a brief notation for Nuclear Instrument Module Standard, a program of the AEC and the National Bureau of Standards, which facilitates compatibility between instruments having different manufacturers. This standards program is outlined in USAEC Report TID-20893 (Rev). This report contains bin and module dimensions, bin and module connections and pin assignments, along with power supply requirements and tolerances.

Such a system utilizing NIM Standard components is illustrated by the flow sheet in figure 7. The thin windowed proportional

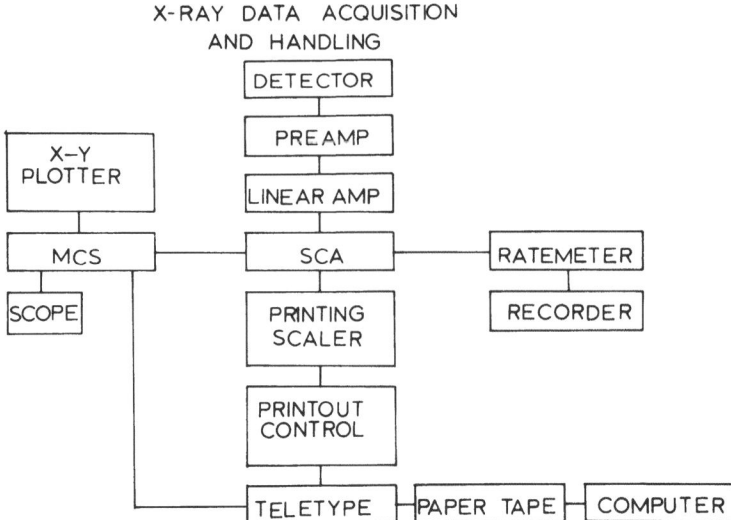

Figure 7.   Data Acquisition Flow Sheet

counter is connected to the preamplifier through a vacuum flange
BNC connection using coaxial cable which is kept as short as
possible. The preamplifier is a versatile solid state FET input
device offering pulse shaping if desired, variable charge gain,
and variable voltage gain. No pulse shaping is used in the preamp
in this work since a pulse shaping linear amplifier follows the
preamp. The signal is then fed from the linear amplifier to a
single channel analyzer (SCA). The signal is split at this point,
going to a ratemeter for recorder output and to a printing scaler
for Teletype output. The Teletype printout control is inter-
connected with the scaler, timer, and stepping motor control so
that it is possible to pre-set either time or counts. When the
pre-set condition has been reached, the scaler and timer stop;
and counts, time, and other information are printed. Simultaneously,
the timer stop pulse has caused the stepping motor controller to
advance the motor by a pre-set increment. When the Teletype has
completed printout, the sequence is repeated. In the system
presently in use, this sequence continues until stopped manually.
In a computer controlled system, the sequence would be stopped by
the computer after a pre-set number of steps and then either
recycled or reset to a new starting angle. One NIM Standard manufacturer has a step scanner and digital count rate meter which
perform many of the functions of a computer and print out time,
count, and angle in 2θ.

Another data acquisition system uses a multichannel analyzer
in a multichannel scaling mode. The multichannel analyzer is
normally used in the nuclear field and simultaneously registers

pulse height distribution data up to 4096 (or even more) channels.
In the mode used here, multiscaling, each channel is sequentially
advanced and data stored. Thus, each channel serves as an individual scaler having a capacity $10^6 - 1$ counts (or more in some
models). These instruments are now available in NIM configurations
and consist of an analog to digital converter, memory and data
handling modules, and special input/output devices. The signal
from the detector is routed through the same components as in
the previous system up to and including the single channel analyzer.
The scanning and timing are essentially the same as in the previous
system except that the multichannel analyzer acts as a data buffer
and readout takes place only after all data are obtained. Some
multichannel analyzers permit observation of accumulation of data
on the oscilloscope as it is being recorded. Each channel may be
advanced by the timer in coincidence with the stepping motor to
allow pre-set time intervals for each channel or the channels may
be advanced one per second (or any pre-set time period) while the
scanning motor is scanning in a continuous mode. When used this
way the multichannel scaler acts only as a recorder and starting
and ending angles must be noted, especially when different scanning speeds are used. The multichannel scaler has several advantages. It may be used to record and store a number of spectra
for visual comparison on the oscilloscope. Only those spectra
for which it is desired to have a permanent record need be dumped
to other readout devices such as the Teletype. Data manipulations
may be performed such as background subtraction, spectrum overlapping, and log presentation.

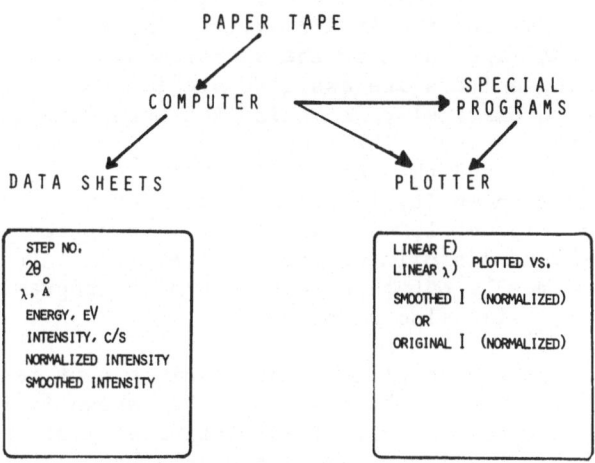

Figure 8.   Data Handling Flow Sheet

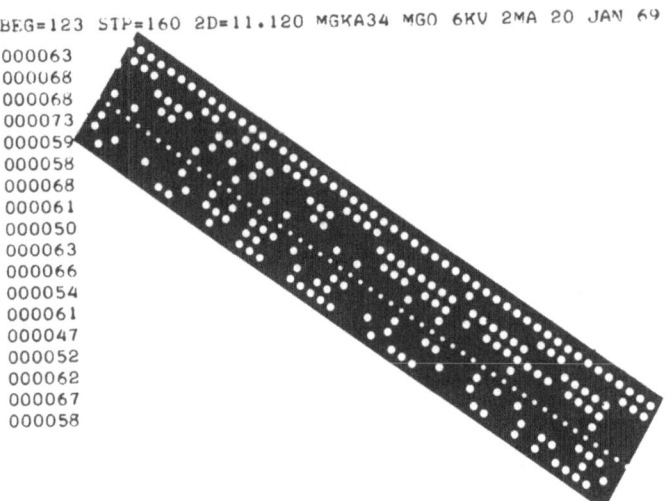

Figure 9.   Teletype Printout And Paper Tape

DATA HANDLING

In either of the data acquisition systems described, Teletype
paper tape output provides a simple, versatile, and inexpensive
method of rough data presentation. The flow diagram in figure 8
shows the route the data points take from tape to computer plotted
spectra. The Teletype provides a printed sheet as it punches the
paper tape. The printed sheet and tape are shown in figure 9. In
this case only the number of counts per step are printed. Other
information such as counting time could be printed. In the
heading is found the starting angle 2θ (BEG=123), the number of
steps per degree 2θ (STP=160), the crystal 2d value in Å (2D=11.120),
the X-ray series designation (MGKA34), the sample (MGO), excitation
conditions (6KV2MA) and the date the spectrum was obtained. The
first three of these items are assigned special positions to allow
the computer to automatically recognize the beginning angle, step
size, and crystal spacing. The computer then calculates the wave-
length (λ =2d sin θ) and the energy (E= $\frac{12398.10}{\lambda}$) for each step.
For each step, the intensity is normalized; that is, the strongest
line is given a value of 100 and all other intensities are based
on 100. Data sheets are printed for each spectrum. In addition,
selected spectra may be put through a smoothing program and the
smoothed intensity will also be printed.

When it is desired to plot the spectrum, a plot tape is
generated which results in plots such as that shown in figure 10,
CuLα (13.3Å) from copper metal. This particular plot has been
computer smoothed. The insert shows the same band as recorded
using multichannel scaling.

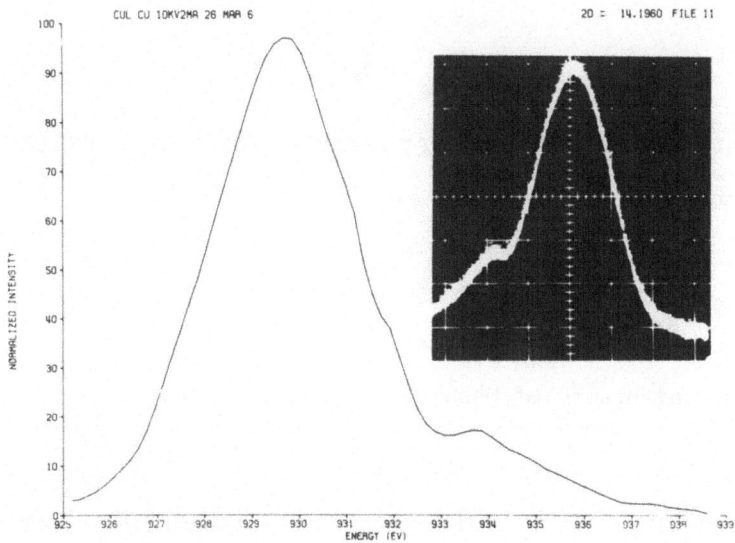

FIGURE 10. COMPUTER PLOT OF CuLα FROM COPPER TARGET.

CONCLUSIONS

This new soft X-ray spectrometer design embodies many novel features that make it particularly well suited to studies and applications of the "chemical effect on X-ray spectra." Initial tests confirm that it is performing as well as anticipated. Modern data acquisition and handling techniques have minimized the tedium previously associated with point by point measurements of fine structure on X-ray spectra. Rapid developments in electronics and computer techniques will further facilitate such measurements in the near future.

REFERENCES

1.   W. L. Baun, accepted for publication, Review of Scientific Instruments.

POINT SCATTERING THEORY OF X-RAY K-ABSORPTION FINE STRUCTURE

D. E. Sayers and F. W. Lytle

Boeing Scientific Research Laboratories, Seattle, Wn.

E. A. Stern

University of Washington, Seattle, Wn.

ABSTRACT

We calculate the extended x-ray absorption fine structure by treating the ejected photoelectron as a spherical wave which expands in the lattice and is partially scattered by neighbors of the absorbing atom. The neighboring atoms are treated as point scatterers and the total scattered wave is summed from the waves scattered by each atom. The fine structure is determined from the dipole transition matrix between the initial K-state and the final photoelectron state. Calculations compare favorably with experimental data.

INTRODUCTION

In spite of many serious attempts an adequate theory of the Kronig or extended x-ray absorption fine structure (EXAFS) has yet to be proposed. (References to several of these theories and a comparison of them can be found in the review article by Azaroff[1].) Most of the theories are based on scattering of the ejected photoelectron by the surrounding atomic potential array. This scattering leads to oscillations in the square of the amplitude of the dipole transition matrix which is directly related to the linear absorption coefficient.

Any successful theory of EXAFS must be flexible enough to include the following properties: (1) EXAFS has been found in crystalline and noncrystalline solids, gases, liquids and polyatomic gases--everything except monatomic gases; (2) EXAFS in all materials is dependent upon the interatomic distance and the symmetry of the neighboring atoms about the absorbing atom; (3) EXAFS reflects the symmetry of the initial electron state[2]. In a given material the structure of the L_{II} and L_{III} edges, which have

initial p electron states are identical, but different from the
structure of the K and L_I edges which have initial s electron
states. Previously, one of us[2,3] proposed a "particle-in-a-box"
theory which identified the primary peaks in the fine structure
spectrum with scattering by the first coordination shell. This
method accounts for only the energy positions of the primary
absorption maxima and its success is due to the fact that the
nearest neighbor atoms are most important in shaping the EXAFS (as
shown herein). This success and its simple formulation has made
it a useful technique for accurate determination of the interatomic
distance using the energy locations of the absorption maxima.[4]

The present theory explains the shape of the EXAFS spectrum
by extending the approach of the "particle in a box" model to
include scattering from the atoms of many coordination shells. It
is in some ways similar to the approach of Shiraiwa, et al.[5] We
consider EXAFS to be the result of partial scattering of the
ejected photoelectron by atoms in the neighborhood of the absorbing atom, i.e. it is a short range order theory. The contribution
of the scattering to the final photoelectron wave function can be
written as a sum over the scattering contributions of each atom.
The oscillations in the absorption coefficient arise when this
scattering contribution is included in the square of the amplitude
of the dipole transition matrix between the initial and final
electron states involved in the absorption process. This result
includes the effects of the thermal motion of the lattice, the
attenuation of the ejected photoelectron as it propagates away
from its origin, and the change in wave vector produced by multiple
scattering. Our theory has considerable potential flexibility in
treating complex crystal structures and in identifying features
of the fine structure spectrum which arise from particular atoms
or coordination shells.

In this paper results are given for the K-absorption spectra
in three cubic crystal structures (copper, face-centered; iron,
body-centered; and germanium, diamond cubic). The shape and
energy position of the EXAFS extrema have been investigated by
retaining just those terms which lead to an oscillating absorption
coefficient. No attempt has been made to calculate the absolute
magnitude of the absorption coefficient and the relative amplitudes
in different materials, nor to calculate the structure near the
absorption edge which is affected by the shielding of the
ionized atom.[6,7,8]

THEORY

During an x-ray absorption event conservation of energy
demands that the energy of the ejected photoelectron be

$$E = h\nu - E_K \tag{1}$$

where $h\nu$ is the x-ray photon energy and E_K is the K-shell binding energy and it is assumed that the ion core is left in an unexcited, singly ionized state. If I_o is the intensity of the incident x-ray beam then the intensity of the beam transmitted through an absorber of thickness x is given by the familiar equation

$$I = I_o \exp(-\mu x) \tag{2}$$

or

$$\mu x = \ln(I_o/I) \tag{3}$$

where μ is the total linear attenuation coefficient. Experimentally I and I_o are measured at a series of photon energies and $\ln(I_o/I)$ is plotted as a function of photoelectron energy. Any observed fine structure is attributed to oscillations in the linear attenuation coefficient. This coefficient is in turn related to the total attenuation cross-section by

$$\frac{\mu}{\rho} = \sigma_T \tag{4}$$

where ρ is the number density of the absorber and

$$\sigma_T = \sigma_A + \sigma_{SC} \tag{5}$$

where σ_A is the photoelectric absorption cross section and σ_{SC} is the total scattering cross section. σ_{SC} will be ignored since it does not contribute to the oscillatory behaviour. We consider only the photoelectric cross section, which may be written using the dipole approximation for radiative transitions[9]

$$\sigma_A = 4\pi^2 \, \alpha h\nu \, |\vec{r}_{if}|^2 \, N(E) \tag{6}$$

where α is the fine structure constant, $N(E)$ is the density of final states, and r_{if} is the dipole matrix element given by

$$\vec{r}_{if} = \int \phi_i^*(\vec{r}_1, \vec{r}_2, \ldots, \vec{r}_N) \cdot \sum_n \vec{r}_n \cdot \phi_f(\vec{r}_1, \vec{r}_2, \ldots, \vec{r}_N) \, d\tau \tag{7}$$

To evaluate this matrix element we must estimate the many electron wave functions ϕ_i and ϕ_f which are given approximately by Slater determinants of one-electron wave functions and N-1 of the N electron states are the same for the initial and final states. The discrete wave functions are assumed to be normalized to unity and the continuum functions normalized per energy interval.

(A good discussion of the evaluation of this matrix element and the assumptions relating to it can be found in the paper by Cooper.)[10] Many body effects are now included as approximations in the one electron wave functions which are solutions of the Hartree-Fock equation. The effect of shielding of the hole in the K-shell is also not included in the wave functions. Since shielding can distort the wave function particularly at energies near the K edge[6,7,8] this theory is not expected to be reliable for photoelectrons of low energy. Under these assumptions the matrix element becomes

$$M = \int \psi_i^*(\vec{r}) \cdot \vec{r} \cdot \psi_f(\vec{r}) \, d\tau \qquad (8)$$

Variations in the absolute square of this matrix element are directly related to the x-ray absorption fine structure spectrum. In the case of K-shell absorption ψ_i may be represented by the atomic 1-s wave function. ψ_f is the final photoelectron wave function including scattering and it is this wave function which must be found.

ψ_f is the state at a given energy which is the solution of the problem of a solid with an atom at the origin ionized by an electron missing from the K-shell while the surrounding atoms are neutral. We assume a muffin-tin type potential -a spherically symmetric potential about each atom smaller in radius than half the distance between atoms with a constant potential outside in the rest of the solid. Of course, the ionized atom at the origin will have a different potential from that of the neutral atoms. We expect the potential of the ionized atom to be shielded by the surrounding conduction electrons justifying its assumed localized nature. The energies of interest are of the order of 100 eV greater than the Fermi energy. These energies are much greater than any of the expected pseudo-potentials or T-matrices that determine the scattering from the neutral atoms. For this reason the scattering should be weak and contributions which correspond to multiple scatterings from the neutral atoms can be neglected. We replace each neutral atom by a delta function potential chosen to give the same back-scattered wave toward the ionized atom as would the actual neutral atom.

The ionized atom is more difficult to treat. It cannot be replaced by an effective potential since the wave function of the final state must be determined where it overlaps with the initial 1-s state, namely inside the atom near the nucleus. Hence the full problem must be solved with the complete potential. The variations in the x-ray absorption spectrum are produced by the interference near the nucleus between the outgoing wave from the absorbing atom and the waves back scattered from the surrounding neutral atoms. When these waves add constructively there is an

absorption maximum and a minimum when they add destructively. We
delay for now solving the final state wave function inside the
ionized atom and consider the form of the wave function just
external to the ionized atom where the potential is a constant.
The wave function will have two parts, one an outgoing wave corresponding to the electron excited from the K-shell and the other
part being back-scattered waves from the surrounding neutral
atoms.[11]   The outgoing wave can be separated into radial and
angular parts. In general the radial part will have the form

$$R = Nh_1'(kr) \tag{9}$$

where

$$h_1'(kr) = \exp(2i\eta_1)h_1^+(kr) \tag{10}$$

h_1^+ is the outgoing Hankel function and may be visualized as an
outgoing spherical wave relative to the ionized atom. The phase
shift η_1 is introduced by the presence of the ionized atom at
the origin and would be zero if there were no such atom present
while N is a normalization constant assumed to be a real number.

For K-shell absorption the initial electron state is an s
state and the dipole selection rule ($\Delta 1 = \pm 1$) for the matrix
element (equation (8)) allows only transitions to states with p
symmetry (1 = 1). The total wave function will be the product of
$h_1'(kr)$ and a linear combination of the three spherical harmonics
for 1 = 1 which represent the x, y, and z components of the wave
function. In a crystal with cubic symmetry these three are all
equivalent and should contribute equally to the final results.
Since we are interested only in the relative magnitude of the
absorption spectrum and not its absolute value only the z component will be considered. This spherical harmonic in the z
direction is $Y_1^0 = \cos\theta$, so that the wave function of the outgoing part of the photoelectron wave function is

$$\psi(\vec{r}) = Nh_1'(kr)\ \cos\theta \tag{11}$$

To obtain the total wave function we must add to this the
contribution from the scattered wave.[11] This contribution is
composed of the scattering from all of the atoms in the solid.
As discussed previously, we are interested in the energy range
where the scattering from a single atom is small. We will therefore neglect the multiple scattering contribution, i.e. scattering
from more than one atom. However, we will include multiple
scattering from the same atom which converts the potential matrix
to the T-matrix. It is only the T-matrix which is small from an
isolated atom. The scattered wave can now be calculated by[11]

$$\psi_{SC} = \sum_i (E-H_o)^{-1} T(\vec{r}_i) \, \psi_f^i \tag{12}$$

where $(E-H_o)^{-1}$ is the Green's function of free space for energy E, $T(r_i)$ is the T-matrix of the neutral atom at r_i and ψ_f^i is the incident wave at point r_i produced by all of the other neutral atoms excluding the atom at r_i and is given by

$$\psi_f^i = \psi + \sum_{m \neq i} (E-H_o)^{-1} T(\vec{r}_m) \psi_f^m \tag{13}$$

In our approximation of small T matrix ψ_f^i is independent of the atom at r_i and is equal to ψ_f. We replace $T(r_m)$ by

$$T(\vec{r}_m) = V_o \, \delta(\vec{r}-\vec{r}_m) \tag{14}$$

V_o is chosen so as to produce the same back-scattered wave as produced by $T(r_m)$. The equation we need to solve is, substituting $\psi_f^i = \psi_f$ and equation (14),

$$\psi_{SC} = V_o \sum_i (E-H_o)^{-1} \delta(\vec{r}-\vec{r}_i) \psi_f \tag{15}$$

It is important to note that ψ_f is significantly different from ψ. As ψ_f moves further from the origin more and more of the scattered waves from surrounding atoms are added to it and it deviates more from ψ. Sufficiently far enough away from the origin ψ_f should be modified to the form of an eigenstate and produce no more scattered waves. In addition, electron-electron and electron-phonon scatterings make ψ_f lose coherence with ψ. Thus, only the atoms within a certain distance of the origin contribute to the scattering. An estimate of the scattering length is given by appendix B. We approximate this effect by assuming that ψ_f decays exponentially away from the origin.

The scattering has two other effects. The scattering from the periodic array of neutral atoms produces energy gaps corresponding to energy regions where no propagating states occur. In these regions the E versus k spectrum and the density of states are distorted from a simple free electron behavior. However, because the scattering is weak, the widths of the gaps are small, and since the experimental data average over several eV, an energy interval large compared to the gaps, the gap effects should not be observable. For that reason we neglect the gap effects on the density of states and assume that it varies as in a free electron gas, proportional to k. The scattering from the periodic array also changes the average k-value of ψ_f. The equation determining this is[11,12]

$$E - \frac{\hbar^2 k'^2}{2m} - nT_{kk} = 0 \qquad (16)$$

where n is the number of neutral atoms per unit volume and T_{kk} is the forward scattering amplitude given by

$$T_{kk} = \int \exp(-i\vec{k}\cdot\vec{r}) T(\vec{r}) \exp(i\vec{k}\cdot\vec{r}) d\tau \qquad (17)$$

Equation (15) can be written as

$$\psi_{SC}(\vec{r}) = -\frac{V_o 2m}{4\pi\hbar^2} \sum_i \int \frac{\exp(ik|\vec{r}-\vec{r}'|)}{|\vec{r}-\vec{r}'|} \delta(r'-\vec{r}_i) \psi_f(\vec{r}') d\tau' \qquad (18)$$

where we have used the spatial representation of the Green's function and

$$E = \frac{\hbar^2 k^2}{2m} \qquad (19)$$

As discussed above we approximate ψ_f by

$$\psi_f(r) = N h_1'(k'r) \cos\theta \exp(-\gamma r) \qquad (20)$$

where k' is determined from equation (16) and γ is a measure of the distance estimated in appendix B. Combining (18) and (20) we obtain

$$\psi_{SC}(r) = -\frac{V_o 2mN}{2\pi\hbar^2} \sum_i \frac{\exp(ik|\vec{r}-\vec{r}_i|)}{|\vec{r}-\vec{r}_i|^2} h_1'(k'r_i)(z-z_i)\exp(-\gamma r_i) \qquad (21)$$

where we have used the fact that the cosine of the angle between r_i and the z axis is (figure 1)

$$\cos\theta_i = \frac{(z-z_i)}{|\vec{r}-\vec{r}_i|} \qquad (22)$$

and z_i is the z-component of r_i. The total wave function outside the ionized atom will be a combination of equations (9) and (21) giving

$$\psi_f(\vec{r}) = N h_1'(kr) \cos\theta - \frac{V_o mN}{2\pi\hbar^2} \sum_i \frac{\exp(ik|\vec{r}-\vec{r}_i|)}{|\vec{r}-\vec{r}_i|^2} h_1'(k'r_i)$$

$$\cdot \exp(-\gamma r_i)(z-z_i) \qquad (23)$$

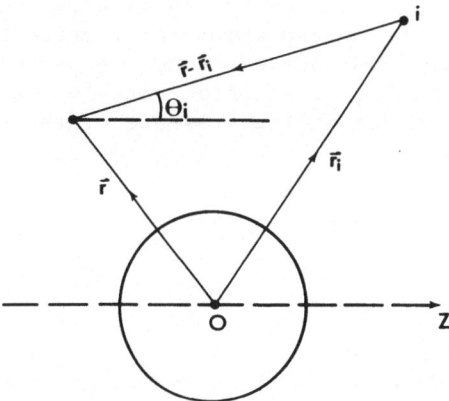

figure 1   The circle about 0 indicates the finite extent of
                the potential of the absorbing atoms, the point
                scatterer is at i and the scattered wave is ob-
                served at P. $\vec{r}_i$ is a lattice vector, $\vec{r}$ is the
                vector from 0 to P and θ_i is the angle between
                the normal to the scattered wave front and the z
                direction. The total wave (scattered plus out-
                going) is evaluated at the origin (P is moved to
                0).

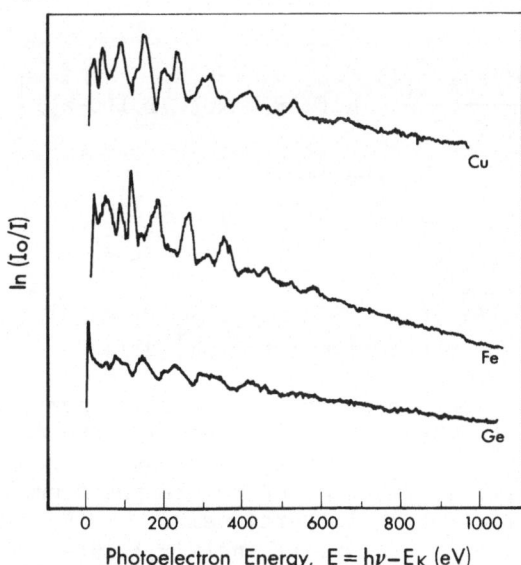

figure 2   Experimental x-ray absorption spectra for copper,
                iron, and germanium.

This expression for ψ_f is expected to be a reasonable approximation just outside the ionized atom. It remains to extrapolate this expression inside the ionized atom. We perform this extrapolation by using the WKB approximation with the Langer correction. The phase shift introduced by the presence of the atom at the origin is given by13

$$\eta_1 = \int_{r_1}^{r_o} \left(k^2 - \frac{2mV(r)}{\hbar^2} - \frac{2.25}{r^2} \right)^{\frac{1}{2}} dr - \int_{r_2}^{r_o} \left(k^2 - \frac{2.25}{r^2} \right)^{\frac{1}{2}} dr \tag{24}$$

where r_1 and r_2 are the zeros of their respective integrands, $V(r)$ is the atomic potential, and r_o is a point on the constant part of the muffin-tin potential (assumed zero). In determining ψ_f near $r=0$ this phase shift must be subtracted from the first term and added to the second term on the right side of equation (23).

For the medium and heavy elements the 1-s function is sharply peaked near the origin where the p wave can be approximated by a straight line passing through the origin and the amount of p character in the wave should be closely approximated by the magnitude of the slope of this line. The matrix element (equation (8)) will therefore be proportional to

$$M \propto \frac{\partial \psi_f(o)}{\partial z} = N \frac{\partial}{\partial z} (h_1^+(0) \cos\theta_o) - \frac{\exp(2i\eta_1) V_o mN}{2\pi\hbar^2}$$

$$\cdot \frac{\partial}{\partial z} \left[\sum_i \frac{\exp(ik|\vec{r}-\vec{r}_i|)}{|\vec{r}-\vec{r}_i|^2} h_1'(k'r_i) \exp(-\gamma r_i)(z-z_i) \right]_{r=0} \tag{25}$$

$$M \propto A + C \exp(2i\eta_1) V_o \sum_i \frac{\exp(ikr_i)}{r_i^2} h_1'(k'r_i) \exp(-\gamma r_i)$$

$$\left(\frac{2z_i^2}{r_i^2} - 1 - \frac{ikz_i^2}{r_i} \right) \tag{26}$$

where A is a constant proportional to $\partial h_1^+(0)/\partial z$.

The x-ray absorption cross-section per unit energy is proportional to $|M|^2 k$ where the k comes from the density of final states (equation (6)). Only the cross product in $|M|^2$ contributes to the oscillatory fine structure spectrum. Therefore, the fine structure spectrum is proportional to

$$\chi = \text{Re} \left[\exp(2i\eta_1) V_o k \sum_i \frac{\exp(-kr_i) h_1'(k'r_i) \exp(-\gamma r_i)}{r_i^2} \right.$$

$$\left. \left(\frac{2z_i^2}{r_i^2} - 1 - \frac{ikz_i^2}{r_i} \right) \right] \tag{27}$$

where   Re [   ]   means the real part of the bracketed term.
Using the relationship

$$h_1^+(kr) = - \exp(ikr) \left(\frac{i}{(kr)^2} + \frac{1}{kr}\right) \tag{28}$$

in equation (27) we can write

$$\chi = V_o k \sum_i \frac{\exp(-\gamma r_i)}{k'r_i^{\,3}} \left[\left(\frac{2z_i^{\,2}}{r_i^{\,2}}-1\right) \alpha_i - \frac{kz_i^{\,2}}{r_i} \beta_i\right] \tag{29}$$

where

$$\alpha_i = \frac{\sin[(k+k')r_i+2\eta_1]}{k'r_i} - \cos[(k+k')r_i+2\eta_1] \tag{30}$$

and

$$\beta_i = \frac{\cos[(k+k')r_i+2\eta_1]}{k'r_i} + \sin[(k+k')r_i+2\eta_1] \tag{31}$$

Expressing equation (27) in terms of a sum over coordination shells we have

$$\chi = V_o k \sum_j \frac{\exp(-\gamma r_j)}{k'r_j^{\,3}} \left[\left(\frac{2SUMZ_j}{r_j^{\,2}} - N_j\right) \alpha_j - \frac{kSUMZ_j\beta_j}{r_j}\right] \tag{32}$$

where N_j is the number of atoms in the jth coordination shell and $SUMZ_j$ is the sum of the squares of the   z   coordinates of the atoms in the jth shell.

   The factor V_o in equation (32) is chosen so as to give the correct backscattering of the photoelectron. In backscattering, the electron changes its wave vector by   2k   and thus $V_o \propto V(2k)$, the $2k^{th}$ fourier component of the scattering potential. Since the energy of the photoelectron is 100 eV or more we expect that $V(2k)$ will approximate the unshielded value. For a point charge $V(2k) \propto 1/k^2$ and we expect

$$V_o k = B/k \tag{33}$$

where   B   is a constant.

Before equation (32) can be compared to experimental spectra corrections must be applied to it due to the thermal motion of the lattice. The temperature correction used is taken from the theory developed by Shmidt[14,15] who proposed that the temperature correction is of the form exp (-I) where

$$I = \frac{12\hbar^2}{k_B m \theta_D} \ k^2 \ \Phi(T/\theta_D) \qquad (34)$$

k_B is the Boltzman constant
m is the mass of the atoms in the absorber
θ_D is the Debye temperature of the absorber
k is the wavevector of the photoelectron and

$$\Phi(T/\theta_D) = \phi_1 \frac{T}{\theta_D} + \phi_2 \frac{\theta_D}{T} + \phi_3 \left(\frac{\theta_D}{T}\right)^3 \qquad (35)$$

ϕ_1, ϕ_2, and ϕ_3 are defined in reference 15 and depend only on the crystal structure. The experimental data in figures 2 and 3 were taken at liquid nitrogen temperature (T=77.4 K); it should be noted that for the materials studied here $T/\theta_D \sim 0.2$, while Shmidt suggests that equation (34) is valid for $T/\theta_D \sim 1/3$. However, equation (34) and the result derived in reference 14, which is valid for $T << \theta_D$ both give similar results and are of the right order of magnitude. Equation (34) was used as the correction since it is somewhat easier to evaluate.

In estimating η_1 in equation (24) the self-consistent Hartree potentials of Herman and Skilman[16] were used. These atomic potentials will be shielded in the solid and will not have as large a tail as that given by Herman and Skilman producing an error in η_1. An additional error is introduced by the fact that the potential of the ionized atom differs from that given by Herman and Skilman because the one electron in the K-shell is missing. Of course, a further error is introduced by the WKB approximation. For these reasons the calculations of η_1 are not expected to be reliable enough to give detailed agreement with experiment, although one expects semi-quantitative agreement. We used the WKB method as a tentative example of phase shift determination. The final values of η_1 were determined from the experimental data and this value used to calculate the fine structure. This was done as follows: the primary peak positions were assumed to come from the first coordination shell and the dominant term in equation (29). The energy positions of these peaks were then used to calculate the phase shifts for these energies. This was done for various choices of T_{kk} in equation (16), and the results compared to the phase shift curve calculated using the WKB approximation. The WKB curve and the experimentally determined curve are compared in figure 6. Techniques similar to this also have been used by D. Mott[17] and Kozlenkov[18].

The value of γ in equation (32) is also determined phenomenlogi-
cally by the best fit to the data. The values of γ found by this
procedure are of the correct magnitude as predicted in appendix B.

EXPERIMENTAL RESULTS

The apparatus used to collect the data consisted of standard
molybdenum or silver target x-ray tubes and a Siemen's single
crystal diffractometer with narrow slits (50 microns). LiF was
used as an analyzing crystal and a scintillation counter with
associated electronics was used to record the x-ray intensity.
The numerical resolving power which corresponds to an energy
resolution of 4-6 eV was $\Delta\lambda/\lambda \sim 2$-3000. The absorber was mounted
in a cryostat and suspended in the x-ray beam in front of the
analyzing crystal. All data were taken at liquid nitrogen
temperature (77.4 K) by automatically recording the intensity for
the absorber in (I) and out (I_o) of the beam at a given setting
of 2θ. A sufficient number of counts was collected to keep the
statistical counting accuracy better than 0.3%. The experiment
was automated to the extent that a control unit moved the absorber
in and out of the beam, stepped the diffractometer by 0.01° after
each (I,I_o) pair was recorded, and punched the data on IBM cards.
The computer program used to analyze the data corrected for
coincidence losses in the counting due to strong emission lines,
accurately located the absorption edges with respect to reference
lines included in the data, and plotted the absorption spectrum as
a function of photoelectron energy. The iron and copper samples
were thin metal foils, well annealed and 6 and 2.5 microns thick,
respectively. The germanium was vapor deposited onto aluminum
foil and annealed to develop the crystal structure and equivalent
to 6 microns thick. The absorption curves for the three materials
are shown in figure 2.

COMPARISON WITH EXPERIMENT AND DISCUSSION

A comparison of the theoretical and experimental fine
structure is shown in figure 3 for copper, iron, and germanium and
the primary and secondary absorption maxima are listed in table I.
The theoretical spectra contain the corrections for the temperature
and electron attenuation and use a linear variation versus k for
the phase shifts. A sample program for the complete calculation
is contained in appendix A. The summation in each case is carried
out to thirteen coordination shells, which is sufficient to
include all shells which give significant contributions to the
fine structure. We have performed the summation to include 20
coordination shells and observed no change in the EXAFS. The
experimental spectra are those from figure 2 with the vertical

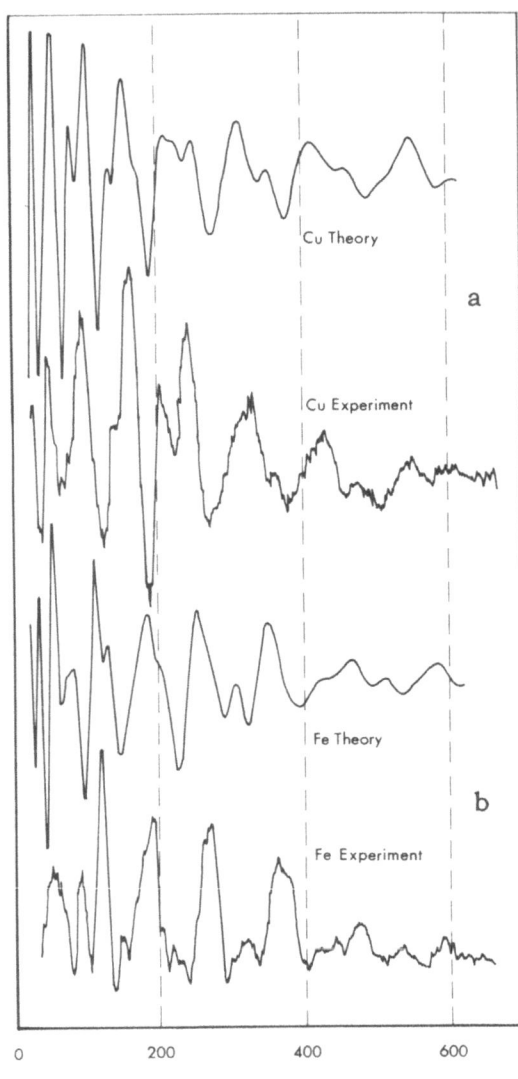

Photoelectron Energy (eV)

figure 3a & b Comparison of theoretical and experimental EXAFS
              for copper and iron. The theoretical spectrum
              is obtained from equations (32), (33), and (34)
              and is summed over 13 coordination shells.

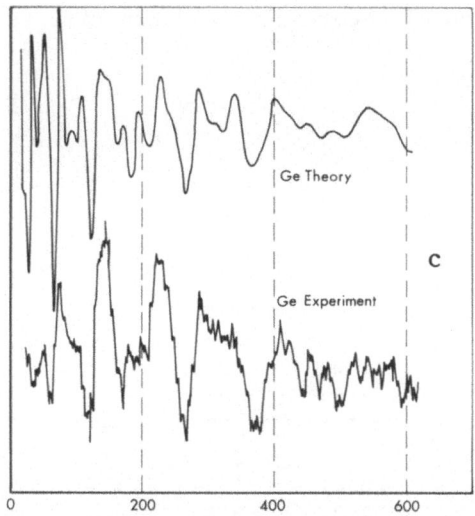

Photoelectron Energy (eV)

figure 3c   Comparison for theoretical and experimental
                 EXAFS for germanium. The theoretical spectrum
                 is obtained from equations (32), (33), and (34)
                 and is summed over 13 coordination shells.

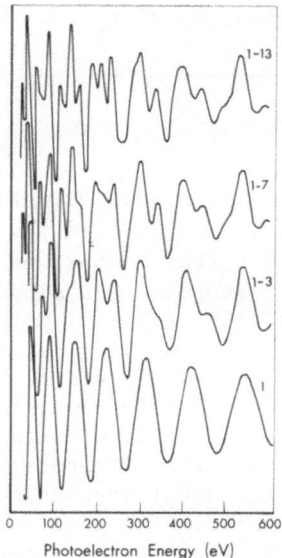

Photoelectron Energy (eV)

figure 4   The accumulative contribution to the EXAFS of
                 copper by successive coordination shells. The
                 numbers on the right refer to the coordination
                 shells included in each spectrum. Temperature
                 and attenuation effects are included.

Table I. Comparison of Theoretical and Experimental EXAFS (in eV)

Cu		Fe		Ge	
Exp.	Theory	Exp.	Theory	Exp.	Theory
25	24	20	22	--	38
--	40	--	33	53	57
49	51	48	52	80	81
66	77	--	67	--	96
95	101	84	80	108	113
135	130	110	109	--	140
157	154	138	128	148	155
202	207	174	182	--	176
--	225	210	205	202	200
237	250	256	250	230	234
322	313	--	270	--	254
356	353	304	305	320	294
417	415	352	349	--	318
467	457	420	425	--	348
537	550	459	466	420	410
		518	512	--	462
		574	586	--	498
				545	553

-- means not resolved experimentally

_____

scale expanded to give a better comparison of the relative
amplitudes and spectral features. The sloping background has
been subtracted graphically.

The agreement between the theoretical and experimental spectra
is generally good. In the case of copper the shape of the theoretical spectrum compares favorably with the experimental spectrum,
particularly at higher energies. The positions of primary absorption maxima are in fair agreement even below 100 eV where our
approximations are less valid. The theoretical spectrum contains
extra structure at 173 eV, and a doublet around 200 eV rather than
the single peak observed experimentally. The detail predicted in
the spectrum below 100 eV can't really be compared with the
experimental results unless the effect of the experimental
resolution is included in the comparison. Peaks may be hidden by
the experimental resolution function (which is not included in the

theoretical curves) and/or a more accurate determination of the
phase shift may move extra peaks into position. The extra features
predicted at 173 eV and 225 eV do not agree with our experimental
data, however a more accurate experimental spectrum of copper
taken by Krogstad[19] contains maxima in the same relative positions
as our theoretical spectrum. A comparison of the iron spectra
shows that, except for minor discrepancies, the theoretical spectrum
agrees very closely with the observed spectrum. The comparison
of theory and experiment for germanium is more difficult because
a lack of absorber uniformity decreased the experimental resolution.
The positions of the major absorption maxima agree very closely
but the additional structure predicted by the theory cannot be
confirmed.

An important aspect of our theory is that particular features
of the absorption spectrum may be identified as originating in
specific coordination shells. This can be seen in the copper
spectra of figure 4 where curves obtained by summing to specific
coordination shells are compared. A good example is the double
peak in the copper spectrum at 320 eV and 356 eV in figure 4 (see
also figure 3a). The first shell fixed the position of the main
peak at 320 eV, with the sub-peak arising mainly from the seventh
shell with a contribution from the third shell. Coordination
shells beyond the seventh contribute primarily to the observed
asymmetry in the main peak.

In figure 5 the effect of the correction terms on the
theoretical spectrum is shown. Figure 5a is equation (32) plotted
for copper and summed over 13 coordination shells; in figure 5b the
temperature correction is applied; and in 5c the attenuation correction is applied. Both corrections lower the amplitude of the
peaks and smooth the spectrum by reducing the fine structure
contributed by the outer coordination shells.

In figure 6 the effect of using the linear approximation to
the phase shift is examined where the best fit for the linear
approximation is compared with the WKB phase shift. The linear
phase is generally below the WKB phase especially at higher
energies because the potential of the absorbing atom in a solid
is more localized than the free atom potential used in calculating
the WKB phase. At lower energies the linear phase is also lower
than the WKB phase which is due in part to the fact that the
linear approximation is not as good in this region. The spectra
calculated using the WKB phase shift and the linear phase shift
are shown in figure 7. For high energies the linear phase gives
results in good agreement with the experimental spectrum and
proves that the linear approximation is the better result for the
phase at higher energies. At low energies both phase shifts give
results which are similar but show differences from the experimental results. Some differences with the experimental spectrum

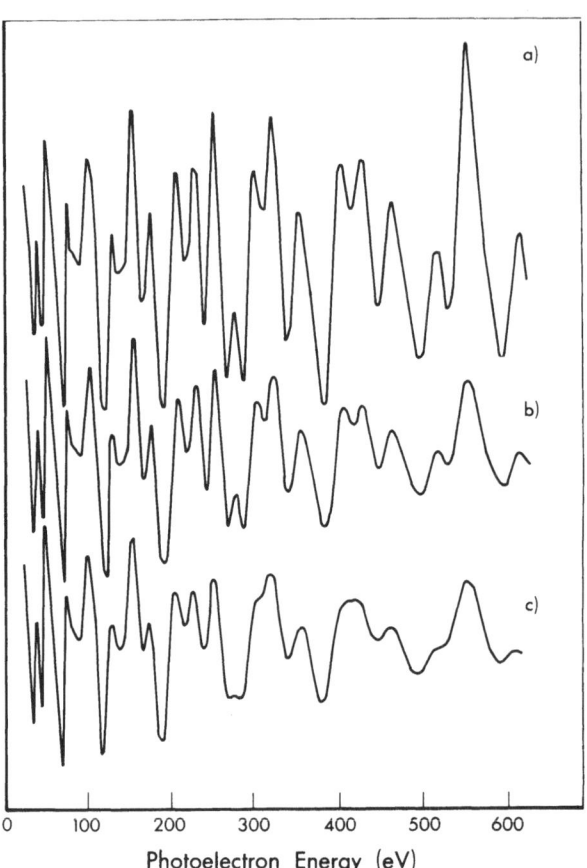

figure 5   a) Equation (32) without the exponential attenuation and the k dependence of equation (33)
plotted for 13 coordination shells for copper.
b) Temperature correction of equation (34) is
applied. c) Exponential attenuation is applied.
Intensities of EXAFS are all plotted to the
same (arbitrary) scale.

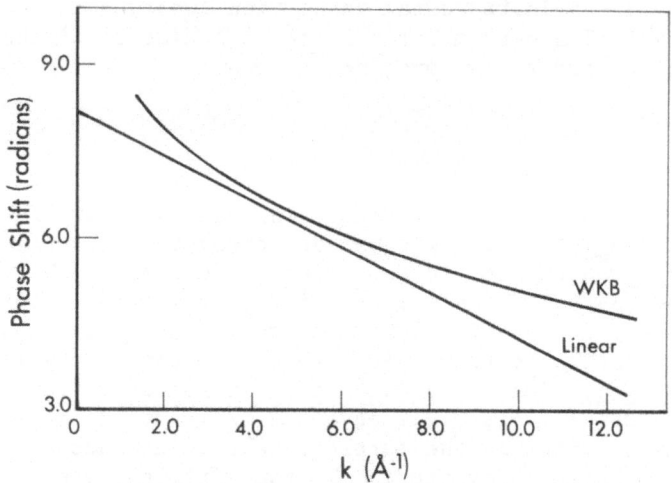

figure 6   Comparison of WKB and linear phase shifts.

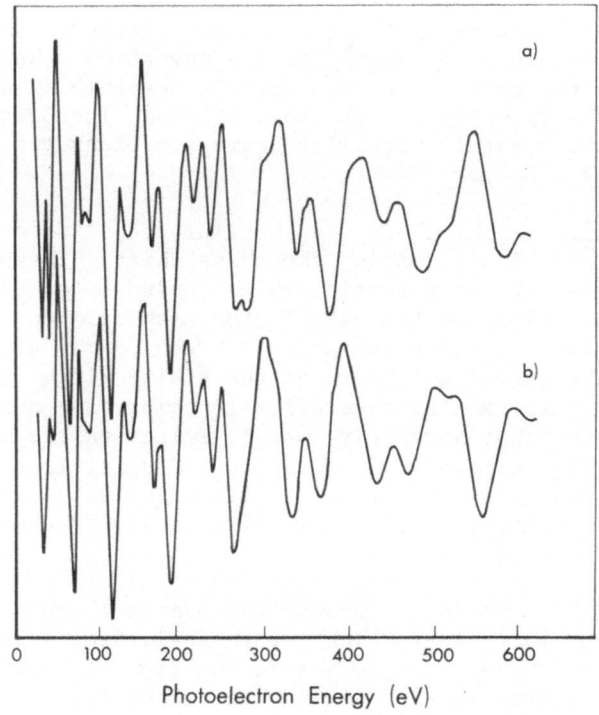

figure 7   Comparison of theoretical copper EXAFS for 1-13
                  coordination shells with   a) linear phase shift
                  and   b) WKB phase shift.

may disappear if an instrumental resolution function is included in the theoretical spectrum but a better calculation of the phase shift is also needed to improve the agreement.

CONCLUSION

The success of this point scattering theory in predicting the EXAFS spectra in materials of cubic structure verifies the importance of short-range order in determining the EXAFS. The contribution to the EXAFS from each succeeding coordination sphere as illustrated by figure 4 clearly demonstrates the build up of each characteristic spectrum. We hope to use the theory for studying ordering, clustering, and defect structures in alloys. In these more complex materials equation (29) must be summed over all the types of atoms in the lattice, taking into account the relative differences in scattering by the different atoms in the pseudo potentials of equation (14).

The calculation of EXAFS in non-cubic materials will involve re-determination of equation (22) where the scattered wave along x, y, and z will not be equivalent and the scattered wave from all three directions must be taken into the summation. Note that for any structure the theory is, in principle, capable of calculating the total K-shell photoelectric cross section, including EXAFS, provided that accurately normalized wave functions are used in the integration of equation (8).

The results of the theory may be improved by using more accurate methods of determining the phase shift. These would include using an atomic potential which includes the solid state effect and the effect of the hole in the K-shell. An improved method of evaluating the attenuation correction may also be desirable. The smearing effects of the finite width of the K state and the instrumental resolution function have not been included because they have only a small effect on the higher energy EXAFS, but probably should be included for investigations of the lower energy structure. The effects of shielding of the K-hole are also important in determining the lower energy structure and would need to be included.

In Appendix B it is shown that the dominant mechanism in determining the range over which the photoelectron back scattering can contribute to EXAFS is the scattering from the initial electron state into an eigen state of the system rather than into inelastic scattering. The shorter range due to this mechanism may also be important in calculating the range of electrons in LEED experiments.

APPENDIX A

PROGRAM TO CALCULATE X-RAY ABSORPTION FINE STRUCTURE
SPECTRUM USING THE POINT SCATTERING THEORY (EXAMPLE FOR IRON)

```
REAL K,K1
DIMENSION ETA(600),CHI1(600),E(600),K(600)
DIMENSION AN(13),D(13),PHI(13),SUMZ(13),RHO(13)
```

LATTICE PARAMETER FOR IRON

```
A=2.87
```

READ IN CRYSTAL STRUCTURE AND TEMPERATURE DATA WHERE AN IS THE
NUMBER OF ATOMS IN THE COORDINATION SHELL, SUMZ IS THE SUM OF
THE SQUARES OF THE Z COMPONENTS IN A SHELL DIVIDED BY A*A, D IS
THE RADIUS OF THE SHELL DIVIDED BY A, AND PHI IS THE TEMPERATURE
PARAMETER FROM EQUATION 35.

```
    READ (5,10)(AN(I),SUMZ(I),D(I),PHI(I),I=1,13)
 10 FORMAT(2F4.0,2F6.0)
    WRITE(6,20)
 20 FORMAT(1H1,6X,1HE,15X,1HK,13X,3HCHI)
```

CALCULATE THE SPECTRUM AS A FUNCTION OF ENERGY IN 1 EV STEPS
FROM 20 EV to 620 EV.

```
    DO 100 I=1,600
    XI=I
    E(I)= XI+20.
    E1=E(I)
    K(I)=SQRT(.263*(E1+15.))
    K1=K(I)
```

CALCULATE THE PHASE SHIFT USING A LINEAR APPROXIMATION.

```
    ETA(I)=-.194*K1+7.96
    CHI=0.0
    DO 200 J=1,13
    ARG=K1*A*D(J)
    ARG2=ARG+ETA(I)
    ALPHA=SIN(2.*ARG2)/ARG-COS(2.*ARG2)
    BETA=COS(2.*ARG2)/ARG+SIN(2.*ARG2)
```

CALCULATE THE RESULT OF EQUATION 32. THIS DOESN T CONTAIN THE
THE EXPONENTIAL ATTENUATION FACTOR.

```
    RHO(J)=-((AN(J)-2.*SUMZ(J)/(D(J)**2))*ALPHA+K1*SUMZ(J)*BETA*A/
```

```
   1 D(J))/(K1*(D(J)*A)**3)/K1

     TEMPERATURE CORRECTION FROM EQUATION 34.

     RHO(J)=RHO(J)*EXP(-.0252*K1*K1*PHI(J))

     EXPONENTIAL ATTENUATION FACTOR FROM EQUATION 32 WITH GAMMA=.3

     RHO(J)=RHO(J)*EXP(-.3*D(J)*A)
     CHI=CHI+RHO(J)
 200 CONTINUE
     CHI1(I)=CHI
     WRITE(6,30) E1,K1,CHI1(I)
  30 FORMAT(1X,F7.1,2(2X,E14.6))
 100 CONTINUE
     PLOT THE SPECTRUM AS A FUNCTION OF ENERGY.
     CALL PLOT(E,CHI1,120,120,50)
     STOP
     END
```

In table II the input data for iron is listed where AN is the number of atoms in the jth coordination shell and corresponds to N_j in equation (32), SUMZ is the sum of the squares of the Z-coordinates in the jth shell and must be multiplied by a^2 to correspond to $SUMZ_j$ in equation (32), D is the radius of the jth shell divided by a, and PHI is the temperature parameter defined by equation (35).

Table II   Input Data for Iron

AN	SUMZ	D	PHI
8	2	0.866	0.2098
6	2	1.000	0.2355
12	8	1.414	0.2598
24	22	1.658	0.2570
8	8	1.732	0.2566
6	8	2.000	0.2601
24	38	2.208	0.2664
24	40	2.236	0.2672
24	48	2.249	0.2676
32	72	2.598	0.2740
12	32	2.828	0.2740
48	140	2.958	0.2736
30	90	3.000	0.2735

APPENDIX B

The distance, d, that the photoelectron travels before it is scattered into an eigenstate of the system can be estimated from

$$d = \frac{\hbar k \tau}{m} \qquad (36)$$

where τ is the time it takes to be scattered into an eigen state and is not related to the usual lifetime due to inelastic scattering. τ can be estimated by considering the expansion of the electron wave function in eigenstates of the solid

$$\psi(r,t) = \sum_n a_n \psi_n \frac{\exp(iE_n t)}{\hbar} \qquad (37)$$

Where the sum is over the states of the system which differ by reciprocal lattice vectors and is subject to the condition

$$\psi(r,0) = \exp(ik \cdot r) \qquad (38)$$

Generally in a solid the two dominant states in the sum will be k and $k+g_1$ where g_1 is the shortest reciprocal lattice vector. Considering only these two states the phase term is $\exp\left[i\left(\frac{E_{k+g}-E_k}{\hbar}\right)t\right]$. This may be rewritten using the definition for the free electron energy (equation (19)) and the fact that $k > g_{min}$ for the electron energies of interest.

$$\exp\left[i\frac{(E_{k+g}-E_k)t}{k}\right] = \exp\left[\frac{i\hbar k g_1 t}{2m}\right] \qquad (39)$$

When this argument is non-zero the plane wave term will be important and the other terms will partially cancel each other; however, when the argument of the exponential is π all terms add directly and change the wave function to an eigenstate of the solid. This gives an estimate of the time, τ.

$$\tau \approx \frac{2m}{\hbar k g_1} \qquad (40)$$

In the case of copper which has a b.c.c. reciprocal lattice

$$g_1 = \frac{2\pi}{a}\sqrt{3} \qquad (41)$$

Therefore from equations (36) and (41)

$$d = \frac{a}{\sqrt{3}} \qquad (42)$$

or d = 2.1 Å in copper. This mechanism then dominates all other scattering mechanisms in determining the range over which the photo-electron back scattering contributes to the EXAFS, since the mean free path due to electron-electron or electron-phonon scattering is expected to be much larger (on the order of several atomic spacings). This result is also independent of k for k >> g_1. Comparing this with the EXAFS experiment we note that the spectrum is essentially complete with the inclusion of 1-3 or perhaps 1-7 coordination shells (figure 4) which have radii of 4.4Å and 6.5Å, respectively. Considering the approximate nature of this argument, the agreement is good.

REFERENCES

1.   L. V. Azaroff, "Theory of Extended Fine Structure of X-ray Absorption Edges," Rev. Mod. Phys. 35:1012-1022, (1963).

2.   F. W. Lytle, "Determination of Interatomic Distances from X-ray Absorption Fine Structure," in G. R. Mallett, M. J. Fay and W. M. Mueller, Editors, Advances in X-ray Analysis, Vol. 9, Plenum Press, New York, 1966, p. 398-409.

3.   F. W. Lytle, "X-ray Absorption Fine Structure in Crystalline and Non-Crystalline Materials," in J. A. Prins. Editor, Physics of Non-Crystalline Solids, North Holland Publishing Co., Amsterdam, 1965, p. 12-29.

4.   P. Chivate, P. S. Damle, N. V. Joshi and C. Mande, "A Model for Extended Fine Structure in X-ray Absorption Spectra," J. Phys. C. (Proc. Phys. Soc.) Ser. 2, 1:1171-1175, (1968).

5.   T. Shiraiwa, T. Ishimura, and M. Sawada, "The Theory of the Fine Structure of the X-ray Absorption Spectrum," J. Phys. Soc. Jap. 13;8:847-859, (1958).

6.   B. Roulet, J. Gavoret, and P. Nozieres, "Singularities in the X-ray Absorption and Emission of Metals. I. First Order Parquet Calculation," Phys. Rev. 178;3:1072-1083, (1969).

7.   P. Nozieres, J. Gavoret, and B. Roulet, "Singularities in the X-ray Absorption and Emission of Metals. II. Selfconsistent Treatment of Divergences," Phys. Rev. 178;3:1084-1096, (1969).

8.   P. Nozieres and C. T. DeDominicus, "Singularities in the X-ray Absorption and Emission of Metals. III. One Body Exact Solution." Phys. Rev. 178;3:1097-1107, (1969).

9.   H. A. Bethe and E. Salpeter, "Quantum Mechanics of One and Two
     Electron Systems," in S. Flügge, Editor, Encyclopedia of
     Physics, Vol. 35, Springer-Verlag, Berlin, 1957, p. 334-338;
     381-385.

10. J. W. Cooper, "Photoionization from Outer Atomic Subshells.
     A Model Study, " Phys. Rev. 128;2:681-693, (1962).

11. M. Lax, "Multiple Scattering of Waves," Rev. Mod. Phys. 23:
     287-310, (1951).

12. E. A. Stern, "Electronic Properties of Alloys." (Submitted
     for publication).

13. L. S. Rodberg and R. M. Thaler, "Introduction to the
     Quantum Theory of Scattering," Academic Press, New York,
     1967, p. 34.

14. V. V. Shmidt, "Contribution to the Theory of the Temperature
     Dependence of the Fine Structure of X-ray Absorption Spectra,"
     Bull. Acad. Sci. USSR, Ser. Phys. 25:988, (1961).

15. V. V. Shmidt, "Contribution to the Theory of the Temperature
     Dependence of the Fine Structure of X-ray Absorption Spectra.
     II. Case of High Temperatures," Bull. Acad. Sci. USSR, Ser.
     Phys. 27:392, (1963).

16. F. Herman and S. Skilman, "Atomic Structure Calculations,"
     Prentice Hall, Englewood Cliffs, 1963.

17. D. L. Mott, "Fine Structure in X-ray Absorption Spectra,"
     Ph.D. Thesis (unpub.), New Mexico State University, 1963.

18. A. I. Kozlenkov, "Theory of the Fine Structure of X-ray
     Absorption Spectra," Bull. Acad. Sci. USSR. Ser. Phys.
     25:968, (1961).

19. Reference 1 p. 1020-1021.

This work sponsored in part by the Air Force Office of Scientific
Research, Office of Aerospace Research, U. S. Air Force under
AFOSR Grant No. AF-AFOSR-1270-67.

A VERSATILE VACUUM SCANNING DOUBLE CRYSTAL SPECTROMETER FOR SOFT X-RAY ABSORPTION EDGE STUDIES

A.S. Bhalla and E.W. White

Pennsylvania State University

University Park, Pennsylvania 16802

ABSTRACT

Studies of the fine structure of absorption edges in the soft x-ray region are becoming increasingly important as a tool for materials characterization. Examples of application includes determination of chemical state of elements, bonding and band structure studies. Intensity and resolving power of the x-ray spectrometer are important experimental considerations. As a rule adjustment of instrumental parameters, such as collimation, to give increased intensity adversely affect resolving power. Optimization of intensity and resolving power must therefore be achieved.

A newly designed double crystal spectrometer has been constructed for high-resolution absorption edge studies in the wavelength region of 5Å to 70Å.

The entire system is enclosed in a vacuum chamber ion pumped to the 10^{-7} torr range. The second crystal motion is obtained by means of an ULTRADEX 360-sided polygon and sine arm that is automatically step-scanned. The spectrometer functions equally well in the $(1, + 1)$ and $(1, - 1)$ orientation and as a precise single crystal spectrometer. Rotation of the second crystal throughout $360°$ and the automatic 2:1 turning of the detector permit x-ray determination of the zero mean position. Conveniently interchangeable crystal mounts are easily interchanged on the temperature-controlled ($\pm .01°C$) crystal mountings. An improved alignment procedure is detailed. A specially built x-ray generator powers demountable x-ray tubes for operation at up to 10 kev and 500 m .

INTRODUCTION

X-ray transitions from the valence band (includes all the electrons from outside of the outermost filled shells) to the outermost filled shells are of very low energy and generally occur in the range 10Å to 700Å. This is known as the soft x-ray region. Grazing incidence grating spectrometers are generally used in the studies. This region is further extended to about 2Å on the high-energy side when transitions from the valence band to the K-shell are considered. In the region 2Å-50Å crystal spectrometers generally provide higher resolution and dispersion than the grating spectrometers. Flat and curved single crystal spectrometers can be used with their limited resolving power, but appreciable intensity. Detailed spectral studies in the high-energy region definitely demand the apparatus with high resolving power without the appreciable loss in the intensity and high line-to-background ratio. Double crystal spectrometers theoretically[1] are the most suitable appliances for these studies. Most of the work has been done with an extra attached second crystal on conventional diffractometers. Good descriptions of the double crystal spectrometers are given by various authors.[2-13] Double crystal spectrometers are now being used for studies for the fine structure[13-20] of x-ray absorption edges. Jaegle[10] has used two concave gratings instead of two flat crystals.

In this paper a very compact vacuum double crystal spectrometer, and its various modes of application and different attachments, is described. Care has been taken to avoid unnecessary deformation in the base plate due to the vacuum. Optimization of line-to-background ratio has also been an important consideration. In all the applications of the instrument precise measurements can be taken from either the first or the second crystal.

DESCRIPTION OF THE INSTRUMENT

Most of the details of the spectrometer are shown in figures 1 and 2. In brief, the description is given as follows; The entire spectrometer mechanism is supported by a surface (A_1) that serves as a reference flat. All other surfaces are references to this which has an overall flatness of 0.001" and leveled with the help of an optical clinometer with an accuracy of better than five seconds of arc.

Motion of Second and First Crystal

The second crystal motion assembly (B) is supported by a tapered roller thrust bearing (Torrington #40-TTHD-015) enclosed in the dust cover (b_3) that also contains the three centering screws for concentric alignment of the precision seven-inch model Ultradex 360-sided polygon (b_1) [supplied by AA Industries, Inc.]. The angular

Figure 1.   External side view of double crystal spectrometer

Figure 2.   Internal top view of the double crystal spectromete

setting of the polygon is read on scale (b_1). The polygon can be
indexed by integral numbers of degrees throughout the entire 360
degrees with an accuracy of ± 1/4 second of arc. The polygon release lever (b_2) serves to disengage the polygon for free rotation
of the second crystal, mounted inside the chamber on an axle (b_4),
which rotates in a vacuum seal in base plate A_2 (19 x 19 x 1 1/2
inch stainless steel plate with top and bottom surfaces ground
parallel to within 0.005 inch). Plate A_2 is set parallel to the
surface A_1.

Precise and accurate rotational motion in steps of one second
of arc is imparted to the polygon via a tangent arm (b_5) of length
20.6265 inches. The tangent arm is motor-driven in step increments
by the stepping motor (m) [Slo-Syn motor #HS-25]. Position of the
tangent arm (b_5) is read from the precision micrometer (T) with a
least count 10^{-4} inch [Brown and Sharpe #599-299-100]. Beside the
step increment feature, a certain range on the drive can be selected anywhere and can be repeated many times automatically back and
forth, either with a variable speed continuous motion or in the
step mode.

The first crystal is translated by a linear motion imparted
externally to the vacuum via feedthrough (M_E). The first crystal
mount itself is placed on a small base plate which has a sine arm
attachment, also, in order to set the contact plane of the three
screws parallel to that of the second crystal. The first crystal
mount assembly moves on a rod (14" long with nicely polished surfaces). Very precise translation of the crystal along the dovetail
slide is given by rotation of a fine pitch screw (100 threads per
inch) with position read outside the chamber from a mechanical
counter and verier dial.

Crystal Mounting Device

A very simple system has been devised for supporting the crystals as shown in figure 3. It features a three-point suspension
mount with fine adjusting screws aligned to bring the plane of the
points to coincide with the spectrometer axis. By having the points
in contact with the front surface of the crystal (or support ring)
it is possible to interchange any preoriented crystals rapidly.
Thus bigger crystals (which have been cut with their plane parallel
to the surface with a misorientation less than 10") can be used
directly and rapidly interchanged for measurements. Very light
spring loading against the back surface of the large crystals or
support ring does not create any measurable stress in the crystals
which would result in shift or broadening of the diffraction peak.
Crystals having a maximum face dimension less than about 3/4 inch
are first mounted in support rings. Each ring has its front surface

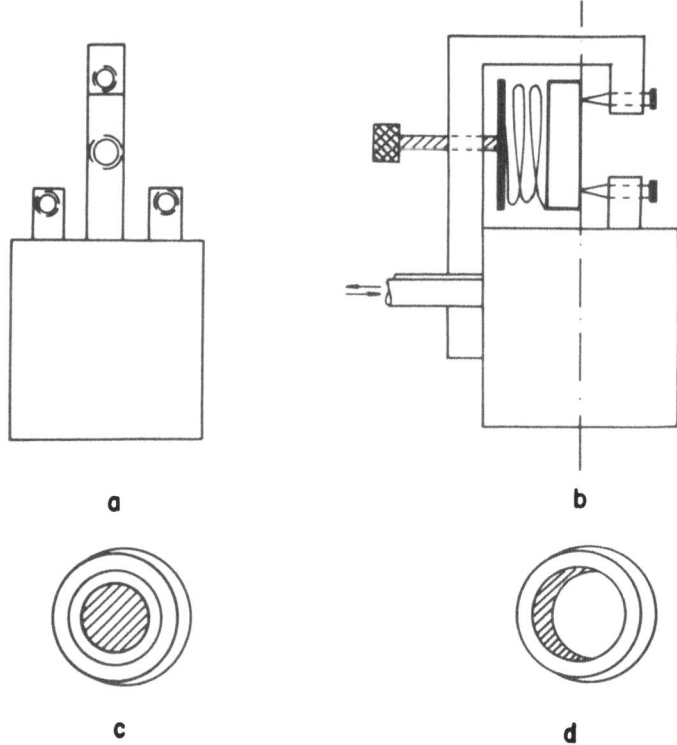

a   b

c   d

Figure 3. (a), (b) Sketch of crystal mounts and
           (c), (d) Support rings

precision-lapped for flatness. By laying the support ring "face
down" on an optical flat and positioning the crystal also "face
down" in the center of a ring of proper inside diameter and then
filling the mold with suitable material, a perfectly prealigned
crystal can be obtained. For crystals in the size range of 1 mm
to 5 mm, two concentric rings may be used as illustrated in figure
3c. The specimen is stuck on the inner ring with either silicone
grease (or highly viscous silicone oil) or colloidal graphite. Thus
no extra stress has been put on the specimen during mounting of the
crystal in the above method. The small differences in the crystal
misorientation can be corrected for by the precise sine arm device
(figure 2) attached to the crystal mounts. The crystal mounts are
provided with coolant-circulating chambers for precise temperature
control. The "precision" constant temperature circulating system
used is Precision Scientific Company Model TS-66600-6 which has a
control sensitivity of ± 0.01°C.

                    Detecting Mount Assembly

    A compact detector and preamplifier are used in the spectrometer.

The detector assembly (D) consisting of proportional detector (C_1)
and solid state preamplifier (C_2) are mounted on the top gear
(C_3) of a simple differential. The bottom gear is not allowed to
rotate but a flat spring keeps it loaded vertically so that the
differential will not disengage when the polygon is indexed. The
middle gear of the differential is clamped to the axle (b_4) so
that the detector rotates at twice the angular velocity of the
second crystal. The detector ring (C_3) can be manually disengaged
and rotated to any desired position. Electrical connections to the
detector are made through connector (C_b). A principal advantage of
this system is that one can go from the parallel to antiparallel
geometry by merely indexing the polygon. All the data are directly
recorded by a Model 33 teletypewriter.

Vacuum System

The whole system may be operated in air, suitable gas or
vacuum. The standard 18" CVC feedthrough ring #AR-005 forms the
vertical wall of the spectrometer chamber. Viton O-rings are used
in all seals. The feedthrough ring V is clamped to plate A_2 by
use of clamps (CL_1, CL_2, CL_3,..). An Ultek model 206-0565 ion pump
and foreline valve have been used for evacuating the chamber. A
molecular sieve foreline trap is used. The current meter for the
power supply serves as the high-vacuum gauge.

X-ray Sources

Two types of x-ray sources are compatible with the double crystal spectrometer, each having its own advantages for particular
applications.

A sealed target, high-intensity tube (Siemens A8Cr64) is used
as an intense source of continuous radiation over the range of 0.5
to about 3.0Å. The chromium K_α lines from the tube are also useful
for air path alignment and tests as described later.

A simple demonstrable high-intensity tube has been built for a
continuous source from 1.0 to 50Å. The relatively large target
(1/4 inch x 1/4 inch) of copper, or other electro-plated targets,
is operated at 0 - 10 kv and 0 - 500 ma. A special power supply
has been designed and built for this wide current range capability.
The filament and target are electrostatically shielded to confine
scattered electrons to the area of the tube.

Alignment Procedure

Many methods for the alignment of double crystal spectrometers have been used by different authors ranging from visual observation on the fluorescent screen[21] to careful optical method[22]

For the classical way of using the double crystal spectrometer one does not need highly complicated precision setting of the first crystal. The method of alignment is much simplified by adopting the present design for the first crystal mount.

The surface plate of the spectrometer is reference-marked for the position of most of the parts of the instrument. The single crystal spectrometer part with its tangent arm attachment and base plate, on its prefixed supports, are set and the two surfaces are made parallel to each other with the help of an optical clinometer.

The spindle attachment (to the single crystal spectrometer) is set within an accuracy of 1-2 sec-arc parallel to the axis of the spectrometer. The clinometer with a mirror attachment is put behind the spindle and an autocollimator is set in front of the spindle. With the help of 180° rotation of the spindle, the clinometer mirror is used as the reference position for the autocollimator. The second crystal is mounted with its detector assembly and the top screw of the crystal mount set parallel to the axis by conventional methods. An optical flat is placed in contact with the three screws and by adjusting the lower two screws and 180° rotation of the axis, and using the position of the autocollimator with reference to the spindle, one can readily set the second crystal mount parallel to the axis of the spectrometer with an accuracy 1 sec-arc.

In a similar way by moving the clinometer and autocollimator the plane of the three screws of the first crystal mount can be set perpendicular to the base plate. In this way no extra trouble is faced for the critical setting of the first crystal.

Tests are made to ensure a straight line translation of the crystal mount along the dovetail slide. Throughout the 8 inches of movement there is no appreciable deviation of the crystal mount from its original setting.

To calibrate the zero position of the first crystal two optical flats have been put in the two crystal mounts and are set in a straight line (second crystal setting at zero degree position). Once the zero position of the first crystal is determined it is very easy then to set it at any angle with respect to the collimator, simply by calculating the linear motion of the crystal mount.

Zero setting of the collimator is done by precise reference mark positions on the collimator and the known geometry of the tube.

The above alignment is true--in practice--only if the crystal planes are parallel to the surface of the crystal. If the misorientation of the crystal is known, then that can be compensated for by proper rotation of the sine arm (S) attached at the axis of the mount of the first crystal. The setting of the plane of the three screws is not at all affected except that the apparent plane of the three screws is tilted equivalent to the misorientation of the diffracting plane with respect to the crystal surface.

The first crystal is placed in its position and set for a particular diffraction peak. A very fine slit is placed in the position of the axis of the second crystal. The detector is moved to its maximum signal intensity position while the spectrometer disc is fixed at its zero position.

Performance of the Double Crystal X-ray Spectrometer

A series of performance tests have been made on a variety of crystals including NaCl, silicon, germanium, ADP and quartz in order to evaluate the spectrometer's resolution, precision and accuracy. A chromium x-ray tube was used as an x-ray source and tests conducted in air. No fine slits have been used either between the source and first crystal or between the first and second crystals. Some of the typical performances of the apparatus using NaCl crystals are shown in figures 4 and 5.

Figure 4 shows the reproducibility of the peak intensity and a positional stability to within 1/4 second of arc. The crystal can be removed and returned to the spectrometer with no observed shift in the peak position or intensity.

Figure 5 shows the successive improvement in the peak width which may be due to slight misorientation in the crystal surface. Thus the results show that the system which has been developed for crystal cutting and mounting in this spectrometer is ideally suited for rapid interchange of crystals and rotational adjustment about the axis perpendicular to the diffracting planes.

After etching the NaCl crystals, a peak width of 25 seconds of arc is achieved. A pair of such crystals has been selected out of some forty crystals. The angular misorientation of these crystals can be calculated[23] as

$$W_m^2 = W_1^2 + W_2^2,$$

(when first and second crystals are of the same material and the vertical divergences are very small.)

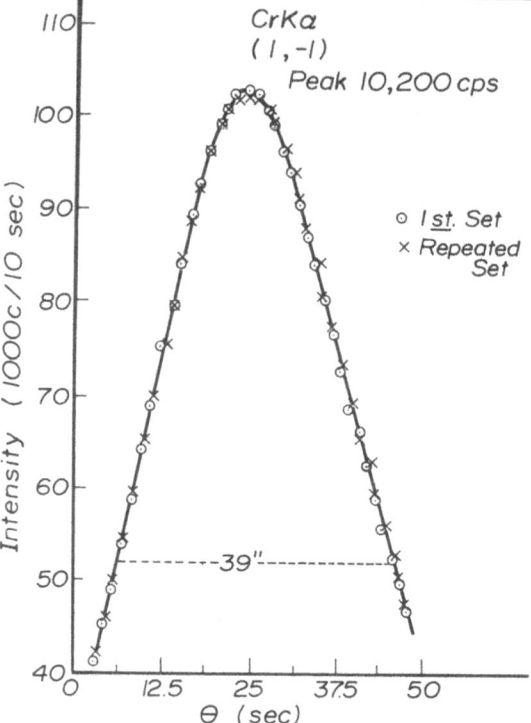

Figure 4. Two repeated sets of observations for CrK_{α_1} peak using (200) plane of NaCl in first and second crystal position. Voltage 20KV, current 6 ma.

W_1, W_2 → angular misorientations of first and second crystals

W_m → observed peak width

When the first and second crystals are interchanged, the same peak width is obtained; thus

$$W_1 = W_2 = W$$

$$\therefore 2W^2 = W_m^2, \quad W_m = 25 \text{ seconds of arc}$$

$$\therefore W \simeq 18 \text{ seconds of arc.}$$

Several other crystals of good quality have also been prepared including ADP (100, 100 cuts) which will be of particular use in the soft x-ray work.

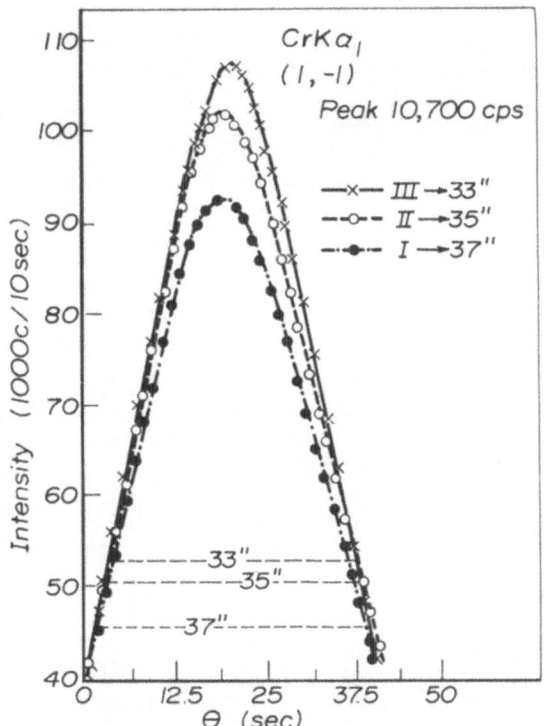

Figure 5. CrK_{α_1} peaks
from three different
settings of the second
crystal of NaCl(200).
Voltage 20KV, current 6 ma

Performance as a Single Crystal Spectrometer

The single-crystal spectrometer part of the instrument (ignoring the first crystal) with tangent arm attachment is an accurate way of measuring Bragg angles. By using it for precise lattice parameter measurements, the Bragg angle should be measured[24] to an accuracy of 0.77 seconds of arc (3.7 x 10^{-6} radians or 2 x 10^{-4} degrees) at $\theta = 75°$ and to 0.36 seconds of arc at $\theta = 30°$.

In order to establish the capability of the tangent arm the critical analysis of the measurements taken by it is essential for two main reasons:

(i) The movement of the tangent drive is linear, whereas that of the center of the ball is along the arc as shown in figure 6. There is a displacement of the center of the ball from line 00" as the tangent drive moves, and hence the effective movement is different from that of the reading on the tangent drive. With the movement of the T.D. the point 0 on the T.A. shifts from the path of the T.D. by a distance 0' - 0". The apparent movement of the T.D. 0 - 0" = d', the effective movement of the T.D. 0 - 0' = d.

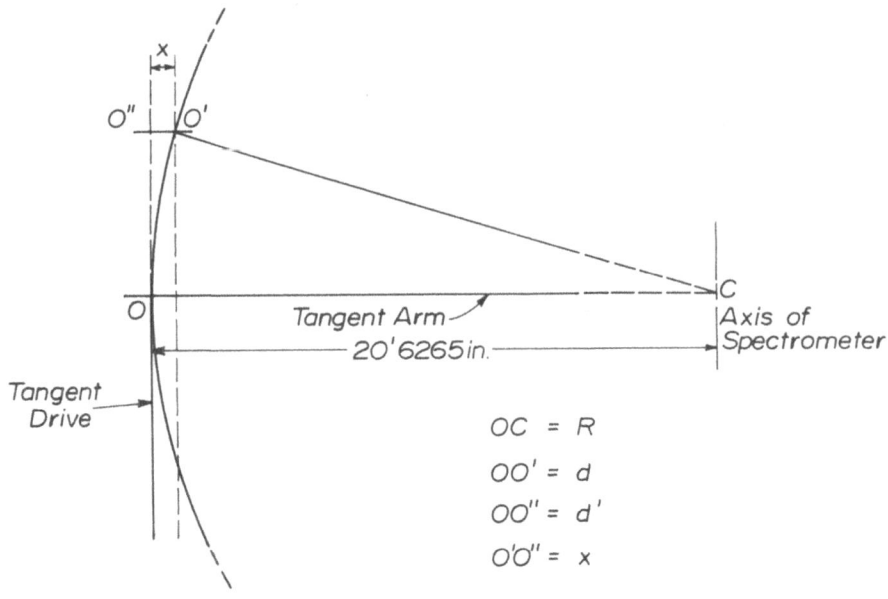

Figure 6.   Motions of the tangent arm and tangent drive. The
            center of the ball 0 moves along arc 00' whereas
            the tangent drive along the path 00".

From the simple geometry we see that

$$(2R-x) \; x = d'^2$$

or

$$x = R \pm \sqrt{R^2 - d'^2}$$

and

$$d^2 = x^2 + d'^2$$

or

$$d = \sqrt{x^2 + d'^2}$$

Figure 7 shows how the systematic error is introduced for various displacements of the tangent drive. The graph shows the safest
region for measuring a small angle with the tangent arm is ± 0.2
inches from the zero setting of the tangent drive.

(ii) The other important factor is the tangent arm length R. A
small change in setting the center of the ball at the distance of

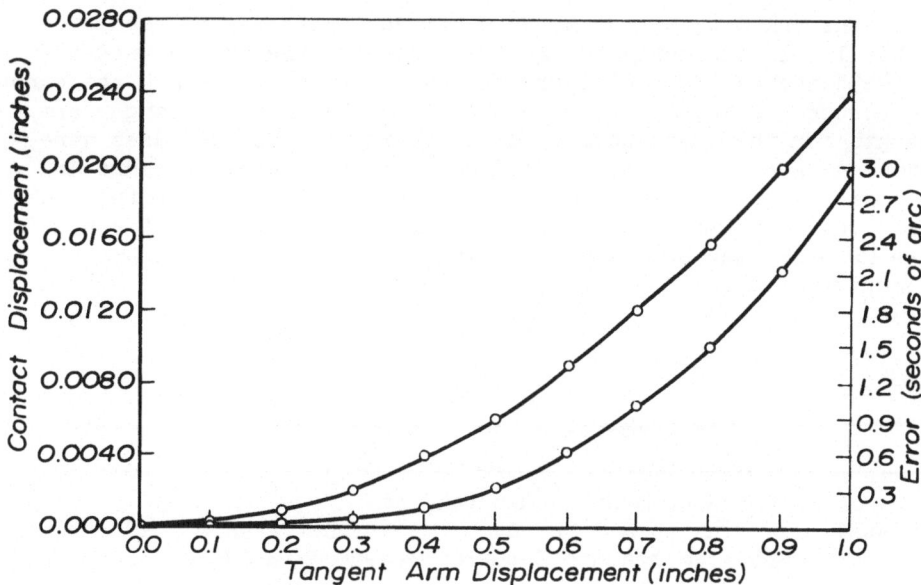

Figure 7.   Successive error introduced in the angle measurement
            due to the motion of tangent arm along the curve as
            shown in Figure 6.

20.6265 inches from the center of the spectrometer axis also affects
the accuracy of the measurement. Simple estimation can be made
from the relation

$$D \frac{\Delta R}{R_1 R_2} = \frac{2\pi}{360} \Delta\theta$$

where D = 0.36 inch is the movement of T.D. for $1°\theta$.

For $\Delta R = \pm 1 \times 10^{-3}$ inch (which can be easily achieved by a
good workshop) the total error is introduced in the measurement of
θ, when the tangent arm used to measure $1°\theta$ is 0.48×10^{14} degrees
($\approx 0.18"$).

Error in the first part can eliminated either by choosing the
proper range of movement of the tangent drive or applying proper
correction. Thus taking into consideration the above errors and
error $\pm 1/4$ second, which may be introduced during the ultradex
disc setting, the accuracy of the angle measurements is quite within
a limit of 1 ppm determination of lattice parameters. The above
facts put the angle measurement device on a level of accuracy equal
to that of other devices used to make angle measurements.

The validity of the above-mentioned facts may be seen from table 1. In the early design the length of the arm was set at 11·4545 inches (± 0.001 inch);for the shorter arm length the error introduced in angle measurement is more than for the longer arm length for the same accuracy of R setting. The two cases were studied under identical conditions of temperature, pressure and alignment. The order of difference in the 2θ measurements in the sets taken on two crystals is the same. The mutual difference in the 2θ error may be due to the ultradex disc setting within an accuracy of ± 0.25".

Table I

Measurements using first-order CrK_{α_1}

Crystal	2θ measurements, when tangent arm is used		Difference in 2θ	2θ error in seconds of arc
	Set 1:to measure 0.2°θ angle	Set 2:to measure more than 1°θ	Set 1∿Set 2	
NaCl (Cleavage)	47·948125 deg.	47·948750 deg.	0·000625 deg. (2θ)	2·25"(2θ)*
Si(III)	42·83370 deg.	42·83440 deg.	0·0007 deg. (2θ)	2·52"(2θ)*

* ± 0·27"

θ-θ Spectrometer Geometry

The described geometry can be used as an ideal θ-θ spectrometer as shown in figure 8.

Crystal C is fixed perpendicular to the dovetail slide OP and parallel to the line DCX. Thus any translation of the crystal along OP will give a θ-θ diffraction. The value of λ and d can be found simply by measuring C as follows:

$$C = \sqrt{a^2 + c^2} \, \sin\theta \qquad \theta = \text{Bragg angle}$$

$$\sin\theta = C / \sqrt{a^2 + c^2}$$

Substituting this in the Bragg equation we get the form

$$\lambda = \frac{2d \quad C}{\sqrt{a^2 + c^2}}$$

$$= \frac{2d(\overset{\circ}{A}) \; C \; (Cm)}{\sqrt{a^2 (CM)^2 + c^2 \; (Cm)^2}}$$

$$= \frac{2dC}{\sqrt{a^2 + c^2}} \quad (\overset{\circ}{A})$$

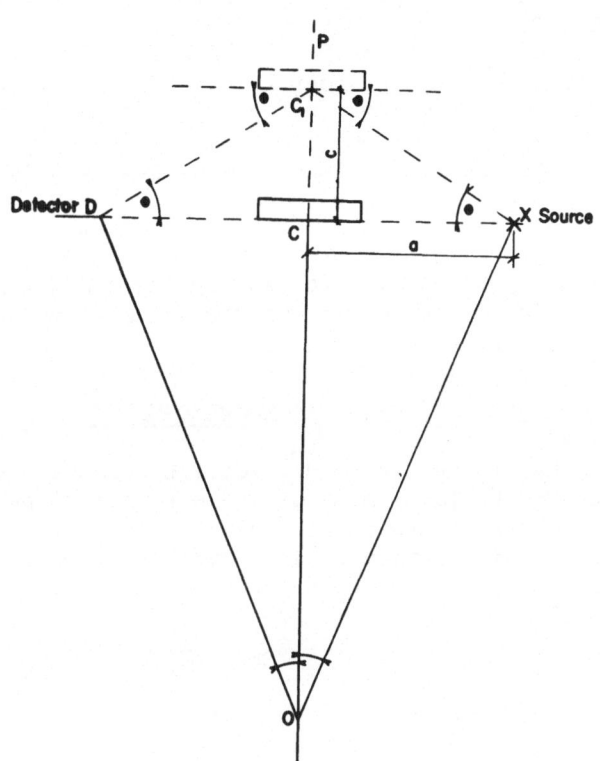

Figure 8. θ-θ spec-
trometer geometry.

SUMMARY

The double-crystal spectrometer described above can be used in
many different ways. Due to its high resolving power, having the
system under vacuum and having the high-gain miniature solid state
preamplifier with the detector, the spectrometer is valuable for
soft x-ray spectroscopy. This includes the K-edge of most of the

lighter elements and the L absorption edges of the first transition series elements. The objective is to study these edges in as great a detail as possible in order to further evaluate the use of x-ray absorption spectra in problems of materials characterization. In addition to studies of absorption by thin films, research continues on the problem of also obtaining reliable absorption edge measurements directly in the emission from thick targets (Hach and White).[25] This approach to x-ray spectroscopy of solids provides the most direct technique for measuring both emission bands and absorption edges in a given material. Besides its versatile use with its different attachments in x-ray spectroscopy (emission and absorption) work, it can be used in several other important fields of materials characterization including: identification of polar[26] surfaces of crystals, finding dislocation-free crystals, precise crystal cutting, choosing the best method for growing crystals, precise lattice parameter measurements; coefficient of thermal expansion, studies of strains and stresses in the crystals, and studies of real surfaces.

Attachments for thin film evaporation device and x-ray fluorescence are under construction. By having the fluorescent attachment one can do the studies of chemical shift, chemical analysis, etc. The thin film evaporation device will provide a method for preparing thin films for absorption studies without appreciable oxidation or the other deteriorating effects due to exposure to atmospheric conditions.

ACKNOWLEDGMENTS

Thanks are due to Mr. Jozef Lebeidzik for his valuable assistance for building most of the electronics parts, to Prof. J.W.Faust, Jr. for providing single crystals of silicon and to Dr. Pfeiffer of NBS for high-purity single crystals of ADP. This work is being sponsored by Air Force Contract F33615-67-C1047.

REFERENCES

1. A.H. Compton and S.K. Allison, "X-rays in Theory and Experiments" p. 709-750, D. Van Nostrand Company, Inc. N.Y. (1935).

2. S.K. Allison and J.H. Williams, "The Resolving Power of Calcite X-rays and Natural Width of the Molybdenum $K\alpha$ doublet." Phys. Rev. 35, 1476 (1930).

3. L.G. Parratt, "Design of Double Crystal Spectrometer." Phys. Rev. 41, 553 (1932).

4. J.A. Bearden, A. Henins, J.G. Marzolf, W.C. Sander and S. Thomsen, "Precision Redetermination of Standard Reference Wavelengths for X-ray Spectroscopy," Phys. Rev. $\underline{135}$, A899-910 (1964).

5. R.D. Deslattes, "Single Axis, Two Crystal X-ray Instrument," Rev. Sci. Inst. $\underline{38}$(6), 815-20 (1967).

6. G. Brogen, "The Width of X-ray Emission Lines," Arkiv. Fysik. $\underline{3}$, 507 (1952).

7. H.W. Schnopper, "The Argon K-absorption Edge-Two Crystal X-ray Spectrometry," Thesis dissertation, Cornell University (1962).

8. L.V. Azaroff, "Plane-Polarized, Two Crystal X-ray Spectrometer," Advances in X-ray Analysis, Vol. $\underline{9}$, 251-258 (1965).

9. P. Jaegle, "Diffraction of Soft X-rays Produced by Two Gratings," (in French) Compt. Rend (French Academy) $\underline{245}$, 1412-15 (1957).

10. P. Jaegle, "Calculation of the Resolving Power of a Spectrograph with Two Concave Gratings at Glancing Incidence," (French Academy) $\underline{250}$, 3620-3621 (1960).

11. M. Sawada and K. Tsutsirmi, "Source Scanning Type X-ray Grating Spectrometer," Jap. J. Appl. Phys. $\underline{40}$(4), 1950 (1969).

12. Y. Gohshi, "Simple Two Crystal Spectrometer and its Application to X-ray Spectrochemical Analysis," Advances in X-ray Analysis, Vol. $\underline{12}$, Plenum Press, N.Y. (1968).

13. Annotated Bibliography on Soft-X-Ray Spectroscopy, NBS Monograph 52.

14. A. Herman, et al, "X-ray Absorption and Emission," Analytical Chem. $\underline{34}$ (5), 282R-294R (1962).

15. J.E. Dowdey, "Fine Structure of the X-ray Absorption Edge of Cubic Ferrites," Univ. Microfilms MIC-59-154, 99 pp (1959).

16. J.O. Porteus, "K Absorption Spectra of Alkali Metal Chlorides," Univ. Microfilms MIC-59-135, 155 pp (1959).

17. J.A. Soules and C.H. Shaw, "X-ray K Absorption Spectra of Solid Argon and Kripton," Phys. Rev. $\underline{113}$, 470-72 (1959).

18. R.J. Liefield, "L Series X-ray Emission and Absorption Spectra in Zirconium," Univ. Microfilms LC card # MIC 60-1197, 99 pp (1960).

19. J.N. Singh, "Fine Structure of the K X-ray Absorption Edge in Germanium," Microfilms MIC 60-5350, 92 pp (1961).

20. L.W. Azaroff and B.N. Das, "X-ray K Absorption Spectra of Cu-Ni Alloy," Phys. Rev. 134, A747-A751 (1964).

21. H. Glaser, "The Absolute Absorption Co-efficient of Germanium and the Fine Structure in the K-edge of Some of its Compounds," Phys. Rev. 82 (5), 616-620 (1951).

22. J.A. Bearden, "Precision X-ray Spectroscopy," International Conference on Physics of X-ray Spectra (Cornell University) June 1968.

23. L.S. Birks, J.W. Hurley and W.C. Sweeney, "Perfection of Ruby-laser Crystals," J. App. Phys. 36 (11) 3562-65 (1965).

24. W.L. Bond, "Precision Lattice Constant Determination," Acta. Cryst. 13 (10), 814-818 (1960).

25. J.T. Hach and E.W. White, "Study of Extended X-ray Absorption Fine Structure with the Use of Thick Targets," Advances in X-rays, Vol. 11, Plenum Press, New York, 1967, pp 339-344.

26. A.N. Mariano and G.A. Wolff, "Polarity Identification of Some AB Compounds of Tetrahedral Coordination," Zeit. Krist. 126, 244-261 (1968).

X-RAY ASTRONOMY

Herbert Friedman

Naval Research Laboratory

Washington, D. C. 20390

ABSTRACT

Although searches so far have been restricted to
a few small rockets and balloons, some 40 discrete
x-ray sources have already been resolved against a
diffuse, nearly isotropic background radiation. The
strongest source is about 2000 times as bright as the
weakest detectable with present rocket instruments.
Nearly all of the discrete sources lie close to the
galactic plane and most likely are members of the
spiral arms of the Milky Way. One x-ray source at high
galactic latitude is identifiable with a distant radio
galaxy, Virgo A, and its x-ray luminosity is 70 times
its radio power. The diffuse background radiation
seems to be resolvable into at least two components:
one may be associated with the interaction of cosmic
rays and the microwave photons of the cosmological 3 K
background; the other with bremsstrahlung from hot,
intergalactic gas.

INTRODUCTION

High energy astronomy is concerned with the most
energetic forms of radiation--x-rays, gamma rays and
cosmic ray particles. The observable energy range
extends from soft x-rays, about 250 eV, to the highest
energy cosmic rays, about 10^{19} eV. In contrast, all of
visible light covers a range of about 2 eV. We can

expect to discover entirely new cosmic phenomena in the
broad spectral range that has become accessible from
space platforms.

We speak of "windows on the universe" encompassing
different frequency ranges. From the ground our windows
are limited to visible light and part of the radio
spectrum. All other rays are degraded or absorbed by
the air above. As balloons, rockets and satellites
surmount the atmospheric veil, the universe comes into
view on all frequencies. In particular, x-ray astronomy
is the story of an entirely new field of celestial observations, totally unsuspected before the advent of
space research. It has revealed the existence of x-ray
stars and x-ray galaxies, whose x-ray powers exceed
their optical and radio powers. It is possible that
most of the mass of the universe can be observed only
in the x-ray spectrum and that important clues to the
origin of cosmic rays will be revealed by the x-rays
that result from interactions of relativistic particles
with matter and fields in the cosmos.

Proof of the existence of cosmic x-ray sources
was first obtained in 1962. Since then, some 40 discrete sources have been discovered against a diffuse
x-ray background that covers the entire sky almost
uniformly.[1] Nearly all are believed to lie in the
galaxy and are characterized by an average x-ray power
of about 10^{8b} erg/sec, which is a thousand times the
luminosity of the sun integrated over all wavelengths
from x-rays to infrared. The brightest x-ray source
is associated with a blue starlike object, but its
x-ray power is 1000 times its visible luminosity. One
x-ray source at high galactic latitudes is identifiable
with a distant radio galaxy, Virgo A (M-87), and its
x-ray power (1-10 Å) is about 70 times its radio power.
Most x-ray sources are still unidentified with visible
or radio counterparts, but there is strong evidence for
associations with supernova remnants, pulsars, and
close binary pairs and weaker evidence for x-rays from
radio galaxies and quasars.

THE CRAB NEBULA

One of the strongest x-ray sources, Tau XR-1, and
certainly one of the most interesting of all celestial
objects is the Crab Nebula in Taurus. It was discovered
in mid-eighteenth century and is located at a position

where, in 1054 A.D., Oriental astronomers observed the
sudden appearance of a "guest star" as bright as
Jupiter. We now recognize that the Crab Nebula is the
expanding debris of that ancient supernova. Only two
others have been similarly documented--the Tycho Brahe
supernova of 1572 and the Kepler supernova of 1604.

Figure 1. This photograph of the Crab Nebula
(Mount Wilson Observatory) was taken in green light,
produced largely by optical synchrotron radiation. The
pulsar is the lower of the two central stars.

The Crab Nebula, figure 1, appears as a whitish
amorphous glow enmeshed in tangled filaments of red,
glowing hydrogen. The entire nebula fills a roughly
ellipsoidal volume that projects an area 4 minutes by
6 minutes of arc. From photographs taken at different
epochs, the expansion is readily evident. Projecting
back in time, the nebula appears to converge to a
point close to a central double star of about 15th
magnitude, and the time of origin is indicated as
1140 A.D., if the rate of expansion has been constant.
To account for the century difference from the histori-
cal event, the expansion must be accelerating.

Both the light of the amorphous core and the radio
emission are highly polarized, which is evidence that

the mechanism of radiation is the synchrotron process.
To account for radio emission, we require electrons of
energy about 10^7 eV gyrating in a magnetic field of about
10^{-4} G (about 1000 times weaker than the earth's field,
but 100 times stronger than interstellar magnetic fields).
The amorphous white light requires electrons of energy
as high as 10^{12} eV. If the x-ray emission is also
synchrotron, the electron energy spectrum must extend
up to 10^{14} eV--the range of high-energy cosmic rays.
Figure 2 (segments E, F, G, H, I) shows the spectral
shape from 10^7 to 10^{20} Hz. The shape of the spectrum
and the high degree of polarization of radio and visible
emission is strongly suggestive of the synchrotron process. However, there is a fundamental problem of the

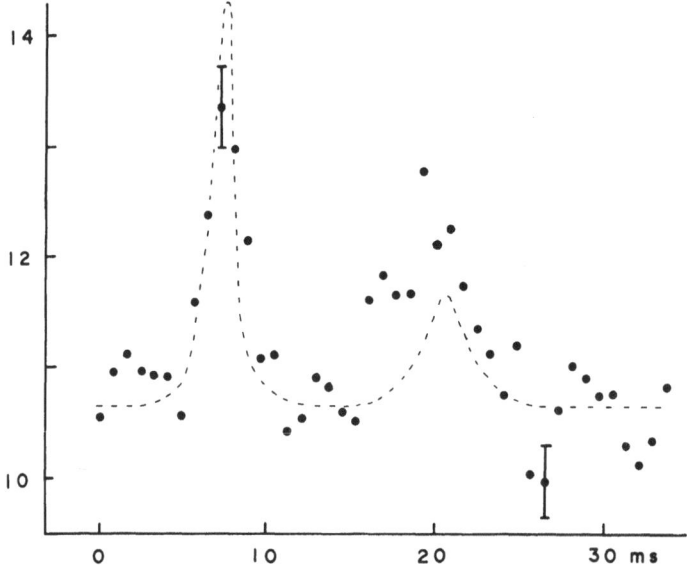

Figure 2.   Spectrum of the Crab Nebula and pulsar
NP 0532. The logarithm of the flux is plotted against
the logarithm of the electromagnetic frequency. Some
lines are labeled with negative slopes (spectral index).
A represents the small radio source in the Crab, which
may be associated with the pulsar. B is the radio pulsar, C the optical pulsar, and D the present x-ray
pulsar observation, which is plotted relative to G, the
x-ray spectrum of the entire nebula as determined by
other workers. H and I are higher energy spectra of the
Crab (as a whole), while E and F are similar spectra in
the radio and optical regions respectively.

lifetime of the electrons.   In the process of producing
synchrotron radiation, the electron energy is depleted.
Although the low energy radio electrons could survive
for thousands of years, the optical electrons should
decay in one or two hundred years and the x-ray electrons in about a year, or less.   Clearly the presently
observed x-rays could not be produced by electrons that
have survived for nearly a thousand years.

The identification of x-ray emission from the Crab
Nebula in 1963 was accompanied by speculation that the
radiating object was a neutron star remnant of the
supernova.   In theory, a supernova results from the
collapse of a highly evolved stellar core.   The infall-
ing matter compacts to nuclear density, about a billion
tons per cubic inch in about one second.   A rebounding
shock wave drives off the outermost layers of stellar
atmosphere to produce the expanding nebula.

The compacted star contains on the order of a
solar mass within a sphere of 10 km radius, and the
core maintains a constant temperature of about 5×10^9 K
out to the last few meters.   Only in the last meter or
so, at the surface of the star, does the temperature
fall to 10^7 K.   Such an object would radiate virtually
all its energy as a black body spectrum concentrated in
the x-ray range of a few angstroms.

In 1964, a lunar occultation of the Crab was observed from a rocket launched at White Sands, and the
x-ray source was found to be about two light-years in
diameter, only slightly more concentrated than the
optical nebula, and certainly not a star-like object.
At about the same time, physicists concluded that cooling by neutrino radiation processes was so efficient
that temperatures above one- or two-million degrees
could not last more than a year.   Thus, the idea of a
neutron star as a black body source of x-rays fitted
neither the observed nor the theoretical evidence.
However, the neutron star has recently reentered the
picture with the discovery of a pulsar in the Crab.

The first pulsar was found barely a year ago, but
nearly 40 more have been detected since.   They shoot
out pulse-like bursts of broad band radio waves, most
of them with periods of the order of a second.   A
pulsar has been found in the Crab that coincides with
the position of the southernmost star of the central
pair, and it produces both radio, optical and x-ray

bursts with a period of 33 msec. It is believed that
the periodicity is associated with a spinning star at
such a high rate. Alternatively, it is possible that
the pulsations are connected with vibration of the star.

The discovery of a pulsar in the Crab has naturally
lent much more credence to the neutron star hypothesis.
Over the short span of a couple of months, it was determined that the period of the Crab radio pulsar is
increasing at a surprisingly large rate, about one part
in 2000 per year. The slowing down of rotation must
release a large amount of rotational energy, between
10^{38} and 10^{39} erg/sec. These numbers fit the entire
energy budget of the Crab remarkably well. About
10^{37} erg/sec go into radiating x-rays and about
10^{38} erg/sec into accelerating relativistic particles
and creating magnetic fields in the nebula.

Just how a neutron star can convert rotational
energy into relativistic particles to fill the nebula
remains a mystery. T. Gold has suggested[2] that a
magnetosphere forms about the neutron star and spins
with it. Over some area of the surface of the star,
gas may be liberated and propelled outward along the
stiff radial magnetic field until it reaches relativistic speed, at which time it is flung out of the
magnetosphere. The crossing of the magnetosphere by
the relativistic gas creates the drag which slows the
spinning star.

At the time of the 1964 occultation of the Crab,
radio astronomers[3] discovered a very bright and very
small low frequency source (36 Mcps) within the nebula.
More refined measurements now place the radio source
close to the central star and its size is less than
0.5 arc second. This source may be identical with the
pulsar even though no pulsation is apparent. Dispersion
of the radio pulse by interstellar electrons at low
frequencies would broaden the pulse so that successive
pulses would overlap.

The latest contribution to the pulsar story has
come from optical and x-ray astronomy.[4,5,6,7,8] Using
a stroboscopic technique, astronomers at the Lick
Observatory succeeded in photographing the central
double star and have shown that the southernmost component blinks on and off at 30 cps. The average power of
the pulsed light is 10^{35} erg/sec and its color is very
blue. An attempt to detect x-ray pulses by the NRL

group with an Aerobee rocket launched on 13 March 1969
succeeded in measuring a double pulse profile similar
to the main pulse and interpulse found in radio and
visible light.

The x-ray interpulse (figure 3) has a lower peak
and is about twice as wide as the main pulse; in these
features it resembles the optical pulsar. However, the
x-ray interpulse actually contains more counts than the
main pulse, and its onset is earlier than the optical

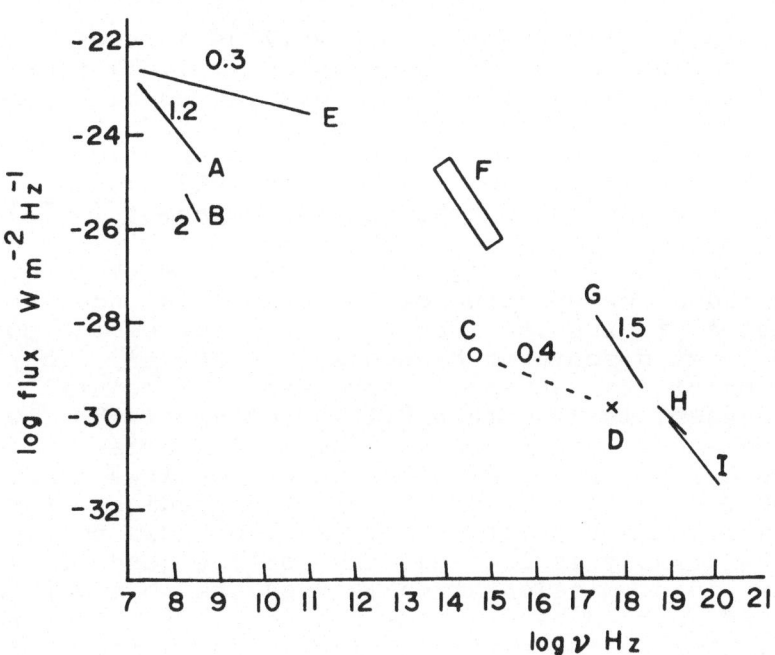

Figure 3. The "light curve" (intensity as a function
of phase) for the Crab Nebula x-ray pulsar (dots).
The vertical scale is hundreds of counts per 0.8 msec
phase interval. Error bars indicate plus and minus
one standard deviation. Phase is in milliseconds. The
dashed line is the optical pulsar adjusted so that the
higher x-ray and optical peaks coincide in phase. The
vertical scale on the optical pulsar has been adjusted
to provide best agreement with the x-ray data, solely
to facilitate relative phase comparisons.

counterpart. Whether these are significant permanent
features or merely reflect time-variability in the
pulsar will only be determined when simultaneous optical
and x-ray measurements are made. The relative phase of
the optical and x-ray data shown in figure 3 is
arbitrary; the peaks are superimposed simply to illustrate the relative similarities.

The x-ray pulsed flux amounts to about 5% of the
integrated x-ray flux of the entire nebula   (3 to 15 Å).
If the total x-ray flux (1 to 10 Å) of the Crab is
about 3×10^{-8} erg/cm^2/sec, then the corresponding x-ray
pulsar flux is 1.5×10^{-9} erg/cm^2/sec. Furthermore, the
pulsar x-ray power is about 200 times the optical power
and about 2×10^4 times the radio power.

Shortly after the NRL observations were made, they
were confirmed in most details by an M. I. T. rocket
observation[7]. The latter group attributed 8% to the
pulsed flux. It may be that the pulsed radiation is
highly variable. A Rice University observation[8] in the
balloon range (30 - 100 keV) made in 1967 has now been
analyzed and also shows the pulsed component ($\sim$ 7%).

Figure 2 shows (segments B, C, D) the shape of the
pulsed radiation spectrum of the Crab. Included
(segment A) is the low frequency spectrum of the point
source first discovered by Hewish and Okoye[3]. The
dimension of this source, determined from scintillation
measurements, is less than 0.1 sec of arc ($\sim 3 \times 10^{15}$ cm
at 2 kparsec). Rees[9] has attributed the emission to
protron-synchrotron radiation, which requires energies
of $\sim$ 50 G. The fact that the spectral indices of both
the low frequency point source and the radio pulsar
are considerably larger than that of the general
nebular radio emission strongly suggests that both concentrated sources have a common origin. Although the
optical flux is indicated in figure 2 by a point,
evidence already exists that the optical spectrum is
comparatively flat[10]. The spectrum of optical
pulsations does not appear to be related to the radio
pulsations in the simple way that all of the nebular
radiation can be attributed to electron synchrotron
radiation. If the optical and x-ray pulsed fluxes are
joined by a straight line, as a preliminary spectral
analysis of the pulsed x-rays suggests may be appropriate, the slope is only 0.4 compared to about 1.5
for the nebular soft x-ray emission.

SCORPIUS XR-1

The brightest x-ray source, Sco XR-1, lies in the constellation Scorpius at a distance of 270-1000 parsec. It differs in almost every essential characteristic from the Crab Nebula. More than 99% of the entire luminosity of Sco XR-1 is concentrated in the x-ray region. The x-ray source has been identified with a faint blue star-like object of 12th magnitude and is barely detectable in the radio spectrum at a level of about 2×10^{-2} flux units (one flux unit = 10^{-26} W/m$^2 \cdot$Hz). Over the x-ray range it is possible to fit the spectrum quite well by a hot plasma (thermal bremsstrahlung) model with temperatures that range from $(4-9) \times 10^7$ K. Some of the difficulty of establishing the thermal or nonthermal character of the spectrum is illustrated[11] in figures 4 and 5. If a bremsstrahlung

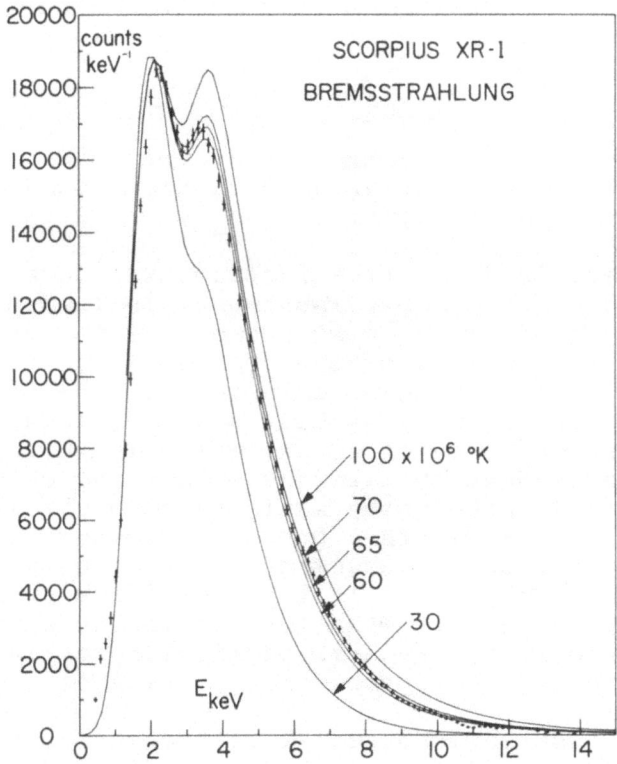

Figure 4. Observed spectrum of Scorpius XR-1 matched to bremsstrahlung models. Courtesy Meekins et al.[11].

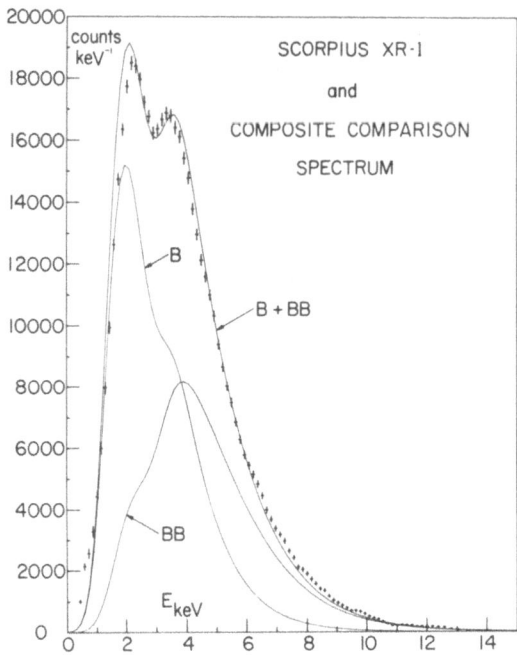

Figure 5. Observed spectrum of Scorpius XR-1 matched
to composite of bremsstrahlung and black body spectra.
Courtesy Meekins et al.[11].

spectrum is folded into the instrumentation spectral
response function, the results shown in figure 4 are
obtained. The decline in response at low energies
represents the opacity of the counter window; at high
energies, the counter gas no longer absorbs the incident
radiation. The saddle at about 4 keV is an effect of
the K-absorption edge of argon, which is the counting
gas. It is evident from figure 4 that the observed
data are well blanketed by a simple exponential spectrum
at $65 \pm 5 \times 10^6$ K. Figure 5, however, shows that black-
body and thin bremsstrahlung spectra can be combined
to give an equally good fit. Manley[12] has argued that
a synchrotron model is a better fit to the entire range
from radio to x-rays and, in particular, explains the
high flux at 0.25 keV, reported by Fritz et al.[13]

Early observations near 1 keV seemed to indicate a
turnover in the spectrum, perhaps attributable to interstellar absorption[14], but more recent observations[15],[16]
show no clear evidence of any deviation from a simple
exponential at 1 keV. Friedman et al.[17] reported a

1965 observation at 0.25 keV, which lay close to the
extrapolated bremsstrahlung with no interstellar
absorption. Since unit optical depth was expected to
be only about 50 parsec, they argued that the source
could not be more than 100 parsec distant without
requiring an implausibly high brightness to compensate
for interstellar absorption. Fritz et al.[13] have
reported a more recent observation 30^{+3}_{-10} keV/keV/cm^2/sec,
which is about one-fifth the earlier value for the flux.
An exponential extrapolation of the spectrum simultaneously observed in the range 1.5 to 13 keV predicts
32 ± 3 keV/keV/cm^2/sec. The apparently small absorption
requires an optical thickness of 0.5 or less. If the
distance to the source is taken at 270 parsec, as
indicated by interstellar Ca II absorption, the interstellar density must be as low as 0.1/cm^3. Alternatively, there could be an enhanced contribution to the
flux from lower temperature plasma within the source.

Efforts have been made to coordinate optical and
x-ray measurements of Sco XR-1, but thus far only meager
results are available from a few rocket flights. The
x-ray and optical emission are both variable and, in
general, increase and decrease together. It has been
observed that the color becomes bluer when the x-ray
intensity increases. The optical object has some of
the variability characteristics of an old nova, but its
spectrum shows significant differences. Briefly summarized, its main optical variations are as follows:
V and B magnitudes vary in the range 12.2 to 13.2, on
the average as rapidly as 0.3 magnitude per hour. Near
maximum brightness, it flickers. The intensity and
duration of the flickers are roughly constant and the
flickers often appear in triplets about 10 minutes
apart. Harvard Observatory plates provide a 70-year
record which shows no secular change in average
brightness level.

If the x-ray emission from Sco XR-1 is attributed
to a plasma at a temperature of about 5×10^7 K,
certain limits can be placed on the plasma models. The
maximum possible diameter of the fireball is 10^{15} cm,
but it may be as small as 10^{10} cm. The corresponding
densities would be 10^6 and 10^{14} particles per cubic
centimeter and the thermal energy contents 10^{47} and
10^{38} erg, respectively. The radiative cooling time for
the thinner plasma would be about ten years, and for
the denser plasma, about 10^{-7} years. As in the case of
the Crab Nebula, a continuous supply of energy seems to

be required to maintain the x-ray luminosity.

Because ex-novae are generally close binary systems, there have been various models proposed, based on a binary configuration, involving one highly condensed star, a white dwarf or a neutron star. In such a close binary system, gas streams to the compact star forming a ring or disk in the plane of the orbit. The gravitational field of the neutron star is so great that the accreted gas acquires tremendous energy in the infall process. In fact, the energy of accretion exceeds that which would be released by thermonuclear conversion of the same material. As a result, a comparatively modest rate of accretion onto the surface of the neutron star could produce the high temperature plasma required for the x-ray emission. The back illumination of the companion red star by the strong x-ray flux could raise the surface temperature to 30,000 degrees or more, and change the color from red to blue, thus possibly accounting for the color variations. While the accretion model is very attractive, there is as yet no clearly demonstrated periodicity in either the light curve or the x-ray emission, but the orbital plane could be perpendicular to the line of sight or the period could be very short. The x-ray emission has been searched for pulsating frequencies in the range from about 50 msec to several seconds with only negative results. It can be stated that no pulsating component in this range greater than 1% of the total x-ray emission is present.

CYGNUS XR-2

Cyg XR-2 strongly resembles Sco XR-1 in its x-ray properties. It has been identified with an optical object which appears to be a spectroscopic binary. Furthermore, it is a very short period binary (5-7 hr) and one of the components seems to be a normal G-type star. Unlike Sco XR-1, however, the optical radiation of the sun-like star is much brighter than the extrapolated bremsstrahlung of the 4×10^7 K x-ray spectrum.

CENTAURUS XR-2

The most variable source thus far observed is Cen XR-2. Sometime between October 1965, when it was not detectable, and April 1967 it became the brightest

x-ray source so far observed in the 1-10 keV range.
Within six months after having reached its peak, it
disappeared. Because this flashing behavior is suggestive of a nova phenomenon, it is surprising that a
companion visible event was not observed. Recently,
x-rays from Cen XR-2 have been detected again and
perhaps a new outburst is developing.

POLARIZATION AND LINE EMISSION

Direct observation of polarization in the x-ray
flux from a cosmic source would be conclusive proof of
the synchrotron process. Equally convincing proof of a
thermal mechanism would be the observation of line
emission from highly stripped atoms such as Fe XXV and
XXVI. Both types of measurement are possible in principle, but only marginally feasible in rockets of the
size of the Aerobee-150 or the Skylark.

An x-ray polarimeter, based on Thompson scattering
in a block of lithium hydride, sandwiched between proportional counters, was flown aboard an Aerobee by
R. Novick[18] and his collaborators to observe Sco XR-1
in July 1968. An apparent polarization of 4.9% ± 5.2%
was indicated, which is essentially a null result. No
polarization has been detected in the optical emission
of Sco XR-1. For the Crab Nebula, the visible polarization is as much as 23%. Novick estimates that with
improved background discrimination, his x-ray polarimeter could achieve a standard deviation of 7.5% on the
Crab. He hopes to attain substantially better performance with a larger instrument being developed for the
Aerobee-350.

Attempts to obtain evidence of line emission from
the Crab and from Sco XR-1 have been limited to the use
of proportional counters in the 1-10 Å range. Theoretically, the most prominent feature would be the contribution from Fe XXV and XXVI at 6.64 and 7.15 keV.
In practice, it is possible to achieve a resolution of
about 15%, or about one keV. Fritz et al.[19] obtained
the results shown in figure 6 which show no statisti-
cally significant evidence of the iron lines. They
conclude that the contribution of iron lines to the
total flux from 2 to 8 Å is much less than 5%, excluding
the low mass supernova model calculation of Tucker[20],
which predicts 5 to 7%. Tucker's estimates of about
one percent for normal cosmic abundance and for a

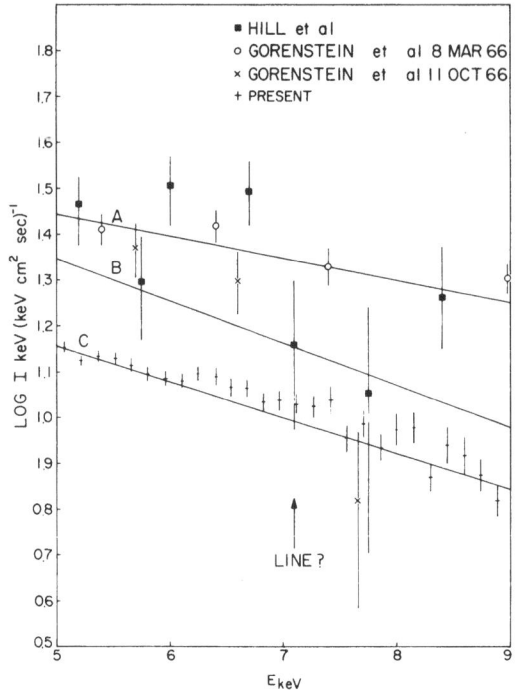

Figure 6. Spectrum of Scorpius XR-1, 5-9 keV. The
points on line C were obtained with 100 cm² counter for
40 sec of observing time. Position of the expected
Fe XXVI line is indicated. Line D is drawn through
average of points in 1 keV bins centered at 6 keV and
8 keV. The 1 keV bin at 7 keV lies several standard
deviations above line D. All three points are higher
than the ungrouped points because the detector efficiency used was that of the center of the bin. Line A is
for the spectrum of 8 March 1966 and line B for that of
11 October 1966. Courtesy Fritz et al.[19].

massive supernova cannot be excluded by the observations. Results obtained by Hill et al.[15] and by
Gorenstein et al.[21] are also shown in figure 6, but are
of poorer statistical significance. Holt et al.[22]
have observed Tau XR-1 with an energy resolution of
about 25% at 6.9 keV, with no positive evidence of line
emission, but their standard deviation was 9%.

X-RAY GALAXIES

At this time only one source has been reliably

identified with a distant galaxy, M-87 (Virgo A), and its x-ray signal is near the limit of detectability with present instrumentation. In further attempts to find x-ray galaxies, the Naval Research Laboratory has conducted two Aerobee rocket experiments to search two of the most powerful radio galaxies, Cygnus A and Centaurus A. No x-ray emission could be detected from Cygnus A, but there is weak evidence of x-rays from Centaurus A. This small sampling of three radio galaxies now permits us to identify certain highly speculative relationships between x-ray and radio galaxies:

(1) X-ray galaxies, like radio galaxies, are produced by violent explosions in the galactic nuclei.

(2) X-ray emission dominates over all other radiation for tens of thousands of years after the explosion.

(3) Subsequently, the x-ray emission dies away, leaving large clouds of relativistic particles and magnetic fields (plasmons). These plasmons may remain strong radio sources for a billion years.

To appreciate the difficulty of detecting x-ray galaxies, compare their x-ray fluxes with that from Sco XR-1, the strongest x-ray star in the Milky Way. The x-ray signal from M-87 is 200 times weaker than that from Sco XR-1; the signal from Centaurus A is at least 2000 times weaker; and the signal from Cygnus A must be well below that from Centaurus A.

Cygnus A is the most powerful of all radio galaxies. It radiates 10^{34} kW of radio power and, as observed from the earth, is ten times as bright as M-87 (Virgo A) and Centaurus A, the latter two being comparable in brightness. All three galaxies exhibit structures typical of the majority of radio galaxies. The radio emission comes from pairs of clouds of relativistic particles and magnetic fields, symmetrically disposed with respect to the parent optical galaxy. These plasmons have been ejected by catastrophic explosions with velocities approaching the speed of light. Each explosion releases an amount of energy comparable to what would be derived from the complete transformation of the mass of a million stars. In M-87, a series of plasmons is strung out along a luminous blue jet emerging from the core of the galaxy.

The farthest plasmon is 3000 light-years from the galactic center. Recent photographs show that the jet extends through the center of the galaxy in the opposite direction. Centaurus A, because it is relatively close--about 14-million light-years--is spread over several degrees of the sky. It is, therefore, easy to resolve at least three pairs of plasmons symmetrically placed along a common line of expulsion passing through the optical galaxy. The most widely separated plasmons reach almost 500,000 light-years on opposite sides of the galaxy. Closest in is a pair of plasmons only 10,000 light-years from the galactic center. Finally, Cygnus A exhibits a simple dumbbell pattern of two plasmons separated by 250,000 light-years, but radiates no detectable x-rays.

These three sources show: that the youngest explosion--that in M-87--radiates the greatest x-ray power; Centaurus A, in which the most recent explosion is at least three times as old as that in M-87, is at most one-tenth as x-ray luminous as M-87; and Cygnus A, in which the last explosion is at least 40 times as old as that in M-87, does not produce detectable x-rays. The x-ray luminosity of M-87 is 70 times its radio luminosity; the x-ray luminosity of Centaurus A may be only twice its radio luminosity.

If we assume that galactic explosions eject clouds of relativistic particles and magnetic fields with nearly the speed of light, the phase during which the x-ray luminosity overwhelms all other radiations must be little more than 10,000 years. Radio emission would continue for several million years. If the radio plasmons slow down in their separation rate after the first few thousand years until they reach speeds more typical of hydromagnetic propagation, they may have ages as great as a billion years (in the more extended radio sources).

We conclude that only a few of the most powerful radio galaxies will exhibit strong x-ray emission. X-ray galaxies may be more abundant among galaxies of the Seyfert type, which exhibit intense activity in extremely bright nuclei, where the explosion phase may be very young.

THE BIG BANG

There exists a strong body of evidence now to support the theory of an evolutionary universe which began about 10-billion years ago with a "Big Bang." At the moment of "creation" all of the matter of the universe was contained in a primordial hydrogen fireball at a temperature in excess of 10^{10} K. About 200 seconds from the beginning of expansion, the fireball had cooled to a billion degrees and was filled with gamma rays. Further expansion and cooling progressed rapidly, while the extremely energetic gamma rays worked against the gravitational forces to support the expansion. Accordingly, the temperature of the radiation fell from 10^9 K to 3 K, and the wavelengths were stretched from gamma rays to infrared and microwaves. At the present time, the universe is filled with a dilute 3-degree background radiation, through which galaxies are speeding away from us in all directions with velocities proportional to distance.

The existence of the 3-degree background radiation has important significance for the question of the origin of cosmic rays, which has remained a puzzle since their discovery more than half a century ago. Relativistic electrons colliding with photons of the 3-degree background field can convert them to x-ray quanta. Such x-rays would be diffusely distributed over the entire sky and the intensity would be proportional to the energy density of cosmic rays in space. The observed x-ray intensity implies that the density of intergalactic cosmic ray electrons is only about a thousandth of the density of cosmic ray electrons in our galaxy.

We have strong evidence that the Crab Nebula can produce relativistic particles with energies approaching the highest ranges of cosmic ray energies. Is it possible that galactic supernovae supply all the cosmic rays that we observe? Since the energy density of cosmic rays near the earth is 10^{-12} erg/cm^3 and the volume of the galaxy is about 10^{63} cm^3, the total energy in the galaxy must amount to about 10^{56} erg. The lifetime of a cosmic ray particle as it wanders through the galaxy is perhaps a few hundred million years. Therefore, to maintain the total energy constant over cosmic time requires that fresh cosmic rays must be produced at a rate of 10^{40} erg/sec. Since each supernova explosion is estimated to release 10^{51} to

10^{53} erg, of which about 10^{49} erg is in the form of
relativistic particles, we would need about one supernova every 30 years to sustain the density of cosmic
rays. The present evidence is that the rate of occurrence of supernovae is about one every 50 years. On
the basis of such rough estimates then, it appears
entirely possible that supernovae can account for
nearly all of the observed cosmic rays.

THE DIFFUSE BACKGROUND

The spectrum of the x-ray background approximates
a differential photon energy distribution $E^{-2.4}$ from
50 keV to 1 MeV and appears to be isotropic to within
about 10%. If the cosmic ray electron spectrum maintained its slope to lower energies, the x-ray background
would follow a simple power law to the soft x-ray
region (0.25 keV). A decrease in slope of the x-ray
spectrum would be explainable in terms of a flattening
of the electron energy spectrum, but no physical mechanism is apparent which would explain an increasing
slope toward lower energies. The spectrum does show a
flattening from 2 to 10 keV where a number of observers[23],[24],[25],[26],[27],[28] have reported power law
(photon) indices of 1.3 to 1.6. The soft component of
the diffuse x-ray background (0.25 keV) has been
measured by Bowyer, Field and Mack[29] and by Henry,
Fritz, Meekins, Friedman, and Byram (NRL)[30]. The observations were made with similar proportional counters,
but the flux obtained by BFM was only 50% of that
derived by HFMFB. Although this discrepancy has not
been resolved, it is possible to conclude from the
shape of the spectrum in the 2-10 keV range that the
soft x-ray flux is not simply an extension of the background observed at higher energies, but is substantially
higher.

The general picture of our current knowledge of
the soft x-ray background is summarized in figure 7.
The dashed line is a straightforward extrapolation of
the spectrum observed above 50 keV. Several observations are illustrated in the 2-10 keV range which is
characterized by a lower slope. Extrapolating the
slopes observed in the 2-10 keV range to 0.25 keV would
give the points B, C, and D in figure 7. They fall
well below the directly observed 0.25 keV fluxes of the
NRL group and BFM. The lower pair of points marked
"NRL" are uncorrected for absorption, whereas the upper

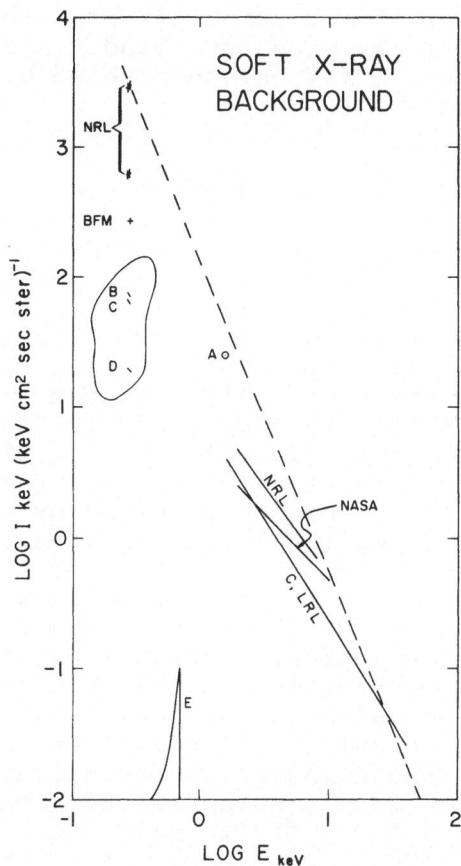

Figure 7. Soft x-ray background. Dashed line is
extrapolation of spectrum 50-150 keV. Lines marked
NRL (Ref. 30), C (Calgary, Ref. 25), LRL (Ref. 27),
and NASA (Ref. 28) extrapolate to points B, C, D at
0.25 keV. Lower points marked NRL (Ref. 30) at 0.25
keV are observed flux; upper points are corrected for
interstellar absorption. BFM is Bowyer, Field and
Mack[29]. Transmission window of Teflon at 0.7 keV is
indicated by E. Future observations are planned in
this window.

pair of points allow for interstellar absorption in the
direction of the galactic pole. The point marked "A"
was derived by the NRL group for the range 0.5 to 3 keV
from the difference in counts obtained with a pair of
counters having a substantially different sensitivity
in the pass band. No error estimate is indicated for

point "A" because of the uncertainty of the spectrum within the relatively wide pass band. A similar rise in background near 1 keV has been noted by Green, Wilson and Baxter[25].

OPEN OR CLOSED UNIVERSE

An expanding universe may expand forever, without limit, or the expansion may eventually come to a halt and the universe then contract again to the original primordial fireball condition. In a limitless or "open" universe, the expansion continues without deceleration; stars eventually burn out and the universe becomes totally black, but the dead universe grows toward infinity. To "close" the universe, i.e., to bring the expansion to a halt, requires the gravitational force that would be provided by a certain critical amount of matter in the universe. This critical mass, if spread uniformly over the entire universe, would amount to an average density of 2×10^{-29} g/cm^3, or 10^{-5} hydrogen atoms/cm^3. The masses of nearby galaxies, including our own, have been determined from observations of their internal rates of rotation and from estimates of their stellar populations. These calculations indicate an average mass density for all the galaxies in the universe of approximately 7×10^{-31} g/cm^3, which is only about 1/30th of the mass required for a finite universe.

The most likely source of the "missing matter" in the universe is intergalactic gas. The gravitational effect of such gas on clusters of galaxies can be inferred from measurements of internal motions of the galaxies within a cluster. When one calculates the mass holding a cluster together gravitationally, it is found to be much greater than the sum of the masses of the individual galaxies, and the discrepancy for large clusters usually exceeds a factor of 30. If this missing mass is present as gas that pervades intergalactic and intra-cluster space with the same average density, it would require 2×10^{-29} g/cm^3 spread throughout the universe. This very thin intergalactic gas would then constitute the glue that holds the universe together and causes it ultimately to collapse back to its original state.

From their observations of the soft x-ray background, Henry et al.[30] conclude that the point at

0.25 keV represents a different emission mechanism from
the high energy spectrum. The observations at 0.25
keV could be attributed to bremsstrahlung from a hot
intergalactic gas at a density of $10^{-6}/cm^3$ and a
temperature of 3×10^5 K to 10^6 K. If the point marked
"A" at 1 keV is valid, figure 7, the associated temperature must be closer to 10^7K. This higher temperature
makes a severe heating problem for any evolutionary
cosmological model. It should also be recognized that
the field of view in the background measurements was
large and that the measurements may contain a strong
contribution from galactic clusters. More spatial
definition will be required in future measurements to
isolate the flux originating in intergalactic gas from
that produced in galactic clusters.

Whereas the spectra of some discrete sources in the
galactic plane appear to show evidence of interstellar
absorption below 3 keV, the diffuse flux shows no such
effect. If the intergalactic gas has cosmic abundance,
the heavy elements should absorb about 50% as much as
the hydrogen and helium at low temperatures. At
temperatures high enough to ionize hydrogen and helium,
oxygen and heavier elements would still be only
partially ionized and the K-shell absorption barely
diminished. Therefore, the observation of 0.25 keV
radiation sets an upper limit of about 2% of the relative cosmic abundance of the heavier elements in intergalactic gas. Photons of energy 1 keV could be
received from a volume confined to a radius of about
0.01 Hubble distance if the intergalactic density is
$10^{-5}/cm^3$.[31]

REFERENCES

1. H. Friedman, E. T. Byram, and T. A. Chubb,
    "Distribution and Variability of Cosmic X-ray
    Sources," Science 156, 374 (1967).

2. T. Gold, "Rotating Neutron Stars and the Nature of
    Pulsars," Nature 221, 25 (1969).

3. A. Hewish, S. J. Bell, J. D. H. Pilkington,
    P. F. Scott, and R. A. Collins, "Observation of a
    Rapidly Pulsating Radio Source," Nature 217, 709
    (1968).

4.   W. J. Cocke, M. J. Disney, and D. J. Taylor,
     "Discovery of Optical Signals From Pulsar NP 0532,"
     Nature 221, 525 (1969).

5.   J. S. Miller and E. J. Wampler, "Television Detec-
     tion of the Crab Nebula Pulsar," Nature 221, 1037
     (1969).

6.   G. Fritz, R. C. Henry, J. F. Meekins, T. A. Chubb,
     and H. Friedman, "X-ray Pulsar in the Crab Nebula,"
     Science 164, 709 (1969).

7.   H. Bradt, S. Rappaport, W. Mayer, R. E. Nather,
     B. Warner, M. MacFarlane, and J. Kristian, "X-ray
     and Optical Observations of the Pulsar NP 0532
     in the Crab Nebula," Nature 222, 728 (1969).

8.   G. J. Fishman, F. R. Harnden, Jr., and R. C. Haymes,
     "Observations of Pulsed Hard X-Radiation from
     NP 0532 From 1967 Data," Astrophys. J. 156, L107
     (1969).

9.   M. J. Rees, "Proton Synchrotron Emission From
     Compact Radio Sources," Astrophys. Lett. 2, 1
     (1968).

10. J. B. Oke, "Photoelectric Spectrophotometry of the
     Crab Pulsating Radio Source NP 0532," Astrophys. J.
     Lett. 156, L49 (1969).

11. J. F. Meekins, R.C. Henry, G. Fritz, H. Friedman,
     and E. T. Byram, "X-ray Spectra of Several Discrete
     Cosmic Sources," Astrophys. J. 157, 197 (1969).

12. O. Manley and S. Olbert, "Models of X-ray Stars,"
     Astrophys. J. 157, 223 (1969).

13. G. Fritz, J. F. Meekins, R. C. Henry, E. T. Byram,
     and H. Friedman, "Soft X-rays from Sco XR-1,"
     Astrophys. J. 153, L199 (1968).

14. R.J. Grader, R. Hill, F. Seward, A. Toor, "X-ray
     Spectra from Three Cosmic Sources," Science 152,
     1499 (1966).

15. R. W. Hill, R. J. Grader, and F. D. Seward, "The
     Soft X-ray Spectrum of Sco XR-1," Astrophys. J.
     154, 655 (1968).

16. S. Rapparport, H. V. Bradt, S. Naranan and G. Spada, "Low-Energy X-ray Spectra of Sco X-1 and Four Sagittarius Sources," Nature 221, 428 (1969).

17. H. Friedman, E. T. Byram, T. A. Chubb, "The Spectrum and Distance of Sco XR-1," Science 153, 1527 (1966).

18. J. R. P. Angel, R. Novick, P. Vanden Bout, and R. Wolff, "Search for X-ray Polarization in Sco X-1," Phys. Rev. Lett. 22, 861 (1969).

19. G. Fritz, J. F. Meekins, R. C. Henry, and H. Friedman, "On X-ray Line Emission from Sco XR-1," Astrophys. J. 156, L33 (1969).

20. W. H. Tucker, "Cosmic X-ray Sources," (Thesis), Univ. Calif., S. Diego, Univ. Microfilm 66-8431 (1966).

21. P. Gorenstein, H. Gursky, and G. Garmire, "The Analysis of X-ray Spectra," Astrophys. J. 153, 885 (1968).

22. S. S. Holt, E. A. Boldt, and P. J. Serlemitsos, "Iron Line Emission from X-ray Sources," Astrophys. J. 154, L137 (1968).

23. S. Hayakawa, M. Matsuoka, and D. Sugimoto, "Galactic X-rays," Space Sci. Rev. 5, 109 (1966).

24. S. L. Mandelshtam and I. P. Tindo, "Measurement of Diffuse X-ray Background of Outer Space in the Energy Region 1-1.5 keV," J. Exptl. Theoret. Phys. 6, 796 (1967).

25. D. W. Green, B. G. Wilson, and A. J. Baxter, "A Spectral Measurement of the Cosmic X-ray Background Down to 2 keV," in Space Research IX, ed. K. S. W. Champion, P. A. Smith and R. L. Smith-Rose, North-Holland Publishing Co., Amsterdam, 1969, p. 222-225.

26. M. Matsuoka, M. Oda, Y. Ogawara, S. Hayakawa, and T. Kato, "Cosmic X-rays," Proceedings 10th Intl. Conference on Cosmic Rays, Calgary, Canada, Can. J. Phys. (In Press).

27. F. Seward, G. Chodil, H. Mark, C. Swift, and
     A. Toor, "Diffuse Cosmic X-ray Background Between
     4 and 40 KeV," Astrophys. J. 150, 845 (1967).

28. E. A. Boldt, U. D. Desai, and S. S. Holt, "2-20
     kev Spectrum of X-rays from the Crab Nebula and
     the Diffuse Background Near the Galactic Anti-
     center," NASA-GSFC Preprint X-611-68-353.

29. C. S. Bowyer, G. B. Field, and J. E. Mack,
     "Detection of an Anisotropic Soft X-ray Background
     Flux," Nature 217, 32 (1968).

30. R. C. Henry, G. Fritz, J. F. Meekins, H. Friedman,
     and E. T. Byram, "Possible Detection of a Dense
     Intergalactic Plasma," Astrophys. J. 153, L11
     (1968).

31. B. G. Wilson, "Cosmic X-ray Background Radiation,"
     Institute of Space and Aeronautical Science,
     Univ. Tokyo Report No. 428, Vol. 33, No. 10.

SOFT X-RAY INSTRUMENTATION FOR SPACE EXPERIMENTS

W. P. Reidy

American Science and Engineering, Inc.*

Cambridge, Massachusetts 02142

This paper will discuss some of the techniques used for soft x-ray astronomy (10 kev - 0.5 kev). A more extensive treatment will be found in review papers by Giacconi, Reidy, Vaiana, Van Speybroeck and Zehnpfennig[1] and by Giacconi, Gursky and Van Speybroeck[2]. The earliest experiments in x-ray astronomy were performed using mechanically collimated small area ($\sim$10 cm^2) geiger tubes flown in open stabilized rockets[3]. Subsequent improvements in sensitivity have resulted from the use of much larger ($\sim$1000 cm^2) counters and improved background rejection techniques[4]. Important spectral information has been obtained by using gas counters which are operated as proportional counters rather than as geiger tubes and the development of the modulation collimator[5] has significantly improved the obtainable angular resolution. The most promising new technique for X-ray astronomy is the development of grazing incidence x-ray optics[1].

GAS COUNTERS

Gas counters[6-8] are particularly suited to use in x-ray astronomy because of their spectral response,

*Presently with Visidyne, Inc., Woburn, Mass. 01801

efficiency, and the background rejection techniques that
can be used with them. Figure 1 shows the calculated
efficiencies of typical counters used in x-ray astronomy.
The construction of the gas tube (figure 2) is such that
the highest electric fields are at the anode. When ionization occurs in the sensitive volumn, electrons drift
towards the anode colliding with gas molecules in the
tube. As the operating voltage is increased, the energy
gained between collisions by the electrons in the vicinity
of the anode will be sufficient to ionize a molecule of
the counter filling gas. At higher voltages, a cascade
effect will occur (electrons formed by collisional ionization will themselves be accelerated, collide, and produce additional ionization) and the charge collected at
the anode will be proportional but substantially larger
than the initial ionization. Typical proportional
counters are operated at gas gains between 10^2 and 10^5.

There is a maximum gas gain at which a particular
tube can be operated in the proportional region. This
maximum is determined by the gas filling and the counter

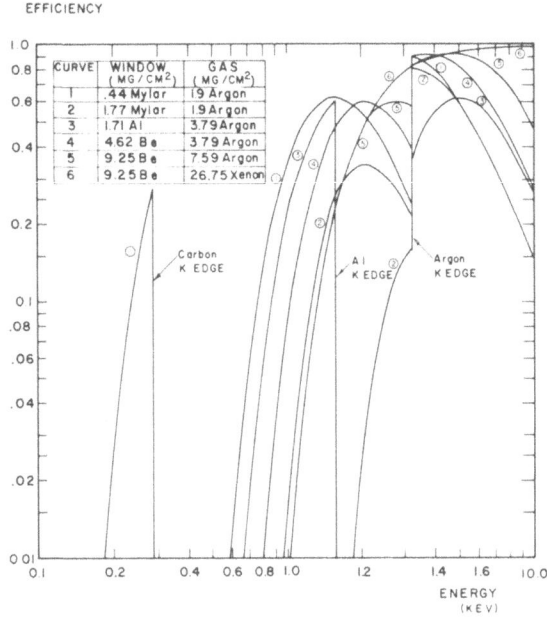

Figure 1. Calculated efficiencies of various gas
proportional counters.

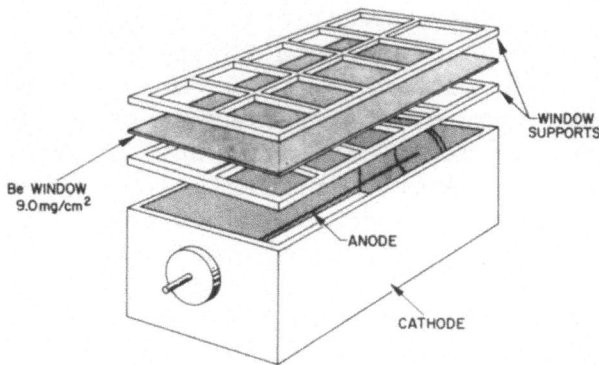

Figure 2.   Thin beryllium window proportional counter.

construction.   At higher operating voltages, secondary
effects in the initial cascade produce additional ionization in the counter which results in a saturation
signal independent of the initial amount of ionization.
The gas counter operated in this mode is a geiger tube.
Because of its large output signal, it can be operated
with relatively simple electronics.   For this reason, it
was used in some of the earliest experiments.

        However, the advantages of proportional counter
operation; energy discrimination, higher time resolution,
and improved background rejection more than compensate
for the increased electronic complexity and for these
reasons the proportional counter is the principal detector
now used in soft x-ray astronomy.

BACKGROUND REJECTION

        Gamma rays and penetrating particles constitute the
principal background in these experiments.   Most of the
energy deposition in a detector due to this background
will be by minimum ionizing particles with an energy
deposition of about 2 Mev/gm/cm^2.   In typical scintillators
(e.g., 1 mm thickness of NaI) the energy deposited will
be in excess of 1 Mev, well outside the x-ray range.
However, in a gas proportional counter (e.g., 5 cm-atm
argon) the energy deposition will be in the kev range
making it impossible to reject background by energy

discrimination.   The conventional background rejection
technique has been to surround the x-ray detector with
either gas or scintillation guard counters and to
reject signals from events traversing both counters.
This technique is extremely efficient in rejecting the
background due to penetrating particles, however, the
rejection of the gamma background is more difficult.   A
promising technique for proportional counter background
rejection is pulse shape discrimination.   This technique,
which was first introduced by the Leicester group, has
been developed by Gorenstein and Mickiewicz[4] at American
Science and Engineering.   It is based on the fact that
x-ray pulses have a fast and almost constant rise time
while background pulses have a longer varying rise time
(figure 3).   The difference in rise time has been
attributed to the difference in collection time for
electrons produced at various points along the ionization
track.   The ionization track for x-ray absorption will
be much shorter and more uniform than that produced by

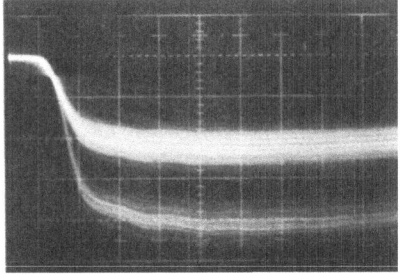

PRE-AMP OUTPUT, X-RAYS

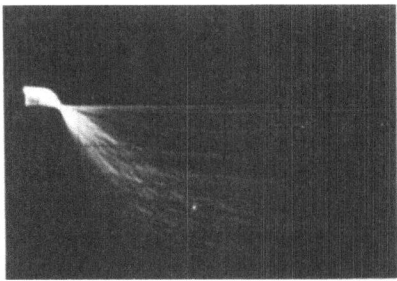

PRE-AMP OUTPUT, Co^{60}

Figure 3.   Comparison of x-ray and gamma ray pulse
shapes.

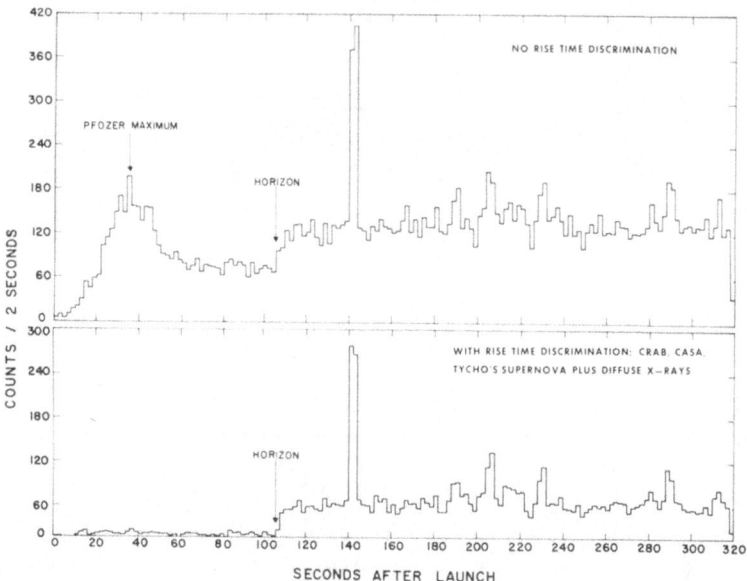

Figure 4. Comparison of rocket data with and without
pulse shape discrimination (Gorenstein and Mickiewicz[4]).

minimum ionizing particles. Figure 4 illustrates the
application of pulse shape discrimination in a rocket
experiment. The upper portion of the figure gives the
total counting rate as a function of time. In the lower
portion of the figure, background events have been rejected using pulse shape discrimination.

MODULATION COLLIMATOR

The simplest collimator consists of a set of hollow
rectangular tubes or pipes which limit the field of view
of the detector. Mechanical collimators can be built
with a field of view as small as 30 arc minutes. Higher
precision and angular resolution can be obtained using
the modulation collimator. This technique which was
devised by Oda[5] is illustrated in figure 5. Its angular
response function comprises a series of narrow transmission
bands covering a wide field of view. For small value
of ψ the individual transmission band is approximately
2d/D and the response function is approximately periodic
with period (2^{n-2}) arc tan 2d/D. Modulation collimators

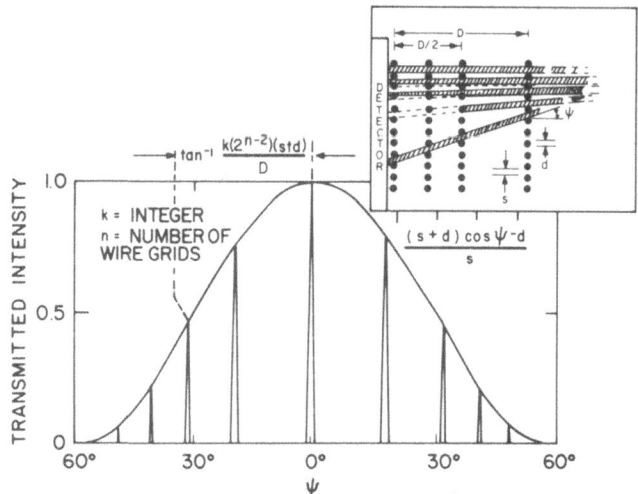

Figure 5. Response of a modulation collimator shown
schematically in the upper right. d is the wire diameter,
s is the wire to wire spacing, and n is the number of
wire planes. In the four-grid modulation collimator used
by the AS&E/MIT group to observe the Crab and Sco X-1,
s = d = 0.005 inches, D = 24 inches, which yielded a
width at the base of each of the triangular transmission
bands of 80 arc sec.

can be practically fabricated with an angular resolution
of less than 1 arc minute. Because of the periodic
response functions, the source position is not uniquely
determined by a measurement with a single collimator.
This ambiguity can be reduced by using two modulation
collimators with slightly different spacings in a manner
analogous to a vernier scale. The use of this technique
resulted in the optical identification of Sco X-1[9].

X-RAY TELESCOPES

X-ray telescopes have been used in several rocket[10-12]
and satellite[13-14] experiments to study solar x-ray emission.
The first x-ray telescope measurement of stellar x-ray
source will be a rocket experiment to be conducted in the
early part of 1970. The application of grazing incidence
x-ray telescopes to x-ray astronomy will provide several

important advantages.

1.   The inherently higher resolution obtainable with
an x-ray telescope will result in a more precise location
of sources. In addition, as the sensitivity of the experiments is increased, this higher angular resolution
will be required to study individual sources in regions
of high source density.

2.   The x-ray telescope is an imaging device and
therefore can be used to study the structure of extended
objects such as the Crab Nebula.

3.   The telescope offers a significant improvement
in signal to noise for a given effective aperture. The
focussing property allows one to use much smaller detectors resulting in a corresponding reduction in the
background counting rate.

The grazing incidence x-ray telescope is a reflection
system. The index of refraction for all materials at
x-ray wavelengths is slightly less than unity and therefore according to Fresnel's equations[15] at sufficiently
shallow angles of incidence the x-rays will be reflected
with an efficiency approaching unity. The calculated
reflection efficiency for several representative materials
is shown in figure 6 as a function of angle wavelengths
for several angles of incidence. The general behavior
is that there is a fairly sharp short wavelength cut off
(λ_c) which is determined by the angle of incidence (θ).
This relation is approximately given by

$$\lambda_c = \theta \left[\frac{\pi}{r_o N} \right]^{\frac{1}{2}} \tag{1}$$

where

r_o is the classical electron radius
N   is the electron density
θ is in radians

W. P. Reidy

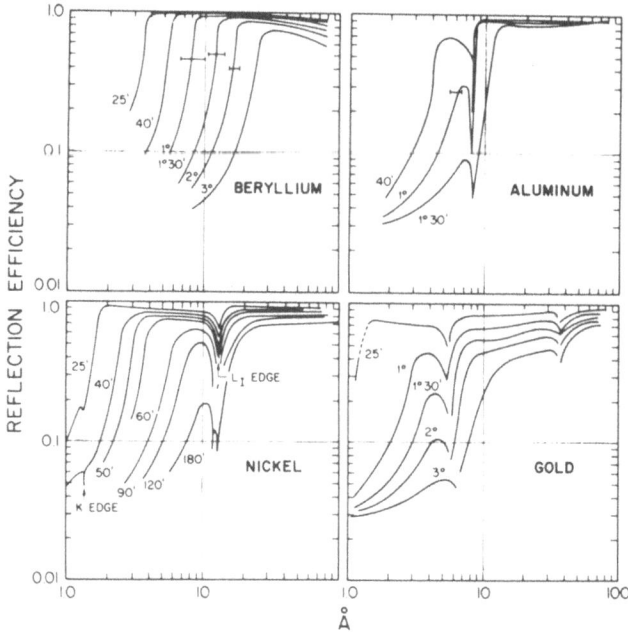

Figure 6.   Theoretical x-ray reflection efficiency.

From figure 6 it is apparent that x-rays cannot be
efficiently reflected at grazing angles of incidence
greater than a few degrees.   Therefore, one cannot use
conventional mirror designs which operate at near normal
x-ray incidence for x-ray imaging.   Several grazing
incidence systems have been proposed for use in x-ray
microscopy.   Kirkpatrick and Baez[16] proposed in 1948 a
system using successive reflection from two perpendicular
mirrors and this system has been successfully demonstrated
by McGee[17].   Wolter[18-19] in 1954 proposed the three
systems of conic sections shown in figure 7.   While it
has not been feasible to fabricate Wolter's systems in
the small sizes proposed for microscopy, the design is
well suited to use as an x-ray telescope.

The x-ray telescopes which have been developed at
our laboratory have all been of the internally reflecting
paraboloid-hyperboloid configuration (figure 7a).   This
design minimizes the difficulties of maintaining alignment between the two reflecting surfaces and reflection
efficiency and short wavelength response for a system

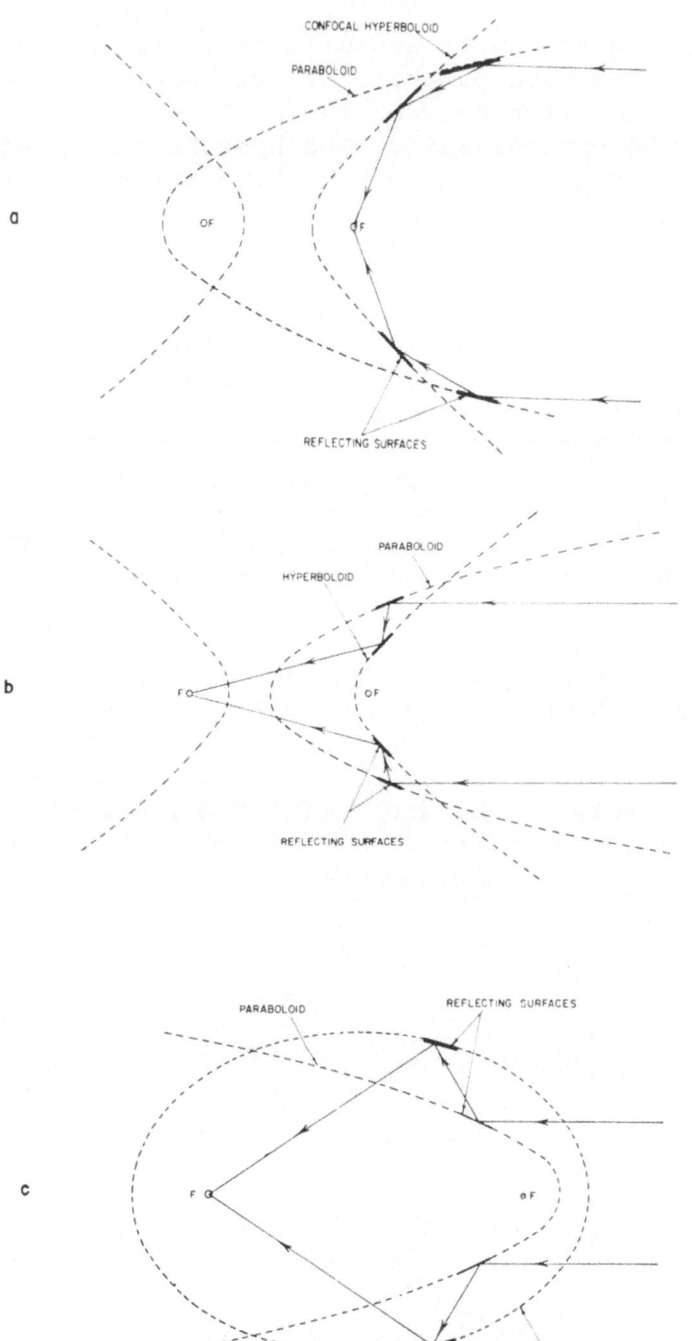

Figure 7. Three configurations for image forming x-ray telescopes (Wolter[18-19]).

with a given focal length and diameter. Typically,
these telescopes have paraboloidal reflecting surfaces
with an average pitch between 40 and 80 minutes with
respect to the optical axis. The hyperboloidal reflecting surface has an average pitch three times that of the
paraboloidal. With this design, paraxial rays strike
both surfaces at the same angle.

   The imaging properties of this type of mirror system
are determined by two parameters; the length of each reflecting surface (paraboloid length equals hyperboloid
length) and the ratio of mirror diameter to focal length.
Figure 8 shows the blur circle diameter calculated from
ray tracing bundles of 60 parallel rays uniformly spaced
over the telescope aperture. In this typical case, the
focal length was ten times the mirror diameter. The
length of the paraboloid was equal to the mirror diameter.
Hence, the intensity distribution sharply peaks at the
center of the blur circle. The blur circle diameter
represents an upper limit on the mirror resolution.
Curve A shows the blur circle diameter obtained with a
flat focal plane at the mirror focus. Curve B shows
that one can improve off axis resolution at the expense
of on axis resolution by displacing the image plane
slightly toward the mirror. This shift corresponds to

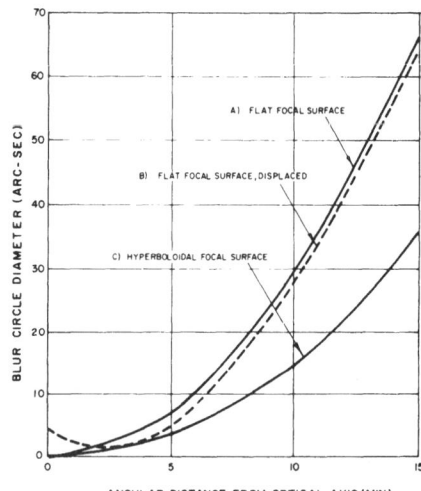

Figure 8. Blur circle diameter as a function of angular
distance from the optical axis.

2×10^{-4} of the focal length. Since the optimum image
distance is a function of angular displacement from the
optical axis, the optimum image surface is curved not
flat. This analysis showed that the optimum image surface could be closely approximated by an hyperboloid.
This gives an improvement of blur circle diameter of
approximately two as shown in Curve C.

The field of view of an x-ray telescope is limited
by both the degradation in angular resolution shown in
figure 8 and by interference from rays reflected from the
paraboloid. This is illustrated in figure 9, which is a
photograph of a pattern of x-ray sources. The individual
dots are separated by 3 arc minutes center to center.
The variation in intensity is due both to vignetting and
to non-uniformities in the source emission. The usable
field of view is about 50 arc minutes. The single reflection rays produce the pattern of arcs which are
evident at the edge of the photograph.

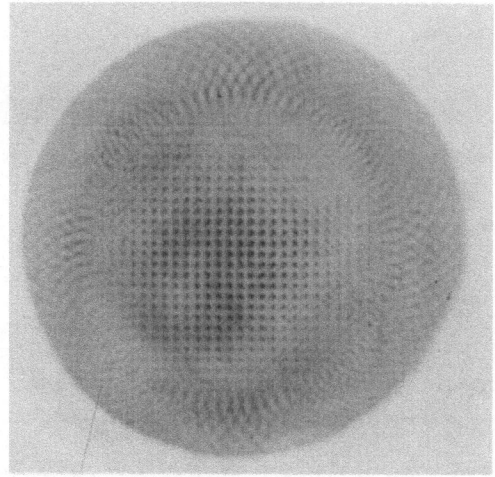

Figure 9. An x-ray exposure illustrating the full extent
of the usable field of view. The real image is the square
array of dots in the central area. The pattern of arcs at
the edge is not part of the real image and is due to rays
that were reflected from the hyperboloid but not the
paraboloid. Since the dots are spaced 3 arc min center-to
-center, the full diameter of the field of view is about
50 min.

Figures 10, 11, and 12 illustrate the evolution in
x-ray telescope fabrication. The telescope in figure 10
is the type used to obtain the first solar x-ray photograph in 1965 and was also used in the OSO-4[13-14]
satellite x-ray telescope experiment. It was fabricated
by electroforming nickel on an optically polished stainless steel mandrel. This method potentially offered
several advantages. The optical finish was performed
on the outside rather than the inside of the conic section
which simplified the polishing technique and the electroformed process offered the possibility of producing a
large number of replicas from a single polished master.
However, since the telescope walls were necessarily thin,
it is difficult to maintain the mechanical tolerances
required for high resolution imaging. Thick walled
telescopes have been fabricated to avoid this problem.
The telescope shown in figure 11 is beryllium coated
with kanigen and has been used in a series of rocket
experiments to study the sun[10]. The beryllium base was
first ground to within a fraction of a mil of the required

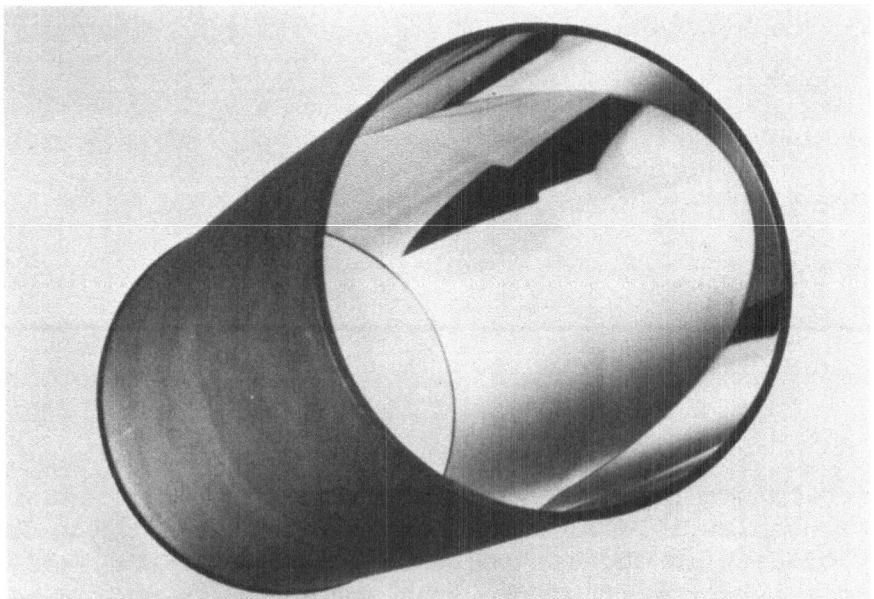

Figure 10. Nickel telescope which was electroformed on
an optically polished stainless steel mandrel and then
removed. The collecting area is 2 cm^2; the diameter is
7.5 cm; and the focal length is 84 cm.

Figure 11. The 23-cm diameter Kanigen telescope which produced the x-ray picture of the sun shown in figure 13. The collecting area is 34 cm^2 and the focal length is 132 cm. The x-ray resolution is illustrated in figure 13.

Figure 12. Recently completed telescope system for ATM. The combined collecting area of the nested confocal 23 and 31 cm diameter mirrors is 44 cm^2.The focal length is 213 cm. The distorted reflection of a small telescope, which will form a separate image on the face of the image dissector, can also be seen.

surface. Five to eight mils of kanigen, a nickel alloy,
was then deposited on the beryllium base and the kanigen
was figured and polished (polishing and figuring was performed by Diffraction Limited, Inc., Bedford, Mass.)
using fairly conventional optical techniques. A more
advanced mirror made in a similar manner is shown in
figure 12. In this model, two mirrors of identical focal
length have been nested to increase the available collecting area inside a given aperture envelope. The rays re-

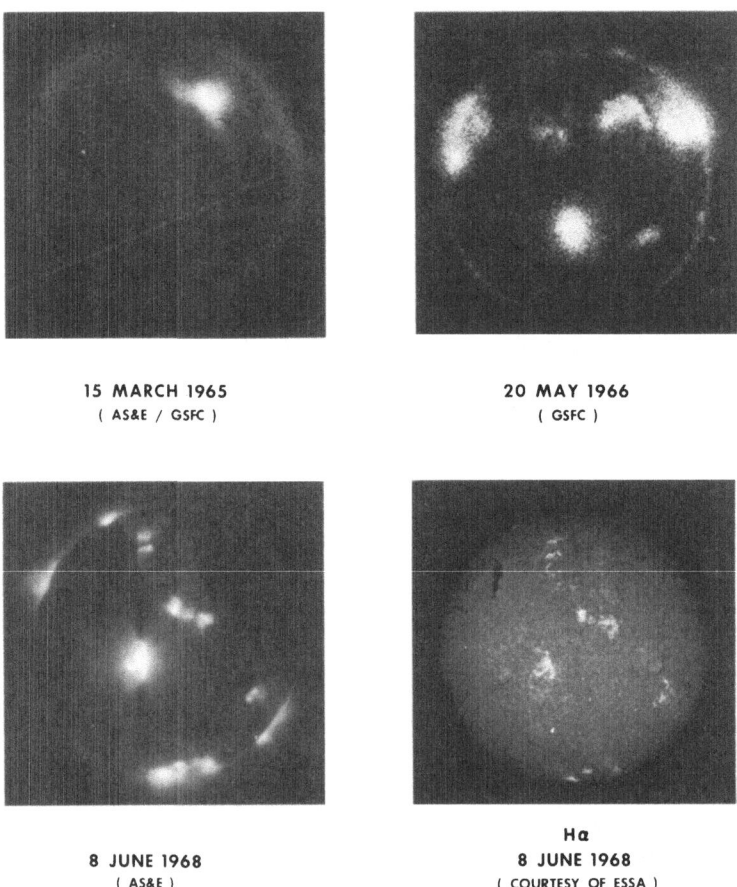

15 MARCH 1965
(AS&E / GSFC)

20 MAY 1966
(GSFC)

8 JUNE 1968
(AS&E)

Hα
8 JUNE 1968
(COURTESY OF ESSA)

Figure 13. Three x-ray photographs of the sun showing
the evolution in x-ray telescope performance over the
past few years. Also shown for comparison is an Hα
picture taken at the time of the 8 June 1968 flight.

flected by the two mirrors are not coherent and the images add in intensity not in amplitude. This mirror will be used on the Apollo Telescope Mount (ATM) solar observatory which will be in orbit in the period 1972 to 1973.

Figure 13 shows the evolution in x-ray telescope performance. The 1965 photographs were obtained with an electroformed telescope of the type which is shown in figure 10. This experiment was performed jointly by the Solar Physics Branch of Goddard Space Flight Center (GSFC) and the group at American Science and Engineering (AS&E). The resolution was about 1 arc minute. The Goddard 1966 photographs were obtained with a polished stainless steel mirror and the resolution was about 20 arc seconds. The latest photograph shown (1968) was taken by American Science and Engineering with the telescope shown in figure 12. In this experiment an image resolution of 3 to 4 arc seconds was achieved in the center of the field of view.

The x-ray telescope has now been developed to the point where it is a major instrument for x-ray astronomy. This development has taken place in the last eight years and continuing efforts should lead to qualitative improvements in performance in the next decade.

REFERENCES

1. R. Giacconi, W. P. Reidy, G. S. Vaiana, L. P. Van Speybroeck and T. F. Zehnpfennig, "Grazing Incidence Telescopes for X-Ray Astronomy", Space Sci. Rev. 9:3-57, (1969).

2. R. Giacconi, H. Gursky and L. P. Van Speybroeck, "Observational Techniques in X-Ray Astronomy", Ann. Rev. Astron. Ap. 6:373-416, (1968).

3. R. Giacconi, H. Gursky, F. R. Paolini and B. B. Rossi, "Evidence for X-Ray Sources Outside the Solar System", Phys. Rev. Letters 9:439-442, (1962).

4.   P. Gorenstein and S. Mickiewicz, "Reduction of
      Cosmic Background in an X-Ray Proportional Counter
      through Rise Time Discrimination", Rev. Sci. Inst.
      39:816-820, (1968).

5.   Minora Oda, "High Resolution X-Ray Collimator with
      Broad Field of View for Astronomical Use", Appl.
      Opt. 4:143, (1965).

6.   B. B. Rossi and H. H. Staub, Ionization Chambers
      and Counters, McGraw-Hill, New York, 1949.

7.   D. H. Wilkinson, Ionization Chambers and Counters,
      Cambridge University Press, 1950.

8.   S. A. Korff, Electron and Nuclear Counters, D. Van
      Nostrand Co., Inc., New York, 1955.

9.   H. Gursky, R. Giacconi, P. Gorenstein, J. R. Waters,
      M. Oda, H. Bradt, G. Garmire, and B. V. Sreekantan,
      "A Measurement of the Angular Size of the X-Ray
      Source Sco X-1", Astrophys. J. 144:1249-1252, (1966).

10.   G. S. Vaiana, W. P. Reidy, T. Zehnpfennig, L. P. Van
      Speybroeck and R. Giacconi, "X-Ray Structures of the
      Sun during the Importance 1N Flare of 8 June 1968",
      Science 161:564-567, (1968).

11.   W. P. Reidy, G. S. Vaiana, T. F. Zehnpfennig and
      R. Giacconi, "Study of X-ray Images of the Sun at
      Solar Minimum", Astrophys. J. 151:333-349 (1968).

12.. J. H. Underwood and W. S. Muney, "A Glancing Inci-
      dence Solar Telescope for the Soft X-Ray Region",
      Solar Phys. 1:129-144, (1967).

13.   F. R. Paolini, G. S. Vaiana, R. Giacconi, W. P. Reidy
      and T. F. Zehnpfennig, "Spectroheliograms of X-Ray
      Flares from OSO-IV", Midwest Cosmic Ray Conference,
      Iowa City, 1968.

14. G. S. Vaiana, F. R. Paolini, R. Giacconi, W. P.
     Reidy and T. F. Zehnpfennig, "X-Ray Plage Studies
     from OSO-IV", Midwest Cosmic Ray Conference, Iowa
     City, (1968).

15. A. H. Compton and S. K. Allison, <u>X-Rays in Theory
     and Experiment</u>, D. Van Nostrand Co., New York,
     (1963).

16. P. Kirkpatrick and A. V. Baez, "Formation of
     Optical Images by X-Rays", J. Opt. Soc. Am.
     <u>38</u>:766-774, (1948).

17. J. F. McGee, "An Introduction to Total Reflection
     X-Ray Microscopy", in V. E. Cosslett, A. Engstrom,
     and H. H. Pattee, Editor, X-Ray Microscopy and
     Microradiography, Academic Press, New York, 1957,
     p. 213-233.

18. H. Wolter, "Spiegelsysteme streifenden Einfalls
     als abbildende Optiken fur Rontgenstrahlen", Ann.
     Physik <u>10</u>:94-114, (1952).

19. H. Wolter, "Verallgemeinerte Schwarzschildsche
     Spiegelsysteme streifender Reflexion als Optiken
     fur Rontgenstrahlen", Ann. Physik <u>10</u>:286-295, (1952).

SYSTEM FOR NON-DISPERSIVE ANALYSIS OF LUNAR X-RAYS FROM APOLLO

P. Gorenstein and H. Gursky

American Science and Engineering, Inc.

Cambridge, Massachusetts

I. Adler and J. Trombka

Goddard Space Flight Center

Greenbelt, Maryland

ABSTRACT

A non-dispersive X-ray detection system consisting of proportional counters plus filters is being prepared for the Command-
Service Module of the Apollo spacecraft as part of a "geochemistry"
package. It will detect solar induced characteristic X-rays from
the abundant elements on the lunar surface during the orbiting
phases of the mission. The objective will be a compilation of a
map of the lunar chemical composition and to detect regional
differences. The system and its theoretical performance are
described.

INTRODUCTION

A collaborative effort involving several institutions has been
undertaken for the purpose of implementing a "geochemical" investigation of the lunar surface. An experimental package is being
prepared for flight aboard the Command-Service Module of the
Apollo spacecraft. Data will be obtained during the orbital phases

of the lunar missions which will be of a few days' duration. An important component of this package is a system of proportional counters plus filters that is designed to be sensitive to the presence of several abundant elements on the sun-lit portion of the Moon whose relative concentrations are an index for specifying chemical variation. Other components of the package include a gamma ray detector that will measure the variation of natural radio-activity plus the most prominent cosmic ray induced nuclear gammas and a matrix of high resolution alpha particle detectors that would detect the presence of radon evolution from the Moon.

Fluorescence following the interaction of solar X-rays is the principal mechanism responsible for the emission of characteristic lunar X-rays. It appears that the typical solar X-ray spectrum is capable of inducing substantial amounts of characteristic X-rays from all the abundant elements up to $Z = 14$. Observations of the gross rate of lunar X-ray emission from the Soviet spacecraft Luna 12 has confirmed this.[1] During brief periods of intense solar activity there will be observable radiation from elements of higher atomic number. A system is being constructed that will look for K_α X-rays of magnesium, aluminum, silicon, potassium, calcium, and iron from lunar orbit. Spatial resolution will be incorporated into the detection system to facilitate the search for regional differences in the elemental abundances. Observations from orbit cannot of course compete in sensitivity or precision with returned samples or in situ measurements but they are complementary in the sense that they have the unique capability of determining the average composition of a large region in a single mission and have the capability of exploring a large portion of the lunar surface and possibly discovering new regions with distinct chemical compositions that might otherwise remain undetected.

Below we describe the expected effect of solar X-rays upon various compositions, the experimental system that is being prepared and its theoretical performance. To an extent the performance has been checked in the laboratory but a comprehensive program of experimental simulation has not yet been completed.

SOLAR X-RAY FLUORESCENCE

Although the Sun does provide sufficient intensity for the production of characteristic lunar X-rays there are several features of the interaction that must be considered in detail if the maximum possible quantitative interpretation is to be given to the observations.

There exists several difficulties which are not normally encountered when X-ray fluorescence analysis spectroscopy is practiced in the laboratory under controlled conditions with standard reference samples. The solar intensity exhibits a great degree of variability on a time scale of minutes to hours. Furthermore, the systematic changes associated with the 11-year solar cycle is reflected in the long-term behavior trend of the average daily intensity. It has been established that the solar spectrum as observed in an instrument of low resolution such as a proportional counter is decreasing with increasing energy at such a rate that, if a thermal mechanism is assumed one finds variable temperatures in the range $\sim 10^6$ to $> 10^{7\circ}$K. It is evident that an increase in temperature, at a constant intensity, will result in a large change in the relative number of fluorescent photons from the various elements. Lines of highly ionized Mg, Si, and Fe are certainly present in the solar spectrum but their importance relative to the continuum has not yet been established [2, 3].

Scattering of the solar X-rays occurs simultaneously with the fluorescence effect. Hence, there is a background spectrum accompanying the fluorescence photons consisting of a continuum plus lines whose wavelengths are not far removed from the K_α radiation of the most abundant elements. This background is relatively insensitive to the details of the lunar composition.

In order to keep track of the variable solar X-ray intensity and spectral hardness during the mission we hope to obtain detailed simultaneous measurements of the solar X-ray spectrum from the various OSO and Solrad satellites that have been especially designed for that purpose. In addition, the detection system will include a small proportional counter that will act as a solar monitor.

The long term behavior trend of the average solar intensity (1-8Å) at the Earth (Moon) is shown in figure 1. A curve of smoothed sunspot number appears on the same axes. A definite correlation exists between the two so that the extrapolated smoothed sunspot number can be taken as an indication of the expected solar intensity for the next few years. It appears that the peak X-ray emission was attained early in 1968. The intensity will decline to about half this value by about January 1972 and be at minimum early in 1976.

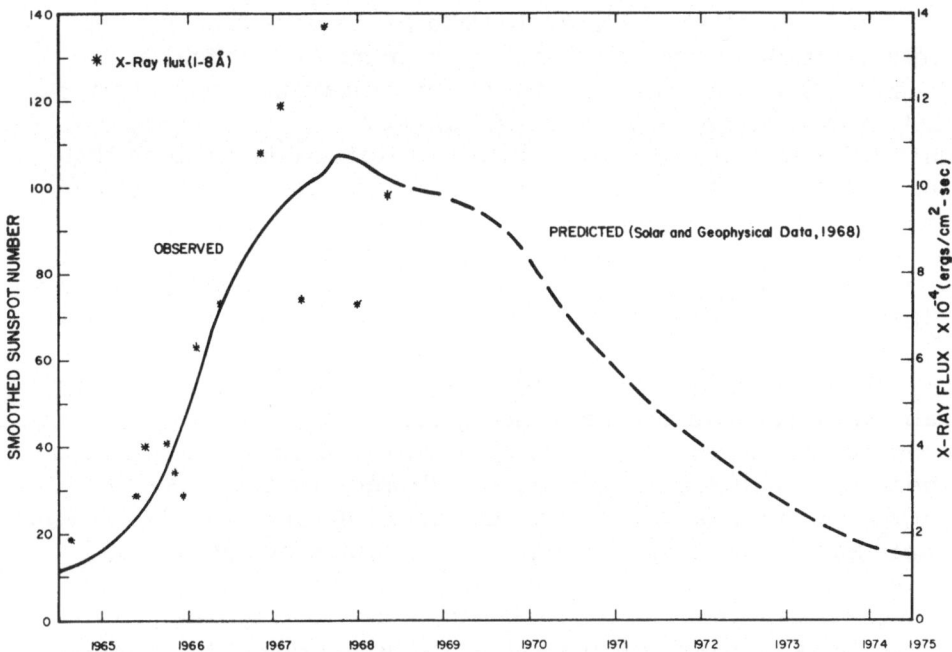

Figure 1. Smoothed sunspot number and observed X-ray intensity as
a function of time during the solar cycle. These data and the extrapolated portion of the smoothed sunspot number curve have been obtained
from various issues of Solar and Geophysical Data.

A very important factor in the determination of the fluorescence
effectiveness is the solar temperature or the spectral hardness.
Assuming a simple "free-free" solar spectrum:

$$I(\lambda) \sim \exp(-\lambda_o / \lambda)/\lambda \qquad \text{photons/cm}^2 - \text{sec-Å}$$

where

$$\lambda_o \text{ (Å)} = 12.4/kT \text{ (keV)}$$

we have estimated the effective solar temperature from the ratio
of the reported daily average intensities of the 0-8 Å to 8-20 Å
wavelength bands for the period 1966-1969[4].   A representative
mean value appears to be $T = 4 \times 10^6$ °K ($\lambda_o = 36$Å) although
there is considerable scatter and a tendency for the temperature
to follow the secular variation of the solar cycle.

There are other factors which add to the difficulty of interpreting complex fluorescent spectra in terms of chemical abundances. One is the well-known chemical matrix effect: the fact
that various elements of a mixture do not act independently of
each other with regard to the number of fluorescent photons that
are emitted from the surface. Another is the geometric relation
involving the angles between the solar radiation, the lunar surface, and the detector which also influences the relative yields.
Also the particle size and the geometrical condition of the surface
will have an effect.

We have written a computer code for a transfer operator that
calculates analytically the return spectrum from an arbitrary
chemical composition that interacts with an arbitrary model solar
spectrum. Included in the program's library are the X-ray attenuation coefficients of Henke et al [5], the fluorescent yields of Fink,
et al[6] and the coherent scattering coefficients of Compton and
Allison[7]. The calculation is not only useful for predicting the
expected yield from the lunar surface under varying solar conditions but is an essential ingredient in the analysis of the data.
With the instantaneous observed solar spectrum as an input the
chemical abundances are left as free parameters that are varied
until the best agreement with the data is achieved on the basis
of a least squares test. Uncertainties in the final results then
emerge quite naturally. As experience is gained in the laboratory,
semi-empirical quantities that improve agreement with experiment
will be incorporated into the transfer operator.

For an assumed solar free-free continuum characterized by
$T = 4 \times 10^6$ °K we have calculated the return spectrum for various
possible lunar compositions (figure 2). The integrated solar
intensity in the (1-8Å) band has been set equal to the average
reported value during April 1967 [4]. Figure 3 illustrates the sensitivity of the relative $K\alpha$ yields from one of the materials to changing
solar temperature at a constant integrated intensity. Figure 4 illustrates the theoretical response to a solar flare in which solar
lines are an important component of the spectrum.

DETECTION SYSTEM

The low value of the lunar X-ray brightness does not permit
the use of dispersive high resolution devices to isolate the characteristic lines. In order to accumulate a sufficient number of
counts, detectors with areas of at least several cm^2 and

LUNAR FLUORESCENCE X-RAYS

SOLAR TEMP = 4×10^6 (NO LINES)

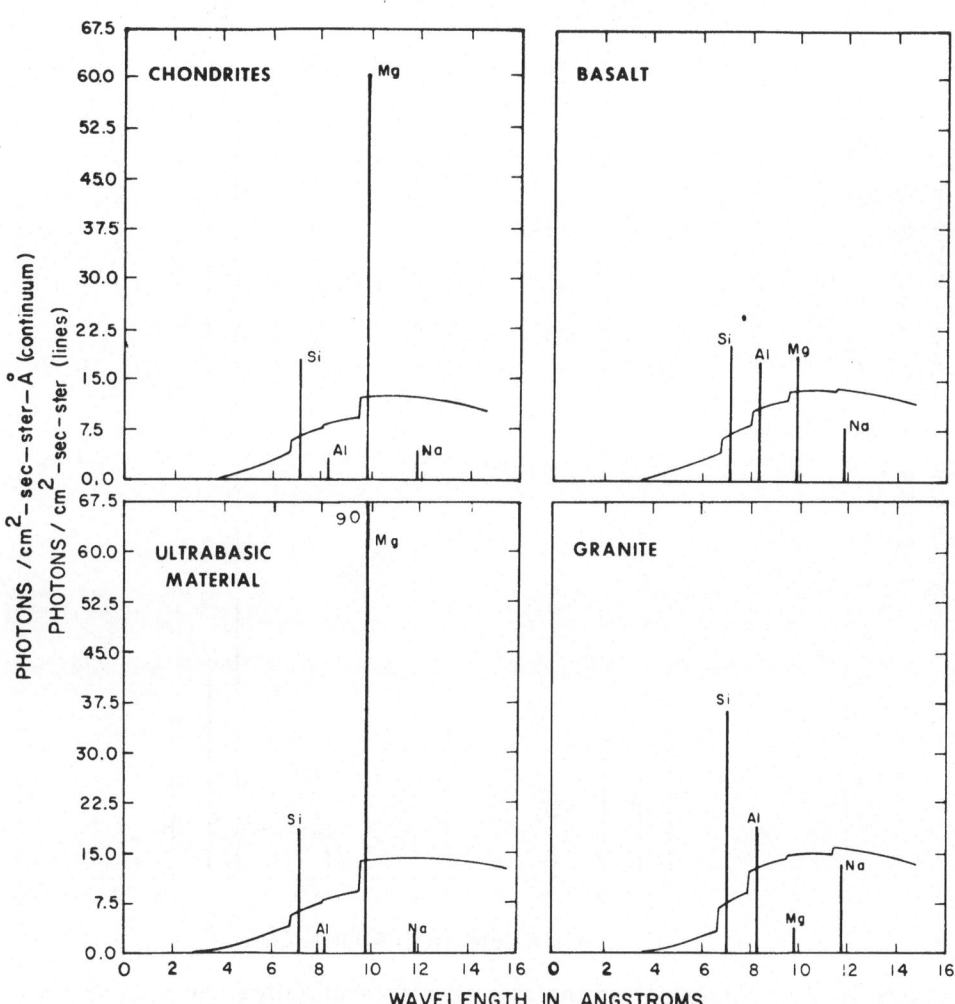

Figure 2. Calculated albedo X-ray spectra of various lunar compositions.
The solar X-ray spectrum is a "free-free" continuum with T = 4 x 10^6°K.
The solar intensity ($1 < \lambda < 8$ Å) is normalized to the average daily
intensity of April 1967. [4] Vertical lines denote K_α fluorescent radiation.
The continuum is the scattered solar spectrum.

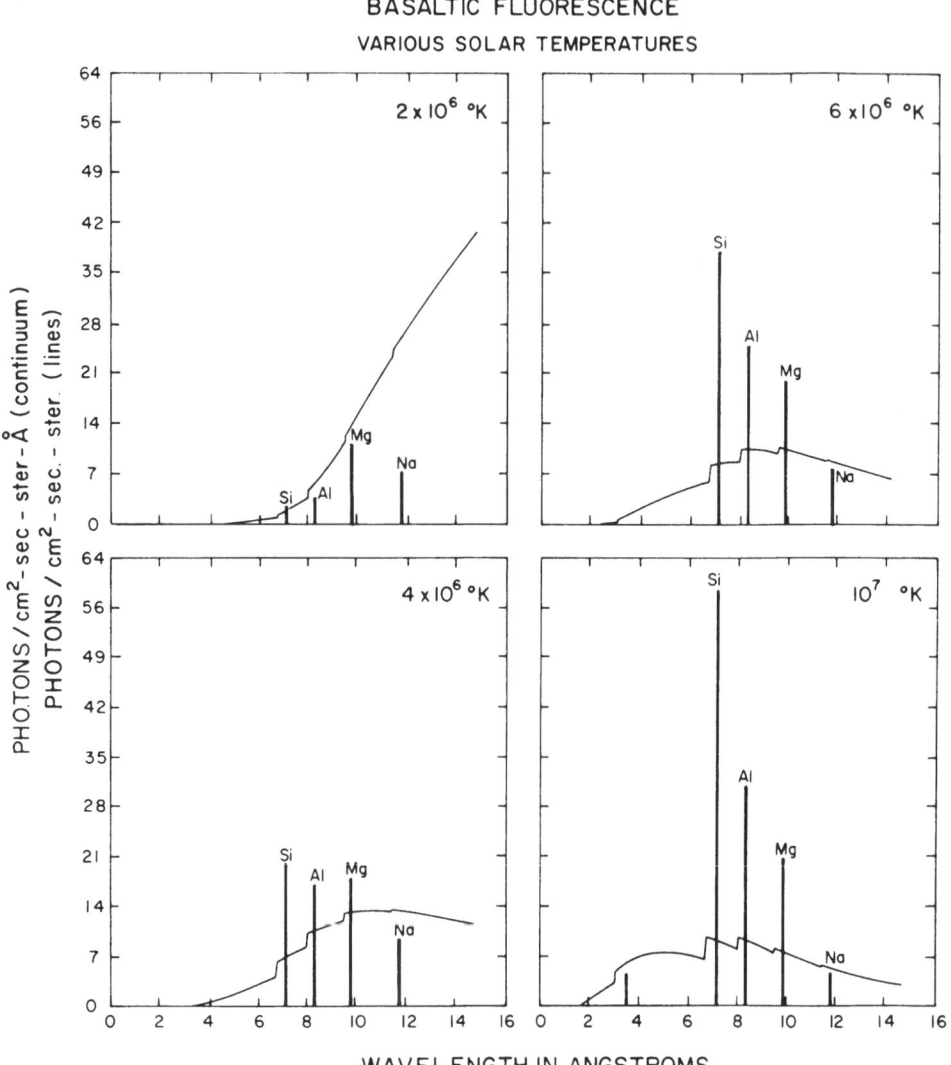

Figure 3. The effect of varying solar X-ray temperature upon the yield of albedo X-rays as calculated (See figure 2 caption). Below about 4 x 10^6°K the characteristic radiation is quite sensitive to variations in solar temperature.

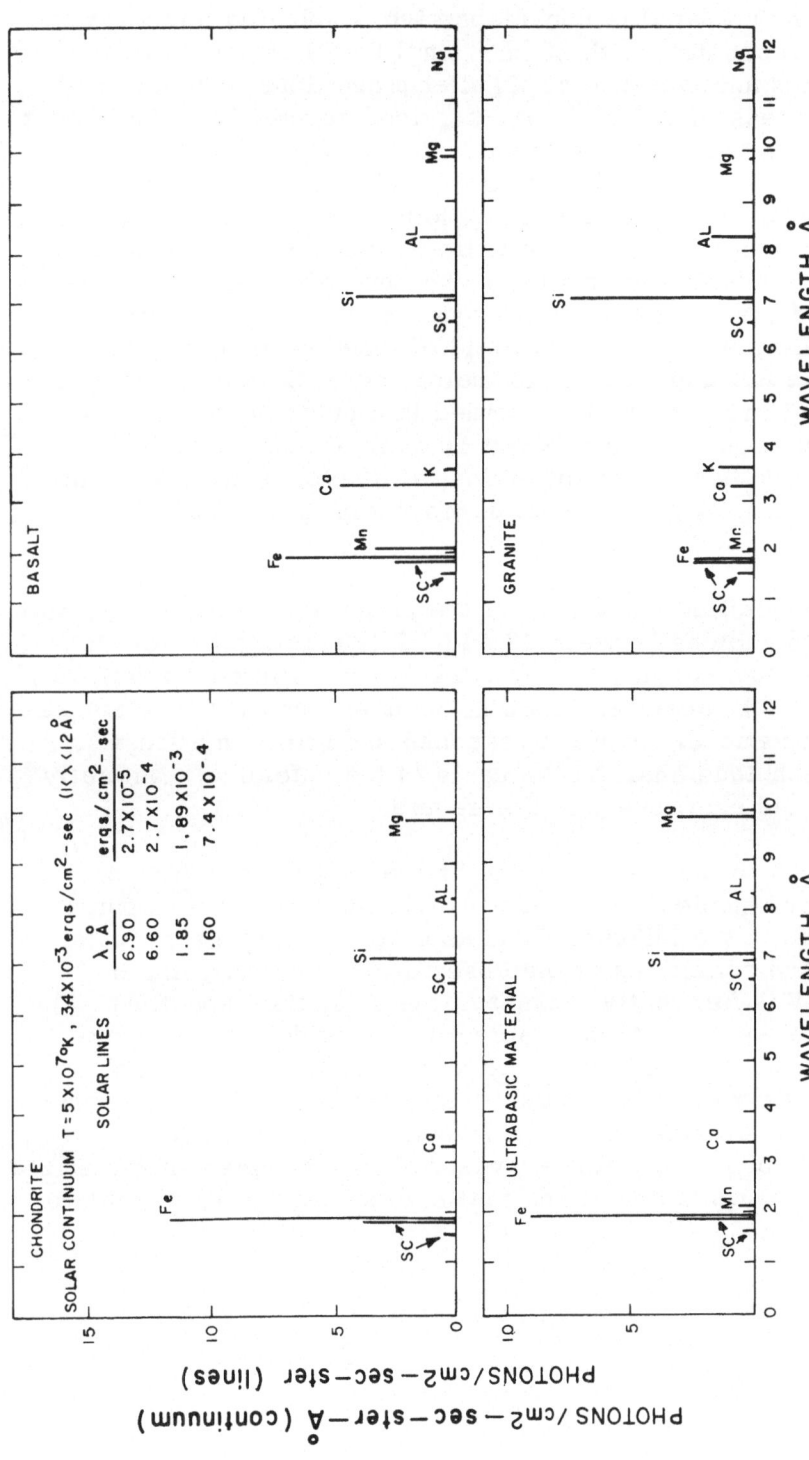

Figure 4. Calculated albedo X-ray spectra during model "solar flare". The solar spectrum includes a continuum plus lines and is normalized to that of a typical flare during April 1967. The vertical lines indicate: K_α fluorescent radiation and scattered lines of the solar spectrum (labeled "SC"). The scattered solar continuum (not shown) is slowly varying from ~ 2 photons/cm²-sec-ster-Å (10 Å) to a peak of about 30 at 1.8 Å.

acceptances of about 1 ster are required. The system (figure 5) that is being prepared is non-dispersive; employing three sealed proportional counters with 25 cm^2 1 mil beryllium windows plus P-10 fillings and two filters. Similar proportional counters with areas in excess of 100 cm^2 have survived several sounding rocket flights over a two-year period with no appreciable change in characteristics. These detectors are 20% efficient at the wavelength of magnesium $K\alpha$ radiation. Rise time discrimination techniques which make the proportional counter highly selective in distinguishing X-rays from a background of cosmic ray interaction products and radioactive gamma sources[8] will be used. A slat collimator will limit the field of view of the device to a cone whose full angle is approximately 60°. Outputs from each proportional counter will be recorded in a pulse height analyzer for accumulation periods of about 10 min. Consequently, from lunar orbit the basic element of spatial resolution will be about 70 x 600 miles. Improved spatial resolution is obtainable by shortening the accumulation time.

Rather detailed knowledge of the proportional counter spectral response to photons between 12 and 1 Å is required in order to extract the fluorescent line strengths (plus scattered X-rays) from the pulse height analyzer spectra. Another computer code simulaing the proportional counter's response and utilizing a least squares technique shall be the basis of the unfolding. Naturally the filters aid enormously in this regard.

Figure 6 shows the pulse height analyzer spectra from a leucite rock sample that is under irradiation from a Fe^{55} source as observed in the laboratory and as calculated by the computer codes for the conditions of the observation. A strong line of about 6 keV is frequently present in the solar flare spectrum. The two spectra are in quite good agreement considering the uncertainties in the composition of the rock sample giving us confidence that our estimate of the lunar rates is correct. The complete program of experimental simulation will include the examination of spectra from a variety of rock samples characterized by different surface conditions that are excited by representative solar spectra.

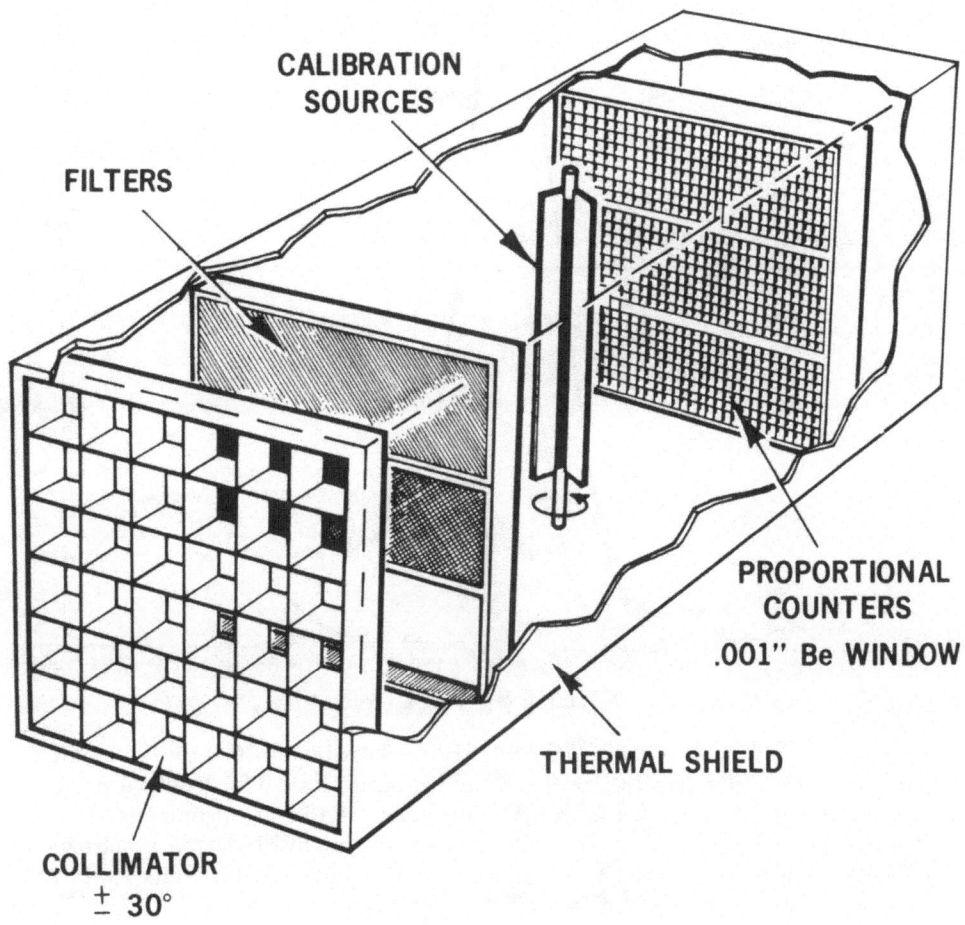

Figure 5.   Diagrammatic sketch of X-ray detection system to be placed aboard Command-Service Module of Apollo spacecraft.

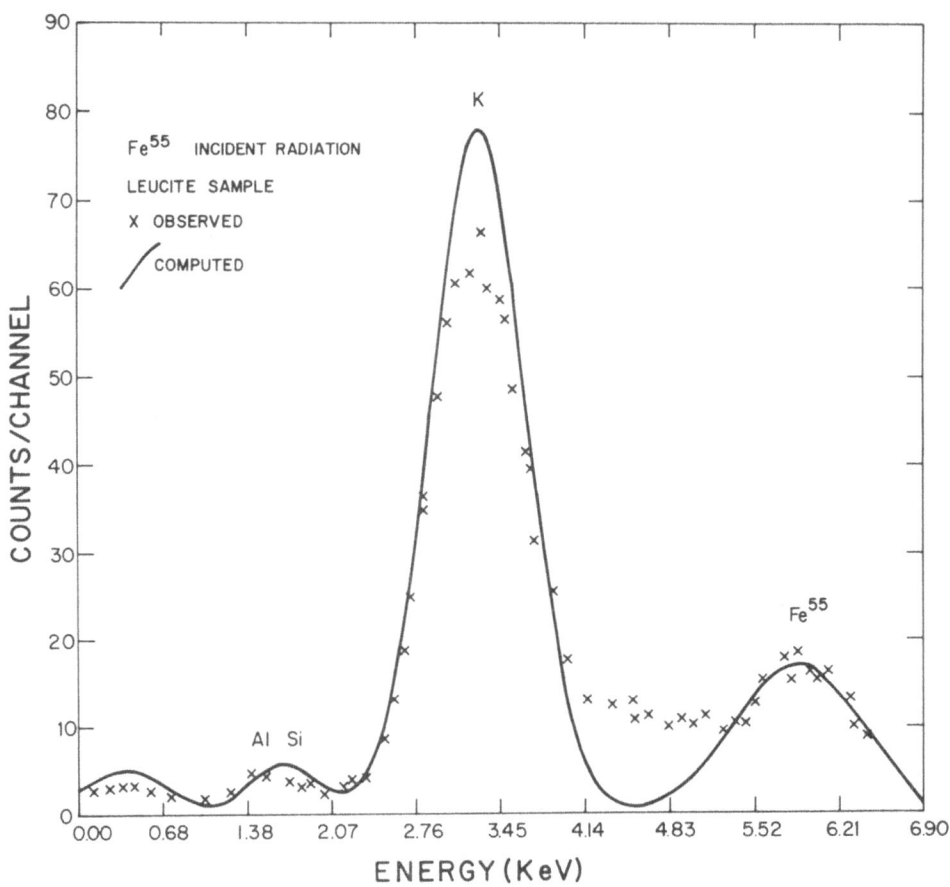

Figure 6. Calculated and observed counting rate spectrum from leucite rock sample under irradiation from Fe^{55} source as observed in a proportional counter and pulse height analyzer. Lack of agreement between the two spectra is consistent with uncertainties in chemical composition of rock sample and strength of source. (Silicon and aluminum lines are suffering considerable air attenuation in this observation.)

ACKNOWLEDGEMENT

We would like to thank Bernard Harris of AS&E for carrying out the calculations of the lunar x-ray spectra.

This work has been supported in part by the National Aeronautics and Space Administration.

REFERENCES

1. S.L. Mandel'shtam, I.P. Tindo, G.S. Cheremukhin,
    L.S. Sorokin and A.B. Dmitriev, "Lunar X-rays and the
    Cosmic X-ray Background Measured by the Lunar Satellite,
    Luna 12", UDC 523:36:629.192.32 (Trans. from Komicheskie
    Issledovaniya 6: 119 (1968)).

2. J.F. Meekins, R.W. Kreplin, T.A. Chubb and H. Friedman,
    "X-ray line and Continuum Spectra of Solar Flares from 0.5
    to 8.5 Angstroms", Science 162: 891 (1968).

3. W.M. Neupert, W. Gates, M. Swartz and R. Young,
    "Observation of the Solar Flare X-ray Emission Line
    Spectrum of Iron from 1.3 to 20Å", Ap. J. Letters 149: L79
    (1967).

4. Solar - Geophysical Data, Solar Radiation Monitoring
    Satellite X-ray. U.S. Department of Commerce, E.S.S.A.
    Environmental Data Service.

5. B.L. Henke, R.L. Elgin, R.E. Lent, and R.B. Ledingham,
    "X-ray Absorption in the 2-200Å Region", Norelco Reporter
    XIV: 112 (1967) Philips Electronics Inst., Mt. Vernon, N.Y.

6. R.W. Fink, R.C. Jopson, Hans Mark, C.D. Swift, "Atomic
    Fluorescence Yields", R.M.P. 38: 513 (1966).

7. A.H. Compton and S.K. Allison, "Atomic Structure or Form
    Factors", X-Rays in Theory and Experiment, App. IV,
    D. Van Nostrand Co., Princeton, Toronto, London and
    New York (1935).

8. P. Gorenstein and S. Mickiewicz, "Reduction of Cosmic
    Background in an X-ray Proportional Counter through Rise
    Time Discrimination", Rev. Sci. I. 39: 816 (1968).

DEVELOPMENT OF A SLITLESS SPECTROGRAPH FOR X-RAY ASTRONOMY

T. Zehnpfennig

American Science and Engineering

Cambridge, Massachusetts 02142

ABSTRACT

The soft x-ray slitless spectrograph is described, and its use-fulness in making rapid surveys of wide spectral regions is discussed. Fabrication of the 1μ thick gold-shadowed transmission gratings used in the instrument is described. Spectra taken in the laboratory and during a solar sounding rocket flight are presented. Because a simple absorbing strip model is inadequate to explain the distribution of intensities among the spectral orders, a phase-grating model must be adopted.

INTRODUCTION

This paper describes the design and capabilities of a slitless or objective grating spectrograph which utilizes, as its dispersive element, an x-ray transmission grating. Development of this device has proceeded in our laboratory over the past few years[1,2] ; and we have twice flown a model of the slitless spectrograph in a sounding rocket payload[3], obtaining soft x-ray spectra of solar active regions. The development effort to date has centered on applications in x-ray astronomy, although there are possible applications in other fields.

The soft x-ray slitless spectrograph, shown schematically in figure 1, consists of the combination of an x-ray transmission grating, an image-forming x-ray telescope[4], and an appropriate

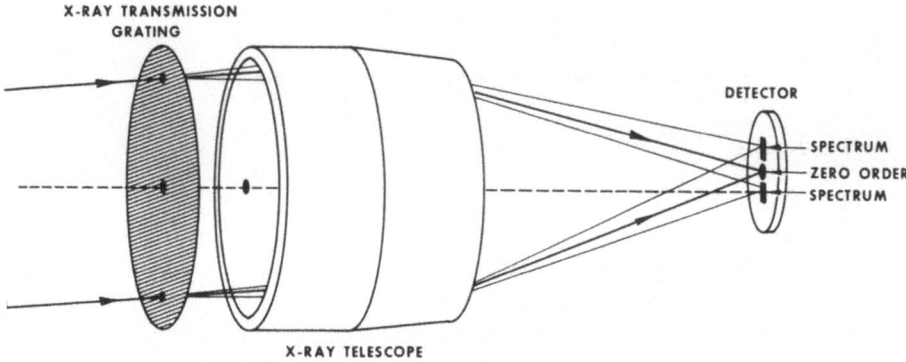

Figure 1.   Diagram of the soft x-ray slitless spectrograph.

filter and detector in the focal plane. The device functions in a
manner similar to the objective grating spectrographs widely used
in the visible range. For each x-ray source in the field of view of
the telescope, the system produces a real image (or zero-order
spectrum), and bracketing it, corresponding spectra of various
orders. Although the grating as shown here is placed in front of the
x-ray telescope, we have found that, because of the relatively small
convergence angle of the ray cone, it can alternatively be placed
just behind the telescope. As an example of the typical dimensions
and parameters involved, table I describes a slitless spectrograph
we are presently building for flight on the Apollo Telescope Mount
in 1972. The useful wavelength range, as determined by the reflection properties of the telescope and the transmission of the grating,
is 3 Å to 15 Å.

SURVEY CAPABILITY

The basic usefulness of a spectrograph of this type is that its
capabilities are intermediate between the extremes of very high efficiency and very high spectral resolution that are characteristic of
the more traditional instruments of soft x-ray spectroscopy. It can
provide surveys of broad regions of the soft x-ray spectrum at greater
speed than the crystal spectrometer. At the same time, a level of
spectral resolution is achieved which, although lower than that of
the crystal spectrometer, is superior to that of proportional counters
and other devices relying on pulse height analysis. The high efficiency compared to the crystal spectrometer arises because the

TABLE I

SLITLESS SPECTROGRAPH PARAMETERS

X-ray telescope

      Focal length: 213 cm.

      Diameter: 30.5 cm.

      Geometrical collecting area: 44 cm^2

      Angular resolution: Better than 5 arc-sec.

      Reflection efficiency for 5 sec. resolution element: 5%

      Effective collecting area for 5 sec. res. element: 2.2 cm^2

         (area x reflection efficiency)

Grating

      Grating constant: 6900 Å

      Dispersion: 0.5 arc-min $Å^{-1}$

      Spectral resolution (dispersion x angular resolution): 0.16 Å

      Nominal grating efficiency (flux sent into combined first order

         spectra): 20%

Detector (microchannel plate image intensifier)

      Nominal efficiency: 10%

      Noise per 5 arc-sec. res. element: $\approx 10^{-3}$ events/sec.

---

slitless spectrograph is not a scanning instrument, but simultaneously examines all wavelengths within its spectral range.

    To illustrate the efficiency of the slitless spectrograph, consider the potential application of the device in producing a medium resolution survey of the spectrum of a strong stellar x-ray source such as Sco X-1 from 3 Å to 10 Å. Using Tucker's model[5] for the spectrum of Sco X-1, we may expect to find several emission lines in this wavelength interval with intensities on the order of 0.5 photons $cm^{-2}sec^{-1}$. The observed continuum intensity is about 7 photons cm^{-2} sec^{-1} $Å^{-1}$. Then in a 1000 second observation period, the spectrograph described in table I could:

1.   Detect the principal emission lines present, and locate them to within one spectral resolution element of 0.16 Å. During the 1000 second period, the detector would accumulate about 22 counts from the typical emission line, and about 50 counts in each spectral resolution element from the continuum radiation. This is sufficient to reduce to 0.08 the probability of observing,

in any of the 42 resolution elements, a false or statistical "emission line" due to fluctuations in the contribution from the continuum.

2.   Measure the intensity of the continuum to an accuracy of 10% at intervals of 0.33 Å, thus determining the general shape of the continuum and to locating principal absorption edges that are present. An average of 100 counts would be accumulated in each 0.33 Å interval.

On the other hand, consider the problem of performing an equivalent 3 Å to 10 Å survey with a crystal spectrometer. Assume the crystal to be ADP, with a reflection coefficient of 2×10^{-5} radians, and that the detector efficiency is essentially 100%. Require that we detect at least 100 counts from each 0.33 Å interval of the continuum and at least 10 counts from each 0.5 photon $cm^{-2} sec^{-1}$ emission line. Let the expected emission line width from Sco X-1 be 10^{-3} radians. Then the effective collecting area required for the crystal spectrometer to complete the survey in 1000 seconds is an enormous 2000 cm^2. (Here, the effective collecting area is just the area of the crystal if a non-focusing, flat crystal spectrometer is used; or it is the effective collecting area of the telescope if a curved crystal is used in the focal plane of an x-ray telescope.)

Thus, as a fast, medium resolution instrument for mapping continuum spectra and for locating emission lines to be studied in further detail with crystal spectrometers, the capabilities of the slitless spectrograph are impressive.

TRANSMISSION GRATINGS

Since the unique component of this spectrograph is the transmission grating, some details of its fabrication should be described. The grating could be considered to be a pickett fence structure. It consists of a microscopic array of parallel absorbing strips, interspersed by regions that are less absorbing to soft x-rays. The absorbing strips can either form a free-standing structure, or be supported on a substrate thin enough to be transparent to soft x-rays. The gratings developed to date have been supported on thin plastic substrates. The plastic layer, which is typically one micron thick parylene[6], is first formed on a thick replica of a conventional ruled grating, and then stripped off, retaining an impression of the grating lines. The absorbing strips, which are the actual dispersive element

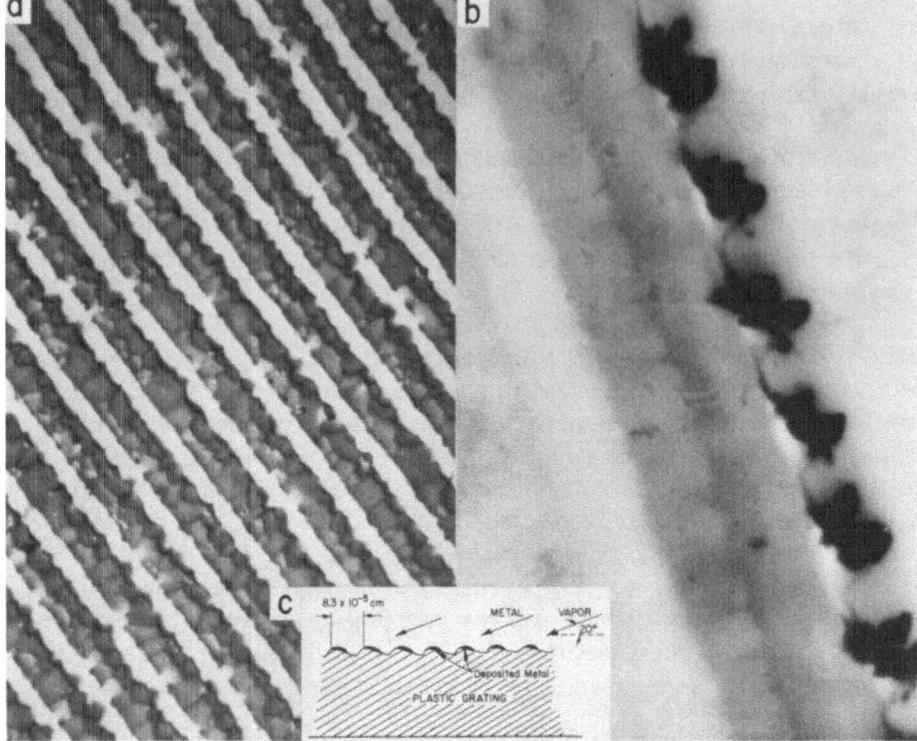

Figure 2. (a) Electron microscope view of the grating lines of a
1440 line/mm gold-shadowed grating; (b) electron microscope view
of the cross section of a similar grating, with the ends of the gold
strips showing as a row of irregular dark areas; and (c) diagram of
the shadowing geometry.

of the grating, are then formed on the plastic layer by a shadowing
process in a vacuum evaporator. Figure 2 shows two electron
microscope views of a typical gold-shadowed grating[7].

The limiting dispersion of transmission gratings intended for
use in the 5 to 15 Å range is determined by the properties of the
absorbing material. Because of the geometry of the shadowing
process, the thickness of the deposition obviously cannot exceed
the grating constant. Thus there is a trade-off between dispersion
and the effectiveness of the absorbing strips. The limiting dispersion
is reached when the strips must become so narrow and thin that they
effectively cease to absorb. The optimum shadowing materials
which permit the highest dispersion and efficiency are those with
the largest x-ray absorption per unit path length, such as gold,
tungsten or platinum. We have fabricated 1440 line/mm gratings

shadowed with 2000 Å of gold having efficiencies of about 15%; but
the efficiency of similar 2880 line/mm gratings was found to be less
than 1%. The low efficiency obtained with the higher dispersion
gratings is believed to be the result of the large inherent grain size
in vapor-deposited gold films, which causes the absorbing and
transmitting areas of the grating to be poorly defined. (The term
efficiency, as used here, is defined as the fraction of the incident
power sent into the combined first-order spectra. The maximum
theoretical efficiency is 20% for a grating with a completely transparent substrate and with completely absorbing strips of width one-
half the grating constant.)

    Alternatively, self-supporting gratings without substrates can
be fabricated in the manner used by the Tübingen group to make
Fresnel zone plates[8] . This technique is of great interest for longer-
wavelength studies, where the grating efficiency is limited by the
x-ray transmission of the substrate.

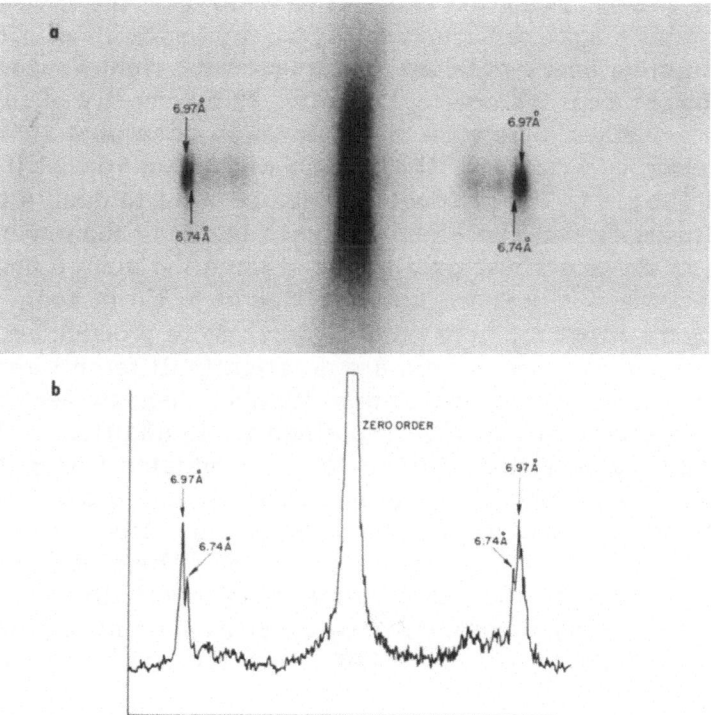

Figure 3.   (a) Spectrum showing the resolution of the tungsten $M\alpha$ and
$M\beta$ lines at 6.97 Å and 6.74 Å, respectively; and (b) microdensitometer
scan of spectrum.

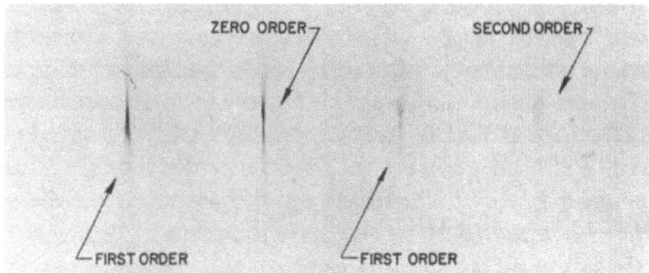

Figure 4.   Aluminum spectrum showing asymmetry in the intensities
of corresponding spectral orders.

SPECTRA

Spectra of laboratory sources made with two versions of the
slitless spectrograph are shown in figures 3 and 4. Figure 3, in
which the tungsten Mα and Mβ lines at 6.97 Å and 6.74 Å are clearly
separated, demonstrates the spectral resolution of the instrument.
On the original negative a curious asymmetry in the two first order
spectra could be seen, with the spectrum on the right appearing
slightly denser than the one on the left. An asymmetry of this kind
cannot be explained purely on the basis of an absorbing strip model
for the grating. In figure 4, the grating was tilted so that its normal
made a 60° angle to the incident x-rays, in order to double the effective dispersion (to 1 arc-min/Å) and to increase the apparent
thickness of the absorbing gold strips. Since the source emitted
mainly aluminum K radiation, only the line at 8.3 Å is seen. Here,
the asymmetry about the zero order is much more pronounced. The
densities of the two first orders are drastically different; and, in
fact, the denser first order is slightly stronger than the zero order
itself. This effect can only be explained if, in addition to the
regular or cyclic variation of the x-ray transmission from point to
point across the plane of the grating, there is also a cyclic variation
of the optical path length. In other words, the effect is, in part,
that of a phase grating. For this explanation to be plausible, it
should be possible to show that optical path length differences on
the order of an x-ray wavelength can occur from point to point across
the grating. The optical path length difference ϵ is given by:

$$\epsilon \ = \ T\delta \ = \ \frac{ne^2\lambda^2}{2\pi mc^2} \ T \tag{1}$$

where δ is the unit decrement of the index of refraction, n is the
electron density, m is the electron mass, and T is the thickness of

the refracting material. For a gold thickness of 2500 Å and with
λ = 8.3 Å, the resulting ϵ is 3.5 Å. Since this is comparable to λ ,
we would indeed expect that there could be phase grating effects in
the diffraction pattern. The asymmetry can be explained if, for instance, the cross sections of the gold strips were tiny refracting
wedges of some appropriate angle. The effect is obviously of
potential use, since it suggests that x-ray transmission gratings
can, in effect, be blazed to concentrate the power into some desired
spectral order.

Figure 5 is a recent slitless spectrogram of the sun made
during a rocket flight of April 8, 1969. This exposure, which has
not yet been analyzed in detail, was made on Pan X film in 25
seconds through an aluminized parylene filter. The dispersion of
the grating was 0.5 arc-min/Å. The 6 Å to 15 Å spectra of several
active regions, as well as the solar limb emission, are evident.
We are making microdensitometer scans of these spectra in order to
determine the general spectral shape and to bring out the emission
lines that are present. The spectral resolution seen here is appreciably
lower than the ultimate resolution of the instrument, because of the
relatively large angular extent of the active regions. As is the case
with any slitless spectrograph, the apparent size of any feature in
the spectrum, such as a discrete emission line, is no smaller than
the angular size of the source itself. The spectral resolution is

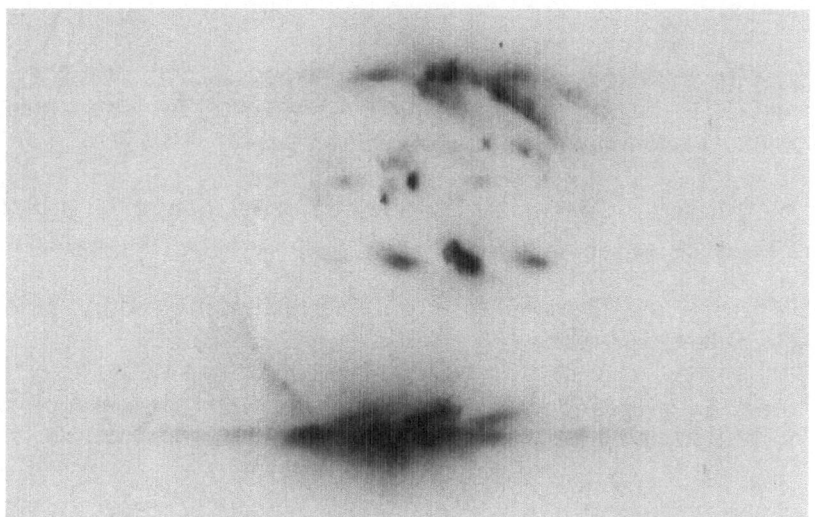

Figure 5. Soft x-ray slitless spectrogram of the sun, showing spectra
of the solar limb emission and of several active regions.

thus degraded by the finite size of the source. The most useful domain of the slitless spectrograph is, probably, with the point-like sources found in stellar x-ray astronomy. We are presently preparing such a stellar payload to be carried in a sounding rocket, which we plan to fly within the next few months.

ACKNOWLEDGMENT

The author wishes to thank Donald Yansen for his contributions to grating development and grating test techniques.

REFERENCES

1.   H. Gursky and T. Zehnpfennig, "An Image-Forming Slitless Spectrometer for Soft X-Ray Astronomy," Appl. Opt. 5:875-876, (1966).

2.   T. Zehnpfennig, "A Functioning Model of the Soft X-Ray Slitless Spectrometer," Appl. Opt. 5:1855-1856, (1966).

3.   G. S. Vaiana, W. P. Reidy, T. Zehnpfennig, L. VanSpeybroeck, and R. Giacconi, "X-Ray Structures of the Sun during the Importance 1N Flare of 8 June 1968," Science 161:564-567, (1968).

4.   R. Giacconi, W. P. Reidy, G. S. Vaiana, L. P. VanSpeybroeck, and T. F. Zehnpfennig, "Grazing Incidence Telescopes for X-Ray Astronomy," Space Sci. Rev. 9:3-57, (1969).

5.   W. Tucker, "Cosmic X-Ray Sources," Astrophys. J. 148:745-765, (1967).

6.   Parylene is a thermoplastic polymer manufactured by Union Carbide Corporation.

7.   Electron microscope pictures of the gratings were supplied by Jeolco (U.S.A.) Inc., Medford, Massachusetts.

8.   G. Möllenstedt, K. H. VonGrote, and C. Jönsson, "Production
      of Fresnal Zone Plates for Extreme Ultraviolet and Soft X
      Radiation" in H. H. Pattee, V. E. Cosslett, and A. Engström,
      Editors, X-Ray Optics and X-Ray Microanalysis, Third
      International Symposium, Acadamic Press, New York, 1963,
      p. 73-79.

X-RAY INTERACTION COEFFICIENTS: EFFECT ON INTERPRETATION OF SOLAR X-RAY DATA

R. L. Blake

The University of Chicago

Chicago, Illinois 60637

ABSTRACT

The interpretation of solar x-ray data obtained from KAP crystal spectrometers requires new laboratory measurements. We used a plasma x-ray source and a KAP spectrometer similar to solar spectrometers and measured the following:

1. The optical reflectivity of KAP for EUV wavelengths $\sim$ 170 - 200Å.

2. "Effective" crystal spacings including effects of anomolous dispersion.

3. Precision wavelengths of a few selected lines.

4. Oxygen spectra in the 23-24Å range with emphasis on the peculiar 23.28Å feature that has been observed previously in electron beam and fluorescent sources.

Results of these measurements are presented. Measurements (1) and (4) were combined with preliminary data from R. Liefeld in an analysis of solar spectra. We report only the result of this analysis, which yields the tentative conclusion that non-thermal processes occur in solar active regions even when no optical flares are observed.

INTRODUCTION

Solar spectra in the wavelength range 1 - 50Å have been recorded since 1963 by plane crystal spectrometers,[1,2] by concave crystal monochromators,[3] and by concave grating spectrographs.[4] The plane crystal instruments have been used more extensively. Because of their simplicity and the fact that the intensity distribution across the source can readily be mapped, these spectrometers will no doubt continue to be prominent in solar studies. They may also find limited application for x-ray stars. However, the measured parameters reported in this paper are applicable to KAP crystal instruments of all kinds.

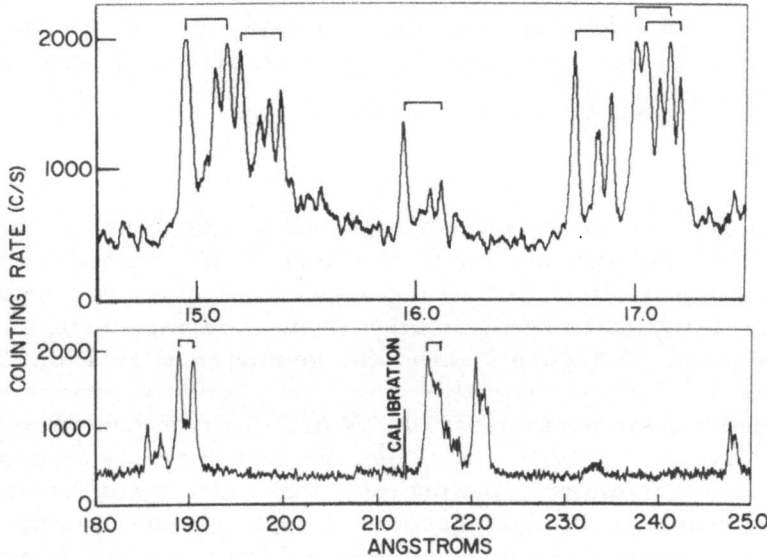

Figure 1. X-ray spectrum of sun recorded 4 October 1966. Printed by permission of the U. S. Naval Research Laboratory.

A set of solar x-ray spectra that demonstrates the rationale for this paper appears in figure 1. These spectra were obtained by the U. S. Naval Research Laboratory group,[2] who used a plane KAP crystal without collimation of the input x-ray beam. Therefore, a one dimensional image of the sun is obtained in each discrete wavelength (line). Brackets above peaks in the figure indicate the same line emitted primarily from two active regions on

opposite limbs of the sun (1/2 degree separation). For example
the strong set of peaks near 19Å belongs to the hydrogen-like Lyman
alpha line of oxygen. Considerable background radiation exists
beneath the lines. Some lines in these spectra and many lines in
shorter wavelength spectra remain unidentified.

Our goal in this paper is to present results of measurements
that bear directly on the interpretation of these and other solar
spectra. We present a curve of the specular reflectivity of KAP
versus angle for extreme ultraviolet (EUV) radiation near 180Å.
This relates to the background radiation in figure 1. Then we give
measured and calculated values of the effective 2d-spacings for
KAP. These will be necessary for future line identifications in
solar spectra. Some wavelength measurements that corroborate
the 2d-spacing values are tabulated. Finally we discuss the way
in which the 23.28Å feature of KAP spectra may be used to interpret solar data.

X-RAY SOURCE AND SPECTROMETER

All laboratory investigations reported here were performed
with Scylla I, a theta-pinch plasma device kindly loaned to us by
the Los Alamos Scientific Laboratory. This pulse-type source
forms an x-ray emitting plasma by magnetic compression of a
pre-ionized gas. Figure 2 shows the geometrical relation of the
hot plasma to the spectrometer. The hot plasma is characterized
by an electron temperature $\approx$ 300 eV and electron density $\approx$
5×10^{16} cm^{-3}. Viewing along the tube axis one sees a plasma of
about 1.5 cm diameter. During the 2 μsec duration of hot plasma
the ions of the spectral gas become stripped of outer electrons,
leaving primarily hydrogen- and helium-like ions for a duration of
about 1μsec (for example O VII and O VIII or Ne IX and Ne X for
oxygen and neon spectral gases).

We chose a plane crystal spectrometer design to permit
direct comparison of laboratory and solar data. A 2 arc minute
FWHM Soller collimator defined the entrance beam to the KAP
crystal as indicated in figure 2. For x-ray spectral measurements
a relatively thick plastic and aluminum window blocked EUV and
visible radiation from the scintillation detector. A line profile
consisted of a point-by-point scan of θ and 2θ with an average
of four Scylla I discharges per spectral point.

Because the source intensity varied from one discharge to
another we had to employ a normalization procedure. In one

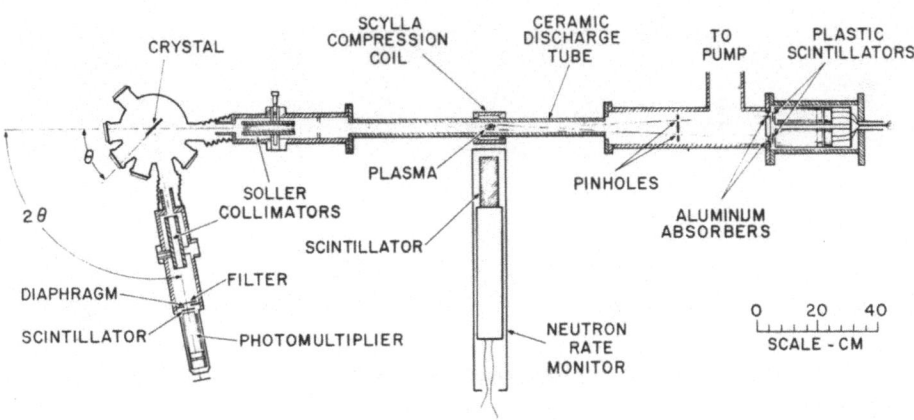

technique illustrated in figure 2 we used two broadband flux monitors on the far end of the tube. With different absorber materials
in front of the two detectors one could monitor the fluxes in two
bands and also calibrate the ratio of fluxes in terms of the plasma
electron temperature. A more reliable technique, which was used
for all measurements on which our effective 2d-spacing and wavelength determinations were based, consisted of adjusting an auxiliary
spectrometer to the peak of the line under study and using this peak
intensity as a reference for all other points on the profile. We are
indebted to Mr. A. E. Unzicker who loaned us one of the NRL rocket
spectrometers for this purpose.

A sample normalized profile appears in figure 3. Our angle
measuring technique was essentially that of Bond[5] with measurements on both sides of the incident beam direction. If θ_1' represents the normal first order Bragg reflection angle, then θ_1''
represents the corresponding angle on the opposite side of the incident beam. Increasing values of θ_1'' correspond to decreasing
values of wavelength. The profile of the helium-like F VIII resonance line in figure 3 is slightly asymmetric toward longer wavelengths (smaller θ_1'') because it is not completely resolved from
the intercombination line $1s-2p^3P$.

Our spectrometer accuracy was about 2 arc sec but other
sources of error limited the final accuracy to 10-15 arc sec. For
a more complete assessment of errors, alignment procedures,
and source diagnostic techniques see the report by Blake.[6]

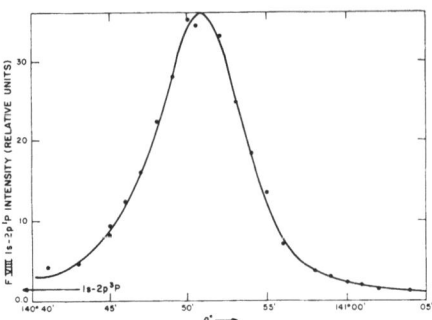

Figure 3. F VIII resonance line profile.

EUV MEASUREMENTS

Scylla I emits radiation over a wide spectral range from the visible to the soft x-ray region, but emission in different wavelength regions depends on time in the capacitor discharge cycle. Accordingly time resolution of the signals proved beneficial in confirming the isolation of EUV flux. Isolation was achieved primarily through control of the plastic filter thickness. In both solar and laboratory spectrometers one uses ~ 2000Å aluminum to block visible light and a thin plastic layer with the dual purpose of supporting the aluminum and blocking EUV radiation in the band λ > 170Å passed by aluminum. The EUV flux dominates for very thin layers of plastic (~ 1000Å) while x-ray flux dominates for thick plastic ($\gtrsim$ 4000Å).

We first measured the reflection coefficient for all the Scylla I EUV radiation passed by a 7000Å thick aluminum filter, which has an EUV passband from 170Å to ~ 400Å. This EUV flux appeared earlier in the discharge cycle than the soft x-rays below 100Å and its reflection coefficient was less steep at large angles. Then we removed half the aluminum and added 1300Å of Parlodion. The Parlodion restricted the EUV passband to ~200Å on the long wavelength side while the remaining aluminum still provided a cutoff below 170Å. One may assume this narrow band to be monochromatic at an effective wavelength of 180Å. Figure 4 shows the measured KAP crystal reflection coefficients versus glancing angle for the Parlodian-aluminum combination. The reflection coefficient is the ratio of EUV flux incident upon the crystal to the reflected flux (energy per unit area per unit time).

By adding more Parlodion it was possible to determine an absorption coefficient at a nominal wavelength of 180Å. When 2800Å more Parlodion was added the attenuation measured at 27.5° and 38° was a factor 90 relative to the filter used for figure 4. This gives a mean absorption coefficient of μ/ρ (180Å) $\approx$ 1.0 x 10^5 cm^2/gm. By comparison Hunter[7] finds a value 7.3 x 10^4 at 180Å and extrapolation of absorption coefficients measured at shorter wavelengths[8] yields a value 9.0 x 10^4. We conclude, from the measured μ/ρ and from the analysis of the time history of signals through different filters, that the data shown in figure 4 contain negligible contribution from radiation outside the 170-200Å passband; except for angles less than 10°. At these small angles specularly reflected soft x-rays begin to dominate the signals. Therefore, the slope at small angles may be wrong. We estimate the maximum ordinate shift of the curve on this account is a factor two. A few representative error bars are shown.

The reflection coefficient data in figure 4 applies directly to solar data obtained with aluminum-plastic windows.

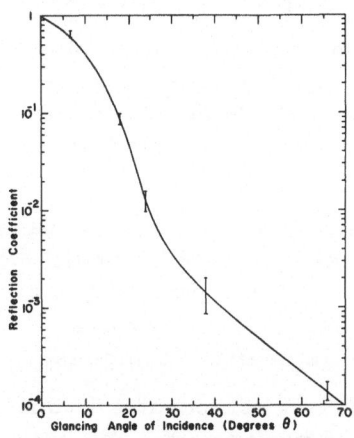

Figure 4. Specular reflectivity of KAP versus glancing angle of incidence. Effective wavelength $\approx$ 180Å.

EFFECTIVE CRYSTAL SPACING OF KAP

Measurements

Let us designate the effective n'th order 2d-spacing as $2d_n$ and write

$$2d_n = (2d_n)_N + \Delta(2d_n) \tag{1}$$

where $(2d_n)_N$ is the crystal interplanar spacing modified by a refraction correction for normal dispersion (N) and $\Delta(2d_n)$ designates the correction for anomolous dispersion (A). In terms of the unit decrement of the refractive index δ we have

$$(2d_n)_N = \left[1 - \left(\frac{2d}{n}\right)^2 \left(\frac{\delta}{\lambda^2}\right)_N \right] 2d \tag{2}$$

and

$$\Delta (2d_n) = \left[- \left(\frac{2d}{n} \right)^2 \left(\frac{\delta}{\lambda^2} \right)_A \right] 2d \tag{3}$$

where 2d is the true interplanar spacing. This notation permits one to use the high precision measurements of 2d and $(2d_n)_N$ from Bearden and Huffman.[9] Since $(2d_n)_N$ depends only on the order of diffraction, one can obtain the effective 2d-spacing for any order by simply adding a correction $\Delta(2d_n)$ that depends only on wavelength.

Our purpose has been to evaluate the correction term $\Delta(2d_n)$ arising from anomolous dispersion. This required reference wavelengths of high accuracy. In the 10-30Å region only the lines of hydrogenic ions approach the requirements for primary standards. For practical reasons only the O VIII Ly α and Ly β lines were used for extensive measurements reported here. Their nominal wavelengths are 19Å and 16Å.

Table I contains all the measured data on peak angles. Each group of lines corresponds to a particular set of system conditions. The collimator setting error is the deviation of a central ray through the collimator from the crystal turntable zero. This error and others were eliminated by measuring and alignment techniques.[5,6]

Since the crystal expands with increasing temperature, one must apply a correction $\Delta \theta$ to the observed peak angle. We have

$$\Delta\theta = -\alpha \tan\theta \, \Delta t \tag{4}$$

where α is the linear expansion coefficient and Δt the variation of temperature from the reference value 26°C selected by Bearden and Huffman.[9] We could not control our room temperature precisely, but continuous monitoring permitted us to establish a rough mean value for each profile. From the general run of expansion coefficients we had estimated $\alpha \sim 2 \times 10^{-5}$. However, from a few profiles selected out of tables I and III (see later for table III) for conditions where temperatures and peak angles seemed least uncertain we made a rough determination of $\alpha \sim 7.6 \times 10^{-5}$ (°C)$^{-1}$. Since we did not have a convenient way to measure α for a soft material like KAP and the value estimated from the run of expansion coefficients was only $\sim 2 \times 10^{-5}$, we adopted a compromise value 4×10^{-5} (°C)$^{-1}$ that would cause minimum systematic error. This value was used to get the peak angles corrected to 26°C. From the reference line peak positions, θ_1' and θ_1'', on opposite

Table I

Measured angles of reference lines for KAP d-spacings

Group	Line	Angle	t̄(°C)	Angle at 26°C	Weight	Soller Collimator	Collimator Setting Error	Mean Angle at 26°C	Notes
1	OVIII Lyα	45° 30' 12" ±06" 134° 29' 30" ±06"	26.5 26.7	45° 30' 16" 134° 29' 24"	1 1	A; 4.3' FWHM "	+ 00' 10" "	θ_1' = 45° 30' 16" θ_1'' = 134° 29' 24" θ_1= 45° 30' 26" ±09"	Normalization against integrated radiation through aluminum–window detector rather than line on rocket spectrometer.
2	OVIII Lyα	45° 33' 12" ±06" 45° 33' 10" ±10" 45° 33' 12" ±06" 134° 32' 06" ±06"	26.5 26.5 26.5 26.5	45° 33' 16" 45° 33' 14" 45° 33' 16" 134° 32' 02"	1/2 1/2 1 1	A; 4.3' FWHM " " "	- 02' 39" " " "	θ_1' = 45° 33' 16" θ_1'' = 134° 32' 02" θ_1 = 45° 30' 37" ±07"	Collimator reset after technical difficulties.
3	OVIII Lyα	45° 30' 50" ±06" 45° 30' 57" ±06" 45° 30' 56" ±04" 134° 30' 48" ±06" 134° 30' 51" ±04" 134° 30' 51" ±06"	26.5 27.0 27.0 27.3 27.5 27.4	45° 30' 54" 45° 31' 05" 45° 31' 04" 134° 30' 37" 134° 30' 41" 134° 30' 41"	1 1 2 1 2 1	B; 1.8' FWHM " " " " "	- 00' 51" " " " " "	θ_1' = 45° 31' 02" θ_1'' = 134° 30' 40" θ_1 = 45° 30' 11" ±05"	
4	OVIII Lyβ	37° 01' 12" ±06" 37° 01' 00" ±06" 37° 01' 09" ±06" 143° 00' 33" ±03" 143° 00' 33" ±03"	25.9 26.1 26.5 26.5 26.4	37° 01' 12" 37° 01' 00" 37° 01' 12" 143° 00' 30" 143° 00' 30"	1 1 1 2 2	B; 1.8' FWHM " " "	- 00' 50" " " "	θ_1' = 37° 01' 08" θ_1'' = 143° 00' 30" θ_1 = 37° 00' 18" ±06"	Same conditions as for Group 3.
5	OVIII Lyβ	37° 02' 57" ±06" 143° 02' 15" ±06" 143° 02' 18" ±04"	26.0 26.6 26.5	37° 02' 57" 143° 02' 11" 143° 02' 15"	1 3/2 2	A; 4.3' FWHM " "	- 02' 35" "	θ_1' = 37° 02' 57" θ_1'' = 143° 02' 13" θ_1 = 37° 00' 22" ±07"	Same conditions as for Group 2.

sides of the incident beam, one obtains the Bragg angle in first
order $\theta_1 = 90° - (\theta_1" - \theta_1')/2$. This appears in column ten of
table I. The errors are estimated limits within which the true
glancing angle falls when only statistical fluctuations are included.
Reference line wavelengths from Garcia and Mack[10] combined with
measured values of θ_1 yield the effective 2d spacings via Bragg's
formula.

Without ambiguity one finds a mean $\theta_1 = 37°00'20"$ for O VIII
Lyβ and the effective first order 2d-spacing at 16Å becomes
$2d_1(16Å) = 26.592_6$Å. Differences among the three determinations
of glancing angle for O VIII Ly α are too large to be attributed to
measurement imprecision. Systematic effects in normalization
probably cause the discrepancies. Group 1 profiles were normalized against continuum flux as indicated by broadband monitors
rather than against the peak line flux. Group 2 profiles were normalized against the line peak flux but the normalization factor showed
an increase with flux amplitude; this implies a non-linearity some-
where in the system. We could not trace the cause of non-linearity,
but it was not in the photomultipliers unless their performances
shifted after calibration.

Data for groups 3 to 5 and all profiles for wavelength determinations (see table III later) were obtained with proper (flat) normalization curves and with line peak normalization. Because of the
questionable normalization in groups 1 and 2, we rejected them and
based the determination of effective 2d-spacing at 19Å upon the more
complete and reliable data of group 3. Thus we state as our best
estimate $2d_1 (19Å) = 26.593_7$Å. For comparison, if we assume all
three groups equally good, the weighted average yields 26.592$_6$Å;
if we use the profiles in group 2 without "correction" for the flux
amplitude dependence and again assume all three groups equally
good, the weighted average gives 26.593$_1$Å.

COMPARISON TO THEORY

Equations (1), (2), and (3) show that the effective 2d-spacing
depends on a single parameter, the unit decrement of the refractive
index. The normal dispersion part $(\delta/\lambda^2)_N$ was evaluated by
Bearden and Huffman,[9] who found its magnitude 2.30 x 10^{-6} in
good agreement with theory. Parratt and Hempstead[11] developed
a useful theory of dispersion. In their notation the anomolous
dispersion part of δ may be written for KAP as

$$(\delta/\lambda^2)_A = 2.21 \times 10^{-8} F \qquad (5)$$

where λ is in Å and

$$F = \sum_s n_s \sum_q G_q \left[Re \left(J_q \right) - 1 \right] . \tag{6}$$

Here n_s is the number of atoms of species s per molecule, G_q the atomic oscillator strength for shell q(K, L_I, L_{II}, etc. . . .), and $\left[Re \left(J_q \right) - 1 \right]$ a universal function tabulated by Parratt and Hempstead. Values of n_s come directly from the KAP formula $KH_5C_8O_4$. The oscillator strength is proportional to the integral of the photoelectric absorption coefficient over all wavelengths shorter than the photo-ionization edge. To suitable approximation in the range of interest for all elements in KAP the absorption coefficients follow simple power law functions of wavelength for each shell. This permits direct evaluation of each G_q from absorption coefficient compilations.[8]

Figure 5 shows the wavelength dependence of F and table II contains numerical values at convenient wavelengths for application to the 2d-spacing corrections. No points were calculated within the interval $0.98\lambda_q < \lambda < 1.02\lambda_q$ because damping cannot be neglected therein. Because each molecule of KAP contains four oxygen and eight carbon atoms, but only one potassium atom, the oxygen and carbon dispersion electrons produce larger effects than potassium at the K edges. However, the potassium L dispersion is comparable to the carbon K dispersion for wavelengths greater than 20Å, and these two effects are so large beyond the oxygen K edge that the curve does not recover. Dispersion by the potassium L electrons has been treated in terms of a single edge at 37Å. The uncertainty of F on this account was estimated to be less than ten percent in the region of interest below 26Å. Additional uncertainty arises from the fact that all absorption edges actually have fine structure whose amplitude may be large when bound excited state absorptions are involved. We do not yet have sufficient knowledge of edge structure in KAP to make the distinction in dispersion theory calculations.

Substitution of the values of F listed in table II into equation (5) provides the desired theoretical values of $(\delta/\lambda^2)_A$. One may then calculate the corrections $\Delta(2d_n)$ directly from equation (3) and insert these into equation (1), along with values of $(2d_n)_N$ measured with high precision by Bearden and Huffman,[9] to obtain the effective 2d-spacing for any order and any wavelength line. The corrections $n^2 \Delta(2d_n)$ are listed in table II along with the effective first order 2d-spacing. In computing $2d_1$ in Å we have

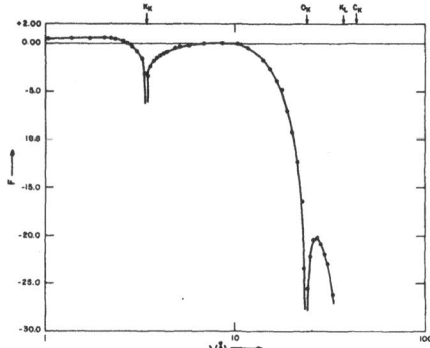

used the conversion factor $\Lambda = 1.002070$ mÅ/xu to convert the $(2d_1)_N$ value of Bearden and Huffman[9] from xu to Å. The uncertainty in Λ corresponds to ± 0.0005Å in $2d_1$.

Comparison of the two measured values to the calculated values at 16 and 19Å shows excellent agreement. We believe the measurements confirm the calculations well enough that table II should be considered the tentative standard for the effective 2d-spacing of KAP at long wavelengths. Future improvements should be concentrated in the range 20 - 26Å. Wavelengths of solar lines may now be determined directly from the Bragg formula with the effective 2d-spacing and measured angles,

$$\lambda = 2d_n \frac{\sin\theta_n}{n} . \quad (7)$$

Table II
Correction to 2d-spacing of KAP calculated from disperson theory

λ (Å)	F	$n^2\Delta(2d_n)$ (Å)	$2d_1$ (Å)
7	− 0.05	0.0000	26.5906
8	+ 0.01	.0000	.5906
9	+ 0.01	.0000	.5906
10	− 0.04	.0000	.5906
11	− 0.22	.0001	.5907
12	− 0.59	.0002	.5909
13	− 1.11	.0005	.5911
14	− 1.77	.0007	.5914
15	− 2.52	.0010	.5917
16	− 3.40	.0014	.5920
17	− 4.43	.0018	.5925
18	− 5.85	.0024	.5931
19	− 7.65	.0032	.5938
20	− 9.75	.0041	.5947
21	−12.35	.0052	.5958
22	−16.0	.0070	.5973
23	−26	.0109	.6015
23.6	−30	.0125	.6032
24	−24.3	.0102	.6008
25	−20.9	.0087	.5994
26	−20.25	.0084	.5990

WAVELENGTH MEASUREMENTS

Several factors prompted the wavelength measurements reported here. First we wanted to evaluate the accuracy attainable in practice with a plane crystal spectrometer and plasma source that effectively simulate solar experiments. Second there exist two sets of precision wavelength measurements[12,13] in this region

that permit a direct confirmation of the accuracy of our 2d-spacing
values. Third we hoped to resolve a disagreement between Tyren[12]
and Flemberg[13] about the helium-like fluorine lines. Finally it was
relatively easy to measure the Ne IX lines with this sytem; these
had not been measured previously.

Table III summarizes the measured data on angles. Only the
upper level configuration is listed for each line, since all transitions
go to the ground state (see table IV). Other columns have the same
interpretation as table I.

The Bragg formula (7) gives wavelengths on an absolute basis
when the effective 2d-spacing is known. On the other hand wavelengths may be determined independently of the crystal spacing
by comparison to a nearby hydrogen-like reference line. Consider
for example the O VII $3p^1P$ line which falls just on the short wavelength side of OVIII Lyα. The ratio of equation (7) for these two
lines gives

$$\lambda \ (3p^1P) = \lambda \ (Ly\alpha) \ \frac{\sin\theta_1 (3p^1P)}{\sin\theta_1 (Ly\alpha)} \tag{8}$$

where λ (Lyα) is the reference wavelength[10] and both angles are
measured. The absolute and relative wavelengths of O VII $3p^1P$
agreed within 0.0002Å. O VIII Lyβ served as a reference for the
F VIII lines, where the largest difference between absolute and
relative wavelengths was 0.0004Å for F VIII $2p^3P$. No reference
wavelengths were measured in close proximity to the Ne IX lines
or the O VII $2p^1P$ line. We used the effective 2d-spacings from
table II for evaluating these wavelengths.

Table IV (below table III) contains all wavelengths measured
in this work together with results of Tyren[12] and Flemberg.[13]
We converted Flemberg's wavelengths from xu to Å with a conversion factor $\Lambda = 1.002070$. The O VII lines agree well with Tyren's
while the Ne IX lines measured here show good agreement with
values extrapolated along the helium isoelectronic sequence. A
bit of irony appears in the tabulation for FVIII; namely, the present
values agree with Tyren on the resonance line but with Flemberg
on the intercombination line. We failed to resolve the discrepancy;
but note that the resonance and intercombination lines have equal
reliability here, whereas in both earlier works the intercombination line must have been considerably weaker than the resonance
line and by inference harder to measure accurately by photographic
methods.

Table III

Measured angles for wavelength determinations

Group	Line	Angle	$\bar{t}$(°C)	Weight	Soller Collimator	Collimator Setting Error	Angle at 26°C	θ'_1	Mean Angle at 26°C	Notes
1	FVIII 2p ^1P	140° 50' 48" ±06"	27.1	1	A; 4.3' FWHM	− 02' 37"	140° 50' 40"	θ'_1 = 140° 50' 37"	θ_1 = 39° 12' 00" ±05"	Collimator correction taken from mean of Groups 2 and 5 of Table 1; system same as for those measurements.
		140° 50' 39" ±06"	26.4	2	"	"	140° 50' 36"			
	FVIII 2p ^3P	140° 27' 06" ±06"	26.8	2	"	"	140° 27' 01"	θ'_1 = 140° 27' 01"	θ_1 = 39° 35' 36" ±07"	
2	Ne IX 2p ^1P	30° 22' 30" ±03"	27	1	A; 4.3' FWHM	+ 00' 10"	30° 22' 35"	θ'_1 = 30° 22' 35"	θ_1 = 30° 22' 45" ±10"	Collimator correction from Group 1 of Table 1. 03" precision increased to ±10" because only one profile was obtained.
	Ne IX 2p ^3P	30° 38' 03" ±03"	27		"	"	30° 38' 08"	θ'_1 = 30° 38' 08"	θ_1 = 30° 38' 18" ±10"	
3	OVII 3p ^1P	44° 28' 39" ±03"	26.5	1	B; 1.8' FWHM	− 00' 46"	44° 28' 43"	θ'_1 = 44° 28' 43"	θ_1 = 44° 27' 57" ±05"	
		135° 32' 57" ±03"	27.0	1	"	"	135° 32' 49"	θ'_1 = 135° 32' 49"		
4	OVII 2p ^1P	54° 19' 39" ±10"	26.5	1	B; 1.8' FWHM	− 00' 43"	54° 19' 45"	θ'_1 = 54° 19' 45"	θ_1 = 54° 19' 02" ±15"	Profiles broad and asymmetric.
		125° 41' 51" ±10"	25.6	2	"	"	125° 41' 56"	θ'_1 = 125° 41' 42"		
		125° 41' 39" ±06"	26.9	2	"	"	125° 41' 29"			

Table IV

Measured wavelengths and comparison to earlier results

	F VIII		Ne IX		OVII	
	$1s^2\,{}^1S_0 - 1s2p\,{}^1P_1$	$1s^2\,{}^1S_0 - 1s2p\,{}^3P_1$	$1s^2\,{}^1S_0 - 1s2p\,{}^1P_1$	$1s^2\,{}^1S_0 - 1s2p\,{}^3P_1$	$1s^2\,{}^1S_0 - 1s3p\,{}^1P_1$	$1s^2\,{}^1S_0 - 1s2p\,{}^1P_1$
This Paper*	16.807 ±0.001	16.948 ±0.001	13.448 ±0.001$_s$	13.551$_s$ ±0.001$_s$	18.628 ±0.001	21.603 ±0.001
Tyrén*	16.807 ±0.001	16.951 ±0.001	-	-	18.627 ±0.001	21.602 ±0.001
Flemberg†	16.8034 ±0.001	16.9478 ±0.001	-	-	-	-
Extrapolated‡	-	-	13.447 ±0.002	-	-	-

*Wavelengths in Å. See text for discussion of error limits.

†Original wavelengths in X.U. have been changed to Å with the conversion factor Λ = 1.002070 mÅ/X.U.

‡The Ne IX resonance line wavelength can be extrapolated with an accuracy nearly equal to that of the measured wavelengths on either side.

PRECISION AND ACCURACY

A detailed account of all known sources of error may be found in reference 6. Let us here simply specify the results. As a consequence of our alignment tolerances and use of the Bond[5] measuring techniques the following geometrical sources of error are negligible or eliminated: turntable zero, collimator horizontal misalignment and tilt, eccentricity, and horizontal divergence. Physical errors and others that depend on crystal angle appear in table V for a first order reflection at a 45 degree crystal angle. Clearly the dominant error is the $\pm$ 6 sec associated with the precision with which the profile peaks could be established from the data. The $\pm$ 5 sec accuracy of our temperature correction can be made negligible when we finally get the expansion coefficient measured. We assume these errors combine in random fashion and calculate the probable total error.

$$\Delta\lambda = \pm\sqrt{(2)^2 + (0.5)^2 + (6)^2 + (1)^2 + (1)^2 + (5)^2} \times 10^{-4}$$

$$= \pm 0.0008 \text{\AA}.$$

We have assumed that the crystals used in this work have the same spacing as those used generally (Isomet Corporation is the standard source for KAP). Moreover, we attributed the differences in O VIII Lyα measurements to faulty normalization but did not prove this. Accordingly we estimate the final accuracy of our wavelength and 2d-spacing measurements to be $\pm$ 0.002Å.

THE 23.28Å FEATURE

Since 1964, when acid phthalate crystals gained common usage in soft x-ray laboratories, many workers have observed the oxygen K emission bands from various oxides and molecular gases. A characteristic feature of all these spectra appears at 23.3Å[14] where earlier workers[12,15] had found little or no emission, but rather they reported the main oxygen absorption edge.[15] In 1963[1] and again in 1966[2] a feature was observed in solar spectra at 23.28Å (see figure 1). Contrary to laboratory observations the solar feature barely rises above the background. Henke[16] suggested that this peak in both laboratory and solar data may be caused by an enhancement of KAP diffraction reflectivity through the influence of anomolous dispersion on the crystal structure factor. Either the weak continuum or the wing of the emission band could be enhanced. There is strong support for this interpretation in

Table V
Error evaluation

Error Source	$\Delta\theta$ (sec)	$\Delta\lambda$ (10^{-4}Å)
1. Angle reading	± 2	± 2
2. Tilt Vertical divergence Vertical misalignment	< 0.5	< 0.5
3. Collimator illumination Source variability Dispersion	± 6	± 6
4. Doppler shift	negligible	negligible
5. Start Effect	negligible	negligible
6. Zeeman Effect	negligible	negligible
7. Absorption	< 1	< 1
8. Lorentz factor Polarization factor	< 1	< 1
9. Temperature	± 5	± 5

the work of Mattson and Ehlert,[14] but their data also showed features that prohibit the exclusion of real emission components.

All spectroscopic studies of oxygen emission and absorption structure have been plagued by the presence of oxygen compounds in parts of the instrument. Consequently there is ambiguity concerning the true emission and absorption structure near 23.3Å for those few compounds that have been studied. This situation left us in doubt as to whether the solar feature was a real line emission or simply a crystal effect. Although the crystal effect has been evaluated through the recent work of Liefeld[17] and his associates, we present here some additional spectra from Scylla I that show the influence of different filters on the 23.28Å feature and also contain new emission features.

Figure 6 shows a composite laboratory spectrum for ten percent oxygen added to the Scylla I plasma. Data points near the bottom edge of the figure represent background (BG) "spectra" with no oxygen added. A peak up to 10 mv sometimes showed up at $61°04'$ (23.28Å) due to impurities released from the discharge

tube walls. The signals with oxygen added are 10 to 100 times the
detector noise level, but still 100 to 1000 times weaker than the
resonance lines reported earlier in this paper. At these low signal levels normalization was poor and some structural detail has
undoubtedly been lost.

The portion of spectrum from $57^\circ 45'$ to $61^\circ 30'$ was obtained
with filtration by $95\mu g/cm^2$ aluminum plus $80\mu g/cm^2$ VYNS. The
peak at $60^\circ 10'$ appeared only with VYNS and sometimes appeared
resolvable, as in the inset, but we did not pursue this portion of
the spectrum further because of the probable dominance of absorption structure.

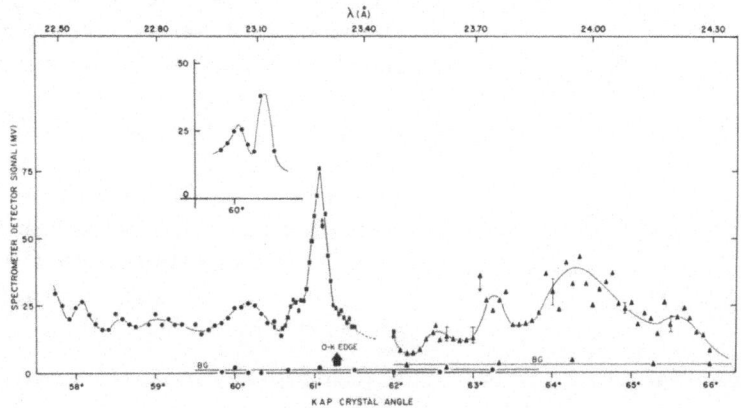

Figure 6. Measurements near 23.3Å with KAP crystal
and oxygen added to Scylla I source.

We tried to minimize the 23.28Å profile uncertainty related
to filter transmission. Unfortunately, all available metal and plastic films without oxygen transmit so much EUV that one cannot use
them on Scylla I. With Parlodion the 23.28Å peak dropped about a
factor five while the spectrum beyond 62° remained about the same.
Profiles obtained through Formvar and VYNS appear in figure 7.
Squares and circles represent VYNS and Formvar, respectively.
For comparison the data were normalized to a common peak amplitude. The profiles coincide near their peaks but exhibit dissimilar
wings. In spite of normalization difficulties, I believe from all the
data collected that the profile differences are real and are caused
by different absorption structures in the two filters. The main O_K
edge in Formvar must fall at a wavelength shorter than 23.28Å;
in Parlodion the main absorption roughly coincides with the 23.28Å
crystal feature.

Our Formvar profile permits a realistic estimate of the actual crystal effect. The peak enhancement is $\approx$ 20 and the FWHM breadth, after a small correction for the 0.02Å beam divergence, is 0.065Å $\pm$ 0.01Å. The peak appears sharper than Gaussian.

Filtration by 140 μg/cm^2 beryllium plus 160 μg/cm^2 Formvar yielded the spectrum from 62° to 66°. Parlodion gave similar results above 62°. The observed feature at 23.61Å (60°35') is weak, but it was observed consistently with both Formvar and Parlodion filters. We know that Scylla I contained all ionization stages from molecular and neutral oxygen to completely stripped ions depending on location in the tube and time during the discharge cycle. Therefore, it seems appropriate to match this peak with the normal oxygen K emission band at 23.6Å. A quantitatively reasonable case can be made for the observed intensity in terms of the relatively few fast electrons that escape across the boundary separating the hot plasma from the cold neutral and molecular gas in the discharge tube.

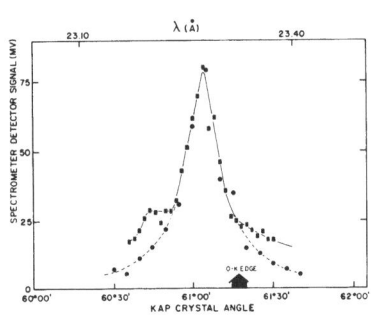

Figure 7. Two profiles of 23.28Å feature with KAP crystal and Scylla I source. Filtration by Formvar (circles) and VYNS (squares).

The range 62° to 66° was scanned several times with filtration by Formvar-beryllium and Parlodion-aluminum. Peaks shown by the solid curve beyond 63° always appeared although some structural detail probably has been lost because of the normalization difficulties. The same fast electrons that ionize neutral and molecular oxygen to produce the 23.6Å multiplet will cause excitations with generally comparable or higher probabilities. For example, in neutral atomic oxygen a 1s → 2p excitation produces the configuration 1s 2s^2 2p^5 which can decay radiatively back to the ground configuration 1s^2 2s^2 2p^4. Hartree-Fock calculations, including multiplet structure and "fitting" shifts estimated from known lines, indicate that some of the peaks between 23.7 and 24.3Å may contain the six multiplets of this OI transition array partially resolved.[6] Excitations in molecular oxygen may also contribute. The present observations are complementary to earlier investigations of both long and short wavelength "satellites" from gases[18] and solids.[12] Future detailed studies with theta-pinch plasmas will have to be done photographically.

APPLICATION TO SOLAR DATA

Direct application of the 2d-spacing and wavelength measurements must await results of planned solar experiments. Here we report the application of the EUV reflectivity data and the 23.28Å data as presently known.

Consider the nominal wavelengths as indicated in figure 8. From our 23.28Å profiles in the previous section plus preliminary data communicated to us by Liefeld we estimated the integrated KAP reflection coefficient R as schematically illustrated, with $R_2/R_1 \sim 20$ and $R_3/R_1 \sim 3$. (This is the standard R - not to be confused with the specular reflection coefficient for EUV.) The transmissions of Parlodion filters used in NRL rocket spectrometers were computed from tables.[8] One gets $T_1/T_3 = 3.1$ for the 1966 experiment. Now T_1/T_2 is unknown. We designate this ratio by the symbol J. Then the rocket spectrometer efficiency RT is accidentally about the same on both sides of the 23.3Å peak;

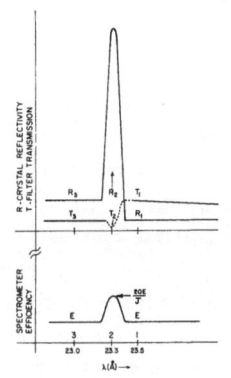

call it E = RT. From the estimated spectrometer efficiency shown in the bottom of figure 8 one expects a flat continuum in this region to yield about equal counting rates on each side of a well defined peak. This is observed in figure 1, but some of the flat background may be specularly reflected EUV. This must be removed before one can evaluate the true continuum.

Figure 8. (Upper) Schematic illustration of KAP integrated reflection coefficient R and Parlodion transmission T versus wavelength. (Lower) Overall spectrometer efficiency.

We set up two equations governing the counting rates caused by EUV and true continuum at positions 1 and 2 of figure 8. These equations involve the unknown parameter J. A lower boundary on J follows from the simple requirement that the EUV counting rate cannot exceed the total rate at any angle. Upon application of the data of figure 4 at all angles one finds J > 7.6. This has two important consequences.

First, J is determined by the Parlodion absorption coefficients at wavelengths 1 and 2. From tables of photoionization absorption coefficients[8] one can compute a maximum value J = 3.1 for the Parlodion window used in the 1966 experiment (figure 1). Since we require a larger J, it follows that the oxygen edge absorption in Parlodion at 23.28Å must represent photoexcitation to a bound excited state rather than photoionization. We estimate the actual absorption coefficient lies between 2.5 and 3.5 x 10^4 cm^2/gm at 23.28Å.

The second consequence of the lower limit on J is that we can, after some computations, set an upper limit of ten percent on the EUV contribution at the wavelength interval around 23.3Å. The contribution at smaller angles is naturally larger, but calculations show that in the 10 - 25Å wavelength band one can use J = 7.6 and be assured that the underestimate of true integrated continuum is less than 40 percent. For both the 1963 and 1966 experiments we find that the true continuum, which is practically flat from 13Å to 21Å, dominates the observed counting rates.

Over the three year interval from near minimum solar activity on 25 July 1963 to a more active sun on 4 October 1966 the active region lines increased by a factor seven while the lines from the quiet corona increased only about a factor two. During the same period the integrated continuum flux increased about a factor seven. This implies that the continuum arises mostly from coronal active regions.

The ratio of integrated line flux to integrated continuum flux serves as an indicator of a hot plasma's thermal character. For solar conditions as encountered in 1963 and 1966 theory predicts a ratio > 0.85 in the 10 - 20Å band for a thermal plasma. The present analysis yields an observed ratio of 0.2 for both experiments. In view of the combined uncertainty of theory and our analysis, perhaps as much as a factor four, we conclude tentatively that the low observed ratio implies the presence of non-thermal excitation processes in solar active regions even when no flares are observed.

ACKNOWLEDGEMENTS

This work was supported by the High Altitude Observatory of the National Center for Atmospheric Research. It was an integral part of the laboratory program of L. L. House, to whom I am indebted for generous support and cooperation. Special thanks are also due to Franz Jahoda and George Sawyer of the Los Alamos Scientific Laboratory for arrangements on the Scylla I source, and to A. E. Unzicker and J. F. Meekins of the Naval Research Laboratory for cooperation on solar data and some equipment.

REFERENCES

1. R. L. Blake, T. A. Chubb, H. Friedman, and A. E. Unzicker, "Spectral and Photometric Measurements of Solar X-Ray Emission Below 60Å, " Ap. J., 142, 1 (1965).

2. G. Fritz, R. W. Kreplin, J. F. Meekins, A. E. Unzicker, and H. Friedman, "Solar X-Ray Spectra From 1.9Å to 25Å, " Ap. J., 148, L 133 (1967).

3. H. V. Argo, J. A. Bergey, W. D. Evans, and S. Singer, "Coronal X-Ray Emission During the 12 November 1966 Eclipse, " Solar Physics, 5, 551 (1968).

4. B. B. Jones, F. F. Freeman, and R. Wilson, "XUV and Soft X-Ray Spectra of the Sun, " Nature, 219, 252 (1968).

5. W. L. Bond, "Precision Lattice Constant Determination, " Acta Cryst., 13, 814 (1960).

6. R. L. Blake, "Plane Crystal Measurements in the Ultrasoft X-Ray Region With Application to Solar Physics, " PhD Dissertation, University of Colorado, (1968), available on request to the author.

7. W. R. Hunter, private communication, (1967).

8. B. L. Henke, R. L. Elgin, R. E. Lent, and R. B. Ledingham, "X-Ray Absorption in the 2- to 200Å Region, " Norelco Reporter, 14, 112 (1967).

9. A. J. Bearden and F. N. Huffman, "Precision Measurement of the Cleavage Plane Grating Spacing of Potassium Acid Phthalate, " Rev. Sci. Instrum., 34, 1233 (1963).

10. J. D. Garcia and J. E. Mack, "Energy Level and Line Tables for One-Electron Atomic Spectra, " J.O.S.A., 55, 654 (1965).

11. L. G. Parratt and C. F. Hempstead, "Anomolous Dispersion and Scattering of X-Rays, " Phys. Rev., 94, 1593 (1954).

12. F. Tyren, "Precision Measurements of Soft X-Rays With Concave Grating, " Nova Acta Regiae Societatis Scientarium Upsaliensis, Ser. IV. Vol. 12, No. 1, (1940).

13. H. Flemberg, "Optical Spectra Within the Ordinary X-Ray Region Recorded with Curved Crystal, " Arkiv for Matematik, Astronomi, och Fysik, Vol. 28A, No. 18, (1942).

14. R. A. Mattson and R. C. Ehlert, "The Application of a Soft X-Ray Spectrometer to Study the Oxygen and Fluorine Emission Lines From Oxides and Fluorides," in G. R. Mallett, M. J. Fay, and W. M. Mueller, Editors, Advances in X-Ray Analysis, Vol. 9, Plenum Press, New York, 1966, p. 471.

15. T. Magnusson, "Investigations Into Absorption Spectra in the Extremely Soft X-Ray Region," Nova Acta Regiae Societatis Scientarium Upsaliensis, Ser. IV, Vol. II, No. 3, (1938).

16. B. L. Henke, in Advances in X-Ray Analysis, Vol. 12, 1969, in press.

17. R. J. Liefeld, S. Hanzely, T. B. Kirby, and D. Mott, "Soft X-Ray Spectrometric Properties of Potassium Acid Phthalate Crystals," in Advances in X-Ray Analysis, Vol. 13, in press.

18. L. Groven, "Etude Des Rayon X Emis Par Les Gaz," Ann. Phys. $\underline{4}$, 62 (1949).

X-RAY SPECTROMETRIC PROPERTIES OF POTASSIUM ACID PHTHALATE CRYSTALS

R. J. Liefeld,[*] S. Hanzély,[†] T. B. Kirby, and D. Mott

New Mexico State University

Las Cruces, New Mexico 88001

ABSTRACT

The results of two crystal measurements of potassium acid phthalate crystal first order parallel position rocking curves, per cent reflections, and reflection coefficients are presented. They cover the 4-24 Å wavelength range and are typical of results with cleaved crystals illuminated over areas of one-half to two square inches. The energy resolution available with these crystals is shown to be nearly constant at about two-thirds of an electron volt over most of the energy range studied and the coefficient of reflection is also nearly constant at about 1×10^{-4} radians. A pronounced line-like reflectivity structure at 23.3 Å is exhibited which is probably associated with oxygen atom K-shell absorption.

INTRODUCTION

Crystals of potassium acid phthalate, (KAP), ($C_6H_4COOHCOOK$), have been used as soft x-ray dispersing elements for a number of years. Their cleavage plane d spacing of 13.3 Å,[1] high reflectivity,[2] and moderate resolving power,[2,3] have made them useful for spectroscopy to about 25 Å. Other welcome characteristics of KAP crystals are their commercial availability (or relatively easy crystallization from water solutions), excellent cleavage, and stability in high vacuum. Offsetting these are the facts that KAP is soft and somewhat hygroscopic so that its reflecting surfaces are easily

[*]Presently on leave at the Los Alamos Scientific Laboratory.
[†]Now at Youngstown State University.

damaged.

This paper presents the spectrometric properties of KAP crystals over the 4-24 Å wavelength range as derived from two-crystal measurements of per cent reflections, reflection coefficients, and first order parallel position rocking curves. The measurements were made over a period of about seven years with several crystals in the course of various spectroscopic studies with a vacuum two-crystal monochromator. At some point in each investigation the instrument was arranged in the first order parallel position for crystal reflectivity and rocking curve measurements at one or more wavelengths of immediate interest.

After a recent suggestion by B. Henke, a study was made of the x-ray reflection characteristics of KAP crystals in the neighborhood of the oxygen K absorption edge. This was to determine if an emission line seen at about 23.3 Å by several observers with these crystals was due to a crystal reflectivity phenomenon or to a real K series emission line from oxygen atoms in the x-ray sources. Spectrum scans, excitation curves, and modified reflectivity determinations all verified that KAP crystals do have a narrow line-like reflectivity structure at 23.3 Å

The collected results of all of the measurements are presented and discussed after a description of some of the experimental considerations in the following section.

EXPERIMENTS

Since monochromatic and plane parallel x-ray beams of sufficient intensity for single crystal reflectivity measurements are not as yet available, it is expedient to make reflectivity measurements with a two-crystal instrument. These are of course directly applicable to spectra obtained with the two crystal monochromator in the dispersive configuration, and they are also directly related to single crystal reflectivity characteristics.

Two-crystal reflectivity measurements are made by arranging the instrument as in figure 1(a) and rocking one of the crystals (usually the second), about the parallel position to obtain a two-crystal rocking curve. Then, as shown in figure 1(b) the second crystal is removed and the detector is positioned to measure I_o, the intensity incident on the second crystal. The ratio of the peak intensity of the rocking curve to I_o is called the per cent reflection P_o. The area under the rocking curve is divided by I_o to get the two-crystal coefficient of reflection R, and the width at half-maximum of the rocking curve, $W_{1/2}$, is recorded as a measure of the spectral resolution attainable at the wavelength of the

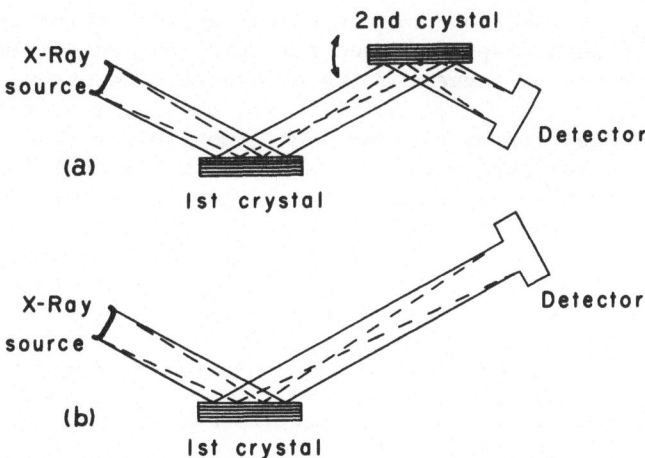

Figure 1. Configurations of the two-crystal monochromator for reflectivity measurements; (a) parallel position rocking curve measurement, (b) direct beam measurement.

rocking curve measurement. These parameters, P_0, R, and $W_{1/2}$, are
related to the corresponding single crystal peak reflectivity, reflection coefficient, and diffraction pattern width.[4]

In the nondispersive parallel position of the two-crystal
monochromator (figure 1(a)) the x-rays diffracted from the first
crystal are collimated in angle per the diffraction pattern of that
crystal. Those which fall on the second crystal (when it is parallel to the first), are all incident within the angular spread of
that diffraction pattern at the correct angle for Bragg reflection.
So all wavelengths incident on the second crystal may be diffracted
into the detector. The range of wavelengths is limited, however,
by the finite horizontal divergence of the x-ray beam. Maximum and
minimum Bragg angles, which determine the limits, are indicated in
figure 1(a) by the dashed lines.

Expositions of the theory of the two-crystal monochromator
usually point out that the shape of the two crystal rocking curve
for parallel positions is independent of the width or height of the
beam defining apertures and of the spectral distribution of the
radiation used for the measurement. This is true only if the crystal diffraction patterns and reflection characteristics are essentially identical for all wavelengths in the range of wavelengths
which may pass through the instrument. If not, some weighted average
of the crystal parameters over the wavelength range limited by the
horizontal divergence will be obtained in a rocking curve and reflectivity measurement. This fact was encountered when attempts
were made to determine the reflection characteristics of the KAP

crystals in the neighborhood of the suspected reflectivity "spike"
at 23.3 Å. Although dispersed spectrum scans and excitation curves
taken at that position suggested the existence of a strong line-like
reflectivity structure there, initial reflectivity measurements made
in the conventional manner yielded much lower values than necessary
to explain the other observations. When the horizontal divergence
of the x-ray beam through the instrument was limited with vertical
slits to a value somewhat less than the width of the crystal diffraction pattern at 23.3 Å, the reflectivity measurements showed a
large maximum at that wavelength. All of the measurements in the
neighborhood of this line-like structure were made with the limited
horizontal divergence modification. All of the others at wavelengths
between 4 and 20 Å were made with the conventional broad beam
technique.

RESULTS AND DISCUSSION

A typical first order parallel position rocking curve obtained
with cleaved KAP crystals is presented in figure 2. This measurement was made with nickel Lα x-rays (14.56 Å) and a central angle
setting of 33.17°. The symmetry of this curve is evidence that the
central portions of the two single crystal diffraction patterns are
similar. It is often assumed that a rocking curve such as this is
a reasonable representation of the central portion of the spectral
window of the two-crystal monochromator when it is in the (1,+1)
dispersive configuration. Note that although the energy resolution
afforded by these crystals at this wavelength is a respectable 0.63
ev, the "spectral window" amplitude is still a few per cent at $\pm$ 7
minutes of arc from the central maximum. Figure 3 displays the
remarkable breadth of these rocking curves somewhat better in a

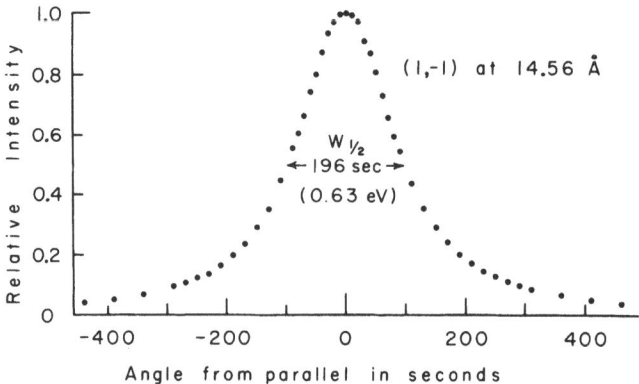

Figure 2. A typical (1,-1) position rocking curve from cleaved KAP
crystals.

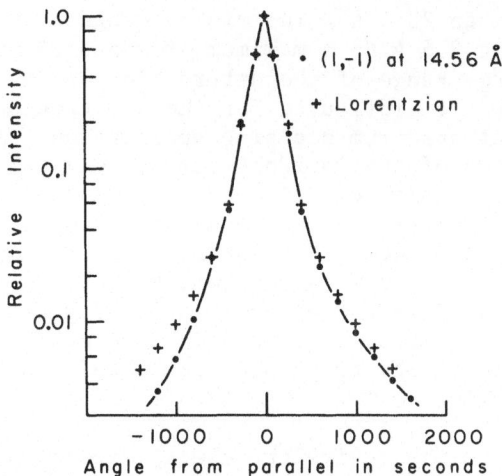

Figure 3. Logarithmic display of a (1,-1) position rocking curve for 14.56 Å.

logarithmic presentation of the same rocking curve. It can also be seen in figure 3 that the tails of the two single crystal diffraction patterns are evidently not identical. They fall below those of a Lorentzian curve fitted at the peak and at the half-maximum points, however, and though not shown in figure 3 this is true to an amplitude of less than 0.0001.

    Figure 4 shows the angular widths at half-maximum of the first order parallel position rocking curves as measured at a number of

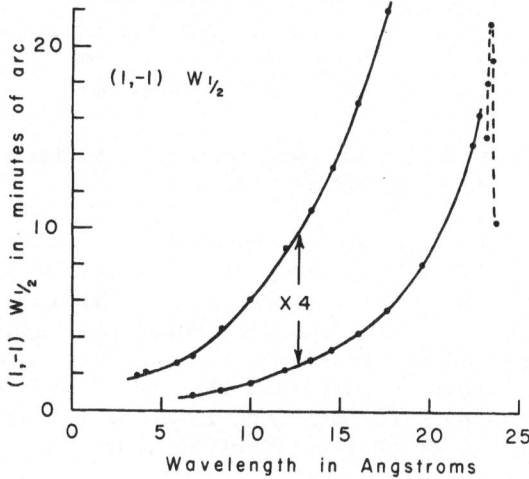

Figure 4. Angular widths at half-maximum of (1,-1) position rocking curves of KAP crystals for wavelengths between 3.6 and 23.6 Å.

wavelengths from 3.6 to 23.6 Å. The widths range from about one-
half minute of arc at 3.6 Å to a maximum of about 20 minutes of arc
at 23.3 Å. This large range of the diffraction widths of KAP crystals causes considerable difficulty for the spectroscopist using
them in a one-crystal spectrum scanning application. A reasonable
mechanical collimation of the incident beam of say 0.1° will neces-
sitate the use of reflection coefficients for wavelengths below
about 15 Å, peak reflectivities for wavelengths above about 20 Å,
and some compromise in between. If KAP is used in a one-crystal
spectrometer which views the x-rays from the sun, where some sources
have an angular width of less than one minute of arc and others have
angular widths of 30 minutes, considerable care must be used in
identifying the reflectivity to be used for obtaining the flux in a
particular observed line.

Figure 5 presents the equivalent energy widths at half-maximum

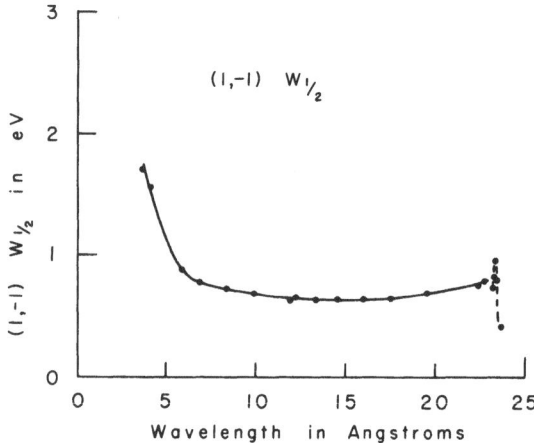

Figure 5. Energy widths at half-maximum of the first order parallel
position rocking curves of KAP crystals.

of the two-crystal rocking curves. Over most of the useful wavelength
range of these crystals this energy width is nearly constant at
about two-thirds of one electron volt. The minimum value of 0.62
ev occurs at about 14 Å. It should be noted that since the two-
crystal rocking curves are in effect the convolution of the two
single crystal diffraction patterns, the energy widths of the single
crystal diffraction patterns are less than the widths of these rocking curves. A single crystal diffraction width of 0.5 ev at 14 Å
is probable.

Both figure 4 and figure 5 show that the rocking curve widths
in the neighborhood of 23.3 Å have a line-like structure. This is

associated with the reflectivity "spike" which exists at that wavelength. It is noteworthy that the rocking curve width at the 23.3 Å
wavelength is twice as large as at 23.6 Å, the position of the normal oxygen Kα line.

In figure 6 the two-crystal per cent reflection numbers are
given as a function of wavelength. The reflectivity "spike" at

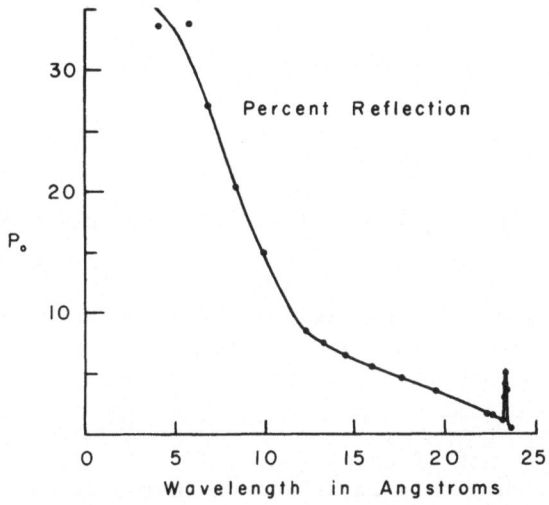

Figure 6. KAP crystal per cent reflection from 3.6–23.6 Å.

23.3 Å is again apparent in this figure as a well defined line-like
structure. Only at the shorter wavelengths where absorption is
small is the per cent reflection fairly large. It decreases with
increasing wavelength to a low of about one-half of one per cent at
23.6 Å, but it is greater than five per cent at 23.3 Å.

The two-crystal coefficient of reflection, calculated as the
area under the two-crystal rocking curve divided by the incident
beam intensity I_O is displayed in figure 7. As the absorption in
the crystals increases (with increasing wavelength), the reflectivity increases smoothly by about a factor of two then drops abruptly
just before exhibiting the pronounced maximum at 23.3 Å. Then it
has fallen by a factor of over 20 at the 23.6 Å position of the
oxygen Kα line. The coefficient of reflection is generally quite
large over the 3–23 Å range which is the principal reason for the
utility of KAP crystals for spectroscopic studies in this region.

In figure 8 we show a two-crystal (1,+1) dispersive configuration spectrum recorded in the neighborhood of the oxygen K absorption threshold with the x-rays from a cobalt anode. The source was
a sheet of high purity cobalt held at a temperature greater than

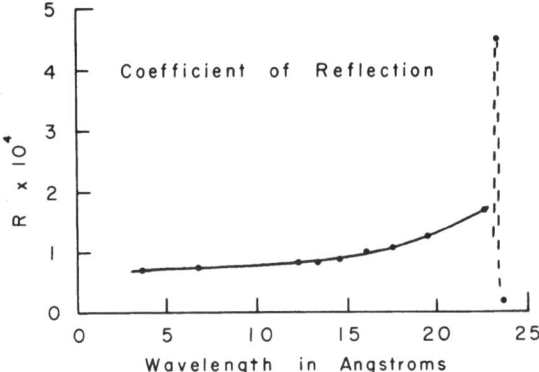

Figure 7.   The two-crystal coefficient of reflection R of KAP crystals.

500°C in an ultra-high vacuum by bombardment with 2 kev electrons. Cobalt has no emission lines in this region and since the anode was essentially clean[*] the expected spectrum is a flat or nearly flat

---

[*]There was no evidence of an oxygen Kα line at 23.6 Å. A line did appear at 23.6 Å when this anode was deliberately oxidized. An approximation to the shape of the oxygen K band in cobalt oxide was then obtained by calculating the point by point ratio of the oxidized anode spectrum to the clean anode spectrum.

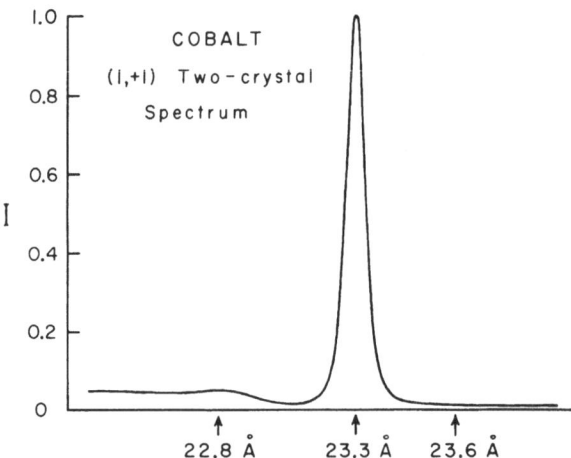

Figure 8.   The continuous spectrum from a clean cobalt anode recorded with the two-crystal (KAP) monochromator.

continuum. All of the structure visible in figure 8 is evidently due to structure in the reflectivity of the KAP crystals. When using two crystals in tandem in a scanning monochromator as was done for this measurement the transmission of the instrument is proportional to the product of the first crystal reflection coefficient and the second crystal per cent reflection so that the line-like reflectivity structure at 23.3 Å is greatly enhanced.

There is a minor maximum in the reflectivity at about 22.8 Å followed by a marked valley between there and the prominent reflectivity maximum at 23.3 Å. This "reflectivity spectrum" is reminiscent of the resonant bound state photon absorption structures usually associated with absorption edges of atoms in the gaseous state and in insulators. This similarity has led to the speculation that the line-like reflectivity maximum is associated with bound state absorption by the oxygen atoms in the carboxyl radicals of the KAP.

The narrow reflectivity maximum at 23.3 Å can have a significant utility. For example, if the reflectivity of a KAP crystal at the 23.3 Å peak and in the 23.5 Å region are known they may be used to determine the fraction of the "background" in an emission spectrum which is true continuum. For only the continuum, not the scattered background, will be modified by the crystal reflectivity. Also, the KAP crystals afford a novel monochromatization at the 23.3 Å wavelength through the combination of the Bragg relation and the sharp reflectivity decrease to both shorter and longer wavelengths.

ACKNOWLEDGMENTS

This work was supported in part by the National Aeronautics and Space Administration under grant NsG—372.

REFERENCES

1. A. J. Bearden and F. N. Huffman,"Precision Measurement of the Cleavage Plane Grating Spacing of Potassium Acid Phthalate," Rev. Sci. Instr. 34: 1233-1234, (1963).

2. R. J. Liefeld, D. Chopra, W. Gray, and D. Mott,"X-Ray Spectrometric Properties of Potassium Acid Phthalate Crystals,"Am. Phys. Soc. 8: 313, (1963).

3. R. J. Liefeld,"The K-Shell Photon Absorption Spectrum of Gaseous Neon,"Appl. Phys. Letters 7: 267-277, (1965).

4. See for example, A. H. Compton and S. K. Allison, "X-Rays in Theory and Experiment" Second Edition, D. Van Nostrand Company Inc., New York, 1935, pp 394-399 and 718-728.

GRATING STUDIES AT X-RAY WAVELENGTHS

R.J. Speer

Imperial College

London, S.W.7.

ABSTRACT

A programme of grazing incidence grating studies is described, undertaken in collaboration with A.W.R.E. Aldermaston, U.K.A.E.A. Culham and N.P.L. Teddington. The wavelength range of interest is from 1Å – 300Å with special emphasis placed on focusing systems for use below 5Å.

INTRODUCTION

In a previous study[1] the ruling and subsequent performance of curved gratings was investigated in detail at one soft X-ray wavelength. The work effectively demonstrated the considerable mutual advantage of a very rapid feedback of information to the ruling engineer concerning surface structure and absolute efficiency at the design wavelength. This experience, gained as part of the calibration analysis of the 304Å grazing incidence monochromator[2], since flown on Orbiting Solar Observatory IV, has led to the fabrication of the more versatile equipment described here. This instrument is a "double" or "tandem" grazing incidence monochromator and is designed to study the fundamental properties of periodic structures throughout the hard and soft X-ray range.

A complimentary study has involved the design and use of a prototype X-ray grating spectrograph at wavelengths below 50Å. In this instrument the emphasis is

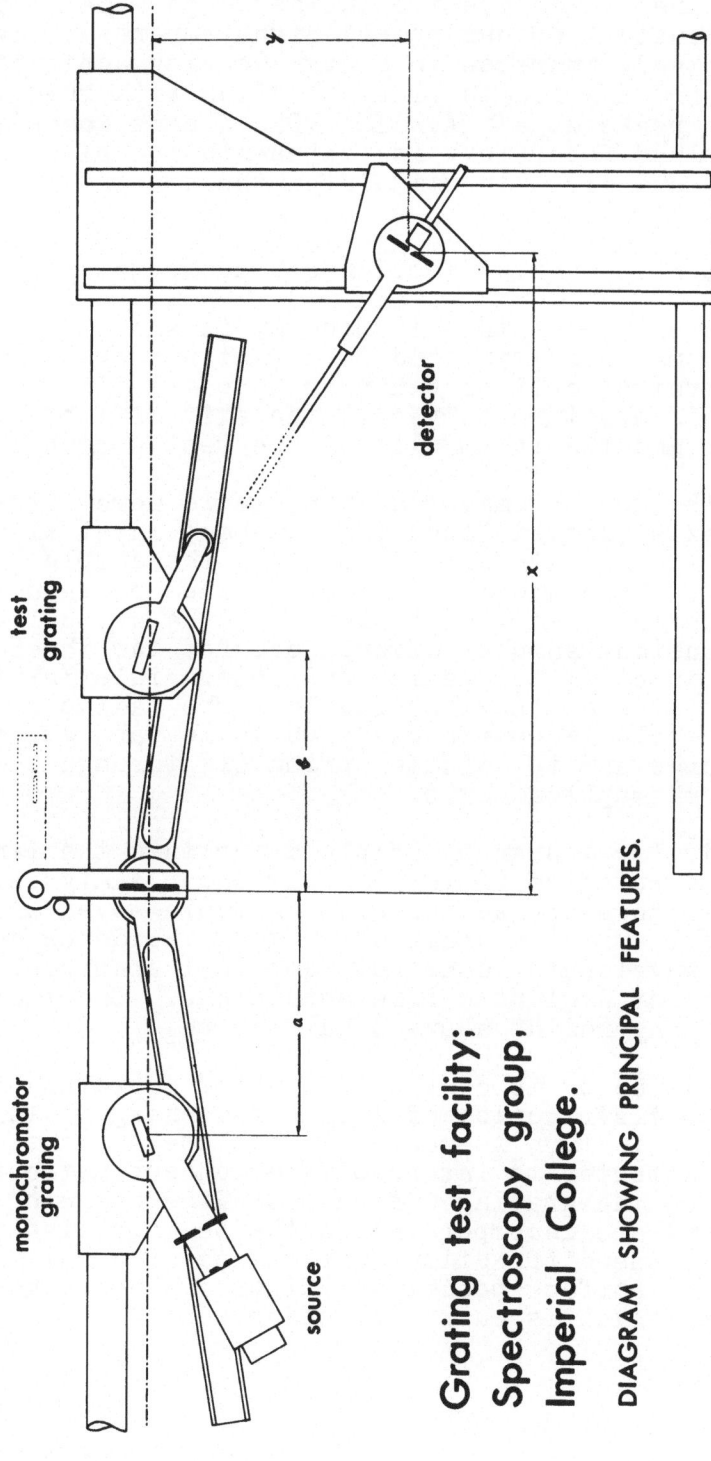

source

monochromator
grating

test
grating

detector

**Grating test facility;
Spectroscopy group,
Imperial College.**

DIAGRAM SHOWING PRINCIPAL FEATURES.

Figure 1.

placed on recording spectra in the range 0.5Å to 50Å
simultaneously from one pulsed high-temperature event:
for the steady increase in energy density achieved in
recent years had led to plasma devices with thermal
radiation peaks at $\lambda \simeq 2$Å. The 70kJ Plasma Focus[3,4,5]
at the Culham Laboratory is representative of this class
of device and directly stimulated the design objectives.

Double Grazing Incidence Monochromator

A schematic of this instrument is shown in figure 1.
A monochromator grating and test grating move along a
common straight optical bench on opposite sides of a common
slit. This aperture serves as the exit slit of the monochromator and the inlet slit of the test grating.

As the monochromator grating table moves along the
optical axis (dotted line) the adjacent inlet slit and
source are constrained to move along the simple linear
cam shown. Monochromatic radiation thus emerges from
the common inlet/exit slit always on the optical axis
and at constant angular divergence, for the three components, inlet slit, grating and common inlet/slit at
all times lie on a Rowland circle. The radius of the
Rowland circle is continuously variable throughout its
design range and is adjusted primarily through the cam
angle to the optical axis.

A similar design principle determines the linear
motion and rotation of the grating under analysis, in
this case, however, as the grating table moves along
the optical axis the angle of grazing incidence changes
at fixed wavelength, constant numerical aperture, and
under correct Rowland circle conditions. The test grating
may be of any radius above a half metre.

The detector is controlled in X-Y coordinates to
record the distribution of diffracted energy. This
measurement in turn is rendered absolute by causing
the same detector to intercept the monochromatic radiation
incident on the grating. A pair of fine scanning jaws
(not shown) are incorporated on the test grating table
to examine the diffraction performance of a ruling as a
function of surface position. Figure 2 shows the completed
instrument which is currently under test.

Figure 2. Assembled Test Facility.

X-ray Grating Spectrograph

This instrument, described in greater detail elsewhere[6], operates at grazing angles of incidence of 10,
20 and 40 arc minutes. The grating, an N.P.L. original
ruling at 300 1/mm[7,8,9] is of 5 metres radius of curvature, and diffracts efficiently at X-ray wavelengths.
The instrument is 17" long, records from 0.5Å to 50Å and
weighs 7.5 kgm.

The spectrum shown in figure 3 was taken from a
commercial X-ray tube using thin Ilford Q2 plates.
Resolution of $CuK\alpha_1, \alpha_2$ at 1.540Å and 1.544Å is achieved
in the third and higher orders. Figure 4 illustrates
order of magnitude calculations for resolving power
Curve A is for detection in a plane of zero thickness
under optimum conditions. Curve B shows the effect of
X-ray film penetration and one micron slit width and
compares the predicted performance with values derived
from measured instrumental line widths. These latter
values are indicated by triangles. Curve C shows the
resolution required to resolve characteristic K radiation
from adjacent elements between bromine and magnesium.

Figure 5 shows the distribution of energy among the
positive diffracted orders for 20 arc minute grazing angle
of incidence. The measurements are normalized to first
order energy and are derived by photographic densitometry.
At $CuK\alpha$ for example, less than 1% of first order energy
is distributed among other orders.

ACKNOWLEDGEMENTS

A considerable number of people have been involved
in the projects outlined above and their continuous
enthusiasm and support has been greatly appreciated,
in particular the advice and assistance of Dr A. Franks
and Mr K. Lindsey of N.P.L. who produced the 5 metre
grating is most gratefully acknowledged. Mr W. Waller,
Mr P.J.H. Osborne and Mr C. Green of A.W.R.E. Aldermaston
have also contributed greatly throughout the design and
manufacture of the equipment. The author is indebted to
Dr N.J. Peacock of the Culham Laboratory for his stimula-
tion and encouragement.

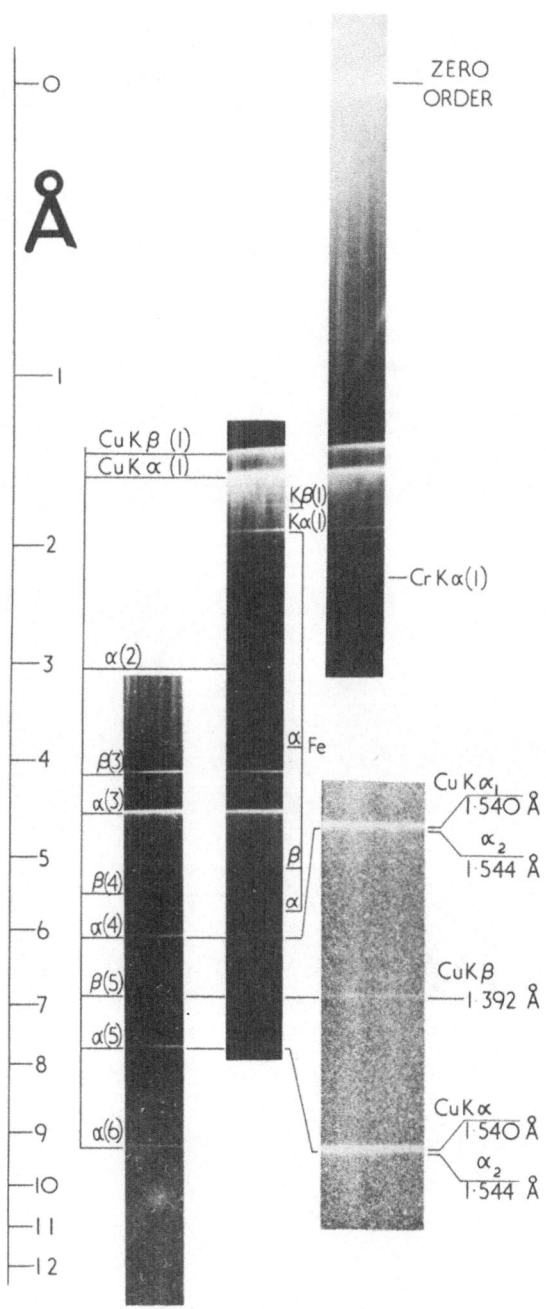

Figure 3. X-ray Test Spectrum

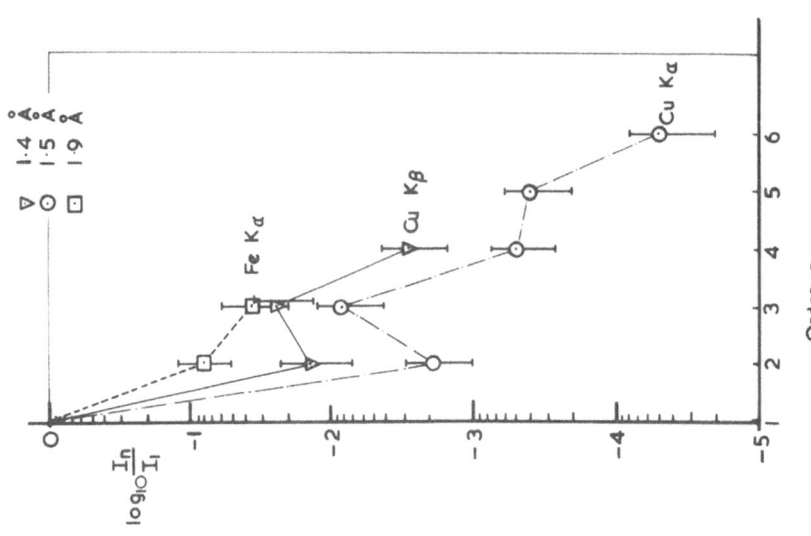

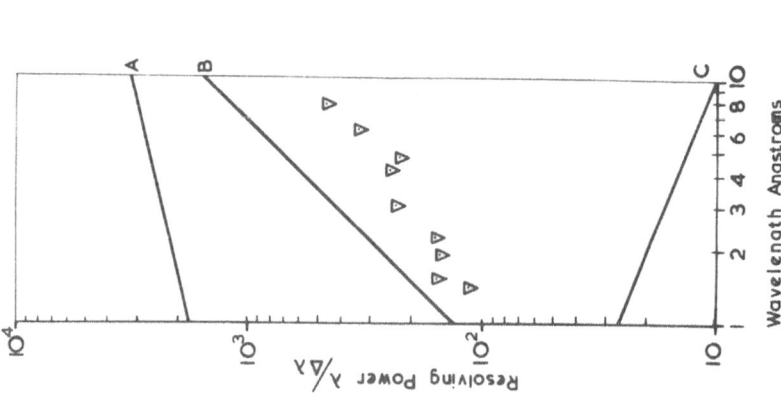

Figure 4. Spectrograph Resolution.   Figure 5. Diffracted Energy Distribution

REFERENCES

1. R.J.Speer, 1966 Ph.D. Thesis. London University.
    "An Extreme Ultra-Violet Monochromator for Solar
    Studies".

2. J.A.Bowles, W.M.Glencross, R.J.Speer, A.F.Timothy,
    J.G.Timothy and A.P.Willmore, "Preliminary Results
    from the Helium II 303.8Å Resonance Line Monochromator
    on the OSO IV Satellite". Ast.J. $\underline{73}$: S56, 1968.

3. J.W.Long, N.J.Peacock, P.D.Wilcock and R.J.Speer,
    "The Formation and Break-up of the Pinch in Plasma
    Focus". Proc.Conf. on Pulsed High Density Plasmas.
    Sept. 1967 Los Alamos. Report 3770. C5-1.

4. N.J.Peacock, P.D.Wilcock, R.J.Speer and P.D.Morgan,
    "Properties of the Dense Plasma Produced in Plasma
    Focus", Plasma Physics and Controlled Nuclear Fusion
    Research. Vol.11, 51-65, 1969.

5. N.J.Peacock, R.J.Speer and M.G.Hobby,
    "Spectra of Highly-Ionized Neon and Argon in a
    Plasma Focus Discharge". J.Phys.B. July 1969.

6. R.J.Speer, N.J.Peacock, W.A.Waller and P.J.H.Osborne,
    "A Focusing X-ray Spectrograph for Use in the Range
    0.5Å - 50Å. Submitted J.Phy

7. L.A.Sayce and A.Franks, "N.P.L. Gratings for X-ray
    Spectroscopy". Proc.Roy.Soc. $\underline{A282}$, 353-357, 1964.

8. A.Franks and K.Lindsey, "Dispersion of 1Å X-rays
    with N.P.L. X-ray Gratings". J.Phys.E. $\underline{1}$. Ser.2,
    144, 1968.

9. N.Bennett, J.Phys.E. 1969, In Press.

ELECTRON SPECTROSCOPY FOR STUDYING CHEMICAL BONDING

Ragnar Nordberg*

Institute of Physics, University of Uppsala

Uppsala, Sweden

The results reviewed in this article were obtained by means of the ESCA technique at the Institute of Physics, University of Uppsala, Uppsala, Sweden and at the Department of Physics, Vanderbilt University, Nashville, Tennessee, USA.

The ESCA technique is basically the study of induced emission of photo and Auger electrons from a sample irradiated with x-rays. If the incident radiation is monochromatic (e.g. an x-ray emission line) the spectrum of these electrons gives precise information about the energy states of the electrons in the sample. To extract this information, high resolution electron spectroscopy is necessary. Instruments for such spectroscopy have therefore been extensively developed during the last decade[1-5].

The binding energy of an electron is defined for solid compounds as the energy required to bring an electron from its orbit to the Fermi level and for gaseous compounds as the energy required to bring an electron from its orbit to infinite distance from the atom or molecule. The electron binding energy is not only a function of the element but also of its chemical environment. This influence on the core electron binding energy can be as large as 10 - 15 eV and is most commonly referred to as a chemical shift.

When a chemical bond is formed the valence electrons are re-arranged. For a pure covalent bond this rearrangement appears as an equal displacement of the valence charge for the two atoms. When atoms with different electronegativities form a bond, the displacement of the valence charge between the atoms is no longer equal. The

*Presently at Department of Physics, Vanderbilt University, Nashville, Tennessee.

bond can then be regarded as partly ionic because the electrons in
the bond have different probabilities of being found in the neighborhood of each of the two atoms. The atoms will thus appear to have
an effective charge. Estimations of this functional charge have been
made in many different ways, most of which are very crude. Oxidation
numbers, charges estimated from partial ionic character of bonds
(calculated from atomic electronegativity differences), as well as
charges obtained from extended Hückel and CNDO calculations have been
used.

As a first approximation the following simple model can be used
as an interpretation of the chemical shift. The changes in the distribution of the valence electrons, mentioned above, change the electrical potential inside the valence shell and thus also the potential
energy of the core electrons.

The chemical shifts are, however, not only a function of the
bonds formed by the studied atom but also of the potential distribution within the molecule. This effect, which can be looked upon as
an induced effect on the electronegativities seems to be a second
order effect and is demonstrated in the electron spectra of carbon
obtained from ethanol, ethyl trifluoroacetate and ethyl chloroformate
(Figure 1). The upper part of the figure shows a double line, which

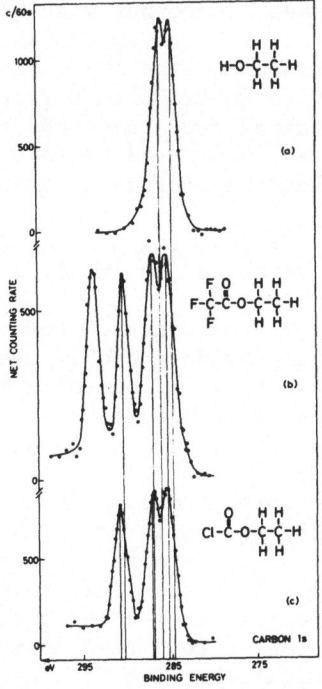

Figure 1. Electron spectrum from (a) ethanol, (b) ethyl trifluoroacetate and (c) ethyl chloroformate.

corresponds to the methyl and the methylene carbons in ethanol. The
line separation, obtained from a graphical resolution of the experimentally obtained line, is 1.4 eV. When the $CF_3C(O)$ group is substituted for the hydrogen atom in the OH group in ethanol, the higher
resulting electronegativity of the $CF_3C(O)$ group affects both the
methyl and methylene carbons. This is observed in the spectrum as a
shift in their binding energies. (See spectrum (b)). The methylene
carbon is influenced more than the methyl carbon, which results in
an increased separation. When the more electronegative $ClC(O)$ group
is substituted for the $CF_3C(O)$ group, the methylene carbon is shifted
towards still higher binding energy, but the effect is not large enough
to induce a measureable shift of the methyl carbon. (See Figure (c)).
The carbonyl carbon is shifted towards higher binding energy since
the Cl atom is more electronegative than the CF_3 group. This indicates that group electronegativities should be used instead of the
atomic electronegativities for the estimation of the partial ionic
character of a bond. Group electronegativities have therefore been
calculated for a series of carbon containing groups from an investigation of about 50 carbon compounds as reported in reference 6.

Another second order effect, very similar to the induced effect,
is the crystal energy contribution to the chemical shift. The size
of this effect has been difficult to establish from the experimental
results since, in many cases, compounds with widely varying material
structures show the same shift. The spread in the established correlations between the charges and the shifts in binding energy has
sometimes been attributed to contributions from the crystal potentials. However, data obtained from compounds studied in the gaseous
phase show the same spread. The cause of this spread is most probably the crude way in which the charges have been calculated.

Another model for interpretation of the chemical shifts is, for
instance, the one described by Fadley et al.[7].

The binding energies can be measured today with an accuracy of
about 0.1 eV and therefore provide a highly useful method for studies
of the chemical bonding as it is reflected in the binding energies.
In addition to this, the ESCA technique has a number of advantages
over established methods. It gives both a qualitiative and quantitative elementary analysis and it can be used to study any element in
the periodic system. The samples may be in any form: gas, liquid
or solid. The measurements are performed in a vacuum of about
10^{-6} torr. Therefore, if a compound is to be studied in the liquid
phase, it must have a vapor pressure lower than 10^{-6} torr at the
measured temperature. Special arrangements, such as a source holder
the temperature of which can be regulated over a wide range, are thus
necessary for most investigations of this type. Compounds, which are
solids at room temperature and normal atmospheric pressure, can therefore be studied in either the solid or gaseous phase and in some

cases even in the intermediate liquid phase. The restrictions for
the latter case are, as mentioned above, that its vapor pressure
must be lower than 10^{-6} torr at the measured temperature and that
its melting point is lower than the measured temperature. Figure 2
shows the electron spectrum from a silicon excited with Mg Kα-
radiation. The spectrum was obtained from the compound in the
liquid phase. This is one of the silicon containing compounds in a
series of about 30 which have been reported[8].

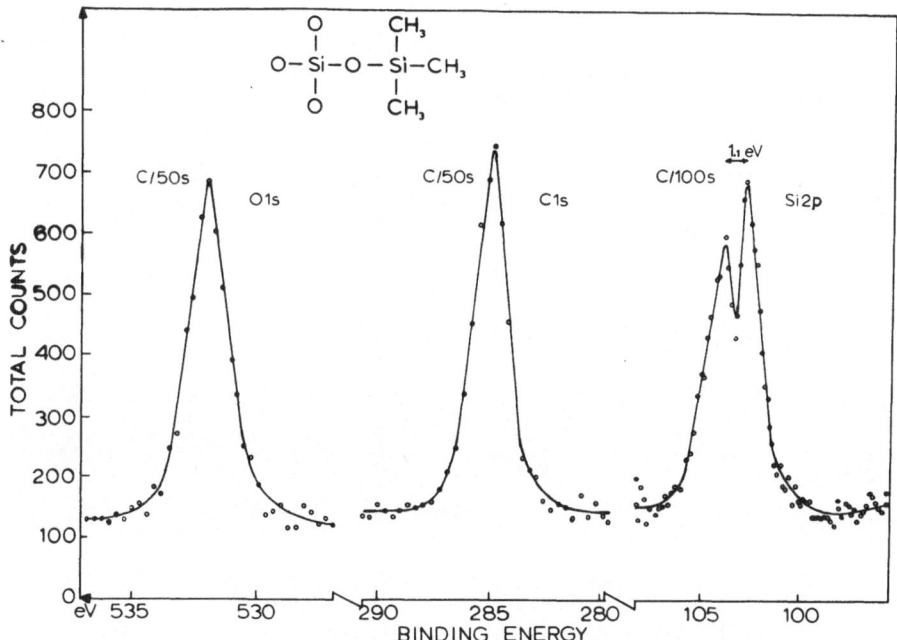

Figure 2.   Electron spectrum from a silicon in the liquid phase.
            The separation of the two lines in the silicon spectrum
            (1.1 eV) is much smaller than expected from compounds
            of the same type studied in the solid phase.

        Studies of the chemical bonding are not necessarily studies
of the valence electrons. The entire electronic structure of an atom
is changed when bonds are formed, so that the spectroscopic data of
core electrons as well as valence electrons give information of the
formed bond.

        The energies of both core and valence electrons are obtainable
with the ESCA technique. The photoelectron spectrum obtained from

N_2 with Mg Kα-radiation is shown in figure 3 as an example of how
all orbitals appear as lines[9]. The width at half maximum of the
nitrogen 1s line is about 0.8 eV. Using x-rays for excitation of
the electron spectra, narrow lines like this have only been obtained
from very simple molecules studied in the gaseous phase. There is
generally a considerable broadening of the lines upon solidification
and broadenings of as much as 50% have been observed. Shifts in the
binding energy have also been observed upon solidification. For
example, a mixture of aminobenzene and nitrobenzene was first studied
in the gaseous phase and then solidified and studied in the solid
phase. The obtained shift in peak position was found to be larger
for aminobenzene (3.1 eV) than for nitrobenzene (2.4 eV). A

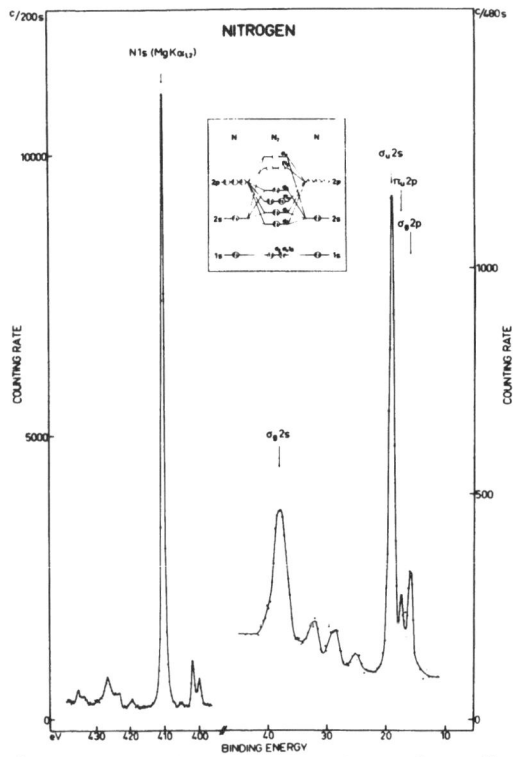

Figure 3. Electron spectrum from N_2.

similar effect seems to appear in silicon when the shifts in binding
energies obtained from compounds in the solid phase are compared with
those obtained in the liquid phase. The possible explanations of
these effects are however beyond the scope of this article. The
reader is referred to reference 9 for further information.

The electron spectrum of N_2 contains line structures on both
sides of the nitrogen 1s line. (Figure 4). Those at higher binding
energy originate from two different processes. The lines denoted A,

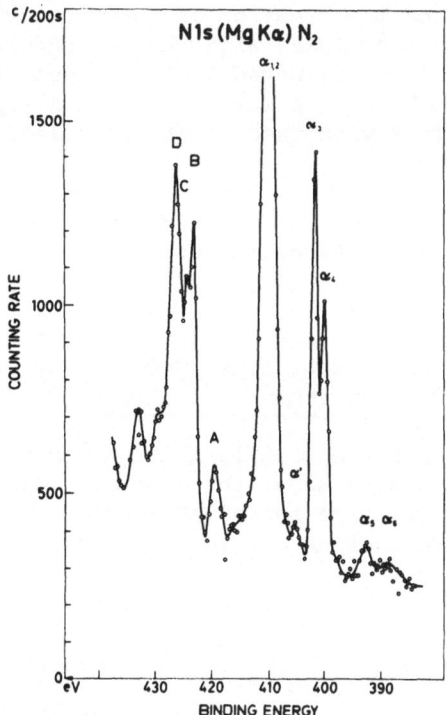

Figure 4. Magnification of the line structure around the N1s line.

B, C and D and a continuum starting at approximately 430 eV, represent electrons expelled from the 1s orbital but which have lost
energy upon excitation or ionization in inelastic collisions with
nitrogen molecules as they passed out of our source chamber. Evidences for the correctness of this interpretation are that the intensity of these peaks varies with the pressure in the source chamber and that the data is consistent with other energy-loss measurements for nitrogen. The second process causes another line structure
and presumably also another continuum partly hidden under the peaks
B, C and D. These peaks originate from the shake-up and shake-off
processes and are thus not pressure dependent. This structure in the
spectrum also consists of electrons expelled from the 1s orbital
with Mg Kα-radiation. The difference in energy between the electrons constituting the main line and thos constituting this structure depends in this case either upon excitation of the molecule,
in which one of the valence electrons is lifted up to one of the
first empty levels (shake-up), or upon ionization of the molecule
in which ejection of one of the valence electrons occurs (shake-
off). These processes are thought to occur as a result of the
sudden change in the electric potential within the atom or

molecule originating from the creation of the vacancy in the 1s
shell. The molecule is therefore, after having undergone the shake-
off process, left in a doubly ionized state. A detailed analysis
of these structures has been made for neon and the experimental data
will be presented together with theoretical calculations of the
process in reference 9.

The structure of the low binding energy side of the main peak
reflects the magnesium x-ray satellite structure and is identified
as the Mg Kα'-, Mg K$\alpha_{3,4}$- and Mg K$\alpha_{5,6}$-radiation.

In the low binding energy part of the spectrum the valence
molecular orbitals appear as four lines and the assignation is made
as shown in the level diagram inserted in figure 3.

The photoelectron spectra contain information not only about
the orbital energies but also about the character of the orbital
which is obtained from relative intensity studies. It has been
found, in separate cross section studies[9], that the Mg Kα photo-
absorption cross section is much higher per electron for the N2s
subshell than for the N2p subshell. These data can thus be used for
estimations of percentage of 2s and 2p character of the molecular
orbitals.

The studies of chemical bonding by means of the ESCA method
have been made for a number of years through investigations of the
energy shifts in the core shells, a technique which even today seems
to be the most rewarding.

Only one core level is involved in the photoelectric process
and the measured chemical shift in the photoelectron energies is
therefore the total influence on the core levels by the valence
electrons. This is in contrast to, for instance, x-ray emission
spectroscopy, where the difference between two levels is measured,
resulting in a much smaller shift.

Chemical shifts of the inner electronic levels in copper and
its oxides were reported as early as 1958[10]. In 1964 the first
photoelectron spectrum was recorded from a molecule containing one
element (sulphur) in two different oxidation states[11]. Two series
of sulphur containing inorganic compounds were then investigated[12],
in which the oxidation state of the sulphur varied from 2- to 6+,
namely, the sulphide, sulphite and sulphate of sodium and potassium.
Sublimated sulphur represented the zero oxidation state. The shifts
obtained in binding energy for the 1s, 2s and 2p electrons were then
plotted as a function of the oxidation state. The difference in
binding energy between the 2- and the 6+ states was found to be about
7 eV and was somewhat larger for the 1s than for the 2s and 2p electrons. This gives a binding energy shift of about 1 eV per oxidation state.

The width at half maximum of the lines in the obtained spectra
was about 6 eV and the accuracy in defining the line position was
well within one tenth of the linewidth. The main contribution to
the linewidth originated from the width of the radiations used to
induce the electron emission, which in this case were Cu $K\alpha_1$- and
Cr $K\alpha_1$-radiations. The need for the relatively hard radiations for
creation of 2s and 2p electron emission results from the fact that
the emitted electrons had to have a kinetic energy of at least 2.5
keV to be detected by the GM-tube, without being absorbed in its
window foil. This relatively large linewidth was, however, a limitation when finer details in the chemical shifts were to be studied.
The contribution to the final linewidth from the impinging radiation has now been successfully reduced by the selection of anode
materials with as low a Z-value as possible. Aluminum[13] and magnesium[14] were obvious choices, as both exhibited rather good thermal
conductivity and a $K\alpha$ spindoublet splitting which is smaller than
the natural line width of the $K\alpha$ emission lines. The width at half
maximum of the $K\alpha_{1,2}$ lines from aluminum and magnesium are < 0.8 eV.
Electrons expelled with this soft X-radiation exhibit a correspondingly lower kinetic energy. Therefore, since for a magnetic spectrometer $\Delta B \rho / B\rho$ is a constant, the spectrometer contribution to
the linewidth, on an absolute scale, decreases. An additional advantage of the low energy radiations is the higher photoelectric
cross section, which generally results in a gain in intensity
although the range of the photoelectrons in the sample is smaller.
Also the ratio between the characteristic radiation and the bremsstrahlung is increased and this is directly reflected in the electron
spectra.

An investigation of a series of nitrogen compounds was performed
with the above mentioned improvements of the art allowing more detailed observations of the chemical shifts[14]. The nitrogen 1s electron spectrum from 2-(4-nitrobenzenesulfonamide)-pyridine is shown
in figure 5. The spectrum was excited with Mg $K\alpha$-radiation. The
three lines in the spectrum originate from the three nonequivalent
nitrogen atoms in the molecule in such a way that the peak with the
highest binding energy corresponds to the nitrogen in the nitro
group and the peak with the lowest binding energy corresponds to the
nitrogen in pyridine. This designation was made after separate
measurements of isolated groups containing only one nitrogen atom,
or several nonequivalent nitrogen atoms of different amounts. In
the latter case the designation was made from the differences in intensity.

The binding energies obtained for nitrogen 1s electrons were
plotted against the charges, q_p, calculated by a method based on the
above mentioned concepts of partial ionic character (figure 6).
Resonance between valence bond structures was taken into account
where the percentage could be obtained from the literature. In

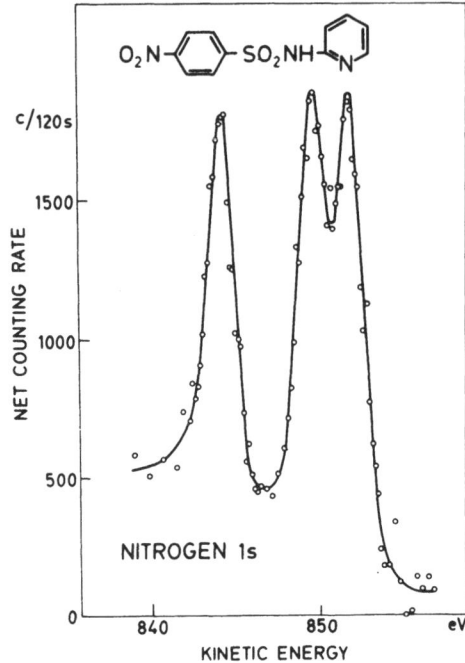

Figure 5.   Electron spectrum from 2-(4-nitrobenzenesulphonamide)-
            pyridine.

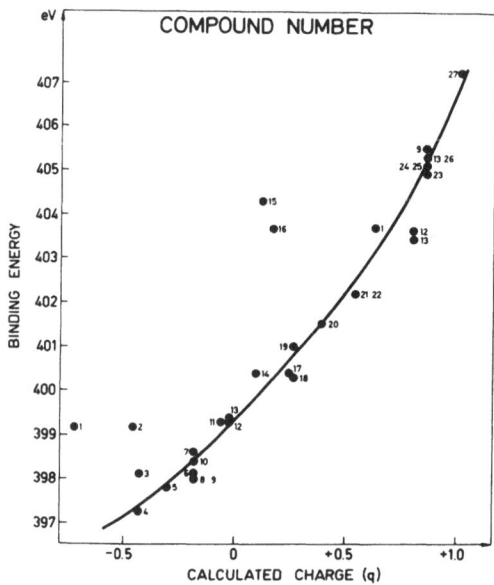

Figure 6.   Binding energy for the nitrogen 1s electrons versus
            calculated charge q_p.

compounds with formal charges on the atoms involved in the bond,
the electronegativities were corrected for the effect of this charge.
The procedure and the underlying theory for these charge calculations has been described in detail[4,14]. The technique will be demonstrated here by an example of how the charge for the nitrogen atom
in nitrobenzene has been calculated (figure 7). χ is the electronegativity of the atoms bonded to the nitrogen. χ is 3.0 for

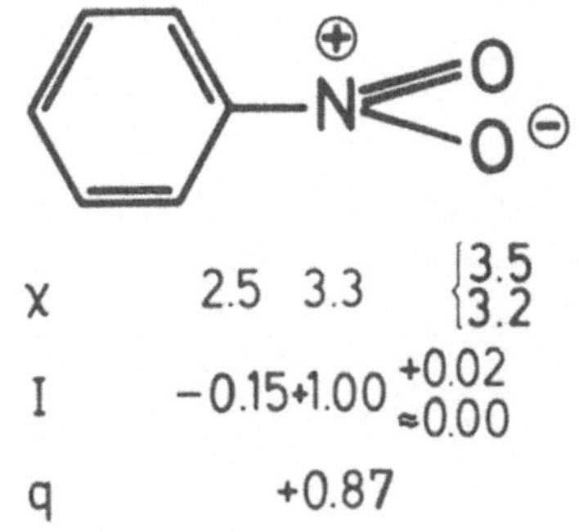

χ	2.5	3.3	$\begin{cases} 3.5 \\ 3.2 \end{cases}$
I	-0.15	$+1.00$	$\begin{array}{l}+0.02 \\ \approx 0.00\end{array}$
q		$+0.87$	

Nitrobenzene

Figure 7.   Calculation procedure of the charge on nitrogen in nitrobenzene.

nitrogen and 3.3 for a nitrogen with a formal plus charge. The
difference in electronegativity between the two oxygen atoms originates also from the formal minus charge on the one oxygen. The
partial ionic character, I, of the bonds is then calculated with
the formula

$$I = 1 - e^{-0.25(\chi_A - \chi_B)^2}$$

empirically derived by Pauling for calculation of the amount of
ionic character from the dipole moments of halides. A and B refer
to the two atoms involved in the bond. The sum of the formal charge,
Q, and the partial ionic character, I, of the bonds, was then taken
as an estimation of the charge, q_B, which gives q_B = +0.87 for the
nitrogen atom. The sign of I is determined by the sign of the electronegativity difference of the elements involved. Despite the very
approximate method for the calculation of the charge parameter and
despite the fact that the series under study was composed of compounds with widely varying structures, the correlation can be regarded as good. A more detailed discussion of the correlation is
given in reference 14.

The charges for most of these compounds have also been calculated by an extended Hückel method and a plot versus the binding
energy gave a rectilinear correlation (figure 8). It should be
observed that this method assigns charges for compounds 1, 2, 15 and
16 that make them fit the line.

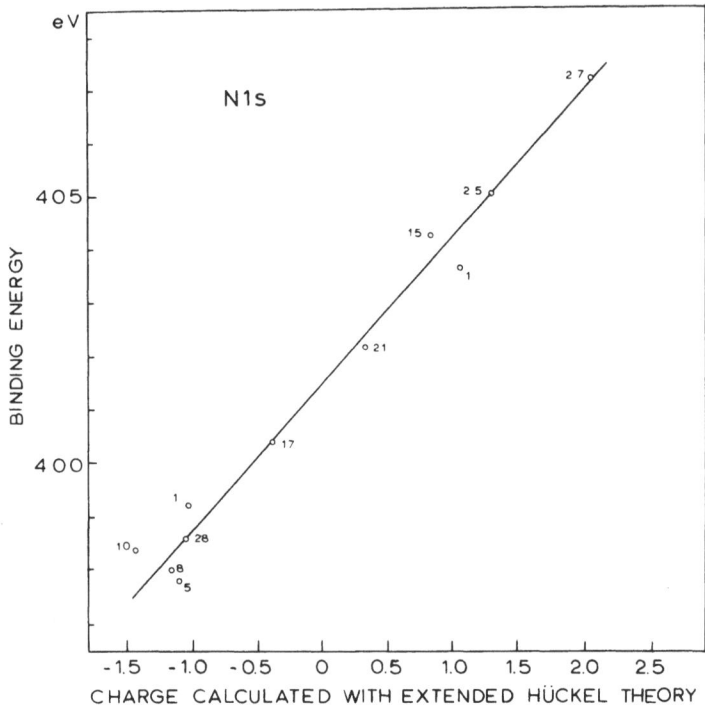

Figure 8.   Binding energy for the nitrogen 1s electrons versus
            charge calculated with the extended Hückel theory.

The charges for nitrogen, in almost the same series of compounds,
have also been calculated from CNDO molecular orbital eigenfunctions
by Hollander et al.[15]. The plot of the measured binding energies as
a function of the charge then forms two straight lines.

The series of sulphur containing compounds has been largely
extended[16] and a maximum binding energy shift of 15 eV has been
obtained for the sulphur 2p subshell. For the sulphur the calculated
charges, obtained by the method of partial ionic character of the
bonds, ranged from about -0.6 to about +2.5. This gives a binding
energy shift of approximately 5 eV per unit calculated charge, which
can be considered as a typical value.

Several different attempts were made to estimate the charge parameter in an analysis of the shift in the carbon ls binding energy[6]. As mentioned earlier, group charges of the groups attached to the studied carbon were estimated by assigning such charges to the groups that a best fit to a parabolic curve was obtained for the measured binding energy values.

The shifts in the binding energy of the carbon ls electrons were also correlated against charges calculated by the partial ionic character method. The good correlation obtained for the nitrogen data, when charges were calculated by the extended Hückel method, were not reproducible in the case of carbon, but instead a rather poor correlation was obtained.

In order to check whether the influence of a molecular potential could be accounted for in terms of charge, a correction term was added to the charges calculated by the extended Hückel method. This term was made proportional to the calculated molecular potential at the center of the atom considered. The proportionality constant was determined by a least square fit of our data to a straight line. This resulted in an improvement of the correlation over that where the uncorrected charges, obtained from the Hückel calculations, were used. This can be taken as an indication that the molecular potential exerts a non-negligible influence on the energy shifts. A plausible conclusion seems to be that the effect of the molecular potential is inherent in the uncorrected electronegativities. Whatever the true meaning of the charges obtained by the simple electronegativity treatment may be, it can be concluded that the expedience of q_b charges in giving useful correlations makes them considerably valuable in the empirical handling of core electron spectroscopy data.

A number of analytical structure problems have been solved with data obtained with the ESCA method. Some problems can be solved easily without the use of a correlation curve as was the case in the determination of whether the cystine S-dioxide had a thiolsulphonate structure or a disulphoxide structure[17]. In the latter case, the two sulphur atoms are equivalent, which means that the electron spectrum obtained would only show one line. In the former case, where the sulphur atoms are nonequivalent, two lines would appear in the electron spectrum. Two lines were obtained and it was therefore concluded that cystine S-dioxide had a thiolsulphonate structure.

In more complicated problems correlation curves are necessary. The procedure is then to construct a spectra for the element (with the help of the correlation curve and the lineshape obtained from the pure element) that would give the best evidence for a certain structure of the possible structures. Carbon has been used in this way for the analysis of two structure problems. The first problem

was to determine whether a compound consisted of the 2- or 6-oxa
analoges of DL-3,5,7-trimethyl-2,4,6,-trithia-1-adamantanol. The
second problem was to find if a compound consisted of 4,4'-thiodi-
(γ-valerolactone) or of 4-oxopentanethiolic anhydride[18]. Chemical
evidences and NMR favored the former structure. The structure
formulas and the constructed spectra are shown in figures 9 and 10.
The obtained electron spectra are shown in the figures 11 and 12.
A comparison between the experimentally obtained spectra and the
constructed spectra shows that in the first problem the compound
consisted of the 2-oxa structure (DL-3,5,7-trimethyl-2-oxa-4,6-dithia-
1-adamantanol) and that in the second problem the compound consisted
of 4,4'-thiodi-(γ-valerolactone).

The above results indicate that the ESCA technique gives information about the charge distribution in molecules and that it constitutes a new tool for a direct study of chemical bonding.

ADAMANTANE

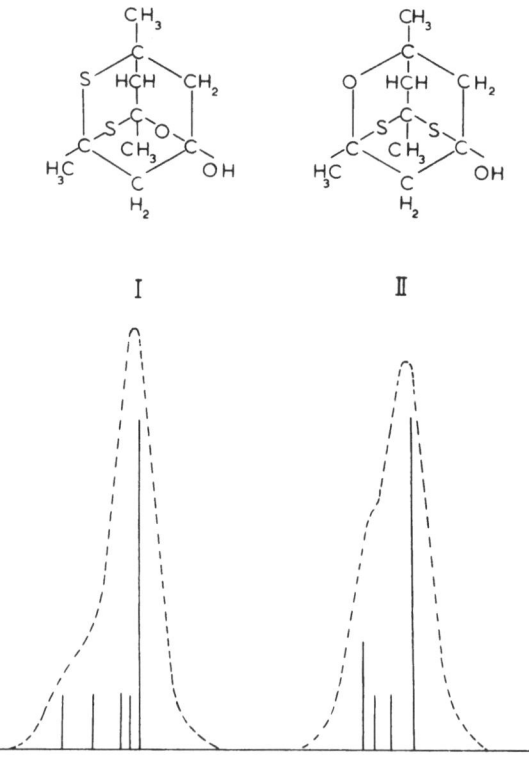

Figure 9.   Estimated lineshapes of the electron spectrum of carbon
            1s from 2- and 6-oxa analoges of DL-3,5,7-trimethyl-2,4,6-
            trithia-1-adamantol.

4,4'-thiodi-(γ-valerolactone)

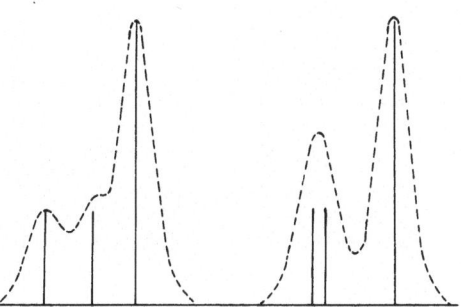

I   II

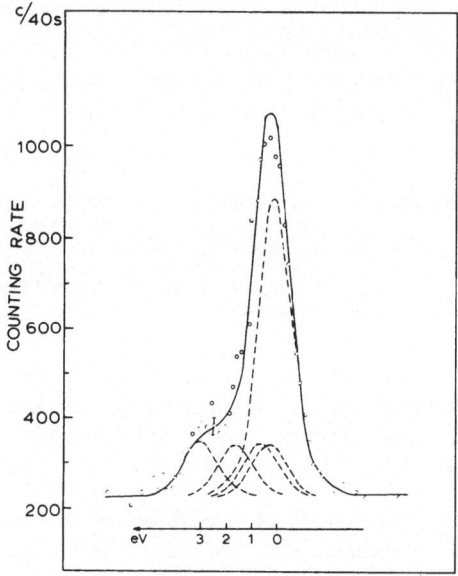

Figure 10. Estimated lineshapes of the electron spectrum of carbon
1s from 4,4'-thiodi-(γ-valerolactone) and 4-oxopentan-
ethiolic anhydride.

Figure 11. Obtained electron spectrum from adamantol. The full
line shows the sum of the obtained components, which
are given with dashed lines.

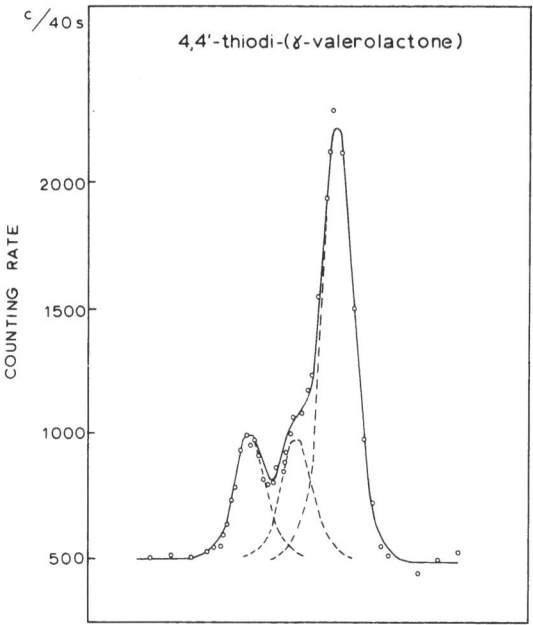

Figure 12. Obtained electron spectrum. The full line shows the
            sum of the obtained components which are given with
            dashed lines. The electron spectrum confirms the
            structure of 4,4'-thiodi-(γ-valerolactone).

REFERENCES

1. E. Sokolowski, C. Nordling, and K. Siegbahn. Arkiv för Fysik
    12, 301, (1957).

2. A. Fahlman, S. Hagström, K. Hamrin, R. Nordberg, C. Nordling
    and K. Siegbahn. Arkiv för Fysik 31, 479 (1966).

3. R. Nordberg, J. Hedman, P. F. Heden, C. Nordling and K. Siegbahn.
    Arkiv för Fysik 37, 489 (1968).

4. K. Siegbahn, C. Nordling, A. Fahlman, R. Nordberg, K. Hamrin,
    J. Hedman, G. Johansson, T. Bergmark, S.-E. Karlsson, I.
    Lindgren and B. J. Lindberg. Nova Acta Regiae Soc. Sci.
    Upsaliensis. Ser. IV. Vol. 20 (1967).

5. A. Fahlman, R. G. Albridge, R. Nordgerg: A Computer-Regulated
    ESCA Spectrometer System. To be published.

6. R. Nordberg, U. Gelius, P. F. Heden, J. Hedman, C. Nordling,
    K. Siegbahn and B. J. Lindberg, UUIP-581 (1968).

7. C. S. Fadley, S. B. M. Hagstrom, M. P. Klein and D. A. S. Shirley. J. Chem. Phys. 48, 3779 (1968).

8. R. Nordberg, R. G. Albridge, A. Fahlman, H. Brecht and J. R. Van Wazer. To be published.

9. K. Siegbahn, C. Nordling, G. Johnasson, J. Hedman, P. F. Heden, K. Hamrin, U. Gelius, T. Bergmark, L. O. Werme, R. Manne and Y. Baer: ESCA Applied to Free Molecules. North Holland Publishing Co. In print.

10. C. Nordling, E. Sokolowski and K. Siegbahn. Arkiv för Fysik 13, 483 (1958).

11. S. Hagström, C. Nordling and K. Siegbahn. Z. Physik 178, 439 (1964).

12. A. Fahlman, K. Hamrin, J. Hedman, R. Nordberg, C. Nordling and K. Siegbahn. Nature 210, 4 (1966).

13. R. Nordberg, K. Hamrin, A. Fahlman, C. Nordling and K. Siegbahn. Z. Physik 192, 462 (1966).

14. R. Nordberg, R. G. Albridge, T. Bergmark, U. Ericson, J. Hedman, C. Nordling, K. Siegbahn and B. J. Lindberg. Arkiv för Kemi 28, 257 (1967).

15. J. M. Hollander, D. N. Hendrickson and W. L. Holly. J. Chem. Phys. 49, 3315 (1968).

16. K. Hamrin, G. Johansson, A. Fahlman, C. Nordling, K. Siegbahn and B. Lindberg. Chem. Phys. Letters 1, 557 (1968).

17. G. Axelson, K. Hamrin, A. Fahlman, C. Nordling and B. J. Lindberg. Spectrochim. Acta. 23A, 2015 (1967).

18. B. J. Lindberg, J. Hedman, P. F. Heden, R. Nordberg and C. Nordling. Spectrochim Acta. In print.

IEE - A NEW TYPE OF X-RAY PHOTOELECTRON SPECTROMETER

N. H. Weichert and J. C. Helmer

Varian Associates

Palo Alto, California 94303

ABSTRACT

A new type of x-ray photoelectron spectrometer has been designed. Although smaller in size, the new instrument shows a considerable improvement in sensitivity compared to other photo-electron spectrometers. This has been achieved through the provision of several new features: (1) The photoelectrons are retarded to low energy (10 eV through 100 eV) before entering the deflecting energy analyzer. (2) A spherical electrostatic analyzer in figure of rotation is used. (3) Annular slits at both source and detector provide a large effective source area. (4) The instrument is shielded against external magnetic fields (earth magnetic field and a.c. fields) by a double layer of mu-metal, thus avoiding the installation of large Helmholtz coils.

Measurements show that counting rates up to 10^4 sec^{-1} are obtained in cases where counting rates from other spectrometers have been in the range of 10^2 to 10^3 sec^{-1}.

INTRODUCTION

Since the recording of the first ESCA spectrum in 1954 and the first observation of chemical shifts in 1957 by K. Siegbahn[1] and his coworkers in Uppsala, X-ray photoelectron spectroscopy has proven to be a very promising analytical tool. The range of possible applications can hardly be overestimated. Until recently the necessary high resolution could be achieved only by the famous magnetic double focusing, iron-free spectrometers, which had been developed by Siegbahn and his colleagues. These instruments are very precise in mechanical design and they rely on a set of large compensating Helmholtz coils to provide a well-defined magnetic environment.

It seemed to us that a smaller instrument, less sensitive to magnetic pertubations, should open a large field of practical applications for X-ray photoelectron spectroscopy. Our new approach, which we present in this paper, appears to fulfill that requirement.

PRINCIPLES

For every type of electron spectrometer, there is a natural relation between its resolution and luminosity. So generally if an instrument is used at high resolution it has a poor luminosity. In an earlier paper[2] we showed that a considerable gain in sensitivity is to be expected, if one uses a spectrometer of low resolution but high luminosity together with an electrostatic retarding field, which slows down the electrons before they enter the spectrometer. When the energy of all electrons of a source is reduced by the same amount upon retardation, a certain spread ΔE in the energy spectrum is of the same absolute value after retardation as before. At the lower energy E, however, the relative spread $\Delta E/E$ is larger, thus allowing the use of a spectrometer of lower resolution* and high luminosity. It is true that the current density of the electrons coming from the source is reduced upon retardation, but this effect is overcompensated by the gain in luminosity. (For a detailed description see ref. 2).

For a system, where these general ideas could be employed, a spectrometer capable of high luminosity had to be designed. High luminosity $L = A\Omega$ requires both a large effective source area A (area given by the entrance slit of the spectrometer) and a large accepting solid angle Ω (solid angle seen from the sample, into which electrons can be emitted and accepted by the spectrometer field). A spectrometer, which can have a very large solid angle, is the spherical condenser type. It was first described by Purcell[3] and later on used in various modifications[4]. In its original design it consists of two concentric spherical electrodes in figure of rotation (figure 1). Electrons emerging from a point S on the axis of rotation a are focused to point F on the axis. With respect to the angular spread of traces in the plane of the figure, focusing is of first order. For reasons of symmetry we get perfect focusing for equivalent traces in different planes containing the axis a. For practical applications the point source S has to be replaced by a

---

*In order to avoid confusion we will call resolution the number, which is given by energy E divided by the energy selectivity $(\Delta E)_S$ of the spectrometer and not its reciprocal, as it is done sometimes in the literature.

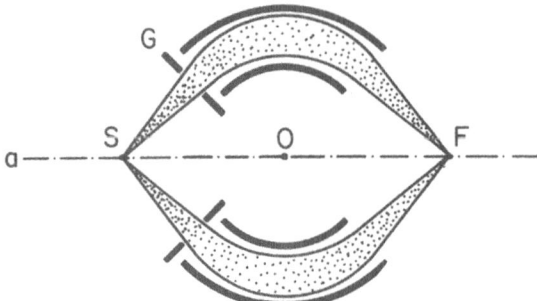

Figure 1. Spherical electrostatic spectrometer, Purcell type.

source of finite size. The image of that source appears at the
area around F, where a suitable detector aperture can be placed.
The angular spread of the electrons, which gives rise to second
order aberrations at the image, can be controlled by an annular
aperture G. The width of the aperture has to be balanced against
the source area in order to obtain the highest possible luminosity
at a given resolution. In the optimum case, the area of a circular
disc source at S turns out to be rather small. So despite the large
solid angle, only moderate luminosity can be achieved.

When the spherical electrodes are not used in figure of
rotation, but only as a sector field (see e.g. K. Siegbahn et al.[1],
page 198), extended line sources can be used. In this arrangement,
however, solid angle is sacrificed and due to the limitation of the
spherical electrodes additional fringing field problems occur.

We found that a combination of both types is indeed possible,
which gives a spectrometer of the desired high luminosity. Again
we use two spherical electrodes in figure of rotation. It is known
(Barber's rule for spherical condenser), that electrons from a
point source S' are focused to a point F', which is given by the

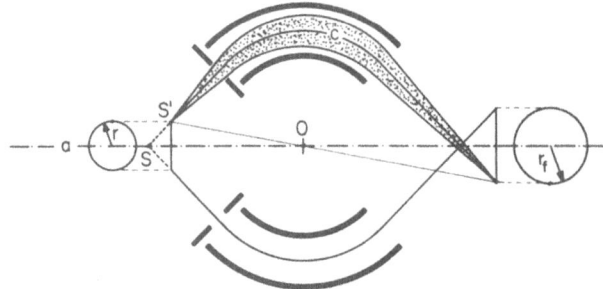

Figure 2. Focusing for an off-axis source point S'.

intersection of the line S'O and the central trajectory c. F' will
be a first order focusing point for all electron trajectories in the
drawing plane. Trajectories leaving S' in other directions are no
longer in a plane equivalent to the drawing plane, as has been the
case in figure 1. They introduce a new kind of second order
aberration at F' depending on the initial inclination of those
trajectories against the drawing plane. The amount of focus
distortion due to this aberration can be controlled by limiting the
angle of inclination. S' can be considered as a representative
point of a circle around the axis with radius r, so obviously an
annular source can be focused onto an annular detector slit (figure 3).
The source can have a finite width comparable to the size of the
disc source in the Purcell type, but as it has a length (2πr) many
times larger than its width, a tremendous increase in source area
has been achieved. The question, under which conditions this
configuration really gives a substantial increase in luminosity,
has to be decided in a detailed analysis, which we give in the
following chapter.

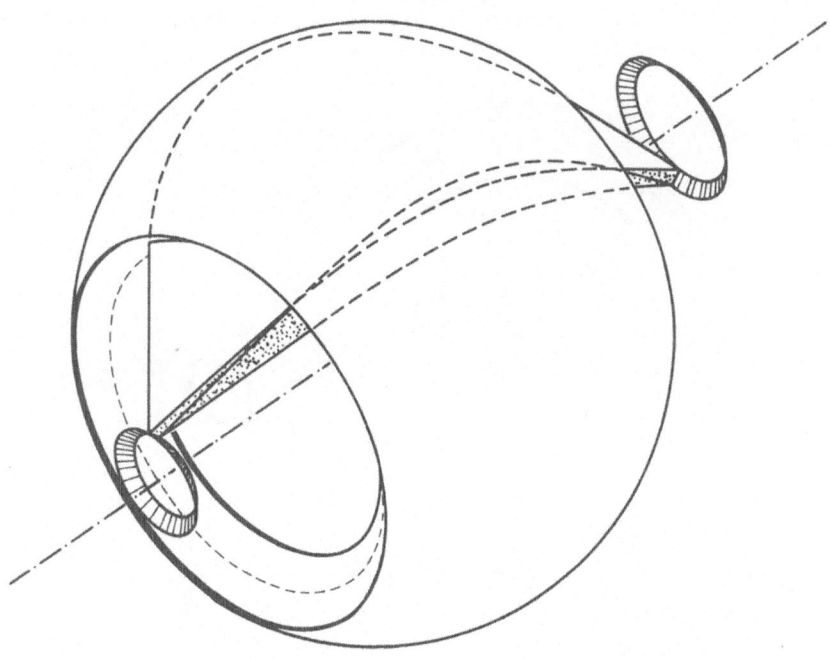

Figure 3. Focusing in the case of an annular source.

ANALYSIS

    The notations can be seen from figure 4 and are basically the
same as used by Purcell[3]. The annular source shall have a mean
radius r and a width w (figure 4). The angle $\pm\varphi$ defines the source
area, from which electrons will be allowed to reach the detector
slit, if they enter the spectrometer field at a point C. Electrons
from source points beyond that area, having traces of higher
inclination at C, will be blocked off by a circular aperture
$X = 2rs\sin\varphi$ at the crossover area. Both for convenience at the final
instrumental arrangement and for simplicity in the calculations an
angle $\emptyset = 90°$ and $\theta_1 = \theta_2 = 45°$ was chosen. Furthermore we assume
$w \ll r \ll r^0$, otherwise the calculations would get very laborious and
complicated.

    The total width of the image produced by a representative part
of the source (shaded areas in figures 4 and 5) is

$$B = (w + 2r^0\alpha^2)/\cos\varphi + r^0(r/r^0)^2 \sin^2\varphi \qquad (1)$$

This is the base width of a line due to monoenergetic electrons, as
it is obtained by scanning the indicated area over a circular
detector slit of radius r_f and zero width. The scanning motion
generated by slightly varying the energy is directed parallel for
all points of the representative area as shown by the arrow (figure 5).

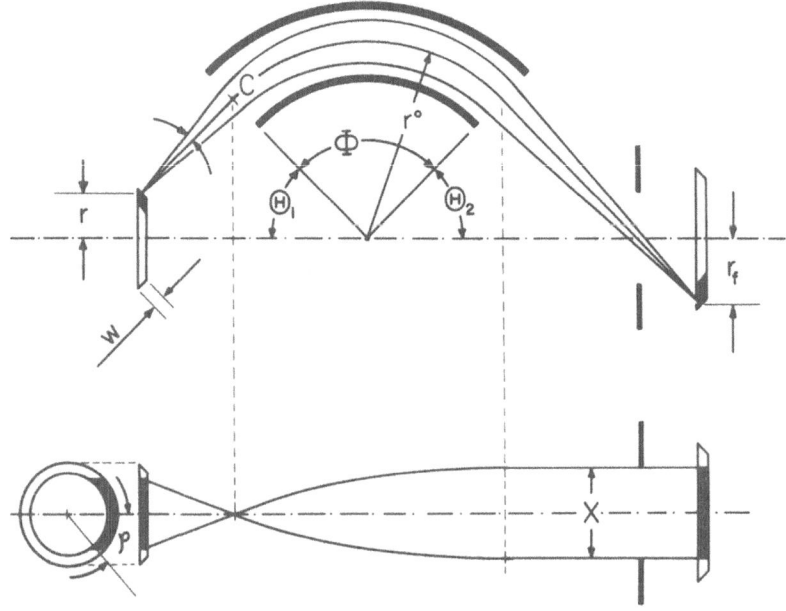

Figure 4. Trajectories and notations in the case of an annular source.

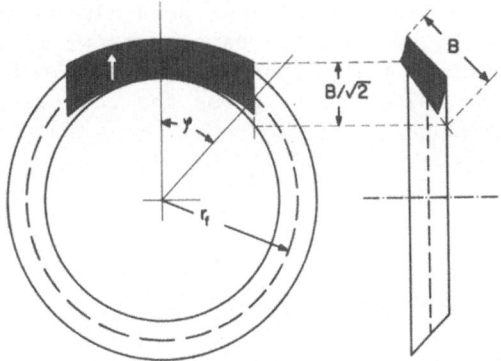

Figure 5. Image of a representative part of the source as defined in figure 4.

In the formula w is the contribution from the finite source width, $2r^o\alpha^2$ gives the well known aberration due to the α-angular spread, the factor $1/\cos\varphi$ taking care of the increased width of a ring-shaped area, which is scanned in a nonradial direction. The last expression in the formula has been derived for the additional image distortion from trajectories with $\varphi \neq o$. These trajectories are passing through a deflecting field, which is slightly longer than $\emptyset = 90°$, thus suffering an additional deflection. A second aberration of the same amount as this deflection has to be included, which is not so obvious. It consists essentially of a parallel offset of such inclined trajectories at the focus.

The formula has been derived assuming $\varphi \leq 45°$ and $w \ll r, r_f$. Again for simplicity, the magnification of the image has been neglected. This should be of no significant influence to the resolution, which is given by $R_B = D/B$, since the magnification affects both the energy dispersion D and the trace width B in a similar way. The energy dispersion in our case ($\theta_1 = \theta_2 = 45°$) will be $D = 2r^o$, so we get for the resolution R_B expressed by its reciprocal

$$\frac{1}{R_B} = \frac{w/2r^o + \alpha^2}{\cos\varphi} + \frac{1}{2}\left|\frac{r}{r^o}\right|^2 \sin^2\varphi \tag{2}$$

No contribution from the finite width of a detector slit is included so far, which will be considered later on.

The source area will be $A = 2r\pi w$, for the solid angle Ω we get $\Omega = 2\sqrt{2}\alpha\varphi$. So the luminosity $L = A\Omega$ is given by

$$L = 4\pi\sqrt{2}\, w\, r\, \alpha\, \varphi \tag{3}$$

The problem now is to find the highest possible value of L for a given resolution R_B by varying the parameters w, r, α, and φ. A solution with respect to the variables w, r, and α by means of Lagrange's method of multipliers, gives the following equations:

$$(w/2r^o)/\cos\varphi = 1/2R_B \tag{4}$$

$$\alpha^2 \cos\varphi = 1/4R_B \tag{5}$$

$$\frac{1}{2}(r/r^o)^2 \sin^2\varphi = 1/4R_B \tag{6}$$

If in this set of equations a special value is assigned to any one of the four parameters w, r, α, and φ, all four will be fixed for a given resolution R_B. If we use φ as the independent parameter, we can express L as a function of φ:

$$L = 2\pi \frac{(r^o)^2}{R_B^2} \frac{\varphi}{\text{tg}\varphi} \sqrt{\cos\varphi} \tag{7}$$

This is a very slowly decreasing function of φ, having its maximum at $\varphi = o$. This value, however, is not allowed, because it implies according to equation (6) an infinite source radius r, which violates our condition $r \ll r^o$. So r may be set as large as possible under the condition $r \ll r^o$, the actual value being not critical with regard to equation (7), since for all corresponding values of φ the luminosity is close to $L = 2\pi(r^o)^2/R_B^2$. For any value of r the other parameters are given by

$$\sin\varphi = \frac{1}{(r/r^o)\sqrt{2R_B}}, \quad \alpha = \sqrt{\frac{\cos\varphi}{4R_B}}, \quad w = \frac{r^o \cos\varphi}{R_B} \tag{8}$$

It might be interesting to compare these results with those for an optimized instrument with a disc source of radius r_0 on the axis:

$$L_d = \frac{32}{125} \sqrt{10} \, \pi^2 \frac{(r^o)^2}{R_B^{5/2}}, \quad \alpha_d = \sqrt{\frac{1}{5R_B}}, \quad r_d = \frac{2r^o}{\sqrt{5R_B}} \tag{9}$$

At a resolution of $R_B = 100$, this gives a luminosity L_d which is about 8 times smaller than L in the case of a ring source.

An accurate treatment of the influence of the detector slit is very complicated, because the actual line shape has to be calculated, to get the most effective slit width. An estimate showed us, that a slit of half the total width B will be a proper choice. The base width of a recorded monoenergetic line thus will be $\frac{3}{2}$ B. Full width at half maximum amounts to about half this value, i.e. $\frac{3}{4}$ B. So the final resolution (FWHM) will be

$$R = (4/3)R_B \qquad (10)$$

In the actual design of the detector slit we consider that a certain magnification of the image appears. Therefore the radius of the detector slit has to be

$$r_f = \frac{r}{1-\sqrt{2}(r/r^o)} \, , \qquad (11)$$

and for the slit width we get

$$w_f = \frac{B}{2} \frac{r_f}{r} = \frac{w}{\cos\varphi} \frac{1}{1-\sqrt{2}(r/r^o)} \qquad . \qquad (12)$$

DESIGN AND CONSTRUCTION

A general schematic of the instrument, which we built based on our theoretical results, is shown in figure 6. The radius r^o of the center trajectory was chosen to about 9 cm, the spherical electrodes being separated by a gap of 2 cm. The spectrometer is designed for a resolution (FWHM) of R = 100, which for example gives an instrumental line width of 1eV at a spectrometer energy of 100eV. The photoelectrons are slowed down to the spectrometer energy by a retarding field between the sample and the annular source slit of the spectrometer. This slit as well as the crossover aperture, the detector slit, and the slit at the edge of the spherical condenser are connected to ground. Voltages of opposite signs are applied to the two spheres, so that the potential at the central orbit equals

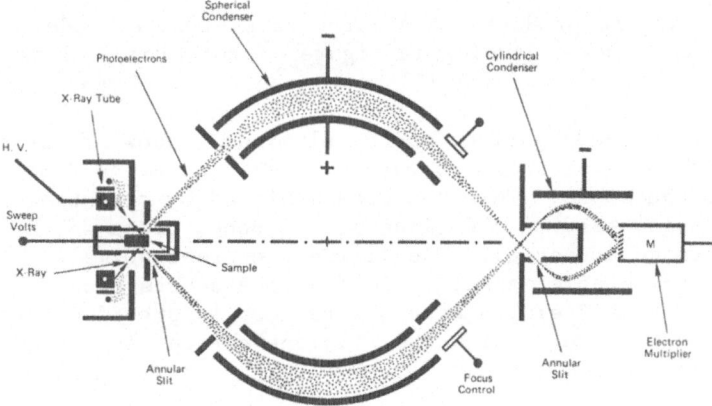

Figure 6.   Schematic of the instrument.

zero. (At an energy of 100eV the voltages are -20V and +25V.) In
order to correct for small, unavoidable misalignments of the whole
arrangement, additional electrodes are provided between the spheres
and the exit slit. These consist of eight segments, to which
different voltages can be applied. In this way we obtain a perfect
annular focus coinciding with the detector slit. Electrons passing
through the detector slit are deflected by a coaxial cylinder
electrode into a windowless electron multiplier (EMI9603). This
cylindrical deflector serves at the same time as a trap for spurious
low energy electrons, which are backscattered from field electrodes
or slits.

The sample, a cylinder of 12 mm in diameter is enclosed in a
housing at the same potential as the sample. The X-rays enter the
housing through a thin aluminum window and eject photoelectrons from
the sample. The photoelectrons escape through an annular slit in
the housing. Retardation takes place between this slit and the
spectrometer source slit. In this way the sample is not exposed to
external electric fields, which is important if insulating samples
are to be used. In the field free environment and under the influence
of the X-radiation, charging up of the sample surface due to the loss
of photoelectrons is almost completely suppressed, since low energy
electrons from the environment of the sample tend to neutralize
surface charges. In an external electric field this balancing
process is strongly disturbed, giving rise to much more severe
charging up effects.

An annular area of about 2 mm width is the active part of the
sample surface. Samples can be cylinders, thin layers deposited on
a substrate cylinder, or powders mounted with double stick adhesive
tape on a base cylinder. The shape of the sample is not critical
and flat samples can be used as well. The sample mount can be cooled
by a cold gas flow or heated by an internal heater.

The X-rays are produced on a large water-cooled anode,
surrounding the sample housing in figure of rotation. It is
bombarded by electrons from a filament loop. The filament is
mounted in such a way, that evaporating filament material cannot
contaminate the anode surface or the aluminum window of the sample
housing. So the electrons are focused on curved paths from the
filament to the anode. This has been achieved by operating the anode
at high voltage, keeping the filament and other parts of the X-ray
source at ground potential. The filament current normally would
produce a considerable magnetic field at the source area. In order
to compensate this field, a second wire loop is provided parallel
to the filament through which the filament current is passed in
opposite direction.

The whole system is incorporated in a cylindrical vacuum
chamber of 33 cm in diameter and 70 cm length. The chamber is

pumped by a titanium sublimation pump, which produces a vacuum of 10^{-7} to 10^{-8} mm Hg. Carbon contamination of samples, as it appears in systems with oil diffusion pumps, is avoided by the use of the sublimation pump.

Since only electric fields are used in the spectrometer, the instrument can be shielded against external magnetic field by mu-metal. Two open ended nested cylinders about 1m in length and just fitting over the vacuum chamber, reduce the earth magnetic field in the critical areas to less than 1 mGauss. The magnetic shielding is not only effective against constant fields, but also against changing, switched and a.c. fields, which are quite common in a normal laboratory. Helmholtz coils are not required.

OPERATION AND ELECTRONICS

In order to make the operation of the instrument as convenient and versatile as possible and to have an effective data acquisition system, a small computer (Varian 620/i) is employed in the electronic part of the instrument. A spectrum is obtained by sweeping the retarding voltage, keeping the spectrometer at a fixed energy. The computer controls a power supply, which sets the lowest retarding voltage of the spectrum. It also generates a ramp voltage, which is superimposed to the initial voltage. As the retarding voltage goes through the sweep, the pulses from the electron multiplier are

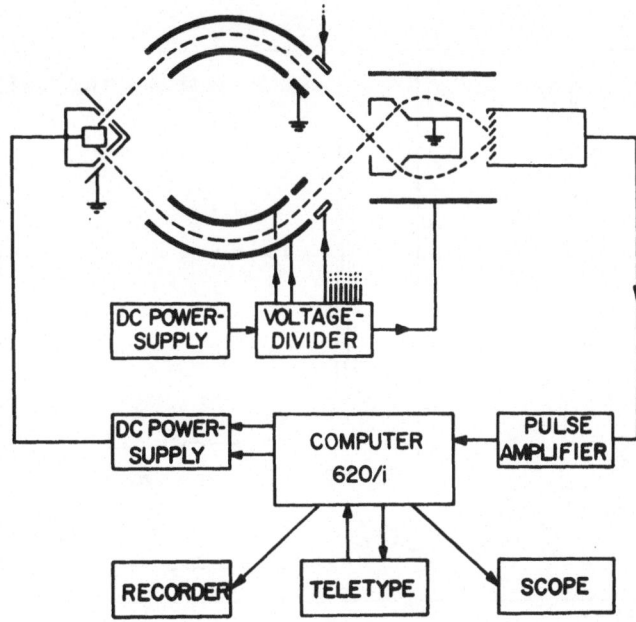

Figure 7. Block diagram of the electronic part of the system.

counted and stored by the computer according to the corresponding
energies. The spectrum can be either displayed on an oscilloscope,
printed as numerical data on a teletype or plotted on a recorder.
Multiscan operation and automatic scanning of several different parts
of the total spectrum is possible. In addition to the basic
procedures, the computer can be used for data processing such as
smoothing of the stored spectra, etc.

The spectrometer energy is defined by the voltage; which is
applied between the two spheres. This voltage and also the voltages
for the eight focus correcting electrodes and for the cylindrical
deflector are derived from a common power supply. So by changing
only one voltage, the spectrometer energy and consequently the
resolution of the total system can be changed. The resolution of
the spectrometer is set for a resolution of 100, so we get an
instrumental contribution to the linewidth of a photoelectron line
of 1eV at a spectrometer energy of 100eV. This value is about equal
to the natural linewidth, when aluminum or magnesium Kα radiation
is used. In cases, where higher resolution is desirable, the
instrumental contribution can be narrowed down to about 0.2eV at the
expense of sensitivity, by reducing the analyzer energy to 20eV.

SPECTRA

We select a few examples of spectra to show the performance of
the instrument. One of our earliest spectra we recorded was from

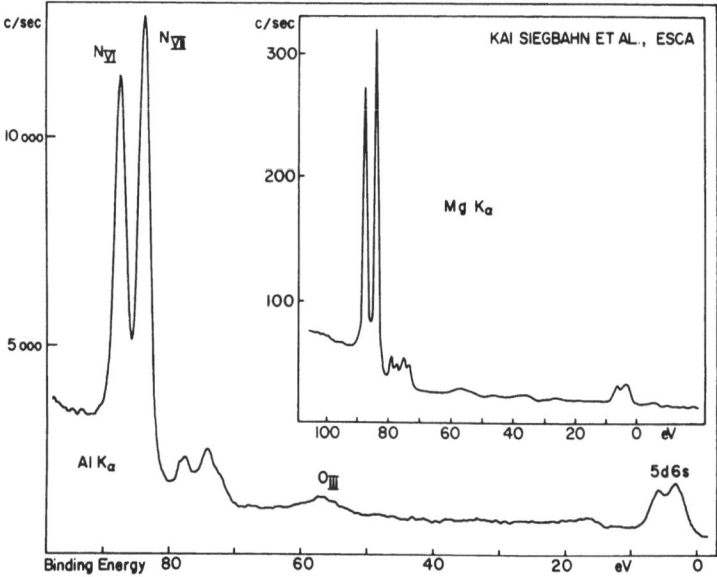

Figure 8. Spectrum of the outer levels in metallic gold.

some outer levels in metallic gold (figure 8). For comparison we
also show the same spectrum as published by the Siegbahn group. At
about the same resolution, it shows a counting rate, which is 30
times lower than in our spectrum. We also recorded the N_{VI}, N_{VII}
doublet of this spectrum at different resolutions, to demonstrate
this feature of the instrument. The two given examples show, that
the new spectrometer although smaller in size than comparable
instruments promises to be a real progress in the field of
instrumentation for X-ray photoelectron spectroscopy.

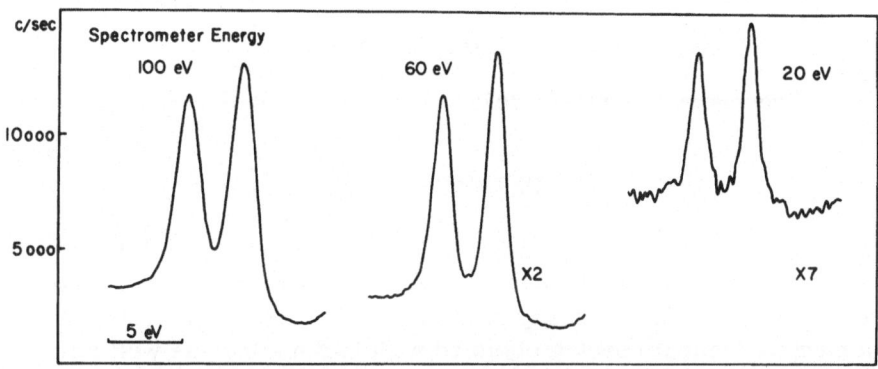

Figure 9. N_{VI}, N_{VII} doublet of gold at different resolutions.
Measuring time for each spectrum 100 sec, number of channels 200.

REFERENCES

1. K. Siegbahn et. al., "ESCA, Atomic, Molecular and Solid State
    Structure Studied by Means of Electron Spectroscopy",
    Almquist & Wicksells Boktryckeri AB, Uppsala 1967.

2. J. C. Helmer and N. H. Weichert, "Enhancement of Sensitivity
    in ESCA Spectrometers", Appl. Phys. Letters 13, 266 (1968).

3. E. M. Purcell, "The Focusing of Charged Particles by a
    Spherical Condenser", Phys. Rev. 54, 818 (1938).

4. See e.g. references 2-10 in J. Kessler and N. Weichert,
    "Untersuchungen an einem Energieanalysator für Relativistische
    Elektronen", Nuclear Instrum. 29, 245 (1964).

α- EXCITED AUGER SPECTRA

L. I Yin, I. Adler and R. E. Lamothe

NASA/Goddard Space Flight Center

Greenbelt, Maryland 20771

ABSTRACT

The possibility of obtaining Auger spectra excited by a Po ^{210} α-source has been investigated. The obvious advantages in the use of radioactive α-sources are simplicity, stability, and high ionization cross section for light elements. Typical spectra obtained with this method, as well as parameters affecting the characteristics of these spectra, are presented and discussed. The following observations have been made: the intense low energy continuum background make it difficult to detect the presence of Auger lines in this region; the abundance of Auger lines in the above region makes identification difficult; the intensity of higher energy Auger lines is too low to be observed with the present α-source and low vacuum system; thus the practicability of α-excited Auger spectroscopy will need further investigation.

INTRODUCTION

There is a rapidly growing interest in the use of Auger and photo-electron spectroscopy for surface studies, chemical analysis, and studies of chemical bonding and valence states, particularly for the light elements. Excitation has commonly been accomplished either with x-rays or with electrons. In the former case both photoelectrons and Auger electrons are produced. In the latter case spectroscopic information is obtained only from Auger electrons. Although the excitation of Auger electrons via nuclear particles exists in theory, to the best of our knowledge the

418

experimental investigation of these spectra is as yet uninitiated. This paper describes our efforts in using, specifically, α-excitation.

OBJECTIVES

Under the bombardment of α-particles the theoretical inner shell ionization cross section has been shown to be proportional to the fourth power of the incident particle energy and inversely proportional to the twelfth power of Z, the atomic charge of the target.[1,2] Judging from this, α-excitation would almost be the ideal mode of excitation for light elements. The usefulness of the Auger spectra of light elements is of course well known and needs no further elaboration. Another obvious advantage of α-excitation is its simplicity. α-sources are easily available, small in size, stable, and require no external power. The geometry of the source, as well as the source-target arrangement, can easily be changed at will. By judiciously choosing the window material for the α-source, the α-excited fluorescence x-rays from the window material can be used to produce photoelectrons from a sample. Thus, in principle, it is possible to study both the Auger electrons and photoelectrons from a sample by using a single α-source.

What are some of the anticipated difficulties? First of all, the theoretical ionization cross section mentioned earlier was arrived at using the Born approximation for inelastic collisions at the low energy limit. One of the assumptions made in the Born approximation is that there be negligible disturbance of the electron orbits by the approaching α-particle. This assumption is valid for high-Z material, where the electron orbits are tightly bound to the nucleus, but it is likely to break down for the light elements. Furthermore, at the energies of the available α-sources, the low energy limit requirement also breaks down for light elements. Therefore the form of the theoretical inner shell ionization cross section is really quite ambiguous for low-Z material. Second, the vast majority of ionizations by the α-particle take place in the outer shells, with low energy-transfers, thus the majority of the emitted electrons for an α-bombarded sample have very low energies. This situation, coupled with the usual energy loss of electrons in emerging from a solid sample, forms an intense continuum background whose shape may be difficult to predict. Such a background would interfere severely with the discrete Auger lines. Third, one faces the often conflicting demands between source strength and health safety that are common to all experiments with radioactive sources. Because of these things, we feel that α-excited Auger spectroscopy holds enough promise and enough ambiguities to warrant experimental investigation.

INSTRUMENTATION

The instrumentation used in this experiment has been reported else-where.[3] An oil-diffusion pump system operating at 10^{-6} torr is used to house the Po^{210} α-source, the sample, and a hemispherical electrostatic spectrometer with 0.6% energy resolution. The 0.5 curie Po^{210} source is sealed in a stainless steel cylinder 3/8" in diameter and 1/4" high with a 0.00015"-thick 302 stainless steel window. The active area is approximately 3/16" in diameter. The source can be mounted in any position. The electrons are pulse-counted by a channel electron-multiplier; a multichannel analyzer operated in the multiscaler mode displays the differential energy spectrum of the electrons. A linear voltage ramp generator supplies the voltages to the spectrometer hemispheres. Source, sample, and the entrance and exit slits of the spectrometer are kept at the same potential (usually ground). The block diagram of a typical experimental arrangement can be seen in figure 1.

RESULTS

X-Rays

The x-rays from the α-source were measured with a proportional counter. As expected, they were shown to be essentially the K radiations

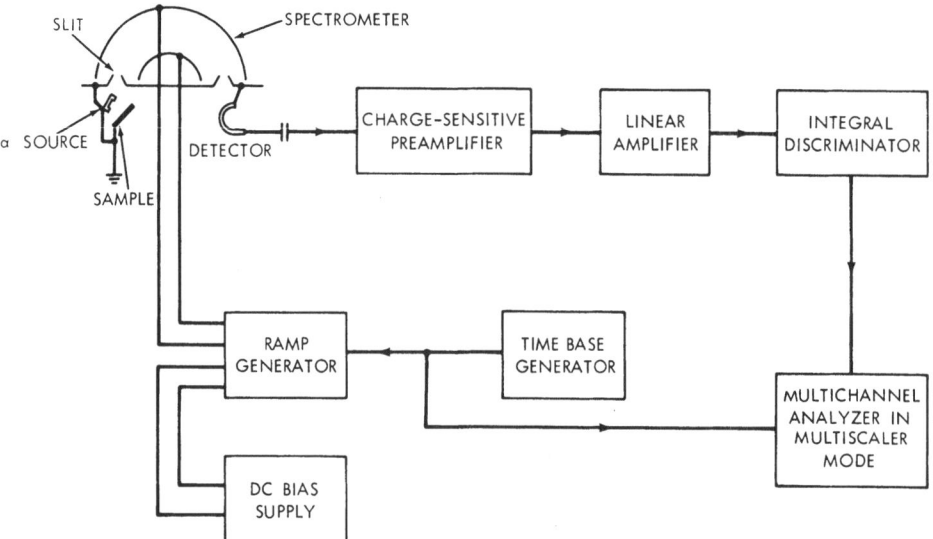

Figure 1. Block diagram of spectrometer and associated electronics
at a typical experimental geometry.

of iron from the stainless steel source window. However, the intensity of the x-rays was so low that we were unable to observe any photoelectrons produced by them from the samples.

Electron Spectrum from the Source

Figure 2 shows the electron spectrum of the α-source itself. This spectrum was obtained by placing the source directly over the entrance slit of the spectrometer. It is evident that, aside from the secondary electron peak at the very low end, the spectrum is essentially continuous. There are, however, small irregularities superposed on the continuum. The majority of the irregularities near the peak region are in fact, quite reproducible. The spectrum failed to show the L Auger lines of iron in the 600 eV to 700 eV region. The peak position of the broad continuum is situated near 120 eV.

Effect of the Angle of Incidence

Figure 2 indicates that the position and the intensity of the continuum will interfere strongly with the discrete Auger lines of the light elements of interest. To what extent can we improve this situation? In high velocity inelastic collisions, most of the ejected electrons are in the direction of the incoming α-particle. Since the main contribution of the continuum is a result of electrons ejected from the outer shells, the directional

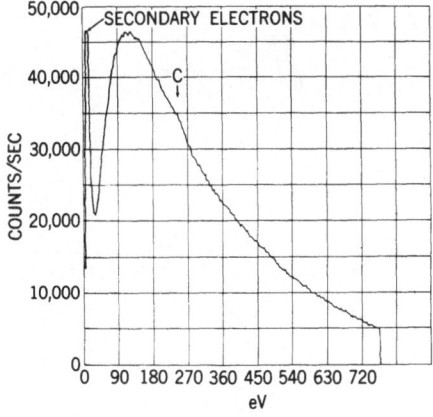

Figure 2. Electron spectrum of Po^{210} source with a stainless steel window.

distribution of these electrons should fall into this category. In contrast,
the emission of Auger electrons should be mostly isotropic. Consequently
one should observe a shift in the shape of the continuum when the angle of
incidence of α-particles on a sample is changed from grazing to normal.
In the latter case, the forward-scattered electrons must change direction
by almost 180° in order to emerge from the sample surface and hence suffer a much greater energy loss. The experimental arrangement to study
this phenomenon is shown in figure 3. A magnesium plate of uncertain
purity was used as a sample in the two geometries sketched in figure 3.
Figure 4 shows the spectrum at grazing incidence. It is rather broad. The
prominent feature is the ever-present carbon Auger peak (probably due to
oil contamination of the sample surface). Figure 5 shows the spectrum at
normal incidence. The change in the shape of the continuum is obvious.
The peak position is now near 45 eV. Furthermore, the structures on top
of the continuum become more easily discernable. Although most of the
excursions at higher energies are statistical fluctuations, those at the low
energy end are quite reproducible. Unfortunately, the intensity at high
energies is so low that we were not able to find the K Auger peaks of
magnesium.

Electron Spectra of Aluminum

Using normal incidence, we obtained the electron spectra of aluminum
in different forms. Figure 6 shows the spectra from a 0.00025"-thick

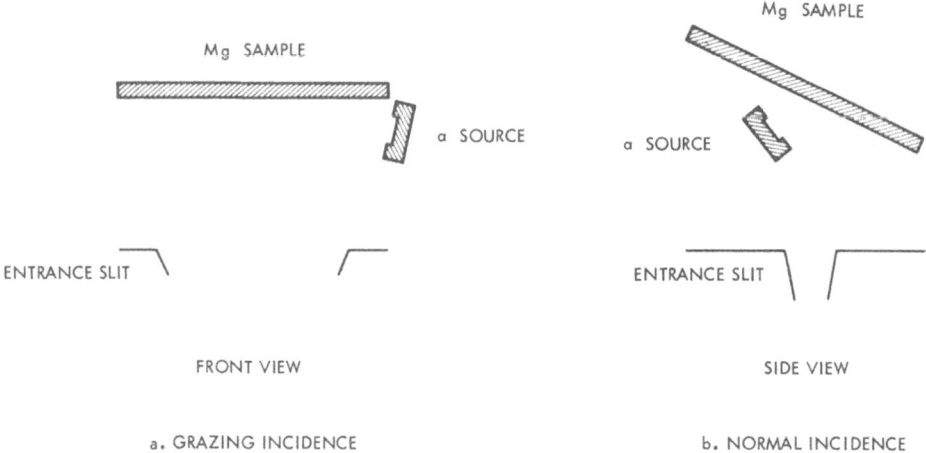

Figure 3. Experimental arrangement to investigate the effect of the angle of incidence.
a. Grazing incidence. b. Normal incidence.

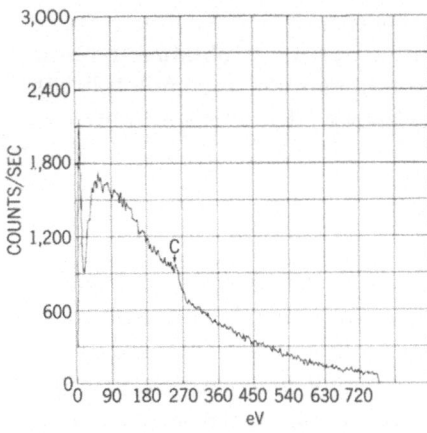

Figure 4. Electron spectrum of Mg
sample at grazing incidence.

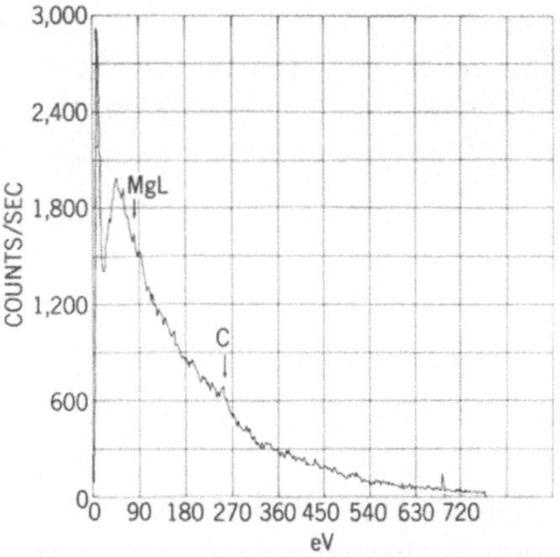

Figure 5. Electron spectrum of Mg
sample at normal incidence.

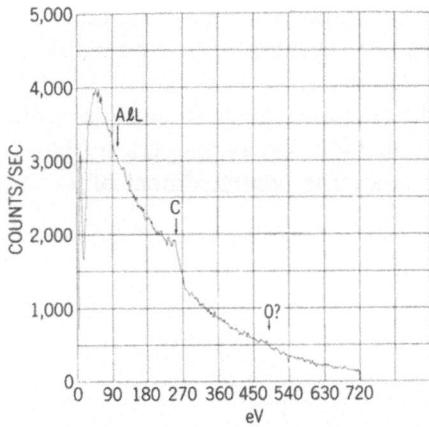

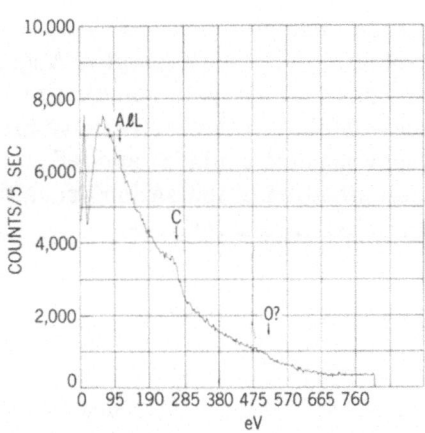

Figure 6. Electron spectrum of mylar
with 500 Å deposition of Al.

Figure 7. Electron spectrum of 500 Å
of Al deposited on 500 Å of plastic
substrate.

mylar sample with about 500 Å deposition of Al. Although the structures
are complicated with varying relative intensities, the positions of the
prominent "lines" at the low energy end are reproducible. It is disappoint-
ing not to be able to see the K Auger peak of oxygen. Figure 7 shows an
extremely thin sample with 500 Å of Al deposited on 500 Å of plastic

substrate. The prominent features of the spectrum are essentially the
same as those of figure 6. Figure 8 shows the spectrum obtained from a
thick anodized aluminum sample. Although this sample is completely different from the previous two, the spectrum shows little change in structure.

DISCUSSION

    The problems involved in obtaining α-excited Auger spectra from
solid samples are twofold. First, the intense low energy continuum makes
the presence of the discrete peaks ambiguous. In addition, since the ionization cross section under α-particle bombardment may be several orders
of magnitude higher for outer shells than it is for inner shells, the low
energy region is also where most of the discrete Auger lines are expected
to occur. This tends to make the identification of lines in this region con-
fusing. Secondly, in our experiment, the intensity of the high energy Auger
lines (such as the K lines of oxygen and magnesium, and the L lines of
iron) has been so low as to evade detection. The first of these two problems is inherent in the physics of inelastic collisions. The second presumably can be improved with stronger, or multiple, sources and better
vacuums. Whether α-excited Auger spectra will eventually prove useful
for chemical analysis remains to be seen. On the other hand, since the
Auger yield is practically unity for the light elements such as carbon, their
Auger spectra could be applied to complement their x-ray spectra in the
determination of ionization cross sections under the bombardment of α-
particles or heavier ions.[4]

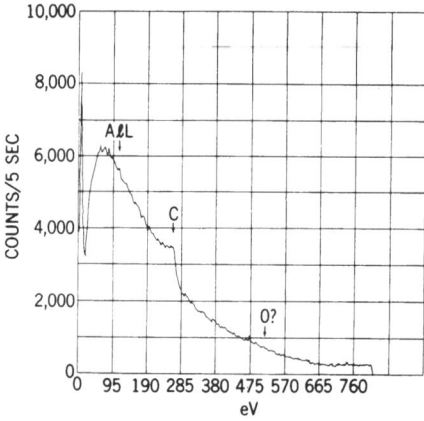

Figure 8. Electron spectrum of thick
anodized aluminum sample.

Other radioactive sources (such as K-capture, internal conversion, and β-sources) also produce Auger electrons; their applicability awaits investigation.

ACKNOWLEDGMENT

We are grateful to K. Omidvar and E. Yellin for many helpful discussions.

REFERENCES

1.   D. R. Bates and G. Griffing, "Inelastic Collisions between Heavy Particles," Proc. Phys. Soc. London, Ser. A 66: 961-971, (1953).

2.   E. Merzbacher and H. W. Lewis, "X-ray Production by Heavy Charged Particles," Handbuch der Physik, 34: 166-192, (1958).

3.   L. I Yin, I. Adler and R. Lamothe, "Some Parameters Affecting Auger and Photoelectron Spectroscopy as an Analytical Tool," Appl. Spect. 23: 41-50, (1969).

4.   R. C. Der, T. M. Kavanagh, J. M. Khan, B. P. Curry and R. J. Fortner, "Production of Carbon Characteristic X-Rays by Heavy-Ion Bombardment," Phys. Rev. Letters 21: 1731-1732, (1968).

THE APPLICATION OF X-RAY DATA TO THE DETERMINATION OF ATOMIC

ENERGY LEVELS

A. F. Burr

New Mexico State University

Las Cruces, New Mexico 88001

ABSTRACT

Atomic energy levels, which represent the energy necessary to remove an electron in a given level from the atom, are used for a variety of purposes-from predicting possible interfering x-ray lines in analysis to serving as experimental values with which to compare the results of complicated computer calculations. X-ray wavelength data is essential to determine energy level differences but cannot be used to place the resulting level scheme on an absolute energy basis. X-ray absorption data can be so used but the results of photoelectron energy measurements are superior. Since there are a very large number of possible x-ray lines from a typical element, one can readily obtain many more energy differences than there are energy levels; hence, to make the best use of the data available, a least squares fitting procedure is used to obtain the most probable value for each energy level.

Recently, excitation curves for selected x-ray lines have been measured with the precision necessary to enable this source to provide useful energy level data. In theory, the minimum voltage across an x-ray tube necessary to just produce a given x-ray line can be used to provide a value for the appropriate energy level. In practice, there are a number of difficulties (including the work function of the x-ray tube filament and the slope of the x-ray excitation curve) which limit the usefulness of this method. However, two excitation curves from the same target can be made to yield energy level differences which have a precision comparable with most of the differences obtainable from the least squares method.

The results of these studies have lead to a number of
interesting observations. A number of errors in observed and
calculated wavelengths have been spotted. The fact that levels
and differences obtained from excitation curves tend to be lower
than levels and differences obtained in the normal way (although
the error bars of the two methods overlap) points to a need for
further study of the excitation curve method; and discrepancies in
the long wavelength region between energy level data and the
wavelength of $L_I L_{II, III}$ transitions clearly point to the existance
of a difficulty in the interpretation of results in that region.

INTRODUCTION

Occupied atomic energy levels are usually labeled by x-ray
physicists with a letter designating the principle quantum number
of the electron and a Roman numeral subscript indicating the j
value of the particular subshell under discussion. Thus when one
electron is removed from an atom it is in an excited state which
is designated by the letter and number of the electron removed.
The amount of energy which it took to remove that electron from the
atom is the value of that energy level. Figure 1 shows schematically
some of the excited states into which a heavy atom can be placed
by the removal of various electrons from different inner atomic
levels. The figure also shows some of the physical processes which
result from the transition of the atom from one state to another.

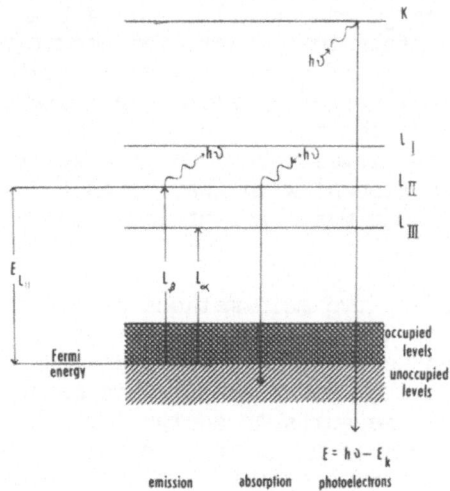

Figure 1. General energy level scheme showing some transitions.

USES OF ENERGY LEVEL VALUES

When precise values have been tabulated for all the energy levels of a number of elements, one is in a position to draw a large amount of additional information from the table. Of basic importance is a knowledge of the structure of the elements. In fact, unambiguous atomic numbers were first assigned to the elements by means of energy level differences. The effect of filled inner orbits (screening) on the outer orbits can be traced from element to element. Energy level tables are very useful to the analyst in selecting transitions and elements to obtain x-rays of specific energy. They can be used to locate possible interfering lines, and to provide the wavelength of lines from hard to handle elements without having to experimentally produce them.

A particularly interesting use of energy level tables is to provide a standard to which one can compare the results of different methods of calculating energy levels. In principle, the Hartree-Fock method will enable one to calculate the value for any energy level. In practice, the method is so complicated that even large computers cannot handle the problem; hence various approximations are made. One can find the effect of these approximations by comparing the calculated energy levels to the experimental ones.

OBTAINING VALUES FOR ENERGY LEVELS

Since there are over 2500 measured characteristic x-ray lines, which represent the energy made available by a transition of an atom from one excited state to another, the amount of information available on energy level differences is very great indeed. When all the characteristic lines of an element are examined, one is able to obtain a good idea of the spacing between levels. However, one cannot place this energy level scheme on an absolute basis or compare the scheme of one element with that of another on the basis of emission lines alone. One must have at least one absolute measurement.

There are at present two main methods of obtaining this absolute measurement. One is to use an x-ray absorption edge measurement - the L_{III} edge is often used for this purpose - the other, better, way to use a photoelectron measurement. This method, which has been described in detail,[1,2] measures an energy level by taking the difference in energy between an incoming absorbed photon and an outgoing ejected electron. Since these absolute measurements are much less precise than the emission line measurements, they do not effect the spacing of the levels but nearly the placing of the whole energy level scheme on an absolute energy scale.

To obtain the energy levels, all the information for a single

element is collected and studied at once. Taking thorium as an
example, one finds that the element has 25 occupied energy levels
and that there are 99 measurements which can be used to determine
these energy levels. In a situation like this (where the number
of unknowns, the 25 levels, are greatly outnumbered by the number
of equations, the 99 measurements) a least squares fit[3] provides a
standard method for obtaining a set of energy levels which best fit
the data.

The above procedure has been carried out for all the elements,[4]
and a detailed analysis of the procedure and difficulties has been
given.[5] The end result is a coherent table of energy levels for
the whole periodic table complete with explicit probable errors.
Some interesting observations were immediately made upon completion
of this table. A number of errors in observed and calculated wavelengths were corrected and an interesting discrepancy in the long
wavelength region between the energy level data and the wavelength
of all L_I $L_{II,III}$ transitions spotted.

ENERGY LEVEL DATA FROM EXCITATION CURVES

There is another method, in addition to the two mentioned above,
for obtaining an absolute energy level measurement. This method
makes use of x-ray emission line excitation curves.

Excitation Curves

An excitation curve is a plot of the voltage across an x-ray
tube vs. the intensity from a given characteristic emission line.
Figure 2 on the next page shows two such curves for cobalt. The
upper curve was obtained with the Lα ($L_{III}M_{IV,V}$) line and the
bottom curve with the Lβ ($L_{II}M_{IV}$) line.

Obtaining Excitation Curves

Figure 3 shows a block diagram of the equipment used to obtain
the curves in Figure 2. Most of the equipment is quite conventional.
The source is an ultra-high vacuum, demountable x-ray tube with
normal electron incidence and normal x-ray take off to minimize self
absorption. The spectrometer is a double crystal vacuum spectrometer
used here to select out the emission line of interest. The detector
is a conventional low pressure, P10 filled, flow proportional counter.
The scalar was a TMC pulse height analyzer used in the multi-scalar
mode. The high voltage supply was a very stable, well regulated,
well filtered variable power supply. The precision voltage measurement was done with a precision voltage divider and a K3 potentio-
meter.

A. F. Burr

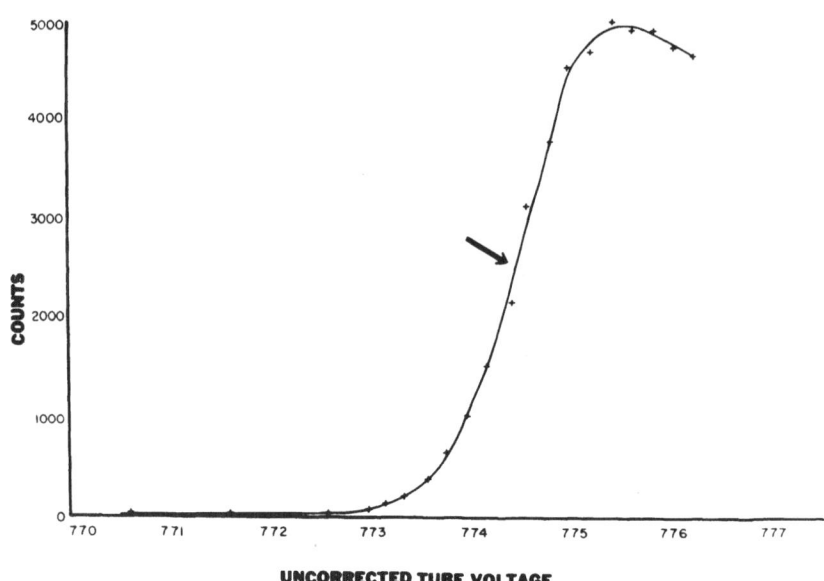

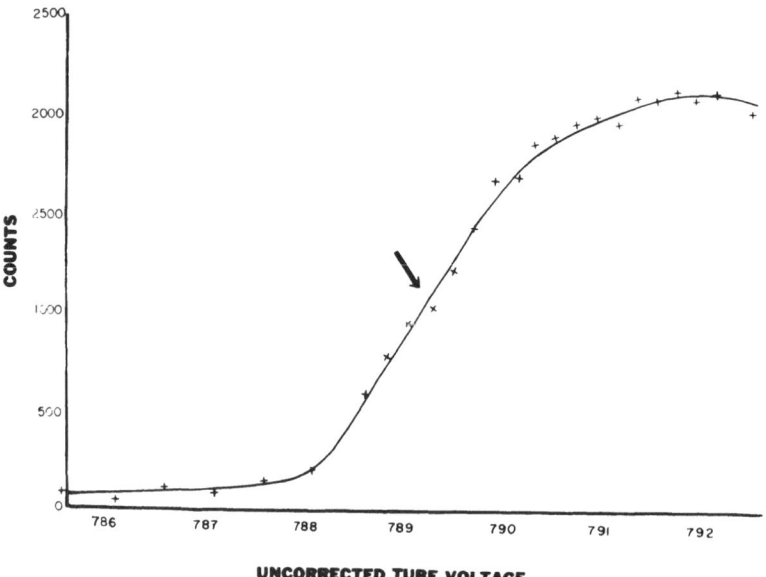

Figure 2. Cobalt excitation curves. The upper curve is for the
Lα line. The lower curve is for the Lβ line.

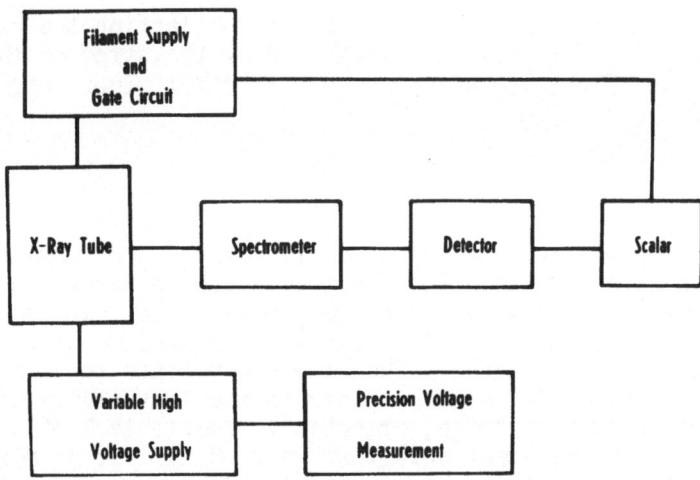

Figure 3. Block diagram of equipment for obtaining excitation curves.

The filament supply was the only unconventional part of the equipment. Alternating current was supplied to the tube through a half wave rectifier. A gate circuit was arranged so that the scalar would accept counts from the detector only during those parts of the AC cycle when current was not flowing through the x-ray tube filaments. Thus the heated filament was always entirely at ground potential whenever the scalar was counting x-rays.

Interpretation of Excitation Curves

One would detect, from a simplified viewpoint, characteristic x-rays only when the bombarding electrons had enough energy to place the target atoms in the upper excited state by removing an L_{III} or L_{II} electron. The observed line would first turn on when the bombarding electrons were given enough energy to just remove the appropriate electron from the target atom. This energy just equals the value for the energy level.

The above interpretation is oversimplified in a number of respects. The most serious oversimplification is the neglect of the work function of the tube. The filament current has to supply the electrons with enough energy to overcome the work function of the filament; hence that much energy has to be added to the energy supplied by the high voltage supply. The next difficulty is deciding just which point on the excitation curve can represent the energy

level. There are a number of reasons for selecting the point of
inflection.[6]   This point is indicated by the arrow on the curves
in figure 2. There are also a number of other minor corrections
to be made.

Results

The preliminary results of experiments run on cobalt and
nickel are given in table I. The middle column gives the results
obtained by the experiment described above. The last column gives
the results obtained by following the procedures detailed in the
first part of this article. The errors given are to be considered
as probable errors. The uncertainty in the last figure of the
middle column is very large, principally due to lack of precision
knowledge of the filament work function. Until better means are
available for measuring and defining the work function, this method
will not enable the precision of an absolute energy level measurement to be increased dramatically.

Much of the uncertainty in these figures can be removed if one
considers energy level differences rather than energy levels. With
level differences the effect of the work function drops out completely. The uncertainty introduced when reading the curve is greatly
reduced. Table II summarizes the results for the L_{II}-L_{III} energy
level difference. Again the middle column gives the results obtained
by the experiment described above. The last column gives the results
obtained by following the procedures detailed in the first part of
this article. The errors given are to be considered as probable
errors. They are much smaller than the errors in Table I because
those errors are highly correlated. The errors in the middle column

Table I. Energy Levels from Excitation Curves

Level	Value (eV)	AEL[a] (eV)
Co L_{II}	791.9	793.6 ± 0.3
Co L_{III}	777.1	778.6 ± 0.3
Ni L_{II}	870.6	871.9 ± 0.4
Ni L_{III}	853.6	854.7 ± 0.4

[a]The energy level value from reference 4.

Table II. Energy Level Differences from Excitation Curves

Difference	Value (eV)	AEL[a] (eV)
Co L_{II}-L_{III}	14.83	15.01 ± 0.05
Ni L_{II}-L_{III}	17.03	17.25 ± 0.05

[a]The energy level difference from reference 5.

are but slightly bigger than the third column errors; hence this
method can yield useful information on energy level differences.

Interpretation of Results

An analysis of the data in tables I and II forces one to conclude that the results of the two methods of measuring energy levels
are in agreement, since the two columns of data differ only by a low
multiple of the probable error. However, one would be better satisfied if this multiple were even lower. Furthermore, in every case
the excitation curve value is lower than the conventional value.
This circumstance makes it desirable to perform more excitation
curve experiments, to attempt to increase their accuracy, and to
try different types of elements.

Energy Level Widths

One can get even more energy level information from these
excitation curve experiments if one considers that the width of the
energy level is related to the slope of the excitation curve in
somewhat the same way that Ritmeyer related the slope of the   *Richtmeyer*
absorption edge to the width of the level.[6] These results are
given in table III. It is difficult to give a realistic evaluation
of the precision of these widths since the theoretical basis for
the measurement is so uncertain.

Table III. Energy Level Widths from Excitation Curves

Level	Width (eV)
Co L_{II}	1.3
Co L_{III}	0.7
Ni L_{II}	1.1
Ni L_{III}	0.8

SUMMARY

One can combine by a least squares method the wavelengths of all the emission lines of a given element with some absolute measurement of any energy level to get a complete energy level scheme for that element which uses all the information available. It appears possible to develop a method based on x-ray line excitation curves which provide an acceptable absolute energy level measurement and which can supply some useful energy level difference information also.

ACKNOWLEDGEMENTS

Dr. R. J. Liefeld of New Mexico State University has contributed much to all aspects of the excitation curve studies.

REFERENCES

1. K. Siegbahn, Alpha-, Beta-, and Gamma-Ray Spectroscopy, North-Holland Publishing Co., Amsterdam 1965, Chapter III.

2. K. Siegbahn, ESCA Atomic Molecular and Solid State Structure Studied by Means of Electron Spectroscopy, Almquist and Wiksells Uppsala 1968.

3. J. A. Bearden and J. S. Thomsen, "A Survey of Atomic Constants", Nuovo Cimento 5: 267 (1957).

4. J. A. Bearden and A. F. Burr, "Reevaluation of X-Ray Atomic Energy Levels", Rev. Mod. Phys. 30:125 (1967).

5. J. A. Bearden and A. F. Burr, Atomic Energy Levels, U. S. Atomic Energy Commission, NYO 2543-1, Oak Ridge, Tenn., 1965.

6. F. K. Richtmyer, S. W. Barnes, and E. Ramberg, "The Widths of the L-Series Lines and of the Energy Levels of Au(79)", Phys. Rev. 46:843 (1934).

ON THE SYMMETRY OF ORIENTATION DISTRIBUTION IN CRYSTAL AGGREGATES*

David W. Baker

University of California

Los Angeles, California 90024

ABSTRACT

In a number of laboratories data from the X-ray pole-figure goniometer are now processed numerically to obtain a suite of complete and normalized pole-figures. From these pole-figures the preferred orientation of the grains in a monomineralic aggregate is determined using spherical harmonic analysis and represented as a frequency distribution of the Euler angles Ψ, θ, ϕ, termed the "orientation distribution function" (ODF). When the frequency density of grain orientations is plotted in a Cartesian coordinate system with Ψ, θ, ϕ as axes, it must have a translation periodicity of 2π or less along each of the axes, because the Euler angles are angles of rotation. The ODF must have a symmetry, corresponding to one of the 230 crystallographic space groups, that reflects both the point group symmetry of the crystal and that of the specimen. The space group symmetry of the ODF has been derived for combinations of specimen and crystal symmetries commonly used. The asymmetric unit in the unit cell, which is the minimum volume of the ODF that must be evaluated, is also listed. Analysis of a naturally deformed quartzite with monoclinic specimen symmetry and trigonal crystal symmetry illustrates the use of these symmetry rules.

*Publication #755, Institute of Geophysics & Planetary Physics, University of California, Los Angeles, California 90024

435

INTRODUCTION

The X-ray pole-figure goniometer is now used in many laboratories to determine preferred orientation of lattice directions in crystal aggregates.[1,2] By combining several scans for each diffraction peak, a suite of complete pole-figures, normalized to multiples of a uniform distribution, is obtained. A smaller number of laboratories use a spherical harmonic analysis to recover the complete orientation of the grains from such a suite of pole-figures.[3-6] In the spherical harmonic analysis of Bunge[7-9] and of Roe[10,11] crystal and specimen symmetry play an important role. Notably lacking, however, is a discussion of the effects of such symmetry on the distribution of grain orientations.

ORIENTATION OF A GRAIN

The orientation of a grain in a specimen of a crystal aggregate can be given by placing a right-handed Cartesian coordinate system XYZ in the grain, a right-handed Cartesian coordinate system ABC in the specimen, and specifying the relative orientation of the two coordinate systems with the three Euler angles Ψ, θ, ϕ (Figure 1). The angles $-\phi$, $-\theta$, $-\Psi$ give the set of rotations necessary to bring XYZ into coincidence with ABC, whereas Ψ, θ, ϕ indicate the inverse operation, i.e., the rotations which take XYZ from coincidence with ABC to a given orientation. From an initial orientation in which XYZ coincides with ABC, XYZ is first rotated through an angle Ψ about an axis parallel to Z to the position X'Y'Z', then through the angle θ about an axis parallel to Y' to the position X"Y"Z", and finally through the angle ϕ about an axis parallel to

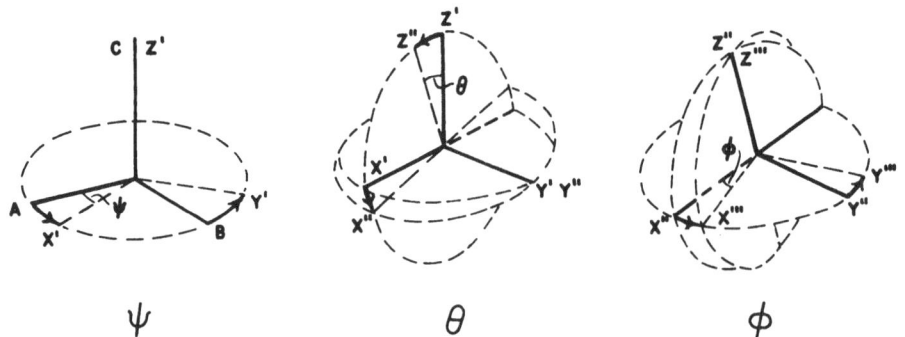

$$\psi \qquad\qquad \theta \qquad\qquad \phi$$

Figure 1. The Euler angles Ψ, θ, ϕ, relating crystal coordinate system XYZ to specimen coordinate system ABC.

Z" to the position X"'Y"'Z"'. (Here a positive rotation about an axis advances a right-handed screw in a
positive direction along that axis.) An alternate way
to attain the same orientation is to start from initial
coincidence of the two coordinate systems and rotate
XYZ through an angle ϕ about C, followed by a rotation
θ about B and a rotation Ψ about C. The inverse operation is accomplished by reversing the order of the
Euler angles and rotating in the negative sense. Equivalent rotations with the Euler angles are listed in
Table I.

ORIENTATION DISTRIBUTION FUNCTION

If the Euler angles for individual grains in an
aggregate are known, one can plot a point for each
grain in a Cartesian coordinate system in which the
Euler angles are the coordinate axes and use an appropriate three-dimensional smoothing technique, normalizing the density of points to multiples of the density in a uniform distribution, to obtain the density
distribution of grain orientations. The continuous
distribution of orientation density with respect to
Euler angles is called the "orientation distribution
function" (ODF) by Bunge.[3]   (It has also been termed
the crystallite distribution function by Roe[10] and the
crystallite orientation distribution by Morris and
Heckler.[4]) The ODF can be recovered from a suite of
pole-figures using a spherical harmonic analysis.[6,8-11]
The ODF's for analyzed specimens have been presented as
a series of serial sections.[3-5]

SYMMETRY OF THE ORIENTATION DISTRIBUTION FUNCTION

Since the addition of 2π to an Euler angle, such as

Table I. Equivalent Rotations with the Euler Angles Ψ, θ, ϕ

Rotated coordinate system	Sense of rotation	Rotation axes parallel to	Order of rotations
XYZ	+	Z,Y,Z	Ψ, θ, ϕ
XYZ	+	C,B,C	ϕ, θ, Ψ
ABC	-	C,B,C	$-\phi, -\theta, -\Psi$
ABC	-	Z,Y,Z	$-\Psi, -\theta, -\phi$

Table II. Symmetry Elements in the Orientation Distribution Function Arising from the Euler Angle Identity and Rotation Axes in the Crystal and Specimen

Symmetry element	Equivalent orientation $w(\psi,\theta,\phi) =$	Symmetry element in ODF
1. Euler angle identity	$w(\psi+\pi, 2\pi-\theta, \phi+\pi)$ $w(\psi+\pi, -\theta, \phi+\pi)$	Glide plane $\perp$ θ axis at $\theta=0,\pi,2\pi$ translation components: $\Delta\psi=\pi$, $\Delta\phi=\pi$
2. U-fold rotation axis parallel Z	$w(\psi,\theta,\phi\pm2\pi/U)$	Unit cell of $w(\psi,\theta,\phi)$ is reduced to $2\pi/U$ in ϕ direction
3. Diad axis parallel Y	$w(\psi,\theta+\pi,-\phi)$	Glide planes $\perp$ ϕ axis at $\phi=0$ translation component $\Delta\theta=\pi$
4. V-fold rotation axis parallel C	$w(\psi\pm2\pi/V,\theta,\phi)$	Unit cell of $w(\psi,\theta,\phi)$ reduced to $2\pi/V$ in ψ direction
5. Diad axis parallel B	$w(-\psi,\theta+\pi,\phi)$	Glide plane $\perp$ ψ axis at $\psi=0$ translation component $\Delta\theta=\pi$

Ψ, leaves the orientation of a grain unchanged, the orientation distribution function w(Ψ+2π,θ,φ) must have the same value as w(Ψ,θ,φ). If the Euler angles are allowed to assume arbitrarily large values, then ODF has a translation periodicity of 2π (or less) with respect to Ψ,θ,φ. Because of this translation periodicity the symmetry of the ODF corresponds to one of the 230 crystallographic space groups. The space group symmetry of the ODF for a given combination of crystal and specimen point group symmetries is considered below.

Symmetry Elements in the ODF

A space lattice can be placed in the ODF that has a cube shaped cell with an edge length of 2π. Because the range of Euler angles required to cover all possible orientations, $0 \leq \Psi \leq 2\pi$, $0 \leq \theta \leq \pi$, $0 \leq \phi \leq 2\pi$, corresponds to a volume only one half the size of this cell, there must be additional symmetry elements present. It can be easily demonstrated that the grain orientations specified by Ψ,θ,φ and Ψ+π, 2π-θ, φ+π are the same, leading to the identities for the ODF of

$$w(\Psi,\theta,\phi) = w(\Psi+\pi, 2\pi-\theta, \phi+\pi)$$

$$= w(\Psi+\pi, -\theta, \phi+\pi).$$

This equivalence of points in the ODF can be described geometrically by a set of glide planes oriented normal to the θ axis in the ODF, located at levels θ = 0, π,2π . . . and with the translation component ΔΨ+Δφ where ΔΨ and Δφ have the value π.

The Euler angles Ψ', θ', φ' for an orientation that is related to an arbitrary orientation Ψ,θ,φ by a rotation axis in the crystal or in the specimen can be deduced in many cases by inspection or with the aid of a Schmidt net. Since the orientations are equivalent the equation w(Ψ',θ',φ')=w(Ψ,θ,φ) must hold. One can determine by inspection whether this equation describes a symmetry element in the ODF. Symmetry elements in the ODF that result from rotation axes present in the specimen and the crystal are listed in Table II. A U-fold axis parallel to Z in the crystal reduces the periodicity of the ODF to 2π/U parallel to the φ axis, whereas a V-fold axis parallel to C in the specimen reduces the periodicity of the ODF to 2π/V parallel to the Ψ axis. Horizontal two-fold axes parallel to Y in the crystal or parallel to B in the specimen generate glide planes

Table III.   Symmetry Elements in the Orientation Distribution Function Resulting from One Mirror Plane in Specimen and One Mirror Plane in Crystals*

		Mirror plane ⊥ crystal axis		
		X	Y	Z
Mirror plane ⊥ specimen axis	A	Inversion center at $(0,0,0)$ $w(-\Psi,-\theta,-\phi)$	Diad parallel θ axis at $(\pi/2,\cdot,0)$ $w(\pi-\Psi,\theta,-\phi)$	Diad parallel ϕ axis at $(0,\pi/2,\cdot)$ $w(-\Psi,\pi-\theta,\phi)$
	B	Diad parallel θ axis at $(0,\cdot,\pi/2)$ $w(-\Psi,\theta,\pi-\phi)$	Diad parallel θ axis at $(0,\cdot,0)$ $w(-\Psi,\theta,-\phi)$	Diad parallel θ axis at $(\pi/2,\pi/2,\cdot)$ $w(\pi-\Psi,\pi-\theta,\phi)$
	C	Diad parallel Ψ axis at $(\cdot,\pi/2,0)$ $w(\Psi,\pi-\theta,-\phi)$	Diad parallel Ψ axis at $(\cdot,\pi/2,\pi/2)$ $w(\Psi,\pi-\theta,\pi-\phi)$	Mirror plane ⊥ θ axis at $\theta=0$ $w(\Psi,-\theta,\phi)$

* Density at equivalent orientation $w(\Psi,\theta,\phi)=$

in the ODF parallel to the θ axis. Although no new
symmetry elements are generated in the ODF by a single
mirror plane in either the crystal or the specimen, new
symmetry elements are generated when the combination of
a mirror plane in the crystal and a mirror plane in the
specimen occurs. These symmetry elements for various
combinations of mirror planes are listed in Table III.

ODF Symmetry for a Quartzite Specimen

As a practical example to illustrate these symmetry rules, the space group symmetry of the ODF for a
quartzite specimen with monoclinic symmetry will be
derived. To do this it is first necessary to define
what the ODF is in the case of quartz. This is because
quartz belongs to one of the seven enantiomorphic crystal
classes. It is the only mineral commonly studied in
petrofabrics that is in such a class. A complete
description of the preferred orientation would require
one ODF for right-handed crystals and another for left-
handed crystals. However, the effect of Friedel's law
for X-ray diffraction is to add an inversion center to
the quartz point group symmetry of 32, giving a diffraction symmetry of 3m. Thus only one ODF is required
because right and left-handed crystals cannot be distinguished using X-ray data obtained with the pole-figure
goniometer.

The convention used here for orienting the coordinate
systems XYZ and ABC in the crystal and specimen, respectively, is that if there are any mirror planes present
in either, the Y or B axis is oriented normal to a mirror plane and Z or C is oriented parallel to the highest
fold rotation axis parallel to the mirror plane. Thus
for quartz Z is parallel to the three-fold axis; and
Y is parallel to a two-fold axis and normal to a mirror
plane (Figure 2). The pole-figures shown in Figure 3
have, to a close approximation, a vertical N-S mirror
plane (and because of centro-symmetry a two-fold axis
or diad normal to this mirror plane). The B axis is
oriented normal to the mirror plane and parallel to the
two-fold axis.

The orientation of axes of the ODF is arbitrary.
In the plots used here, the θ axis is vertical whereas
Ψ and φ are horizontal. The trigonal axis parallel to
Z in the crystal reduces the unit cell of the ODF to
$2\pi/3$ in the φ direction (Table II). The two-fold axis
parallel to Y generates vertical glide planes normal to

Figure 2. Orientation of
XYZ with respect to crystal
axes of quartz. Poles for
reflections used in analysis
are shown in asymmetric unit
of diffraction group 3̄m --
a 60° sector. Equal area
projection on <u>upper</u> hemisphere.

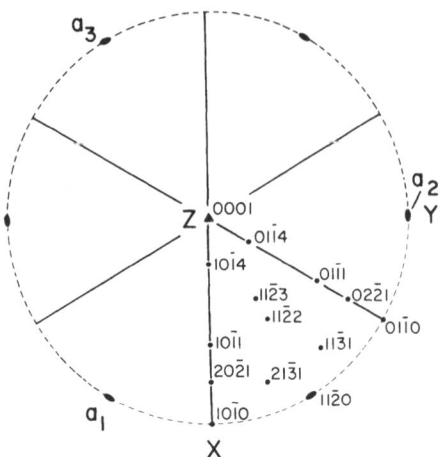

the φ axis in the ODF and the two-fold axis in the
specimen parallel to B generates vertical glide planes
perpendicular to the Ψ axis (Table II). The combination of a mirror plane normal to B and a mirror plane
normal to Y creates two-fold rotation axes in the ODF
parallel to the θ axis (Table III). The identity for
Euler angles with θ greater than π leads to glide planes
normal to the θ axis at levels θ=0,π,2π . . .(Table II).
Combining these symmetry elements leads to screw axes
parallel to the Ψ axis and to the φ axis and also to
inversion centers as shown in Figure 4. The space
group containing this arrangement of symmetry elements
is listed by Buerger[12] as P 2_1/c 2_1/c 2/n. By substituting the axes of the ODF Ψ,θ,φ for the crystallographic axes a,b,c used in the international space
group symbols and rearranging the symbols into the order
Ψ,θ,φ, one obtains the space group symbol for the ODF
of P 2_1/θ 2/n 2_1/θ.

     The ODF calculated for the pole-figures in Figure
3 is shown in Figure 5.[13] (As has been noted by Bunge
and Haessner[3] regions of negative density -- a physical
impossibility -- frequently appear in harmonic expansions
of the ODF and are an artifact of the computation.) The
unit cell spans 0≤Ψ≤2π, 0≤θ≤2π, 0≤φ≤2π/3 whereas the
asymmetric unit of the unit cell spans only one eighth
of that volume, i.e., 0≤Ψ≤π, 0≤θ≤π/2, 0≤φ≤2π/3.

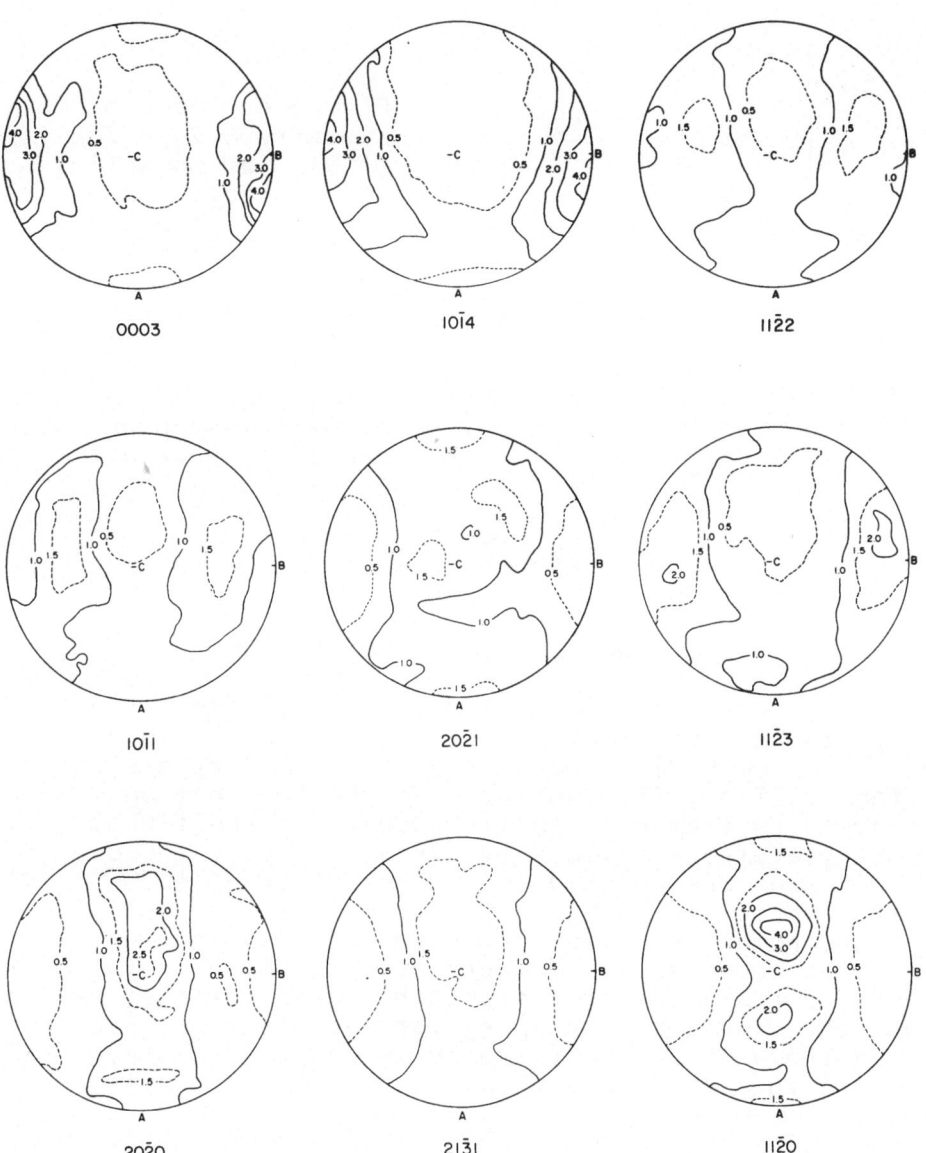

Figure 3. Pole-figures for quartzite specimen determined with X-ray pole-figure goniometer. Pole density
given in multiples of density in a uniform distribution.
Equal area projection on lower hemisphere. Pole-figures
are triclinic but approximate monoclinic symmetry with
mirror plane normal to B and two-fold axis parallel to
B.

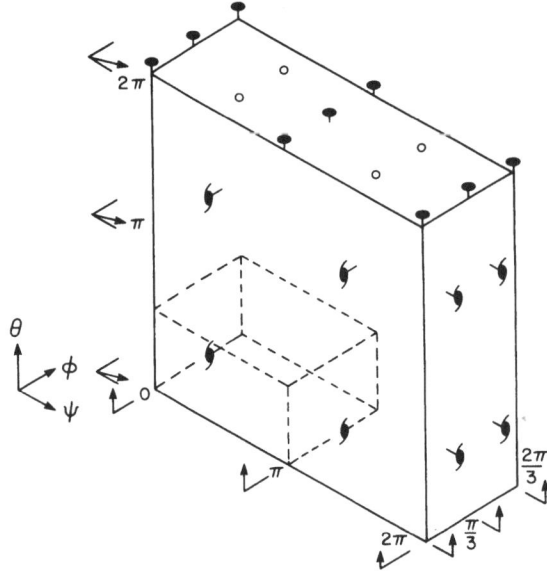

Figure 4. Space group symmetry P $2_1/\theta$ $2/n$ $2_1/\theta$ of ODF for monoclinic (2/m) specimen and trigonal ($\bar{3}$m) crystal. Solid line--unit cell, dashed line--asymmetric unit, open circle--inversion center, ●--diad axis, ●--two-fold screw axis, right angle with arrow--glide plane with translation parallel to arrow.

Space Group Symmetry of the ODF

The space group for the ODF for any combination of crystal and specimen point group symmetry can be deduced by combining the appropriate symmetry elements in Tables II and III, noting that the combination of two symmetry elements usually results in additional symmetry elements. Table IV lists the space group of the ODF when the holohedral crystal class of each crystal system is combined with triclinic, monoclinic or orthorhombic specimen symmetry. The convention, mentioned previously, of wherever possible orienting B and Y normal to mirror planes has been followed here. Among the nine space group symmetries listed in Table IV and shown in Figure 6 are three pairs (b-d,c-e,g-h) in which the same space group is shown in two orientations. Axial specimen symmetry is discussed elsewhere.[6]

Size of Unit Cell and Asymmetric Unit

The unit cell of the ODF spans $0 \leq \Psi \leq 2\pi/V$, $0 \leq \theta \leq 2\pi$, $0 \leq \phi \leq 2\pi/U$ where Z is parallel to a U-fold symmetry axis in the crystal and C is parallel to a V-fold symmetry axis in the specimen. Only a portion of the unit cell need by evaluated. The ranges of Euler angles for the asymmetric unit of the unit cell for the space groups in Table IV are listed in Table V. It should be noted

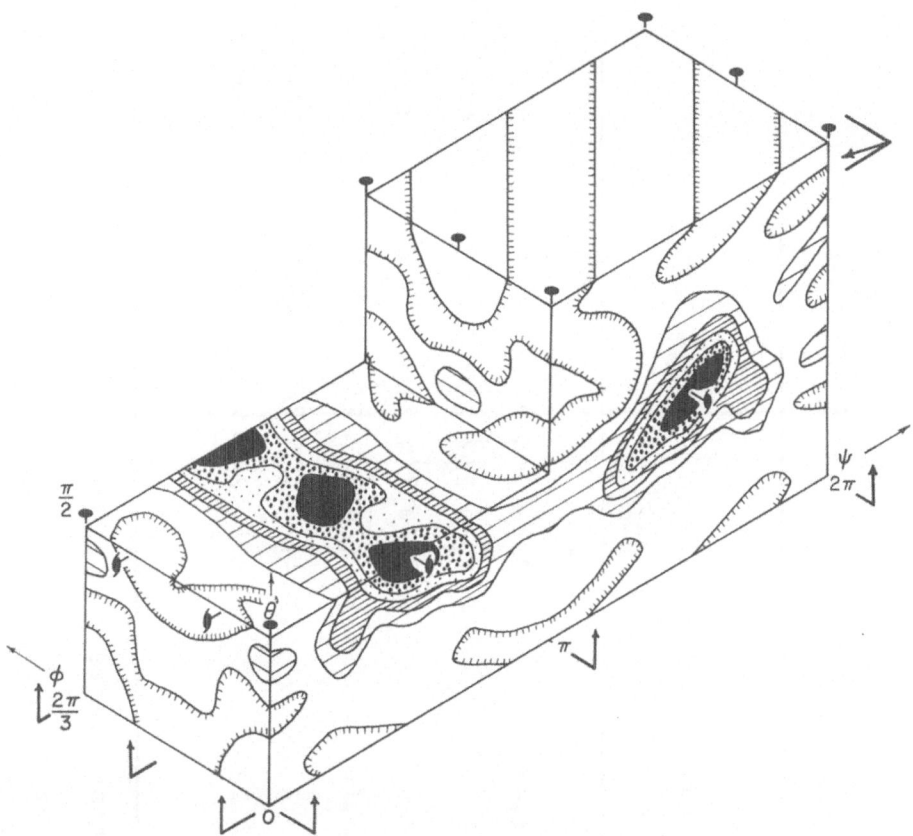

Figure 5. Block diagram of ODF calculated from pole-
figures shown in Figure 3. Orientation density given
as multiples of density in a uniform distribution. Con-
tours at 5,4,3,2,1,0X. Solid black-maximum. Hachures
- negative density, an artifact of calculation that
has no physical meaning.

that in the majority of cases other choices for the
range of Euler angles are possible. The remainder of
the unit cell is obtained from the asymmetric unit by
the symmetry operations of the space group.

        The size of the asymmetric unit is a function of
the symmetry elements in the space group of the ODF and
these in turn are due to symmetry elements in the point
groups of the crystal and the specimen. Orientations
of a crystal that are indistinguishable because of symmetry leads to points that are equivalent in the ODF.

Table IV.   Space Group of the ODF for Combinations of Specimen and Crystal Point Group Symmetries*

Specimen Symmetry

Crystal symmetry	$\bar{1}$	2/m	mmm
$\bar{1}$	Pn	$P\theta n 2_1$	$P\theta\phi 2_1$
2/m	$P 2_1 n \theta$	$P 2_1/\theta\ 2/n\ 2/\theta$	$P 2/\theta\ 2/\phi\ 2_1/\theta$
mmm	$P 2_1 \Psi \theta$	$P 2_1/\theta\ 2/\Psi\ 2/\theta$	$P 2/\theta\ 2/m\ 2/\theta$
$\bar{3}$m	$P 2_1 n \theta$	$P 2_1/\theta\ 2/n\ 2_1/\theta$	$P 2/\theta\ 2/\phi\ 2_1/\theta$
4/mm	$P 2_1 \Psi \theta$	$P 2_1/\theta\ 2/\Psi\ 2/\theta$	$P 2/\theta\ 2/m\ 2/\theta$
6/mmm	$P 2_1 \Psi \theta$	$P 2_1/\theta\ 2/\Psi\ 2/\theta$	$P 2/\theta\ 2/m\ 2/\theta$
m3m	$P 2_1 \Psi \theta$	$P 2_1/\theta\ 2/\Psi\ 2/\theta$	$P 2/\theta\ 2/m\ 2/\theta$

*B and Y are normal to mirror planes if present.

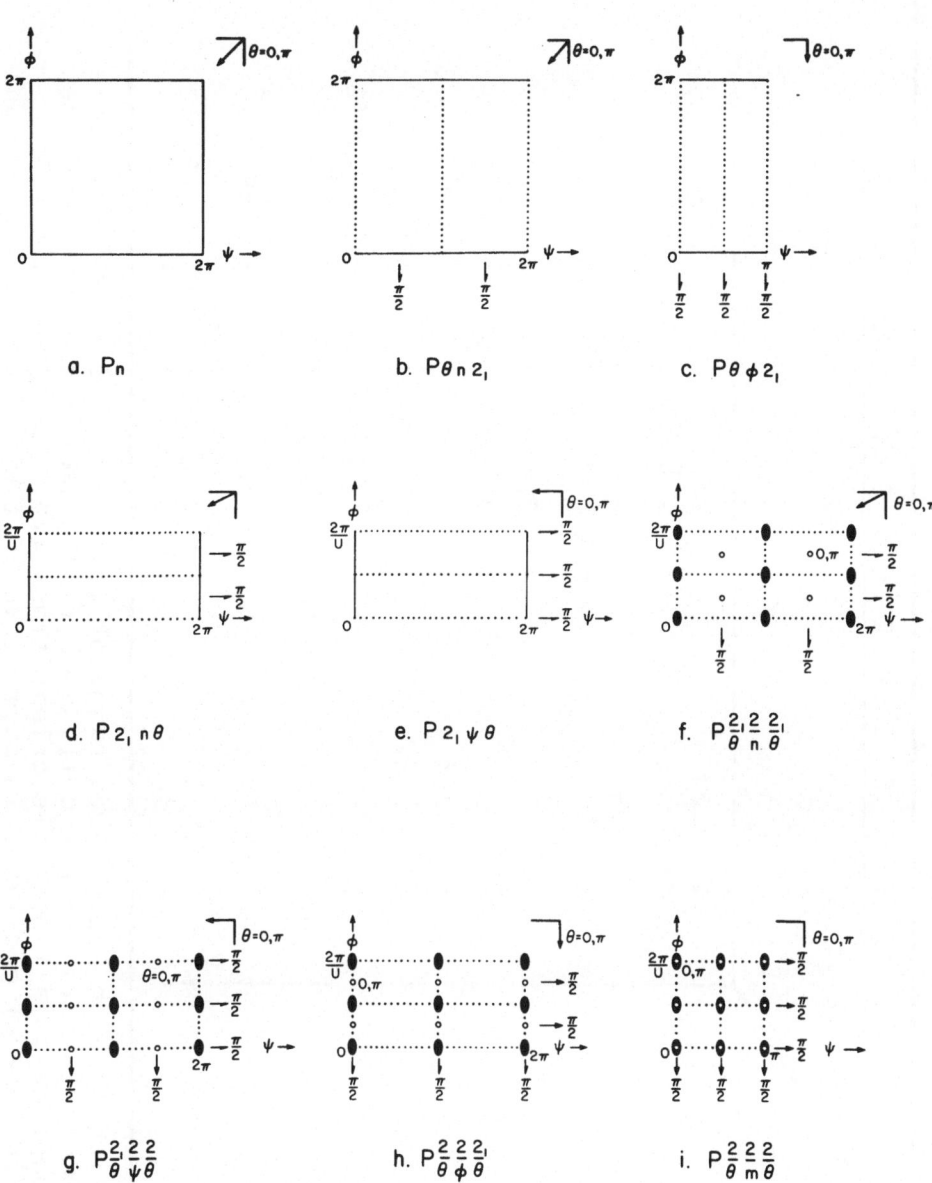

Figure 6.   Space groups for the orientation distribution function.   International symbols[14] for glide planes, mirror planes, rotation axes, screw axes and inversion centers are used.

Table V. Asymmetric Unit of Unit Cell of ODF for Space Groups Listed in Table IV*

CRYSTAL SYMMETRY Point group	U	SPECIMEN SYMMETRY Point group: 1̄ (V=1)	2/m (V=1)	mmm (V=2)
1̄	1	$2\pi,\pi,2\pi$ 2, 1, 1	$2\pi,\pi,\pi$ 4, 2, 1	π,π,π 4, 4, 1
2/m	1	$\pi,\pi,2\pi$ 4, 2, 1	π,π,π 8, 4, 1	$\pi,\pi/2,\pi$ 8, 8, 1
mmm	2	π,π,π 4, 4, 1	$\pi,\pi/2,\pi$ 8, 8, 1	$\pi/2,\pi/2,\pi$ 8, 16, 1
3̄m	3	$\pi,\pi,2\pi/3$ 4, 6, 1	$\pi,\pi/2,2\pi/3$ 8, 12, 1	$\pi/2,\pi/2,2\pi/3$ 8, 24, 1
4/mmm	4	$\pi,\pi,\pi/2$ 4, 8, 1	$\pi,\pi/2,\pi/2$ 8, 16, 1	$\pi/2,\pi/2,\pi/2$ 8, 32, 1
6/mmm	6	$\pi,\pi,\pi/3$ 4, 12, 1	$\pi,\pi/2,\pi/3$ 8, 24, 1	$\pi/2,\pi/2,\pi/3$ 8, 48, 1
m3m	4	$\pi,\pi,\pi/2$ 4, 8, 3	$\pi,\pi/2,\pi/2$ 8, 16, 3	$\pi/2,\pi/2,\pi/2$ 8, 32, 3

* First line in each entry gives range of Ψ,θ,ϕ. Second line lists number of asymmetric units in unit cell, number of asymmetric units in volume corresponding to all possible orientations $(2\pi,\pi,2\pi)$, and number of points in general positions in asymmetric unit which are equivalent due to crystal and specimen symmetry.

Since orientations are given as rotations specified by
the Euler angles, the number of orientations that are
equivalent due to crystal symmetry is equal to the
number of faces in the general form for the point group
consisting of only rotation axes in the crystal. For
example the diffraction symmetry of quartz is 3̄m and
the general form, the trigonal scalenohedron, has 12
faces. Consider one face on a wooden model of a perfect
scalenohedron to have a given orientation. Only five
of the faces may be substituted for this face by rotating the model. The remaining six faces cannot be
brought into congruency by rotating the model because
they are mirror images of the first face. The subgroup
of 3̄m that contains only rotation axes is 32. The
general form of 32, the trigonal trapezohedron, has six
faces and each of the six faces can assume a given
orientation by applying the appropriate rotations.

    The number of equivalent orientations due to speci-
men symmetry is equal to the number of faces in the
general form of the point group formed by rotation axes
in the specimen. The number of equivalent orientations
due to combined crystal and specimen symmetry, which is
given by the product of the values derived above, is
the number of points which are equivalent in the volume
of the ODF corresponding to all possible orientations
$(2\pi,\pi,2\pi)$. (Due to a center of symmetry imposed by
Friedel's law combinations of the type listed in Table
III do not have to be considered here, i.e., m + i=2/m
and the two-fold axes are treated here.) For quartz
crystals in a monoclinic (2/m) specimen there are
$6\cdot2=12$ equivalent orientations. In Table V the number
of asymmetric units in the volume $2\pi,\pi,2\pi$ listed is
also 12. This corresponds to one point per asymmetric
unit.

                    Cubic Crystal Symmetry

    Cubic crystal symmetry requires special considera-
tions because the Euler angles cannot express in a
simple manner the diagonal three fold axes. Furthermore
the horizontal four-fold axes appear in the ODF the
same as two-fold axes, i.e., glide planes parallel to the
θ axis. In the combinations for non-cubic symmetries
listed in Table IV all symmetry operations in either
the crystal or specimen which result in equivalent
orientations are expressed in ODF by additional symmetry
elements and a reduction in the size of the asymmetric

unit. This is not the case for cubic crystal symmetry
as the space group of the ODF and the size of the asymmetric unit are the same as for tetragonal crystal symmetry. The three fold axes are expressed in an indirect
manner in that there are three points in the asymmetric
unit that are equivalent instead of one (cf. Table V).

     Consider the combination of m3m crystal symmetry
with mmm specimen symmetry that occurs in many sheet
metals. From the discussion in the previous section one
obtains 24·4=96 orientations or 96 equivalent points in
the volume $2\pi \cdot \pi \cdot 2\pi$ radians[3] that are related by crystal
and specimen symmetry. However, the asymmetric unit
for P $2/\theta$ $2/m$ $2/\theta$ is 1/32 of this volume (Table V),
so that there are three points which are equivalent per
asymmetric unit. To check whether these three points
are related by a symmetry operation of translation,
rotation or reflection in the ODF an array of three
points, arranged in the form of a right angle, is used
(circle, cross and triangle joined by solid line in
Figure 7). If the crystal in the orientation 30°, 35°,
40°, shown by the open circle in Figure 7, is rotated
through 120° about its [$\bar{1}$11] axis, it attains the orientation of 154°, 68°, 29°. Rotating the point corresponding
to this orientation about the diad axis parallel to the
θ axis in the ODF at $\pi/2,\cdot,\pi/4$ (cf. Figure 6i) gives
the point 26°, 68°, 61° shown as an open circle in the upper left of the asymmetric unit. A rotation of the
crystal of 240° instead of 120° about [$\bar{1}$11] gives the

Figure 7. Asymmetric
unit of P $2/\theta$ $2/m$ $2/\theta$
showing three arrays of
points (circle,cross,
triangle) which are
equivalent because of
crystal symmetry (m3m)
and specimen symmetry
(mmm). Numbers give
θ coordinate in degrees.

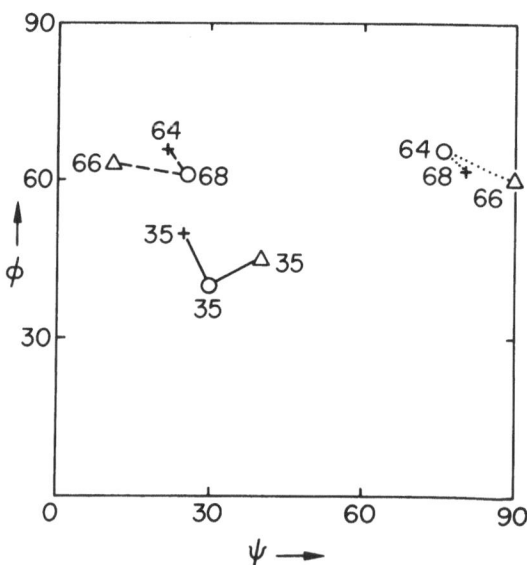

orientation 256°,64,66°. Translating parallel to the
Ψ axis 180° yields the point 76°,64°,66°, shown as an
open circle in the upper right of Figure 7. When these
operations are applied to the points 25°,35°,50° (cross)
and 40°,35°,45° (triangle) the resulting arrays of
points (dashed line and dotted line) do not form right
angles. Thus no symmetry operation can bring the array
shown with the solid line into congruence with either
of the other two arrays in Figure 7.

It should be noted that the space groups derived
for cubic crystals are also valid for the Euler angle
convention used by Bunge[7] in which the second rotation
is about X instead of Y as used here.

Symmetry on Special Sections of the ODF

Up to this point the symmetry on special sections
through the ODF has not been considered. For example,
on the planes θ=0,π or 2π, Z is constrained to be parallel to C whereas the orientation of X is given by the
sum of Ψ and φ. This means that contour lines, w(Ψ,0,φ)=
constant,are straight lines inclined at 45° to the Ψ and
φ axes (Figure 8). In the direction perpendicular to
the contour lines the values of w(Ψ,0,φ) must repeat
with a periodicity of $1/\sqrt{2}$ times the factor listed in
Table VI. Furthermore if the crystal has a mirror plane
perpendicular to Y and the specimen has a mirror plane
perpendicular to B, the contour line through the origin
will be a line of reflection as will be contour lines
at the repeat intervals from the origin and those at one
half the repeat interval (Figure 8). This is a consequence
of the two-fold axis parallel to the θ axis in the space
group (Figure 6i).

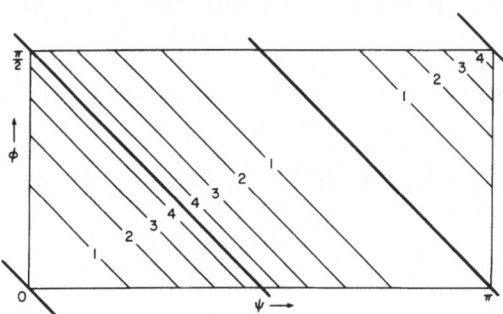

Figure 8. Section through
crystallite orientation
distribution at the level
θ=0°. Density given in
multiples of a uniform
distribution. Poly-
ethylene (mmm) in biaxially stretched sheet (mmm)
(redrawn from Adachi[5]).
Heavy lines are lines
of reflection.

Table VI. Periodicity at Level θ=0 in ODF for Crystals
          with an U-fold Axis Parallel Z and Specimens
          with an V-fold Axis Parallel C.

		U				
		1	2	3	4	6
V	1	2π	π	$2\pi/3$	$\pi/2$	$\pi/3$
	2	π	π	$\pi/3$	$\pi/2$	$\pi/3$

SUMMARY AND CONCLUSIONS

Expressions for the Euler angles of orientations
which are equivalent due to crystal or specimen symmetry can be deduced in many cases by inspection using
a wooden crystal model or with the aid of a stereonet.
Using such expressions the space group symmetry of the
orientation distribution function (ODF), which is a function of crystal and specimen point group symmetry, can
be determined without resort to higher mathematics.
The space group of the ODF is listed for combinations of
crystal and specimen symmetry commonly encountered[15] in
studies of preferred orientation. Also listed is the
asymmetric unit in the unit cell, which is the smallest
volume of the ODF that must be evaluated in numerical
determinations of the ODF. It is sufficient to present
the density in the asymmetric unit (usually in the form
of serial sections) and to refer to a space group shown
in Figure 6. The density for any orientation then can
be obtained by using the symmetry operations in that
space group. The space group symmetry of the ODF plays an
important role in the spherical harmonic analysis of Bunge
where the ODF is expanded with a set of symmetrical generalized spherical harmonics that has the same space group
symmetry as the ODF.

Although the space group of the ODF for cubic crystals
is the same as for tetragonal crystals, there are three
points which are equivalent in the asymmetric unit instead of one.

ACKNOWLEDGEMENTS

The author is grateful to D.T. Griggs and J.M. Christie for the continued encouragement and support for this work. H.R. Wenk collected the X-ray data for the quartzite specimen which was supplied by J.M. Christie. In addition to the above named persons I am indebted to G. Oertel, D. Schwarzenbach and W. Dollase for constructive comments on the manuscript. The work was supported by National Science Foundation grant GA 1389. Computations were made with the IBM 360 computer of the U.C.L.A. Campus Computing Network.

REFERENCES

1. Azaroff, L.V., Elements of X-ray Crystallography, McGraw-Hill, New York, 1968, p. 544-548.

2. Barrett, C.S. and Massalski, T.B., Structures of Metals, 3rd Ed., McGraw-Hill, New York, 1966, p. 200-203.

3. Bunge, H.J.and Haessner,F.,"Three Dimensional Orientation Distribution Function of Crystals in Cold-Rolled Copper," Jour. Appl. Physics, $\underline{39}$: 5503-5514, (1968).

4. Morris, P.R. and Heckler, A.J., "Crystallite Orientation Analysis for Rolled Cubic Materials," in J.B. Newkirk, Editor, Advances in X-ray Analysis, Vol. 11, Plenum Press, New York, 1968, p. 454-472.

5. Adachi, T., Crystallite Orientation Distribution for Biaxially Stretched Polyethylene, Ph.D. Thesis, Duke University, Durham, N.C., University Microfilms, Ann Arbor, Mich., 1967, 174 p.

6. Baker, D.W., Wenk, H.R. and Christie, J.M., "X-ray Analysis of Preferred Orientation in Fine-Grained Quartz Aggregates," Jour. Geology, $\underline{77}$:144-172,(1969).

7. Bunge, H.J., "Die dreidimensionale Orientierungsverteilungsfunktion und Methoden zu ihrer Bestimmung," Kristall und Technik, $\underline{3}$:439-454, (1968).

8. Bunge, H.J., "Zur Darstellung allgemeiner Texturen,"
    Zeitschr. Metallkunde, 56:872-874, (1965).

9. Bunge, H.J., "Einige Bemerkungen zur Symmetrie
    verallgemeinerter Kugelfunktionen," Monatsber.
    Deutsche Akad. Wiss., 7:351-360, (1965).

10. Roe, R.J., "Description of Crystallite Orientation
    in Polycrystalline Materials. III. General Solution
    to Pole Figure Inversion," Jour. Appl. Physics,
    36: 2024-2031, (1965).

11. Roe, R.J., "Inversion of Pole Figures for Materials
    Having Cubic Crystal Symmetry," Jour. Appl. Physics,
    37: 2069-2072, (1966).

12. Buerger, M.J., Elementary Crystallography, McGraw-
    Hill, New York, 1963, p. 337.

13. Baker, D.W. and Wenk, H.R., "Spherical Harmonic
    Analysis of X-ray Pole-Figure Data for Specimens
    with Low Symmetry," (abs.), Am. Geophys. Union
    Trans., 50: 323,(1969).

14. Henry, N.F.M. and Lonsdale, K., International
    Tables for X-ray Crystallography, Vol. 1, Kynoch
    Press, Birmingham, 1952, p. 49-50.

15. Paterson, M.S. and Weiss, L.E., "Symmetry Concepts
    in the Structural Analysis of Deformed Rocks,"
    Geol. Soc. America Bull., 72: 865-868, (1961).

AUTOMATED LATTICE PARAMETER DETERMINATION ON SINGLE CRYSTALS

Armin Segmüller

IBM Watson Research Center

Yorktown Heights, New York 10598

ABSTRACT

A commercial powder diffractometer linked to an IBM 1800 time sharing computer is used for the precision determination of lattice parameters on single crystals with Bond's method of measuring the angle between two diffracting positions of the crystal, symmetric to the incident x-ray beam. A remote x-ray tube and collimator allows a high angle Bragg reflection to be measured easily in these two positions with only one detector. After scanning the two diffraction line profiles, the peak positions and the lattice parameter are determined on line by the computer. Several methods for determining the peak position are discussed. Using these techniques, the lattice parameter of silicon has been determined on two crystals of high purity and perfection to $a_0 = 5.43093 \pm 0.00002$ Å and $a_0 = 5.43095 \pm 0.00002$ Å at 25°C. These values are in excellent agreement with Bond's result, $a_0 = 5.430935 \pm 0.000019$ Å, and they differ slightly from the value $a_0 = 5.43074 \pm 0.00017$ Å, the mean value of lattice parameters measured on silicon powder in 16 laboratories under the I.U.Cr. precision lattice parameter project. The wavelengths of the Cu-Kα radiation used, $\lambda_1 = 1.540562$ Å and $\lambda_2 = 1.544390$ Å, were taken from Bearden's recent precision measurements which are based on the lattice parameter of silicon, $a_0 = 5.430946 \pm 0.000018$ Å, as determined from density measurements. Determination of a_0 from λ_1 and λ_2 indicates that $\Delta\lambda_{12}$ may not be known as precisely as believed.

INTRODUCTION

A precision method for determining lattice parameters on single crystals has been developed by Bond.[1] By means of a one-circle goniometer the angle between two positions of the single crystal is accurately measured. In these two positions, symmetric to the incident x-ray beam, x-rays are diffracted from the desired crystal planes and the two rocking curves are measured by two detectors with large windows, mounted at two fixed positions, also symmetric to the incident beam. Since the angle difference of two crystal positions is measured, rather than that of the two detector positions, eccentricity, absorption and zero errors are completely eliminated, whereas other errors are shown to be very small and negligible by careful adjustment.[1] Several authors have used this method for precision measurement of lattice parameters[2] and of the coefficient of thermal expansion.[3,4] Baker et al.[5] have used an especially designed high precision goniometer under computer control for measuring lattice parameters on single crystals with Bond's technique. This paper describes the use of a standard powder diffractometer linked to an IBM 1800 time sharing computer and operated completely under program control for the automatic determination of lattice parameters on single crystals by means of a modified Bond technique.

EXPERIMENTAL

Implementation of Bond's Technique

On a powder diffractometer, the rotation of the sample (Θ) is linked to that of the detector (2Θ) by a 1:2 ratio. For the measurement of single crystal reflections the receiving slits are normally removed to allow the whole diffracted beam to be received by the detector. If the crystal surface is fairly well aligned to the diffractometer axis rocking curves can be measured in a Θ -2Θ -scan over several orders of one reflection. However, it is impossible to measure one reflection in two positions symmetric to the incident beam, as is required by Bond's technique, because the crystal would have to be rotated 180° when moving the detector from $+2\Theta$ through zero to -2Θ. This would introduce an inconvenience and a serious angle reading error. Also, higher Bragg angles could not be measured on one side, because the x-ray tube would block the detector motion. Principally, the angular difference Θ_{mn} between the m. and n. order of one reflection can be measured and Θ_m can be calculated using the relation

$$\tan \Theta_m = \sin\Theta_{mn}/(n/m-\cos\Theta_{mn})$$

However, the deviation from Bragg's law due to the refraction changes with 1/n rendering this possibility impractical when highest precision is required. The other possibility chosen in this paper is to locate the x-ray tube and the incident beam collimator outside the diffractometer in order to allow the detector to be moved from +2Θ through 180° to −2Θ , with the sample motion linked to it.

Diffractometer

Figure 1 shows a drawing of the Picker Biplane Diffractometer with a second remote x-ray tube. Since the powder tube remains mounted on the diffractometer no realignment is necessary when switching from one tube to the other. With this arrangement Bragg reflections of single crystals can be measured on both sides of the incident beam from Θ = 65° up. If lower Bragg angles are to be used the powder tube and its mounting bracket have to be removed.

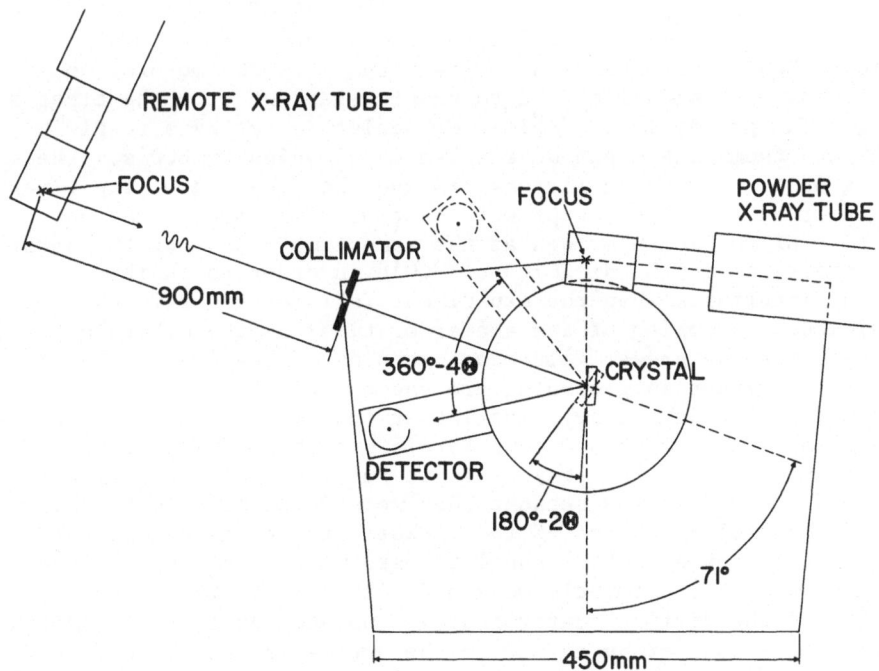

Figure 1.   Picker Biplane Diffractometer with remote x-ray tube.

On one end of the 2Θ -shaft a SLO-SYN ac motor is mounted
allowing the detector to be driven with a high speed (72°/min)
between the two diffracting positions. On the other end of the
2Θ -shaft a 6-digit shaft angle encoder is mounted enabling the
angle 2Θ to be read absolutely with 0.001° precision. A stepping
motor, as used by Baker et al.[5] for measuring angles, may lose
counts if driven with high speed and it reads relative angles
only. The Picker step scan cam mounted on the shaft of the slow
2Θ-motor has been replaced by a cogwheel with 16 cogs cutting the
stepping time from 8 to 1/2 sec. With this cogwheel from 60 to
960 step/° (2Θ) can be chosen depending on the gear ratio. The
detector preamplifier is raised over the detector to enable the
detector to be moved as close as possible to the position 2Θ =
180° without intersecting the incident beam.

                          Computer Control

     The IBM 1800 system as described by Cole[6] was available under
time sharing for controlling the diffractometer and logging the
data. A Canberra scaler and timer with 6-digit parallel BCD output
is used to measure the diffracted intensity. The output of the
shaft angle encoder, scaler and timer is multiplexed in the interface and read into 1 1/2 input words (24 bits) of the computer upon
selection of the unit by an Electronic Contact Operate (ECO)
command. ECO's are also used to start and stop the motors, to
select slow or fast drive, and to reset, start or stop the timer and
scaler. The preset mode of timer or scaler as set by a toggle
switch and thumbwheels can be enabled or disabled by ECO's. The
closing of the step scan microswitch and the stopping of the timer
and scaler by reaching the preset count or time, each cause an
interrupt of the computer and at the same time a stop of the motor.
Both interrupts can be disabled by ECO's in order to inhibit
spurious interrupts when the experiment is inactive or in the
manual mode. Stopping of the scaler or timer by the other unit
upon reaching the preset value and stopping of the 2Θ-motor at the
time the interrupt is generated are the only logic functions
performed in the interface. The latter function is necessary to
avoid loss of data when the computer is not able to accept
interrupts. Certain unnecessary logic functions of the scaler and
timer had to be disabled because they were interfering with the
computer control. Instead of the Picker diffractometer control
panel a very simple control panel is used with a main switch for
choosing manual or automatic mode and with a relay for every
function of the diffractometer that is operated by a toggle switch
in the manual mode or by an ECO in the automatic mode. In a
typical run, first the detector is set by the computer to
the desired starting angle by the fast and slow motor, then a
specified range is scanned slowly either in steps or continuously,

and the intensity is measured either in preset time or preset count
mode or in a combination of both. Upon a timer-scaler interrupt,
the 2Θ-motor is stopped if scanning is continuous, the shaft angle
encoder, timer and scaler are read, the intensity is computed and
stored together with the angle in the INSKEL COMMON area in the
computer and the scanning is continued. Upon a step interrupt in
the step scan mode, the 2Θ-motor is stopped and scaler and timer
are reset and started. In the continuous mode scaler and timer
are reset and started simultaneously with the 2Θ-motor. After 100
data have been accumulated in the COMMON area, they are stored in
a disk file and data acquisition is continued. After a line has
been scanned the detector is set to the next starting angle and
scanning is initiated, or the experiment is terminated according to
the instructions.

A printer and a numerical keyboard at the side of the diffractometer allow a convenient communication with the remote computer.
Simple numerical codes are used to set experimental parameters and
angles, to start the experiment and to inquire its status.

X-Ray Measurements

Unfiltered Cu-Kα radiation emerging from the spot focus of
the remote tube under a take-off angle of 3° is collimated by slits,
1 mm wide, 2 mm high and 900 mm distant from the focus. The values
of the wave lengths used, λ_1 = 1.540562 Å and λ_2 = 1.544390 Å,
have been taken from Bearden's recent tables.[7] No receiving slit
is used in front of the about 4 mm wide proportional counter window.
The single crystal is placed on a barrel-shaped adjustable holder[8]
and its reflecting lattice planes are aligned perpendicular to the
barrel axis and parallel to the diffractometer axis with 0.01°
precision. If the reflecting crystal planes are not parallel to the
crystal surface, a wedge is used to mount the crystal on the barrel.
The crystal surface is aligned to the diffractometer axis within a
few hundredths of a mm using the fine collimated incident x-ray
beam from the powder tube. The crystal angle has to be aligned
relative to the detector angle so that maximum intensity is received
right in the center of the detector window. Within one run
detector and crystal are moved in the same direction on both sides
of the incident beam in order to eliminate any backlash in the
driving gears. The measurement is repeated usually ten times in an
oscillating movement in order to compute a reliable mean value and
standard deviation. A range of 2.5° (2Θ) is step-scanned on both
sides with 120 step/$^\circ$ (2Θ). The intensity of the α_1 peak is about
10,000 cts/sec with the tube operated at 50 KV and 5 mA. Scaler and
timer are both preset to 10,000 cts and 2 sec. The time needed for
one run is about 1/2 hr at the end of which a pair of lattice
parameters is printed out.

TABLE I: Lattice parameters of silicon, as determined with different peak search methods. Sample FZ 2. Values corrected for Refraction and for 25° C. $\lambda_1=1.540562$ Å, $\lambda_2=1.544390$ Å ; a_o is the meanvalue of a_1 and a_2, Δa the standard deviation.

| | Centroid, truncated at | | | | | Parabola | Midchord | Tangent |
	10%	20%	33%	50%	67%			
a_1	5.43085	5.43088	5.43091	5.43094	5.43095	5.43097	5.43098	5.43097
a_2	5.43084	5.43084	5.43086	5.43088	5.43090	5.43093	5.43094	5.43093
a_o	5.43084	5.43086	5.43088	5.43091	5.43093	5.43095	5.43096	5.43095
Δa	0.00001	0.00002	0.00003	0.00003	0.00003	0.00002	0.00003	0.00002

DATA EVALUATION

Peak Position

Immediately after a line has been scanned the peak is examined and its position is determined by computing the tangent of the line profile and searching for its change from positive to negative slope. Fitting a parabola

$$y = a + bx + cx^2 \qquad (1)$$

to the measured points (x_i, y_i) by the least-squares method, the tangent or derivative of this parabola at $x = x_o$ is given by

$$b = \sum y_i (x_i - x_o) \; / \; \sum (x_i - x_o)^2 \qquad (2)$$

where the summations are carried out over an equal number of equi-distant data points on both sides of x_0.[9] Since the denominator in equation (2) is a constant if the same number of data points is used for the parabola fit all over the line profile, the division in equation (2) need not be carried out. The use of double word integer (32 bits) arithmetic allows the calculation of the tangent to be executed very fast. The peak position is calculated by fitting a least-squares line to the derivatives in the neighborhood of their sign change. The number of data points used for the fit of the parabola has to be adjusted to the real structure of the line profile. If it is chosen too small too many closely spaced peaks are found due to statistical fluctuations. If it is chosen too large, it takes unnecessary time and closely spaced real peaks are not resolved. In this study 4 to 10 data points on each side have been found to give satisfactory results.

This tangent method has the advantage that no prior knowledge of the approximate peak position is necessary. Peak search and determination of peak position is done in only one step, simplifying the program. Little satellite peaks, as for instance in the study of hetero-epitaxial growth, are easily recognized, provided the tangent changes its sign. Other methods for determining the peak position have been tried on the same data sets but off line. The midchord method[10] and the parabola method[11] give practically the same results as the tangent method (table I). The results obtained by the centroid method[12] depend on the truncation and they approach the results of the tangent method if the profile is truncated at increasing intensity levels (table I).

Since the background has been found to be constant no background subtraction is necessary before the peak search.

Errors and Corrections

Errors due to the eccentricity, absorption or zero shift are eliminated by Bond's technique. In the following the other errors are listed and estimated in the same order as they are treated by Bond.[1]

Crystal and Collimator Tilt Error. A deviation of the reflecting crystal planes from the position parallel to the diffractometer axis (tilted crystal) or the deviation of the incident beam from the plane perpendicular to the diffractometer axis (tilted collimator) cause both a shift of the measured Bragg angle to larger values. For an error in the lattice parameter of less than 1 ppm the angle of deviation should be less than $0.08°$ [1] which is easily fulfilled for the crystal tilt error by means of the barrel-shaped holder. The collimator tilt error can be minimized by shifting the tube in the vertical direction with the collimator kept in place until the measured Bragg angle has a minimum. This method is equivalent to the "method of the maximum angle" used by Baker et al.[5]

Axial Divergence Error. With a collimator slit height of 2 mm, a focus spot height of 1.3 mm and a collimator length of 900 mm the axial (or vertical) divergence of the incident beam of $2\Delta_0$ = 3.3/900 rad causes the lattice parameter a to be measured to low by

$$\Delta a_{AD} = a\Delta_0^2/6 = a \cdot 0.6 \cdot 10^{-6} \qquad (3)$$

which can be neglected.

Refraction Correction. The refraction correction for the hkℓ reflection of a cubic crystal is given by

$$\Delta a_R = (4.48 \cdot 10^{-6} \, N \, \lambda^2/a^3) \cdot a/\sin^2\theta \; Å \qquad (4)$$

$$= 17.9 \cdot 10^{-6} \, N/(h^2 + k^2 + \ell^2) \; Å$$

where N is the number of electrons per unit cell. For the silicon 444 reflection we obtain

$$\Delta a_R = 4.2 \cdot 10^{-5} \; Å. \qquad (5)$$

Lorentz-Polarisation Correction. The Lorentz-polarisation factor for perfect crystals

$$LP = (1 + |\cos 2\theta|)/\sin 2\theta \qquad (6)$$

TABLE II. Lattice parameter of silicon, at 25° C and corrected for refraction.

λ_1 = 1.540562 Å, λ_2 = 1.544390 Å, a_o = meanvalue of a_1 and a_2, Δa = standard deviation

Sample	LOPEX 1	LOPEX 2	FZ 1	FZ 2	Bond*	Smakula-Kalnajs**	Henins-Bearden***
a_1	5.43094	5.43095	5.43097	5.43097	5.430935	5.43090	
a_2	5.43091	4.43091	5.43094	5.43093			
a_o	5.43092	5.43093	5.43095	5.43095			5.430946
Δa	0.00002	0.00002	0.00002	0.00002	0.000019	0.00002	0.000018

*Reference 1, value of a_1 converted from kX to Å by multiplication of original value with 1.002060 = 1.537395.

**Reference 14, value of a_1 converted to our Å by multiplication of original value with 1.540562/1.54051.

***Reference 15, value of a_o calculated from density measurements.

causes a peak shift to higher angles for $45^{\circ} < \Theta < 90^{\circ}$. Approximation of the line profile by a dispersion curve[1] with the half maximum width $2\Delta_H$ allows the lattice parameter correction to be estimated to

$$\Delta a_{LP} = a\Delta_H^2/2 \sin^2\Theta = a\cdot 10^{-6} \qquad (7)$$

for a measured value $2\Delta_H = 0.156^{\circ}(\Theta)$ of the silicon 444 reflection.

Temperature Correction. All lattice parameters are given for 25° C. Extrapolation from the measuring temperature of $21 \pm 0.5^{\circ}$C, using the thermal expansion coefficient of silicon $\alpha_{Si} = 2.33 \cdot 10^{-6}$ centigrade^{-1}, [13] gives the temperature correction to

$$\Delta a_T = a\cdot(9.3 \pm 1.2)\cdot 10^{-6} \qquad . \qquad (8)$$

Dispersion Error. Since Bearden's wavelengths have been determined with a peak method[7,10] as are our lattice parameters no disperion correction should be necessary.

Adding up all the corrections we arrive at the total correction

$$\Delta a = (1.01 \pm 0.06)\cdot 10^{-4} \overset{\circ}{A} \qquad . \qquad (9)$$

for the measured lattice parameter of silicon. The uncertainty in equation (9) is due to the uncertainty of the temperature.

RESULTS

Two Texas Instruments silicon crystals, LOPEX and Float Zone (FZ), both P-type, with (111) orientation, with a dislocation density less than 500 cm^{-2} and with a resistivity higher than 500 ohm-cm have been used for the measurements. The 444 reflection was chosen with a Bragg angle $\Theta_1 = 79.31^{\circ}$ and $\Theta_2 = 80.10^{\circ}$. Two ten-run cycles have been taken on each sample, rotating the barrel and checking the alignment between the two cycles. Table II shows the lattice parameters corrected for all errors according to equ. (9) together with Bond's result.[1] Also shown is the result obtained by Smakula and Kalnajs[14] using a Philips diffractometer and an extrapolation method. The agreement is within the limits of the rms deviation. Henins and Bearden[15] determined the wavelength of the copper $K\alpha$-radiation from double crystal spectrometer studies and density measurements on 17 selected silicon crystals. Using their mean density $\rho = 2.329004$ g/cm^3 $\pm$ 3.4 ppm, Avogadro's number $N = 6.02252\cdot 10^{23}$ (g mole)$^{-1}$ $\pm$ 11 ppm and the atomic weight of silicon $A = 28.0857$ amu $\pm$ 10 ppm one obtains the density based lattice parameter of silicon, $a_0 =$

(5.430946 ± 0.000018) $\overset{\circ}{A}$ ± 15 ppm, where the error within the parentheses is due to the standard deviation of the density measurements, and the second error is due to that of the atomic constants. Our values of a_0 also agree within the limits of the rms deviation with the density based lattice parameter which is of course the basis of the wavelengths used by us.

Lattice parameters of silicon powder samples have been measured under an I.U.Cr. project. At its conclusion,[16] the meanvalue of lattice parameters measured in 16 laboratories has been given to a = 5.43074 ± 0.00017 $\overset{\circ}{A}$,* whereas a weighted meanvalue has been suggested to a = 5.43084 ± 0.0002 $\overset{\circ}{A}$.[17] In this project, the only reliable diffractometer determination[18] using a not identified peak method gave a = 5.43087 ± 0.00003 $\overset{\circ}{A}$ which is closer to the single crystal values than the average powder value. However, as Parrish[16] pointed out, a difference between powder and single crystal measurements may be due to the thin layer of SiO_2 the powder particles are covered with.

A systematic difference of about 8 ppm can be observed between the lattice parameter as determined from the α_1 and α_2 peak by peak methods (Table I and II). This difference has been observed also measuring single crystals of germanium and gallium-arsenide. It is larger than the standard deviation of lattice parameters calculated from one component of the doublet (1-3 ppm) and it cannot be explained by the overlap of the two peaks since that would cause a_2 to be larger than a_1 contrary to the observation.

The rms deviations of the wavelengths as calculated from the measurements of Henins and Bearden[15] are 5.5 ppm and 4.1 ppm for the α_1 and α_2, respectively.** This gives a possible error for the difference $\Delta\lambda_{12}$ of 7 ppm which is indeed of the order of our observed discrepancy. However, the precision of our set-up is limited by the precision of the angle reading and by the precision of the computer in floating point computations to a few ppm. Therefore, this discrepancy should be investigated by a configuration capable of higher precision, like the one described by Baker et al.[5]

CONCLUSION

An automatic x-ray diffractometer has been described, capable of measuring lattice parameters on single crystals with an error

*This and the following lattice parameters are converted to Bearden's $\overset{\circ}{A}$ by multiplication of the original values with the factor 1.000036 = 1.002056/1.00202.

**The probable errors in the means given by the authors (0.6 and 0.5 ppm) appear too optimistic for measurements taken on different samples. We use the standard deviations in the measurement.

of a few ppm. This achievment is due to the use of Bond's method and to the use of a computer to evaluate the data. The computer allows the processing of much more data than could be done manually, and therefore is able to detect systematic errors. The precision is less than that claimed by Baker et al.[5] for their configuration. However it offers a middle way between that and the powder diffractometer, in precision, price and effort as well.

ACKNOWLEDGEMENTS

The author is idebted to J. Angilello for assembling and modifying the electronics, to R. V. Dobransky for designing and implementing the interface, to R. Ryniker for his valuable assistance in programming and to C. G. Wood for changes and attachments of the diffractometer.

REFERENCES

1.   W. L. Bond, "Precision Lattice Parameter Determination", Acta Cryst. 13:814,1960.

2.   A. S. Cooper, "Precise Lattice Constants of Germanium, Aluminum, Gallium Arsenide, Uranium, Sulphur, Quartz and Sapphire", Acta Cryst. 15:578 , 1962.

3.   F. M. d'Heurle, R. Feder and A. S. Nowick, "Equilibrium Concentration of Lattice Vacancies in Lead and Lead Alloys", J. Phys. Soc. Japan 18, Suppl. II: 184, 1963.

4.   R. Feder and H. P. Charbnau, "Equilibrium Defect Concentration in Crystalline Sodium", Phys. Rev. 149:464, 1966.

5.   T. W. Baker, J. D. George, B. A. Bellamy and R. Causer, "Fully Automated High-Precision X-Ray Diffraction", in J. B. Newkirk, G. R. Mallet and H. G. Pfeiffer, Editors, Advances in X-Ray Analysis, Vol. 11, Plenum Press, New York, 1968, p.359.

6.   H. Cole, "Computer-Operated X-Ray Laboratory Equipment", IBM Journal 13:5, 1969.

7.   J. A. Bearden, "X-Ray Wavelengths", Rev. Mod. Phys. 39: 78, 1967.

8.   J. G. Walker, H. J. Williams and R. M. Bozorth, "Growing and Processing of Single Crystals of Magnetic Materials", Rev. Sci. Instr. 20: 947, 1949.

9.   L. G. Parratt, Probability and Experimental Errors in Science, John Wiley and Sons, New York, 1961, p. 132.

10. J. A. Bearden, "The Wavelengths of the Silver, Molybdenum, Copper, Iron and Chromium $K\alpha_1$ Lines", Phys. Rev. 43: 92,1933.

11. W. Parrish, J. Taylor and M. Mack, "Dependence of Lattice Parameters on Various Angular Measures of Diffractometer Line Profiles", in W. M. Mueller, G. Mallet and M. Fay, Editors, Advances in X-Ray Analysis, Vol. 7, Plenum Press, New York, 1964, p.66.

12. J. Ladell, W. Parrish and J. Taylor, "Interpretation of Diffractometer Line Profiles", Acta Cryst. 12: 561, 1959.

13. D. F. Gibbons, "Thermal Expansion of Some Crystals with the Diamond Structure", Phys. Rev. 112: 136, 1958.

14. A. Smakula and J. Kalnajs, "Precision Determination of Lattice Constants with a Geiger-Counter X-Ray Diffractometer", Phys. Rev. 99: 1737, 1955.

15. I. Henins and J. A. Bearden, "Silicon-Crystal Determination of the Absolute Scale of X-Ray Wavelengths", Phys. Rev. 135: A890, 1964.

16. W. Parrish, "Results of the I.U.Cr. Precision Lattice Parameter Project", Acta Cryst. 13: 838, 1960.

17. H. Weyerer, "Discussion of Error in Lattice Parameter Measurements", Acta Cryst. 13: 821, 1960.

18. R. A. Coyle and R. I. Garrod, Code No. 15 in Reference 16.

CORRELATION OF RESIDUAL STRESS LEVEL AND FATIGUE DAMAGE IN B. C. C. METALS

G Koves

IBM Systems Development Division

Rochester, Minnesota 55901

ABSTRACT

The development of materials with optimum fatigue characteristics hinges on a better understanding of the fatigue mechanism (i.e., fatigue crack nucleation and propagation) under specific loading conditions. A correlation was established between the residual macroscopic stress level--as measured by x-ray diffraction analysis, the fatigue damage in the form of slip-band formation and intensification, and the fatigue life at various applied stress levels of pure iron and low-carbon steel.

Residual stresses were measured by an automated diffractometer technique (Fe 211 K_α reflection, Cr radiation) and computer calculation of both macroscopic (line shift) and microscopic (line broadening) stresses. Surface fatigue damage was observed by optical microscopy.

Rectangular bar specimens with a center notch were subjected to cyclic impact-bending fatigue loading at applied stresses both above and below the endurance limit of the material. Residual stresses were measured, and slip-band formation observed at the point of maximum fatigue stress periodically during cycling.

A consistent pattern was established for specimens cycled at applied stress levels above the fatigue limit. Residual macrostress increased to a peak value, which appeared at about 25% of the total fatigue life. The magnitude of the peak increased with increasing applied stress level. The first visible fatigue damage occurred as residual stresses reached their peak value. This damage (i. e., slip-band formation) intensified at a constant residual stress level until gross fatigue cracking commenced, at which point the residual stress level dropped precipitously.

In specimens cycled below the established fatigue limit, the residual stresses never rose above the inherent level, and slip-band formation was not observed during cycling up to 3×10^7 load cycles.

INTRODUCTION

Fatigue of structural metals and alloys is one of the key problems affecting the performance and reliability of electromechanical data processing machines. The development of materials with optimum fatigue characteristics hinges on a better understanding of the fatigue mechanism, i.e. fatigue crack nucleation and propagation, under specific loading conditions.

The literature dealing with the mechanism of metal fatigue is quite voluminous. However, progress has been rather modest. Among those attempting a truly comprehensive approach, Grosskreutz[1,2] concludes that detailed knowledge and data are not yet available to clarify the complex phenomenon and that the statistical treatment is still the recommended practical tool. Stulen et. al.[3], approaching the question of basic fatigue mechanism through the study of one specific alloy, came to a similar conclusion.

Among the numerous factors affecting the fatigue characteristics of metals, internal (or residual) stress has been recognized and discussed widely. Textbooks on fatigue, such as the one by Forrest[4], usually include a chapter on the effects of internal stresses on the fatigue strength of metals. Major symposia on fatigue generally include papers on the subject. Horger and Neifert[5] and Rowland[6] provide good overviews and point out that compressive residual stresses exert a favorable influence on the fatigue strength of metals, while tensile residual stresses have the opposite effect. A unique compilation of papers on internal stresses and fatigue in metals, edited by Rassweiler and Grube[7], contains much valuable information on measurement methods, effects, and microstructure correlations.

Several papers[8-12] follow in the footsteps of the classical treatment by Bühler and Buchholtz[13] in establishing quantitative correlation between internal stresses--usually generated by cold work or surface treatment--and the resulting change in fatigue strength of specific structural metals.

The study of the opposite phenomenon, namely the effect of cyclic fatigue loading on the internal residual stress condition of metals, could be of considerable interest. This interest is twofold. First, the changes in residual stress level due to cyclic straining could be indicative of the mechanism by which fatigue damage initiates and spreads in metals. Furthermore, if the change of residual stresses follows a regular pattern, the measurement of such stresses at an early stage of fatigue cycling could predict the behavior of the specimen during its complete fatigue life.

Among the relatively few published studies in this area, Pattinson and Dugdale[14] indicate that residual stresses--introduced into specimens by cold deformation--fade quickly with fatigue cycling in mild steel, but persist in aluminum alloy. The most comprehensive effort is that by Hartmann and Macherauch[15-17], who measured residual microscopic stresses in various metals with x-ray line-broadening techniques. They found considerable increase in microstresses with fatigue cycling in nickel but no change in iron and steel.

The present study involves the measurement of both macroscopic and microscopic residual stresses due to fatigue straining and their correlation with fatigue life, as well as structural damage in ferrous alloys.

EXPERIMENTAL PROCEDURE

The impact-fatigue loading of the test specimens was done in an experimental arrangement described by Koves.[18] Figure 1 shows the specimen configuration used in all tests. The notch and its vicinity was electropolished at the beginning of the test sequence, but was not repolished thereafter. The specimens were loaded as cantilever bars at a rate of 1800 cpm so that the maximum impact-bending stress developed at the notch. The specimens were cycled at a broad range of constant stress levels both above and below the endurance limit of the material. The highest stress load produced fracture after about 8×10^3 cycles. Fatigue cycling was interrupted after stops of 10^n load cycles (where $n = 0, 1, 2, 3, 4...$) and the specimens were removed from the fatigue machines for residual stress measurement and microscopic observation. Fatigue cycling was continued to specimen fracture or to 3×10^7 load cycles. Figure 2 shows the actual test setup illustrating the relative positions of specimen and striker head.

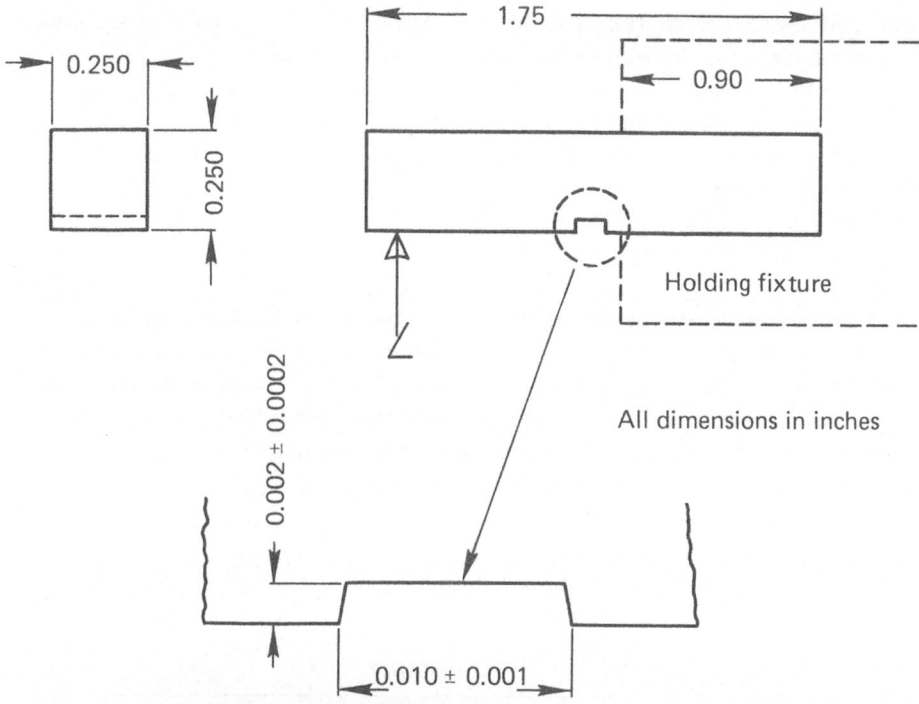

Holding fixture

All dimensions in inches

Figure 1 Specimen configuration

Figure 2 Impact fatigue test setup

Two types of body centered cubic ferrous alloys were tested.
AISI 1018 low carbon steel served as a typical structural alloy frequently
used in electromechanical data processing machines. The impact-fatigue
characteristics of this material were well established in previous studies.
The basic correlation between residual stresses and fatigue life was established on this material, thereby proving the feasibility for practical
applications. A commercial grade of pure iron was utilized for observing surface damage development during fatigue cycling and correlating
it with residual stress level. The homogeneous ferritic microstructure
of this material is better suited for unobstructed formation of slip-bands.
Table I lists relevant materials parameters.

Residual stresses were measured by the x-ray diffraction method
using a diffractometer technique and computer automation for data collection and evaluation, described by Koves and Ho.[19] Both macroscopic
stresses, as indicated by line-shift, and microscopic stresses, measured
by line-broadening, were evaluated. The x-ray beam (unfiltered Cr radiation) covered a narrow rectangle on the tension side of the specimen
immediately adjacent to the notch. The area of the rectangle covered

Table I. Materials parameters

			1018 Steel	Iron
Nominal chemical				
composition (%)	C	Max	0.20	0.06
	Mn	Max	0.90	0.20
	P	Max	0.04	0.02
	S	Max	0.05	0.04
	Si	Max	–	0.04
	Cu	Max	–	0.15
	Fe	Min	–	99.60
Final heat treat			Stress relieve at 1000°F in vacuum for 1/2 hour; cool 100°F/hour	Anneal at 1700°F in vacuum for 1/2 hour; cool 100°F/hour
Impact-fatigue endurance limit			25.000 psi	8,000 psi*

*Fatigue strength at
10^8 cycles

the entire width of the specimen with a depth of about 1/16 in. The Fe 211 Kα reflection was used for both line-shift and line-broadening determinations. Table II summarizes the diffraction parameters used in all stress measurements.

Table II. X-Ray diffraction stress measurement parameters

Tube target	Chromium
Filter	None
Power	30KV, 10 ma
Detector	Sealed proportional counter
Pulse height discriminator	10-V window over 10-V base
Divergence slit system	1-deg slit, masked laterally, plus soller slit
Receiving slit	0.006 in.
Anti-scatter slit system	4 deg, plus soller slit
Angle of inclined scan	ψ = 45 deg
Scanning steps	0.125 deg 2θ

Surface damage due to fatigue cycling was monitored by optical microscopy. The bottom surface of the notch was scanned for evidence of slip-bands. Such evidence usually appeared simultaneously at several spots of the specimen. At the time of the first visible evidence (at X500 magnification on the as-electro-polished surface) three representative areas were selected on the specimen. These areas were observed and photographed during each subsequent test stage. The final fatigue cracks always propagated through these observation areas.

RESULTS

All specimens were annealed before final machining and stress relieved after final machining (see table I) so that residual macroscopic stress measured before fatigue cycling was always below 2500 psi. Accuracy of the x-ray diffraction stress measurement technique is about ± 1000 psi with reproducibility somewhat better than that. The residual microstress ratio (see reference 19 for definition) before fatigue cycling was between 3.5 and 6.4. While macrostresses were quite uniform after final stress relief throughout each specimen, microstresses varied between areas in one specimen approximately within the above limits. The initial macroscopic stresses, while always less than 2500 psi in absolute value, could be either tensile (+) or compressive (−).

The 1018 steel specimens, when cycled with applied stresses
significantly exceeding the endurance limit of the material, showed a
characteristic change in residual macroscopic stress level when measured adjacent to the notch. Figure 3 illustrates this pattern of change
in a specimen cycled at a nominal applied stress of 72, 000 psi, which
caused fracture after 87, 000 cycles.

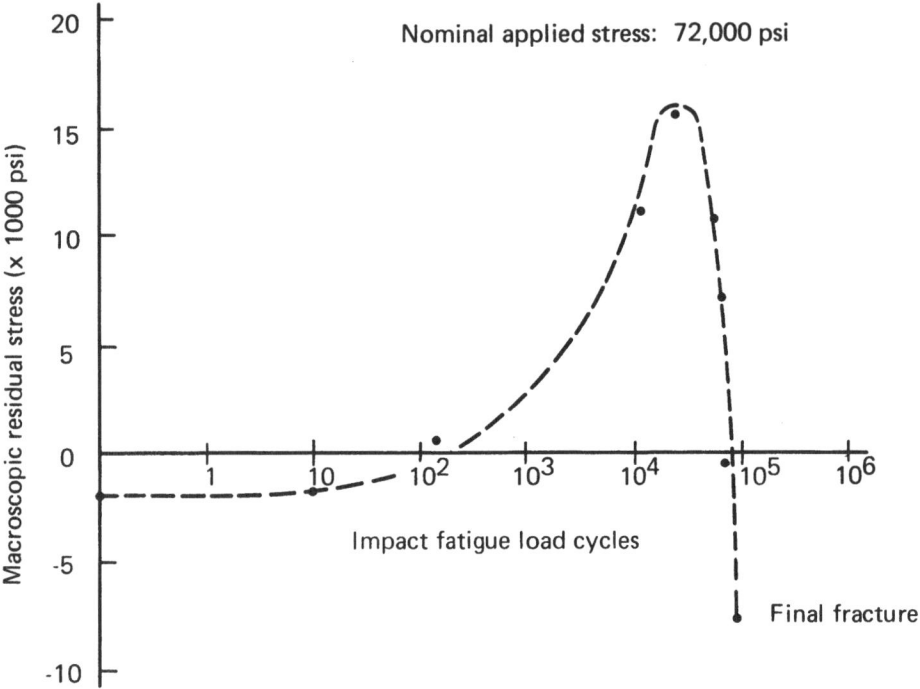

Figure 3 Residual macrostress as a function of impact-fatigue load cycles in 1018 steel.

The residual macrostress level stayed within the inherent range
of ± 2500 psi for about 1000 cycles. After that the stress level rose
steadily until it reached a maximum value of + 16, 200 psi at 20, 000 cycles. From this point on the stress level dropped consistently, reverting
to a compressive value at 80, 000 cycles, which was the last measurement
before final failure. The residual macrostress level measured on the
fractured specimen--after 87, 000 cycles--was 8400 psi in compression.

This pattern was quite closely repeated in other specimens cycled
at the same applied stress level. The peak residual stress varied between
approximately 15, 000 and 20, 000 psi in tension; however, the peak appeared quite consistently between 20, 000 and 25, 000 load cycles, while
fatigue life to fracture ranged from approximately 80, 000 to 100, 000

cycles. The residual stress after fracture was generally close to zero, although usually on the compression side.

A subsequent series of 1018 steel specimens was cycled at constant applied stress levels ranging from 102, 000 psi (causing failure at about 8000 load cycles) to 15, 000 psi, well below the established endurance limit of the material. Figure 4 shows a representative series of residual macrostress versus load cycles curves with applied stress as a parameter. A clear pattern of behavior emerges: The peak value of the residual macrostress increases with increasing applied stress level. At 102, 000 psi applied stress the residual stress peak in tension always exceeded 25, 000 psi and its range approached 30, 000 psi. However, at 35, 000 psi applied stress the residual stress--still in tension--peaked at only about 5000 psi. At the 15, 000 psi applied stress level the residual macrostress never rose above the inherent stress range and no perceptible stress peak developed in 3 x 10^7 load cycles.

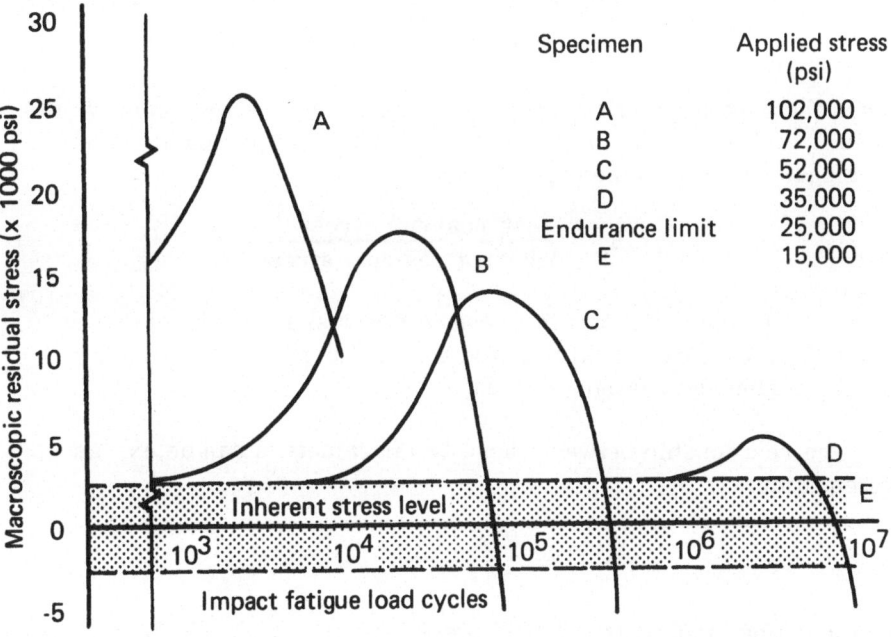

Figure 4   Residual macrostress--load cycle correlation as a function of applied fatigue stress.

A very consistent trend becomes apparent from figure 4 in the relative positions of residual stress peaks as compared to the fatigue life of the specimens. In specimen A, which failed after 8000 cycles, the stress peak appeared at about 2000 load cycles. In specimen B, as mentioned before, the residual stress peak was measured at about 20, 000 cycles, while failure commenced at 87, 000 cycles. Specimen C failed at 270, 000 cycles with the residual stress peak appearing at about 70, 000 cycles. Specimen D peaked at about 2.4×10^6 cycles and failed at 9.5×10^6 load cycles. Finally, specimen E exhibited no residual stress peak and fatigue failure did not occur as the test was terminated after 3×10^7 load cycles.

It is apparent that the residual stress peak appears quite consistently at close to 25% of the specimen's fatigue life. During the investigation using 1018 steel, 28 specimens were tested at 7 different applied stress levels. The 25% correlation between residual stress peak position and fatigue life in these specimens held with an accuracy of ± 3%.

Another correlation is illustrated in figure 5. Two stress parameters can be defined. The applied stress parameter,

$$S_A = \frac{\text{Endurance limit}}{\text{Applied stress}} \quad ,$$

indicates the magnitude of the applied nominal fatigue stress relative to the endurance limit of the material in the impact-fatigue loading mode. The residual stress parameter,

$$S_R = \frac{\text{Peak residual stress}}{\text{Inherent residual stress}} \quad ,$$

is a measure of the maximum increase in residual stress level due to impact-fatigue cycling, relative to the residual stress in the specimen before the beginning of fatigue cycling.

The relationship between these two parameters can be expressed by the equation

$$S_r = ae^{-bS_A}$$

where for the particular use of 1018 steel

$$a \quad = 20.9$$

$$\text{and} \quad b \quad = -3.1$$

Residual macroscopic stress measurements have also been carried out in two other configurations. When measured on the side surface of the notched bar specimen immediately below the notch, the residual stress level did not increase appreciably above the inherent stress level; the occasional measured increase did not show any coherent pattern. Measurements were also taken on the top side of unnotched bars near the area of maximum impact-bending tensile stress. While definite increases in residual stress level were obtained, the peaks were not well defined and valid correlations in the pattern of figures 3-5 could not be established.

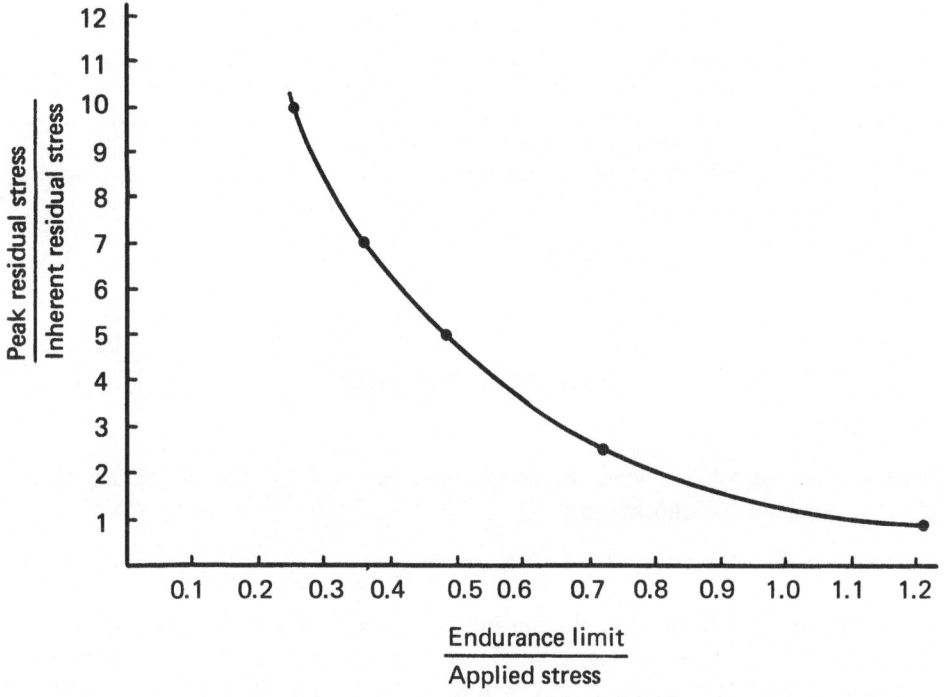

Figure 5   Correlation between residual stress parameter and applied stress
parameter in impact-fatigue loaded 1018 steel

The measurement of residual microscopic stresses brought quite different results. The residual microstress ratio (line-broadening compared to that measured on zone refined, annealed, ultra-high purity iron) stayed essentially constant in all specimens throughout their entire fatigue life. This is shown in figure 6, which depicts the stress band for all specimens. A feasible explanation for this phenomenon is the location of the spot where the residual stresses were measured relative to the occurrence of fatigue damage. This damage was concentrated in the notch, while the stress measurement was carried out on the flat top surface of the specimen, immediately adjacent to, but still not in, the notch.

The small dimensions of the notch precluded valid x-ray diffraction stress measurement in that area. Nevertheless, this observation agrees with earlier studies.[15-17]

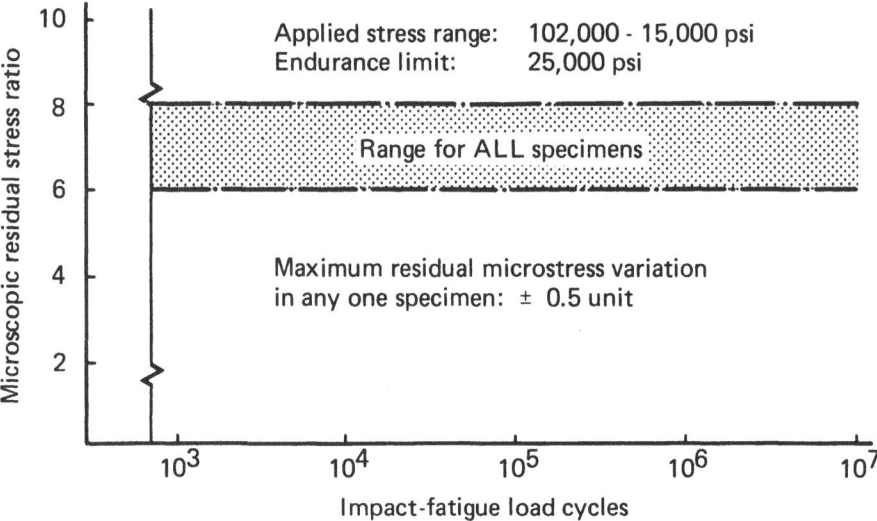

Figure 6   Residual microstress--load cycle correlation as a function of applied fatigue stress.

As the pattern of correlation between residual macrostresses and fatigue life in 1018 steel emerged, an attempt was made to observe the formation and spreading of surface damage in the form of slip bands in and around the notch. However, the frequent pearlite regions prevalent in the microstructure of this material rendered the observation difficult and unreliable. Hence, the commercially pure iron was tried and proved to be quite satisfactory.

First a series of control tests was run to establish the existence of a correlation between residual stress level and fatigue life similar to that found in the steel. When such correlation was indeed apparent, the subsequent test series was run at well established applied stress levels. The fatigue behavior in this material was different from steels inasmuch as it did not exhibit a clear-cut endurance limit. Therefore, the tests were carried out to 10^8 load cycles, and fatigue strength at 10^8 cycles was used as a limiting value. This value was 8000 psi.

The homogeneous ferritic microstructure of this material proved quite suitable for microscopic observation and recording of surface fatigue damage. Although bright field, dark field, as well as phase contrast illumination were all used to detect slip bands, bright field illumination was used for photomicrography.

The graph in figure 7 and the photographic series in figure 8 illustrate a typical test sequence that demonstrates the correlation between fatigue surface damage and residual stresses during impact-fatigue cycling. Figure 7 depicts the change in residual macrostresses as a function of load cycles similar to figures 3 and 4. However, instead of connecting the individual data with a continuous, smooth curve, figure 7 is divided into four phases. Residual stresses in phase I stay at the inherent stress level. In phase II, between approximately 10^2 and 10^5 load cycles, residual stresses increase and then level off between 10^5 and 10^7 cycles (phase III). At the end of phase III a sudden peak appears, after which the residual stress level drops precipitously in phase IV to the terminal value after final fracture.

The photomicrographic series of figure 8 illustrate the gradual progress of surface fatigue damage at one typical area of the notch bottom. The transverse notch is vertical on all photographs. The letters A through F in the graph of figure 7 relate to the respective micrographs in figure 8. No slip-band formation was observable at any point of the specimen until 10^5 load cycles. At that stage, arrows in figure 8A indicate the first observed slip-bands. Similar slip formation occurred simultaneously at other points of the notch.

Significant intensification of slip damage is seen in figure 8B, taken after 5×10^5 load cycles. The three separate areas of figure 8A are starting to merge and outline a grain boundary to their right. Slip-bands are just becoming visible in the adjacent grain. Figure 8C, taken after 2×10^6 cycles, shows further intensification of the slip systems. The grain boundary is now completely outlined by dense slip formation. Within the large grain at center right, the first signs of cross-slip appear. New areas of slip-band start to form in the upper and lower left corner areas. Note that the progressive intensification of slip damage takes place in phase III (see figure 7) at a constant residual stress level.

Figure 8D was taken after 1.1×10^7 load cycles as the residual macrostress rose to a slight peak above the plateau of phase III. Further intensification of all previous damage areas is obvious. The most significant change, however, took place inside the large grain at right center with the already clearly delineated cross-slip pattern. These slip-

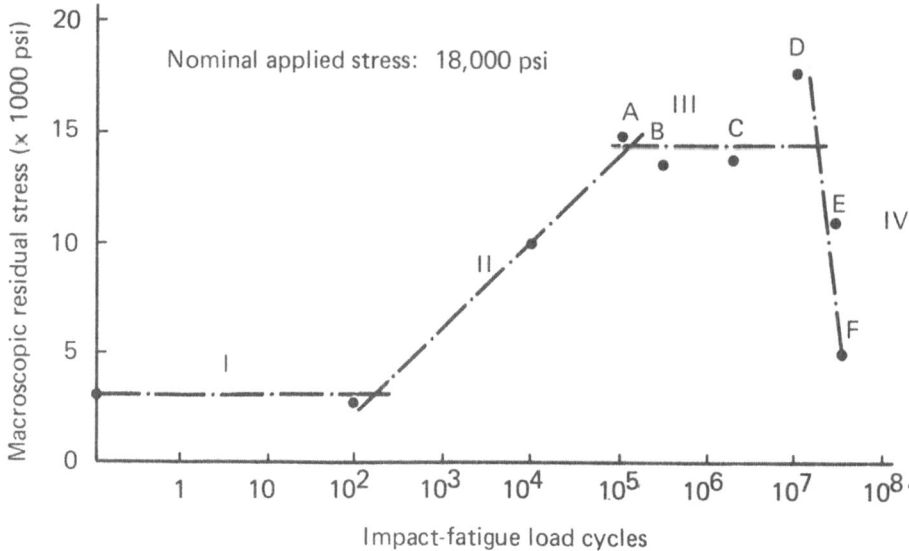

Figure 7 Residual macrostress as a function of impact–fatigue load
cycles in commercially pure iron.

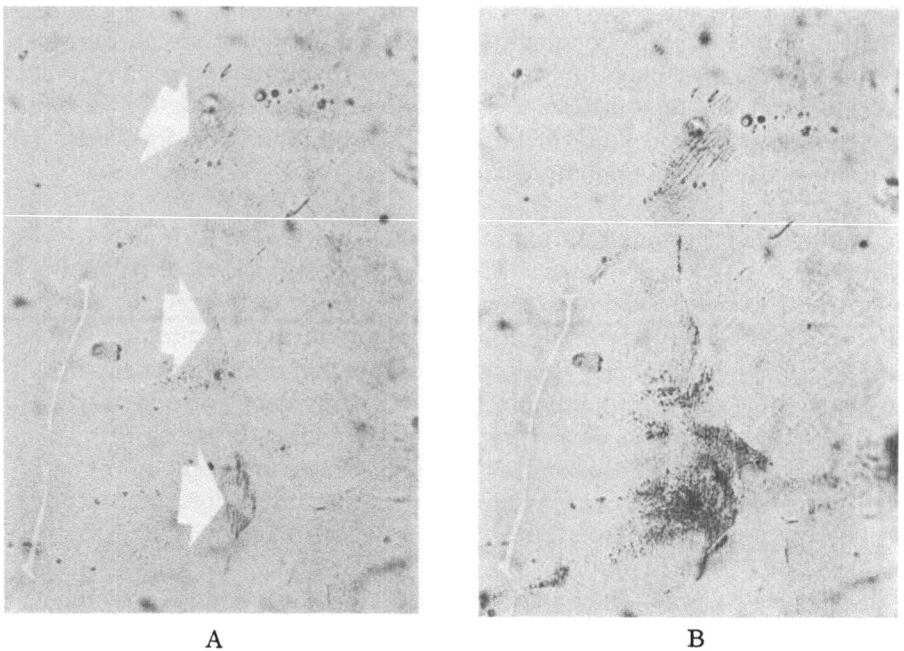

A   B

Figure 8 (A–F) Intensification and spreading of surface fatigue damage
during the fatigue test sequence depicted in figure 7.

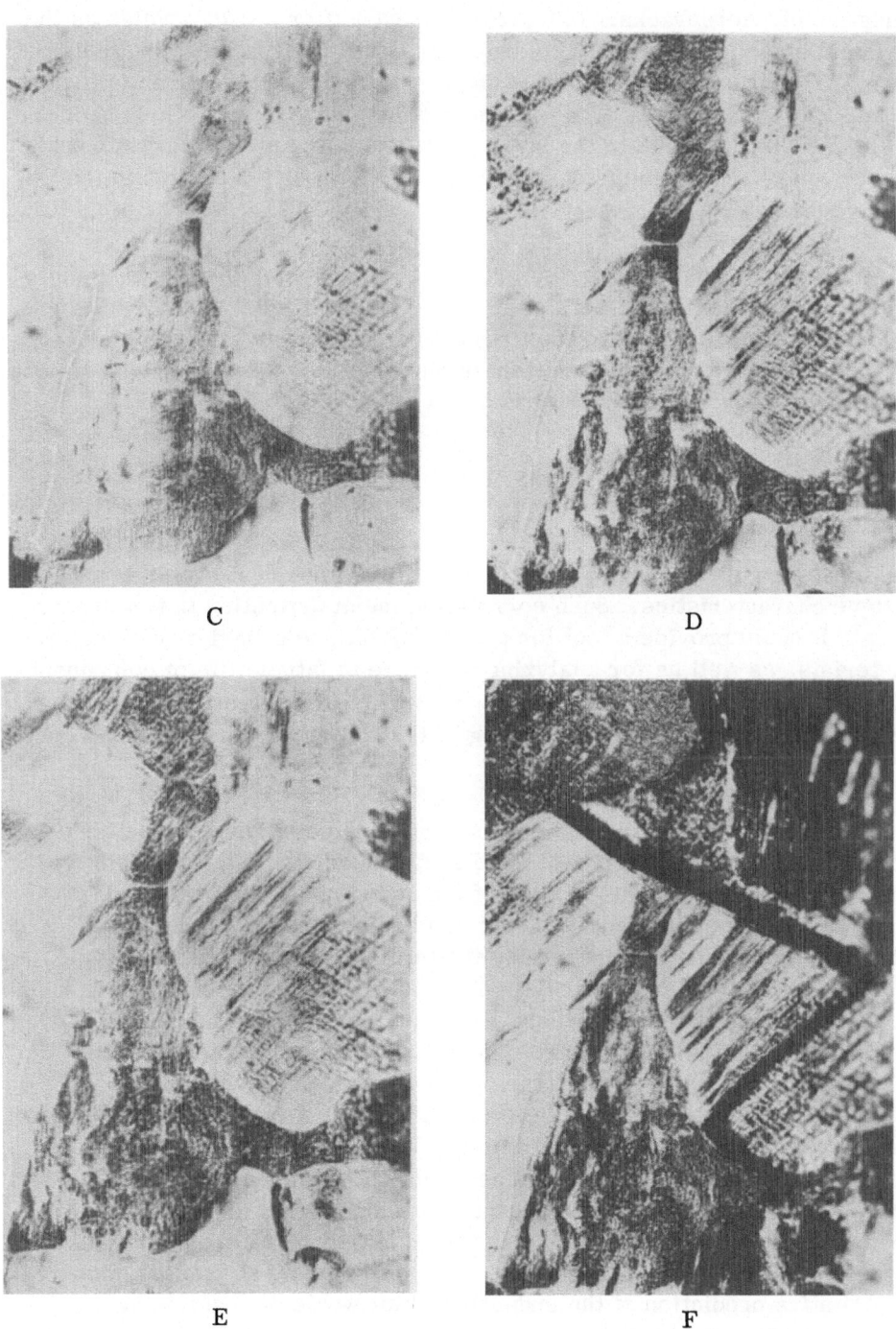

C

D

E

F

Figure 8 (Continued).

bands became much wider and deeper with some approaching the characteristics of microcracks. In figure 8E, taken after 2×10^7 cycles as the residual stress level started to drop, no significant changes can be detected. Finally, figure 8F shows the same area after specimen failure at 4.5×10^7 load cycles. The zigzag pattern of the crack follows the main slip direction within the large grain and reverts to the grain boundary at bottom center before taking up the vertical slip direction in the next grain.

The residual microstress ratio followed the same pattern in the iron specimens as in the 1018 steel. The inherent level was considerably lower in the iron, probably due to the lower level of interstitial elements. That level stayed virtually constant during fatigue cycling in all specimens. All readings fell within a band width of 1.95 - 2.40.

DISCUSSION

The results of both b.c.c. metals confirm the assumption about the existence of meaningful correlation between residual stresses and fatigue characteristics. Such correlation merits attention in two areas. First, it could provide a tool for predicting fatigue behavior of structural materials, as well as for analyzing the stage in fatigue life of components already in service. Secondly, proper interpretation of such correlation could contribute to the understanding of the basic fatigue mechanism operative in these materials.

Figure 4 indicates some possibilities in using the correlation between residual macrostresses and fatigue load cycles as a predictive tool. By determining the location of the stress peak, the expected fatigue life at any applied stress level can be determined in only about one-fourth of the time required for the full conventional fatigue test. For materials with such correlation already completely established, periodic control diffraction tests could catch unexpected increases in residual stress level as danger signals for impending fatigue failure. Finally, for materials with a true but unknown endurance limit, it should be comparatively simple to check whether a certain fatigue load is above or below the endurance limit. One residual stress measurement after a few million load cycles would show whether the stress level stays within the inherent stress band, or a significant increase above that band took place. If figure 5 and the corresponding equation holds true for a broad range of materials (i.e. different steel grades), then quantitative prediction of the endurance limit would be feasible by a comparatively simple and quick test sequence.

The results obtained with the commercial grade of pure iron material permits some insight into the fatigue crack nucleation mechanism in b.c.c. metals. From figure 7 we can see that, for a short initial period of fatigue cycling (phase I) no microstructural change takes place. In the next stage (phase II), a continuous buildup of internal stresses occurs; this stage can be identified with the strain (work) hardening which is known to take place in soft metals undergoing cyclic straining. This process, however, seems to take place entirely on a submicroscopic level with the strain energy completely absorbed by the lattice distortion, resulting in increased internal stresses. This is indicated by the lack of appearance of slip-bands during this phase.

Apparently as the ability of the material to absorb strain energy by lattice distortion is exhausted, additional straining results in the formation and gradual spreading as well as intensification of slip bands. Phase III in figure 7 comprises this stage, which is illustrated by figures 8A through 8D. Irreversible surface fatigue damage takes place during this phase as some slip bands intensify to the extent where they essentially become microcracks. Wood et al[20] describe a fatigue mechanism for b.c.c. metals which is quite compatible with the above observations. As a matter of fact, the formation of pores in bands of concentrated slip can be detected in figure 8D in the large grain at center right. The final coalescence of pores and microcracks into the macroscopic crack leading to final fatigue failure, takes place during phase IV.

Two important tasks remain to be done in future investigations. The general applicability of the correlations found valid for 1018 steel will have to be checked with a number of other structural steels. Beyond that, an attempt should be made to establish similar correlations for face centered cubic and also, perhaps, for hexagonal metals. Some of these studies are presently in progress.

CONCLUSIONS

1. A correlation was established between the residual macroscopic stress level, the fatigue damage in the form of slip-band formation, and the fatigue life at various applied stress levels of low carbon steel and commerically pure iron.

2. The residual macrostress level increased to a peak value proportional to the relative applied cyclic stress above the endurance limit. This peak value appears at about 25% of the total fatigue life.

3.   Residual stresses do not increase beyond the inherent level in
     specimens cycled with applied stresses below the endurance limit.

4.   A four-phase fatigue mechanism in iron is demonstrated in which
     surface fatigue damage by slip-band formation occurs only after
     residual macrostresses reached a plateau of maximum value.

5.   The correlation between residual stresses and fatigue characteris-
     tics could provide a tool for predicting the fatigue performance of
     structural materials.

ACKNOWLEDGEMENT

The author acknowledges with thanks the significant contribution
of P A Talcott, of the Advanced Materials Technology Department,
IBM Systems Development Division, Rochester, Minnesota, who performed most of the tests.

REFERENCES

1.   J. C. Grosskreutz, "Research on the Mechanisms of Fatigue,"
     Wright Air Development Center Technical Report, September
     1959, p. 59-192.

2.   J. C. Grosskreutz, "Research on the Mechanisms of Fatigue,"
     Wright Air Development Center Technical Report, April 1960,
     p. 60-313.

3.   F. B. Stulen, J. H. Redfern, and W. C. Schulte, "An Approach
     to Metal Fatigue," NASA Contractor Report No. CR-246, June
     1965.

4.   P. G. Forrest, "Fatigue of Metals," Pergamon Press, New
     York, N. Y., 1962.

5.   O. J. Horger and H. R. Neifert, "Internal Stresses and Fatigue"
     in Fatigue and Fracture of Metals, John Wiley and Sons, New
     York, N. Y., 1952, p. 103.

6.   E. S. Rowland, "Effect of Residual Stresses on Fatigue," in
     Fatigue, An Interdisciplinary Approach, Syracuse University
     Press, Syracuse, N. Y., 1964, p. 229.

7. J. M. Rassweiler and W. L. Grube, "Internal Stresses and
    Fatigue in Metals," Elsevier Publishing Co. (absorbed by D.
    Van Nostrand Co., Princeton, N. J.), 1950.

8. D. Rosenthal and G. Sines "Effect of Residual Stress on the
    Fatigue Strength of Notched Specimens," ASTM Proceedings,
    51:593, (1951).

9. S. Taira and Y. Murakami, "Effect of Residual Stresses on
    Fatigue Strength," Proceedings of the Fifth Japan Congress on
    Testing Materials, 1962, p. 27.

10. "Influence of Residual Stress on Fatigue of Steel" SAE J 783,
    SAE Handbook Supplement, TR-198, July 1962.

11. H. Sigmart, "Influence of Residual Stresses on the Fatigue
    Limit," Proc. of the International Conf. on Fatigue of Metals,
    the Institution of Mechanical Engineers, London, p. 272, (1956).

12. I. V. Kudryavtsev, "The Influence of Internal Stresses on the
    Fatigue Endurance of Steel," Proc. of the International Conf.
    on Fatigue of Metals, the Institution of Mechanical Engineers,
    London, p. 317, (1956).

13. H. Bühler and H. Bucholtz, "Die Wirkung von Eigenspannungen
    auf die Biegeschwingungfestigkeit," Stahl und Eisen, Vol. 53,
    No. 51, Dec. 21, 1933, p. 1330.

14. E. J. Pattinson and D. S. Dugdale, "Fading of Residual Stresses
    Due to Repeated Loading," Metallurgia, Vol. 66, No. 397,
    November 1962, p. 228-230.

15. R. J. Hartmann and E. Macherauch, "Die Veränderung von
    Röntgeninterferenzen, Hysterese und Oberflächenbild bei Ein
    und Wechselinniger Beanspruchung von Messing, Nickel und
    Stahl," Vol. 54, No. 3, March 1963, p. 161.

16. R. J. Hartman, Op. Cit. No. 4, April 1963, p. 161.

17. R. J. Hartman, Op. Cit. No. 5, May 1963, p. 282.

18. G. Koves, "The Applicability of AISI Free-Machining Steel to
    Complex Fatigue-Shock-Wear Load," Trans. of Metalurgical
    Soc. of AIME 230:58, (September 1964).

19.   G. Koves and C. Y. Ho, "Computer-Automated X-ray Stress
       Analysis," Norelco Reporter, Vol. XI, No. 3, July-September
       1964, p. 99.

20.   W. A. Wood, W. H. Reimann, and K. R. Sargent, "Comparison
       of Fatigue Mechanisms in BCC Iron and FCC Metals," AIME
       Trans., 230:511 (April 1964).

APPLICATION OF THE X-RAY TWO-EXPOSURE STRESS MEASURING TECHNIQUE TO A CARBURIZED STEEL

Bruce A. MacDonald

Member of the Technical Staff, Research Center

Ingersoll-Rand Company, Princeton, New Jersey

ABSTRACT

X-ray stress factors for a carburized steel were determined experimentally as a function of tempering temperature from loaded cantilever specimens. The standard X-ray stress expressions were modified to account for both the applied calibrating stresses and the biaxial residual stresses present in the surface of these specimens. The resulting equations permitted calculation of the {211} stress-free interplanar spacing, elastic modulus and Poisson's ratio from the X-ray data.

Tempering to higher temperatures reduced significantly the surface residual compressive stresses and stress-free spacing. The change in spacing with tempering of these specimens must be considered when determining elastic stresses by the one-exposure X-ray technique. Tempering also produced a linear relationship between the tempered hardness and {211} interplanar spacing.

The ratios, ν/E and $(1 + \nu)/E$, and the stress factors were independent of tempering temperature between 200 to 1000°F. Averaging the experimental X-ray stress factors in this range gives $K_{45} = 44.5 \times 10^6$ and $K_{60} = 29.5 \times 10^6$ psi/A. The values of X-ray stress factor for the as-quenched condition are about seven percent lower. Otherwise, microstructural changes accompanying tempering have, at best, only minor effects on the stress factor. The X-ray stress factors for the tempered condition are about 11 percent higher than calculated stress factors using bulk elastic constants. This is attributed to elastic anisotropy and preferred orientation.

INTRODUCTION

Various X-ray methods are employed to measure the residual stresses present in a metal. For example, the "one-exposure" technique measures the sum of the surface biaxial stresses, and the "two-exposure" and the "$\sin^2 \psi$" methods determine the stresses in any given direction in the surface[1-4]. The residual stresses are calculated from X-ray measured interatomic spacings by using a multiplying factor called the stress factor. The accuracy in the resulting stress values depends primarily on the diffractometer alignment, the specimen material and its geometry, the X-ray counting procedure and appropriate data corrections, and the choice of this stress factor[1].

The stress factor in the two-exposure technique incorporates the elastic modulus E, Poisson's ratio ν, the interplanar spacing of the diffracting planes, and an angle based on the diffraction geometry. Unfortunately, error can be introduced in calculating this factor if the values of E and ν are taken from mechanical measurements of the bulk material[2,3,5]. Any error is then carried directly into the residual stress value. It is preferred, therefore, to determine the stress factor by experimental X-ray calibration using a specimen as similar in composition and processing to the metal of interest as is practical. For example, Donachie and Norton[6] have determined calibrated stress factors in iron and aluminum, Esquivel[7] in deformed steels, aluminum and titanium alloys, and Macherauch[2] recently summarized calibrated stress factors for a number of metals and alloys. It is surprising, therefore, that calibrated stress factors for carburized steels are not presented in the literature; this occurs in spite of the numerous studies of residual stresses developed by carburizing and hardening.

Because of the important role that residual stresses play in carburized steels, this paper reports results of an investigation on the stress factors in a series of carburized specimens tempered from ambient temperature to 1000°F. The tempering temperature is considered an important variable because of its marked effects on the residual stresses, microstresses and mechanical properties. From a thorough analysis of the data, X-ray values of the stress factor, ν/E, $(1 + \nu)/E$, the residual stress, and the stress-free, interatomic spacing are found as functions of the tempering temperature.

THEORETICAL CONSIDERATIONS

The Basic Stress Equations

To relate X-ray measured interatomic spacings to elastic strains and stresses, it is assumed that the material is elastic, homogeneous, and isotropic. From the theory of elasticity the following expression, basic to X-ray stress analysis, is developed:[1-4]

$$e_\psi = \frac{1 + \nu}{E} (\sin^2\psi)\, \sigma_\varphi - \frac{\nu}{E} (\sigma_1 + \sigma_2). \qquad (1)$$

The terms are clear with reference to figure 1; e_ψ is the strain at a direction ψ to the surface; ψ is the angle between the specimen surface normal N and this strain direction; ν and E are material properties, Poisson's ratio and elastic modulus respectively; σ_φ is a stress acting in the plane of the surface in the φ direction; φ is the azimuth angle with respect to σ_1; and σ_1 and σ_2 are the principal stresses in the specimen surface.

Since the stress normal to the free surface is zero, the strain normal to the surface, $e_\perp$, is

$$e_\perp = -\frac{\nu}{E} (\sigma_1 + \sigma_2), \qquad (2)$$

provided X-ray penetration is small. The strains e_ψ and $e_\perp$ are available from X-ray measurements of d_ψ and $d_\perp$, the interplanar spacings for sets of planes that lie normal to the ψ and N directions respectively.

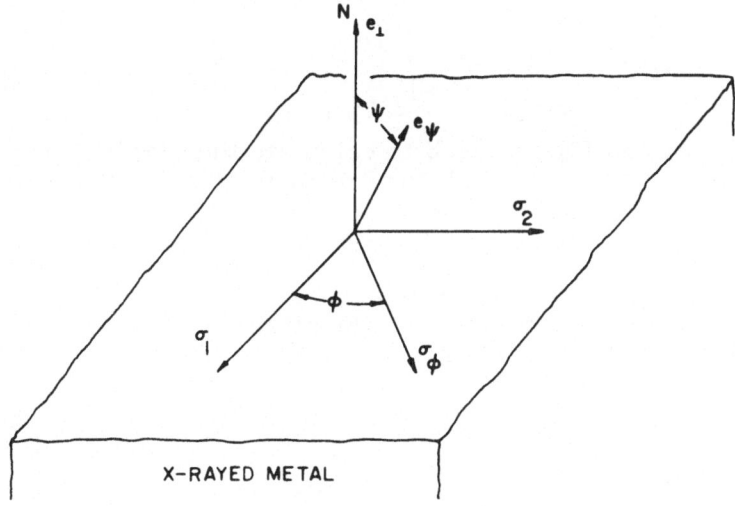

Figure 1: Illustration of Symbols for the Two-Exposure Method

$$e_\psi = \frac{d_\psi - d_o}{d_o} \quad ; \quad e_\perp = \frac{d_\perp - d_o}{d_o} \quad . \tag{3}$$

d_o is the spacing of stress-free planes.

Substituting the above expression for $e_\perp$ in equation (2) and rearranging gives

$$(\sigma_1 + \sigma_2) = -\frac{E}{\nu} \left(\frac{d_\perp - d_o}{d_o} \right) . \tag{4}$$

Equation (4) is used in the one-exposure technique in which only one measurement of interatomic spacing, $d_\perp$, is required to find the sum of the biaxial stresses. However, to avoid serious errors, an accurate value of d_o must be used; for example, in steel, if d_o is known to $\pm 0.00005\text{\AA}$, the sum of the biaxial stresses is, at best, known to $\pm 4,400$ psi.

The two-exposure method is usually preferred for it permits determination of stresses in any direction in the plane of the surface. Substituting equations (2) and (3) into (1) gives the basic expression of the two-exposure method.

$$\sigma_\varphi = \frac{E}{1 + \nu} \cdot \frac{1}{\sin^2 \psi} \cdot \frac{d_\psi - d_\perp}{d_o} \tag{5}$$

An approximate value of d_o can be included in the stress factor K_ψ, as given below:

$$\sigma_\varphi = K_\psi (d_\psi - d_\perp), \tag{6}$$

where

$$K_\psi = \frac{E}{1 + \nu} \cdot \frac{1}{\sin^2 \psi} \cdot \frac{1}{d_o} \quad . \tag{7}$$

Equations (6) and (7) are used to calculate the elastic stress from measurements of the interplanar spacings at two orientations of the specimen to the X-ray beam.

Carburizing introduces a residual stress component, σ_R, which remains constant as the elastic stress, σ_A, is applied mechanically during calibration. Modifying equation (6) to account for both stress components gives

$$\sigma_\varphi = \sigma_A + \sigma_R \tag{8}$$

and

$$\sigma_A = K_\psi (d_\psi - d_\perp) - \sigma_R . \tag{9}$$

In the present calibration experiments on carburized specimens, values of $(d_\psi - d_\perp)$ are determined at many steps of the applied stress. K_ψ and σ_R are then easily found from the experimental data since the slope of σ_A versus $(d_\psi - d_\perp)$ is the stress factor and the residual stress is equal, but opposite, to the applied stress at $(d_\psi - d_\perp) = 0$.

Theoretical Expressions for E, ν, and σ_R

Interplanar spacings $d_\perp$ and d_ψ vary linearly with applied stress. Taking the slope as M_ψ and the intercept at $\sigma_A = 0$ as d_ψ^*,

$$d_\psi = M_\psi \, \sigma_A + d_\psi^* \qquad (10)$$

M_ψ is determined experimentally from the X-ray calibration data. A theoretical expression for d_ψ can be developed from equation (1) by substituting equations (3) and (8) for e_ψ and σ_ϕ, respectively. Noting that $(\sigma_1 + \sigma_2) = (\sigma_A + \sigma_R + \sigma_R)$ for carburizing residual stresses that are approximately in an equal biaxial state,

$$d_\psi = d_o \cdot \frac{1+\nu}{E} (\sin^2\psi)(\sigma_A + \sigma_R) - \frac{\nu}{E} (\sigma_A + 2\sigma_R) + 1 \qquad (11)$$

Differentiating equation (11) with respect to σ_A gives

$$M_\psi = \frac{dd_\psi}{d\sigma_A} = \frac{d_o}{E} \left[(1+\nu) \sin^2\psi - \nu \right] \qquad (12)$$

M_ψ is shown below for each value of ψ (0, 45°, and 60°) used in the calibration experiments:

$$M_\perp = -\frac{\nu}{E} \cdot d_o \; ; \qquad (13)$$

$$M_{45} = \frac{1-\nu}{E} \cdot \frac{d_o}{2} \; ; \qquad (14)$$

$$M_{60} = \frac{1-\nu}{E} \cdot \frac{d_o}{4} \; . \qquad (15)$$

X-ray expressions for E and ν come from combining equation (13) with (14) or (15).

For $\psi = 0$ and 45°,

$$E = \frac{d_o}{2M_{45} - M_\perp} \; ; \qquad \nu = \frac{-M_\perp}{2M_{45} - M_\perp} \qquad (16)$$

For $\psi = 0$ and 60°,

$$E = \frac{3d_o}{4M_{60} - M_\perp} \; ; \qquad \nu = \frac{-3M_\perp}{4M_{60} - M_\perp} \qquad (17)$$

Application of the above expressions to the calibration experiments on 4820 carburized steel is the subject of the remaining paper.

PROCEDURES

Material, Heat Treatment, and Specimen Design

The stress calibration specimens A-F were machined out of hot rolled, AISI 4820 steel bar stock, heat treated in the manner described in table I.

Table I: Specimen Heat Treatment

1. Carburized at 1700°F to 20 to 25 mils case depth.
2. Oil quenched from 1600°F.
3. Hardened at 1400°F, 1-1/2 hours, and oil quenched.
4. Refrigerated at -100°F.
5. Tempered in steps of 200°F from ambient temperature to 1000°F for 2 hours, as shown below:

Specimen	Tempering Temperature, °F
A	75
B	200
C	400
D	600
E	800
F	1000

After carburizing and hardening, specimens were tempered for two hours at the indicated temperatures and then electropolished 2.5 mils over a 2.25 inch length to remove any decarburized surface. The specimens and fixture shown in figure 2 were especially designed to permit X-ray measurements during application of compressive or tensile stresses. Strain gages attached in the polished areas on either side of the X-rayed area verified that the longitudinal surface strains were equal in that region of the specimen taper length; consequently, strain gradients were not a factor. X-ray measurements were made for each 300μ-inch/inch step in applied strain from a maximum compressive strain of 2100μ-inch/inch to a maximum tensile strain of 1800μ-inch/inch. This represented an applied stress range of 63 KSI compression to 54 KSI tension, assuming $E = 30 \times 10^6$ psi, and 14 applied stress steps on each specimen.

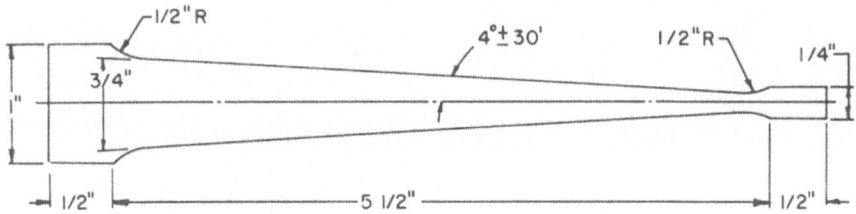

Figure 2A: Tapered Beam, 1/8 Inch Thick, Used for Measuring the Stress Factor.

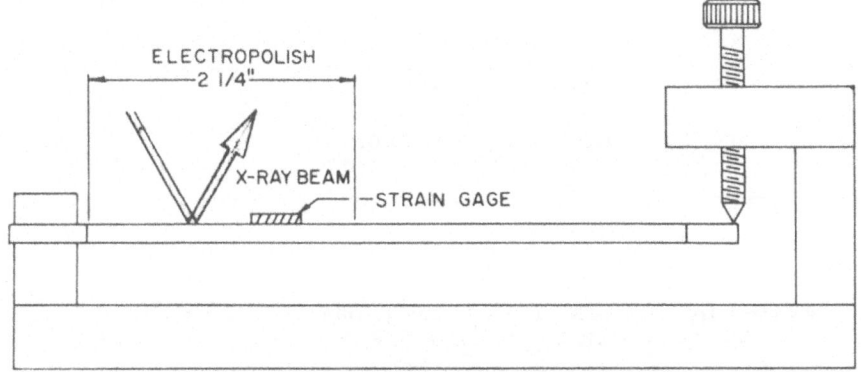

Figure 2B: Fixture Used to Apply a Bending Stress to the Beam.

X-ray Stress Measurement

X-ray stress measurements were made on a Siemens Krystalloflex IV diffractometer with CrKα radiation filtered through a one-mil vanadium filter. A flow proportional counter, pulse height discrimination, and counter movement were employed for good peak resolution and intensity. The times to obtain 200,000 intensity counts were recorded at three angles of 2θ equally spaced within 85 percent of the maximum intensity of the {211} steel diffraction peak at $\psi = 0$, 45° and 60°. The recorded times were corrected for the Lorentz, polarization and absorption factors and the 2θ angle for maximum intensity determined by a parabolic fit of the data[8]. Values of $d_\perp$, d_{45}, and d_{60} were calculated from these 2θ angles at each value of applied stress. A least squares analysis of the data gave the constants in equations (18) to (22) for specimens A to F.

$$\sigma_A = K_{45} (d_{45} - d_\perp) - \sigma_R \qquad (18)$$

$$\sigma_A = K_{60} (d_{60} - d_\perp) - \sigma_R \qquad (19)$$

$$d_\perp = M_\perp \sigma_A + d_\perp^* \qquad (20)$$

$$d_{45} = M_{45}\ \sigma_A + d^*_{45} \tag{21}$$

$$d_{60} = M_{60}\ \sigma_A + d^*_{60} \tag{22}$$

Peak Breadth and Microhardness Measurements

An estimate of the {211} diffraction peak breadth at half-maximum intensity, B, was obtained directly from the parabolic fit of the X-ray stress data using the method of Marburger and Koistinen[9]. There was no correction for the X-ray background. Microhardness measurements were made on a Tukon tester using 300 grams load for specimens A to E and 200 grams for specimen F.

Retained Austenite

Retained austenite was found from

$$V_A = \frac{100}{1 + 2.4\ (\dfrac{A_M}{A_A})}\ .$$

The integrated intensities of the {200} martensite and {220} austenite peaks, A_M and A_A respectively, were measured by planimeter from chart traces using CrKα radiation.

RESULTS

Microstructure and Hardness

Carburizing produced a carbon gradient that varied from about 0.65% at the surface to 0.24% at 35 mils depth, figure 3.

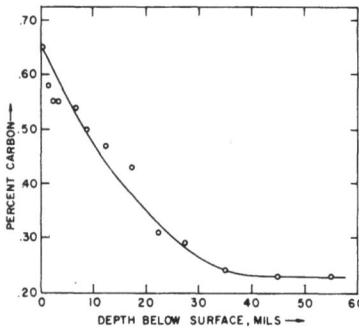

Figure 3: Carbon Gradient in the Carburized 4820 Steel Specimens

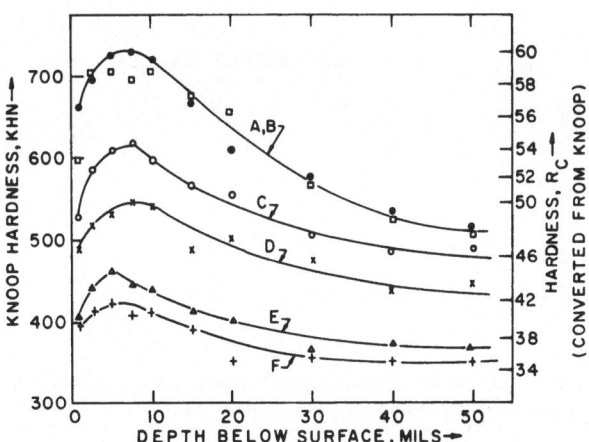

Figure 4: Variation in Hardness with Depth in Specimens A to F.

After hardening, the specimens were martensitic in case and core with a case depth of about 20 mils as estimated by the 50 Rc point in figure 4 for 400°F tempering. The specimens developed a maximum hardness at about 7 mils below the surface.

Hardness at 2.5 mil depth, corresponding to the layer removed by electropolishing, is shown as a function of tempering temperature in figure 5. The hardness fell rapidly on tempering above 200°F as carbides precipitated in the first and third stages of tempering. This softening was interrupted between 400° to

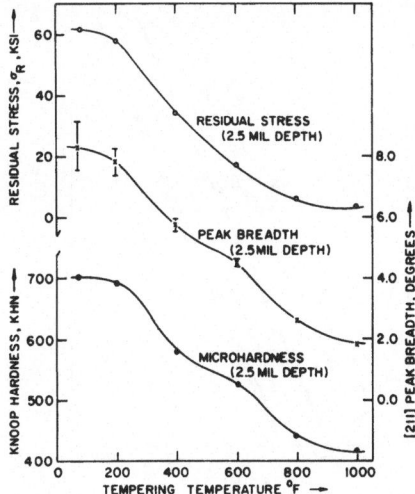

Figure 5: The Effects of Tempering Temperature on Residual Stress, Diffraction Peak Breadth, and Microhardness of Carburized 4820 Steel.

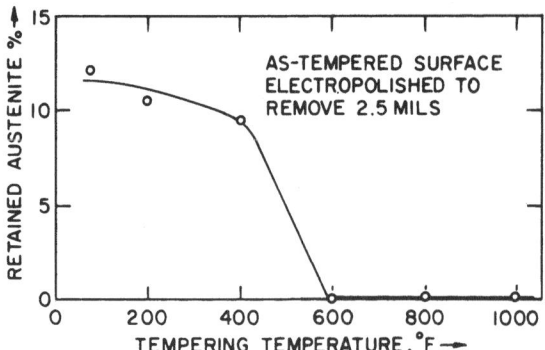

Figure 6: The Effect of Tempering Temperature on Retained
Austenite in Carburized 4820 Steel.

600°F by transformation of the retained austenite, figure 6, to low
temperature, high strength products.

X-ray Stress Results and Analysis

The variations in applied stress produced linear shifts in
$(d_{45} - d_\perp)$ and $(d_{60} - d_\perp)$, figures 7 and 8. The slopes of these
curves were the stress factors K_{45} and K_{60}, which are plotted in
figure 9 versus tempering temperature. With the exception of the
untempered martensite, it is clear that these factors were insensitive to the tempering temperature up to 1000°F.

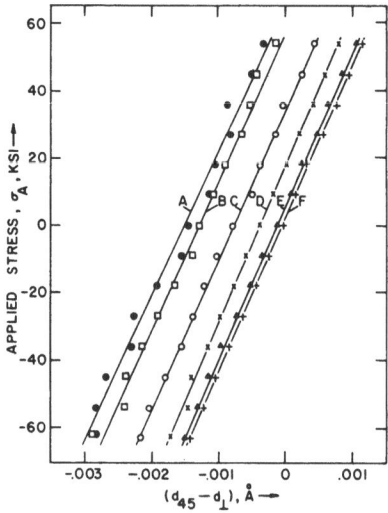

Figure 7: Variation in $(d_{45} - d_\perp)$ with Applied Stress for
Specimens A to F.

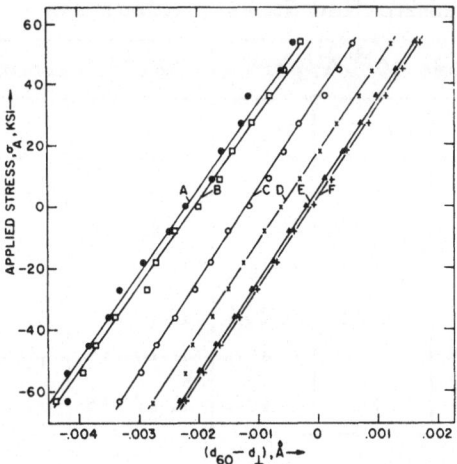

Figure 8: Variation in $(d_{60} - d_{\perp})$ with Applied Stress for Specimens A to F.

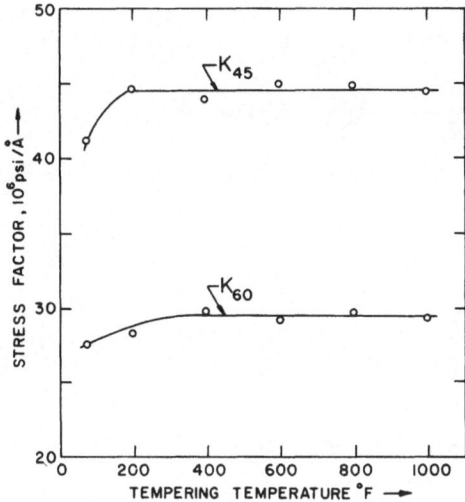

Figure 9: Influence of Tempering Temperature on the Stress Factors at $\Psi = 45°$ and $60°$.

As forecast by equations (20) to (22), the $\{211\}$ interplanar spacings $d_{\perp}$, d_{45}, and d_{60} also varied linearly with the applied stress. The equation constants determined from the X-ray data

Specimen	Table II Least-Squares Results for Interplanar Spacings Versus Applied Stress, Equations (20) to (22)					
	Slopes 10^{-8} Å/psi			{211} Spacing at $\sigma_A = 0$, Å		
	$M_\perp$	M_{45}	M_{60}	$d_\perp^*$	d_{45}^*	d_{60}^*
A	-1.025	1.384	2.585	1.17261	1.17112	1.17038
B	-0.946	1.270	2.563	1.17240	1.17110	1.17034
C	-0.946	1.324	2.402	1.17171	1.17094	1.17056
D	-0.986	1.231	2.435	1.17078	1.17040	1.17021
E	-0.959	1.276	2.410	1.17043	1.17031	1.17025
F	-0.939	1.312	2.468	1.17048	1.17042	1.17037

are listed in table II and a typical example of the effect of applied stress on $d_\perp$, d_{45}, and d_{60} is shown in figure 10 for specimen C tempered at 400°F. The three lines intersect at a single point which denotes σ_{AB}, the applied stress which exactly balances the residual stress, and d_{oB}, the corresponding interatomic spacing. At the intersection $d_{oB} = d_\perp = d_{45} = d_{60}$; thus, d_{oB} can be estimated directly from the graph, or calculated from

$$d_{oB} = \frac{M_\perp}{M_\psi - M_\perp} (d_\perp^* - d_\psi^*) + d_\perp^* \tag{23}$$

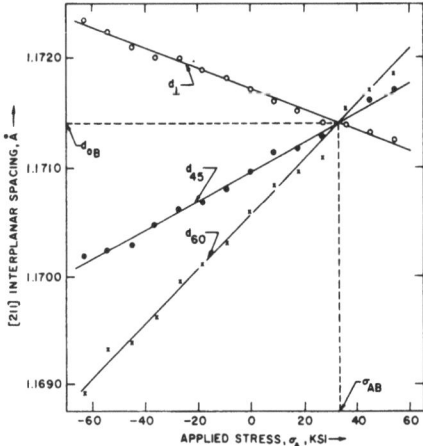

Figure 10: Variation in Interplanar Spacings $d_\perp$, d_{45}, and d_{60} with Applied Stress for Specimen C

In a material free of elastic stresses, the stress-free spacing d_o is equal to d_{oB}. Since a balanced biaxial state of stress exists in the carburized specimens, when σ_A equals σ_R in the longitudinal direction, the spacing is still modified by σ_R in the transverse direction through the Poisson effect; i.e.,

$$d_o = \frac{d_{oB}}{1 - \frac{\nu}{E}(\sigma_R)} \quad . \tag{24}$$

d_o was calculated for specimens A to F from the X-ray values of d_{oB}^o and σ_R. Figure 11 summarizes the effects of tempering temperature on both d_o and d_{oB}.

Using equations (16) and (17) values of E and ν were calculated from experimentally determined slopes M_L, M_{45}, and M_{60} and the spacing d_o. These results are listed in table III.

Specimen	E, 10^6 psi		ν	
	$\Psi = 0, 45°$	$\Psi = 0, 60°$	$\Psi = 0, 45°$	$\Psi = 0, 60°$
A	30.9	30.9	.270	.271
B	33.6	31.4	.271	.253
C	32.6	33.2	.263	.269
D	33.9	32.8	.286	.276
E	33.4	33.1	.273	.271
F	32.9	32.5	.263	.263

Table III: Experimental X-ray Values of Elastic Modulus and Poisson's Ratio

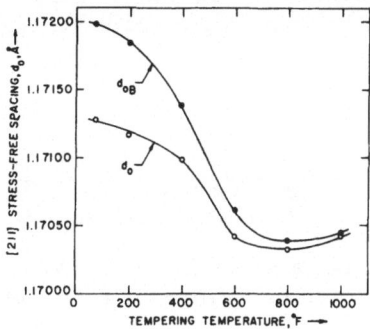

Figure 11: Effect of Tempering Temperature on the Stress-Free {211} Spacing

DISCUSSION

The X-ray Stress Factor

The results in figures 7, 8, and 10 confirm that a linear relationship exists between the applied elastic stresses and {211} interplanar spacings in 4820 carburized steel. Furthermore, the tempering has negligible effect on the stress factor between 200 - 1000°F as shown by figure 12. Averaging the measured X-ray stress factors over this temperature range gives $K_{45} = 44.5 \times 10^6$ and $K_{60} = 29.5 \times 10^6$ psi/A . To generalize from these results, it appears that only one calibration of the X-ray stress factor is required to measure residual stresses in steel specimens which differ only by their degree of tempering.

Let us now look at the common alternative method of determining the stress factor, i.e., by calculation from equation (7) using the bulk elastic constants. This gives $K_{45} = 39.6 \times 10^6$ and $K_{60} = 26.4 \times 10^6$ psi/A, or values which are about 11 percent lower than the X-ray values*. Similar deviations have been recorded for iron[6] and other metals[2]. The difference in the X-ray and calculated stress factors apparently arise from two sources, which can be seen by examining the calibration relationship in equation (25).

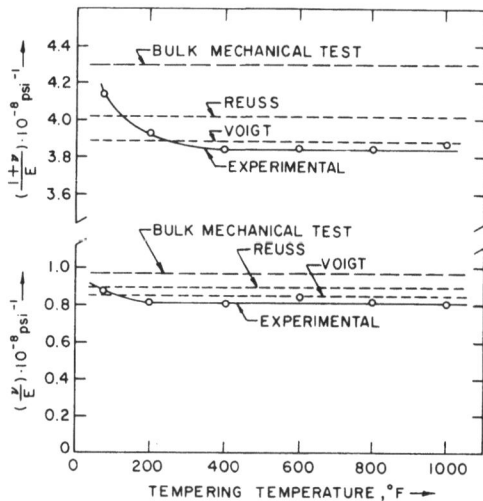

Figure 12: Experimental and Theoretical Values of ν/E and $(1 + \nu)/E$ versus Tempering Temperature

---

*In this calculation the {211} diffraction planes are used as the strain indicators; $d_o = 1.1710A$; $E = 30 \times 10^6$ psi; $\nu = 0.29$.

$$K_{X\text{-}ray} = \frac{\Delta \sigma_A}{\Delta (d_\psi - d_\perp)} = \frac{E_A \, \Delta e}{\Delta (d_\psi - d_\perp)} \qquad (25)$$

The applied stress, σ_A, is calculated from a strain gage output, e, using an assumed elastic modulus, E_A. In our experiments E_A was taken as 30×10^6 psi, typical of the bulk material. If this chosen E_A is incorrect, then $K_{X\text{-}ray}$ is also in error. Furthermore, the strain gage indicates applied strain in the longitudinal direction of the beam while $d_\perp$ and d_ψ are measured at $\psi = 90°$ and 45° or 60° respectively to the surface. If the elastic constants are anisotropic and preferred orientation exists, $\Delta e / \Delta (d_\psi - d_\perp)$ in equation (25) will not give the expected ideal ratio. These points are examined below, in turn.

There are at least two factors which could influence the surface elastic modulus of these specimens. Since the carbides formed by tempering have lower elastic modulus than the ferrite[2], tempered martensite might be expected to have a lower average modulus than untempered martensite. On the other hand, Winchell and Cohen[10] have shown that the elastic modulus of untempered steel decreases with increasing amounts of interstitial carbon. The untempered martensite of specimen A has, at maximum, 0.6 percent carbon interstitially dissolved in the surface, and this could accordingly decrease the elastic modulus by three to four percent. However, for 0.6 percent carbon, the surface martensite-start temperature is about 440°F and the 90 percent martensite point is reached at about 360°F; hence, autotempering should occur tending to push the interstitial carbon level back toward 0.25 percent and the influence of carbon on the modulus can probably be neglected. Furthermore, any change in the overall surface elastic modulus affects equally both the X-ray stress factor through σ_A and the calculated stress factor from equation (7). Hence, changing the value of E_A cannot rectify the disagreement in the X-ray and calculated stress factors.

In view of the above argument, it is best to remove any effect of an assumed modulus by examining instead the ratio K_ψ / E_A which is equal to $\Delta e / \Delta (d_\psi - d_\perp)$. The ideal ratios calculated from equation (7) are 1.32 and 0.88 for $\psi = 45°$ and 60° respectively. The X-ray ratios are 1.48 and 0.98 for the tempered and 1.37 and 0.92 for the untempered steel, which indicates that $\Delta (d_\psi - d_\perp)$ measured by X-rays does not respond as much to applied stress as does the strain gage. This suggests that anisotropy in the elastic constants is the major cause producing the high X-ray values of stress factor and explains the high values of E in table III, as well.

It should be pointed out that the calculated stress factors represent only 11 percent error in the measured stress, while a measurement accuracy of 10 to 15 percent is sufficient for many purposes. It is nevertheless pertinent that attempts have been made to estimate the stress factor by using theoretical values of $(1 + \nu)/E$ obtained from the elastic properties of single crystals[2, 5] A major obstacle to this approach is developing the model to transmit stress or strain from any single grain to its neighbors. Briefly, Voigt[11] assumed that each grain in a metal of small, randomly oriently grains underwent the same strain. By averaging stresses for all orientations of the grains, he derived expressions for the elastic constants in terms of the elastic stiffness coefficients. In a different approach, Reuss[12] assumed that all grains underwent the same stresses and derived elastic constants in terms of the elastic compliances by averaging the strains for all grains oriented at the proper diffraction angle. The resulting elastic constants were dependent on crystallographic direction in the "constant stress" approach, but not in the "constant strain" approach.

Since the ratios ν/E and $(1 + \nu)/E$ are used directly in the $\sin^2\psi$ stress method, these ratios are calculated using the X-ray elastic constants in table III, the theories of Voigt and Reuss, and the bulk elastic constants. As shown in figure 12 the X-ray values for tempered steel are approximated closest by the Voigt result. However, the increase observed for the untempered specimen is not predicted by either theory.

The Tempering Behavior

This X-ray investigation has provided an unusually close look at the tempering behavior of 4820 carburized steel. As expected, tempering reduced both the hardness and $\{211\}$ diffraction peak breadth. Marburger and Koistinen[13] have previously demonstrated that the peak breadth provides a useful measure of the hardness in tempered steels. For specimens A to F, the plot of Knoop hardness versus the $\{211\}$ peak breadth is a straight line, figure 13, and a first-order least-squares analysis gives

$$H_{KHN} = 40.9\ B + 362 \tag{26}$$

The calculated hardness using the experimental values of peak breadth deviates from the measured hardness by less than three percent.

In view of the relationship between hardness and peak breadth, the author has indicated in figure 5 the mean average and the range of observed peak breadths for the 14 measurements at each tempering temperature. The variation about the mean results from fluctuations in the X-ray background intensity for which a

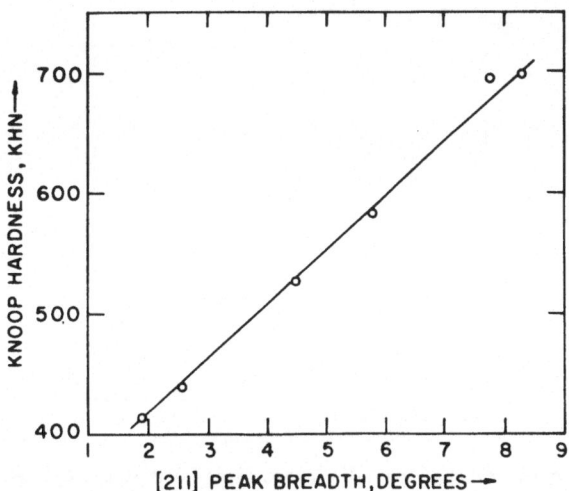

Figure 13: Knoop Hardness versus Peak Breadth for Carburized
          and Tempered 4820 Steel

correction is not made in this approximate analysis. For tempering temperatures below 600°F the variation is noticeable
because background is 10 to 20 percent of the maximum peak intensity. Above 600°F the effect is smaller than the data points in
the figure.

        Carbide precipitation during tempering depletes the inter-
stitial carbon with a concomitant reduction in the average interatomic spacing.[10, 14]   The associated decrease in the average {211}
stress-free spacing shown in figure 11 must be considered when
using the one-exposure stress method with equation (4). Using
the experimentally determined values of d_o and d_i^* , the sum of
the biaxial residual stresses were calculated by the one-exposure
method. The results indicate that the transverse residual stresses
were about equal to the longitudinal values shown in figure 5.

        Considering the significant effects that residual stresses
can have on the behavior of steel, let us now examine the heat-
treating stresses developed by the carburizing. Hardening of the
carburized 4820 steel produced high compressive stresses in the[15, 16]
case, figure 5, as observed before in other carburized steels.
The subsequent tempering severely reduced the residual compressive stresses such that, after the 1000°F tempering, only 2 KSI
compression remained. This observed stress relief is compared
to results by Brown and Cohen[17] in tempered 52100 steel, figure
14. Although the as-quenched 52100 steel supported mechanically
applied stresses, the degree of stress relief occurring at the indicated temperatures is remarkably similar to the 4820 steel.
The stress relief accompanying tempering is undoubtedly related
to carbide precipitation, but a descriptive theoretical mechanism
has not yet been developed.

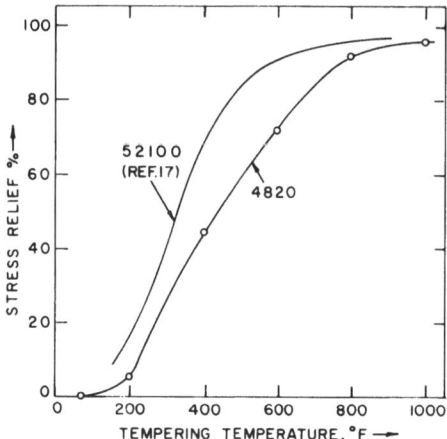

Figure 14: Stress Relief on Tempering Carburized 4820 Steel
and 52100 Steel

CONCLUSIONS

1.   This investigation has clearly demonstrated the suita-
bility of the X-ray two-exposure method for measuring elastic
stresses in a series of tempered 4820 carburized steel specimens.
X-ray stress factors found by the linear shifts in the {211} interplanar spacings with applied stress were independent of the tempering temperature from 200 to 1000°F; averaging gave K_{45} =
44.5 x 10^6 and K_{60} = 29.5 x 10^6 psi/Å.   The calibrated stress
factors observed for the untempered steel were about seven
percent lower.

2.   X-ray values for ν/E and $(1 + \nu)/E$ in the tempered
steel compared favorably with estimates from Voigt's constant
strain theory.   However, the increase in these ratios for the untempered condition was not predicted by the theory.

3.   X-ray values of the stress factors in the tempered
steel were about 11 percent above the calculated values using bulk
elastic constants.   This is attributed to elastic anisotropy and
preferred orientation in the specimens.

4.   The progress of tempering was closely followed by
measuring the microhardness, {211} peak breadth and residual
stress.   Tempering reduced the microhardness and peak breadth
such that a linear relationship resulted between these parameters.
Tempering also produced marked stress relief, the 60 KSI compression on the surface falling progressively as a function of temperature to only 2 KSI after the 1000°F treatment.

ACKNOWLEDGEMENTS

The author thanks J. R. Kunkel who kindly supplied the cantilever specimens, heat treatment, and carbon analyses. Acknowledgement is also due Messrs. F. Tufano and C. Warznak for their assistance in the measurements.

REFERENCES

1. A. L. Christenson, D. P. Koistinen, R. E. Marburger, M. Semchyshen, and W. P. Evans, "The Measurement of Stress by X-ray", TR-182, Society of Automotive Engineers, New York, 1960, p. 1-36.

2. E. Macherauch, "X-Ray Stress Analysis", Proceedings of 2nd SESA Institute Congress on Experimental Mechanics, Washington (1965).

3. C. S. Barrett and T. B. Massalski, Structure of Metals, McGraw Hill, New York, 1966, p. 466-485.

4. B. D. Cullity, Elements of X-Ray Diffraction, Addison-Wesley Publishing Co., Reading, Mass., 1959, p. 431-453.

5. G. B. Greenough, "Quantitative X-Ray Diffraction Observations on Strained Metal Aggregates", Progress in Metal Physics, Vol. 3, Interscience Publishers, New York, 1952, p. 176-219.

6. M. J. Donachie, Jr. and J. T. Norton, "X-Ray Studies of Lattice Strains Under Elastic Loading", Trans. ASM 55:51, (1962).

7. A. L. Esquivel, "X-Ray Diffraction Study of the Effects of Uniaxial Plastic Deformation on Residual Stress Measurements", 17th Annual Denver X-Ray Conf., Estes Park, Colorado (1968).

8. D. P. Koistinen and R. E. Marburger, "Simplified Procedure for Calculating Peak Position in X-Ray Residual Stress Measurements on Hardened Steel", Trans. ASM 51:537, (1959).

9. R. E. Marburger and D. P. Koistinen, "X-Ray Measurement of Residual Stresses in Hardened Steel", in Internal Stresses and Fatigue in Metals, G. M. Rassweiler and W. L. Grube, Editors, Elsevier Publishing, New York, 1959, p. 98-119.

10. M. Cohen, "The Strengthening of Steel", Trans. AIME 224: 638-656, (1962).

11. W. Voigt, Lehrbuch der Kristallphysik, B. G. Teubner, Editor, Leipzig, Berlin (1928).

12. A. Reuss, Z. angnew Math. u. Mech., 9:49, (1929).

13. R. E. Marburger and D. P. Koistinen, "The Determination of Hardness in Steels from the Breadth of X-Ray Diffraction Lines", Trans. ASM 53:743, (1961).

14. S. G. Fletcher and M. Cohen, "The Effect of Carbon on the Tempering of Steel", Trans. ASM 32:333, (1944).

15. D. P. Koistinen, "The Distribution of Residual Stresses in Carburized Cases and Their Origin", Trans. ASM 50:227, . (1958).

16. W. B. Bond, "X-Ray Diffraction Studies of the Residual Stress Levels in an AMS 6260 (SAE 9310) Carburized Case", Norelco Reporter XIII, No. 1:18-20, (1966).

17. R. L. Brown and M. Cohen, "Stress Relaxation of Hardened Steel", Metals Progress 81:66, (1962).

X-RAY DIFFRACTION FROM VIBRATING QUARTZ PLATES

W. J. Spencer and G. T. Pearman

Bell Telephone Laboratories, Inc.

Allentown, Pennsylvania

ABSTRACT

X-ray diffraction topographs of modes in rectangular AT-cut quartz plates are shown. The change in the resulting patterns when different Bragg planes are used is demonstrated. Diffracted x-ray intensity was measured as a function of strain in a vibrating quartz plate. The Lame solutions for quartz plates are briefly described and some experiments suggested for x-ray topography. Finally modes in quartz plates with multiple electrodes are shown indicating the complexity of this type of vibration problem.

INTRODUCTION

Nearly forty years ago Fox and Carr[1] discovered that the diffracted x-ray intensity from piezoelectrically vibrating quartz crystals was significantly enhanced. Nishikawa[2] and his coworkers found the same effect almost similtaneously and showed that the nature of vibration patterns in the quartz could be related to the diffracted intensity. Barrett and Howe[3] correctly related this intensity change to a reduction in extinction due to the inhomogeneous strains associated with the piezoelectrically excited resonant vibrations. Cady[4] summarized this work in 1946 and based on the results of Jauncey and Bruce[5] associated the intensity change to a reduction of secondary extinction. Although the diffraction effects are not completely understood, it now appears that primary extinction effects are more important in the cases reviewed by Cady and shown here. White[6] first obtained complete pictures of vibration patterns in quartz resonators in the Bragg geometry and measured the change in intensity as a function of a static deformation

of the quartz plates. Transmission x-ray topographs of vibrating
quartz plates were obtained more recently at Bell Telephone Laboratory[7] and Georgia Institute of Technology[8]. These experiments use
scanning x-ray cameras similar to those first proposed by Lang[9] for
the study of defects in nearly perfect single crystals. A review
of the work in this area through 1967[10] has been recently published.
Usual camera arrangements, methods of mounting and piezoelectrically
exciting the quartz plates were presented along with a qualitative
x-ray theory.[*] None of these results will be covered in this paper.
In addition it is assumed that Lang's technique of x-ray topography
is well known to the reader.

The first section covers the change in x-ray intensity as a
function of vibration amplitude and shows the effect of using different Bragg planes to study the same resonant mode. The second
section shows some results obtained from rectangular AT-cut quartz
plates. The next section briefly describes Lame' modes and some
experiments which might be used to study these particularly simple
vibrations. The last section shows some recent patterns obtained
in the study of AT-cut quartz plates with several coupled resonators
on a single substrate.

EFFECT OF VIBRATION AMPLITUDE AND DIFFRACTING PLANE
ON X-RAY INTENSITY

Ideally one would like to find a single simple vibration
which could be used to relate the diffracted x-ray intensity to
vibration amplitude for a given set of Bragg planes. However, in
any finite body with stress free faces the displacements are exceedingly complex. In general it requires an infinite set of modes to
satisfy the boundary conditions and the displacement at any position
in the vibrating body is a sum of the displacements due to each of
these modes. In some cases it is possible to find dimensions of a
given plate or bar where one or two modes predominate and displacements at a particular resonance are much simpler. This is true of
vibrations in rectangular AT-cut quartz plates[12] where the principal modes are thickness-shear and flexure. AT-cut quartz plates
and many of the resonances which are strongly excited in these
plates have been studied in some detail by Mindlin[13].

The AT-cut quartz plates studies have been motivated by the
wide use of this resonator in communications. The orientation of
AT-cut plates with respect to the optic (or c-axis) axis is shown
in figure 1. The c-axis in quartz is an axis of trigonal symmetry.

---

[*]A more rigorous theory relating diffracted x-ray intensity to
vibration amplitude has been published by Kuriyama.[11]

Normal to this are three axis of digonal symmetry generally noted
as a-axis. In most of the results shown here the x-ray topographs
are made by diffraction from the $(2\bar{1}{\cdot}0)$ plane which is normal to
the a or x_1 axis and to the major faces of the quartz plate.

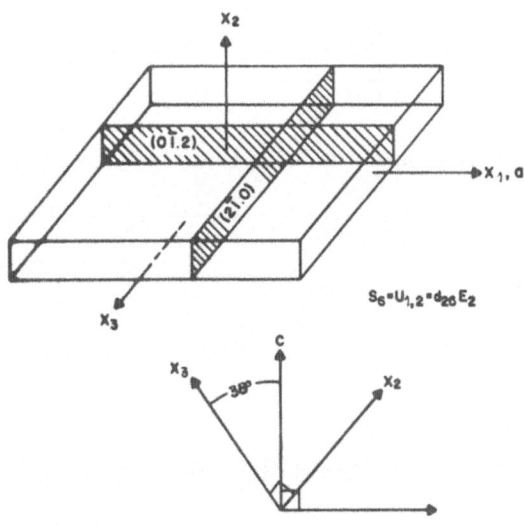

Figure 1.   Orientation of AT-cut quartz plates and principal Bragg
            planes normal to plate surfaces.

By applying an electric field E_2 along the x_2 axis an S_6
strain may be induced. When the electric field is sinusoidal the
strain alternates in phase with the field and at certain frequencies
resonant vibrations are set up in the crystal plate. In AT-cut
quartz plates, the strongest modes are thickness-shear and coupled
strongly to these the flexural vibrations. The displacements associated with thickness-shear and flexural vibrations are shown in
figure 2. The thickness-shear displacement is sinusoidal through
the thickness and antisymmetric about the center of the plate.
These coupled modes are discussed in more detail in the next section.

When flat AT-cut plates are contoured the coupling between
thickness-shear and flexure is reduced[14]. The shear displacement
in the x_1 direction through the plate thickness is then given to a
good approximation by[15]

$$u_1 = A \sin \frac{m\pi x_2}{2b} \tag{1}$$

where 2b is the plate thickness and m = 1, 3, 5.... is the number of half wave lengths along x_2. The frequency of these modes is then given by

$$\omega = \frac{m\pi}{2b} \sqrt{\frac{c'_{66}}{\rho}} \tag{2}$$

where c'_{66} is the rotated elastic constant for the AT-cut plate and ρ is the density.

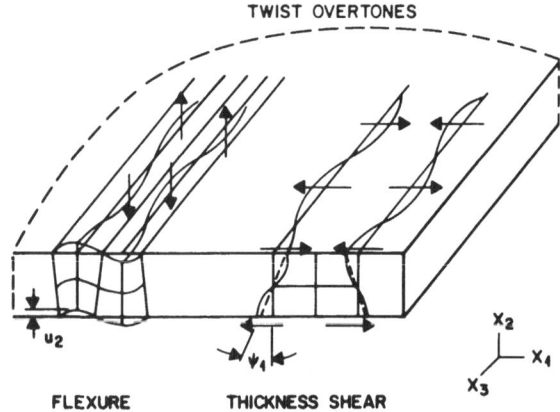

Figure 2.   Displacements associated with twist overtones of thickness-shear and flexural vibrations.

An x-ray topograph of a circular plane convex AT-cut plate vibrating on the third overtone of thickness-shear is shown in figure 3. The dark circle at the center is due to the resonant vibration. This vibration was driven by applying a sinusoidal electric field across the quartz plate. The field is obtained by voltage put on thin metallic electrodes vacuum deposited on the

major faces of the quartz plate. The x-ray intensity is sensitive
to displacement normal to the diffracting plane. In figure 2 this
is the displacement along x_1. Defects in the quartz[16] are apparent
running diagonally across the plate. The ($2\bar{1}\cdot0$) plane is vertical.
The diffraction vector is to the right. The topograph was made
using Mo-Kα radiation on Ilford type G film with an exposure time
of one minute per millimeter. The displacement through the thickness is shown just below the topograph. In this case m = 3 in
equation 1 and 2b $\cong$ 1.67 mm. The frequency was about 3 MHz.

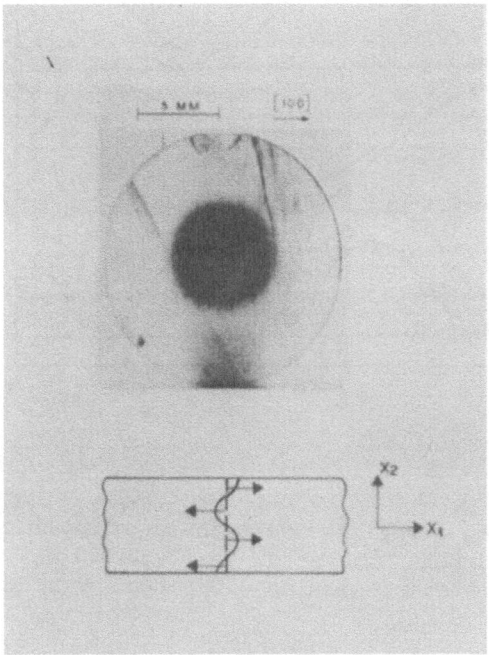

Figure 3. X-ray topographs of the third overtone of thickness-
shear in a circular contoured AT-cut quartz plate.

Using a very narrow slit on the diffracted beam and scanning
along the plate thickness as described by Young, et. al.[8], it is
possible to determine the intensity variation through the plate.
Figure 4 shows the experimentally measured intensity for the same
vibration shown in figure 3. Plotted on the same scale are u_1 and
$S_6 = u_{1,2}$ calculated from equation (1). The x-ray intensity does
not go to zero at $\pm$ 2b/3 and 0 due to the finite slit width. This
also accounts for the two outer peaks having lower intensity than
the inner peaks. The x-ray intensity is in phase with the displace-
ment u_1 and proportional to either $|u_1|$ or $|u_{1,22}|$. Physically one

would expect the diffracted intensity to be related to the strain gradient, $u_{1,22}$. The dark circle in figure 1 is thus seen to be an overlay of four diffraction maxima projected onto the x-ray film. In general for any value of m in equation 1, the pattern on the topograph will represent m+1 patterns projected one on top of another.

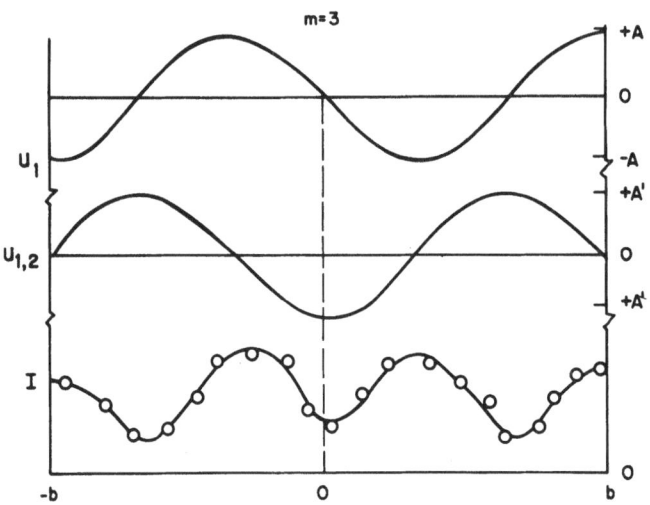

Figure 4. Diffracted x-ray intensity through the thickness of a third overtone thickness shear vibration. The displacement and strain calculated from equation (1) are shown for comparison.

Reducing the slit used to obtain the intensity curves in figure 4 to a small aperture only large enough to integrate the intensity over the entire thickness and replacing the x-ray film with an x-ray counter, the diffracted intensity as a function of strain amplitude can be measured. The strain due to the thickness-shear vibration can easily be changed by changing the electric field used to piezoelectrically excite the vibration.

The results of such an experiment are shown in figure 5. The intensity has been normalized to that measured from the quartz plate in an unstrained condition. As the amplitude of the vibration is increased, the diffracted intensity increases until it reaches a maxima at strains above 1.5×10^{-5}. The strain can be approximated by

$$S_6 = d'_{26} \, E_2 Q \qquad\qquad (3)$$

where d'_{26} is the piezoelectric constant for an AT-cut quartz plate and E_2 is the electric field along x_2. At resonance, S_6 must be multiplied by the effective Q of the vibration mode, which in this case is about 10^5. The lower and upper horizontal lines are calculated from James[17] equations for the diffraction from ideally perfect and imperfect crystals. For the $(2\bar{1}\cdot0)$ plane in quartz this ratio of intensities is about eighteen. At low strain levels the diffracted intensity is parabolic. This behavior is predicted by both Katos[18] and Kuriyamas[11] theory. Over most of the range, the x-ray intensity is linearly related to the strain. This makes it possible to measure relative amplitudes of various displacements from photodensitometer traces of x-ray topographs or by directly counting the x-ray intensity. At higher levels of strain, the diffracted intensity approaches that predicted for imperfect crystals.

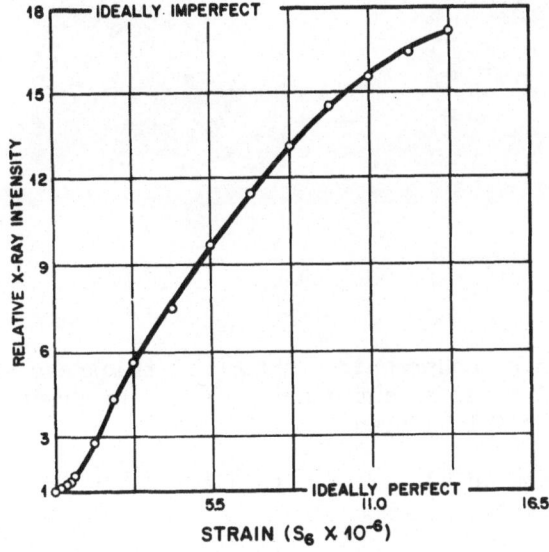

Figure 5. Variation of x-ray intensity diffracted from an AT-cut quartz plate as a function of strain.

The mode shapes determined by x-ray topography depend on the particular Bragg diffraction planes. Figure 6 shows topographs of an anharmonic overtone of thickness-shear[12] taken from $(2\bar{1}\cdot0)$ planes and the $(01\cdot2)$ planes. The relation of these planes to the AT-cut plate was shown in figure 1. The patterns bear little resemblance to one another which is the usual case. The complexity of the pattern in figure 6b indicates the magnitude of the problem associated with analytically treating resonant vibrations in contoured circular plates. In some cases, diffraction from one set of planes will give a sharp well defined pattern while diffraction from other planes will produce little or no pattern. Haruta[19] has given three conditions which must be met to be able to observe vibrations by x-ray topography.

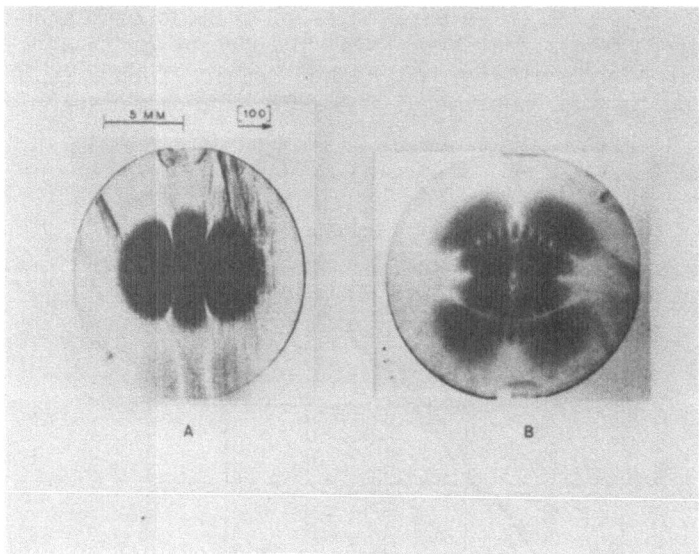

Figure 6. a) Third anharmonic overtone of thickness-shear in a circular AT-cut quartz plate. Diffraction is from the $(2\bar{1}\cdot0)$ plane.

   b) Same as a) except with diffraction from the $(01\cdot2)$ plane.

Modes in Rectangular AT-cut Quartz Plates

The principal (although not be any means the only) vibrations found in AT-cut plates are thickness-shear, flexure and their twist overtones. The displacement associated with these modes was shown in figure 2. The thickness-shear motion produces considerable distortion of the $(2\bar{1}\cdot0)$ planes. Several examples of vibrations in

AT-cut quartz plates both with[20] and without electrodes[12] have been
published. Figure 7 shows the theoretical frequency of several
twist families of thickness-shear and flexural modes for plates
with length-to-thickness variations of 16 to 19. The width of all
these plates was twenty times their thickness. The points are
experimentally determined frequencies where topographs were obtained
to show mode shapes. The theory and experimental techniques have
been published elsewhere[12]. The frequencies have been normalized
by dividing each by the infinite plate frequency given in equation 1
with m = 1. In the flat portions of the curves just above $\Omega = 1$
the motion is principally thickness-shear. Above and below the
plateaus the motion is principally flexural. A family of twist
overtones are nearly parallel to the shear-flexure branches. These
twist branches correspond to 0, 2, 4..[12] antinodes along x_3. The
displacement is given approximately by[12]

$$\psi = K \left[\cos \delta_1 x_1 + f(\delta_1, \delta_2)\cos \delta_2 x_1\right]\cos \frac{n\pi}{2c} x_3 \qquad (5)$$

where δ_1 is the wave number for flexure, δ_2 the wave number for
thickness-shear, 2c is the plate width, $f(\delta_1, \delta_2)$ gives the relative
amplitude of the two modes and n corresponds to a particular twist
overtone. δ_2 is imaginary when $\Omega < 1$ and real when $\Omega > 1$. The
mode shapes for n = 0 and for two frequencies below $\Omega = 1$ and one
above are shown in figure 8.

The displacements in figure 8a are the sum of two cosine
functions, one with a short wave length, the flexural mode, and one
with a longer wave length, the thickness-shear mode. Below $\Omega = 1$,
figure 8b and c, the thickness-shear displacement becomes hyperbolic with a maximum at the plate edges. The absolute value of the
displacements has been plotted since x-ray intensity does not depend
on the sign of the displacement as was shown in figure 4.

The modes shown in figure 9 were obtained at the experimental
points shown in figure 7. The patterns indicate the nature of
coupled thickness-shear and flexure and their twist overtones. The
displacements correspond closely to those compiled from equation 5.
There are however many modes which do not fit on the curves shown
in figure 7. One of these is shown at $\Omega = 1.035$ and a/b = 18.6.

The mode shapes at the points in figure 7 can be calculated
and compared with photodensitometer traces of the mode patterns in
figure 9. This was done for a single mode and is shown in figure 10.
There was a change made in equation (5) to obtain the agreement
shown. Equation (5) gives only the variation of ψ with x_1 and
assumes the variation through the thickness to be linear[12]. This
of course is not the case even though it gives an excellent approximation for frequency calculations. Qualitatively one may argue that
the variation through the thickness is sinusoidal for the shear

branch and hyperbolic for the flexural branch. Since the x-ray
intensity is proportional to the strain gradient and not the displacement a second derivative of equation (5) would change the sign
of one term and not the other. This has been assumed to be the case
for the plot shown in figure 10. The effect of the sign change on
the mode shape is shown in figure 11.

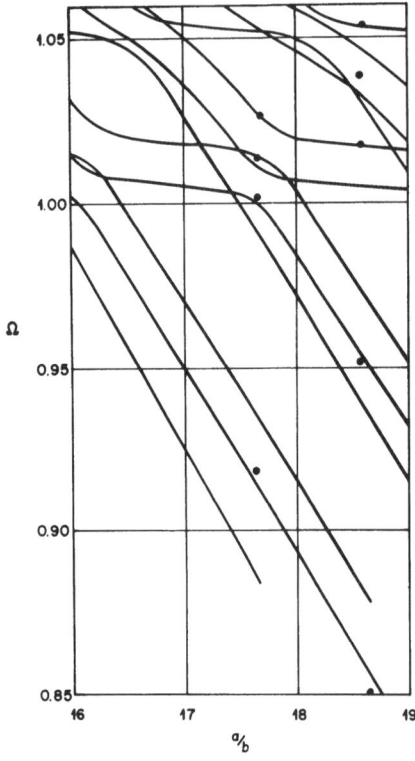

Figure 7. Frequency of twist overtones of thickness-shear and
flexure as a function of plate length-to-thickness ratio
a/b. Points are experimentally measured frequencies.

Lame' Modes

Exact closed form solutions for the vibrations described in
the last section do not exist. The frequencies and displacements
used are calculated from approximate equations which provide a good
deal of insight into the physical nature of the vibrations as well
as very closely predicting the actual measured resonances. In isotropic solids and certain orientations of anisotropic solids very
simple solutions in closed forms may be obtained. These modes first
studied by Lame' for the isotropic case were generalized by Mindlin[21]
for several different crystal classes.

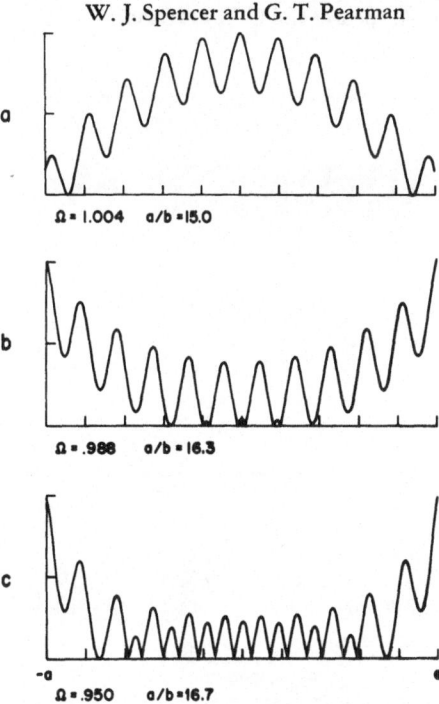

Figure 8.   Displacements in unplated rectangular AT-cut quartz
            plates.

The Lame' modes are sinusoidal solutions of the equations of
motion which can satisfy stress free boundary conditions when the
dimensions of a solid are related by integral multiples. The exact
dimensions of course depend on the elastic constants. Mindlin was
able to show that for certain orientations of single crystals, simple vibrations could be obtained which would satisfy stress free
boundary conditions. In quartz, for example there are two orientations in which stress free boundaries on four of the six faces of a
rectangular parallelepiped can be satisfied by very simple vibrations.
In calculating these orientations, Mindlin used elastic constants
which were in error. These incorrect constants change the orientations of quartz which can be used to obtain simple vibrations. For
the details, Mindlin's paper provides an excellent reference.

Following his procedure and using the following elastic constants for quartz (in units of 10^{10} dynes/cm^2)

$$c^{\circ}_{11} = 86.74 \qquad\qquad c^{\circ}_{13} = 11.90$$

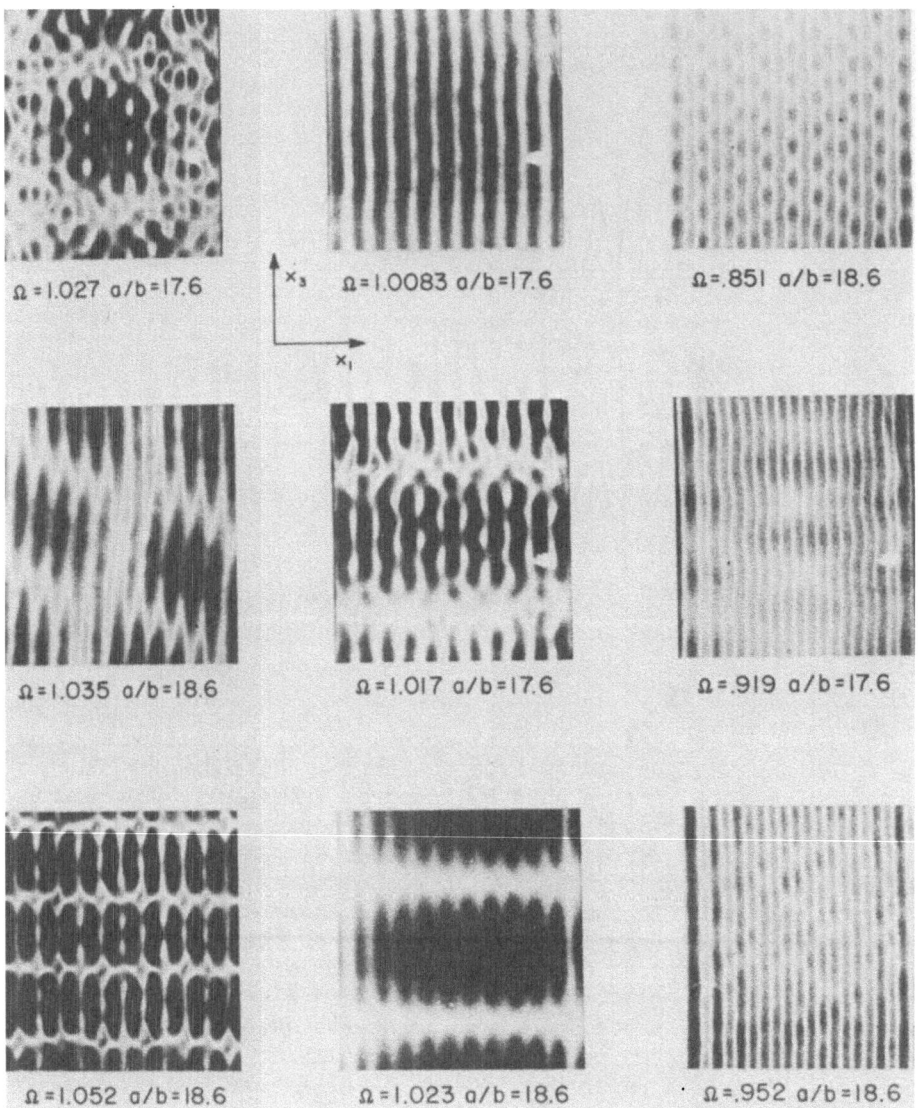

Figure 9:   X-ray Topographs of the Resonances Shown in Figure 7.

$$c^o_{12} = 6.99 \qquad c^o_{44} = 57.94$$

$$c^o_{33} = 107.2 \qquad c^o_{14} = -17.91,$$

the orientations for simple modes were recalculated. Figure 12 shows the orientation of rectangular parallelpipeds which satisfy the conditions for simple modes with stress free faces normal to x_1 and x_2. The bars are assumed infinite along x_3. The ratios of the length along x_1 to that along x_2, a/b is given by

$$\frac{mb}{na} = \begin{cases} 0.97 \\ 1.26 \end{cases} \qquad \begin{array}{l} \theta = 26°45' \\ \theta = -41°20'. \end{array}$$

The frequency is given by

$$\omega = \frac{m\pi}{2a} \sqrt{\frac{c}{\rho}}$$

where

$$c = (c_{11}c_{22} - c_{12}{}^2)/(c_{22} + c_{12}). \tag{6}$$

The c_{11} in equation (6) must be calculated for the two angles shown in figure 13.

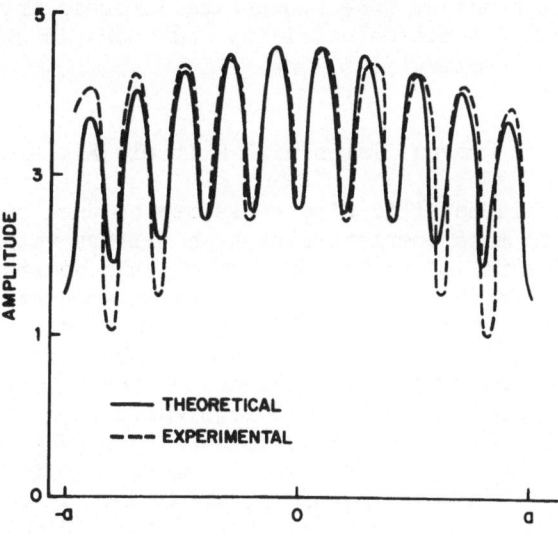

Figure 10.   Comparison of measured and computed displacements in rectantular AT-cut quartz plates.

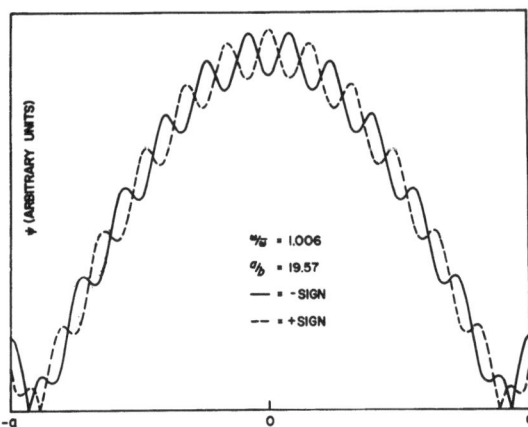

Figure 11. Effect on displacement of sign change in equation (5).

These modes could be driven by applying electrodes to the
faces normal to x_1. Both the d_{11} and the d_{12} piezoelectric constants are non-zero for these orientations. The modes driven in
quartz prisms with these orientations should provide extremely simple vibration patterns and permit verification of some of the aspects
of dynamical diffraction. Even better would be the modes which
satisfy all three traction free boundaries in cubic crystals. Some
cubic crystals exhibit piezoelectricity and modes in others could
be driven electrostrictively.

Vibrations in Quartz Plates with Multiple Electrodes

Arrays of thin metallic electrodes may be used on single piezo-
electric plates to make complete band pass filters without any
additional components. A sketch of a monolithic quartz crystal filter is included in figure 13. In this particular piezoelectric
plate the width is along x_1 - the digonal axis of quartz - and the
gold electrodes are identical. Electrically the center six electrode pairs are shorted to ground and one of the end electrodes is
driven through a properly terminated line while the electrode on
the opposite end (same termination) is used to input a detector.
The example illustrated here is obviously more complicated than a
single electrode resonator. Figure 13 also shows a frequency scan
of the plate and many resonances are seen. Identification of the
vibrational modes is possible using x-ray topography. The characterization of resonant modes has helped develop a piezoelectric
vibration theory which is now used to predict many of the resonances

in multi-electroded plates.   Topographs of some of the modes are
shown in figure 13.   It should be pointed out that the topographs
show only a portion of the plate.   Since the electrode geometry is
symmetrical with respect to the center line indicated by the arrows,
the mode patterns are also symmetrical about this line.   A little
more than half the plate is shown as proof of the symmetrical behavior.   The center line is indicated in each topograph as well as
in the sketch of the plate.   A portion of the end resonator is
shadowed by a ceramic base which supports the plate.   The fundamental resonance (or pass band region), is readily identified without x-ray topographs.   However, resonances around 8.206 MHz, 8.266 MHz
and 8.408 MHz are not so easily identified but the topographs indicate the true nature of the vibrational modes:

> 8.206 MHz is the first shear anharmonic (break-up along
> x_1),
>
> 8.266 MHz is the first twist anharmonic (break-up along
> x_3),

and   8.408 is a series of modes showing various amounts of
coupling between the second shear anharmonic and the
first twist anharmonic (break-up in two directions under
each electrode).   These patterns are extremely complicated due to the break-up in two directions as well as
interactions between the eight resonators.

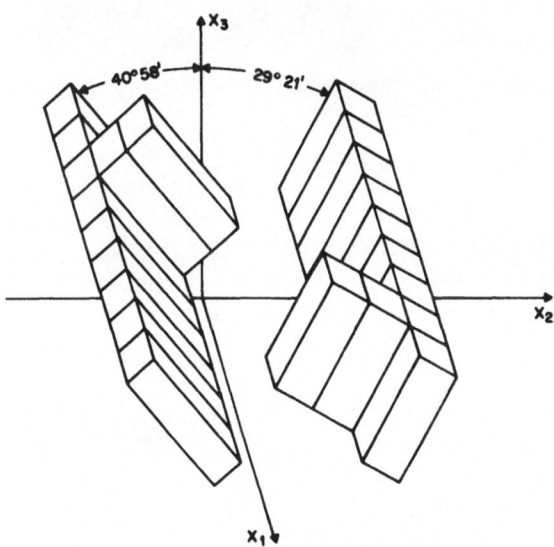

Figure 12.   Orientations of quartz prisms for simple vibrations
with stress free faces.

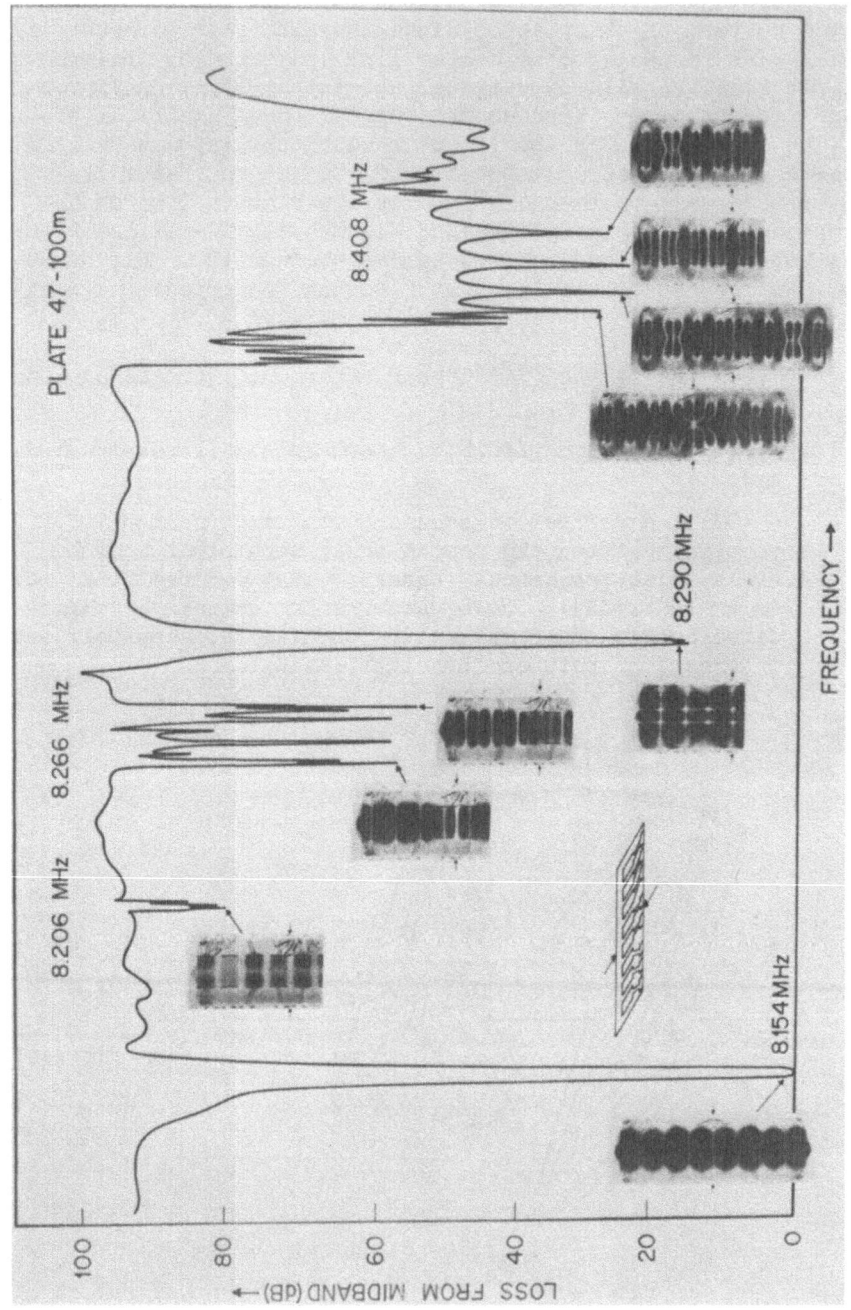

Figure 13: Frequency Response of an Eight Resonator Plate (including X-ray topographs).

Seen in all the topographs are flexure components appearing as short wavelength "ripples" surrounding the primarily shear motion.

This example illustrates the fairly complicated mode patterns existing in an 8 resonator monolithic crystal filter with 8 identical electrodes. In cases where the various electrodes have assorted dimensions, the patterns can be expected to be even more complicated.

By using topography to determine the areas of the plate which vibrate methods of preferentially damping out certain unwanted modes can be devised. It is well known that the flexure component which tends to be spread out over the whole plate can be damped somewhat by clamping the edge of the plate without serious effects on the thickness shear modes which are largely contained under the electrodes as shown in the topographs. Similarly unwanted anharmonic modes can be suppressed by proper dimensioning of the electrodes.

The point is to be stressed again that only with the knowledge of exactly what kind of motion is present can an intelligent decision regarding possible means of suppressing unwanted resonances be made. X-ray topographs have proven to be an invaluable tool in such work.

REFERENCES

1. G. W. Fox and P. H. Carr, "The Effect of Piezoelectric Oscillation on the Intensity of X-ray Reflections from Quartz", Phys. Rev. 37: 1622-1625, 1931.

2. S. Nishikawa, Y. Sakisaka and I. Sumoto, "X-ray Investigation of the Mode of Vibration of Piezoelectric Quartz Plates," Proc. Phys.-Math. Soc. Japan 25, 20-30, 1934.

3. C. S. Barrett and C. E. Howe, "X-ray Reflection from Inhomogeneously Strained Quartz," Phys. Rev. 39, 889-897, 1932.

4. W. G. Cady, "Piezoelectricity," Dover Publications, Inc. New York, 1964.

5. G. E. M. Jauncey and W. A. Bruce, "Diffuse Scattering of X-rays from Piezoelectrically Oscillating Quartz," Phys. Rev. 54, 163-165, 1938.

6. J. E. White, "X-ray Diffraction by Elastically Deformed Crystals," J. Appl. Phys. 21, 855-859, 1950.

7. W. J. Spencer, "X-ray Diffraction Study of Acoustic Mode Patterns in Crystalline Quartz," Appl. Phys. Letters 2, 133-135, 1963.

8. A. L. Bennett, R. A. Young and N. K. Hearn, Jr., "X-ray Diffraction Topography of Vibrating Quartz Crystals," Appl. Phys. Letters $\underline{2}$, 154-156, 1963.

9. A. R. Lang, "The Projection Topograph: A New Method in X-ray Diffraction Microradiography," Acta. Cryst. $\underline{12}$, 249-250, 1959.

10. W. J. Spencer, "Observation of Resonant Vibrations and Defect Structure in Single Crystals by X-ray Diffraction Topography," in Physical Acoustics, Vol. 5, Edited by W. P. Mason, Academic Press, Inc., New York, 1968, p. 111-161.

11. M. Kuriyama and T. Miyakawa, "Theory of X-ray Diffraction by a Vibrating Crystal," J. Appl. Phys. $\underline{40}$, 1697-1702, 1969.

12. W. P. Mason, "Piezoelectric Crystals and their Application to Ultrasonics," D. Van Nostrand Co., New York, 1950.

13. R. D. Mindlin and W. J. Spencer, "Anharmonic, Thickness-Twist Overtones of Thickness-Shear and Flexural Vibrations of Rectangular, AT-Cut Quartz Plates," J. Acoust. Soc. Am. $\underline{42}$, 1268-1277, 1967.

14. P. C. Y. Lee and S. S. Chen, "Vibrations of Contoured and Partially Plated, Contoured, Rectangular, AT-Cut Quartz Plates," (to be published).

15. R. D. Mindlin, "An Introduction to the Mathematical Theory of Vibrations of Elastic Plates," (U.S. Army Signal Corps Engineering Laboratories, Ft. Monmouth, N. J., 1955).

16. W. J. Spencer and K. Haruta, "Defects in Synthetic Quartz," J. Appl. Phys. $\underline{37}$, 549-553, 1966.

17. R. W. James, "The Optical Principles of the Diffraction of X-rays," G. Bell and Sons Ltd., London, 1962.

18. W. Kato, "Pendellösung Fringes in Distorted Crystals III," J. Phys. Soc. Japan $\underline{19}$, 971-985, 1964.

19. K. Haruta, "New Method of Obtaining Steroscopic Pairs of X-ray Diffraction Topographs," J. Appl. Phys. $\underline{36}$, 1789-1790, 1965.

20. P. C. Y. Lee and W. J. Spencer, "Shear-Flexure-Twist Vibrations of Rectangular AT-Cut Quartz Plates with Partial Electrodes," J. Acoust. Soc. Am. $\underline{45}$, 637-645, 1969.

21. R. D. Mindlin, "Simple Modes of Vibrations of Crystals," J. Appl. Phys. <u>27</u>, 1462-1466, 1956.

22. R. A. Sykes, W. L. Smith and W. J. Spencer, "Monolithic Crystal Filters," 1967 IEEE International Convention Record, 1967, p. 78-93.

X-RAY TOPOGRAPHIC STUDY OF VIBRATING DISLOCATIONS IN ICE UNDER AN AC ELECTRIC FIELD

K. Itagaki

USA Cold Regions Research and Engineering Laboratory

Hanover, New Hampshire 03755

INTRODUCTION

The behavior of charged dislocations in alkali-halide crystals has been drawing attention in connection with the charge transfer which occurs during plastic deformation.[1-9] Recently, Itagaki proposed a charged dislocation mechanism to account for the dielectric properties of ice.[10] His theory is in part supported by the dielectric measurements of strained ice made by Ackley and Itagaki.[11] Brantley and Bauer[12] derived similar equations for the dielectric constant based on charged dislocation motion. They also proposed a new mechanism for apparent piezoelectricity based on moving charged dislocations in an electric field.

An estimate of the charge concentration can be made if the amplitude of vibrating charged dislocations is measured under a known electric field. X-ray topography is the most promising method to make direct observations of vibrating dislocations in ice. Electron microscopy could not be used because ice would sublime in the high vacuum. An etch pit method would not reveal the vibrating dislocation and the surface can affect the movement of dislocations. The intent of the present study is to establish the charge concentration on the dislocation line by X-ray topography.

THEORY

The response of a stretched string in a viscous medium to the local electric field E' exp $i\omega t$ is the model for charged dislocation motion and the equation of motion is:

$$m \frac{\partial^2 \eta}{\partial t^2} + B \frac{\partial \eta}{\partial t} - T \frac{\partial^2 \eta}{\partial x^2} = E' \sigma \exp i\omega t \qquad (1)$$

where m, B and σ are mass, coefficient of drag and electric charge on unit length of dislocation respectively, T is the line tension of the dislocation and η is the amplitude of oscillation of the dislocation line. The solution is:

$$\eta = \frac{E' \sigma \exp i\omega t}{k^2 T} \left(\frac{\cos kx}{\cos \frac{1}{2}kl} - 1 \right) \qquad (2)$$

where $k^2 = (\omega^2 m - iB\omega)/T$. Expanding the bracket in a power series and dropping the k^2 terms in the low frequency range, the maximum amplitude is

$$\eta_{max} = l^2 E' \sigma / 8T \qquad (2')$$

η_{max} and l can be measured from the topograph and T is roughly $Gb^2/2 = 2.5 \times 10^{-11}$ nt. Assuming that the local field E' is the Mossoti field, $E' = E(\varkappa' + 2)/3$ where the specific permittivity $\varkappa'$ is about 90 for an unstrained sample at 60 Hz, one can expect a local field within the crystal about thirty times larger than the applied field. The charge concentration, σ, is thus:

$$\sigma = \frac{8T \eta_{max}}{l^2 E'} \qquad (3)$$

The amplitude of vibration and the distance between pinning points can be obtained from the X-ray topograph of ice under a known electric field.

EXPERIMENTAL APPARATUS AND PROCEDURE

Specimens used in these experiments were single crystals produced in the Mendenhall Glacier. Conductivity of molten water was about 5×10^{-7} mho/cm indicating that the amount of ionic impurities was quite low. However, minute amounts of mineral fragments were sometimes found. Dislocation density was about $10^4/cm^2$. The specimen was oriented to have a diffraction plane of $(10\bar{1}0)$ and was mounted on a holder as shown in figure 1 with a small amount of water. It was then sliced without introducing strains by a hot

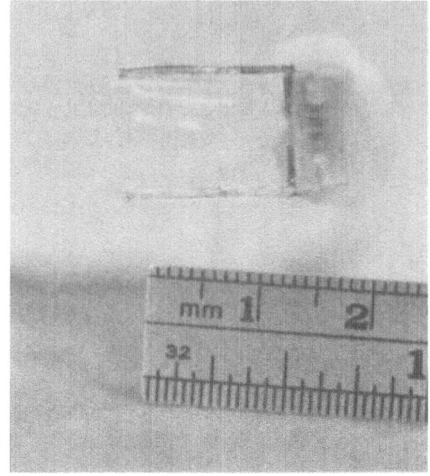

Figure 1. Ice specimen slab mounted
           on the holder.

wire cutter parallel to the (0001) plane to a thickness of about
2mm. Both surfaces of the specimen were sublimed by a clean air
stream to reduce the thickness and to provide a final finish. The
final thickness was generally about 0.5mm or less. The holder was
placed into a socket on a plate and mounted on the goniometer head
as shown in figure 2. Further sublimation from the specimen was
prevented by encasing the specimen holder with saran film and
placing a small amount of powdered ice in the goniometric holder

Figure 2. Ice specimen mounted
           in the "cave" of
           the goniometer.

with the sample. Lang diffraction topography was made using a
Jarrell-Ash Lang camera with Norelco X-ray equipment. A fine focus
copper target in a sealed-off tube with a nickel K-beta filter at
40kv and 10ma was used as the source. Ilford L4 nuclear emulsion
plates were used and developed using Kodak D19 developer at 18°C
for about 15 min. A point source port (0.4mm by 0.84mm) was
placed about 40cm from the specimen and an adjustable slit (3.0mm
by ~10mm) was placed 5cm from the specimen. The vertical height
of the slit was adjusted to expose the major portion of the specimen. With this configuration, reasonably sharp, high contrast
topographs were obtained with relatively short exposure times (30
min). The fields applied to the electrode supports were a 60 Hz
sine wave, and 1 1/3 Hz and 1/30 Hz square waves. The field
strength ranged from 3 V/cm to 600 V/cm. These frequencies were
used because the drag and mass effects on the amplitude and wave
shapes of the vibrating dislocations can be considerable at higher
frequencies. No specific frequency effect was detected within this
frequency range.

RESULTS

The following precautions were taken to separate the dislocation motion under the electric field from the motion caused by the
other forces. About one-fourth of the total exposure of a specimen
was made without any electric field when the 60 Hz AC field was
applied. Only spindle-shaped diffused lines with a prominent
center core were selected as the dislocations vibrating under the
electric field (figure 3). Those lines were further confirmed by
comparing them with the corresponding lines in topographs taken
without any electric field. Lower frequency square waves (1 1/3 Hz
and 1/30 Hz) were produced by a cam driven microswitch. A shutter
was rotating with the cam to reduce the exposure during the negative cycle to one-half that of the exposure during the positive
cycle. This arrangement made it possible to distinguish the dislocations vibrating under the electric field as well as the sign
of their charge (figure 4). The measurement of length and amplitude of diffused segments was required as shown in equation (3) to
obtain the charge density. However, when the dislocation density
was high it was difficult to make accurate measurements. A further
difficulty was that the dislocation images overlapped and smeared
out when the higher electric fields were applied while the displacement under the low electric fields was limited by the resolution and was indistinguishable from motion produced by other forces.
The optimum electric field was between 10V/cm and 50V/cm for individual dislocations. Bundles of dislocations became mobile under
higher electric fields in some cases (540 V/cm). No motion was
observed for straight, sharp, well-defined dislocations when up to
a 160 V/cm electric field was applied while curved lines were
vibrating (figure 5). Presumably those dislocations were trapped

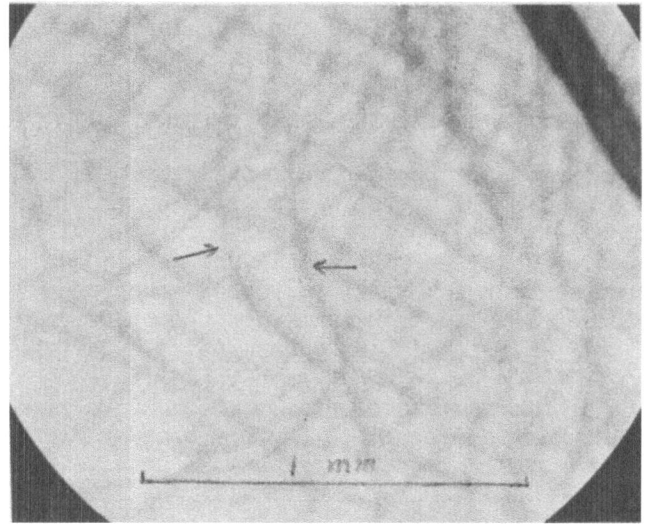

Figure 3. Vibrating dislocations with center core is
           shown by the arrows. Electric field ←→ .
           Diffraction vector ←→ field strength
           14.5 V/cm. Frequency 60 Hz sine wave.

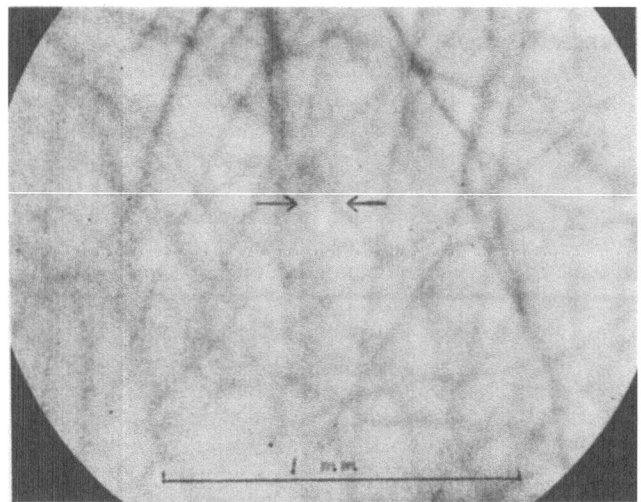

Figure 4. A spindle-shaped shadow of vibrating disloca-
           tion is indicated between two arrows. Note
           left side is darker than right side. Exposure
           during the field direction → is one-half of
           ←. Field strength 29.5 V/cm. Diffraction
           vector ←→ . Frequency 1/30 Hz.

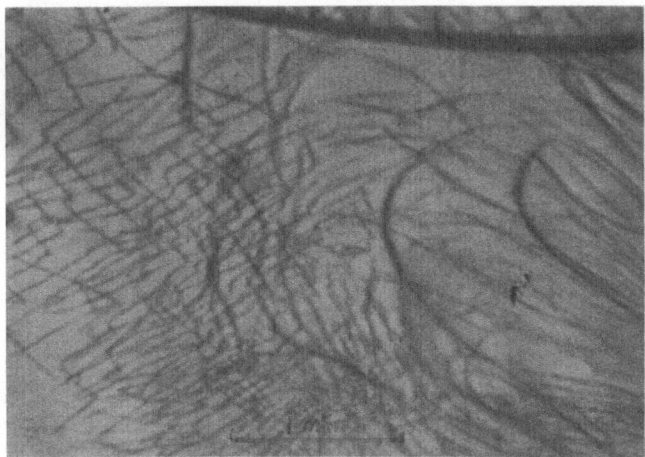

Figure 5.   Straight dislocations under the higher
            electric field do not vibrate while
            curved lines vibrate (160 V/cm, 1 1/3 Hz).
            Direction of electric field ↕   .
            Diffraction vector ←→ .

in the trough of the Peierls potential. Measurements were made
only on the segments which conformed to the following standards:
(1) length and amplitude of the segment could be established; (2)
no motion was detected in the topographs taken without an applied
field; and (3) a center core was seen when the 60 Hz field was
applied or (3') spindle-shaped shadow was darkly outlined and one
side of the outline was darker than the other when the 1 1/3 Hz or
1/30 Hz square wave was applied. Although most of the dislocations
lying perpendicular to the electric field were found to vibrate
under the influence of the field, only a few lines conformed to
the standards mentioned above. The measured results are shown in
table I, columns 2 and 3. As derived from equation (2) η_{max}/l^2 is
proportional to the applied field as shown in figure 6 although
there is considerable scatter. The scatter may be due to difficulty
in the measurement of the diffuse image. The charge density may
possibly be different depending on the direction of the dislocations.

     There are several problems in estimating the local field.
Even assuming a value for the dielectric constant does not allow
accurate calculation of the local field for the following reasons:
(1) The Mosotti field cannot be used for accurate calculation because it is based on the spherical cavity surrounding the dipole
considered. In the present case, however, the cavity should be
cylindrical with the center line in common with the dislocation
line. These considerations would change some numerical factors.

Table I

No.	Segment length l $\times 10^{-5}$ m	Amplitude η max $\times 10^{-5}$ m	η max$/l^2$ $\times 10^2$/m	Sign of Charge	σ max $\times 10^{-10}$ Cb/m	σ min $\times 10^{-12}$ Cb/m
		4.75×10^3 V/m 60 Hz				
1	44	6	3.1		1.5	4.9
2	60	8	2.2		1.1	3.5
3	68	8	1.7		0.8	2.7
4	60	8	2.2		1.1	3.5
5	60	4	1.1		0.5	1.7
6	60	8	2.2		1.1	3.5
7	52	6	2.2		1.1	3.5
8	100	12	1.2		0.6	1.9
9	88	6	0.8		0.4	1.3
10	80	6	0.9		0.4	1.4
11	52	8	3.0		1.4	4.8
		1.45×10^3 V/m 60 Hz				
12	88	8	1.0		1.6	5.2
13	123	10	0.7		1.1	3.7
		2.95×10^3 V/m 1/30 Hz				
14	128	12	0.7	+	0.5	1.8
15	116	8	0.6	+	0.5	1.5
16	48	6	2.6	+	2.0	6.7
17	64	8	2.0	?	1.5	5.1
Mean	75.9				1.0	3.3

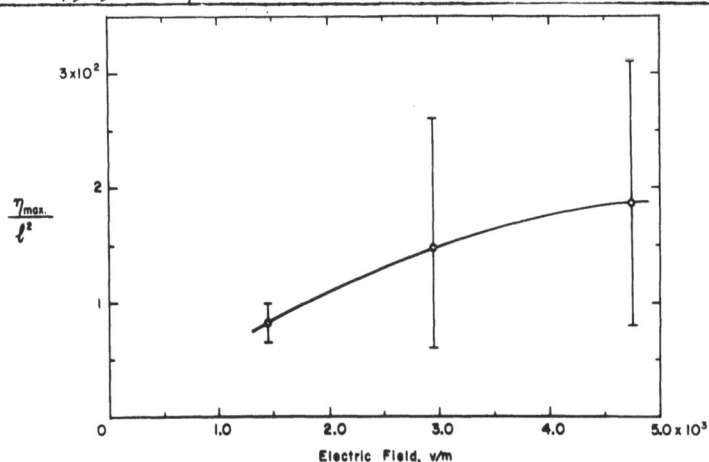

Figure 6. η max$/l^2$ vs. field strength. $\circ$ denotes mean values at the field strength.

(2) The distribution of dislocation lines is not uniform so that the local field produced by the dislocation lines and their displacement is not uniform. (3) The specimen is quite thin so that the integral of the field cannot be made on an infinite volume but only on the finite slab of the specimen. Moreover, the dielectric constants are highly dependent on the strain, presumably because of the changing dislocation structure with increasing strain especially in the frequency range in which the present work was done. Therefore, the exact strength of the local field is in considerable doubt. Probable highest and lowest limits were used in the present paper to define the range. The highest limit of the local field was calculated as the Mosotti field using the dielectric constant of unstrained ice ($\chi' = 90$) while the lowest limit is the external field. The charge density σ calculated using both values are shown in columns 5 and 6 of table I. Only three segments were found appropriate by the mentioned standards to identify the sign of the charge. These had a positive charge. The possibility still remains that dislocation lines of negative sign can exist under certain conditions.

DISCUSSION

Several possible causes seem to prevent dislocation line motion in these experiments.

1. Surface effect. The portions of dislocation lines which lay parallel and near to the surface and which terminated on the surface seemed to be affected by the surface. Dislocations in the thinner samples were immobilized, presumably by this effect.

2. Local field effect. As the dislocation density decreased during the thinning by sublimation or by annealing out of dislocations, the effective dielectric constant became lower because it is directly proportional to the dislocation density. This effect will modify the Mossoti field acting on the individual dislocation lines. A lower dislocation density portion is surrounded by a higher density part as shown in figure 7. The amplitude of dislocation motion in the low density portion is apparently smaller than in the high density portion. The smaller amplitude can be attributed to the surface or the local field effect or both.

3. Charge cloud effect. According to Eshelby, et al.[13], charged dislocations in alkali halide crystals are gradually surrounded by a charge cloud of opposite sign. Coulomb interaction between the charged dislocation and the slowly moving charge cloud would suppress the dislocation motion.

4. Impurity effect. Various types of impurities diffuse from the atmosphere into the ice crystals, through the crystal lattice,

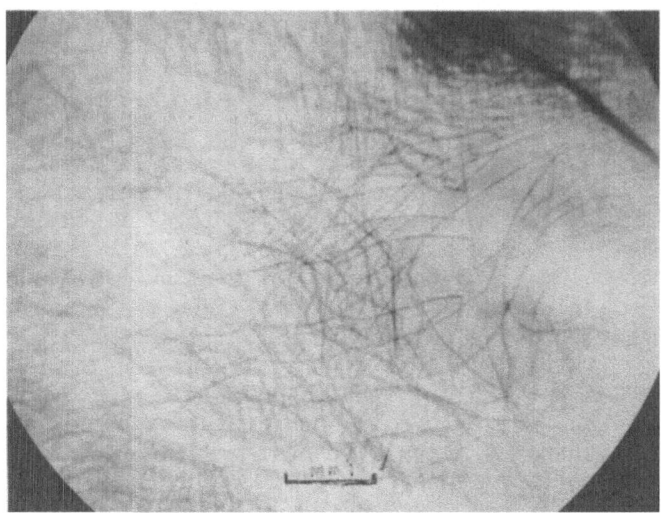

Figure 7. Vibrating dislocations in lower dislocation
density portion show less vibration than the
higher dislocation density portion. Note
that some lines remain sharp in high density
portion. Field strength 29.5 V/cm. Direction↕.
Diffraction vector ←→.

and along the dislocation lines. Although hydrofluoric acid, HF,
seems to accelerate the dislocation motion, most types of impurities
would tend to constrain the motion. It has been observed that the
dislocation lines in newly prepared specimens are curved and mobile
while dislocations in older specimens are rather straight and immobile even under the highest electric field applied (600 V/cm).
The difference in character of the dislocations can be attributed
to the charge cloud or the effect of impurities which diffuse into
the ice from the atmosphere.

5. <u>X-ray effect</u>. It is known that, in the case of alkali
halides, point defects produced by X-ray irradiation at color
centers pin down the dislocations.[14] A similar effect seems to
exist in ice although no color center in ice is known to the
present author. The dielectric constant of ice shows a very rapid
increase followed by a gradual decrease after X-ray irradiation[15].
The initial increase is presumably induced by charge carrier production and the gradual decrease may be related to the pinning of
charged dislocations by point defects produced during X-ray irradiation. In the case of rock salt, it is known that irradiation by
visible light (6300 Å) can unpin the dislocations.[14] Visible light
irradiation by a small incandescent light was used in an effort to
unpin the dislocations in ice but the results were not clear.

6. <u>Trough of the Peierls potential</u>. A dislocation which lies in the "Peierls' trough" running along certain crystallographic orientations requires a higher energy to move out of the trough than a dislocation which does not lie in the trough. It was frequently observed that straight dislocations are generally in the $\langle 11\bar{2}0 \rangle$ directions and were immobile (figure 5). Presumably the dislocations were trapped in the trough of energy minima after prolonged annealing. The X-ray topographic study on dislocation structure in ice made by Webb and Hayes[16] showed that all burgers vectors are of the $\langle 11\bar{2}0 \rangle$ type. Another possibility is that the pure screw dislocation in ice may not have any electric charge as was found in some of the alkali halide crystals.[3]

All of the above factors tend to suppress the dislocation motion. Numbers 1, 3, 4, and 6 can be avoided by using thick, freshly prepared specimens. To avoid the local field effect one must observe the dislocations in the higher density regions which makes it difficult to distinguish the pinning points of the dislocations. The X-ray effect is unavoidable as an X-ray topography method is used. Fortunately, this effect does not seem prohibitively strong although it may affect the results.

Brantley recently suggested that piezo-electric deformation is a possible mechanism to drive dislocations in the crystal.[17] However, Tippe[18] found no detectable piezo-electric effect in ice, so piezo-electric deformation seems unlikely. Deubner, et al.[19], found some piezo-electric effect in a newly prepared sample but the effect disappeared after three to four days. The disappearance of piezo-electricity can be more easily explained by the effects of charge clouds, impurities, or troughs in the Peierls potential as discussed previously in this paper.

A possible temperature rise produced by passing current through the specimen does not appear responsible for the dislocation vibration. The distinctive differences between figure 8a (made with no applied field) and figure 8b (made with a 300 V/cm field) is the support of this notion. Most of the dislocation lines lying perpendicular to the electric field which appeared in figure 8a are smeared out, of low contrast, or are spread to a spindle-shape in figure 8b while little change is observed in the lines parallel to the electric field. The vibration of dislocations due to a thermal effect would not depend on orientation.

There are several papers on the charge density of dislocation lines.[1-9] The method based on charge generation during deformation requires information on mobile dislocation density. If, however, some dislocations are immobilized by factors such as the charge cloud effect, the impurity effect, or the "Peierls' trough" effect, an estimate of mobile dislocation density based on etching (which reveals all dislocations whether mobile or not) may be an over-

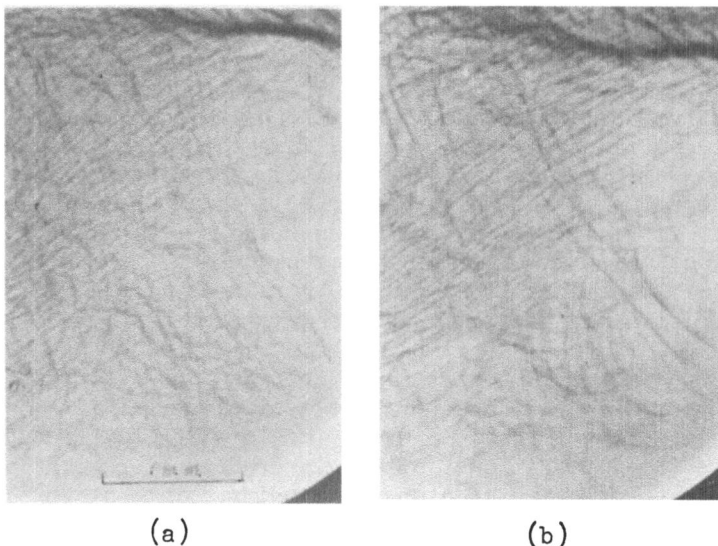

(a)   (b)

Figure 8.   Comparison of dislocation image without (a)
             and with (b) electric field. Direction of
             field ◄─►. Field strength in (b) 320 V/cm,
             1 1/3 Hz square wave. Diffraction vector ↗ .

estimate causing the charge concentration to be underestimated.

Zagoruiko[5] observed etch pits moving under a static field.
Several sources of error may appear when this method is used to
estimate the charge concentration. For example, the local field
near the surface may be affected by fringing, and be different
than the applied field strength, thus, from his equation (1'),
$E = 0.8 \frac{\sigma^2 K}{\epsilon}$, affecting the estimate of charge concentration, σ.
The concentration of divalent impurities required to calculate K
in this same equation may be different near the surface. Also, the
motion of a dislocation would be hampered near the surface, especially at the end attached by the etchant. So, his equation to
calculate the charge density contains quite a few assumptions which
are quite difficult to verify.

CONCLUSIONS

The present author estimated the charge density, σ, from the
dielectric constant values of undeformed ice as 5.16×10^{-11} coulomb/m.
This estimate was based on the segment length $L = 10^{-5}$m which is
considerably shorter than the segment length measured in this experiment since only the longest lengths satisfied the experimental
criteria in the present experiment. Even so, the segment length

previously assumed can be adjusted with little change in result to agree with the present work and the results of other researchers. The value for charge density seems to agree reasonably well with that obtained for other materials such as sodium chloride or cesium iodide considering the amount of error possible in most calculations. The charging mechanism on dislocations in ice crystals may be considerably different from that in ionic crystals and nothing is conclusive on this point yet. The agreement with the alkali halide work at this stage seems satisfactory. Although there are several difficulties in applying X-ray topographs for the measurement of dislocation charge density in a crystal, the assumptions required to determine the charge density by this method seem less complex and more direct. Future refinement may prove quite fruitful.

ACKNOWLEDGEMENTS

The author wishes to express sincere thanks to SP5 S. F. Ackley for his help in preparing this paper and to Mrs. Jonilee Lange for typing it. He also appreciates the advice on X-ray topography technique given by Dr. R. G. Wolfson.

REFERENCES

1. R. L. Sproull, "Charged Dislocations in Lithium Fluoride", Phil. Mag. 5:815-831, (1960).

2. G. Remaut and J. Vennik, "Observations on an Electrical Effect Obtained during Deformation of Sodium Chloride Crystals," Phil. Mag. 6:1-8, (1961).

3. R. W. Davidge, "The Sign of Charged Dislocations in NaCl," Phil. Mag. 8:1369-1377, (1963).

4. R. W. Whitworth, "The Production of Electrostatic Potential Differences in Sodium Chloride Crystals by Plastic Compression and Bending," Phil. Mag. 10:801-816, (1964).

5. N. V. Zagoruiko, "Effect of an Electrostatic Field and a Pulsed Magnetic Field on the Movements of Dislocations in Sodium Chloride," Soviet Physics-Crystallography, 10:63-67, (1965).

6. R. W. Whitworth, "A Measurement of the Charge on Edge Dislocations in a Sodium Chloride Crystal," Phil. Mag. 15:305-319, (1967).

7. R. J. Schwensfeir, Jr. and C. Elbaum, "Electric Charge on Dislocation Arrays in Sodium Chloride," J. Phys. Chem. Solids, 28: 597-606, (1967).

8. R. M. Turner and R. W. Whitworth, "Movement of Dislocations in Sodium Chloride Crystals in an Electric Field", Phil. Mag. 18: 531-539 (1968).

9. R. De Batist, E. von Dingen, Yu. N. Martyshev, I. M. Sil'vestrova and A. A. Urusovskaya, "Charged Dislocations in Cesium Iodide," Soviet Physics-Crystallography, 12:881-888, (1968).

10. K. Itagaki, "Contribution of Charged Dislocation Motion on Dielectric Behavior of Ice," Bulletin Amer. Phys. Soc. 14: 411, (1969) Abstract.

11. S. Ackley and K. Itagaki, "Strain Effect on the Dielectric Properties of Ice," Bulletin Amer. Phys. Soc. 14:411, (1969) Abstract.

12. W. A. Brantley and C. L. Bauer, "Effect of Charged Dislocations on Dielectric Piezoelectric and Elastic Properties," J. Mat. Sci. Eng. 4:29-38, (1969).

13. J. D. Eshelby, C.W.A. Newey, P. L. Pratt, and A. B. Lidiard, "Charged Dislocations and the Strength of Ionic Crystals," Phil. Mag. 3:75-89, (1958).

14. C. L. Bauer and R. B. Gordon, "Mechanism for Dislocation Pinning in the Alkali Halides," J. Appl. Phys. 33:672-682, (1962).

15. K. Itagaki, "X-ray Effect on Dielectric Properties of Ice," to be published.

16. W. W. Webb and C. E. Hayes, "Dislocations and Plastic Deformation of Ice," Phil. Mag. 16:909-925, (1967).

17. W. A. Brantley, Private Communication.

18. A. Tippe, "Zum Piezoeffekt bei Eis I," ZS, Naturwissenschaften, 3:1, (1967).

19. A. Deubner, R. Heise and K. Wenzel, "Nachweis des Piezoeffektes am Eis," Naturwissenschaften, 47:600-601, (1960).

AN APPROACH TO THE SOLID SOLUTION PROBLEM USING A COMPUTERIZED

IDENTIFICATION TECHNIQUE

Gerald G. Johnson, Jr. and Frank L. Chan

Pennsylvania State University, University Park, Pa.

Wright-Patterson Air Force Base, Ohio

ABSTRACT

Since for most real systems, solid solution effects influence the position and intensity of the x-ray powder diffraction pattern, it is desirable and necessary to have an automatic system which will identify standard reference phases regardless of the amount of solid solution. Using the system CdS-ZnS, where the lattice parameter a_o changes from 4.136 to 3.820Å, with complete solid solution over the entire range of composition, an illustrative study was made. This work presents the results obtained from a computer analysis of the powder pattern obtained. It has been found that 1) if the starting chemistry is known and 2) the end members of the series are in the ASTM Powder Diffraction File, that the solid solution can be identified. Once the phases present are identified, a plot following Vegard's law yields the approximate composition of the sample under consideration. These two methods of compositional determination agree quite well. Examples of the computer system and description of the program input and output with interpretation of the results will be discussed.

This work is primarily sponsored by a grant by the Joint Committee on Powder Diffraction Standards.

INTRODUCTION

The compiling of reference x-ray diffraction patterns by ASTM dated back to early 1941. The Joint Committee on Powder Diffraction Standards at that time undertook the task to publish a card file of powder diffraction data for use in the identification of

539

unknown crystalline materials by x-ray powder diffraction methods.
In those days there were only one thousand compounds for which the
x-ray diffraction data were known. These data were organized
according to an indexing system developed by Hanawalt, now Professor at The University of Michigan, Rinn and Frevel at Dow Chemical
Company, Midland, Michigan[1].

In the 1950 edition of the ASTM Powder Diffraction File, the
Davey alphabetical index appeared first, followed by the Hanawalt
numerical index[2]. Organic anions such as the acetates and oxa-
lates were listed under "General Inorganic and Organic Index."
All other organic compounds were listed under "Organic Index"
followed by index for "Mineral."

In the early stages of the compilation of the powder diffraction file, the question was often asked as to the reliability of
the diffraction data as to the "d" values and their relative intensities. Instead of throwing out some of the diffraction data,
the earlier editions listed multiple cards of the same compound.

At the outset, a fellowship was established at the National
Bureau of Standards to work out the powder diffraction data[3].
Stars placed on ASTM data furnished by the Bureau indicated the
high reliability. Some of the questionable quality powder patterns
appearing in the earlier editions were deleted from the file in
later editions. Since 1941 the file has grown in size and the
quality has increased. At the end of 1969 this file had 18,000
inorganic and organic powder patterns. These data are used not
only in the United States but throughout the world. Incidentally,
the ASTM Powder Diffraction File has recently been put on Micro-
fiche to reduce storage space.

As the data in the file became larger, other indexing systems
have been introduced for effective use of these data. The identification of x-ray powder diffraction patterns has been made possible by a published collection of standard reference patterns. A
series of book indexes to this ASTM File are available (Hanawalt-
Davey Index, Fink Index, and KWIC Index)[4,5,6,7] to facilitate the
identification of crystalline unknown samples. Furthermore, the
basic data of the Powder Diffraction File have been computerized
and stored on magnetic tape. They can be sorted into any desired
classification or configuration.

The Fink Index is one of the newer indexing systems. The
primary purpose of developing the Fink Index in the last decade
was to extend the existing file for both x-ray and electron diffraction techniques. In most cases the d values by both methods
agree to within one percent; however, the intensities differ considerably for the two types of patterns. Thus the Hanawalt system
of identifying a compound based upon the three most intense lines

as recorded in the existing file may not apply to data obtained
from the electron diffraction. Therefore, in order to accommodate
the existing ASTM Powder Diffraction File to facilitate the identification of x-ray and electron diffraction patterns, the relative
intensity factor is not emphasized. The system of listing the
eight strongest lines and the cyclic permutation are fully explained in this index for the users.

Unlike the Davey Index[5,6] which is presently based only on
cations, Drs. V. Vand and G. G. Johnson, Jr. have recently developed the KWIC Index (Key-Word-in-Context Index)[7] for inorganic
substances. The KWIC Index is based on major chemical fragments.
By this sytem, patterns of a chemical compound can then be per-
muted and therefore it is more comprehensive than the Davey Index.
In theory, it is possible for the researchers to solve any problems
made up from standards in the reference file. However, in practice,
due to different experimental procedures and errors by the diffractionist, the problem can become extremely difficult. In order to
eliminate the necessity that anyone who wishes to use this reference File become a professional diffractionist, the Joint Committee
has sponsored the development of a complete identification system.

Perhaps the most reliable and rapid method for the detection
of crystallization phases is by x-ray diffraction, provided the
data have been stored in the ASTM Powder Diffraction File. With
the data already computerized, it is not necessary to rely on
human pattern recognition. Based on data input with chemical information, one can derive from a computer print-out definite,
unbiased conclusions as to the presence of different species
present. This can further be confirmed by other physical and
chemical methods.

The magnetic tape with the powder diffraction data can be
used to search unknown materials containing one or more components.
The computer program and a full description of the ASTM computerized tape with user's instruction can be obtained from that
organization[8].

THE PROBLEM

The original computerized identification system was developed
at the Materials Research Laboratory of The Pennsylvania State
University by the late Prof. V. Vand and by G. G. Johnson, Jr.[9].
Although two other programs exist for the same purpose[10,11] this
paper will discuss the results obtained with the Johnson-Vand
program.

Since the 1967 meeting (at which time the use of computer
identification of phases in a multicomponent unknown was pointed
out), these computer programs have undergone various improvements

such as the simplification of file preparation and the reduction
of the parameters needed to be entered by the diffractionist. In
addition to the simplification of the running of the programs,
certain extensions have been included in the revised program
(Version 10). This paper will discuss the applicability of this
program for the identification of phases present in solid solution.

The solid solution problem is the changing of the unit cell
lattice parameters due to the presence of a second phase which
"shifts" the spectra of both the host phase and the second phase.
Although in the high symmetry case the shift in line position, due
to solid solution is the same for all lines (either toward/from
higher angle from/toward "normal position"), the general case must
be able to solve more complex patterns. In the low symmetry case,
solid solution will remove multiplicities or degeneracies, and the
direction of the Δθ shift is quite complex. This is a problem
which is solved by the proper indexing of the powder pattern.
However, for a pattern to be indexed, the components must be known.
In the problem proposed, we do not know the phases present, hence
we cannot index the pattern until we solve the multiphase unknown.

Since we do not know the amount or direction of each Δθ shift
(or even if there will be a degeneracy removed), and we do not know
the number of phases present, the problem seems quite insurmount-
able. We do know, however, one component part - the chemistry -
and have another hope that the substance is present in the Powder
Diffraction File. It is there the assumptions which, if true, will
allow the identification of the unknown phases, even if they are
present with a large amount of solid solution and large changes in
the lattice parameters.

METHOD OF SOLUTION

The method of computer identification is quite different than
the one used with the book indices. In the manual approach, lines
of relatively high intensity are chosen in pairs and these are
used to enter a book index. If an "identification" is made, all
these lines of that phase are marked and the process is repeated
using other relatively intense peaks to hunt for other phases. If
an "identification" fails, other combinations of relatively intense
peaks are chosen and that entire problem is repeated until success
or human exhaustion prevails.

The method of the computer search[8,9] involves the opposite
(inverted) logic. Every standard is compared with the unknown,
with the most intense line of the standard first, and with lines
of lesser intensity in each step. Since this method involves no
subtraction, each phase can be accounted for independently.

Since there is a finite error on both the line position and

and relative intensity of each line, the computer system must be
able to account for these errors. The error or misjudgment in
relative intensity is minimized by the using of $\log_{10} I/I_0$ while
the entire powder diffraction pattern of each standard is stored
in d^* (reciprocal). The errors of measurement in d^* are linear
and the "band-pass" in error tolerance of each line becomes quite
simple.

For the identification of a solid solution phase it is easy
for the computer to allow a larger shift between the measured line
and the position for the standard. In the view for the computer
program, the diffractionist has seemingly made a very poor measurement, but in reality the measurement is good (we hope), and the
line shift has been caused by the solid solution.

This seems simple enough; but as we increase the "band-pass"
around each line of the unknown diffraction pattern, we allow more
and more patterns to fit the criteria for an identification. Let
us look at a table of "band-pass," or windows, for various d
values:

	±2	±4	±10	±20
1 A	±.002Å	±.004Å	±.010Å	±.020Å
2 A	±.008Å	±.016Å	±.040Å	±.080Å
3 A	±.018Å	±.036Å	±.090Å	±.180Å
4 A	±.032Å	±.064Å	±.160Å	±.320Å

It can thus be seen that for a very large window, corresponding to
a large change in lattice parameter with solid solution, we must
reduce the number of possibilities which "fit" the identification
criteria. This criteria is chemistry.

The mode of the running of other Johnson-Vand SEARCH programs
must now be discussed. For major and minor components the logic
is to first check the d-match (how well the lines of the standard
match the unknown). The second is to check the agreement of the
relative intensity of each line of standard with the corresponding
relative intensity line of the unknown. If the standard possesses
these criteria, the chemistry (both positive and negative elements)
and functional groups (positive and negative) are checked. A
standard reference pattern possessing all of these criteria is then
considered in the MATCH section of the program. In the MATCH section of the program the interdependence of phases to make up the
unknown powder diffraction pattern is considered.

In contrast to the above procedure, the solid solution technique introduces the chemical criteria (elemental and groups)
immediately. Since this is such a stringent requirement, very few
standards pass this test. There are now only a small number of
patterns to be viewed, and the d and I matches are not nearly as
selective, since wide tolerance can be placed on them to account

for the shifts in position, and for reduction of intensity which
are characteristic of solid solution.

EXPERIMENTAL

Three solid solutions were standard: ZnS-CdS, CdS-CdTe and Zn,
the olivines between forsterite and fayalite. This paper will
discuss only the results of ZnS-CdS since the conclusions for all
these solid solutions are the same.

The method of growing single crystals for ZnS-CdS has been
previously discussed at this meeting by Chan. Using the technique
which he discussed with the standard Debye-Scherrer powder diffraction method the data were collected.

The instrument used for taking the powder patterns is the
Norelco 114.7 mm diameter powder camera. Other instruments such as
the XRD6 diffractometer, the G.E. powder camera with an effective
circumference of 45.00 cm and the Guinier camera are available[14,15].
For rapid identification, a 57.35 mm diameter Norelco powder camera
was also used[16].

Normally the exposure time was one hour with the target tube
operated at 50 kV and 20 ma. using copper target having the K_β radiation removed by nickel filter. The spectra positions of the exposed films were carefully determined and the intensity of each
spectra line were visually determined by calibrated strips[17].

The entire list of observed lines and intensities for the
series will not be presented here. Rather, a summary of the lattice parameters for these experiments is given as follows:

		a_o	c_o	c
CdS-ZnS	(0% ZnS)	4.135	6.713	1.623
CdS-ZnS	(32.5 ZnS)	4.031	6.566	1.629
CdS-ZnS	(53.0 ZnS)	3.967	6.474	1.632
CdS-ZnS	(68.0 ZnS)	3.920	6.406	1.634
CdS-ZnS	(84.5 ZnS)	3.867	6.330	1.637
CdS-ZnS	(100 ZnS)	3.819	6.260	1.639

COMPUTER SEARCH RESULTS

After preparing different species of ZnS and CdS, x-ray diffraction powder patterns were then taken on a single phase and on
mixtures of two or more phases. The diffraction powder patterns
containing a single phase prepared at ARL checked very closely with
those reported in the ASTM File. A computer search was performed
by the Penn State University's IBM 360/67 on single phase combination with several species to correlate the pattern recognition.

FORTRAN IV Version 10, which has not been published before,
was used in this phase of the work.

The card for this program is as follows:

(A) Title or identification card
(B) Parameter card
(C) Positive elements
(D) Negative elements
(E) Positive Functional Groups
(F) Negative Functional Groups
(G) Data - 8 sets of d and I per card (as many as needed).

The various parameters, which are somewhat involved, as well as the system of punching the cards can be found in FORTRAN IV Version 10, compiled under the auspices of the ASTM Joint Committee on Powder Diffraction Standards. If the user does not specify the desired parameter, there are parameters already written in the program which conform to normal practice.

From the experience of one of the authors (Johnson) with this program, the earlier version proved to be effective if properly used. The use of this program by approximately 80 groups throughout the world has brought comments, suggestions and criticism which were used to improve the later versions. The continued development of the program is being brought about by means of this interaction. The present version not only searches the existing File of Standard Reference Patterns (which can consist of both ASTM and preparatory patterns) but in addition it matches the "best" results with the unknown in d tabular form and attempts to give a relative concentration of each standard in the unknown.

There are four sections of program output:

(1) Input
(2) Output
(3) Report
(4) Subtraction of intensities and scaling factor.

The results on the data presented in this paper prove quite interesting. Although this problem is not difficult to carry out by manual searching technique, the analysis by the computer took less than one minute on an IBM 360/67 at Penn State, and cost less than $5.00. At The University of Michigan, on this dual processor IBM 360/67, the time was 20 seconds with a cost of $4.00. Although the time will vary inversely with the speed of the computer, the cost remains approximately the same, since the speed and cost per second product is usually constant. The cost of $5.00 per complete identification is quite reasonable when we consider the necessary cost of trained technicians and their overhead rate.

The results from the running of the same problem under various windows shows how the solid solution problem is approached. The system CdS-ZnS at the 32.5 mole % ZnS (by weight) was run at windows of ±20, ±10 and ±5. This corresponds to approximately 20, 10 and 5 σ (standard deviations) from the standard reference file without solid solution. The normal window of ±2 means if solid solution were not encountered and a pure phase of the standard were analyzed by the powder method, that 95% of the line analysis would fall within the window.

In order to reproduce computer output, many, many long pages would be required, and if the results were photographically reduced the numbers would be impossible to read. To avoid both of these situations, the important results will be briefly summarized here, and those interested in seeing the entire printout should write to ASTM for a complete 88-page booklet containing these results.

SUMMARY OF RESULTS

The following titles marked with a double star indicate each of the sections of the computer printout.

**Input

The parameters, such as number of line matches, range of experiment, chemistry, error window and measured pattern are as follows:

Line matches	3	
Percentage line matches	33	
D_{max}	3.883	
D_{min}	1.093	
Positive elements	Cd, Zn, S	
Negative elements	OTHERS	
Error window	±10	
Pattern	3.4796	90
	3.2813	100
	3.0841	90
	2.3890	30
	2.0166	70
	1.8536	50
	1.7430	10
	1.7172	50
	1.6865	10
	1.6399	1
	1.5418	5
	1.4860	1
	1.3654	10
	1.3211	5
	1.2945	1
	1.2740	5

```
1.2302        5
1.2299        5
1.1644        5
1.1638        1
1.1311        5
```

**Output

All patterns which pass the chemical tests, and the d and I tests are printed here. The 8 results presented were:

```
4-831        Zn
5-566        ZnS
5-674        Cd
6-314        CdS
8-247        S
10-434       ZnS
10-454       CdS
12-688       ZnS
```

**Report

The entire powder pattern for each of the above results are listed in order of decreasing intensity for viewing by the diffractionist.

**Matchd

The standard reference pattern and the unknown are listed side by side for the comparison of every line of unknown. Thus every line of the unknown should be accounted for from standard reference pattern.

**SubstI

The standard reference patterns are subtracted one by one from the unknown in order of "closeness" of fit. The scaling factor is calculated to put each pattern on the same basis for the subtraction.

The Standard ZnS has scaling factor .202.
The Standard CdS has scaling factor .851.
The Standard Cd has scaling factor .063.
The Standard S has scaling factor   .100.

CONCLUSIONS

By varying the size of the "error window" it is possible to both identify phases and tell the effects of solid solution. When the "error window" is decreased so that a known phase is no longer found, the results automatically present an upper limit on the change in lattice parameters with solid solution. However, the aim of this program is not to displace Vegard's Law but to allow rapid identification of phases present with solid solution. The supplementary use of Vegard's Law to get a lattice constant and indexing of the identified phases presents the most likely use of the program.

Thus it can be seen that the results presented by the computer analysis still must be judged by the investigator. Since the computer cannot make judgments on quality of input data or know information which is not in the data base, the final analysis and conclusion still rests with the researcher. Evidences from other sources such as chemical analysis[16,17] and other physical testing may at times complement the computer search.

ACKNOWLEDGMENTS

The authors wish to acknowledge the help of their colleagues at ARL and Penn State University. The sincere appreciation of the authors is also given to the Joint Committee on Powder Diffraction Standards for the use of their data base for this undertaking.

REFERENCES

1. J. D. Hanawalt, H. W. Rinn and L. K. Frevel, Ind. Eng. Chem., Anal. Ed. 10, 457 (1938).

2. Alphabetical and Grouped Numerical Index of X-ray Diffraction Data, Special Tech. Pub. No. 48B, American Society for Testing and Materials, Philadelphia, Pa. (1950).

3. National Bureau of Standards Monograph 25, Section 6, p. 10, (1968).

4. Fink Inorganic Index to the Powder Diffraction File, ASTM Pub. PDIS-19-f, American Society for Testing and Materials, Philadelphia, Pa. (1969).

5. Inorganic Index to the Powder Diffraction File, ASTM Pub. PDIS-19-i, American Society for Testing and Materials, Philadelphia, Pa. (1969).

6. Organic Index to the Powder Diffraction File, ASTM Pub. PDIS-19-o, American Society for Testing and Materials, Philadelphia, Pa. (1969).

7. KWIC Guide to the Inorganic Patterns of the Powder Diffraction File, ASTM Pub. PDIS-19-k, American Society for Testing and Materials, Philadelphia, Pa. (1969).

8. V. Vand and G. G. Johnson, Jr., Fortran IV Programs (Version 7) for the Identification of Multiphase Unknown Powder Diffraction Patterns, American Society for Testing and Materials, Philadelphia, Pa. (1968).

9. G. G. Johnson, Jr. and V. Vand, A Computerized Powder Diffraction Identification System, Ind. Eng. Chem. 59, 19 (1967).

10. L. K. Frevel, Anal. Chem. 37, 471 (1965).

11. M. Nichols, Twenty-Fourth Pittsburgh Diffraction Conference, Paper No. B-3 (1966).

12. G. G. Johnson, Jr. and V. Vand, Computerized Multiphase X-ray Powder Diffraction Identification System, Denver X-ray Conference, Vol. 11, Advances in X-ray Analysis, pp. 376-384 (1968).

13. F. L. Chan, Determination of Zinc Sulfide and Cadmium Sulfide in Solid Solutions of Small Single Crystals Used for Semiconductors by Xray and Chemical Methods, W. M. Mueller, Editor, Advances in X-ray Analysis, Vol. 5, Plenum Press, New York, 1962, pp. 142-152.

14. F. L. Chan and R. W. Moshin, X-ray Powder Patterns and Index, Metal Compounds with n-Benzoyl - N-Phenylhydroxylamine, WADC Tech. Report 59-533, Sept. 1959.

15. F. L. Chan and L. Sprialter, Perarylated Silanes, Identification by X-ray Diffraction Powder Patterns, WADC Tech. Report 59-512, 1959.

16. F. L. Chan, Some Modifications of the X-ray Instruments and Their Utilization to the Study of Analytical Problems, Norelco Reporter, Vol. X, pp. 133, 1963.

17. H. P. Klug and L. E. Alexander, X-ray Diffraction Procedures for Polycrystalline and Amorphous Material, Chapman and Hall, Ltd., London (1954).

A VERSATILE BRAGG-BRENTANO/SEEMAN-BOHLIN POWDER DIFFRACTOMETER

H.W. King, C.J. Gillham and F.G. Huggins

Imperial College

London, S.W.7 , England

ABSTRACT

An interchangeable Bragg-Brentano:Seeman-Bohlin diffractometer, with variable specimen setting, has been constructed. A study of the influence of the geometry on the resolution, angular displacement and intensity of diffraction profiles clearly demonstrates that the standard Bragg-Brentano method is to be preferred when recording Bragg reflections below $40^{\circ}, 2\theta$. At higher Bragg angles, which permit the specimen to be set further away from the X-ray source, it is demonstrated that the diffracted intensity of the Seeman-Bohlin method can be increased above that of the Bragg-Brentano because the use of an increased divergent angle does not reduce the resolution. A study of the effect of specimen setting on the systematic errors associated with the Seeman-Bohlin geometry indicates that a setting of 60° (half-angle from the X-ray source) gives optimum resolution with minimum displacement error. An extrapolation function to minimise systematic errors is tabulated in an Appendix, and its application to the measurement of precision lattice parameters is demonstrated using the I.U.Cr. specimens of silicon and tungsten.

INTRODUCTION

When using a conventional Bragg-Brentano (B-B) para-focusing X-ray diffractometer, an increase in the intensity of a diffracted beam can only be obtained at the expense of its angular resolution [1,2]. The Seeman-Bohlin (S-B) method, on the other hand, uses a polycrystalline or powdered specimen curved to the diffractometer radius R as shown in figure 1, so that the various Bragg reflections come to a series of sharp foci, D, along the diffractometer circle. With this geometry, the aperture slits can be enlarged to increase the diffracted intensity with little or no loss in resolution. The latter technique also has the advantage that service connections, such as electrical leads, water hoses, vacuum lines, etc., do not interfere with the collection of diffraction data, since the specimen remains stationary while the counter-tube is tracked around the focusing circle.

It is thus rather surprising to find that manufacturers of X-ray diffractometers do not offer S-B attachments as standard optional equipment. The reason for this, no doubt, is that the advantages of the Seeman-Bohlin diffractometer are more than offset by a lack of versatility resulting from the usual practice of mounting the specimen in a fixed position on the diffractometer circle, so that the measurable range of Bragg angles is severely restricted. Thus, if the specimen is located as near as possible to the X-ray source, as in the instruments of Parrish and Mack[3] and Das Gupta et al[4], the physical dimensions of the X-ray tube and counter-tube rotational mechanism limit the scanning range to $23^{\circ} - 108^{\circ}, 2\theta$. This design also suffers from the disadvantage that the alignment of the zero angle position cannot be achieved with certainty, so that standard substances must be used to calibrate the angular scale when precision measurements are required. Alternatively, if the specimen is mounted diametrically opposite the X-ray source, to overcome the need to rotate the X-ray tube from the B-B position and thus assist the zero angle calibration, as proposed by Pike[5] and incorporated into the instrument of Baun and Renton[6], the counter-tube can only scan the back reflection region above $120^{\circ}, 2\theta$, as indicated in figure 1(b). The specimen position can be varied in the S-B instrument described by Wassermann and Wiewiorowsky[7]. This is achieved by rotating the diffractometer bodily about the fixed X-ray source, by a complex rotational and sliding mechanism which appears to make zero point alignment, using the B-B method, very difficult.

To overcome the limitations of existing instruments, a Seeman-Bohlin attachment has been constructed to fit on a standard horizontal diffractometer so that the X-ray specimen can be clamped to a curved dovetail track, running parallel to the focusing circle, at any angular position from 34° to 215° from the X-ray source. The operator thus has the option of choosing a specimen setting to give him the most convenient angular scale to suit a specific problem. The design of the attachment also permits the diffractometer to be readily converted from one geometry to the other, without disturbing the zero point of the angular scale which can thus be accurately aligned by a procedure which makes use of both geometries.

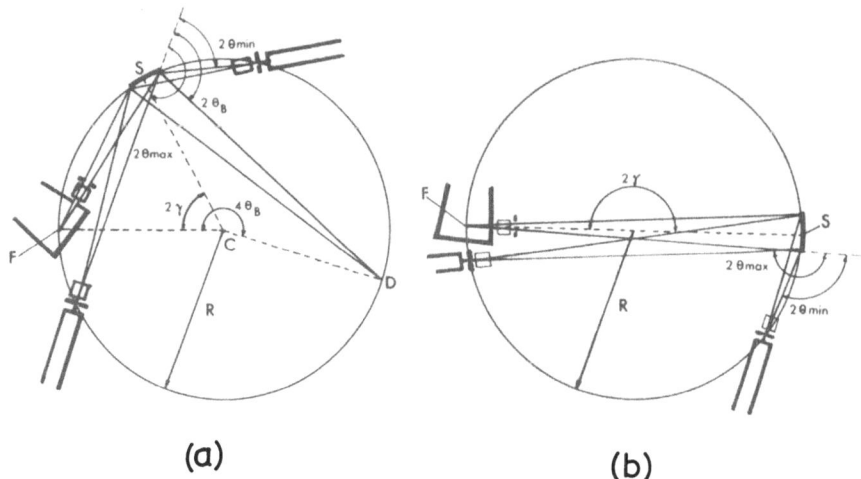

(a)   **(b)**

Figure 1. S-B geometry at various specimen settings

The aim of the present paper is two-fold: (1) to compare the
relative merits of the two types of geometry and (2) to examine
the effect of specimen setting on the geometrical errors and
instrumental misalignments associated with the S-B geometry.
The latter is intended to be a complementary study to an earlier
paper[8], presented at the Denver X-ray Conference in 1962, on the
errors associated with the conventional Bragg-Brentano method.
Following the pattern of the earlier paper, systematic errors
which affect the angular position of S-B reflections are discussed
in terms of the resultant error in the lattice parameter of a cubic
crystal, and values of an extrapolation function to minimise such
errors are tabulated in an Appendix.

AN INTERCHANGEABLE B-B:S-B DIFFRACTOMETER

An attachment to enable a powder diffractometer to be used at
will with either the B-B or the S-B geometry, is shown schematically
in figure 2.   A sealed-off X-ray tube, with vertical line focus
F, is mounted horizontally so that it can be adjusted both longi-
tudinally (ΔF) and transversely (ΔR_F) to align its focus with the
axis of a pivot point set in a base-plate which supports the
entire apparatus.   The X-ray tube can be rotated about this pivot
point, through an angular range of -10° to +90° from the standard
B-B setting, to direct the beam at various positions along the
diffractometer circle.   A modified Wooster-Martin diffractometer
is mounted on a set of parallel slides, A, which enable its zero
angle position to be adjusted with respect to the axis of rotation

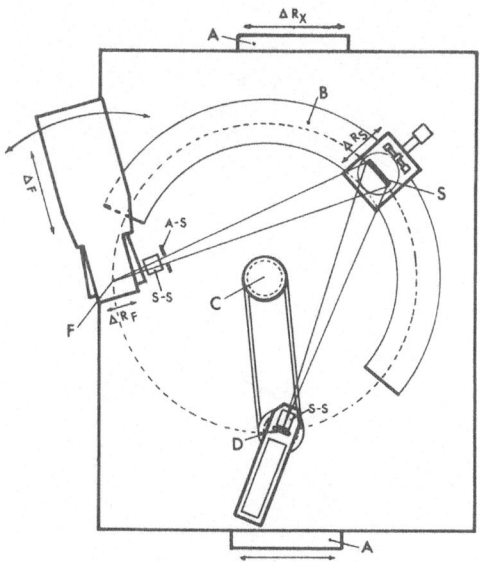

Figure 2.   Schematic diagram of B-B:S-B diffractometer

of the X-ray tube.   The distance from the axis of the diffractometer to the centre of the receiving slit (CD), is fixed at
175 mm and this defines the radius of the focusing circle of the
S-B geometry.   The specimen S is mounted on a curved dovetail
track, B, which lies concentric with the focusing circle.   The
specimen holder can be clamped to this track at any angular posi-
tion, 2γ, between 34° and 200° from the X-ray source, and the specimen surface can be aligned coincident with the focusing circle with
the aid of transverse and rotational micrometer adjustments.

A unique feature of the attachment is that the rotational
mechanism which maintains the counter-tube directed towards the
centre of the specimen, is a 1:1 direct chain drive from a gear
locked on to the θ drive of the standard B-B diffractometer.   Thus
when it is required to alter the setting of the specimen, the X-ray
tube is rotated through the required angle and the rotational setting
of the counter-tube is simply re-adjusted by locking the chain drive
at a new position with respect to the θ drive.   To convert the
instrument back to B-B geometry, the gear is removed from the θ
drive, and replaced by the standard flat specimen holder, while
the counter-tube is directed towards the centre of the diffractometer and locked in this position with a dowel pin.   It is not
necessary to remove the dovetail track during this operation, as
it does not interfere with the scan of the counter-tube.   When
both the zero angle and the X-ray line focus have been accurately
aligned to coincide with the axis of rotation of the X-ray tube,

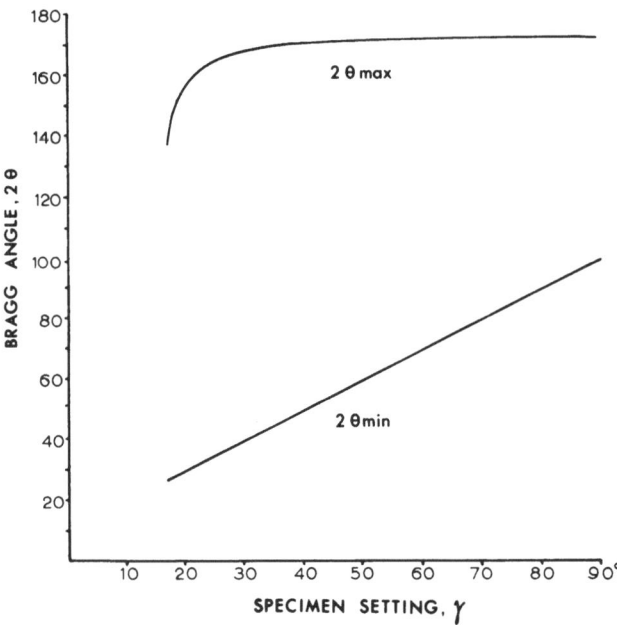

Figure 3. Maximum and minimum measurable angles with
S-B geometry, as a function of specimen setting, γ

the specimen setting or the geometry can be altered in matter of
minutes.

Reference to figure 1 shows that the angle FCD, subtended at
the centre of the diffractometer by the X-ray source and the centre
of the receiving slit, is equal to four times the Bragg angle θ_B,
regardless of where the specimen is positioned along the focusing
circle.   Since the angular scales of conventional B-B diffractometers are calibrated in $^{\circ}2\theta$, with the zero angle position diametrically opposite the X-ray source, the Bragg angle of a diffraction profile recorded with a S-B attachment is given by the
relationship:

$$4\theta_B \quad = \quad 180^{\circ} \quad + \quad 2\theta' \tag{1}$$

$$\text{or} \quad 2\theta_B \quad = \quad 90^{\circ} \quad + \quad \theta' \tag{2}$$

where $2\theta'$ is the angle indicated on the B-B diffractometer scale.
If however, the B-B diffractometer is equipped with an independent
specimen drive (omega drive) and thus provided with a second odo-
meter which measures θ', the Bragg angle $2\theta_B$ can be obtained directly by re-setting the θ' scale to read 90° when the $2\theta'$ scale
reads zero, in which case $\theta' = 2\theta_B$.   Since all analyses of

systematic errors are expressed as trigonometric functions involving 2θ, and this angle can be read directly with the present diffractometer, all further references to Bragg angles are given in $^{\circ}2\theta$. For the same reason, specimen settings are expressed in terms of γ, defined as half the angle FCS, subtended at the centre of the diffractometer by the X-ray source and the centre of the specimen.

When using S-B geometry the range of measurable Bragg angles depends primarily on the specimen position and secondarily on the physical dimensions of the X-ray tube, specimen holder and counter-tube rotational mechanism. The maximum and minimum $2\theta_B$ angles which can be measured with the present instrument, for various specimen settings γ, are shown in figure 3. The lowest measurable Bragg angle is $28^{\circ},2\theta$, which means that d-spacings greater than 3.2 Å cannot be measured with CuK_{α} radiation. At the other end of the scale Bragg angles can be measured up to $172.5^{\circ},2\theta$, over a wide range of γ settings from 55° - 90°. This versatility of angular scale follows as a direct result of the design of the counter-tube rotational mechanism. Apart from the prime need to measure a specific Bragg reflection, the factors which govern the selection of the optimum specimen position, or the particular form of geometry, are concerned with the intensity, resolution and positional accuracy of the diffraction profile, and these are considered in detail in the following section.

SOURCES OF ERROR IN THE S-B METHOD

It is a general rule that the more sensitive a technique, the more it is prone to incidental errors. The X-ray diffractometer method is no exception to this rule, but has the inherent advantage that the various sources of error can be readily identified and, where necessary, studied both analytically and experimentally. Thus, in order to obtain the maximum benefit from the use of S-B geometry, an operator should be fully aware of the sources and significance of errors, and be familiar with the methods available for their minimization, or complete elimination. The relevant sources of error may be conveniently summarized under the following headings:

1. Errors due to Physical Effects

    a. Errors in definition of wavelength units
    b. Refraction
    c. Filters and monochromators
    d. Dispersion, Lorentz factor and polarization

2. Errors Arising from S-B Geometry

    a. Width of X-ray source and take-off angle

    b. Aperture slits

    c. Axial divergence

    d. Specimen curvature, surface roughness and particle size

    e. Absorption by specimen and air path

3. Errors Inherent in the Instrument

    a. Eccentricity of the main gear

    b. Backlash in gears

    c. Rotational following of the counter-tube

    d. Eccentricity of receiving slit

    e. Eccentricity of specimen track

    f. Diffractometer slides not parallel to $0°-180°$ line.

4. Errors Associated with Alignment

    a. X-ray beam not in diffractometer plane

    b. $0°$ point displaced from X-ray tube axis

    c. X-ray line focus displaced from rotational axis

    d. Rotational mis-setting of X-ray tube

    e. Specimen surface displaced from focusing circle

    f. Rotational mis-setting of specimen surface

    g. Tilt of specimen surface

    h. Rotational mis-setting of receiving slit

A number of the errors due to physical effects and instrumentation are common to both the B-B and S-B methods and have thus been discussed at some length in standard review articles on the diffractometer technique[8,9]. Systematic errors arising from S-B geometry have been rigorously analysed by Segmuller[10] and by Kunzel[11] and studied experimentally at low γ angles by Parrish and Mack[3]. This material will not be repeated except to emphasize the role of the specimen setting and to draw the reader's attention to the pertinent references. Errors arising from alignment, however, have not received as much attention and are thus discussed in some detail and, where necessary, demonstrated experimentally.

Errors Due to Physical Effects

The effect of refraction[12] and of the use of filters[13] and monochromators[14] is to cause an effective change in the wavelength of the characteristic X-radiation. Since these sources of error are independent of the geometry or technique, they are best eliminated by applying correction factors to the experimentally determined lattice constants at the time when the results are corrected to standard units of wave length[15,16].

The combined effects of dispersion, Lorentz factor and polarization cause a shift in the centroid of a diffraction profile, which has $\tan^3\theta_B$ dependence when using B-B geometry[17], and thus becomes critical at Bragg angles greater than $155°, 2\theta$ [8].   The shift in the peak of the profile due to these physical effects is uncertain, but it is generally considered to be significantly smaller than that of the centroid.   Because of the obvious importance of these errors when using S-B geometry, which enables Bragg angles to be measured up to $172°, 2\theta$, they are the subject of a special study to be published later[18].

Errors Arising from S-B Geometry

The size of the X-ray source is a fundamental parameter, since the angular aperture of its effective width governs the breadth of the diffraction profile and hence the ultimate angular resolution of the technique.   The effective source width W_F is a function of the take-off angle ψ and is given by the following expression:

$$W_F = W_o \cdot \sin \psi \qquad (1)$$

where W_o is the true width of the focal spot.   In the B-B method the source-specimen distance is constant and hence the angular aperture E_{B-B} of the source is given by:

$$E_{B-B}(°2\theta) = 2 \text{ arc tan } W_F/2R \qquad (2)$$

for all diffraction angles.   In the S-B method, the source-specimen distance varies with specimen setting angle γ, as shown in figure 1, and the angular aperture of the source is now given by:

$$E_{S-B}(°2\theta) = \text{arc tan } W_F/(2R \sin \gamma) \qquad (3)$$

A comparison of equations (2) and (3) reveals that for a given X-ray tube and take-off angle, the source aperture of the S-B diffractometer becomes smaller than that of the B-B instrument for γ greater than $30°$, and falls to half $E_{B-B}(°2\theta)$ as $\gamma \to 90°$. The S-B diffractometer is thus intrinsically capable of twice the angular resolution of the B-B instrument, in the back reflection region.

A direct consequence of the fully-focusing S-B geometry is that the width of the divergent slit does not affect the angular resolution of the diffracted beam, provided of course, that the divergence does not exceed twice the take-off angle ψ.   The angular divergence, α, of this slit is given by:

$$\alpha = 2 \text{ arc tan } W_{DS}/2a \qquad (4)$$

where W_{DS} is the width of the divergent slit and a its distance from the X-ray source. The divergence slit controls ℓ, the length of the specimen irradiated by the X-ray beam, since

$$\ell = 2\alpha R \tag{5}$$

and thus has a direct influence on the intensity of the diffracted beam. It is important to note that ℓ is independent of the specimen setting γ, and hence the aperture of the divergent slit can be set to irradiate the entire length of the curved specimen and remains unaltered throughout a scanning experiment.

In the present S-B diffractometer the receiving slit is maintained normal to the central diffracted ray, as the counter-tube is scanned around the focusing circle. Although the width of this slit remains constant, the specimen-slit distance changes continuously during scanning and the angular aperture of the slit ε_{S-B} is thus a function of both the Bragg angle $2\theta_B$ and the specimen setting γ, i.e.

$$\varepsilon_{S-B}(^{O}2\theta) = \arctan \frac{W_{RS}}{2R \sin(2\theta_B - \gamma)} \tag{6}$$

The equivalent expression for the angular aperture of the receiving slit on a B-B diffractometer is

$$\varepsilon_{B-B}(^{O}2\theta) = 2 \arctan W_{RS}/2R \tag{7}$$

A comparison of these equations shows that for a given slit width W_{RS}, $\varepsilon_{S-B}(^{O}2\theta)$ becomes equal to $\varepsilon_{B-B}(^{O}2\theta)$ when $2\theta_B = \gamma + 30^{\circ}$, and falls to one half $\varepsilon_{B-B}(^{O}2\theta)$ when $(2\theta_B - \gamma) = 90^{\circ}$; i.e. when the counter-tube lies diametrically opposite the specimen. Thus the high resolution of the S-B instrument, derived from the γ-dependence of the X-ray source aperture (equation 3) can only be utilized over certain ranges of $2\theta_B$. Since the integrated intensity of a diffraction profile is also proportional to the angular aperture of the receiving slit, it also follows that relative intensity measurements at different Bragg angles should be corrected for ε_{S-B} changes, as discussed by Kunze[11].

Errors due to axial divergence were not included in the analyses of Segmuller[10] and Kunze[11]. Pike's analysis[19] of the equivalent effect for B-B geometry indicates that profile shifts should occur at high and low Bragg angles. A similar analysis for the S-B geometry is now in progress.

An error in the curvature of the specimen surface causes an

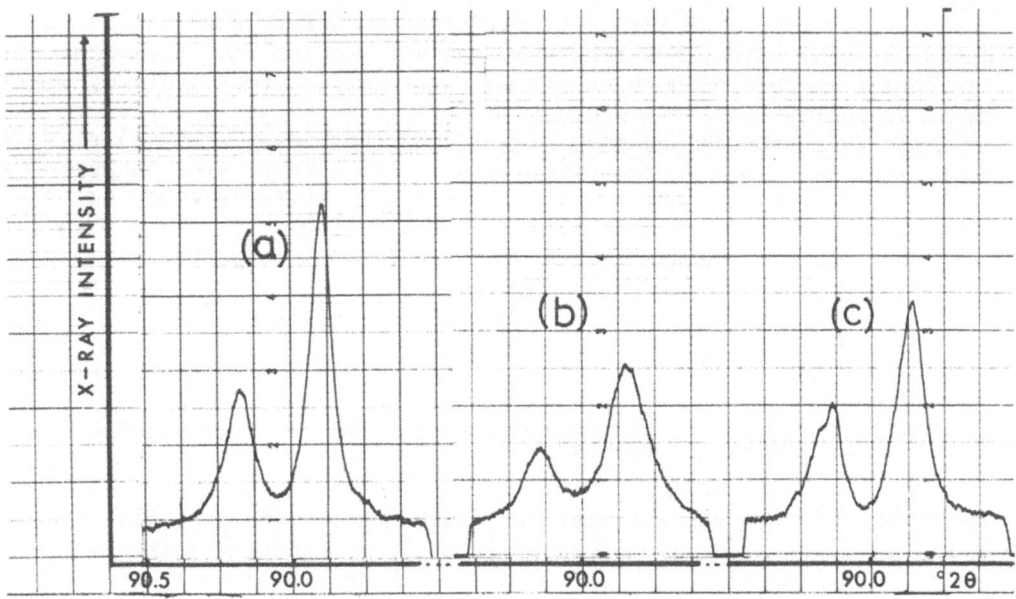

Figure 4.   311 S-B profiles from polycrystalline copper.
(a) R = 175mm, (b) R = ∞, (c) R = 125mm.

asymmetrical broadening of the diffraction profile.   The peak of
the profile is relatively unaffected by this error, but a significant shift occurs in the centroid.   According to Segmuller[10] and
Kunze[11] the centroid shift is related to the angular aperture α of
the divergent slit, the specimen setting γ and the Bragg angle 2θ,
according to the following relationship:

$$\Delta 2\theta_{\text{(curv.)}} = \frac{\alpha^2}{3} \frac{\sin 2\theta}{\sin \gamma.(\sin 2\theta - \gamma)} \tag{8}$$

As shown in the figure 4(b) a flat specimen causes appreciable
broadening, but a small error in curvature, figure 4(c), has much
less effect on the resolution.   Hence when measuring $2\theta_B$ from
peak positions, the exact radius of curvature of the powdered
specimen is not critical, provided it approximates to R.

Parrish and Mack[3] have pointed out that large particle size
or surface roughness of the specimen may cause a loss of diffracted
intensity at low values of γ and $2\theta_B$ because of the shallow inclination of the incident and diffracted beams, but these factors should
not cause a shift in the profile position.

The absorption of X-rays by different materials also introduces an asymmetrical broadening to the diffraction profile. The
broadening is more pronounced the more transparent the specimen,
as layers below the surface can then contribute to the diffraction
profile, the centroid and peak of which are both displaced towards
lower angles. This source of error has also been studied analytically by Segmuller[10] and by Kunze[11], who give the following two
expressions for the centroid shifts, one for highly absorbent thick
specimens ($\mu t \rightarrow \infty$) and another for transparent thin specimens
($\mu t \rightarrow 0$).

$$\Delta 2\theta_{(\mu t \rightarrow \infty)} = \frac{1}{2\mu R} \frac{\sin 2\theta}{\sin \gamma + \sin(2\theta - \gamma)} \tag{9}$$

$$\Delta 2\theta_{(\mu t \rightarrow 0)} = \frac{t}{4R} \frac{\sin 2\theta}{\sin \gamma \cdot \sin(2\theta - \gamma)} \tag{10}$$

Absorption by the specimen also decreases the integrated intensity
of the profiles. Parrish and Mack[3] have combined the available
analyses of this effect to derive a comprehensive correction factor,
which also takes into account the effects of air scatter and the
chord length of the receiving slit.

Errors Inherent in the Instrument

Errors arising from inaccurate machining, backlash or calibration of the main or subsidiary gears of the B-B diffractometer can
be measured by calibrated polygons[20]. The techniques available
for eliminating or minimizing these errors in the B-B diffractometer[8] apply equally to the S-B attachment. The significant
difference between the two instruments is that the $2\theta{:}\theta$ following
mechanism of the B-B diffractometer is now used to maintain the
counter-tube directed towards the centre of the stationary curved
specimen. This possible source of error in the S-B technique
turns out to be insignificant, since as demonstrated in the following section, a counter-tube rotational error of $\pm 10^{0}$ can be
tolerated without causing a peak or centroid shift greater than
$\pm 0.005^{0}, 2\theta$. The latter experiment will also determine any eccentricity of the rotational axis with respect to the receiving slit.

A manufacturing error unique to the S-B diffractometer is that
the curved specimen track may not be concentric with the focusing
circle, so that as the specimen is set at different positions along
the track its surface will be systematically displaced from the
focusing circle. As discussed later, part of the alignment procedure for the S-B diffractometer involves adjusting the X-ray
source until a Bragg reflection appears at the same angle ($\pm 0.005^{0}$,
2θ) for all settings of the specimen. The presence of a track

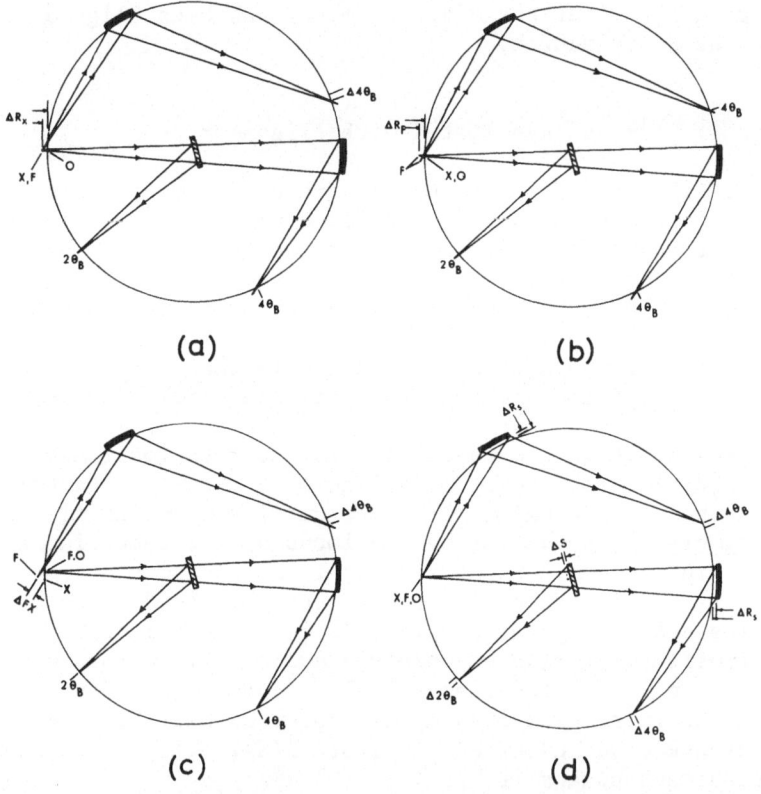

Figure 5. Alignment errors associated with S-B geometry
(a) ΔR_X $(X=F{\neq}0^o)$ (b) $\Delta R_F \Delta (X=0^o{\neq}F)$
(c) $\Delta F X$ $(F=0^o{\neq}X$ at $\gamma = 90^o)$ (d) ΔR_S $(F=X=0^o)$
        $(F{\neq}0^o{\neq}X$ at $\gamma < 90^o)$

Note: (1) ΔF is a general translation of the X-ray tube;
       (2) $\Delta X(F=X{\neq}0^o)$ causes an equal shift of all angles.

eccentricity error will thus be indicated by an inability to align
the instrument. In this event the eccentricity of the track
should be checked with an optical technique, and successive adjustments made to its setting so that the alignment procedure can be
effected.

     In theory, another possible source of instrumental error will
arise if the alignment slides of the diffractometer (A in figure 2)
do not lie parallel to the 0° - 180° diameter of the focusing circle.
Such an error is overcome in practice by re-setting the zero of the

B-B angular scale, after the diffractometer has been aligned with respect to the X-ray source.

Errors Associated with Alignment

A small tilt, ρ, of the focusing circle of the diffractometer with respect to the plane of diffraction causes some loss in intensity, but contributes a negligible error to the angular measurements and hence the instrument may be levelled adequately with a bubble-gauge. The various adjustments required to bring the instrument into alignment in the plane of diffraction are illustrated in figure 5, for various specimen settings. In the following discussion, it will be assumed that these mis-alignments occur individually.

The fixed point on the instrument is the rotational axis of the X-ray tube, referred to as X in figure 5. The diffractometer must be adjusted on its parallel slides until this point lies on the focusing circle traced out by the locus of the receiving slit. An error ΔR_X in this adjustment results in de-focusing of the divergent X-ray beam for the B-B diffractometry, leading to broadened profiles but no angular displacement[8,9]. A symmetrically broadened profile will also be observed due to this error, when the S-B specimen is located diametrically opposite the source at $\gamma = 90^\circ$, but an increasing angular displacement will be observed as γ is decreased, as indicated in figure 5(a). It can be shown that the resultant peak shift is

$$\Delta 2\theta = \frac{\Delta R_X}{2 R} \cot \gamma \qquad\qquad (11)$$

A tangential displacement, ΔX, of the zero angle position from the point X, in an otherwise well aligned instrument, is equivalent to a rotational error in the position of the main gear and hence causes all diffraction angles to be in error by the same amount. It can thus be eliminated, as the last step in alignment, by re-calibrating the zero angle position.

A displacement of the X-ray source along the direction of the primary beam of the B-B geometry ΔR_F again causes a de-focusing of the B-B diffracted X-ray beam leading to a symmetrically broadened profile, the effect being indistinguishable from ΔR_X, for this geometry. Reference to figure 5(b) shows that a similar broadening, i.e. without an angular shift, will also occur for all settings of the specimen when using S-B geometry. A displacement of the X-ray source tangentially to the focusing circle, ΔFX, on the other hand causes a significant shift in the position of a B-B diffracted profile, but does not cause it to become broadened (figure 5(c).

TABLE I

Theoretical and Observed Peak Shifts for ΔR_S = 1.25 mm, for various Specimen Settings, γ.

γ^0	$\Delta 2\theta^0$ Obs.	$\Delta 2\theta^0$ Theor.
20	0.653	0.647
25	0.528	0.538
35	0.434	0.437
45	0.390	0.409
55	0.446	0.432
65	0.535	0.523
70	0.628	0.617
75	0.790	0.781
80	1.103	1.106

When using this geometry, the tangential displacement of the focal spot from X, cannot be distinguished from an equivalent displacement of the zero point, i.e. $\Delta FX \equiv \Delta X$, and the profile shift may thus be used as the basis of a precision method for aligning the X-ray source at the 180^0 point of the B-B angular scale[21].   A displacement ΔFX causes a similar profile shift for $\gamma = 90^0$ in the S-B geometry, but at lower γ angles the shift is accompanied by broadening of the profile, as shown in figure 5(c).   The equivalent peak shift is given by:

$$\Delta 2\theta = \frac{\Delta FX}{2R \sin \gamma} \tag{12}$$

A rotational mis-setting of the X-ray tube $\Delta \phi$ causes a change in diffracted intensity, for a narrow specimen, because of an effective change in take-off angle[3].   The intensity is also affected when the narrow divergent beam does not irradiate the entire specimen. In the latter event, an angular shift may also be observed, if the specimen surface is not concentric with the focusing circle, and thus provides a sensitive technique for checking this alignment of the specimen ($\Delta \tau$).

A displacement, ΔR_S, of the specimen surface from the focusing circle causes an asymmetrical broadening of the diffraction profile, As shown in figure 5(d), both the peak and the centroid of the profile are displaced, the direction of the shift depending on whether the specimen lies inside or outside the focusing circle and its magnitude increasing with decreasing γ.   This source of error was also analysed by Segmuller[10] and Kunze[11], when considering the other

geometrical errors of the S-B method, and the dependence of the
profile shift on Bragg angle and specimen setting was found to be
similar to that for the flat specimen and transparency errors, i.e.

$$\Delta 2\theta_{Spec.} = \pm \frac{\Delta R_S}{2 R} \cdot \frac{\sin 2\theta}{\sin \gamma \cdot \sin(2\theta - \gamma)} \qquad (13)$$

This analysis has been confirmed for various γ settings as shown
in table I, which gives the observed and theoretical peak shifts
associated with ΔR_S = 1.25 mm, for the 132 profile ($\sim 90°, 2\theta$) of
permaquartz.

A rotational mis-setting, τ, of the specimen surface with
respect to the focusing circle causes a broadening of the diffraction profile but no angular displacement, provided the instrument
is otherwise well aligned. A tilt of the specimen surface with
respect to the plane of diffraction has much the same effect as a
tilt in the focusing circle, discussed above.

It can be easily shown, by direct experiment, that no change
in the shape or position of a profile is observed for rotational
mis-settings of the counter-tube up to $\Delta\sigma = \pm 10°$. Beyond this
limit, the diffracted beam is abruptly extinguished by the bevelled
edge of the slit or housing. It is thus quite unnecessary to
design a sophisticated drive mechanism for this rotation, particularly as this tends to restrict the scanning range of the counter-
tube[3,5].

DISCUSSION

Alignment Procedure for B-B:S-B Diffractometer

A study of the differing responses of the diffraction profile
to the various alignment errors, and their dependence on geometry
and/or specimen setting, has led to the following procedure for
progressively aligning the combined B-B:S-B diffractometer. The
alignment effected at each step is indicated by the symbols in
brackets at the end of each instruction.

1. The radius of the focusing circle is fixed at 175 mm by the
distance between the centre of the diffractometer and the axis of
rotation of the receiving slit. The diffractometer is set with
its axis of rotation ~175 mm from the axis of rotation of the X-ray
tube, using a scale attached to the base-plate, (ΔR_X).

2. With the B-B $2\theta'$ scale set at zero and a 0.05 mm receiving slit

pointed accurately towards the axis of the goniometer, using the dowel pin which locks the rotational movement of the receiving slit/counter-tube bracket:- the receiving slit assembly is adjusted into co-linearity with two needles previously set accurately on the axes of rotation of the X-ray source and diffractometer, thereby establishing that the axis of rotation of the X-ray tube lies approximately on the 0° - $180^o,2\theta'$ diameter of the B-B diffractometer, ($\sim\Delta X$).

3. With a CuK$_\alpha$ X-ray tube inserted in the holder and the take-off angle set at 6°, an $\alpha = 1^o$ divergent beam is directed towards the centre of a standard flat gold specimen mounted in the B-B holder and aligned for $2\theta:\theta$:- the X-ray source is adjusted tangentially (ΔF) until the 111 peaks ($\sim39^o,2\theta$) of the specimen become superimposed when scanned on either side of the beam[21].   A slight rotation $\Delta\phi$ of the X-ray tube is required after each ΔF adjustment to re-direct the beam towards the centre of the diffractometer, ($\sim\Delta FX$).

4. The S-B specimen holder is mounted at the selected γ position on the curved track and the receiving slit is aligned normal to the central diffracting ray by setting the counter at $2\theta_B=(\gamma+90^o)$ and locking the chain drive to the drive shaft, before removing the dowel pin which locks the receiving slit/counter-tube rotational movement, ($\Delta\tau$).

5. Using a sample of G.E. Permaquartz, with its surface accurately ground to a radius of 175 mm, in the S-B specimen holder:- adjustments of the type ΔF and ΔR_X are made alternately until the 132 profile ($\sim90^o,2\theta$), scanned at specimen settings of $\gamma = 20^o$ and 70°, become superimposed, and maximum peak intensity is obtained.   From equations 8, 9, 10 and 13 all systematic errors are equal for these two γ settings, when measuring a profile at $90^o,2\theta$.   During this procedure the specimen is located in accurately machined and dowelled positions, to ensure that no relative ΔR_S errors occur when moving the specimen between the two settings, (ΔFX and ΔR_X).

6. ΔR_S errors are now eliminated by making adjustments until the 132 permaquartz peak occurs at the same angle for all settings of γ, (ΔR_S).

7. Using an $\alpha = 0.25^o$ aperture slit:- the angular rotation τ of the specimen is adjusted until no peak shifts are observed when the beam is directed at different parts of the specimen, ($\Delta\tau$).

8. Transverse adjustments to the X-ray tube of the type ΔR_F are made until maximum peak intensity is obtained from the 132 permaquartz reflection, (ΔR_F)

9. Procedures 5-8 are reiterated until the 132 profile occurs at the same angle over the whole range of γ.   If the profiles match, the instrument is aligned within the required accuracy.

10. The movement ΔF in steps 5 or 9 disturbs the zero angle setting
of the B-B method, which was aligned in step 3 on the assumption that
ΔX = zero. This is corrected by repeating step 3 but re-setting
the zero angle on the calibrated drum attached to the end of the
main worm gear, instead of altering ΔF, to bring the zero angle
position into coincidence with the rotational axis of the X-ray
tube. This procedure also calibrates the zero angle of the S-B;
scale. Since this alters the indicated angular positions of all
diffraction profiles by a fixed amount, the alignment achieved in
steps 5-8 is not affected, (ΔX).

    Failure to align the instrument after several cycles of steps
5-8 indicates that the eccentricity of the specimen track should
be checked, or that the specimen has too high a transparency.
Once the alignment has been accurately established, the specimen
setting or geometry can be altered without error, since the rotational movements involved in this operation Δϕ and Δτ, do not influence
the angular positions of diffraction profiles.

                    Selection of Instrumental Settings

    The selection of the peak position or the centroid of a pro-
file as the indicator of the Bragg angle, follows the same reason-
ing as in the B-B method. In general, the peak is easier to
measure and less prone to error, but the centroid is more amenable
to analysis[1,2,9].

    The resolution, and hence the ultimate accuracy, of a dif-
fractometer measurement of a Bragg angle is governed by the angular
aperture of the X-ray source. To take full advantage of this
resolution, a receiving slit of the same angular aperture should
be selected. This presents no problem for the B-B method[1,2], but
when the S-B geometry is used, the source aperture is found to vary
with specimen setting (equation 3) while the receiving slit aperture
depends on both $2\theta_B$ and γ (equation 6). To provide a direct comparison between these two quantities, the source apertures of standard 2Kw, 1Kw and fine-focus X-ray tubes for various take-off
angles and specimen settings are listed in table II, while the
angular apertures for a standard 0.1 mm receiving slit, for different specimen settings and Bragg angles are given in table III,
for R = 175 mm. Also for purposes of comparison, the equivalent
values of $E_{B-B}(^{o}2\theta)$ are included in table II and horizontal lines
are drawn in table III to indicate the position of the equivalent
value of $\varepsilon_{B-B}(^{o}2\theta)$. Since the arc tan function is equal to the
argument at very small angles, the angular apertures of other standard slits such as 0.05 mm, 0.2 mm and 0.4 mm can be obtained by
simply multiplying the values in table III by a factor 0.5, 2.0 or
4.0, respectively. Both tables may also be used to calculate the

TABLE II

Angular Aperture of X-ray source $E_{S-B}(^{o}2\theta)$, as a function
of Source Width W_o, Take-off angle ψ, and specimen
Setting γ, for R = 175 mm

W_o (mm)	ψ	$E_{S-B}(^{o}2\theta)$ at various values of						E_{B-B} $(^{o}2\theta)$
		15°	30°	45°	60°	75°	90°	
2.0	3°	0.066	0.034	0.024	0.020	0.018	0.017	0.034
	6°	0.132	0.068	0.048	0.040	0.036	0.034	0.068
1.0	3°	0.033	0.017	0.012	0.010	0.009	0.009	0.017
	6°	0.066	0.034	0.024	0.020	0.018	0.017	0.034
0.4	3°	0.013	0.007	0.005	0.004	0.004	0.003	0.007
	6°	0.026	0.014	0.010	0.008	0.007	0.007	0.014

source and slit apertures for S-B attachments mounted on commercial
diffractometers of similar radius, e.g. Hilger and Watts (180 mm),
Philips (170 mm) and Siemens (172.5 mm).

The proper receiving slit for maximum resolution is thus sel-
ected by scaling the appropriate ε_{S-B} (0.1 mm, $^{o}2\theta$) value given in
table III so that it equals the experimental source aperture derived
from table II. For example for W_o = 1 mm, ψ = 6° and a specimen
setting γ = 60°, the source aperture from table II is 0.02^o,2θ,
and the ε_{S-B} (0.1 mm, $^{o}2\theta$) values in table III are found to be
0.094^o, 0.033^o, 0.019^o and 0.017^o for Bragg angles occurring at
70°, 90°, 120° and 160°,2θ, respectively. The correct width of
receiving slit for precision measurements of these Bragg angles
should thus be 0.02 mm at 70°, 0.06 mm at 90° and 0.1 mm at 120°
and 160°,2θ. Larger slit widths would, of course, be used for
less precise measurements, such as a rapid scan to locate the
approximate positions of the profiles.

The variation in the aperture of the receiving slit also provides a complication when selecting the appropriate time constant
for the electronic smoothing circuits of the rate meter. For
minimum distortion of a diffraction profile, the electronic time
constant, in seconds, is set at one half the time width of the
receiving slit, which must also be expressed in seconds[1,2]. The
latter quantity depends on the scanning speed ω, which is measured
in $^{o}2\theta$,/min (or perhaps $^{o}4\theta$/min). Thus, to obtain the required
time width of the slit, its angular aperture $\varepsilon_{S-B}(^{o}2\theta)$ = scaling
constant ε_{S-B}(0.1 mm, $^{o}2\theta$), should be multiplied by 60/ω and hence

TABLE III

Angular Aperture of 0.1 mm Receiving Slit $\varepsilon_{S-B}(0.1\ mm, ^{\circ}2\theta)$ as a Function of Bragg Angle $2\theta_B$ and specimen setting γ, for $R = 175$ mm.

$2\theta_B$	$_{S-B}$ (0.1 mm $^{\circ}2\theta$) for various values of γ					
	15°	30°	45°	60°	75°	90°
20°	0.188					
30°	0.063					
40°	0.039	0.094				
50°	0.029	0.048	0.188			
60°	0.023	0.033	0.063			
70°	0.020	0.025	0.039	0.094		
80°	0.018	0.021	0.029	0.048	0.188	
90°	0.017	0.019	0.023	0.033	0.063	
100°	0.016	0.017	0.020	0.025	0.039	0.094
110°	0.016	0.017	0.018	0.021	0.029	0.048
120°	0.017	0.016	0.017	0.019	0.023	0.033
130°	0.018	0.017	0.016	0.017	0.020	0.025
140°	0.020	0.017	0.016	0.017	0.018	0.021
150°	0.023	0.019	0.017	0.016	0.017	0.019
160°	0.029	0.021	0.018	0.017	0.016	0.017
170°	0.039	0.025	0.020	0.017	0.016	0.017

Notes: 1. The horizontal lines indicate the position of the equivalent value of $\varepsilon_{B-B}(0.1mm, ^{\circ}2\theta) = 0.033^{\circ}$.

2. To obtain $\varepsilon_{S-B}(^{\circ}2\theta)$ for other slit sizes multiply by $10 \times W_{RS}(mm)$.

the proper time constant is given by

$$\text{Time constant (S-B)} = \varepsilon_{S-B}(^{\circ}2\theta) \cdot 30/\omega(^{\circ}2\theta/min) \qquad (14)$$

Thus for $\omega = \frac{1}{8}^{\circ}2\theta/min$, the proper time constant for all the receiving slits in the example quoted above is 4.8 sec.

N.B.   The above equations, and all previously quoted analyses of systematic errors are based on the direct measurement of Bragg angles in terms of $°2\theta$,.   If this facility is not available and the Bragg angle is measured in $°4\theta$ using equation (1), the error functions in equations (8-13) and the source and receiving slit apertures in tables II and III should all be multiplied by a factor of x 2 to convert them into $°4\theta$.   When using $°4\theta_B$, it should also be noted that indicated B-B scanning speeds given in $°2\theta$, represent $°4\theta$ for the S-B geometry.   Thus both ε_{S-B} and ω in equation (14) must be doubled, when working in $°4\theta$, which leaves the numerical value of the time constant unaltered.

Selection of Optimum Geometry and Specimen Setting

The availability of an interchangeable Bragg-Brentano/Seeman-Bohlin diffractometer poses the practical problem of selecting the appropriate geometry and specimen setting to obtain maximum resolution and intensity, with minimum angular displacement, when recording a specified diffraction profile.

It is evident from the data listed in table II that the basic angular resolution of the S-B diffractometer falls below that of the B-B instrument at specimen settings below $\gamma = 30°$.   Since the minimum measurable Bragg angle ($°2\theta$) at any specimen position is equal to $(\gamma + 10°)$ (figure 3), it follows that all measurements of $2\theta_B$ less than $40°$ are subject to a loss of resolution, due to the increased angular aperture of the X-ray source.   Thus at low Bragg angles the "doubled" $2\theta_B$ scale associated with the S-B geometry is really empty magnification, since the profiles are broader than those recorded with the B-B diffractometer.   Further, as demonstrated by Parrish and Mack[3], at low Bragg angles the S-B geometry is more susceptible than the B-B method to systematic errors due to wrong curvature, absorption and specimen displacement.   At $\gamma = 15°$ the systematic errors are greater than the equivalent B-B errors [22], over the entire range of $2\theta_B$.   The cross-over point between the two methods occurs at $\gamma = 30°; 2\theta_B = 60°$.

The relatively short source-specimen and specimen-receiving slit distances, required by the S-B method at low $2\theta_B$ values, also cause intensity losses due to micro-absorption arising from surface irregularities and particle size roughness[3].   Thus on all three accounts, of resolution, angular error and diffracted intensity, the Bragg-Brentano diffractometer is definitely to be preferred for precision work at $2\theta_B$ values below $60°$.   This conclusion is much the same as that reached by Parrish and Mack[3], on the basis of their experience with a S-B diffractometer with γ fixed at $15°$.   It does not, of course, rule out the use of the S-B diffractometer for qualitative studies at low angles, e.g. when indexing a known pattern

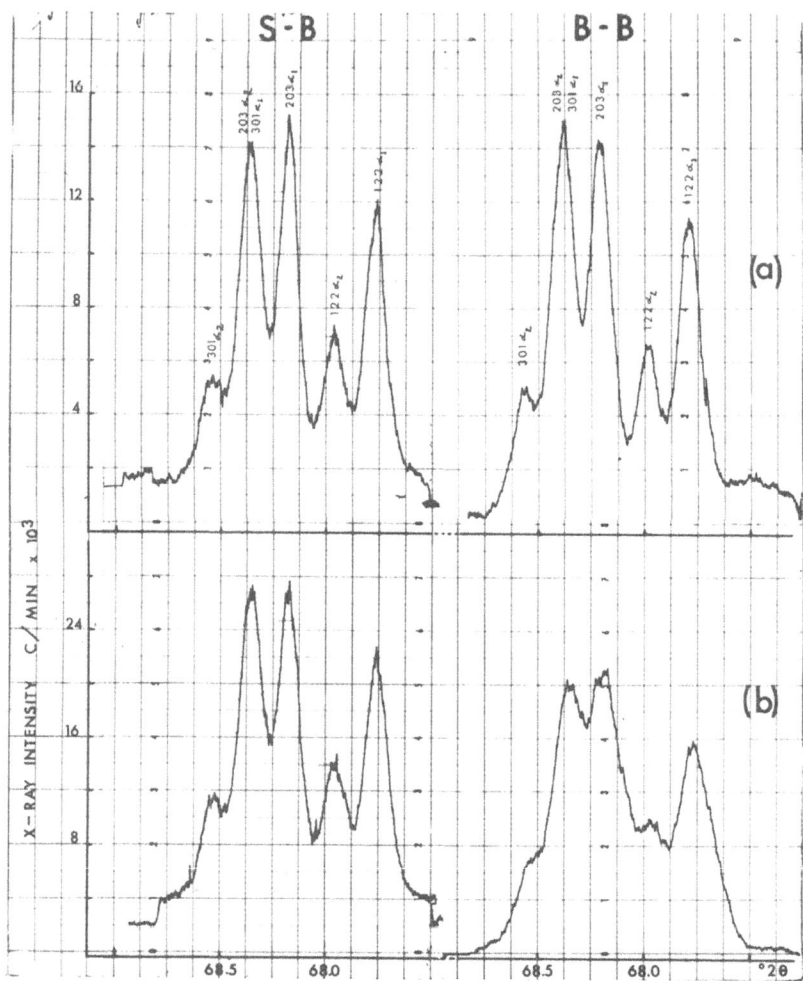

Figure 6. Overlapping 122, 203 and 301 profiles from
Permaquartz, recorded with S-B and B-B geometries.
(a) $\alpha = 2^\circ$, (b) $\alpha = 4^\circ$

prior to making precision measurements in the back-reflection region.

To compare the relative advantages of the two geometries at intermediate γ values, i.e. between 30° and 45°, the overlapping 122, 203 and 301 profiles of permaquartz, which occur near $68^\circ, 2\theta$, were studied using a flat specimen in the B-B position and a specimen ground to a radius of 175 mm set at $\gamma = \frac{1}{2}(2\theta_B) = 34^\circ$. This particular setting approximates to the special condition, $\gamma = 30^\circ, 2\theta_B = 60^\circ$,

at which both the source-specimen and specimen-receiving slit distances are equal to the diffractometer radius R, so that the angular
aperture of the source and the receiving slit are the same for both
geometries (tables II and III).   The length of specimen irradiated,
and hence the diffracted intensities, should also be approximately
the same for the two methods.   When using a divergent aperture of
α = 2°, the fine structure of the three overlapping profiles is
clearly revealed by both geometries as shown in figure 6(a).   On
increasing α to 4° the diffracted intensity is doubled in both
cases, but while the profiles recorded with S-B geometry remain
equally well resolved, the B-B profiles show considerable broadening and loss of resolution.   The S-B geometry is thus undoubtedly
superior to the B-B even in the intermediate region just beyond
the theoretical cross-over point.   The advantage of being able
to increase the area of specimen irradiated without loss of resolution, is of particular importance when it is required to make precision back reflection measurements on coarse grained powders.

    Although the present S-B method has the advantages of high
resolution and intensity, it is not convenient for quantitative
measurements of relative intensities of widely spaced diffraction
profiles because of the corrections required as a result of the
variable angular aperture of the receiving slit.   This limitation
is best overcome by using the variant of the geometry suggested by
Pike[5] in which the counter-tube is directed towards the axis of
the diffractometer instead of at the specimen.   This geometry
results in a constant angular aperture for the receiving slit[3] but
is not easy to achieve because it involves the efficient detection
of oblique irradiation.

    The selection of the optimum γ setting for precision measure-
ments using S-B geometry depends on a number of slightly contra-
dictory factors.   From the aspect of the angular resolution of
the X-ray source, and of the systematic errors contributing to a
shift in angular position, γ should be set at 90°.   This position
is also convenient because it demands no rotation of the X-ray
source from the B-B position, as pointed out by Pike[5] and Baun and
Renton[6], and it is certainly to be recommended for precision measurements of individual profiles in the far back reflection region.
When determining precision lattice parameters, however, it is
usually necessary to use Bragg angles down to 90°,2θ, in order to
obtain sufficient points to make an accurate extrapolation plot.
At γ = 90° this is prevented by the lower limit of (γ+30°) set for
$2\theta_B$, below which the angular aperture of the receiving slit falls
below the equivalent B-B resolution (table III).   The use of γ =
90° also means that the finest resolution of the receiving slit
lies in the inaccessible region of $2\theta_B$ above 172° (table III).

    A careful examination of the data listed in tables II and III

indicates that the optimum specimen setting for precision lattice
parameter determination lies between γ = 45° and 60°. At γ = 60°,
the angular aperture of the X-ray source is less than two thirds
that of the B-B method, and this degree of resolution can be maintained over the important range of Bragg angles from 115° - 172°,
2θ, without requiring a change in receiving slit. Maximum resolution is achieved from 100° - 170°, 2θ, at γ = 45°, but this is
partly off-set by the cut-off of 2θ max at 170°, (figure 3) and by
the slightly larger source aperture (0.7 E_{B-B}) at this setting.

Elimination of Systematic Errors

The sources of error contributing to an angular displacement
of a diffraction profile are mis-alignments of the X-ray source or
specimen, wrong specimen curvature and specimen transparency. As
indicated in equations (11 and 12), transverse or tangential displacements of the X-ray source cause a profile displacement, the
magnitude of which is independent of 2θ, but decreases as γ approaches 90°. Thus when lower values of γ are used, these sources of
error should be eliminated by careful alignment of the instrument.
Although the remaining sources of error have a similar γ-dependence,
(equations 8-10, 13), they all tend to zero as 2θ goes to 180°, and
their effect on the lattice constants of a crystal can therefore be
eliminated by extrapolation.

The error in the lattice parameter $\underline{a}$ of a cubic crystal is
given by

$$\frac{\Delta a}{a} = \frac{\Delta d}{d} = -\frac{1}{2} \cot \theta . \Delta 2\theta \qquad (15)$$

substituting $\Delta 2\theta$ from the expressions in equations (8, 10 and 13)
for errors due to specimen curvature or displacement of a thin
transparent specimen ($\mu t \rightarrow 0$), it follows that

$$\frac{\Delta a}{a} = const. \frac{1 + \cos 2\theta}{\sin \gamma . \sin(2\theta - \gamma)} \qquad (16)$$

If, on the other hand, the specimen is thick and highly absorbent
($\mu t \rightarrow \infty$), and the specimen curvature and displacement errors are
insignificant, equation (10) becomes a more appropriate expression
for $\Delta 2\theta$, and hence the resultant error in lattice parameter is
given by

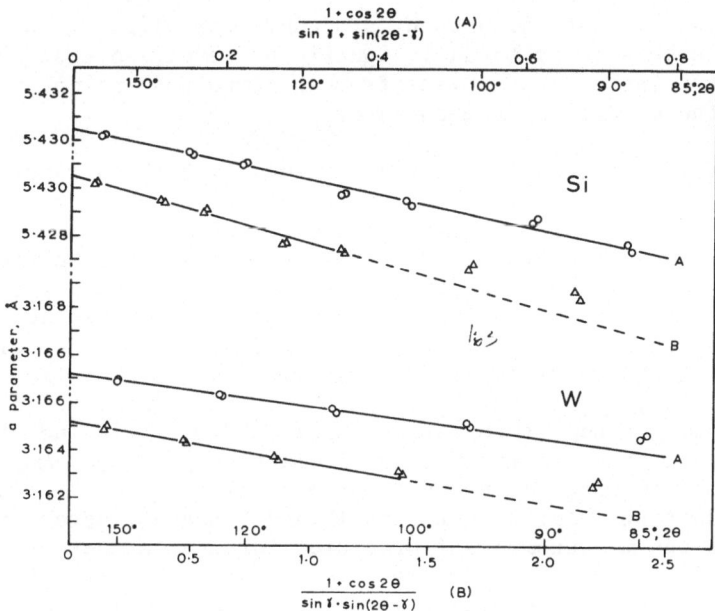

Figure 7. S-B extrapolation plots for I.U.Cr. specimens of silicon and tungsten.

$$\frac{\Delta a}{a} = \text{const.} \frac{1 + \cos 2\theta}{\sin \gamma + \sin(2\theta - \gamma)} \qquad (17)$$

To evaluate the instrument, the technique and the extrapolation procedure, and to provide a comparison between the S-B and B-B diffractometers, the lattice parameters of the I.U.Cr. specimens[23] of silicon and tungsten were determined. Filtered CuKα radiation was used with the specimens at a γ-setting of 45°. The instrument was carefully aligned according to the procedure described above and profiles were scanned at $1/8°(2\theta)$/min. using the proper slits and time constants, in accordance with the data listed in tables II and III. Lattice parameters, calculated from all Bragg reflections which occur above 87°,2θ, are plotted against the functions $(1 + \cos 2\theta)/(\sin \gamma . \sin(2\theta - \gamma))$ and $(1 + \cos 2\theta)/(\sin \gamma + \sin(2\theta - \gamma))$ in figure 7. The extrapolation function for thin transparent specimens (μt → 0), which also takes account of specimen curvature and displacement errors, is found to be effective only at Bragg angles greater than 110°,2θ. The results however, conform to a straight line plot over the entire range of 2θ above 90° when plotted against the absorption function for thick specimens (μt → ∞). As the effectiveness of the specimen displacement error function (equation 12) has been demonstrated by deliberately mis-setting the specimen, as shown in table I, the results in figure 7 confirm that the present procedure for aligning the instrument reduces this error to an insignificant level, so that the dominant systematic error arises from specimen absorption, which is out of the control of the operator.

To avoid needless calculation, values of both extrapolation functions have been computed, to three decimal places, over the measurable ranges of $2\theta_B$ for specimen settings from $\gamma = 90°$ down to $\gamma = 15°$, in increments of $15°$. These values are tabulated in tables IV and V in the Appendix. The operator can thus plot his results against both functions, as in figure 7, and then select the most appropriate function for his particular experimental conditions. If the results fail to conform to a straight line with either function, errors may be present in the alignment of the X-ray source. Since these cause a shift $\Delta 2\theta$ which is independent of Bragg angle, the resultant error in lattice parameter may be minimized by plotting against cot θ(see equation 15). Although the low γ settings listed in tables IV and V may not necessarily be used for the measurement of precision lattice parameters, they are useful for calculating errors which occur in instruments with fixed low γ settings, such as that described by Parrish and Mack[3], and for use with flat or curved specimen Seeman-Bohlin cameras.

Extrapolation of the lattice parameter results in figure 7 to $2\theta = 180°$, give the lattice parameters of silicon and tungsten at 25°C as 5.43050 Å + .00005 Å and 3.16520 Å + .00005 Å respectively, (λ CuKα_1 = 1.54051 Å, λ CuKα_2 = 1.54433 Å and corrected for refraction[23]). These results fall well within the limits of the mean I.U.Cr. values[23] and lie within + 0.00008 Å of both the subsequent precision B-B determinations of Delf[24], Parrish, Taylor and Mack[25], and King and Russell[26] and the precision focusing powder camera results of Franks[27], confirming that the present S-B method yields results on a par with other precision powder techniques.

ACKNOWLEDGEMENTS

The Authors are grateful to Professor J.G. Ball for his helpful advice and encouragement, to Messrs. G.J. Green and A. Lloyd for their assistance with the construction of the instrument and to Mr. M. Losty for his help with the computing. This work was supported in part by the Ministry of Technology, R.R.E., Great Malvern.

REFERENCES

1. H.P. Klug and L.E. Alexander, "X-ray Diffraction Procedures",
    John Wiley and Sons., Inc., New York, 1954.

2. L.F. Vassamillet and H.W. King, "Diffractometer Techniques",
    in the Handbook of X-rays, Ed. Kaelble, McGraw-Hill Book, Co.,
    Ltd., New York, 1967.

3. W. Parrish and M. Mack, "Seeman-Bohlin X-ray Diffractometry,
    I Instrumentation; II Comparison of Aberrations and Intensity
    with Conventional Diffractometer", Acta Cryst. 23, 687+693,
    1967.

4. K. Das Gupta, H.W. Schnopper, A.E. Metzer and A.R. Shields,
    "A Combined Focusing X-ray Diffractometer and Non-dispersive
    X-ray Spectrometer for Lunar and Planetary Analysis", Advances
    in X-ray Analysis, Plenum Press, 9, 221, 1966.

5. E.R. Pike, "Focusing Geometry in X-ray Diffractometer", J.
    Sci. Instrum. 39, 222, 1962.

6. W.L. Baun and J.J. Renton, "The Design and Use of Special-
    Purpose Attachments for the Horizontal Diffractometer",
    Advances in X-ray Analysis, Plenum Press, New York, 7, 302,
    1964.

7. G. Wassermann and J. Wiewiorosky, "Uber ein Geiger-Zahlrohr-
    Goniometer nach dem Seeman-Bohlin Prinzip", Z. Metallk. 44,
    567, 1953.

8. L.F. Vassamillet and H.W. King, "Precision X-ray Diffractometry
    using Powder Specimens", Adv. in X-ray Analysis, Plenum Press,
    New York, 6, 142, 1963.

9. W. Parrish and A.J.C. Wilson, International Tables for X-ray
    Crystallography, Vol. 11, 216, 1959, Kynoch Press, Birmingham.
    England.

10. A. Segmuller, "Die Bestimmung von Glanzwinkeln, Linienbreiten
    und Intensitaten der Rontgen-Interferenzen mit einem Geiger-
    Zahlrohr-Goniometer nach dem Seemann-Bohlin-Prinzip", Z. Metallk.
    48, 448, 1957.

11. G. Kunze, "Korrekturen hoherer Ordnung fur die mit Bragg-
    Brentano- und Seemann-Bohlin-Systemen gewonnenen Messgrossen
    etc.," Z. Angew. Phys. 17, 412, and "Intensitats-, Absorptions-
    und Verschiebungsfaktoren von Interferenzlinien bei Bragg-
    Brentano- und Seemann-Bohlin-Diffraktometern I and II", Z.
    Angew. Phys. 17, 522 and 18, 28, 1964.

12. A.J.C. Wilson, "Correction of Lattice Spacings for Refraction", Proc. Camb. Phil. Soc., 36, 485, 1960.

13. A.J.C. Wilson, "Effect of Absorption on Mean Wave-Length of X-ray Emission Lines", Proc. Phys. Soc., (London), 72, 924, 1958.

14. J. Cermak, "The Intensity Distribution in the Faces of Curved Crystal Monochromators and an Estimate of Its Influence on Precision Measurements of Lattice Parameters", Acta Cryst. 13, 832, 1960.

15. K. Lonsdale, International Tables for X-ray Crystallography, Vol. 111, 1962, Kynoch Press, Birmingham, England.

16. K. Beardon, "X-ray Wavelengths", U.S. Atomic Energy Commission, Division of Technical Information Extension, Oak Ridge, Tenn., 1964.

17. E.R. Pike, "Counter Diffractometer - The Effect of Dispersion, Lorentz and Polarization Factors on the Position of X-ray Powder Diffraction Lines", Acta Cryst. 12, 87, 1959.

18. C.J. Gillham and H.W. King, "Seeman Bohlin Diffractometer - Effect of Dispersion, Lorentz factor and Polarization on the Position of Diffraction Profiles", to be published.

19. E.R. Pike, "Counter Diffractometer - The Effect of Vertical Divergence on the Displacement and Breadth of Powder Diffraction Lines", J. Sci. Instrum. 34, 355, 1957.

20. J.C. Evans and C.O. Taylerson, "Measurement of Angle in Engineering", Nat. Phys. Lab. Notes on Appl. Sci. No. 26, 1961, H.M. Stationery Office, London.

21. H.W. King and L.F. Vassamillet, "Precision Lattice Parameter Determination by Double-Scanning Diffractometry", Advances in X-ray Analysis, Plenum Press, New York, 5, 78, 1962.

22. A.J.C. Wilson, "Geiger-Counter X-ray Spectrometer - Influence of Size and Absorption Coefficient of Specimen on Position and Shape of Powder Diffraction Maxima", J. Sci. Instrum., 27, 321, 1950.

23. W. Parrish, "Results of the I.U.Cr. Precision Lattice Parameter Project", Acta Cryst. 13, 838, 1960.

24. B.W. Delf, "The Practical Determination of Lattice Parameters using the Centroid Method", Brit. J. Appl. Phys. 14, 345, 1963.

25. W. Parrish, J. Taylor and M. Mack, "Dependence of Lattice Parameters on Various Angular Measurements of Diffraction Line Profiles", Advances in X-ray Analysis, Plenum Press, New York, 7, 66, 1964.

26. H.W. King and C.M. Russell, "Double-Scanning Diffractometry in the Back-Reflection Region", Advances in X-ray Analysis, Plenum Press, New York, 8, 1, 1965.

27. A. Franks, "The Precision Measurement of Lattice Parameters using a Back Reflection Focusing Camera", N.P.L. Report No., M.3072, 1963.

APPENDIX

Table IV

S-B Extrapolation Function for Absorption when $\mu t \to \infty$

$$\frac{1 + \cos 2\theta}{\sin \gamma + (\sin 2\theta - \gamma)}$$

$\mu t \to \infty$　FUNCTION FOR　GAMMA=90

θ	0.0	0.1	0.2	0.3	0.4	0.5	0.6	0.7	0.8	0.9
50	.704	.699	.694	.689	.684	.680	.675	.670	.665	.660
51	.656	.651	.646	.642	.637	.633	.628	.624	.619	.615
52	.610	.606	.602	.597	.593	.589	.585	.580	.576	.572
53	.568	.564	.560	.556	.552	.548	.544	.540	.536	.532
54	.528	.524	.520	.516	.513	.509	.505	.501	.498	.494
55	.490	.487	.483	.479	.476	.472	.469	.465	.462	.458
56	.455	.452	.448	.445	.441	.438	.435	.431	.428	.425
57	.422	.419	.415	.412	.409	.406	.403	.400	.397	.394
58	.390	.387	.384	.381	.378	.376	.373	.370	.367	.364
59	.361	.358	.355	.353	.350	.347	.344	.341	.339	.336
60	.333	.331	.328	.325	.323	.320	.318	.315	.312	.310
61	.307	.305	.302	.300	.297	.295	.292	.290	.288	.285
62	.283	.280	.278	.276	.273	.271	.269	.266	.264	.262
63	.260	.257	.255	.253	.251	.249	.246	.244	.242	.240
64	.238	.236	.234	.232	.230	.228	.225	.223	.221	.219
65	.217	.215	.214	.212	.210	.208	.206	.204	.202	.200
66	.198	.196	.195	.193	.191	.189	.187	.185	.184	.182
67	.180	.178	.177	.175	.173	.172	.170	.168	.167	.165
68	.163	.162	.160	.158	.157	.155	.154	.152	.150	.149
69	.147	.146	.144	.143	.141	.140	.138	.137	.135	.134
70	.132	.131	.130	.128	.127	.125	.124	.123	.121	.120
71	.119	.117	.116	.115	.113	.112	.111	.109	.108	.107
72	.106	.104	.103	.102	.101	.099	.098	.097	.096	.095
73	.093	.092	.091	.090	.089	.088	.087	.086	.084	.083
74	.082	.081	.080	.079	.078	.077	.076	.075	.074	.073
75	.072	.071	.070	.069	.068	.067	.066	.065	.064	.063
76	.062	.061	.060	.059	.059	.058	.057	.056	.055	.054
77	.053	.052	.052	.051	.050	.049	.048	.048	.047	.046
78	.045	.044	.044	.043	.042	.041	.041	.040	.039	.038
79	.038	.037	.036	.036	.035	.034	.034	.033	.032	.032

Table IV continued

$\mu t \rightarrow \infty$ FUNCTION FOR GAMMA=90

θ	0.0	0.1	0.2	0.3	0.4	0.5	0.6	0.7	0.8	0.9
80	.031	.030	.030	.029	.029	.028	.027	.027	.026	.026
81	.025	.025	.024	.023	.023	.022	.022	.021	.021	.020
82	.020	.019	.019	.018	.018	.017	.017	.016	.016	.016
83	.015	.015	.014	.014	.013	.013	.013	.012	.012	.011
84	.011	.011	.010	.010	.010	.009	.009	.009	.008	.008
85	.008	.007	.007	.007	.006	.006	.006	.006	.005	.005
86	.005	.005	.004	.004	.004	.004	.004	.003	.003	.003
87	.003	.003	.002	.002	.002	.002	.002	.002	.001	.001
88	.001	.001	.001	.001	.001	.001	.001	.001	.000	.000
89	.000	.000	.000	.000	.000	.000	.000	.000	.000	.000

$\mu t \rightarrow \infty$ FUNCTION FOR GAMMA=75

θ	0.0	0.1	0.2	0.3	0.4	0.5	0.6	0.7	0.8	0.9
40	1.114	1.108	1.101	1.094	1.087	1.080	1.074	1.067	1.060	1.054
41	1.047	1.041	1.034	1.028	1.022	1.015	1.009	1.003	.996	.990
42	.984	.978	.972	.966	.960	.954	.948	.942	.936	.931
43	.925	.919	.913	.908	.902	.896	.891	.885	.880	.874
44	.869	.864	.858	.853	.848	.842	.837	.832	.827	.822
45	.816	.811	.806	.801	.796	.791	.786	.782	.777	.772
46	.767	.762	.757	.753	.748	.743	.739	.734	.729	.725
47	.720	.716	.711	.707	.702	.698	.694	.689	.685	.680
48	.676	.672	.668	.663	.659	.655	.651	.647	.643	.639
49	.635	.630	.626	.622	.618	.615	.611	.607	.603	.599
50	.595	.591	.588	.584	.580	.576	.573	.569	.565	.561
51	.558	.554	.551	.547	.543	.540	.536	.533	.529	.526
52	.523	.519	.516	.512	.509	.506	.502	.499	.496	.492
53	.489	.486	.483	.479	.476	.473	.470	.467	.464	.461
54	.457	.454	.451	.448	.445	.442	.439	.436	.433	.430
55	.427	.424	.422	.419	.416	.413	.410	.407	.404	.402
56	.399	.396	.393	.391	.388	.385	.383	.380	.377	.375
57	.372	.369	.367	.364	.361	.359	.356	.354	.351	.349
58	.346	.344	.341	.339	.336	.334	.332	.329	.327	.324
59	.322	.320	.317	.315	.313	.310	.308	.306	.303	.301
60	.299	.297	.294	.292	.290	.288	.286	.283	.281	.279
61	.277	.275	.273	.271	.269	.266	.264	.262	.260	.258
62	.256	.254	.252	.250	.248	.246	.244	.242	.240	.238
63	.236	.235	.233	.231	.229	.227	.225	.223	.221	.220
64	.218	.216	.214	.212	.211	.209	.207	.205	.204	.202

Table IV continued

$\mu t \to \infty$ FUNCTION FOR GAMMA=75

θ	0.0	0.1	0.2	0.3	0.4	0.5	0.6	0.7	0.8	0.9
65	.200	.198	.197	.195	.193	.192	.190	.188	.187	.185
66	.183	.182	.180	.178	.177	.175	.174	.172	.171	.169
67	.167	.166	.164	.163	.161	.160	.158	.157	.155	.154
68	.152	.151	.150	.148	.147	.145	.144	.142	.141	.140
69	.138	.137	.136	.134	.133	.132	.130	.129	.128	.126
70	.125	.124	.122	.121	.120	.119	.117	.116	.115	.114
71	.112	.111	.110	.109	.108	.106	.105	.104	.103	.102
72	.101	.099	.098	.097	.096	.095	.094	.093	.092	.091
73	.089	.088	.087	.086	.085	.084	.083	.082	.081	.080
74	.079	.078	.077	.076	.075	.074	.073	.072	.071	.070
75	.069	.068	.067	.067	.066	.065	.064	.063	.062	.061
76	.060	.059	.059	.058	.057	.056	.055	.054	.054	.053
77	.052	.051	.050	.050	.049	.048	.047	.047	.046	.045
78	.044	.044	.043	.042	.041	.041	.040	.039	.039	.038
79	.037	.037	.036	.035	.035	.034	.033	.033	.032	.031
80	.031	.030	.030	.029	.028	.028	.027	.027	.026	.025
81	.025	.024	.024	.023	.023	.022	.022	.021	.021	.020
82	.020	.019	.019	.018	.018	.017	.017	.016	.016	.016
83	.015	.015	.014	.014	.013	.013	.013	.012	.012	.011
84	.011	.011	.010	.010	.010	.009	.009	.009	.008	.008
85	.008	.007	.007	.007	.007	.006	.006	.006	.005	.005
86	.005	.005	.004	.004	.004	.004	.004	.003	.003	.003
87	.003	.003	.002	.002	.002	.002	.002	.002	.002	.001
88	.001	.001	.001	.001	.001	.001	.001	.001	.000	.000
89	.000	.000	.000	.000	.000	.000	.000	.000	.000	.000

$\mu t \to \infty$ FUNCTION FOR GAMMA=60

θ	0.0	0.1	0.2	0.3	0.4	0.5	0.6	0.7	0.8	0.9
35	1.291	1.283	1.276	1.269	1.261	1.254	1.247	1.240	1.233	1.226
36	1.219	1.212	1.205	1.198	1.191	1.185	1.178	1.171	1.165	1.158
37	1.151	1.145	1.138	1.132	1.125	1.119	1.113	1.106	1.100	1.094
38	1.088	1.082	1.076	1.069	1.063	1.057	1.051	1.046	1.040	1.034
39	1.028	1.022	1.016	1.011	1.005	.999	.994	.988	.983	.977
40	.972	.966	.961	.955	.950	.944	.939	.934	.929	.923
41	.918	.913	.908	.903	.898	.893	.888	.883	.878	.873
42	.868	.863	.858	.853	.848	.844	.839	.834	.829	.825
43	.820	.815	.811	.806	.802	.797	.793	.788	.784	.779
44	.775	.771	.766	.762	.758	.753	.749	.745	.740	.736

Table IV continued

$\mu t \to \infty$ FUNCTION FOR   GAMMA=60

θ	0.0	0.1	0.2	0.3	0.4	0.5	0.6	0.7	0.8	0.9
45	.732	.728	.724	.720	.716	.711	.707	.703	.699	.695
46	.691	.687	.683	.680	.676	.672	.668	.664	.660	.656
47	.653	.649	.645	.641	.638	.634	.630	.627	.623	.620
48	.616	.612	.609	.605	.602	.598	.595	.591	.588	.584
49	.581	.578	.574	.571	.567	.564	.561	.558	.554	.551
50	.548	.544	.541	.538	.535	.532	.528	.525	.522	.519
51	.516	.513	.510	.507	.504	.501	.498	.495	.492	.489
52	.486	.483	.480	.477	.474	.471	.468	.465	.463	.460
53	.457	.454	.451	.449	.446	.443	.440	.438	.435	.432
54	.429	.427	.424	.421	.419	.416	.414	.411	.408	.406
55	.403	.401	.398	.396	.393	.390	.388	.385	.383	.381
56	.378	.376	.373	.371	.368	.366	.364	.361	.359	.357
57	.354	.352	.350	.347	.345	.343	.340	.338	.336	.334
58	.331	.329	.327	.325	.322	.320	.318	.316	.314	.312
59	.310	.307	.305	.303	.301	.299	.297	.295	.293	.291
60	.289	.287	.285	.283	.281	.279	.277	.275	.273	.271
61	.269	.267	.265	.263	.261	.259	.257	.255	.254	.252
62	.250	.248	.246	.244	.242	.241	.239	.237	.235	.233
63	.232	.230	.228	.226	.225	.223	.221	.219	.218	.216
64	.214	.213	.211	.209	.208	.206	.204	.203	.201	.199
65	.198	.196	.195	.193	.191	.190	.188	.187	.185	.184
66	.182	.181	.179	.178	.176	.175	.173	.172	.170	.169
67	.167	.166	.164	.163	.161	.160	.158	.157	.156	.154
68	.153	.151	.150	.149	.147	.146	.145	.143	.142	.141
69	.139	.138	.137	.135	.134	.133	.131	.130	.129	.128
70	.126	.125	.124	.123	.121	.120	.119	.118	.117	.115
71	.114	.113	.112	.111	.110	.108	.107	.106	.105	.104
72	.103	.102	.100	.099	.098	.097	.096	.095	.094	.093
73	.092	.091	.090	.089	.088	.087	.085	.084	.083	.082
74	.081	.080	.079	.078	.078	.077	.076	.075	.074	.073
75	.072	.071	.070	.069	.068	.067	.066	.065	.065	.064
76	.063	.062	.061	.060	.059	.058	.058	.057	.056	.055
77	.054	.053	.053	.052	.051	.050	.050	.049	.048	.047
78	.046	.046	.045	.044	.043	.043	.042	.041	.041	.040
79	.039	.039	.038	.037	.036	.036	.035	.035	.034	.033
80	.033	.032	.031	.031	.030	.029	.029	.028	.028	.027
81	.027	.026	.025	.025	.024	.024	.023	.023	.022	.022
82	.021	.021	.020	.020	.019	.019	.018	.018	.017	.017
83	.016	.016	.015	.015	.014	.014	.014	.013	.013	.012
84	.012	.012	.011	.011	.011	.010	.010	.009	.009	.009

Table IV continued

$\mu t \to \infty$　FUNCTION FOR GAMMA=60

θ	0.0	0.1	0.2	0.3	0.4	0.5	0.6	0.7	0.8	0.9
85	.008	.008	.008	.007	.007	.007	.007	.006	.006	.006
86	.005	.005	.005	.005	.004	.004	.004	.004	.003	.003
87	.003	.003	.003	.003	.002	.002	.002	.002	.002	.002
88	.001	.001	.001	.001	.001	.001	.001	.001	.001	.000
89	.000	.000	.000	.000	.000	.000	.000	.000	.000	.000

$\mu t \to \infty$　FUNCTION FOR GAMMA=45

θ	0.0	0.1	0.2	0.3	0.4	0.5	0.6	0.7	0.8	0.9
25	2.068	2.056	2.044	2.032	2.019	2.007	1.996	1.984	1.972	1.960
26	1.949	1.938	1.926	1.915	1.904	1.893	1.882	1.871	1.860	1.849
27	1.839	1.828	1.818	1.807	1.797	1.787	1.776	1.766	1.756	1.746
28	1.736	1.727	1.717	1.707	1.698	1.688	1.679	1.669	1.660	1.651
29	1.641	1.632	1.623	1.614	1.605	1.596	1.588	1.579	1.570	1.561
30	1.553	1.544	1.536	1.528	1.519	1.511	1.503	1.494	1.486	1.478
31	1.470	1.462	1.454	1.446	1.439	1.431	1.423	1.416	1.408	1.400
32	1.393	1.385	1.378	1.371	1.363	1.356	1.349	1.342	1.334	1.327
33	1.320	1.313	1.306	1.299	1.293	1.286	1.279	1.272	1.265	1.259
34	1.252	1.246	1.239	1.232	1.226	1.220	1.213	1.207	1.200	1.194
35	1.188	1.182	1.176	1.169	1.163	1.157	1.151	1.145	1.139	1.133
36	1.127	1.122	1.116	1.110	1.104	1.098	1.093	1.087	1.081	1.076
37	1.070	1.065	1.059	1.054	1.048	1.043	1.037	1.032	1.027	1.021
38	1.016	1.011	1.006	1.001	.995	.990	.985	.980	.975	.970
39	.965	.960	.955	.950	.945	.940	.936	.931	.926	.921
40	.916	.912	.907	.902	.898	.893	.888	.884	.879	.875
41	.870	.866	.861	.857	.853	.848	.844	.839	.835	.831
42	.826	.822	.818	.814	.810	.805	.801	.797	.793	.789
43	.785	.781	.777	.773	.769	.765	.761	.757	.753	.749
44	.745	.741	.737	.733	.730	.726	.722	.718	.715	.711
45	.707	.703	.700	.696	.692	.689	.685	.682	.678	.674
46	.671	.667	.664	.660	.657	.653	.650	.647	.643	.640
47	.636	.633	.630	.626	.623	.620	.616	.613	.610	.607
48	.603	.600	.597	.594	.591	.587	.584	.581	.578	.575
49	.572	.569	.566	.562	.559	.556	.553	.550	.547	.544
50	.541	.538	.536	.533	.530	.527	.524	.521	.518	.515
51	.512	.510	.507	.504	.501	.498	.496	.493	.490	.487
52	.485	.482	.479	.477	.474	.471	.468	.466	.463	.461
53	.458	.455	.453	.450	.448	.445	.442	.440	.437	.435
54	.432	.430	.427	.425	.422	.420	.418	.415	.413	.410

Table IV continued

$\mu t \to \infty$ FUNCTION FOR GAMMA=45										
θ	0.0	0.1	0.2	0.3	0.4	0.5	0.6	0.7	0.8	0.9
55	.408	.405	.403	.401	.398	.396	.394	.391	.389	.387
56	.384	.382	.380	.377	.375	.373	.371	.368	.366	.364
57	.362	.359	.357	.355	.353	.351	.348	.346	.344	.342
58	.340	.338	.336	.333	.331	.329	.327	.325	.323	.321
59	.319	.317	.315	.313	.311	.309	.307	.305	.303	.301
60	.299	.297	.295	.293	.291	.289	.287	.285	.283	.281
61	.280	.278	.276	.274	.272	.270	.268	.267	.265	.263
62	.261	.259	.257	.256	.254	.252	.250	.248	.247	.245
63	.243	.241	.240	.238	.236	.235	.233	.231	.229	.228
64	.226	.224	.223	.221	.219	.218	.216	.215	.213	.211
65	.210	.208	.207	.205	.203	.202	.200	.199	.197	.196
66	.194	.192	.191	.189	.188	.186	.185	.183	.182	.180
67	.179	.177	.176	.174	.173	.172	.170	.169	.167	.166
68	.164	.163	.162	.160	.159	.157	.156	.155	.153	.152
69	.151	.149	.148	.147	.145	.144	.143	.141	.140	.139
70	.137	.136	.135	.134	.132	.131	.130	.128	.127	.126
71	.125	.123	.122	.121	.120	.119	.117	.116	.115	.114
72	.113	.112	.110	.109	.108	.107	.106	.105	.103	.102
73	.101	.100	.099	.098	.097	.096	.095	.094	.092	.091
74	.090	.089	.088	.087	.086	.085	.084	.083	.082	.081
75	.080	.079	.078	.077	.076	.075	.074	.073	.072	.071
76	.070	.069	.068	.068	.067	.066	.065	.064	.063	.062
77	.061	.060	.059	.059	.058	.057	.056	.055	.054	.054
78	.053	.052	.051	.050	.049	.049	.048	.047	.046	.046
79	.045	.044	.043	.042	.042	.041	.040	.040	.039	.038
80	.037	.037	.036	.035	.035	.034	.033	.033	.032	.031
81	.031	.030	.029	.029	.028	.027	.027	.026	.026	.025
82	.024	.024	.023	.023	.022	.022	.021	.021	.020	.020
83	.019	.018	.018	.017	.017	.016	.016	.016	.015	.015
84	.014	.014	.013	.013	.012	.012	.012	.011	.011	.010
85	.010	.010	.009	.009	.008	.008	.008	.007	.007	.007
86	.006	.006	.006	.006	.005	.005	.005	.004	.004	.004
87	.004	.003	.003	.003	.003	.003	.002	.002	.002	.002
88	.002	.002	.001	.001	.001	.001	.001	.001	.001	.001
89	.000	.000	.000	.000	.000	.000	.000	.000	.000	.000

Table IV continued

$\mu t \to \infty$ FUNCTION FOR   GAMMA=30

θ	0.0	0.1	0.2	0.3	0.4	0.5	0.6	0.7	0.8	0.9
20	2.622	2.605	2.589	2.572	2.556	2.540	2.524	2.509	2.493	2.478
21	2.462	2.447	2.432	2.417	2.403	2.388	2.374	2.360	2.345	2.331
22	2.317	2.304	2.290	2.276	2.263	2.250	2.236	2.223	2.210	2.198
23	2.185	2.172	2.160	2.147	2.135	2.123	2.111	2.099	2.087	2.075
24	2.063	2.052	2.040	2.029	2.017	2.006	1.995	1.984	1.973	1.962
25	1.951	1.940	1.930	1.919	1.909	1.898	1.888	1.878	1.867	1.857
26	1.847	1.837	1.827	1.818	1.808	1.798	1.789	1.779	1.770	1.760
27	1.751	1.742	1.733	1.724	1.715	1.706	1.697	1.688	1.679	1.670
28	1.662	1.653	1.644	1.636	1.628	1.619	1.611	1.603	1.594	1.586
29	1.578	1.570	1.562	1.554	1.546	1.538	1.531	1.523	1.515	1.508
30	1.500	1.492	1.485	1.478	1.470	1.463	1.456	1.448	1.441	1.434
31	1.427	1.420	1.413	1.406	1.399	1.392	1.385	1.378	1.371	1.365
32	1.358	1.351	1.345	1.338	1.332	1.325	1.319	1.312	1.306	1.300
33	1.293	1.287	1.281	1.275	1.268	1.262	1.256	1.250	1.244	1.238
34	1.232	1.226	1.220	1.214	1.209	1.203	1.197	1.191	1.186	1.180
35	1.174	1.169	1.163	1.158	1.152	1.147	1.141	1.136	1.130	1.125
36	1.120	1.114	1.109	1.104	1.099	1.093	1.088	1.083	1.078	1.073
37	1.068	1.063	1.058	1.053	1.048	1.043	1.038	1.033	1.028	1.023
38	1.019	1.014	1.009	1.004	.999	.995	.990	.985	.981	.976
39	.972	.967	.963	.958	.954	.949	.945	.940	.936	.931
40	.927	.923	.918	.914	.910	.905	.901	.897	.893	.889
41	.884	.880	.876	.872	.868	.864	.860	.856	.852	.848
42	.844	.840	.836	.832	.828	.824	.820	.816	.813	.809
43	.805	.801	.797	.794	.790	.786	.782	.779	.775	.771
44	.768	.764	.760	.757	.753	.750	.746	.743	.739	.736
45	.732	.729	.725	.722	.718	.715	.711	.708	.705	.701
46	.698	.695	.691	.688	.685	.681	.678	.675	.671	.668
47	.665	.662	.659	.655	.652	.649	.646	.643	.640	.637
48	.633	.630	.627	.624	.621	.618	.615	.612	.609	.606
49	.603	.600	.597	.594	.591	.588	.586	.583	.580	.577
50	.574	.571	.568	.565	.563	.560	.557	.554	.551	.549
51	.546	.543	.540	.538	.535	.532	.530	.527	.524	.521
52	.519	.516	.513	.511	.508	.506	.503	.500	.498	.495
53	.493	.490	.488	.485	.482	.480	.477	.475	.472	.470
54	.467	.465	.463	.460	.458	.455	.453	.450	.448	.446
55	.443	.441	.438	.436	.434	.431	.429	.427	.424	.422
56	.420	.417	.415	.413	.410	.408	.406	.404	.401	.399
57	.397	.395	.393	.390	.388	.386	.384	.382	.379	.377
58	.375	.373	.371	.369	.366	.364	.362	.360	.358	.356
59	.354	.352	.350	.348	.346	.343	.341	.339	.337	.335

Table IV continued

θ	0.0	0.1	0.2	0.3	0.4	0.5	0.6	0.7	0.8	0.9
				$\mu t \to \infty$	FUNCTION FOR		GAMMA=30			
60	.333	.331	.329	.327	.325	.323	.321	.319	.317	.315
61	.314	.312	.310	.308	.306	.304	.302	.300	.298	.296
62	.294	.292	.291	.289	.287	.285	.283	.281	.279	.278
63	.276	.274	.272	.270	.269	.267	.265	.263	.261	.260
64	.258	.256	.254	.253	.251	.249	.247	.246	.244	.242
65	.241	.239	.237	.235	.234	.232	.230	.229	.227	.225
66	.224	.222	.221	.219	.217	.216	.214	.212	.211	.209
67	.208	.206	.205	.203	.201	.200	.198	.197	.195	.194
68	.192	.191	.189	.187	.186	.184	.183	.181	.180	.178
69	.177	.176	.174	.173	.171	.170	.168	.167	.165	.164
70	.163	.161	.160	.158	.157	.155	.154	.153	.151	.150
71	.149	.147	.146	.144	.143	.142	.140	.139	.138	.136
72	.135	.134	.132	.131	.130	.129	.127	.126	.125	.123
73	.122	.121	.120	.118	.117	.116	.115	.114	.112	.111
74	.110	.109	.107	.106	.105	.104	.103	.102	.100	.099
75	.098	.097	.096	.095	.094	.092	.091	.090	.089	.088
76	.087	.086	.085	.084	.082	.081	.080	.079	.078	.077
77	.076	.075	.074	.073	.072	.071	.070	.069	.068	.067
78	.066	.065	.064	.063	.062	.061	.060	.059	.058	.057
79	.057	.056	.055	.054	.053	.052	.051	.050	.049	.048
80	.048	.047	.046	.045	.044	.043	.043	.042	.041	.040
81	.039	.039	.038	.037	.036	.035	.035	.034	.033	.032
82	.032	.031	.030	.030	.029	.028	.028	.027	.026	.026
83	.025	.024	.024	.023	.022	.022	.021	.020	.020	.019
84	.019	.018	.018	.017	.016	.016	.015	.015	.014	.014
85	.013	.013	.012	.012	.011	.011	.010	.010	.010	.009
86	.009	.008	.008	.008	.007	.007	.006	.006	.006	.005
87	.005	.005	.004	.004	.004	.004	.003	.003	.003	.003
88	.002	.002	.002	.002	.001	.001	.001	.001	.001	.001
89	.001	.000	.000	.000	.000	.000	.000	.000	.000	.000

Table IV continued

$\mu t \to \infty$ FUNCTION FOR GAMMA=15

θ	0.0	0.1	0.2	0.3	0.4	0.5	0.6	0.7	0.8	0.9
10	5.606	5.547	5.489	5.432	5.376	5.322	5.268	5.215	5.163	5.112
11	5.062	5.013	4.965	4.918	4.871	4.825	4.781	4.736	4.693	4.650
12	4.608	4.567	4.526	4.486	4.447	4.408	4.370	4.332	4.295	4.259
13	4.223	4.188	4.153	4.119	4.085	4.052	4.019	3.986	3.955	3.923
14	3.892	3.862	3.832	3.802	3.773	3.744	3.715	3.687	3.659	3.632
15	3.605	3.578	3.552	3.526	3.500	3.475	3.450	3.425	3.401	3.377
16	3.353	3.329	3.306	3.283	3.260	3.238	3.216	3.194	3.172	3.151
17	3.130	3.109	3.088	3.068	3.048	3.028	3.008	2.988	2.969	2.950
18	2.931	2.912	2.894	2.876	2.857	2.840	2.822	2.804	2.787	2.770
19	2.753	2.736	2.719	2.703	2.686	2.670	2.654	2.638	2.623	2.607
20	2.592	2.576	2.561	2.546	2.532	2.517	2.502	2.488	2.474	2.459
21	2.445	2.432	2.418	2.404	2.391	2.377	2.364	2.351	2.338	2.325
22	2.312	2.299	2.287	2.274	2.262	2.250	2.238	2.225	2.214	2.202
23	2.190	2.178	2.167	2.155	2.144	2.133	2.121	2.110	2.099	2.088
24	2.077	2.067	2.056	2.045	2.035	2.024	2.014	2.004	1.994	1.984
25	1.974	1.964	1.954	1.944	1.934	1.925	1.915	1.905	1.896	1.887
26	1.877	1.868	1.859	1.850	1.841	1.832	1.823	1.814	1.805	1.796
27	1.788	1.779	1.771	1.762	1.754	1.745	1.737	1.729	1.721	1.712
28	1.704	1.696	1.688	1.680	1.672	1.665	1.657	1.649	1.641	1.634
29	1.626	1.619	1.611	1.604	1.596	1.589	1.582	1.574	1.567	1.560
30	1.553	1.546	1.539	1.532	1.525	1.518	1.511	1.504	1.497	1.491
31	1.484	1.477	1.471	1.464	1.458	1.451	1.445	1.438	1.432	1.425
32	1.419	1.413	1.407	1.400	1.394	1.388	1.382	1.376	1.370	1.364
33	1.358	1.352	1.346	1.340	1.334	1.329	1.323	1.317	1.311	1.306
34	1.300	1.294	1.289	1.283	1.278	1.272	1.267	1.261	1.256	1.250
35	1.245	1.240	1.234	1.229	1.224	1.219	1.213	1.208	1.203	1.198
36	1.193	1.188	1.183	1.178	1.173	1.168	1.163	1.158	1.153	1.148
37	1.143	1.138	1.133	1.129	1.124	1.119	1.114	1.110	1.105	1.100
38	1.096	1.091	1.087	1.082	1.077	1.073	1.068	1.064	1.059	1.055
39	1.051	1.046	1.042	1.037	1.033	1.029	1.024	1.020	1.016	1.012
40	1.007	1.003	.999	.995	.991	.986	.982	.978	.974	.970
41	.966	.962	.958	.954	.950	.946	.942	.938	.934	.930
42	.926	.922	.919	.915	.911	.907	.903	.900	.896	.892
43	.888	.885	.881	.877	.873	.870	.866	.863	.859	.855
44	.852	.848	.845	.841	.837	.834	.830	.827	.823	.820
45	.816	.813	.810	.806	.803	.799	.796	.793	.789	.786
46	.783	.779	.776	.773	.769	.766	.763	.760	.756	.753
47	.750	.747	.744	.740	.737	.734	.731	.728	.725	.721
48	.718	.715	.712	.709	.706	.703	.700	.697	.694	.691
49	.688	.685	.682	.679	.676	.673	.670	.667	.664	.661

Table IV continued

$\mu t \to \infty$ FUNCTION FOR   GAMMA=15

θ	0.0	0.1	0.2	0.3	0.4	0.5	0.6	0.7	0.8	0.9
50	.658	.656	.653	.650	.647	.644	.641	.638	.636	.633
51	.630	.627	.624	.622	.619	.616	.613	.610	.608	.605
52	.602	.600	.597	.594	.591	.589	.586	.583	.581	.578
53	.576	.573	.570	.568	.565	.562	.560	.557	.555	.552
54	.550	.547	.544	.542	.539	.537	.534	.532	.529	.527
55	.524	.522	.519	.517	.514	.512	.509	.507	.505	.502
56	.500	.497	.495	.493	.490	.488	.485	.483	.481	.478
57	.476	.474	.471	.469	.467	.464	.462	.460	.457	.455
58	.453	.450	.448	.446	.444	.441	.439	.437	.435	.432
59	.430	.428	.426	.424	.421	.419	.417	.415	.413	.410
60	.408	.406	.404	.402	.400	.397	.395	.393	.391	.389
61	.387	.385	.383	.381	.378	.376	.374	.372	.370	.368
62	.366	.364	.362	.360	.358	.356	.354	.352	.350	.348
63	.346	.344	.342	.340	.338	.336	.334	.332	.330	.328
64	.326	.324	.322	.320	.318	.316	.314	.312	.310	.308
65	.307	.305	.303	.301	.299	.297	.295	.293	.291	.290
66	.288	.286	.284	.282	.280	.279	.277	.275	.273	.271
67	.269	.268	.266	.264	.262	.260	.259	.257	.255	.253
68	.251	.250	.248	.246	.244	.243	.241	.239	.237	.236
69	.234	.232	.231	.229	.227	.225	.224	.222	.220	.219
70	.217	.215	.214	.212	.210	.209	.207	.205	.204	.202
71	.200	.199	.197	.196	.194	.192	.191	.189	.188	.186
72	.184	.183	.181	.180	.178	.176	.175	.173	.172	.170
73	.169	.167	.166	.164	.163	.161	.159	.158	.156	.155
74	.153	.152	.150	.149	.148	.146	.145	.143	.142	.140
75	.139	.137	.136	.134	.133	.131	.130	.129	.127	.126
76	.124	.123	.122	.120	.119	.117	.116	.115	.113	.112
77	.111	.109	.108	.107	.105	.104	.103	.101	.100	.099
78	.097	.096	.095	.093	.092	.091	.090	.088	.087	.086
79	.085	.083	.082	.081	.080	.078	.077	.076	.075	.074
80	.072	.071	.070	.069	.068	.067	.065	.064	.063	.062
81	.061	.060	.059	.058	.056	.055	.054	.053	.052	.051
82	.050	.049	.048	.047	.046	.045	.044	.043	.042	.041
83	.040	.039	.038	.037	.036	.035	.034	.033	.032	.032
84	.031	.030	.029	.028	.027	.026	.026	.025	.024	.023
85	.022	.022	.021	.020	.019	.018	.018	.017	.016	.016
86	.015	.014	.014	.013	.012	.012	.011	.011	.010	.009
87	.009	.008	.008	.007	.007	.006	.006	.005	.005	.005
88	.004	.004	.003	.003	.003	.002	.002	.002	.002	.001
89	.001	.001	.001	.001	.000	.000	.000	.000	.000	.000

H. W. King, C. J. Gillham, and F. G. Huggins

Table V

S-B Extrapolation Function for Specimen Curvature and
Displacement for Thin Transparent Specimens with $\mu t \to 0$

$$\frac{1 + \cos 2\theta}{\sin \gamma . \sin(2\theta - \gamma)}$$

$\mu t \to 0$ FUNCTION FOR GAMMA=90

θ	0.0	0.1	0.2	0.3	0.4	0.5	0.6	0.7	0.8	0.9
50	4.759	4.647	4.540	4.436	4.337	4.241	4.148	4.059	3.973	3.890
51	3.810	3.732	3.657	3.584	3.514	3.445	3.379	3.315	3.253	3.192
52	3.134	3.077	3.021	2.967	2.915	2.864	2.814	2.766	2.719	2.673
53	2.628	2.584	2.542	2.500	2.460	2.420	2.382	2.344	2.307	2.271
54	2.236	2.202	2.168	2.135	2.103	2.072	2.041	2.011	1.981	1.952
55	1.924	1.896	1.869	1.842	1.816	1.790	1.765	1.741	1.716	1.693
56	1.669	1.647	1.624	1.602	1.581	1.559	1.538	1.518	1.498	1.478
57	1.459	1.439	1.421	1.402	1.384	1.366	1.349	1.331	1.314	1.298
58	1.281	1.265	1.249	1.233	1.218	1.203	1.188	1.173	1.158	1.144
59	1.130	1.116	1.103	1.089	1.076	1.063	1.050	1.037	1.025	1.012
60	1.000	.988	.976	.964	.953	.942	.930	.919	.908	.898
61	.887	.877	.866	.856	.846	.836	.826	.817	.807	.798
62	.788	.779	.770	.761	.752	.743	.735	.726	.718	.710
63	.701	.693	.685	.677	.669	.662	.654	.646	.639	.632
64	.624	.617	.610	.603	.596	.589	.582	.575	.569	.562
65	.556	.549	.543	.537	.530	.524	.518	.512	.506	.500
66	.494	.489	.483	.477	.472	.466	.461	.455	.450	.445
67	.440	.434	.429	.424	.419	.414	.409	.404	.400	.395
68	.390	.386	.381	.376	.372	.367	.363	.359	.354	.350
69	.346	.341	.337	.333	.329	.325	.321	.317	.313	.309
70	.305	.302	.298	.294	.290	.287	.283	.280	.276	.272
71	.269	.266	.262	.259	.255	.252	.249	.246	.242	.239
72	.236	.233	.230	.227	.224	.221	.218	.215	.212	.209
73	.206	.203	.201	.198	.195	.192	.190	.187	.184	.182
74	.179	.177	.174	.172	.169	.167	.164	.162	.159	.157
75	.155	.152	.150	.148	.146	.143	.141	.139	.137	.135
76	.133	.130	.128	.126	.124	.122	.120	.118	.116	.115
77	.113	.111	.109	.107	.105	.103	.102	.100	.098	.096
78	.095	.093	.091	.090	.088	.086	.085	.083	.082	.080
79	.079	.077	.076	.074	.073	.071	.070	.068	.067	.066

Table V continued

$\mu t \to 0$ FUNCTION FOR GAMMA=90

θ	0.0	0.1	0.2	0.3	0.4	0.5	0.6	0.7	0.8	0.9
80	.064	.063	.062	.060	.059	.058	.056	.055	.054	.053
81	.051	.050	.049	.048	.047	.046	.045	.043	.042	.041
82	.040	.039	.038	.037	.036	.035	.034	.033	.032	.032
83	.031	.030	.029	.028	.027	.026	.025	.025	.024	.023
84	.022	.022	.021	.020	.019	.019	.018	.017	.017	.016
85	.015	.015	.014	.014	.013	.012	.012	.011	.011	.010
86	.010	.009	.009	.008	.008	.008	.007	.007	.006	.006
87	.006	.005	.005	.004	.004	.004	.004	.003	.003	.003
88	.002	.002	.002	.002	.002	.001	.001	.001	.001	.001
89	.001	.000	.000	.000	.000	.000	.000	.000	.000	.000

$\mu t \to 0$ FUNCTION FOR GAMMA=75

θ	0.0	0.1	0.2	0.3	0.4	0.5	0.6	0.7	0.8	0.9
40	13.94	13.37	12.84	12.34	11.88	11.45	11.05	10.68	10.32	9.99
41	9.677	9.381	9.101	8.836	8.584	8.345	8.118	7.901	7.695	7.498
42	7.310	7.130	6.957	6.792	6.634	6.482	6.335	6.195	6.060	5.930
43	5.804	5.683	5.567	5.454	5.345	5.240	5.138	5.040	4.945	4.852
44	4.763	4.676	4.592	4.510	4.431	4.354	4.279	4.207	4.136	4.067
45	4.000	3.935	3.871	3.809	3.749	3.690	3.633	3.577	3.523	3.469
46	3.417	3.367	3.317	3.269	3.221	3.175	3.130	3.085	3.042	3.000
47	2.958	2.917	2.878	2.839	2.801	2.763	2.726	2.691	2.655	2.621
48	2.587	2.554	2.521	2.489	2.458	2.427	2.397	2.367	2.338	2.309
49	2.281	2.253	2.226	2.199	2.173	2.147	2.122	2.097	2.072	2.048
50	2.024	2.001	1.978	1.955	1.933	1.911	1.889	1.868	1.847	1.827
51	1.806	1.786	1.767	1.747	1.728	1.709	1.691	1.672	1.654	1.636
52	1.619	1.602	1.584	1.568	1.551	1.535	1.519	1.503	1.487	1.471
53	1.456	1.441	1.426	1.411	1.397	1.382	1.368	1.354	1.341	1.327
54	1.313	1.300	1.287	1.274	1.261	1.249	1.236	1.224	1.212	1.200
55	1.188	1.176	1.164	1.153	1.141	1.130	1.119	1.108	1.097	1.086
56	1.076	1.065	1.055	1.045	1.035	1.025	1.015	1.005	.995	.985
57	.976	.967	.957	.948	.939	.930	.921	.912	.903	.895
58	.886	.878	.869	.861	.853	.845	.837	.829	.821	.813
59	.805	.798	.790	.783	.775	.768	.761	.753	.746	.739
60	.732	.725	.718	.711	.705	.698	.691	.685	.678	.672
61	.665	.659	.653	.647	.640	.634	.628	.622	.616	.611
62	.605	.599	.593	.587	.582	.576	.571	.565	.560	.554
63	.549	.544	.539	.533	.528	.523	.518	.513	.508	.503
64	.498	.493	.489	.484	.479	.474	.470	.465	.461	.456

Table V continued

$\mu t \to 0$　　FUNCTION FOR GAMMA=75

θ	0.0	0.1	0.2	0.3	0.4	0.5	0.6	0.7	0.8	0.9
65	.451	.447	.443	.438	.434	.430	.425	.421	.417	.413
66	.408	.404	.400	.396	.392	.388	.384	.380	.376	.373
67	.369	.365	.361	.358	.354	.350	.346	.343	.339	.336
68	.332	.329	.325	.322	.318	.315	.312	.308	.305	.302
69	.298	.295	.292	.289	.286	.283	.279	.276	.273	.270
70	.267	.264	.261	.258	.255	.253	.250	.247	.244	.241
71	.238	.236	.233	.230	.228	.225	.222	.220	.217	.214
72	.212	.209	.207	.204	.202	.199	.197	.194	.192	.190
73	.187	.185	.183	.180	.178	.176	.173	.171	.169	.167
74	.165	.162	.160	.158	.156	.154	.152	.150	.148	.146
75	.144	.142	.140	.138	.136	.134	.132	.130	.128	.126
76	.124	.123	.121	.119	.117	.115	.114	.112	.110	.108
77	.107	.105	.103	.102	.100	.098	.097	.095	.094	.092
78	.091	.089	.088	.086	.085	.083	.082	.080	.079	.077
79	.076	.075	.073	.072	.070	.069	.068	.067	.065	.064
80	.063	.061	.060	.059	.058	.057	.055	.054	.053	.052
81	.051	.050	.049	.047	.046	.045	.044	.043	.042	.041
82	.040	.039	.038	.037	.036	.035	.034	.033	.033	.032
83	.031	.030	.029	.028	.027	.027	.026	.025	.024	.023
84	.023	.022	.021	.020	.020	.019	.018	.018	.017	.016
85	.016	.015	.015	.014	.013	.013	.012	.012	.011	.011
86	.010	.010	.009	.009	.008	.008	.007	.007	.007	.006
87	.006	.005	.005	.005	.004	.004	.004	.003	.003	.003
88	.003	.002	.002	.002	.002	.001	.001	.001	.001	.001
89	.001	.001	.000	.000	.000	.000	.000	.000	.000	.000

$\mu t \to 0$　　FUNCTION FOR GAMMA=60

θ	0.0	0.1	0.2	0.3	0.4	0.5	0.6	0.7	0.8	0.9
35	8.924	8.729	8.542	8.362	8.189	8.022	7.861	7.705	7.555	7.410
36	7.270	7.134	7.003	6.876	6.753	6.634	6.518	6.406	6.297	6.191
37	6.089	5.989	5.892	5.797	5.706	5.616	5.529	5.444	5.362	5.281
38	5.203	5.126	5.051	4.979	4.907	4.838	4.770	4.704	4.639	4.576
39	4.514	4.453	4.394	4.336	4.279	4.223	4.169	4.116	4.064	4.012
40	3.962	3.913	3.865	3.818	3.772	3.726	3.682	3.638	3.595	3.553
41	3.511	3.471	3.431	3.392	3.353	3.315	3.278	3.242	3.206	3.170
42	3.136	3.102	3.068	3.035	3.002	2.970	2.939	2.908	2.877	2.847
43	2.818	2.789	2.760	2.732	2.704	2.677	2.650	2.623	2.597	2.571
44	2.545	2.520	2.496	2.471	2.447	2.423	2.400	2.377	2.354	2.332

Table V continued

θ	0.0	0.1	0.2	0.3	0.4	0.5	0.6	0.7	0.8	0.9
45	2.309	2.288	2.266	2.245	2.224	2.203	2.182	2.162	2.142	2.122
46	2.103	2.084	2.065	2.046	2.027	2.009	1.991	1.973	1.956	1.938
47	1.921	1.904	1.887	1.870	1.854	1.838	1.822	1.806	1.790	1.775
48	1.759	1.744	1.729	1.714	1.699	1.685	1.670	1.656	1.642	1.628
49	1.615	1.601	1.587	1.574	1.561	1.548	1.535	1.522	1.509	1.497
50	1.484	1.472	1.460	1.448	1.436	1.424	1.413	1.401	1.389	1.378
51	1.367	1.356	1.345	1.334	1.323	1.312	1.302	1.291	1.281	1.270
52	1.260	1.250	1.240	1.230	1.220	1.210	1.201	1.191	1.182	1.172
53	1.163	1.154	1.144	1.135	1.126	1.117	1.108	1.100	1.091	1.082
54	1.074	1.065	1.057	1.048	1.040	1.032	1.024	1.016	1.008	1.000
55	.992	.984	.976	.969	.961	.953	.946	.938	.931	.924
56	.916	.909	.902	.895	.888	.881	.874	.867	.860	.853
57	.847	.840	.833	.827	.820	.814	.807	.801	.795	.788
58	.782	.776	.770	.764	.758	.752	.746	.740	.734	.728
59	.722	.717	.711	.705	.700	.694	.688	.683	.677	.672
60	.667	.661	.656	.651	.645	.640	.635	.630	.625	.620
61	.615	.610	.605	.600	.595	.590	.585	.581	.576	.571
62	.566	.562	.557	.552	.548	.543	.539	.534	.530	.525
63	.521	.517	.512	.508	.504	.499	.495	.491	.487	.483
64	.479	.475	.471	.466	.462	.458	.455	.451	.447	.443
65	.439	.435	.431	.428	.424	.420	.416	.413	.409	.405
66	.402	.398	.395	.391	.387	.384	.380	.377	.374	.370
67	.367	.363	.360	.357	.353	.350	.347	.344	.340	.337
68	.334	.331	.328	.325	.321	.318	.315	.312	.309	.306
69	.303	.300	.297	.294	.291	.289	.286	.283	.280	.277
70	.274	.272	.269	.266	.263	.261	.258	.255	.252	.250
71	.247	.245	.242	.239	.237	.234	.232	.229	.227	.224
72	.222	.219	.217	.214	.212	.210	.207	.205	.203	.200
73	.198	.196	.193	.191	.189	.187	.184	.182	.180	.178
74	.176	.173	.171	.169	.167	.165	.163	.161	.159	.157
75	.155	.153	.151	.149	.147	.145	.143	.141	.139	.137
76	.135	.133	.132	.130	.128	.126	.124	.122	.121	.119
77	.117	.115	.114	.112	.110	.109	.107	.105	.104	.102
78	.100	.099	.097	.096	.094	.092	.091	.089	.088	.086
79	.085	.083	.082	.081	.079	.078	.076	.075	.073	.072
80	.071	.069	.068	.067	.065	.064	.063	.062	.060	.059
81	.058	.057	.055	.054	.053	.052	.051	.049	.048	.047
82	.046	.045	.044	.043	.042	.041	.040	.039	.038	.037
83	.036	.035	.034	.033	.032	.031	.030	.029	.028	.027
84	.027	.026	.025	.024	.023	.022	.022	.021	.020	.019

$\mu t \to 0$ FUNCTION FOR GAMMA=60

Table V continued

$\mu t \to 0$ FUNCTION FOR GAMMA=60

θ	0.0	0.1	0.2	0.3	0.4	0.5	0.6	0.7	0.8	0.9
85	.019	.018	.017	.017	.016	.015	.015	.014	.013	.013
86	.012	.012	.011	.010	.010	.009	.009	.008	.008	.007
87	.007	.006	.006	.006	.005	.005	.004	.004	.004	.003
88	.003	.003	.003	.002	.002	.002	.002	.001	.001	.001
89	.001	.001	.001	.000	.000	.000	.000	.000	.000	.000

$\mu t \to 0$ FUNCTION FOR GAMMA=45

θ	0.0	0.1	0.2	0.3	0.4	0.5	0.6	0.7	0.8	0.9
25	26.66	25.59	24.61	23.69	22.84	22.04	21.30	20.60	19.95	19.33
26	18.75	18.20	17.68	17.19	16.72	16.28	15.85	15.45	15.07	14.70
27	14.35	14.02	13.70	13.39	13.10	12.82	12.54	12.28	12.03	11.79
28	11.56	11.33	11.11	10.90	10.70	10.51	10.32	10.13	9.96	9.78
29	9.618	9.457	9.300	9.148	9.000	8.857	8.717	8.581	8.449	8.321
30	8.196	8.074	7.956	7.840	7.728	7.618	7.511	7.407	7.305	7.205
31	7.108	7.013	6.920	6.830	6.741	6.654	6.569	6.486	6.405	6.326
32	6.248	6.172	6.097	6.024	5.953	5.882	5.814	5.746	5.680	5.615
33	5.551	5.489	5.428	5.367	5.308	5.250	5.193	5.137	5.082	5.028
34	4.975	4.923	4.872	4.821	4.772	4.723	4.675	4.628	4.581	4.536
35	4.491	4.447	4.403	4.360	4.318	4.276	4.235	4.195	4.155	4.116
36	4.078	4.040	4.002	3.965	3.929	3.893	3.858	3.823	3.788	3.755
37	3.721	3.688	3.656	3.623	3.592	3.560	3.530	3.499	3.469	3.439
38	3.410	3.381	3.353	3.324	3.297	3.269	3.242	3.215	3.189	3.162
39	3.136	3.111	3.086	3.061	3.036	3.012	2.987	2.964	2.940	2.917
40	2.894	2.871	2.848	2.826	2.804	2.782	2.761	2.740	2.718	2.698
41	2.677	2.657	2.636	2.616	2.597	2.577	2.558	2.538	2.519	2.501
42	2.482	2.464	2.445	2.427	2.410	2.392	2.374	2.357	2.340	2.323
43	2.306	2.289	2.273	2.256	2.240	2.224	2.208	2.192	2.177	2.161
44	2.146	2.131	2.116	2.101	2.086	2.071	2.057	2.042	2.028	2.014
45	2.000	1.986	1.972	1.959	1.945	1.932	1.918	1.905	1.892	1.879
46	1.866	1.853	1.841	1.828	1.816	1.803	1.791	1.779	1.767	1.755
47	1.743	1.731	1.720	1.708	1.697	1.685	1.674	1.663	1.652	1.641
48	1.630	1.619	1.608	1.597	1.587	1.576	1.565	1.555	1.545	1.535
49	1.524	1.514	1.504	1.494	1.484	1.475	1.465	1.455	1.446	1.436
50	1.427	1.417	1.408	1.399	1.389	1.380	1.371	1.362	1.353	1.344
51	1.336	1.327	1.318	1.310	1.301	1.292	1.284	1.276	1.267	1.259
52	1.251	1.243	1.234	1.226	1.218	1.210	1.202	1.195	1.187	1.179
53	1.171	1.164	1.156	1.148	1.141	1.133	1.126	1.119	1.111	1.104
54	1.097	1.090	1.082	1.075	1.068	1.061	1.054	1.047	1.040	1.034

Table V continued

FUNCTION FOR GAMMA=45

θ	0.0	0.1	0.2	0.3	0.4	0.5	0.6	0.7	0.8	0.9
55	1.027	1.020	1.013	1.007	1.000	.993	.987	.980	.974	.967
56	.961	.954	.948	.942	.936	.929	.923	.917	.911	.905
57	.899	.893	.887	.881	.875	.869	.863	.857	.852	.846
58	.840	.834	.829	.823	.817	.812	.806	.801	.795	.790
59	.785	.779	.774	.769	.763	.758	.753	.748	.742	.737
60	.732	.727	.722	.717	.712	.707	.702	.697	.692	.687
61	.682	.677	.673	.668	.663	.658	.654	.649	.644	.640
62	.635	.630	.626	.621	.617	.612	.608	.603	.599	.595
63	.590	.586	.582	.577	.573	.569	.564	.560	.556	.552
64	.548	.543	.539	.535	.531	.527	.523	.519	.515	.511
65	.507	.503	.499	.495	.491	.488	.484	.480	.476	.472
66	.469	.465	.461	.457	.454	.450	.446	.443	.439	.435
67	.432	.428	.425	.421	.418	.414	.411	.407	.404	.400
68	.397	.394	.390	.387	.383	.380	.377	.374	.370	.367
69	.364	.361	.357	.354	.351	.348	.345	.341	.338	.335
70	.332	.329	.326	.323	.320	.317	.314	.311	.308	.305
71	.302	.299	.296	.293	.290	.288	.285	.282	.279	.276
72	.273	.271	.268	.265	.262	.260	.257	.254	.252	.249
73	.246	.244	.241	.238	.236	.233	.231	.228	.226	.223
74	.221	.218	.216	.213	.211	.208	.206	.203	.201	.199
75	.196	.194	.191	.189	.187	.184	.182	.180	.178	.175
76	.173	.171	.169	.166	.164	.162	.160	.158	.156	.153
77	.151	.149	.147	.145	.143	.141	.139	.137	.135	.133
78	.131	.129	.127	.125	.123	.121	.119	.117	.116	.114
79	.112	.110	.108	.106	.105	.103	.101	.099	.098	.096
80	.094	.092	.091	.089	.087	.086	.084	.082	.081	.079
81	.078	.076	.075	.073	.071	.070	.068	.067	.066	.064
82	.063	.061	.060	.058	.057	.056	.054	.053	.052	.050
83	.049	.048	.046	.045	.044	.043	.042	.040	.039	.038
84	.037	.036	.035	.033	.032	.031	.030	.029	.028	.027
85	.026	.025	.024	.023	.022	.022	.021	.020	.019	.018
86	.017	.016	.016	.015	.014	.013	.013	.012	.011	.011
87	.010	.009	.009	.008	.008	.007	.006	.006	.005	.005
88	.005	.004	.004	.003	.003	.003	.002	.002	.002	.001
89	.001	.001	.001	.001	.000	.000	.000	.000	.000	.000

Table V continued

$\mu t \to 0$ FUNCTION FOR GAMMA=30

θ	0.0	0.1	0.2	0.3	0.4	0.5	0.6	0.7	0.8	0.9
20	20.34	19.92	19.52	19.13	18.75	18.39	18.04	17.71	17.38	17.07
21	16.77	16.48	16.19	15.92	15.65	15.39	15.14	14.90	14.66	14.44
22	14.21	14.00	13.79	13.58	13.39	13.19	13.00	12.82	12.64	12.47
23	12.30	12.13	11.97	11.81	11.66	11.51	11.36	11.22	11.07	10.94
24	10.80	10.67	10.54	10.42	10.29	10.17	10.06	9.94	9.83	9.72
25	9.606	9.500	9.395	9.292	9.192	9.093	8.996	8.901	8.808	8.716
26	8.626	8.538	8.451	8.365	8.281	8.199	8.118	8.038	7.960	7.883
27	7.807	7.733	7.660	7.588	7.517	7.447	7.378	7.310	7.244	7.178
28	7.114	7.050	6.987	6.926	6.865	6.805	6.746	6.687	6.630	6.573
29	6.518	6.463	6.408	6.355	6.302	6.250	6.199	6.148	6.098	6.049
30	6.000	5.952	5.905	5.858	5.811	5.766	5.721	5.676	5.632	5.589
31	5.546	5.504	5.462	5.420	5.380	5.339	5.299	5.260	5.221	5.183
32	5.144	5.107	5.070	5.033	4.996	4.961	4.925	4.890	4.855	4.821
33	4.787	4.753	4.720	4.687	4.654	4.622	4.590	4.558	4.527	4.496
34	4.465	4.435	4.405	4.375	4.346	4.317	4.288	4.260	4.231	4.203
35	4.176	4.148	4.121	4.094	4.067	4.041	4.015	3.989	3.963	3.938
36	3.913	3.888	3.863	3.838	3.814	3.790	3.766	3.742	3.719	3.696
37	3.673	3.650	3.627	3.605	3.583	3.560	3.539	3.517	3.495	3.474
38	3.453	3.432	3.411	3.391	3.370	3.350	3.330	3.310	3.290	3.270
39	3.251	3.231	3.212	3.193	3.174	3.156	3.137	3.119	3.100	3.082
40	3.064	3.046	3.029	3.011	2.993	2.976	2.959	2.942	2.925	2.908
41	2.891	2.875	2.858	2.842	2.826	2.809	2.793	2.778	2.762	2.746
42	2.731	2.715	2.700	2.685	2.669	2.654	2.639	2.625	2.610	2.595
43	2.581	2.566	2.552	2.538	2.524	2.510	2.496	2.482	2.468	2.454
44	2.441	2.427	2.414	2.400	2.387	2.374	2.361	2.348	2.335	2.322
45	2.309	2.297	2.284	2.272	2.259	2.247	2.235	2.222	2.210	2.198
46	2.186	2.174	2.162	2.151	2.139	2.127	2.116	2.104	2.093	2.081
47	2.070	2.059	2.048	2.036	2.025	2.014	2.004	1.993	1.982	1.971
48	1.960	1.950	1.939	1.929	1.918	1.908	1.898	1.887	1.877	1.867
49	1.857	1.847	1.837	1.827	1.817	1.807	1.797	1.788	1.778	1.768
50	1.759	1.749	1.740	1.730	1.721	1.712	1.702	1.693	1.684	1.675
51	1.666	1.657	1.648	1.639	1.630	1.621	1.612	1.603	1.595	1.586
52	1.577	1.569	1.560	1.552	1.543	1.535	1.526	1.518	1.510	1.501
53	1.493	1.485	1.477	1.469	1.461	1.452	1.444	1.437	1.429	1.421
54	1.413	1.405	1.397	1.389	1.382	1.374	1.366	1.359	1.351	1.344
55	1.336	1.329	1.321	1.314	1.307	1.299	1.292	1.285	1.277	1.270
56	1.263	1.256	1.249	1.242	1.235	1.228	1.221	1.214	1.207	1.200
57	1.193	1.186	1.179	1.173	1.166	1.159	1.152	1.146	1.139	1.133
58	1.126	1.119	1.113	1.106	1.100	1.094	1.087	1.081	1.074	1.068
59	1.062	1.055	1.049	1.043	1.037	1.031	1.024	1.018	1.012	1.006

Table V continued

μt → 0   FUNCTION FOR   GAMMA=30

θ	0.0	0.1	0.2	0.3	0.4	0.5	0.6	0.7	0.8	0.9
60	1.000	.994	.988	.982	.976	.970	.964	.958	.952	.947
61	.941	.935	.929	.923	.918	.912	.906	.901	.895	.889
62	.884	.878	.873	.867	.862	.856	.851	.845	.840	.834
63	.829	.824	.818	.813	.808	.802	.797	.792	.787	.781
64	.776	.771	.766	.761	.756	.751	.746	.740	.735	.730
65	.725	.720	.716	.711	.706	.701	.696	.691	.686	.681
66	.677	.672	.667	.662	.657	.653	.648	.643	.639	.634
67	.629	.625	.620	.616	.611	.606	.602	.597	.593	.588
68	.584	.579	.575	.571	.566	.562	.557	.553	.549	.544
69	.540	.536	.532	.527	.523	.519	.515	.510	.506	.502
70	.498	.494	.490	.486	.481	.477	.473	.469	.465	.461
71	.457	.453	.449	.445	.441	.438	.434	.430	.426	.422
72	.418	.414	.410	.407	.403	.399	.395	.392	.388	.384
73	.380	.377	.373	.369	.366	.362	.359	.355	.351	.348
74	.344	.341	.337	.334	.330	.327	.323	.320	.316	.313
75	.309	.306	.303	.299	.296	.293	.289	.286	.283	.279
76	.276	.273	.270	.266	.263	.260	.257	.254	.250	.247
77	.244	.241	.238	.235	.232	.229	.226	.223	.220	.217
78	.214	.211	.208	.205	.202	.199	.196	.193	.190	.188
79	.185	.182	.179	.176	.174	.171	.168	.165	.163	.160
80	.157	.155	.152	.150	.147	.144	.142	.139	.137	.134
81	.132	.129	.127	.124	.122	.119	.117	.115	.112	.110
82	.108	.105	.103	.101	.099	.096	.094	.092	.090	.088
83	.086	.083	.081	.079	.077	.075	.073	.071	.069	.067
84	.065	.063	.062	.060	.058	.056	.054	.052	.051	.049
85	.047	.046	.044	.042	.041	.039	.038	.036	.035	.033
86	.032	.030	.029	.027	.026	.025	.023	.022	.021	.020
87	.019	.018	.016	.015	.014	.013	.012	.011	.010	.010
88	.009	.008	.007	.006	.006	.005	.004	.004	.003	.003
89	.002	.002	.001	.001	.001	.001	.000	.000	.000	.000

Table V continued

$\mu t \to 0$ FUNCTION FOR GAMMA=15

θ	0.0	0.1	0.2	0.3	0.4	0.5	0.6	0.7	0.8	0.9
10	85.99	82.64	79.54	76.66	73.97	71.47	69.13	66.93	64.87	62.93
11	61.10	59.37	57.73	56.18	54.71	53.32	51.99	50.72	49.52	48.36
12	47.26	46.21	45.20	44.23	43.31	42.42	41.56	40.74	39.95	39.18
13	38.45	37.74	37.06	36.40	35.76	35.14	34.54	33.97	33.41	32.87
14	32.34	31.83	31.34	30.86	30.39	29.94	29.50	29.07	28.66	28.25
15	27.86	27.47	27.10	26.73	26.38	26.03	25.69	25.36	25.04	24.73
16	24.42	24.12	23.83	23.54	23.26	22.99	22.72	22.46	22.20	21.95
17	21.71	21.47	21.23	21.00	20.77	20.55	20.33	20.12	19.91	19.71
18	19.50	19.31	19.11	18.92	18.73	18.55	18.37	18.19	18.02	17.85
19	17.68	17.52	17.35	17.19	17.04	16.88	16.73	16.58	16.43	16.29
20	16.15	16.01	15.87	15.73	15.60	15.47	15.34	15.21	15.08	14.96
21	14.84	14.71	14.60	14.48	14.36	14.25	14.14	14.03	13.92	13.81
22	13.70	13.60	13.49	13.39	13.29	13.19	13.09	13.00	12.90	12.81
23	12.71	12.62	12.53	12.44	12.35	12.26	12.18	12.09	12.01	11.92
24	11.84	11.76	11.68	11.60	11.52	11.44	11.37	11.29	11.21	11.14
25	11.07	10.99	10.92	10.85	10.78	10.71	10.64	10.57	10.51	10.44
26	10.37	10.31	10.24	10.18	10.12	10.05	9.99	9.93	9.87	9.81
27	9.748	9.689	9.631	9.573	9.515	9.459	9.402	9.347	9.291	9.237
28	9.183	9.129	9.076	9.023	8.971	8.919	8.868	8.817	8.767	8.717
29	8.667	8.618	8.570	8.522	8.474	8.427	8.380	8.333	8.287	8.241
30	8.196	8.151	8.107	8.062	8.019	7.975	7.932	7.889	7.847	7.805
31	7.763	7.722	7.681	7.640	7.600	7.559	7.520	7.480	7.441	7.402
32	7.364	7.325	7.287	7.250	7.212	7.175	7.138	7.102	7.066	7.030
33	6.994	6.958	6.923	6.888	6.853	6.819	6.785	6.751	6.717	6.683
34	6.650	6.617	6.584	6.552	6.519	6.487	6.455	6.424	6.392	6.361
35	6.330	6.299	6.268	6.238	6.208	6.178	6.148	6.118	6.089	6.060
36	6.031	6.002	5.973	5.945	5.916	5.888	5.860	5.832	5.805	5.777
37	5.750	5.723	5.696	5.669	5.643	5.616	5.590	5.564	5.538	5.512
38	5.486	5.461	5.435	5.410	5.385	5.360	5.336	5.311	5.286	5.262
39	5.238	5.214	5.190	5.166	5.143	5.119	5.096	5.072	5.049	5.026
40	5.003	4.981	4.958	4.936	4.913	4.891	4.869	4.847	4.825	4.803
41	4.782	4.760	4.739	4.717	4.696	4.675	4.654	4.633	4.612	4.592
42	4.571	4.551	4.530	4.510	4.490	4.470	4.450	4.430	4.411	4.391
43	4.371	4.352	4.333	4.313	4.294	4.275	4.256	4.237	4.219	4.200
44	4.181	4.163	4.144	4.126	4.108	4.090	4.071	4.053	4.036	4.018
45	4.000	3.982	3.965	3.947	3.930	3.912	3.895	3.878	3.861	3.844
46	3.827	3.810	3.793	3.777	3.760	3.743	3.727	3.710	3.694	3.678
47	3.661	3.645	3.629	3.613	3.597	3.581	3.566	3.550	3.534	3.519
48	3.503	3.487	3.472	3.457	3.441	3.426	3.411	3.396	3.381	3.366
49	3.351	3.336	3.321	3.307	3.292	3.277	3.263	3.248	3.234	3.219

Table V continued

$\mu t \to 0$ FUNCTION FOR GAMMA=15

θ	0.0	0.1	0.2	0.3	0.4	0.5	0.6	0.7	0.8	0.9
50	3.205	3.191	3.176	3.162	3.148	3.134	3.120	3.106	3.092	3.078
51	3.065	3.051	3.037	3.024	3.010	2.996	2.983	2.969	2.956	2.943
52	2.929	2.916	2.903	2.890	2.877	2.864	2.851	2.838	2.825	2.812
53	2.799	2.786	2.774	2.761	2.748	2.736	2.723	2.711	2.698	2.686
54	2.673	2.661	2.649	2.637	2.624	2.612	2.600	2.588	2.576	2.564
55	2.552	2.540	2.528	2.516	2.505	2.493	2.481	2.469	2.458	2.446
56	2.434	2.423	2.411	2.400	2.389	2.377	2.366	2.355	2.343	2.332
57	2.321	2.310	2.298	2.287	2.276	2.265	2.254	2.243	2.232	2.221
58	2.211	2.200	2.189	2.178	2.167	2.157	2.146	2.135	2.125	2.114
59	2.104	2.093	2.083	2.072	2.062	2.051	2.041	2.031	2.021	2.010
60	2.000	1.990	1.980	1.969	1.959	1.949	1.939	1.929	1.919	1.909
61	1.899	1.889	1.879	1.870	1.860	1.850	1.840	1.830	1.821	1.811
62	1.801	1.792	1.782	1.772	1.763	1.753	1.744	1.734	1.725	1.715
63	1.706	1.697	1.687	1.678	1.669	1.659	1.650	1.641	1.632	1.622
64	1.613	1.604	1.595	1.586	1.577	1.568	1.559	1.550	1.541	1.532
65	1.523	1.514	1.505	1.496	1.487	1.479	1.470	1.461	1.452	1.443
66	1.435	1.426	1.417	1.409	1.400	1.392	1.383	1.374	1.366	1.357
67	1.349	1.340	1.332	1.324	1.315	1.307	1.298	1.290	1.282	1.273
68	1.265	1.257	1.249	1.240	1.232	1.224	1.216	1.208	1.200	1.191
69	1.183	1.175	1.167	1.159	1.151	1.143	1.135	1.127	1.119	1.111
70	1.104	1.096	1.088	1.080	1.072	1.064	1.057	1.049	1.041	1.033
71	1.026	1.018	1.010	1.003	.995	.987	.980	.972	.965	.957
72	.950	.942	.935	.927	.920	.912	.905	.897	.890	.883
73	.875	.868	.861	.853	.846	.839	.832	.824	.817	.810
74	.803	.796	.788	.781	.774	.767	.760	.753	.746	.739
75	.732	.725	.718	.711	.704	.697	.690	.684	.677	.670
76	.663	.656	.650	.643	.636	.629	.623	.616	.609	.603
77	.596	.589	.583	.576	.570	.563	.557	.550	.544	.537
78	.531	.524	.518	.512	.505	.499	.493	.486	.480	.474
79	.467	.461	.455	.449	.443	.437	.430	.424	.418	.412
80	.406	.400	.394	.388	.382	.376	.371	.365	.359	.353
81	.347	.341	.336	.330	.324	.319	.313	.307	.302	.296
82	.291	.285	.280	.274	.269	.263	.258	.253	.247	.242
83	.237	.232	.226	.221	.216	.211	.206	.201	.196	.191
84	.186	.181	.176	.171	.167	.162	.157	.153	.148	.143
85	.139	.134	.130	.126	.121	.117	.113	.109	.104	.100
86	.096	.092	.088	.084	.081	.077	.073	.070	.066	.062
87	.059	.056	.052	.049	.046	.043	.040	.037	.034	.032
88	.029	.026	.024	.022	.019	.017	.015	.013	.011	.010
89	.008	.007	.005	.004	.003	.002	.001	.001	.000	.000

MEASUREMENT OF LONG RANGE ORDER IN THE γ' PHASE OF NICKEL-BASE SUPERALLOYS

J.R. Mihalisin
The International Nickel Company, Inc.
Paul D. Merica Research Laboratory
Sterling Forest
Suffern, New York

ABSTRACT

The long range order parameter (S) has been measured at room temperature on the γ' phase extracted from IN-731 and alloys 713C and 713LC by x-ray diffraction techniques. Measurements were obtained from specimens of IN-731 and alloy 713LC in the as-cast condition and after long time rupture testing. The alloy 713C specimens were in the as-cast and high temperature heat treated conditions.

It was found that long range order in the γ' phase of IN-731 and alloy 713LC was changed very little after long time rupture testing, and after high temperature heat treatment in the case of alloy 713C. It was also found that chromium and molybdenum most probably distribute themselves approximately in the ratio of 4 to 1 on nickel sites in the γ' phase of these alloys.

INTRODUCTION

The γ' phase in nickel-base superalloys is based on the compound Ni_3Al, an ordered B_3A type structure of the LI_2 type, where B is usually a transition element. Extensive solubility of other elements can take place in the γ' phase as demonstrated by Guard and Westbrook[1]. Some elements prefer either B or A sites while others show intermediate behavior. A previous paper[2] reported the chemical composition of γ' extracted from various commercial nickel-base

superalloys. In that same work, the long range order
parameter of the γ' phase was measured to establish the
formula of the γ' phase for density determinations for
subsequent electron vacancy calculations.

These long range order measurements were not
reported in this earlier study[2] since they were
incidental to the objective of that work. Because
data of this type are of importance to investigators
interested in strengthening mechanisms in nickel-base
superalloys, these results are presented here.

EXPERIMENTAL PROCEDURE

Extracted residues of the γ' phase were obtained
from IN-731, alloy 713LC and alloy 713C and chemically
analyzed as described in[2]. Integrated X-ray intensity
measurements were made on the powder residues which
had been compacted into Lucite* holders. The (100)
superlattice reflection and (200) fundamental reflection
of the γ' phase were used in order to eliminate
preferred orientation effects. Iron filtered CoKα
radiation was used with a diffractometer scan rate of
$1/8°$/minute. A $1°$ slit and 0.3 receiving slit was
used. The (100) superlattice peak was scanned from
$2\theta = 28-30°$, while the (200) peak was scanned from
$2\theta = 59-61°$. The total counts were accumulated by a
timer-scaler. The background correction was made by
averaging the counts at both ends of the counting
ranges for times equivalent to that for scanning the
peaks. The integrated X-ray intensity of each peak is
then simply the total counts obtained from scanning
each peak minus the average background count.

The relationship between the long range order
parameter S for a B_3A compound and the integrated
intensities (A_{100} and A_{200}) of the (100) and (200)
reflections is as follows:

$$A_{100}/A_{200} = \frac{(f_A-f_B)^2}{(f_A+3\ f_B)^2} \times S^2 \times \frac{L.P._{100}}{L.P._{200}} \times \frac{\exp -2M(100)}{\exp -2M(200)} \quad (1)$$

where f_A is the atomic scattering factor for element A
and f_B is the atomic scattering factor for element B
and f_A-f_B corresponds to the (100) superlattice
reflection and $(f_A + 3\ f_B)$ to the (200) fundamental
reflection. In this study, use was made of previous
investigations to establish the tendency for solubility
of various elements on either A or B sites in the B_3A

*Trademark of the E.I. DuPont de Nemours and Co., Inc.

compound. From this information it was assumed that
f_A and f_B was then simply the sum of the atomic
scattering factors (multiplying atom fraction of each
element by its scattering factor) for elements on A and
B sites. Thus f_A and f_B become the sum of the atomic
scattering factors for elements on A and B sites
respectively. L.P. is the Lorentz Polarization Factor
for both reflections. In the experimental arrangement
used, L.P. is equal to $\dfrac{1 + \cos^2 2\theta}{\sin^2\theta \cos\theta}$. exp -2M is the
temperature factor for each reflection.
The 2M factor is equal to $\dfrac{B \sin^2\theta}{\lambda^2}$ where

B, the Debye parameter, $= \dfrac{6 h^2}{mkT_D} \left[\dfrac{\phi(x)}{x} + \dfrac{1}{4}\right]$

which can be evaluated from tables in Smithells[3] and
h = Planck's constant, m = mass of atom, k = Boltzmann's
constant, T_D = characteristic temperature, $x = \dfrac{T_D}{T}$.

This factor could only be estimated from the available
tables for pure elements. In this case, the values for
pure nickel were chosen. This is a valid assumption
since the B_3A compound in these alloys is rich in
nickel and aluminum (mostly nickel) and the
characteristic temperature (T_D) for nickel and aluminum
are not very different, so that B would not vary greatly
and the factor exp -2M would mainly depend on $\dfrac{\sin^2\theta}{\lambda^2}$.

Table I

Values of X-Ray Constants Used in Equation (1)

	(hkl)	
	(100)	(200)
Lorentz Polarization Factor	29.18	5.80
exp -2M	.992	.969
Atomic Scattering Factors (f)	$\sin\theta/\lambda$ for (100)	$\sin\theta/\lambda$ for (200)
Ni	23.87	18.65
Al	10.18	7.99
V	17.93	13.43
Cb	35.67	28.68
Mo	36.59	29.46
Cr	19.66	14.84
Ti	16.75	12.75
Co	22.97	17.90

Table I lists the values of the X-ray constants used in the computations for equation (1). The long range order parameter S is evaluated by using the constants in Table I and the experimentally determined values of the integrated intensity ratio of the (100) superlattice and the (200) fundamental peaks.

RESULTS AND DISCUSSION

The chemical compositions of the three alloys studied here are given in Table II. Table III lists the values of the long range order parameter for the γ' of IN-731 as-cast and after rupture testing. As mentioned earlier, previous work[1] had shown that aluminum, titanium, and vanadium will occupy A sites, and nickel and cobalt, B sites in the ordered B_3A structure. Chromium and molybdenum show intermediate behavior (i.e., they can occupy both sites). This is important in the determination of the long range order parameter S. It is obvious from equation (1) that S depends upon which atoms occupy A and B sites. In Table III there is no problem in this respect with the γ' phase extracted from the rupture tested IN-731 specimens. In that case, all the chromium and molybdenum is needed to occupy B sites. It also shows that vanadium can enter B sites as well as A sites in order to satisfy stoichiometry. In the case of the as-cast specimen, there is a dilemma. Chromium and molybdenum have to be distributed between A and B sites to satisfy the formula B_3A. An inspection of equation (1) will show that two extremes are found for the value of S with all vanadium on A sites:

(a) S will have a minimum value when molybdenum is placed on all the remaining B sites.

(b) S will have a maximum value when chromium is placed on all the remaining B sites.

For (a) S has the value .82, for (b) S has the value .88.

Table II

Chemical Composition (Wt./%) of Alloys Studied

IN-731

Alloy	C	Cr	Co	Mo	V	Ti	Al	B	Si	Fe	Ni
1	.16	9.50	9.72	2.46	.85	4.66	5.60	.014	.045	.23	66.94

Alloy 713LC

Alloy	C	Cr	Mo	Cb+Ta	Al	Ti	B	Si	Fe	Ni
2	.06	12.32	4.46	2.13	5.90	.72	.009	.05	.20	74.40
3	.03	12.52	4.41	2.10	5.90	.60	.009	<.10	.15	74.43

Alloy 713C

Alloy	C	Cr	Mo	Cb+Ta	Al	Ti	B	Si	Fe	Ni
4	.11	13.23	4.46	2.09	5.86	.79	.009	.04	.10	73.45

Table III

γ′ Formulas and Long Range Order Parameters S for IN-731 As-Cast and After Rupture Testing (Alloy 1)

Treatment	γ′ Formula	Measured Integrated Intensity Ratio A_{100}/A_{200}	Long Range Order Parameter S
As-Cast	$(\mathrm{Ni}_{.880}\ \mathrm{Co}_{.079}\ \mathrm{Cr}_{.034}\ \mathrm{Mo}_{.009})_3\,(\mathrm{Al}_{.613}\ \mathrm{Ti}_{.331}\ \mathrm{V}_{.023}\ \mathrm{Cr}_{.023}\ \mathrm{Mo}_{.010})$	.099	.84
1350°F/ 80,000 psi 2711.3 Hrs Life	$(\mathrm{Ni}_{.892}\ \mathrm{Co}_{.012}\ \mathrm{Cr}_{.033}\ \mathrm{Mo}_{.011}\ \mathrm{V}_{.003})_3\,(\mathrm{Al}_{.641}\ \mathrm{Ti}_{.336}\ \mathrm{V}_{.023})$	.103	.83
1500°F/ 50,000 psi 1925.6 Hrs Life	$(\mathrm{Ni}_{.890}\ \mathrm{Co}_{.070}\ \mathrm{Cr}_{.030}\ \mathrm{Mo}_{.007}\ \mathrm{V}_{.003})_3\,(\mathrm{Al}_{.682}\ \mathrm{Ti}_{.312}\ \mathrm{V}_{.006})$	.106	.82
1800°F/ 15,000 psi 773.7 Hrs Life	$(\mathrm{Ni}_{.875}\ \mathrm{Co}_{.070}\ \mathrm{Cr}_{.031}\ \mathrm{Mo}_{.007}\ \mathrm{V}_{.007})_3\,(\mathrm{Al}_{.590}\ \mathrm{Ti}_{.410})$	.100	.84

It will be noticed for the rupture specimens where all the chromium and molybdenum is required on B sites that the Cr/Mo ratio is approximately 4 to 1. Using this ratio (as was done in Table III) for the as-cast specimen yields a value of .84 for S which is in excellent agreement with the other values for S in the table. However, considering the assumptions made in the calculations, experimental error and the accuracy of the chemical analysis indicates these measured values of S to be no better than $\pm$.03 units.

The same situation holds true for all the specimens of alloy 713C and 713LC in Tables IV and V. Table IV compares long range order in the γ' phase from as-cast and rupture tested alloy 713LC (2 heats). Table V shows the effect of high temperature heat treatment on long range order in the γ' phase from alloy 713C. Here again one is faced with the problem of apportioning chromium and molybdenum to B and A sites. The values of S shown in Tables IV and V are calculated on the basis that the Cr/Mo ratio on B sites is 4/1 as for IN-731. It will be seen that this produces values of S for alloy 713LC that are in excellent agreement with one another (comparing as-cast and rupture tested specimens). With alloy 713C the agreement is not as good (Table V). It is possible here that the higher temperatures to which alloy 713C was subjected might have influenced the degree of order to some extent. Note that S tends to be lower after a 2150°F/1 hr treatment.

Table IV

γ' Formulas and Long Range Order Parameters S for Two Heats of
Alloy 713LC As-Cast and After Rupture Testing

Treatment	γ' Formula	Measured Integrated Intensity Ratio A_{100}/A_{200}	Long Range Order Parameter S
As-Cast	(Alloy 3) (Ni .990 Cr .008 Mo .002)₃(Al .706 Cb .098 Ti .042 Cr .114 Mo .041)	.078	.95
1500°F/ 40,000 psi 1862.7 Hrs Life	(Ni .966 Cr .027 Mo .007)₃(Al .721 Cb .116 Ti .048 Cr .065 Mo .050)	.074	.96
As-Cast	(Alloy 2) (Ni .976 Cr .019 Mo .005)₃(Al .716 Cb .106 Ti .045 Cr .094 Mo .039)	.079	.95
1500°F/ 35,000 psi 4108.6 Hrs Life	(Ni .973 Cr .022 Mo .006)₃(Al .737 Cb .090 Ti .045 Cr .073 Mo .055)	.081	.95

Table V

γ' Formulas and Long Range Order Parameters S for Alloy 713C
After Heat Treatment (Alloy 4)

Treatment	γ' Formula	Measured Integrated Intensity Ratio A_{100}/A_{200}	Long Range Order Parameter S
As-Cast	$(Ni_{.982} Cr_{.014} Mo_{.004})_3 (Al_{.717} Cb_{.087} Ti_{.050} Mo_{.039} Cr_{.107})$	.080	.93
2150°F/1 Hr	$(Ni_{.978} Cr_{.018} Mo_{.004})_3 (Al_{.712} Cb_{.097} Ti_{.053} Mo_{.034} Cr_{.104})$	.074	.91
1900°F/4 Hrs	$(Ni_{.978} Cr_{.018} Mo_{.004})_3 (Al_{.718} Cb_{.100} Ti_{.051} Mo_{.037} Cr_{.097})$	.080	.95
2150°F/1 Hr + 1900°F/4 Hrs	$(Ni_{.976} Cr_{.019} Mo_{.005})_3 (Al_{.722} Cb_{.098} Ti_{.052} Mo_{.032} Cr_{.096})$	.074	.90

Using the same criteria as with IN-731 for extreme values of S yields a maximum and minimum range for S for the γ' phase from Tables IV and V of .88-1.0 where either molybdenum or chromium is apportioned entirely to B sites. It appears unlikely that the high value could be correct since a recent measured value of S for annealed pure Ni_3Al yielded a value of S = .94[4] and there is no reason to believe that long range order is increased by alloying. On the contrary theoretical calculations indicate that the reverse should be true[5]. An average of the values of S for Tables IV and V (Cr/Mo ratio = 4/1) gives a value of $\bar{S}$ = .94. This could mean that the value of S for alloys 713C and 713LC should be somewhat less than that obtained for pure Ni_3Al because of solution of other elements in the γ'. This could be accomplished by using a smaller Cr/Mo ratio. In view of the accuracy of these measurements such refinements can hardly be made with the data here with any degree of confidence. On the other hand, the measured S values of IN-731 are significantly smaller than for alloy 713LC or alloy 713C. In the case of IN-731 there is heavier substitution of alloy elements in the γ' phase particularly cobalt and titanium. This may indicate certain elements have a greater effect on long range order than others in the γ' phase.

From the above it therefore does not appear unreasonable to assume that chromium and molybdenum distribute themselves in the ratio of approximately 4/1 on B sites in the γ' phase of these alloys. Assuming this to be the case, there is very little variation in long range order in the γ' phase in IN-731 and alloy 713LC after long time rupture testing or after high temperature heat treatment in the case of alloy 713C. Regardless of how chromium or molybdenum is distributed the maximum variation in S could only be from 7-12 percent.

It should be kept in mind, however, that these measurements were made at room temperature and it is possible that under conditions of stress and high temperatures there may be a decrease in the order parameter S and subsequent increase in order on cooling. However, the work of Corey and Potter[6] on pure Ni_3Al indicates no substantial increase in order after isothermal annealing and subsequent cooling to room temperature.

CONCLUSIONS

1.   There is a very little variation in long range
order (measured at room temperature) in the γ' phase of
IN-731 and alloy 713LC after long time rupture testing
and after high temperature heat treatment in alloy 713C.

2.   It is most probable that chromium and
molybdenum distribute themselves approximately in the
ratio of 4 to 1 on nickel sites in the γ' phase of
IN-731, alloy 713LC and alloy 713C.

3.   The long range order parameter of IN-731 is
significantly smaller than that of alloy 713LC and
alloy 713C. Vanadium, like chromium and molybdenum,
can occupy both nickel and aluminum sites in the γ'
phase of this alloy.

REFERENCES

1.   R.W. Guard and J.H. Westbrook, "Alloying Behavior
     of Ni₃Al (γ' Phase)", Trans. AIME, Vol. 215,
     October 1959.

2.   J.R. Mihalisin and D.L. Pasquine, "Phase
     Transformations in Nickel-Base Superalloys",
     International Symposium on Structural Stability
     in Superalloys, Vol. I, Sept. 4-6, 1968.

3.   Smithells, "Metals Reference Book", 4th Edition,
     Vol. 1, New York Plenum Press, 1967, pp. 120-123.

4.   C.L. Corey, Private Communication, Wayne State
     University.

5.   H.K. Hardy, "The Phase Diagram A-BC (Ordered
     Compound), Acta Metallurgica, Vol. 1, 1953.

6.   C.L. Corey and D.I. Potter, "Recovery Processes
     and Ordering in Ni₃Al", Journal of Applied
     Physics, Vol. 38, No. 10, Sept. 1967.

MEASUREMENT OF THE MOLECULAR SIZE OF A SODIUM HUMATE FRACTION*

R. L. Wershaw, S. J. Heller and D. J. Pinckney

U. S. Geological Survey

Denver, Colorado 80225

INTRODUCTION

Many of the chemical and physical interactions that take place in natural soil-water systems are strongly influenced by the presence of natural organic poly-electrolytes. The most common of these organic poly-electrolytes is humic acid, which is defined as the alkali soluble, acid insoluble fraction of soil organic matter. Over a hundred years of experimental data have demonstrated that a wide variety of different materials having different physical and chemical properties fit this definition. If we are to make any progress in understanding the role of humic materials in soil-water systems, then we must have a classification system which, as far as possible, provides a unique taxonomic defini-tion for each separate molecular species that falls into the category of humic acid. This in turn requires that we isolate the various molecular species found in dif-ferent humic acids, and measure their physical and chemical properties and elucidate their chemical structures. As a first step in developing such a clas-sification system we have isolated a fraction of sodium

---

*Publication authorized by the Director, U.S. Geological Survey.

humate from a North Carolina soil and have measured
its molecular size by small angle x-ray scattering.

As our work progresses we plan to isolate the
other fractions of the North Carolina soil sodium
humate and to measure their molecular sizes. We will
then have a means of characterizing this particular
sodium humate and of comparing it with other sodium
humates. In this way a classification system can be
built up.

Mehta, et al[1], and a number of subsequent workers
have been able to successfully fractionate humic acids
by gel permeation chromatography on Sephadex. These
workers however, have not been able to demonstrate that
they have isolated monodisperse fractions of humic acid.
Our work takes up where these workers left off. We have
used Sephadex chromatography to isolate the various humic
acid fractions, and small angle x-ray scattering and
infrared spectrometry to characterize the fractions.
Small angle x-ray scattering gives a direct measure
of the sizes and shapes of macromolecules in solution
and in this respect it provides information that is
not readily available by any other method. By combining
these measurements with density measurements it is then
possible to calculate the molecular weights of macromolecules.

A wide variety of different molecular weights have
been published for humic acid. Piret, et al[2], Mehta,
et al[1], Gjessing[3], Bailly[4], Wershaw, et al[5] and others
have shown that unfractionated humic acid is polydis-
perse; as we shall show in the following we have isolated a fraction which is either monodisperse or is
made up of molecules having no more than two different
sizes.

SAMPLE PREPARATION AND RESULTS

The humic acid used in this study was extracted
with 0.1N NaOH from a North Carolina sandy soil. After
16 hours of extraction the solution was centrifuged for
2 hours at 7000 revolutions per minute in a 12.25 inch

28° rotor, and the supernatent was filtered through a
Reeve Angel ultra glass fiber filter. After filtration
the humic acid was precipitated by lowering the pH to
one. The precipitate was washed twice and then redis-
solved in 0.1N NaOH, dialyzed against distilled water
and lyophilized.

One percent solutions of the lyophilized sodium
humate were prepared in water and were fractionated on
Sephadex G-50 chromatographic columns using distilled
water as an eluant. Sephadex gel filtration column
chromatography has been widely used for fractionation
and molecular weight estimation of natural products
and biological materials. The remarkable success that
has been achieved in the fractionation of these materials
has resulted in a number of attempts to fractionate
sodium humate on Sephadex columns. (See Lindqvist[6] for
a brief review of this work). A variety of different
techniques have been used, some of the workers have
employed eluents of high ionic strength while others
have used eluents of low ionic strength or a combination
of the two. Posner[7] and Lindqvist[6] have pointed out
however, that sodium humate is adsorbed by Sephadex
and therefore the degree of fractionation obtained on
Sephadex is controlled by the ionic strength of the
eluents. They have shown that fractionation is completely absent with some eluent systems and only
partially effected by others. Therefore, different
fractionation patterns on Sephadex columns will be
obtained for the same sodium humate sample when different eluents are used.

Each of the various grades of Sephadex will fractionate a particular molecular-weight range of proteins
or polysaccharides, and molecules above this range will
pass through the column without fractionation. The
elution curves from columns of different grades of
Sephadex have proved to be particularly successful in
estimating the molecular weights of globular proteins
and polysaccharides. However, as we have pointed out
previously (Wershaw and Burcar[9]) because sodium humate
is sorbed by Sephadex, the mechanism of its fractionation
by Sephadex is apparently different from that of proteins

and polysaccharides; therefore, it is not possible to
use the elution curves for estimating molecular weights
of the sodium humate fractions. The Pharmacia Fine
Chemical Company which manufactures Sephadex has also
pointed out that the relationships between molecular
weights and elution volumes are different for different
types of molecules, and therefore since very little is
known about the size, shape, and functional group distribution in humic acids there is no way to predict the
behavior of these compounds on Sephadex. However, as
our knowledge of the physical chemical properties of
sodium humate grows it should be possible to devise
chromatographic systems that can be calibrated for
particular ranges of molecular weight.

The different sodium humate fractions obtained by
Sephadex chromatography have distinctly different infrared absorption spectra (figure 1). These differences
in infrared spectra indicate that the fractions differ
either in chemical structure, in degree of polymerization,
or in both of these properties.

After fractionation each of the fractions was
lyophilized and 0.5 percent solutions of each were
prepared for small angle x-ray scattering studies.
Approximately 0.2 milliter of solution was put into a
1 millimeter diameter thin-walled quartz capillary tube
and the tube was mounted in the Kratky small angle x-ray
scattering camera (Kratky[8]). We have described previously
(Wershaw, et al[9]) the procedure used for making the
scattering measurements. Beeman, et al[10] and Guinier
and Fournet[11] have very complete treatises on the interpretation of small angle x-ray scattering data. As
they have shown, the intensity of scattered radiation,
I, from a large number of randomly oriented particles
having a radii of gyration of R is given by the Guinier
equation

$$\text{Log } I(h) = - h^2 R^2 \log e + \text{constant}$$

where

$$H = \frac{4\pi \sin \theta}{\lambda},$$

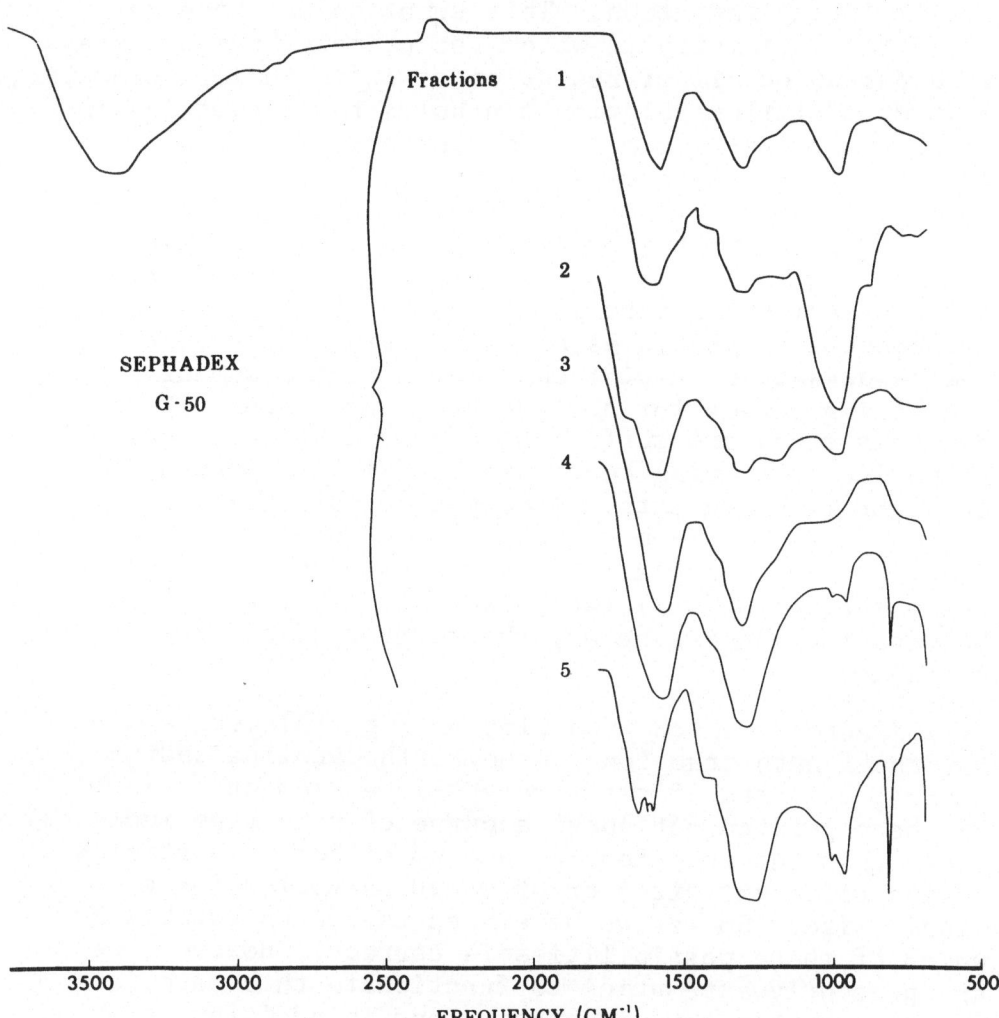

figure 1. Infrared absorption spectra of North Carolina
          soil Sodium humate fractions separated on
          Sephadex G-50

2θ is the scattering angle and λ is the wavelength of the impinging radiation. This equation has been derived for a geometry in which the primary beam of x-rays is a narrow pencil of radiation of infinitesimal diameter, however a similar relationship holds for geometries in which the primary beam is long and narrow, as is the case in the Kratky camera.

It is apparent from the Guinier equation that a plot of log I(h) versus h^2 will yield a straight line for a system of uniform particles and that if several different size particles are scattering the radiation the resultant scattering curve will be the sum of scattering curves for the separate particles. If the particle sizes are sufficiently separated the resultant composite scattering curve will consist of several straight line segments. A composite scattering curve will also result from a group of particles of a single density. For example in a particle having two zones, x-ray scattering will take place both at the interface between the inner zone and the outer surface of the particle.

Figure 2, which is a plot of log I(h) versus h^2 for the fourth fraction of the North Carolina sodium humate, consists of two straight-line segments. As has been pointed out above a curve of this type indicates that the fraction is made up either of particles of two different sizes or of zoned particles for a single size. So far we have been unable to establish which of these possibilities is correct. However, we are presently attempting to fractionate this material by ion exchange chromatography, and in addition we plan to make measurements at different pH's, ionic strengths, and with different counter ions present.

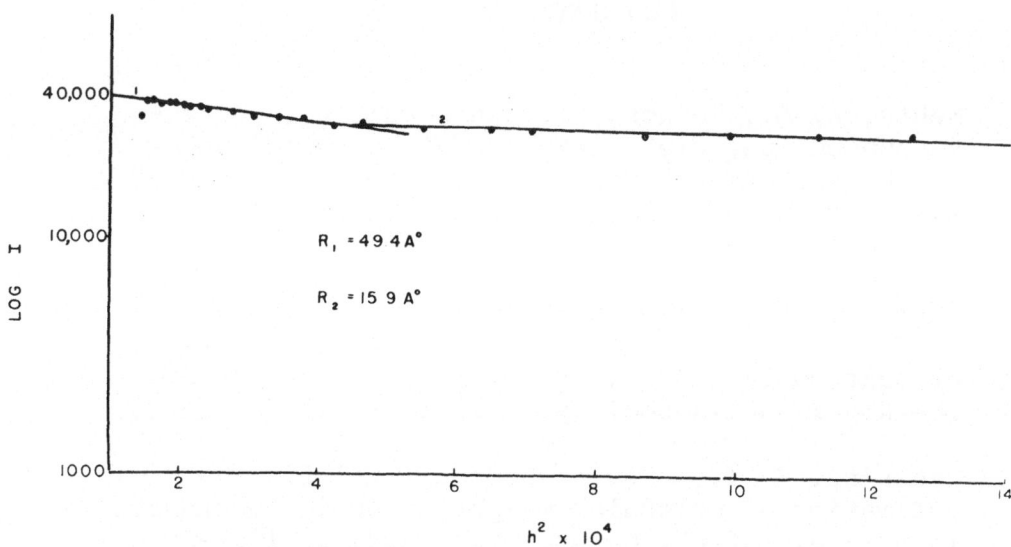

figure 2. Guinier plot (log of scattered intensity
          versus h², where h = $\frac{4\pi \sin \theta}{\lambda}$) for fraction 4

CONCLUSIONS

Most of the sodium humate fractions that we have
isolated by ion exchange and gel permeation chromatography have distinctly different infrared absorption
spectra, which indicates that these fractions differ
from one another in either their chemical structure
or degree of polymerizations. Size measurements of
one of these fractions by small angle x-ray scattering
indicates that the fraction is made up either of
molecules of two different sizes or of zoned molecules
of a single size.

REFERENCES

1. Mehta, N. C., Dubach, P., and Deuel, H., "Unter-
   suchungen uber die molekulargewichtsverteilung
   von huminstoffen durch gelfiltration an Sephadex",
   Z Pflanzenernahr, Dung., Bodenkunde Vol. 102,
   p. 128-137, 1963.

2. Piret, E. L., White, R. G., Walther, H. C. and
   Madden, A. J., "Some physico-chemical properties
   of peat humic acids", Scientific Proceedings of
   the Royal Dublin Society, Vol. 1A p. 69-79, 1960.

3. Gjessing, E. T., "Use of Sephadex gel for the
   estimation of molecular weight of humic substances
   in natural water", Nature, Vol. 208, p. 1091-1092,
   1965.

4. Bailly, J. R., "Fractionnement des acides humiques
   suivant leur encombrement moleculaire. Appreciation
   sommaire de quelques poids moleculaires privilegies",
   Acad. Sci. (Paris) Compte Rendus, Vol. 264, p. 564-
   566, 1967.

5. Wershaw, R. L., Burcar, P. J., Sutula, C. L., and
   Wiginton, B. J., "Sodium humate solution studied
   with small-angle x-ray scattering", Sci. Vol. 157,
   no. 3795, p. 1429-1431, 1967.

6. Lindqvist, Ingvar, "Adsorption effects in gel
   filtration of humic acid", Acta Chem. Scand.,
   Vol. 21, p. 2564-2566, 1967.

7. Posner, A. M., "Importance of electrolyte in the
   determination of molecular weights by Sephadex gel
   filtration with especial reference to humic acid",
   Nature, Vol. 198, p. 1161-1163, 1963.

8. Kratky, O., "X-ray small angle scattering with substances of biological interest in dilute solutions",
   Progress in Biophysics, Vol. 13, p. 105-173, 1963.

9. Wershaw, R. L. and Burcar, P. J., "Physical-Chemical properties of naturally occurring poly-electrolytes -- I Sodium humate", Proceedings of the Third Annual American Water Resources Conference, p. 351-364, 1967.

10. Beeman, W. W., Kaesberg, P., Anderegg, J. W. and Webb, M. B., "Size and Particles and Lattice Defects, Handbuch der Physik, Vol. 32, p. 321-442, 1957.

11. Guinier, A. and Fournet, G., "Small-Angle Scattering of X-rays", Wiley, New York, 1955.

A NEW ABSOLUTE-SCALE SMALL-ANGLE X-RAY SCATTERING INSTRUMENT

H. Pessen, T. F. Kumosinski, S. N. Timasheff,
R. R. Calhoun, Jr., and J. A. Connelly

Eastern Utilization Research and Development Division
U. S. Department of Agriculture
Philadelphia, Pennsylvania 19118

ABSTRACT

A small-angle X-ray scattering apparatus is described which is based on a Guinier focusing arrangement, utilizing a fine-focus tube, a Johann-type curved-crystal monochromator, two goniometer-mounted beam-defining slits and proportional detection. It differs from previously described instruments in important respects. Using a horizontal goniometer, it affords ease of access and mechanical stability to mounted parts, such as slits and beam stop. The major instrument assemblies--the horizontal tube housing producing the vertical line-shaped beam, and the goniometer table--are mounted on a granite surface plate for stability. For ease, precision and re-producibility in alignment, fine adjustments, with micrometer heads or dial indicators where advisable, are provided for all pertinent rotational and translational motions of the various subassemblies. In particular, provision is made for fine adjustment of the goniometer as a whole with respect to the X-ray source, to facilitate threading the monochromator-focused beam through the slit system and the center of rotation of the detector arm. Maximum distance between the beam-defining slits compatible with monochromator geometry yields increased resolution. Use of a pulse-height analyzer eliminates harmonic radiation from the detection system and thus allows operation of the X-ray tube at greater intensity. Absolute measurements are made by means of calibrated beam-attenuators to refer scattered and direct-beam intensities to the same scale. An accessory is a motor-driven horizontal slit for scanning the vertical beam profile to validate assumptions used in analysis of the data, i.e., slit-desmearing to refer the slit-produced scattering to a theoretical point source. Automatic step scanning with

punched-tape output is provided for automatic data processing.
The facility for symmetrical scanning provides a check on proper
alignment as well as on source stability and system geometry during
the extended scans required for weakly scattering systems, such as
dilute solutions of biological macromolecules. With the use of
this instrument, radii of gyration were determined for several globular proteins (β-lactoglobulin, ribonuclease, lysozyme) and values
in agreement with those in the literature were obtained.

INTRODUCTION

Small-angle X-ray scattering has been employed for many purposes and in the study of many types of system, e.g., for particle
and pore sizes in catalysts, particle sizes and clustering in alloys,
glasses and ceramics, observation of critical phenomena, colloidal
micelles in solution, synthetic polymers and biopolymers. The variety of parameters obtainable by the method (radius of gyration,
particle size, pore size, molecular weight, mass per unit length in
linearly ordered systems, surface-to-volume ratio or specific surface, particle volume, internal hydration, and others) has made it
sufficiently attractive to lead to a considerable number of instruments constructed for it. These instruments, as well as much of the
theory, have been reviewed by Guinier and Fournet,[1] Beeman et al.[2]
and others, and very complete bibliographies of the literature have
been given by Yudowitch in Guinier and Fournet[1] and by Goldmann.[3]
More recent instruments have been described by Luzzati at al.,[4]
Brumberger and Deslattes,[5] Renouprez et al.,[6] Bonse and Hart,[7]
Koffman,[8] and Kavesh and Schultz.[9]

Experimental Obstacles

In the study of proteins in dilute solution in particular, one
is faced with difficult experimental requirements. Since scattered
intensities generally are weak, counting statistics require long
counting times, in the case of step scanning, or very slow scans,
in the case of continuous scanning. Either way, the time for a complete scan may range from 8 to 24 hours, during which the primary-
beam intensity is expected to remain constant. This is generally
attempted by the use of very stable sources, including highly stabilized power supplies, air conditioning and cooling-water temperature control; even so, primary-beam intensity fluctuations are not
always eliminated (cf. Baker et al.,[10] who correlated intensity variations with barometric pressure changes). To the extent that
high-intensity primary beams (by virtue either of high-intensity
sources or of collimating systems passing a relatively wide beam)
will give increased scattered intensities, the time requirements
will be ameliorated.

Another approach would be to utilize a record of monitored
primary-beam intensities to apply point-by-point normalization to
the measured scattered intensities. There are practical difficulties in this approach which appear to have prevented it from being
applied routinely so far.

Under these circumstances it would seem desirable to have some
other check on the stability during a scan. Those instruments which
allow scanning on either side of the zero angle, assuming they are
in perfectly symmetrical alignment, afford this possibility and provide an immediate duplicate determination as well, although of
course at the cost of approximately doubling the time required.
A check for symmetry at the end of each complete scan then requires
no more than a translucent chart record paper and an illuminated
viewing box.

Other experimental difficulties relate to the need for measurements at the very small angles corresponding to large repeating
units or large molecular weights. Although collimating systems
yielding high primary-beam intensities will result in better
counting statistics and shorter scanning times, a broader primary
beam generally will have tails which may be high compared to the
scattered intensities at somewhat wider angles, particularly when
it is remembered that scattered intensities are lower than direct-
beam intensities by a factor, typically, of 10^6. Provisions for
precise and reproducible alignment of all the collimating system
components and for elimination of parasitic scattering due to various causes are indispensable. Otherwise it is not feasible to
find a suitable compromise, for a given sample, between the conflicting demands of high intensity and high resolution.

Further requirements result from the need to measure primary-
beam intensities on the same scale as scattered intensities, if the
instrument is to give the absolute-scale measurements required for
determination of molecular weights and other parameters. The problem has been summarized by Kratky et al.[11] Methods used for this
purpose fall into two broad categories: 1. Use of standard scatterers whose scattering either is calculated from basic data
(Beeman et al.,[12] Weinberg[13]) or is determined by calibration with
reference to one of the other methods (Kratky[11]); 2. Use of attenuators, either by fractional-time sampling of the primary beam (rotating disk with hole, Kratky[14]) or by filter-type foil attenuation
(Luzzati,[15] Weinberg,[13] Damaschun and Müller[16]).

Problems with Existing Instruments

Instruments described in the literature sometimes achieve very
high performance with respect to one or more of the desirable requirements by accepting severe restrictions with respect to others.

Some of the most attractive-appearing designs may be described in detail, but beyond that first description in the literature one often finds little further publication of data obtained by their use.

Instruments frequently fall short of the ideal in one or more of the following respects.

1.   The advantages of a symmetrical primary beam and a symmetrical scan are relinquished, in return for the possibility of eliminating parasitic scattering from the beam-defining slit edges to a very high degree; more precisely, the slit-produced penumbra of the source focal spot is eliminated almost entirely on one side of the zero angle, in return for covering up the beam on the other side of zero angle.

2.   Instruments use a goniometer on a horizontal axis, with the arm moving in a vertical plane. This may result in greater compactness and thus possibly greater rigidity, as well as less space

Figure 1.   Small-angle X-ray scattering instrument.

required, but it may also result in lessened accessibility and may
make mounting of various accessories more difficult and more tenu-
ous. A vertical-axis goniometer, with the arm moving in a horizontal plane, lends itself to a more open construction, with ample room
for access and for the attachment of components to the stationary
horizontal goniometer table.

3. Instruments using pure slit collimation have to discard
most of the divergent beam issuing from the source in order to
obtain a well-defined beam. To increase intensity, high-intensity
rotating-anode X-ray tubes are sometimes employed.

4. Instruments using a curved crystal for primary-beam monochromatization change the divergent beam coming from the source into
a converging beam passing through the sample and do not need to
suffer a great loss of intensity from collimating slits, but the
reflection at the crystal face itself is accompanied by substantial
losses.

5. Most instruments are not adapted to make absolute measurements routinely, if at all. The use of gaseous standard scatterers
is not a simple procedure; solid or liquid secondary standards can
be used routinely but depend on standardization in an instrument of
identical geometry; the rotating-disk method of Kratky is too complex to be used routinely; foil attenuators require a set of carefully calibrated high-grade foils, and provisions to accommodate
them.

6. Some instruments are suitable primarily for very small
angles, and their range is correspondingly small ($\pm$ 3° or less),
whereas an all-purpose instrument may be required to take measurements to $\pm$ 8°.

APPARATUS

On the basis of the above considerations, the instrument shown
in figures 1 and 2 was designed, constructed, aligned and tested.

Construction

The two major assemblies of the instrument, the horizontal
tube housing with attached monochromator, and the goniometer with
slits and detector, are mounted on a 24" x 48" granite surface
plate (SP) supported by a steel frame. Some inconvenience in se-
curing components to this massive slab was felt to be a small price
to pay for obtaining the utmost in rigidity and insuring maintenance of alignment between these two assemblies.

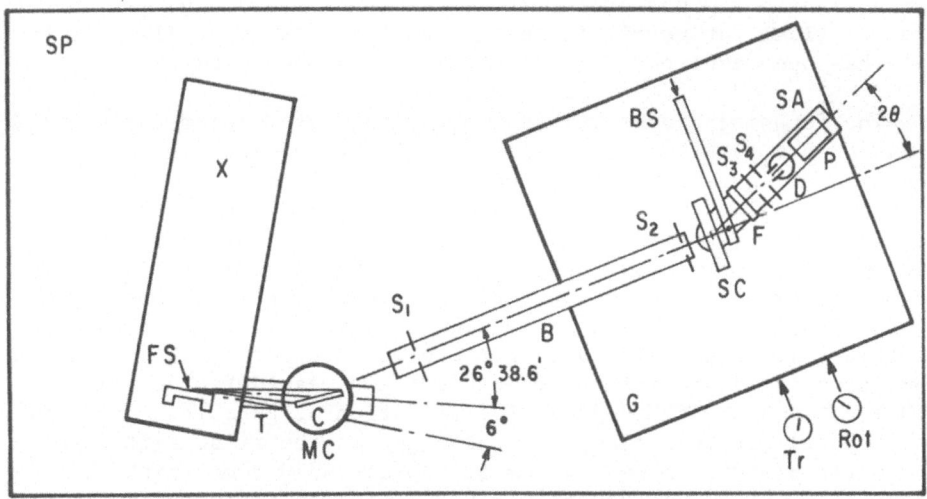

Figure 2. Diagram of scattering instrument.

The X-ray source (X) is a fine-focus copper-target Philips (Eindhoven)* four-window diffraction tube of 1200 watt capacity (PW 2113), mounted in a horizontal tube housing in such a way that a vertical beam issues from one of the line-focus windows and, viewed at a 6° take-off angle, the focal spot (FS) has effective dimensions of 0.04 mm x 8 mm. The ordinarily cantilevered end of the tube housing is further supported by a rigid welded-steel column attached to the foot of the L-shaped housing brace.

Attached to this column also is the monochromator support assembly, which includes a set of four screw motion tracks: circular, vertical and horizontal tracks in planes transverse to the 6° take-off direction, and a longitudinal track (T) at the 6° angle, whose axis is centered on the tube focal spot. This track carries the monochromator housing (MC), which has a rotatable support table with a lockable coarse adjustment (Compagnie Générale de Radiologie). Once this adjustment is locked, a fine motion by means of a differential tangent screw allows adjustment to within 0.1 minute of arc. This table was originally designed for a magnetic crystal-bending clamp of the same manufacturer; since a non-magnetic demountable

---

* Mention of specific firms or products does not imply endorsement by the U. S. Department of Agriculture to the possible exclusion of others not mentioned.

clamp of another manufacturer (Etablissements Beaudouin) was found
more practical for permitting a choice of special crystals, the
table has been equipped with a suitable hold-down clamp.

   The focusing geometry (essentially that of Guinier-Luzzati[1,4])
is based on a flat-cut circularly bent quartz crystal in_the
Johann[17] configuration, utilizing reflection from the $10\bar{1}1$
planes.[18,19] The crystal faces are cut at an 8° angle to the reflecting planes; the crystal (C) is elastically bent to a radius of
1300 mm. This asymmetry results in a favorable geometry in which
the monochromator is very close to the source, thus subtending a
relatively large angle and collecting an intense divergent beam; at
the same time it is relatively far from the detector, thus allowing
room for a large separation between the two beam-defining slits,
with a consequent high degree of elimination of parasitic scattering
from crystal, crystal clamp and the edges of the first slit itself.
The quartz plates, 13 mm x 40 mm in aspect and 0.3 mm thick, supplied by Dr. Steeg and Reuter G.m.b.H., must be individually selected, since their quality is crucial, and must be carefully
mounted in the bending clamp by trial and error. The clamp, in
turn, must be carefully mounted so that the center of the crystal
face passes through the center of rotation of the monochromator
table. Properly adjusted, the monochromator will isolate the α_1
line from the Cu-K$_\alpha$ doublet at a Bragg angle of 13° 19.3'. The
criteria for crystal quality, correct mounting and proper adjustment are the intensity and homogeneity of the monochromatic beam
as viewed on a fluorescent screen, and the shape of its scanned
profile as recorded on a chart.

   The two beam-defining slits (S_1 and S_2) are mounted on an optical bench (B) supported on the goniometer table (G). This bench is
rotatable about the center of rotation of the scanning arm (which is
also the center of the sample cell), and the median line of each of
these slits is adjustable to lie in the vertical plane passing
through this center of rotation.

   The two slit assemblies, especially designed and supplied by
W. W. Beeman, are compact devices embodying transverse horizontal
and rotational adjustments. The 35-mm long tantalum slit jaws open
and close symmetrically by means of a micrometer adjustment. The
slit assemblies are mounted on carriages slidable on the bench.
The fine adjustment and the quality of the slit jaws, particularly
of the second slit, are extremely critical to the apparatus.

   The two scanning slits (S_3 and S_4) mounted on the scanning arm
(SA), define the scattering angle, 2θ, at which intensities are observed. The first of these (S_3) is the receiving slit and determines resolution as well as intensity seen by the detector (D),
also mounted on the arm. The second slit (S_4), an anti-scatter

slit, limits the radiation seen by the detector to the direction
coming from the sample. Without it, parasitic scattering originating from various components and from irradiated air in the vicinity of this direction would pass through the receiving slit and
reach the detector. Because of space limitations the original,
very compact Philips slits are used. These are exchangeable fixed-
width slits with molybdenum jaws. Adjustments originally were
limited to transverse quasi-translations of each (actually small
rotations about a vertical axis), as well as a coarse longitudinal
translation of the two slits and the detector as a unit along the
scanning arm. This latter motion has been modified to a fine adjustment. A trimming adjustment has been added which allows these
same three components (scanning slits and detector) to be rotated
as an entity with respect to the scanning arm.

    A transverse fifth slit (shown in figure 1, but not in fig-
ure 2) is a motor-driven vertical-scanning slit, used for measuring
the vertical beam profile. This information is useful in determining the proper weighting function to be used in desmearing the

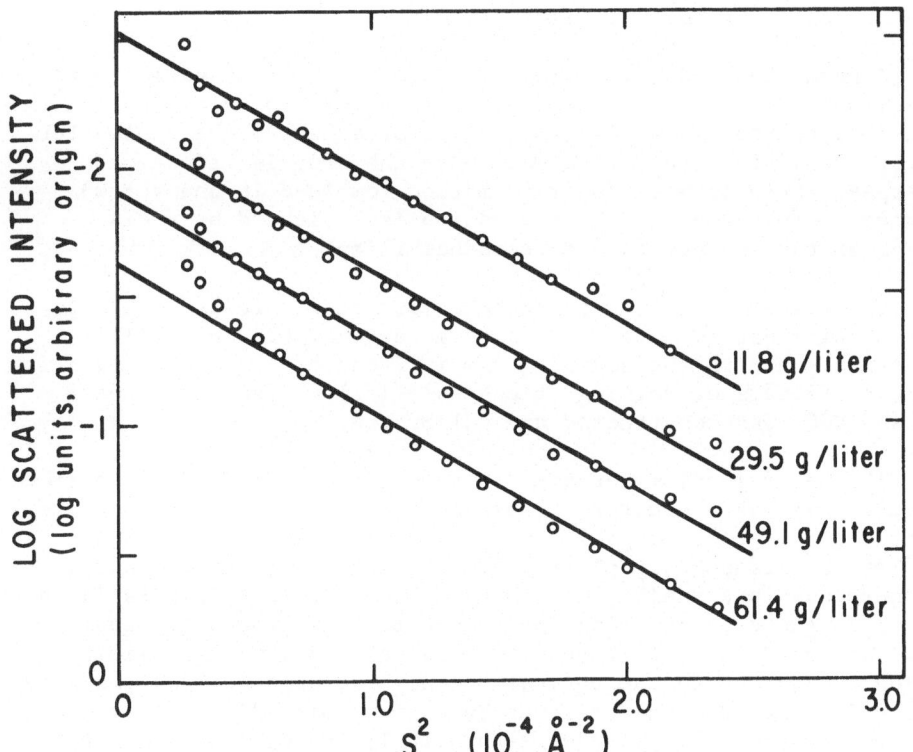

Figure 3. Guinier plots for β-lactoglobulin (pH 5.2 sodium
acetate buffer, Γ/2 = 0.1).

measured scattered intensities, i.e., in deconvoluting observed
intensities derived from a line source to theoretical intensities
ascribable to the point source on which scattering theory is based.
(If the assumption of infinite slit height holds (Guinier,[1] Luzzati,[15]
Kratky et al.[20]) this correction is particularly simple.) This
measurement, furthermore, supplies a check on parallelism of all
four longitudinal slits with one another and with the monochromatized beam. The preferred location for this slit would have been
between sample and receiving slit; because of space limitations,
however, it is removably mounted on the optical bench.

The sample cell (SC) is held by a removable sample holder
having translational adjustments to permit centering on the axis
of rotation. A θ-motion or a fixed position may be selected. The
liquid-sample cell, with window height of 30 mm and volume of
1/6 ml, consists of a 1 mm Teflon spacer (approximately the optimum
sample thickness) clamped between two approximately 0.0007" thick
mica windows by means of the cell frame, together with access pas-
sages for introducing and removing samples. The only materials in
prolonged contact with sample solutions are Teflon and mica. Window thickness represents a compromise between minimizing X-ray
attenuation and window fragility.

A beam stop (BS), consisting of one of a set of gauged wires,
is held by clamps in the C-frame of the beam-stop holder, which in
turn is supported by a dove-tail slide attached to the goniometer
table. Sensitive transverse motion of the stop is effected by a
two-speed screw (coarse and fine motion combined in one screw)
bearing against the slide. The motion is indicated by a dial
indicator for the purpose of reproducibility.

Filters are built up of high-quality nickel foils of the required thickness, mounted in frames held removably in a filter
holder attached to the scanning arm in front of the receiving slit.
A set of filters of varying thicknesses (filter factors between
1.2 and 800) was constructed and calibrated.

The entire goniometer box, supporting the components described
above as well as the motor, gears and controls inside the box, is
mounted on a set of two 5/16"-thick ground steel plates and a
1/2"-thick aluminum jig plate. The aluminum plate is clamped to
the surface-plate table. The lower of the two steel plates is supported on the aluminum plate by a dove-tail slide and by steel
balls in the form of ball-bearing parallels, allowing a sliding
motion transverse to the optical bench. The upper of the two steel
plates is supported on the lower by a pivot pin adjustable to be
located perpendicularly underneath slit 1, as well as by steel
balls like the other plate, allowing rotation about the median line
of this slit as an axis. The translational adjustment is controlled

by a differential screw, the rotational adjustment by another two-speed screw, and both motions are indicated by dial indicators (Tr, Rot) to facilitate aligning the goniometer. By these means the vertical plane passing through the center lines of the slits and the sample may be adjusted to make the correct angle (26° 28.6', twice the Bragg angle) with the monochromator track and may thus be centered on the monochromatized beam.

The detector (D) is a sealed-window Amperex proportional counter tube with a Hamner NB-19 preamplifier, mounted on the scanning arm. Electronics consist of a Wanlass Parax power conditioner, a Fluke 412B high-voltage power supply, a Hamner NA-11 linear amplifier and a Hamner NC-11 pulse-height analyzer feeding into the ratemeter and display portions of a Philips 12206 electronic circuit panel, and from there into a Honeywell Electronik strip chart recorder. The goniometer may be used in the continuous mode or, under the control of the circuit panel, in a step scanning mode. With step scanning, data may be fed to a Victor Digit-Matic tape printer and, through an interface, into a Friden SP-2 tape punch.

Alignment

The monochromator is aligned with the aid of a fluorescent screen. The criterion of longitudinal alignment (cf. Guinier[21]) is the sharp disappearance of the beam without shifting, upon a slight rotation of the crystal in either direction away from the position of maximum intensity and homogeneity. At any but the correct distance between tube and monochromator, rotating the crystal

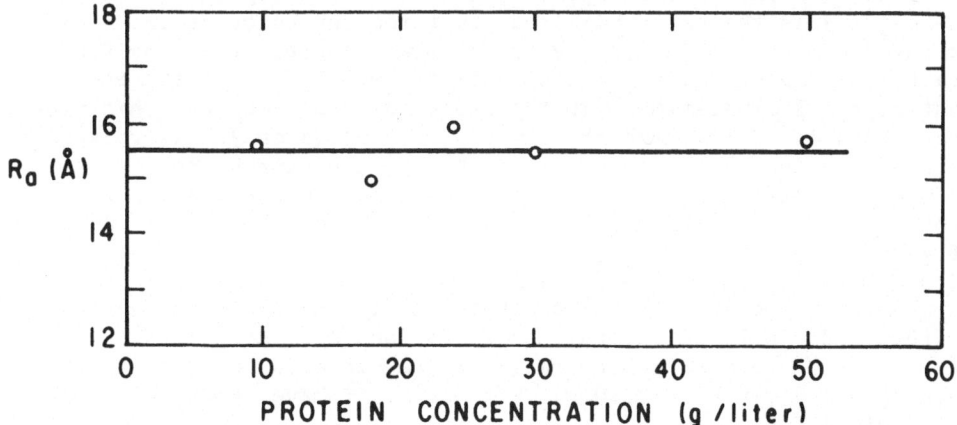

Figure 4. Radius of gyration vs. concentration, for lysozyme (pH 6.6 sodium phosphate buffer, $\Gamma/2 = 0.1$).

either way produces some vertical motion of the fluorescent spot
before it disappears.

The goniometer is lined up with the monochromatized beam by
the use of an auxiliary slit in the sample position. This slit is
first precisely adjusted to coincide with the scanning center of
rotation. With slit 1 on the bench but slit 2 removed, successive
translational and rotational adjustments of the goniometer platforms guide the beam first through slit 1 and then through the
scanning center. With the scanning slits removed, the goniometer
can now be zeroed.

The receiving and anti-scatter slits are next replaced and
aligned in turn, and the auxiliary slit is then removed. Slit 2
is replaced last and is centered and opened so as to barely graze
the direct beam. Exceedingly fine adjustments of this slit are
required to produce a symmetrical direct beam.

RESULTS

Radii of gyration were determined for the globular proteins
β-lactoglobulin B, lysozyme and ribonuclease, each at a series of
concentrations. The β-lactoglobulin B was prepared according to
the method of Aschaffenburg and Drewry[22] from the milk of homo-
zygous B/B cows. Egg-white lysozyme, 3x crystallized, was obtained
from Pentex Incorporated. The ribonuclease was the salt-free
bovine pancreatic product, 5x crystallized, supplied by Mann Research Laboratories, Inc.

The apparent radius of gyration, R_a, at any concentration can
be obtained from the corresponding Guinier plot,[1] log $j_n(s)$ vs. s^2,
where $j_n(s)$ is the net normalized scattered intensity at the angle
measured by $s = (2 \sin \theta)/\lambda$, θ is the Bragg angle, or one-half the
scattering angle, and $\lambda = 1.5405$ Å is the wavelength of the incident copper K_{α} radiation. In the region of small angles, this plot
approaches linearity, and the limiting slope, $(slope)_0$, is related
to the apparent radius of gyration, R_a, by the expression
$(slope)_0 = -(4/3)\pi^2 R_a^2$. Extrapolation to zero concentration of a
plot of R_a as a function of concentration yields the actual radius
of gyration, R_0.

Guinier plots for β-lactoglobulin B (in pH 5.2 sodium acetate
buffer, $\Gamma/2 = 0.1$) at several concentrations are shown in figure 3.
Extrapolation to zero concentration of values of R_a derived from
these plots gives $R_0 = 21.0$ Å. Witz et al.[23] have reported a
value of 21.7 Å.

Figure 4 is a R_a vs. concentration plot for lysozyme (in pH 6.6 phosphate, $\Gamma/2 = 0.1$), where the R_a values were calculated from slopes of Guinier plots similar to those of figure 3. Extrapolation yields $R_0 = 15.5$ Å, compared with 15.2 Å reported by Luzzati et al.[24]

A similar study for ribonuclease (in pH 5.2 acetate buffer, $\Gamma/2 = 0.1$) gave a value of $R_0 = 14.6$ Å. From the crystallographic data of Kartha et al.[25] approximating the shape of the ribonuclease molecule by a right parallelepiped with dimensions of 38 x 28 x 22 Å and using the expression $\sqrt{(a^2 + b^2 + c^2)/12}$ for the radius of gyration, one arrives at $R_0 = 15.03$ Å.

It is seen, therefore, that results obtained with this instrument for three different proteins are in good agreement with comparable values from the literature.

REFERENCES

1. A. Guinier and G. Fournet, "Small Angle Scattering of X-Rays," John Wiley & Sons, Inc., New York, 1955, pp. 83-110, 120-123; pp. 111-120; p. 127; pp. 217-259.

2. W. W. Beeman, P. Kaesberg, J. W. Anderegg, and M. B. Webb, "Size of Partcles and Lattice Defects," in S. Flügge, Editor, Handbuch der Physik, Vol. 32, Springer-Verlag, Berlin, 1950, 1957, pp. 359-371.

3. J. B. Goldmann, Small Angle X-Ray Scattering: an Annotated Bibliography, Lockheed Missiles & Space Co., Sunnyvale, California, 1962, (AD 286156, Clearinghouse, U. S. Department of Commerce, Springfield, Va.).

4. V. Luzzati, J. Witz, and R. Baro, "Description d'un appareil de diffusion centrale des rayons X destinée à la mesure des intensités à une échelle absolue," Journal de Physique--Physique Appliquée, 24:141A-146A, (1963).

5. H. Brumberger and R. Deslattes, "A New High Resolution Small-Angle X-Ray Camera," J. Res. Nat. Bur. Std., 68C:173-175, (1964).

6. A. Renouprez, H. Bottazzi, D. Weigel, and B. Imelik, "Description et réglage d'un appareillage adapté à la diffusion centrale des rayons X," J. Chim. Phys., 62:131-136, (1965).

7. U. Bonse and M. Hart, "A New Tool for Small-Angle X-Ray Scattering and X-Ray Spectroscopy: the Multiple Reflection Diffractometer," in H. Brumberger, Editor, Small-Angle X-Ray Scattering, Gordon & Breach, New York, 1967, pp. 121-130.

8. D. M. Koffman, "An X-Ray Small-Angle Scattering Instrument,"
    in J. B. Newkirk, G. R. Mallett, and H. G. Pfeiffer, Editors,
    Advances in X-Ray Analysis, Vol. 11, Plenum Press, New York,
    1968, pp. 332-338.

9. S. Kavesh and J. M. Schultz, "High Resolution Small and Wide
    Angle X-Ray Diffractometer," Rev. Sci. Instr., 40:98-101, (1969).

10. T. W. Baker, J. D. George, B. A. Bellamy, and R. Causer,
    "Fully Automated High-Precision X-Ray Diffraction," in J. B.
    Newkirk, G. R. Mallett, and H. G. Pfeiffer, Editors, Advances
    in X-Ray Analysis, Vol. 11, Plenum Press, New York, 1968,
    pp. 359-375.

11. O. Kratky, I. Pilz, and P. J. Schmitz, "Absolute Intensity
    Measurement of Small Angle X-Ray Scattering by Means of a
    Standard Sample," J. Colloid Interface Sci., 21:24-34, (1966).

12. W. W. Beeman, "Structural Studies with Small-Angle X-Ray
    Scattering," in M. A. Stahmann, Editor, Polyamino Acids,
    Polypeptides and Proteins, U. of Wisconsin Press, Madison,
    1962, p. 253.

13. D. I. Weinberg, "Absolute Intensity Measurements in Small-
    Angle X-Ray Scattering," Rev. Sci. Instr., 34:691-696, (1963).

14. O. Kratky, "Die Messung der Absolutintensität der diffusen
    Röntgenkleinwinkelstreuung--ein Verfahren zur ,,Wägung" in
    makromolekularen Systemen," Z. anal. Chem., 201:161-194, (1964).

15. V. Luzzati, "Interprétation des mesures absolues de diffusion
    centrale des rayons X en collimation ponctuelle ou lineaire:
    Solutions de particles globulaires et de bâtonnets," Acta
    Cryst., 13:939-945, (1960).

16. G. Damaschun and J. Müller, "Die Messung der Absolutintensität
    der Röntgen-Kleinwinkelstreuung durch Schwächung des
    Primärstrahls mit Absorptionsfiltern nach vorheriger
    Monochromatisierung durch Totalreflexion," Z. Naturforsch.,
    20a:1274-1279, (1965).

17. H. H. Johann, "Die Erzeugung lichtstarker Röntgenspektren
    mit Hilfe von Konkavkrystallen," Z. Physik, 69:185-206, (1931).

18. D. R. Chipman, "Monochromation of X-rays by Crystals, for
    Use in Diffuse Scattering Measurements," in F. H. Herbstein,
    Editor, Methods of Obtaining Monochromatic X-Rays and
    Neutrons, Int. Union of Cryst., Utrecht, The Netherlands,
    1967, pp. 55-58.

19. J. Witz, "Focusing Monochromators," Acta Cryst., A25:30-42. (1969).

20. O. Kratky, G. Porod, and Z. Skala, "Verschmierung und Entschmierung bei Röntgen-Kleinwinkeldiagrammen," Acta Physica Austriaca, 13:76-128, (1960).

21. A. Guinier, Théorie et Technique de la Radiocristallographie, Dunod Cie., Paris, 1945, 1956, 1964, p. 192.

22. R. Aschaffenburg and J. Drewry, "Improved Method for the Preparation of Crystalline β-Lactoglobulin and α-Lactalbumin from Cow's Milk," Biochem. J., 65:273-277, (1957).

23. J. Witz, S. N. Timasheff, and V. Luzzati, "Molecular Interactions in β-Lactoglobulin. VIII. Small-Angle X-Ray Scattering Investigation of the Geometry of β-Lactoglobulin A Tetramerization," J. Am. Chem. Soc., 86:168-173, (1964).

24. V. Luzzati, J. Witz, and A. Nicolaieff, "Détermination de la masse et des dimensions des protéines en solution par la diffusion centrale des rayons X mesurée a 1'échelle absolue: Exemple du lysozyme," J. Mol. Biol., 3:367-378, (1961).

25. G. Kartha, J. Bello, and D. Harker, "Tertiary Structure of Ribonuclease," Nature, 213:862-865, (1967).

MASS ABSORPTION COEFFICIENT MEASUREMENTS USING THIN FILMS

P. Lublin, P. Cukor and R. J. Jaworowski

General Telephone & Electronics Laboratories Incorporated

Bayside, New York 11360

ABSTRACT

For quantitative electron probe analysis, the raw intensity ratios must be corrected to take into account deviations due to absorption, fluorescence and electron beam penetration. The major correction is usually the absorption correction, so that for best results, accurate mass absorption coefficients are required. Many tables of absorption coefficients are calculated by interpolation or extrapolation from available measured values, and therefore new measurements are required for increased reliability. The region which requires the most attention for present-day probe analysis is the 2 to 10 Å range.

Thin foils of the lighter metals are available for mass absorption coefficient measurements, but heavy metal foils, which must be extremely thin, are not obtainable. A method has been developed to prepare thin films of heavy metals on a suitable substrate by pyrolytic decomposition of metal organic compounds. Microscopical examination of the films showed that they were satisfactory for the measurements in respect to density and continuity.

Measurements were made in a standard Philips vacuum spectrograph. Deadtime correction procedures were employed. Measurements were made for aluminum, nickel, tungsten and gold absorbers. Good agreement in comparison with other measurements was obtained in most cases.

INTRODUCTION

In electron probe analysis, one of the obstacles to accurate quantitative analysis is the lack of reliable data for mass absorption coefficients.

Subsequent to the publication of Heinrich's[1] comprehensive tables, the two major articles dealing with absorption data are those of Hughes, Woodhouse and Bucklow[2] and Henke, et al.[3] Other recent publications of interest are by Ogier, et al.[4], Cooke and Stewardson[5] and McMaster, et al.[6] The latter reference is a set of calculated tables based on recent measured · values; the data, however, are tabulated in electron volts rather than the customary selected wavelengths.

The principal region of interest for new measured values for probe analysis is in the range 2 to 10 Å. For radiation wavelengths shorter than 2 Å existing measured and calculated absorption values are generally accurate to 5%, whereas in the 2 to 10 Å range the present data may be in error by as much as 20%.

PREPARATION OF FILMS

Standard specimens for determination of mass absorption coefficients should be of high purity, uniform in thickness, and free of pinholes. Most previous measurements have been made on thin foils prepared by rolling. For the heavier elements the foils for measurements in the 2 to 10 Å wavelength range are in the thickness range from 1000 to 10,000 Å, which presents problems in the preparation of suitable foils. Thin films prepared by evaporation have been used by Cooke[5] and Hughes[2] (the latter by rotation of the substrate under the source) with satisfactory results.

There are many metal-organic compounds used to prepare thin films of metal by pyrolitic decomposition, many of which are commerically available. These compounds are either soaps of fatty acids or alcoholates or mercaptides.

$$RC \overset{O}{\diagup} - O - M \qquad \text{Soap}$$

R -O - M   Alcoholate

R - S - M   Mercaptides

An attempt was made to use these materials for deposition of films for mass absorption coefficient measurements. In our experiments 0.001 -in. Be foils were employed as the substrate. Specimens of foil were subjected to the same heat treatment that the films would undergo in pyrolysis. The attenuation of these foils for silicon Kα was measured. Two groups of foils were obtained where the variation among the foils in attenuation for silicon Kα for each group was within 1% of the mean. One foil from each group was selected as a blank, and the others were used as substrates for the respective blank.

Solutions of the metal-organic compounds were applied dropwise onto substrates and then spun at 1900 rpm. The samples were dried on a hot plate and then fired in hydrogen at temperatures ranging from 400°C to 700°C. A number of successive coats could be applied, depending on the thickness desired. Microscopical examination indicated uniform coatings and continuity. Films of nickel, tungsten, and gold were prepared. The absorbers were weighed before and after deposition and the area was measured to obtain ρx.

EXPERIMENTAL MEASUREMENTS

For the absorption measurements a standard Philips vacuum x-ray spectrograph was employed. A special holder was constructed to hold the absorber flat and perpendicular to the Soller slit assembly in front of the proportional counter. Pulse height analysis was not employed, and the voltage and current were adjusted for each set of measurements. In all cases the potential was well above the excitation voltage. A shutter was also designed so that the blank and film could be measured without shutting off the x-ray beam for better precision. The counting rates were corrected for dead time according to the method of Hon.[7] To check our techniques, measurements were made on aluminum foil absorbers. Four measurements were made on each sample and averaged. Reproducibility was within 1 to 2%. The uniformity of the nickel film was checked by using a quarter sector in front of the absorber and measuring the film by quarters. The results, which follow, indicate an rms deviation less than 0.002.

Area 1	0.237
Area 2	0.240
Area 3	0.235
Area 4	0.239
Total Area	0.238

RESULTS

Aluminum Foil

Table I compares the present results on aluminum foil with other measured and calculated values. The measurements of Hon and Heinrich[8] were also made with a vacuum x-ray spectrograph. The column marked L^6 contains values from the work at Livermore and was obtained by interpolation of graphs of their calculations. Chemical analysis of the foil used in our measurements revealed about 100 ppm of Fe, and this contamination was ignored in the calculations. The dashed lines in the tables indicate absorption edges. The agreement in comparison with the other measured values is excellent.

Tungsten Film

The measured values of tungsten absorption coefficients were initially very low when compared with available calculated values. A

TABLE I -- ALUMINUM FOIL ABSORBER

λ (Å)	Absorption Coefficient (cm²/g)				
	GT&E	H&W&B[2]	H&H[8]	H[1]	L[6]
CuKα 1.54	51.5	50.6	49.8	49.6	50.2
NiKα 1.66	63.6	63.1	61.8	60.7	62.2
FeKα 1.94	99.5	99.7	95.3	93.4	97.5
CrKα 2.29	160	160	153	149	158
TiKα 2.75	270	267	260	247	265
SnLα 3.60	566	567		523	550
AgLα 4.15	816	813		779	818
PdLα 4.37	928	936		895	955
SKα 5.37	1601		1595	1593	1620
SiKα 7.13	3390	3450	3410	3493	3240
		3370 (5)			
———					
AlKα 8.34	399	408	386	386	418
		396 (4)			

subsequent chemical analysis of the film revealed only about 72% tungsten in the film. Previous investigations had indicated that these films may be contaminated with carbon. The coefficients were recalculated using the combined attenuation of tungsten and 28% carbon. Heinrich's[1] tables were used for the carbon absorption values. The contribution of the carbon attenuation to the total attenuation was only 10% in the most severe case aluminum and even a 20% error in the absorption coefficient by carbon represents about a 2% variation in the final figure for tungsten. The possible error is even less for most of the other values. The results are given in Table II. In this case there are no measured values available for comparison.

Nickel Films

In the case of nickel a chemical analysis was made for nickel and carbon. Only 0.4% carbon was found. The calculations were made on the basis of nickel only, and the contribution of the carbon neglected. The results are given in Table III. Results on two foil measurements are also included. Our results are about 5% lower than the other measured values.

P. Lublin, P. Cukor, and R. J. Jaworowski

TABLE II -- TUNGSTEN ABSORBER (FILM)

λ (Å)	Absorption Coefficient (cm^2/g)		
	GT&E	H[1]	L[6]
TiKα 2.75	714	759	753
CaKα 3.36	1276	1279	1325
KKα 3.74	1673	1693	1740
ClKα 4.73	2721	2680	2820
SKα 5.37	3214	3096	3610
SiKα 7.13	1313	1533	1260
AlKα 8.34	2016	2172	1860

TABLE III -- NICKEL ABSORBER (FILM)

λ (Å)	Absorption Coefficient (cm^2/g)				
	GT&E	H&W&B[2]	H&H[8]	H[1]	L[6]
ZnKα 1.44	302		309	309	306
TiKα 2.75	229 233(foil)	240	239	234	240
SnLα 3.60	487 498(foil)	510		489	500
ClKα 4.73	998		1045	1028	1070
SKα 5.37	1390		1472	1458	1435
PKα 6.16	2024			2113	2080
SiKα 7.13	2878	2990		3152	3050

Gold Films

The gold film on analysis contained about 2% carbon. The calculations were again made using values for the carbon absorption from Heinrich's tables. The values given in Table IV are lower in most cases than both the measured and calculated data. Some measurements were made on evaporated foils for comparison purposes, but these were neglected when the foils were found to contain pinholes and cracks. Heinrich's absorption values for gold are high in comparison with all others. The major error in Heinrich's tables for gold is the value for sulfur. In the case of sulfur, the sulfur Kα is right on the M IV absorption edge and both Hon's[8] value and ours are considerably lower than the calculated values.

TABLE IV -- GOLD ABSORBER (FILM)

λ (Å)	GT&E	Absorption Coefficient (cm^2/g)			
		H&W&B[2]	H&H[8]	H[1]	L[6]
CrKα 2.29	573	570		584	568
TiKα 2.75	829	870	900	937	906
SbLβ 3.23	1277	1335		1421	1400
SbLα 3.44	1469	1563		1687	1630
SnLα 3.60	1588	1735		1890	1800
AgLβ 3.94	1820	1966		2050	1930
AgLα 4.15	1973	2150		1958	2150
ClKα 4.73	2221		2422	2373	2230
SKα 5.37	1885		2074	3309	3020
PKα 6.16	1158		1128	1271	1125
SiKα 7.13	1591	1614	1597	1760	1580
AlKα 8.34	2221	2300	2330	2490	2340

CONCLUSIONS

Another method for preparing thin films for mass absorption co-efficients has been demonstrated. The uniformity and thickness range are good, and the accuracy of our measurements is estimated at 3 to 5%. These metal organic compounds can be synthesized for most of the metals in the periodic table. Improvements in chemical purity and applicability to other substrates are presently being investigated. Acknowledgement is made to J. L. Kranick who carried out the measurements.

REFERENCES

1.   K. F. J. Heinrich, "X-Ray Absorption Uncertainty" - The Electron Microprobe, John Wiley & Sons, New York, 1966, 296.

2.   G. D. Hughes, J. B. Woodhouse and I. A. Bucklow, "The Determination of Some X-Ray Mass Absorption Coefficients, " Br. J. Appl. Phys. D, Ser. 2 1, 695, (1968).

3.   B. L. Henke, R. L. Elgin, R. E. Lent, R. B. Ledingham, "X-Ray Absorption in the 2 - 200Å Region " Norelco Reporter XIV, No. 3, 1967, 112.

4.   W. T. Ogier, G. J. Lucas, and R. J. Park, "Ultrasoft X-Ray Absorption Coefficients of Al, Be, C, O, and F, " Appl. Phys. Letters, 5, 146, (1964).

5.   B. A. Cooke and E. A. Stewardson, "The Absorption of X-Rays in the Range 7 - 17Å by Be, Mg, Al, Cu and Ag, " Br. J. Appl. Phys., 15, 1315, (1964).

6.   W. H. McMaster, N. Kerr Del Grande, J. H. Mallett and J. H. Hubbell, "Set 2 Compilation of X-Ray Cross Sections," UCRL-50174, University of California, Livermore, 1969.

7.   P. K. Hon, "Dead Time in Electronic Counting Systems, " Private Communication (to be published).

8.   P. K. Hon and K. F. J. Heinrich, Private Communication (to be published).

X-RAY ABSORPTION TABLES FOR THE 2-to-200 Å REGION

B. L. Henke and R. L. Elgin*

University of Hawaii

Honolulu, Hawaii 96822

Physical and chemical analysis, X-ray astronomy and high temperature plasma diagnostics which utilize the ultrasoft X-radiations have made evident a strong need for filling the gap in measured absorption coefficient data for the radiations between the conventional X-rays and the extreme ultraviolet. More than one hundred new coefficients have been measured in this laboratory on the gas state, atomic or molecular, containing He, C, N, O, F, Ne, S, Cl, Ar, Kr and Xe using eleven fluorescent, characteristic wavelengths Al-K_α (8.34 A) through Be-K (113.8 A). The radiations were isolated by Bragg reflection from multilayer analyzers of the Langmuir-Blodgett type and by pulse height discriminating proportional counter intensity measurements. Using these data and data previously published, a complete table has been determined for He through Ar and for wavelengths below the L_{III} edges and in the region 2-to-200 A. Absorption cross sections have been calculated for many compound materials which are commonly encountered in low energy X-ray analysis. The transmission of X-rays from a source above the earth has been tabulated as a function of altitude and wavelength.

The experimental apparatus, the measurement methods and the data handling for this work have been described previously.[1,2] A schematic of the experimental system is shown here in Figure 1.

In Table I are listed the measurements which have been made in this laboratory, indicating the particular eleven wavelengths employed and the gas samples measured. In italics are printed the absorption coefficients for the elements which have been calculated

---

*California Institute of Technology.

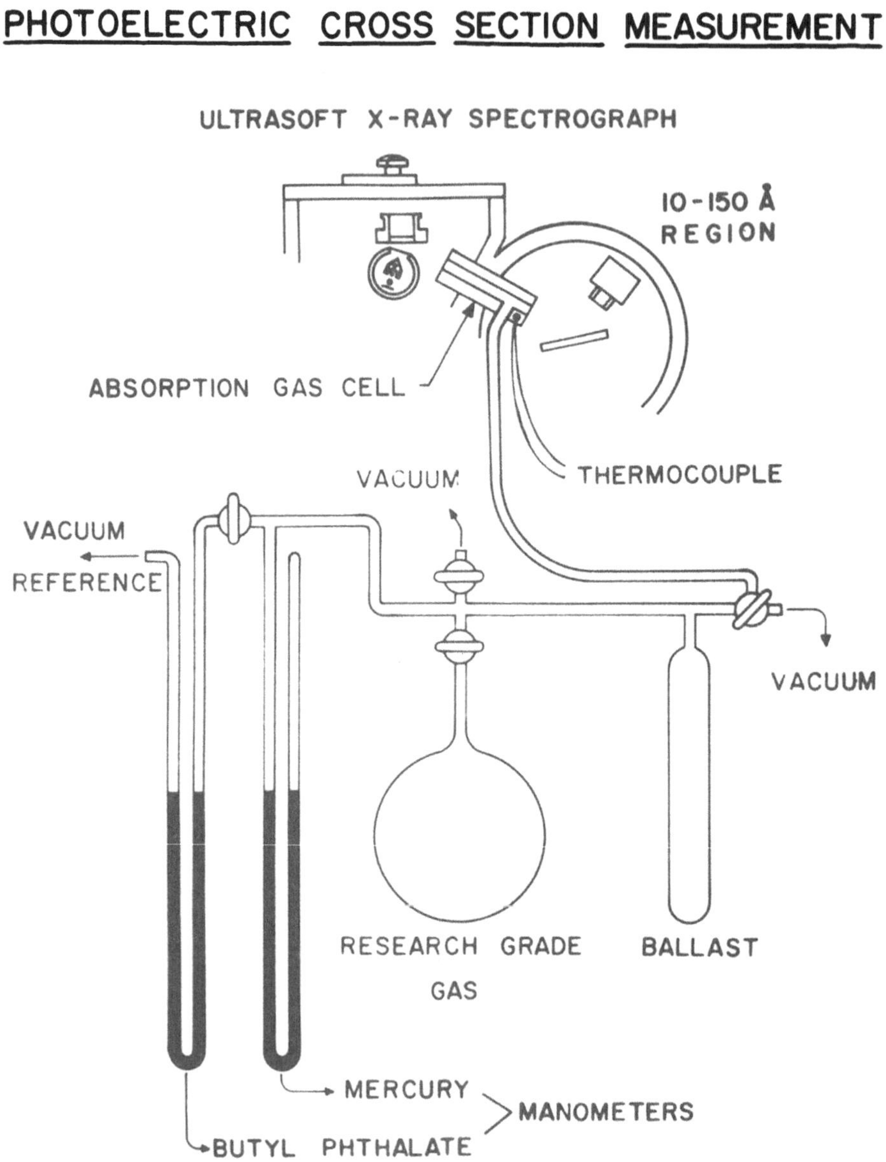

PHOTOELECTRIC CROSS SECTION MEASUREMENT

ULTRASOFT X-RAY SPECTROGRAPH

10-150 Å REGION

ABSORPTION GAS CELL

VACUUM

THERMOCOUPLE

VACUUM REFERENCE

VACUUM

RESEARCH GRADE GAS

BALLAST

MERCURY

BUTYL PHTHALATE

MANOMETERS

Figure 1. Schematic of System as Used for the Measurement of Ultrasoft X-ray Attenuation Coefficients.

TABLE I

MASS ABSORPTION COEFFICIENTS MEASURED FROM GAS STATE

Wave-length	Helium	C_2H_6	Carbon	Nitrogen (N_2)	Oxygen (O_2)	C_2F_6	Fluorine	Neon	H_2S	Sulfur	CCl_4	Chlorine	Argon	Krypton	Xenon
8.34		575	720	1121	1604	1800	2030	2780	820	868	984	1010	1180	1090	4500
9.89		933	1167	1800	2540	2800	3140	4310	1320	1390	1580	1610	1850	1710	6200
13.35		2190	2740	4040	5560	6140	6850	9770	2920	3090	3250	3300	4070	3550	8800
16.00		3570	4470	6550	8850	9800	11000	897	4670	4940			6390	5450	10900
18.32		5100	6400	9100	12620	1830	860	1310	6600	6990			8940	7270	3580
23.57		9700	12200	17200	1440	3520	1700	2600	12200	13000			15900	12300	4250
31.68		20300	25400	1730	2550	7500	3700	5540	24000	25400			30200	21500	6200
44.6	3320	1830	2280	3940	6250	7650	8780	13630	45000	47500			45600	31400	7130
67.9	11300	5230	6550	10270	16500	19700	22500	35900	65000	69000			9170	35800	4000
84.2	21500	8300	10350	17000	26500	32000	36600	56800	8000	8400			12700	32400	10200
113.8	51000	17000	21200	36500	56000	59000	67000	102000	14000	15000			19500	13000	87000

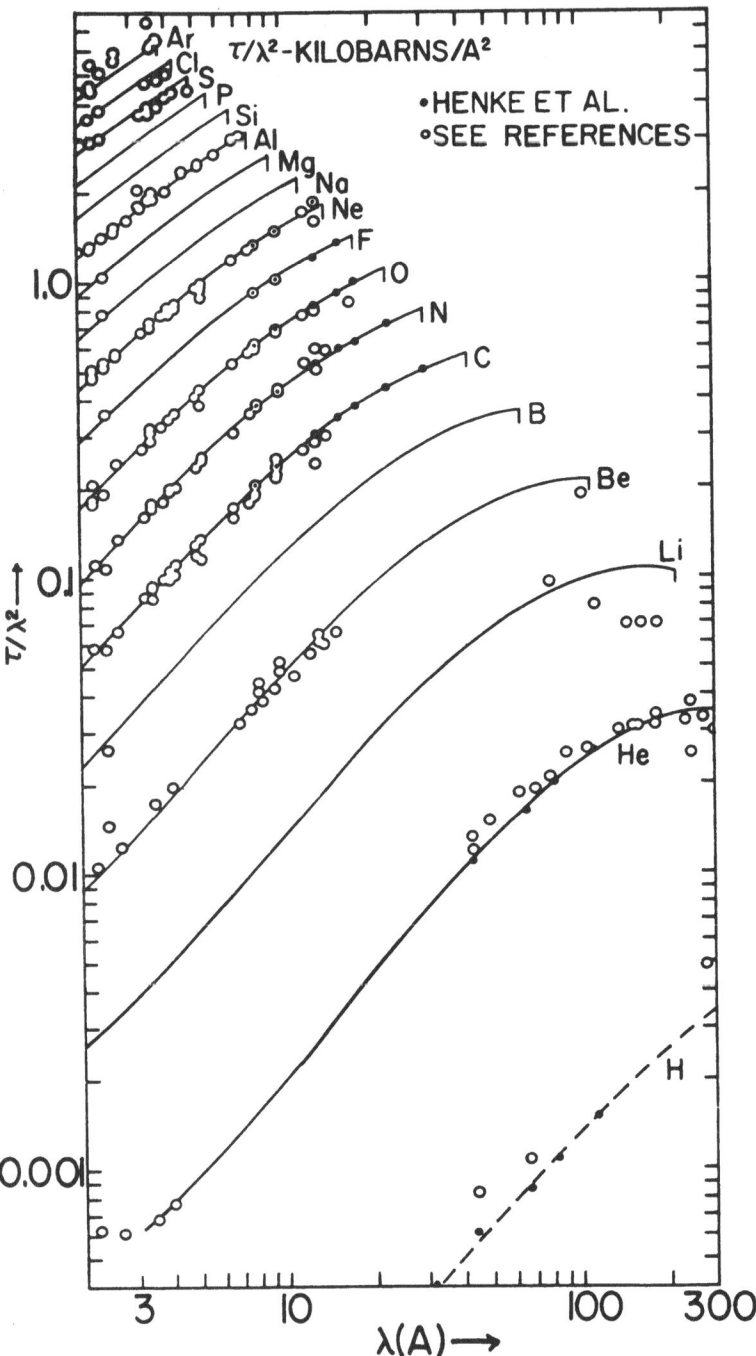

Figure 2. Plots of "Best Fit" Functions and Experimental Points
for the Light Elements and for Wavelengths Below the K-Edges.

from those measured for compounds (assumed to represent a simple sum of elementary cross sections).

In Figures 2 and 3 are shown plots of absorption cross sections as measured previous to this work by others along with these measurements as given in Table I. Because of the strong dependence upon wavelength, the cross sections are plotted here as divided by the square of the wavelength (units of kilobarns/A^2).

A more complete and up-to-date listing of published experimental mass absorption coefficients is presented in Table II. For this wavelength region and for the elements involved, coherent as well as incoherent scattering has a negligible effect on the

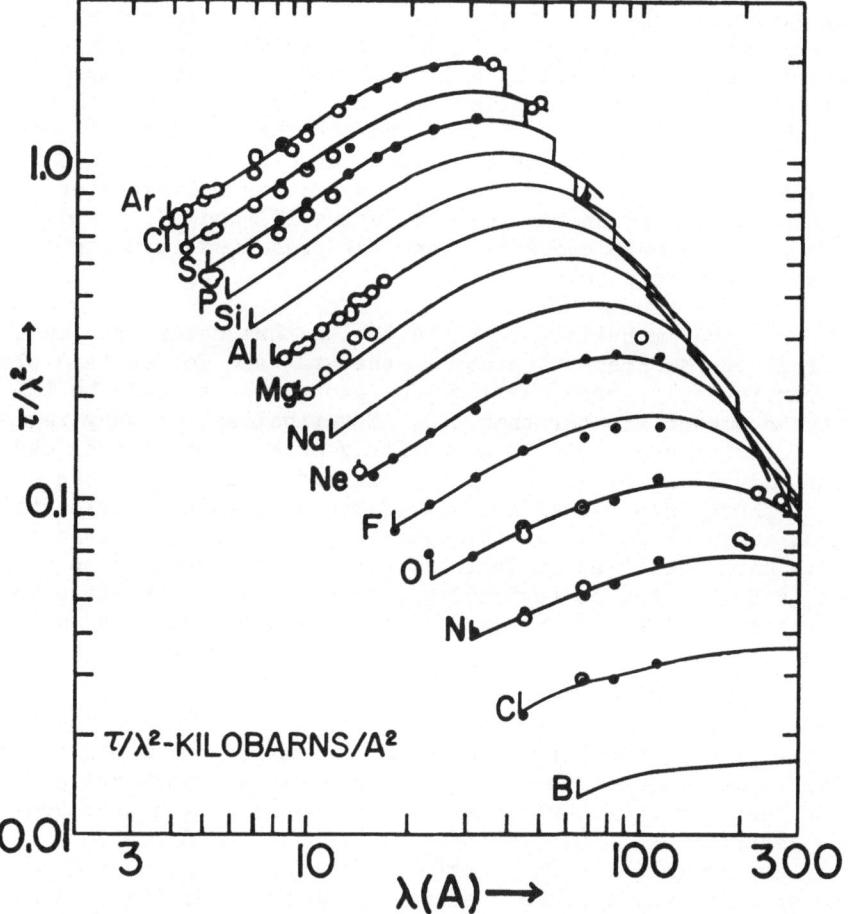

Figure 3. Plots of the "Best Fit" Functions and Experimental Points for the Light Elements and for Wavelengths Between the K and the L_{III} Edges. (o Henke et al.; • See references.)

measured attenuation, and the calculated cross sections can be considered to be essentially photoelectric. Because many of the photoelectric absorptions which are measured here are involved with the valence electron levels, it might well be expected that the particular molecular state will change somewhat the total photoelectric cross section from simply the sum value of atomic or "free atom" cross sections. As a preliminary test for the magnitude for such an effect the atomic cross sections-vs-wavelength were deduced for carbon by simple subtraction from measurements on methane, carbon monoxide, carbon dioxide as well as for ethane. Internal consistency upon neglecting possible chemical effects on such data was within a few percent. Large effects would not be expected because the photon energies which are involved here are still large as compared to the first ionization energies.

By assigning what the authors felt to be appropriate weights to the experimental cross section data for all but the most recent data listed in Table II and by using polynomial-least squares fitting (described in references 1 and 2), "best fit" functions were determined. These are plotted against the experimental data used in Figures 2 and 3. These photoelectric cross section functions have been tabulated in units of cm^2/gm for intervals of wavelength (A) and of energy (eV) in Tables III and IV, respectively. In Table V, the atomic cross sections are presented in units of barns and for energy (eV) intervals.

It is to be emphasized that the tables thus determined must simply reflect the present "state of the art" and for certain elements, particularly among the lightest group, these could be in error by an amount greater than 10%. Nevertheless, in many regions the predicted values could be appreciably more accurate than certain individual experimental values. In order to suggest the precision of these interpolations, all of the available experimental data that could be found for this wavelength region of interest here have been tabulated in Table II along with reference to their sources, and with the corresponding predicted values by these "best fit" functions. The percentage deviation of the predicted values from the individual experimental values are also presented.

Note added in proof: The authors have received, too late to enter into Table II, the tabulations of experimental measurements of mass absorption coefficients for the rare gases neon-through-xenon obtained recently by F. Wuilleumier.[3] These are given from continuous measurement in 0.05 A intervals in the 2-8 A region and 0.1 A intervals in the 8-15 A region. Interpolating these data for the specific wavelengths used by the authors for the rare gas measurements presented here in Table I, we find fairly good agreement between the two sets of data (mostly within one percent, the largest deviation being about 4 percent).

Because of the strong possibility of enhanced chemical effects or other anomalous effects being present for the absorption of wavelengths near an absorption edge wavelength, such data were not used in the polynomial fitting, but are listed in Table II. By simple polynomial extrapolation, the values for the predicted cross sections at the K absorption edge were determined along with their associated absorption jump ratios. These have been presented in Table VI.

Ignoring possible chemical combination effects, the mass attenuation coefficients for many compound samples which are often encountered in the application of ultrasoft X-ray analysis have been tabulated in Table VII. Here the percentage composition values for the gases are given, as conventional, in percentage by volume.

For the wavelengths of interest here, the effective atmospheric composition with respect to X-ray absorption is nearly constant below 150 km altitude, where most of the absorption takes place. Above 150 km, the relative numbers of nitrogen, oxygen, and argon atoms begin to change because of diffusion and the mean molecular weight decreases. Because the measurements reported here were directly on the gas state of N_2, O_2, and Ar, it is particularly appropriate to apply these data to the prediction of atmospheric absorption. In Table VIII is presented the transmission of X-radiations from a source directly above the effective absorption region of the atmosphere as a function of wavelength, of the mass per unit area thickness above the given position, and of the associated altitude. The relation between the mass thickness, m, and the altitude, h, is that deduced from the U. S. Standard Atmosphere Tables.[4] The assumed, effective composition, by volume, is 78% N_2, 21% O_2, and 1% Ar, and with negligible water vapor.

Table VIII should be considered as very approximate for altitudes above 150 km, where the assumption of sea level atmospheric composition no longer obtains and the actual composition is not known with sufficient accuracy at this time.

SUPPLEMENTARY TABLES

Except for the rare gases krypton and xenon, there are very few measurements reported for the heavy element absorption in the long wavelength region. As described in the introductory paper of this volume by one of the authors, the theoretical cross sections which have been calculated recently by McGuire[5] match the experimental data presented here for the light elements rather well. It can be expected that his cross sections for the heavy elements should be representative of the best estimates for the long

wavelength region that are available at this time. Also, the authors feel that McGuire's calculated values for the light element
cross sections for the wavelengths lying between the L_I and the L_{III}
edges are probably more accurate than those presented here (due to
the scarcity of experimental data for this absorption edge region).

For the absorption cross sections of all elements and for
wavelengths shorter than those of principal interest here (<5 A),
the reader is referred to a recent and extensive compilation by
Hubbell, McMaster, Del Grande, and Mallett.[6]

ACKNOWLEDGEMENTS

The authors gratefully acknowledge the assistance of John H.
Hubbell of the Center for Radiation Research, National Bureau of
Standards, in completing and editing the compilation of experi-
mental data.

This work is supported by the Air Force Office of Scientific
Research under AFOSR Grant No. 1262-67.

REFERENCES

1. B. L. Henke, R. L. Elgin, R. E. Lent and R. B. Ledingham,
    "X-Ray Absorption in the 2 to 200 A Region," Technical Report,
    AFOSR 67-1254, June 1967.

2. B. L. Henke, R. L. Elgin, R. E. Lent and R. B. Ledingham,
    "X-Ray Absorption in the 2 to 200 A Region," Norelco Reporter,
    XIV, No. 3-4, 1967, p. 112.

3. F. Wuilleumier, "Contribution a L'Etude de la Photoionisation
    des Gaz Rares par Analyse Continue Entre 1,5 et 15 A," Thesis,
    University of Paris, 1969.

4. United States Committee on Extension to the Standard Atmos-
    phere, U. S. Standard Atmosphere, 1962, Superintendent of
    Documents, U. S. Government Printing Office, Washington, D. C.

5. E. J. McGuire, "Photo-Ionization Cross Sections of the Elements
    Helium to Xenon," Phys. Rev. 175:20-30 (1968).

6. W. H. McMaster, N. K. Del Grande, J. H. Mallett, and J. H.
    Hubbell, "Compilation of X-Ray Cross Sections," Lawrence Radia-
    tion Laboratory Report UCRL-50174 Section II, University of
    California, Livermore (1969).

TABLE II

COMPARISON OF REFERENCED EXPERIMENTAL DATA
WITH VALUES PREDICTED BY THE WEIGHTED
"BEST FIT" FUNCTION EXPRESSED IN TABLE III
(UNITS: CM^2/GM)

The following designations are used:

X Experimental values, the letters referring to the sources listed
   in the references below.
A An average, predicted value.
%D The percentage deviation of the predicted value from the
   experimental value.
* Values read from another author's graphical presentation.
** Values corrected for impurities.

References

A - S. J. M. Allen, reported in Compton and Allison, X-Rays in Theory and
       Experiment, pp. 800-806.
An - C. L. Andrews, Phys. Rev. 54, 994 (1938).
B - A. J. Bearden, J. Appl. Phys. 37(4), 1681 (1966).
Ba - G. B. Bandopadhyaya and A. T. Maitra, Proc. Roy. Soc., 21, 869 (1936).
Bi - H. H. Biermann, Ann. Physik 26, 740 (1936).
C - B. A. Cooke and E. A. Stewardson, Brit. J. Appl. Phys. 15, 1315 (1964).
Cl - W. W. Colvert, Phys. Rev. 36, 1619 (1930).
CO - J. A. Crowther and L. H. H. Orton, Phil. Mag. 10, 329 (1930); 13, 505
       (1932).
CW - M. J. Cole, J. B. Woodhouse, et al., Final Report, U. S. Dept of Army
       Contract DA-91-591-EUC-3094, Oct. 1964.
D - E. Dershem and M. Schein, Phys. Rev. 37, 1238 (1931).
De - R. D. Deslattes, AFOSR-TN-58-784; Dissert.; Johns Hopkins Univ.,
       Baltimore (1959).
DO - Del Grande, Oliver, and Stinner, unpublished (1966-69), quoted in
       LRL Source #39.
Eh - C. E. Ehrenfried and D. E. Dodds, AFSWC-TN-59-33 (1960).
Er - O. A. Ershov and A. P. Lukirskii, Fiz. Tver. Tela 8, 2137 (1966);
       transl. in Sov. Phys. - Solid State 8, 1699 (1967).
G - K. Grosskurth, Ann. Physik 20, 197 (1934).
H - B. L. Henke, et al., AFOSR 67-1254.
He - K. F. J. Heinrich, NBS (1962), private communication.
Hi - R. D. Hill, Proc. Roy. Soc. (London) A161, 284 (1937).
HN - B. L. Henke and J. Nippa (1969), University of Hawaii.
Ho - J. I. Hopkins, J. Appl. Phys. 30:2, 185 (1959).
Hu - W. R. Hunter and R. Tousey, J. Physique 25, 148 (1964).
HW - G. D. Hughes, J. B. Woodhouse and I. A. Bucklow, Brit. J. Appl. Phys.
       1 (Ser. 2), 695 (1968).
J - E. Jonsson, Dissert., Uppsala (1928).
K - H. Kurtz, Ann. Physik 85, 529 (1928).
L - A. P. Lukirskii, I. A. Brytov and T. M. Zimkina, Optika i Spektrosk.
       17, 438 (1964), transl. in Opt. and Spectrosc. 17, 234 (1964)(He);
       A. P. Lukirskii and T. M. Zimkina, Izv. Akad. Nauk SSSR 27, 817
       (1963), transl. in Bull. Acad. Sci. USSR, Phys. Ser. 27, 333
       (1963)(Ar).
M - R. H. Messner, Z. Physik 85, 727 (1933).
O - W. T. Ogier, G. J. Lucas and R. J. Park, Appl. Phys. Letters 5, 146
       (1964).
S - J. A. R. Samson, et al., J. Opt. Soc. Am. 54:7, 876 (1964); 54:12,
       1491 (1964); 54:420, (1964); 54:6, 842 (1964); 55:1035 (1964);
       55:8, 935 (1965); 56:526 (1966); Applied Optics 4, 8 (1964).
Si - S. Singer, J. Appl. Phys. 38, 2897 (1967).
Sp - R. G. Spencer, Phys. Rev. 38, 1932 (1931).
T - D. H. Tomboulian, Phys. Rev. 83:6, 1196 (1951); 94:6, 1585 (1954);
       124:5, 1471 (1961); 128:2, 677 (1962); 133:6A, 1525 (1964).
W - B. Woernle, Ann. Physik 5, 475 (1930).
We - W. Weisweiler, Proc. 5th Internat. Conf. on X-Ray Optics and Micro-
       analysis, Tubingen, July-Aug. (1968).
Wi - P. R. Wise, Johns Hopkins University Thesis NP-12661 (1961).
Wr - W. Wrede, Ann. Phys. 36, 681 (1939).
Wu - F. Wuilleumier, J. Phys. Radium 26, 776 (1965).

λ	X		A	%D
HYDROGEN				
1.00	0.44	A		
1.25	0.45	A		
1.39	0.47	A		
1.54	0.48	A		
	0.55	B		
1.94	0.50	A		
	0.72	B		
2.29	0.86	B		
2.50	0.55	A		
2.75	1.17	B		
3.60	2.15	B		
4.15	2.63	B		
31.68	241.	H		
44.60	1000.	M		
	729.	H		
67.90	2980.	M		
	2375.	H		
84.20	4610.	H		
113.80	11660.	H		
239.6	131000.	S		
283.5	236000.	S		
314.9	335000.	S		
345.1	450000.	S		
374.4	610000.	S		
452.2	1010000.	S		
HELIUM				
1.00	0.25	A		
1.54	0.28	B		
1.94	0.34	B		
2.29	0.49	B		
2.75	0.69	B	0.7	-9.9
3.60	1.33	B	1.4	0.3
4.15	2.	B	2.1	1.1
44.60	3600.	D	3300.	-8.4
	3960.	L		-16.7
	3320.	H		-0.6
51.20	6000.	T	5010.	-16.5
64.35	11700.	L	9880.	-15.6
67.90	11300.	H	11600.	2.3
72.20	15100.	L	13800.	-8.6
81.98	21300.	L	19800.	-6.8
84.20	21500.	H	21400.	-0.5
93.00	33000.	T	28200.	-14.4
108.70	46600.	L	43300.	-7.2
113.80	51000.	H	48900.	-4.1
139.50	89000.	L	83200.	-6.5
157.0	116000.	T	112000.	-3.5
164.6	128000.	L	126000.	-1.8
189.0	170000.	T	175000.	3.0
190.0	185000.	L	177000.	-4.2
239.0	280000.	T	298000.	6.3
250.0	354000.	L	328000.	-7.4
	240000.	S*		36.6
275.0	380000.	T	401000.	5.4
300.0	400000.	S*	478000.	19.4
325.0	540000.	T	560000.	3.4
350.0	580000.	S*	640000.	10.7
356.0	650000.	T	660000.	2.0
395.0	790000.	T	800000.	1.1
400.0	750000.	S*	820000.	8.8
443.0	990000.	T	970000.	-2.0
450.0	940000.	S*	1000000.	5.9
504.0	1260000.	T	1190000.	-5.5
	1120000.	S*		6.3

λ	X		A	%D
LITHIUM				
1.00	0.43	A	0.2	-68.0
1.25	0.35	DO	0.3	<10
1.54	1.10	A	0.5	-61.2
	0.50	DO		<10
1.76	0.66	DO	0.7	<10
1.94	2.10	A	0.9	-61.5
2.08	0.98	DO	1.0	<10
2.50	4.	A	1.7	-58.2
	1.65	DO		<10
83.3	58000.	T	53500.	-7.7
117.9	97000.	T	122000.	26.0
151.5	138000.	T	211000.	53.0
171.6	178000.	T	272000.	53.0
192.8	225000.	T	342000.	52.0
215.2	203000.	T	420000.	107.0
231.2	119000.	T	5450.	-95.4
272.2	139000.	T	8290.	-94.0
BERYLLIUM				
1.00	0.55	A	0.4	-45.4
1.11	0.45	DO	0.5	<10
1.25	0.60	DO	0.6	<10
1.39	1.25	A	0.8	-38.5
1.48	0.96	DO	1.0	<10
1.54	1.60	A	1.1	-35.2
	1.06	B		-2.3
	1.06	DO		<10
1.76	1.48	A	1.6	8.
1.94	3.05	A	2.1	-32.7
	2.29	B		-10.4
	2.00	DO		5.
	2.1	CW		0
2.07	2.32	DO	2.5	8.
2.29	3.4	CW	3.4	0
	3.70	B		-8.8
2.50	6.10	A	4.5	-27.8
	3.97	DO		13.
2.75	5.9	CW	5.9	0
	6.25	B		-5.9
3.52	11.7	DO	12.5	7.
3.60	13.5	CW	13.5	0
	15.00	B		-10.6
4.15	22.80	B	21.0	-8.2
	21.0	CW		0
7.13	110.	C	111.	0.4
	125.7	CW		-11.
7.85	148.	C	149.	0.3
8.34	206.	O	179.	-13.2
	199.	CW		-10.
	192.	Q**		-6.9
8.60	189.	C	197.	3.9
9.45	252.	C	262.	3.9
9.89	340.	O	301.	-11.5
	327.	CW		-8.
	318.	O**		-5.4
11.00	379.	C	416.	9.6
12.50	578.	C	611.	5.7
13.35	770.	O	744.	-3.4
	722.	O**		3.0
14.05	792.	C	867.	9.4
15.50	1050.	C	1160.	10.4
60.0	110000.	T*	45500.	-58.7
105.0	140000.	T*	157000.	12.2
115.0	15000.	T*	5430.	-63.8
200.0	80000.	T*	19000.	-76.3

λ	X		A	%D	λ	X		A	%D
BORON					**CARBON** (cont.)				
1.00	0.76	A	0.7	-19.1	9.89	1063.	A	1170.	10.4
1.25	1.35	A	1.2	-15.9		1156.	W**		1.6
1.39	1.87	A	1.7	-12.4		1090.	B		7.7
1.54	2.45	A	2.3	-9.0		1235.	O		-4.9
1.94	4.70	A	4.5	-4.5		1167.	H		0.6
2.50	9.10	A	9.8	7.1	11.9	2082.	We	1900.	-9.
CARBON					12.25	2030.	B	2150.	5.8
					13.35	2756.	We	2730.	-1.
1.00	1.36	A	1.3	-11.7		2170.	A		25.6
1.04	1.39	G	1.4	-2.4		2550.	B		6.9
	1.43	WR		-5.1		2580.	O		5.7
1.12	1.70	WR	1.7	-5.5		2740.	H		-0.5
	1.69	DO		1.	14.56	3200.	B	3450.	7.9
1.18	1.90	G	2.0	1.8	16.00	4470.	H	4470.	0
	2.01	WR		-3.8	17.59	5591.	We	5800.	4.
1.25	2.42	A	2.4	-4.7	18.32	6400.	H	6420.	0.3
	2.26	DO		6.	23.57	12200.	H	12300.	1.0
	2.40	WR		-1.6		12993.	We		-5.
1.34	2.93	WR	2.9	-2.7	31.68	25400.	H	25600.	0.7
1.39	3.35	A	3.3	-2.7		23803.	We		8.
1.44	3.52	G	3.6	0.9	44.6	2300.	M**	2350.	2.3
	3.53	WR		0.7		2280.	O		3.2
1.48	3.72	DO	4.0	7.		2280.	H		3.2
1.54	4.52	A	4.5	-1.5		2535.	We	2350.	-7.
	4.30	CO**		3.5	67.9	6800.	M**	6530.	-4.0
	4.33	B		2.8		6550.	H		-0.3
	4.32	DO	4.5	4.		6592.	We	6600.	0
1.66	5.57	G	5.5	-1.4	84.20	10350.	H	10700.	3.1
1.76	6.34	DO	6.7	5.	113.80	21200.	H	20900.	-1.6
1.94	8.77	A	9.0	2.6					
	8.83	CO**		1.9	**NITROGEN**				
	8.62	G		4.4	1.00	2.11	A	2.1	-4.7
	8.79	B		2.3	1.25	3.95	A	3.9	-2.3
	8.41	DO		7.	1.39	5.50	A	5.5	-0.8
2.07	10.3	DO	11.0	7.	1.54	7.40	A	7.5	0.6
2.29	15.20	B	15.0	-1.8		7.33	CO		1.6
2.48	18.6	DO	19.6	5.		7.29	B		2.1
	18.00	A		8.4	1.94	14.00	A	15.1	7.2
2.75	25.00	B	26.1	4.3		15.00	B		0
3.10	35.7	DO	37.3	4.	2.29	25.10	B	24.8	-1.2
3.38	49.60	W**	48.9	-1.5	2.50	29.00	A	32.4	11.4
3.52	53.4	DO	55.	3.	2.75	44.20	B	43.1	-2.6
3.60	55.20	A	59.2	7.2	3.38	79.50	W	80.0	0.6
	59.70	W**		-0.9	3.60	96.30	W	96.6	0.2
	60.90	B		-2.9		99.50	B		-3.0
3.93	77.40	W**	77.2	-0.4	3.93	120.70	W	125.	3.8
4.15	92.00	W**	91.	-1.2	4.15	144.30	W	147.	2.0
	85.00	B		7.0		149.	B		-1.2
4.37	97.80	A	106.	7.9	4.37	166.	W	170.	2.6
	106.40	W**		-0.8	5.18	273.	W	281.	3.0
5.18	160.	A	176.	9.9	5.41	312.	W	319.	2.3
	174.	W**		1.1		320.	B		-0.3
5.41	201.	W**	201.	0.2	6.98	645.	W	667.	3.4
	170.	B		18.5	7.98	980.	W	973.	-0.7
6.98	390.	A	426.	9.3	8.34	1109.	W	1110.	0.1
	422.	W**		1.0		1150.	B		-3.5
7.98	570.	B	628.	10.2		1121.	H		-1.0
8.34	656.	A	719.	9.6	9.89	1796.	W	1790.	-0.5
	711.	W**		1.2		1825.	B		-2.1
	670.	B		7.4		1800.	H		-0.7
	725.	O		-0.8	12.25	3430.	B	3210.	-6.5
	720.	H		-0.1					
	806.	We		-11.		(continued)			

NITROGEN (cont.)

λ	X		A	% D
13.35	3836.	A	4040.	5.3
	4530.	B		-10.9
	4040.	H		-0.1
14.56	5300.	B	5070.	-4.3
16.00	6550.	H	6500.	-0.7
18.32	9100.	H	9210.	1.2
23.57	17200.	H	17300.	0.3
31.68	1730.		1640.	-5.1
44.60	3850.	D	3870.0	0.5
	3790.	M		2.1
	3800.	K		1.9
	3940.	H		-1.7
67.90	10900.	M	10600.0	-2.6
	10270.	H		3.3
84.20	17000.	H	17500.0	3.2
113.80	36500.	H	34900.0	-4.5
209.80	189000.	S	127000.0	-8.8
247.20	210000.	S	174000.0	-17.3
297.60	247000.	S	239000.0	-3.2
345.10	318000.	S	300000.0	-5.6
387.40	400000.	S	351000.0	-12.2
452.20	486000.	S	420000.0	-13.6
508.20	490000.	S	467000.0	-4.6
537.00	542000.	S	487000.0	-10.2

OXYGEN

λ	X		A	% D
1.00	3.15	A	3.2	0.7
1.25	5.70	A	6.1	6.5
1.39	8.10	A	8.6	5.8
1.54	11.16	A	11.7	4.6
	11.6	Sp		0.6
	11.5	GA		1.5
	11.1	CO		5.1
	11.8	B		-1.1
1.94	22.2	CO	23.4	5.2
	25.3	B		-7.7
2.29	36.4	W	38.4	5.4
	35.5	Sp		8.0
	41.1	B		-6.7
2.50	45.5	A	49.9	9.5
2.75	69.1	B	66.1	-4.4
3.38	116.8	W	122.	4.0
3.60	150.0	A	146.	-2.6
	141.0	W		3.6
	156.0	B		-6.3
3.93	188.8	W	189.	0
4.15	222.	W	221.	-0.4
	221.	Sp		0.1
	220.	B		0.5
4.37	258.	W	255.	-1.1
5.18	413.	W	417.	0.9
5.41	476.	W	472.	-0.8
	420.	B		12.4
6.98	976.	W	970.	-0.6
	971.	Sp		-0.1
7.98	1380.	B	1400.	1.6
8.34	1585.	W	1590.	0.6
	1615.	B		-1.3
	1540.	O		3.5
	1604.	H		-0.6

OXYGEN (cont.)

λ	X		A	% D
9.89	2540.	W	2530.	-0.2
	2520.	B		0.6
	2480.	O		2.2
	2540.	H		-0.2
12.25	4340.	B	4480.	3.1
13.35	5456.	A	5600.	2.6
	5500.	B		1.8
	5340.	O		4.8
	5560.	H		0.7
16.00	8850.	H	8900.	0.5
17.59	10000.	A	11400.	14.1
18.32	12620.	H	12500.	-1.1
23.57	1440.	H	1190.	-17.3
31.68	2550.	H	2530.	-0.6
44.60	5800.	A	5980.	3.1
	5765.	D		3.7
	5650.	M		5.8
	6000.	K		-0.4
	6150.	O		-2.8
	6250.	H		-4.4
67.90	16250.	M	16600.	2.3
	16500.	H		0.8
84.20	26500.	H	27500.	3.8
113.80	56000.	H	53800.	-3.9
225.2	200000.	S	195000.0	-2.3
266.3	264000.	S	250000.0	-5.3
314.9	309000.	S	308000.0	-0.3
374.4	337000.	S	366000.0	8.6
428.2	365000.	S	404000.0	10.7
522.0	393000.	S	440000.0	11.9
537.0	399000.	S	369000.0	-7.6

FLUORINE

λ	X		A	% D
1.00	4.80	A	4.5	-7.3
1.04	4.94	WR	5.1	1.3
1.11	5.86	WR	6.0	1.1
1.18	7.06	WR	7.2	0.9
1.25	9.50	A	8.7	-8.6
	8.56	WR		1.5
1.34	10.22	WR	10.5	2.3
1.39	12.50	A	12.0	-4.6
1.44	12.59	WR	13.0	3.1
1.54	17.	A	16.2	-4.7
1.94	35.	A	32.2	-8.1
2.50	71.	A	67.9	-4.4
8.34	2035.	O	2040.	0.1
	2030.	H		0.5
9.89	3140.	O	3200.	2.1
	3140.	H		2.1
13.35	6850.	H	6960.	1.6
16.00	11000.	H	11000.	-0.4
18.32	860.	H	874.	1.6
23.57	1700.	H	1680.	-1.3
31.68	3700.	H	3620.	-2.2
44.60	8780.	H	8670.	-1.3
67.90	22500.	H	24000.	6.8
84.20	36600.	H	39100.	6.7
113.80	67000.	H	72900.	8.8

λ	X		A	%D
NEON				
1.00	6.5	A	6.6	1.1
1.25	12.4	A	12.5	0.4
1.39	17.0	A	17.5	2.7
	16.0	Cl		9.1
1.54	24.0	A	23.7	-1.6
	22.0	B		7.3
1.94	49.0	A	46.5	-5.2
	44.7	B		4.0
2.29	74.7	W	75.4	0.8
	75.5	Cl		-0.2
	76.0	B		-0.9
	79.6	Wu		-5.4
2.50	100.0	A	97.1	-2.9
	96.4	Wu		0.7
2.75	130.	B	128.	-1.7
	127.	Wu		0.6
3.38	231.	W	231.	-0.3
3.60	275.	A	276.	0.2
	270.	Wu		2.0
	282.	B		-2.3
3.93	356.	W	353.	-0.9
4.15	416.	W	411.	-1.1
	421.	B		-2.3
	401.	Wu		2.6
4.37	478.	W	472.	-1.2
	459.	Wu		2.9
5.18	763.	W	758.	-0.7
5.41	865.	W	859.	-0.7
	831.	Wu		3.3
	770.	B		11.5
6.98	1727.	W	1710.	-1.1
7.98	2420.	B	2440.	0.6
8.34	2750.	W	2750.	0.2
	2760.	B		-0.2
	2780.	H		-0.9
9.89	4310.	W	4300.	-0.3
	4320.	B		-0.5
	4310.	H		-0.3
12.25	7700.	B	7430.	-3.5
13.35	8500.	A	9230.	8.5
	9800.	B		-5.9
	9770.	H		-5.6
14.56	750.	B	705.	-6.0
16.00	897.	H	919.	2.4
18.32	1310.	H	1310.	0.2
23.57	2600.	H	2570.	-1.2
31.68	5540.	H	5630.	1.6
44.60	13630.	H	13500.	-1.1
67.90	35900.	H	35900.	0.1
80.00	75000.	T*	50700.	-32.4
84.20	56800.	H	56200.	-1.1
100.00	90000.	T*	77900.	-13.5
113.8	102000.	H	97500.	-4.4
150.0	167000.	T*	148000.	-11.3
200.0	220000.	T*	208000.	-5.3
	196000.	S*		6.3
250.0	260000.	T*	253000.	-2.8
	230000.	S*		9.9
300.0	253000.	S*	248000.	-1.9
	270000.	T*		-8.1
350.0	266000.	S*	269000.	1.1
400.0	270000.	S*	281000.	4.1
	280000.	T*		0.4

λ	X		A	%D
NEON (cont.)				
450.0	258000.	S*	287000.	11.1
500.0	238000.	S*	287000.	20.8
	250000.	T*		15.0
550.0	206000.	S*	285000.	38.3
560.0	220000.	T*	284000.	29.1
576.0	188000.	S	283000.	50.3
SODIUM				
1.00	8.80	A	8.7	-1.3
1.54	32.10	A	30.8	-4.1
2.50	128.00	A	124.0	-2.9
MAGNESIUM				
1.00	11.50	A	11.9	3.0
	10.65	De		11.2
1.10	14.71	De	15.7	6.3
1.25	21.40	A	22.7	5.9
	21.54	De		5.2
1.39	30.00	A	30.9	2.7
	29.32	De		5.1
1.54	40.80	A	41.4	1.5
	39.33	De		5.2
1.94	77.2	A	79.	2.3
2.50	161.	A	164.	2.1
10.00	510.	C	506.	-0.9
11.50	767.	C	740.	-3.5
13.00	1100.	C	1040.	-5.9
14.05	1470.	C	1280.	-12.9
15.50	1830.	C	1680.	-8.5
ALUMINUM				
1.00	14.12	A	14.9	5.3
	13.50	B		10.1
	13.64	De		9.0
	14.70	Ho		1.1
1.04	15.30	He	16.6	8.
1.11	18.77	De	19.6	4.2
	19.7	J	20.0	1.
	18.82	He		6.
1.13	20.07	De	20.7	2.7
1.14	20.5	HW	21.6	5.
1.18	23.48	G	23.4	-0.5
	23.1	J	24.0	4.
1.25	26.30	A	27.6	4.9
	29.80	Ho		-7.4
	29.90	B		-7.7
	27.12	De		1.8
1.31	31.8	HW	29.0	-9.
1.39	36.80	A	38.3	4.0
	37.12	De		3.1
	36.40	He		5.1
1.44	41.44	Eh	42.6	2.7
	45.6	Ho		-7.0
	41.8	J		1.8
	40.47	He		5.0
1.54	49.00	A	51.3	4.6
	51.15	G		0.2
	50.70	Eh		1.1
	50.28	De		1.9
	52.00	B		-1.5

(continued)

ALUMINUM (cont.)

λ	X		A	%D
1.54 (cont.)	49.10	He	51.3	4.4
	51.2	J		0
	49.7	An		3.
	50.6	HW		1.
	49.8	He		3.
1.66	68.70	Ho	61.3	-10.8
	62.25	Eh		-1.6
	61.1	J		5.
	61.3	An		4.
	63.1	HW		1.
	61.75	He		4.
1.79	76.34	G	79.	3.1
	72.0	J		8.6
	76.3	He		3.1
1.94	93.50	A	99.	5.6
	97.12	Eh		1.9
	92.4	B		6.7
	92.1	He		6.9
	93.9	J		5.2
	93.5	An		5.5
	99.7	HW		-0.7
	95.33	He		3.7
2.10	122.4	G	123.	0.5
	123.0	Eh		0.
	120.9	He		1.7
2.29	152.3	Bi	157.	0.5
	149.5	He		2.4
	151.9	G		0.8
	149.1	Eh		2.7
	149.0	B		2.8
	153.	An		3.
	160.3	HW		-2.
	153.4	He		3.
2.50	193.0	A	201.	3.8
	196.3	Bi		2.1
	182.	J		10.
	194.	An		4.
	205.	HW		-2.
	200.4	He		0
2.75	255.0	Bi	261	2.3
	252.3	Eh		3.3
	244.0	B		6.5
	248.	J		5.
	268.	An		-2.7
	267.	HW		-2.3
	260.4	He		0.2
3.03	320.	J	342.	7.
	346.	An		-1.
	350.	HW		-2.
3.08	333.0	Bi	345.	3.5
3.29	430.	HW	425.	-1.
3.36	450.0	Bi	452.	0.4
	506.2	Eh		-12.
	433.	J		4.2
	459	An		-1.5
	476.0	He		-5.3
	447.	He		1.1
3.60	500.	A	545.	8.2
	541.	B		0.7
	535.	He		1.8
	567.	HW		-4.

ALUMINUM (cont.)

λ	X		A	%D
3.74	613.	Bi	608.	-0.8
	581.	Eh		4.4
	595.	He		2.1
	639.	HW		-5.
	601.3	He		1.
3.93	710.	An	695.	-2.
4.15	774.	B	806	3.9
	771.	J		4.3
	822.	An		-2.
	813.	HW		-1.
4.37	880.	A	928	5.1
	894.	J		3.7
	941.	An		-1.4
	936.	HW		-1.
4.73	1167.	Bi	1140	-2.3
	1142.	Eh		-0.2
	1108.	An		2.8
	1128.5	He		1.
5.12	1410.	HW	1410.	0
5.18	1370.	A	1450.	6.1
	1430.	An		1.
5.37	1567.	Bi	1600.	2.1
	1598.	Eh		0.1
	1600.	B,Wi		0
	1594.9	He	1610.	1.
5.41	1480.	J	1630.	10.
	1630.	An		0
	1644.	HW		-1.
5.49	1691.	He	1690.	0
5.72	1902.	HW	1890.	-1.
	1880.	He		1.
6.05	2206.	HW	2180.	-1.
6.07	2281.	HW	2210.	-3.
	2130.	An		4.
6.16	2247.	Bi	2310.	2.7
	2271.	He		1.7
6.45	2700.	HW	2590.	-4.
6.98	2800.	A	3200.	12.8
	2900.	Ba		9.6
	2800.	J		12.8
	3299.	HW		-2.7
7.13	3429.	Bi	3390.	-1.1
	3370.	C		0.6
	3170.	An		7.
	3450.	HW		-2.
	3410.	He		-1.
7.25	3640.	HW	3530.	-3.
7.50	3890.	C	3870.	-0.5
7.95	4300.	Ba	4500.	4.
	304.	Ba	355.	14.
8.34	330.	A	404.	22.3
	396.	Bi		1.9
	390.	B,Wi		3.5
	396.	O		1.9
	344.	J		18.
	459.	An		-12.
	408.	HW		-1.
	386.1	He		4.
8.60	430.	C	439.	2.0
9.45	556.	C	567.	1.9

(continued)

λ	X	A	%D
ALUMINUM (cont.)			
9.89	632.	Bi	642.	1.5
	632.	·W1		1.5
	630.	O		1.9
	630.	B		1.9
	650.	HN		-1.
	553.	J		16.
	642.	HW		0
11.00	862.	C	859.	-0.4
12.25	1150.	B,W1	1150.	0.3
12.50	1200.	C	1220.	1.6
13.35	1410.	O	1460.	3.5
	1440.	B,W1		1.3
	1468.	HN		-1.
14.05	1700.	C	1680.	-1.3
14.56	1830.	W1	1850.	0.8
	1840.	B		0.3
	1740.	Ba		6.
	2290.	H1		-19.
15.50	2200.	C	2190.	-0.4
16.00	2330.	Ba	2390.	2.
17.00	2840.	C	2810.	-1.0
17.59	3070.	Ba	3100.	1.
	3520.	H1		-12.
19.45	4050.	Ba	4010.	-1.
21.64	5430.	Ba	5300.	-2.
23.57	6920.	Ba	6700.	-3.
	7330.	H1		-8.
44.60	33000.	Si	29900.	-10.
150.00	115000.	Hu	131000.	13.8
SILICON			
1.00	17.	A	19.3	13.4
1.39	44.	A	49.2	11.7
1.94	111.	A	126.	13.0
PHOSPHORUS			
1.00	21.2	A	23.0	8.4
1.39	54.	A	58.2	7.6
1.94	134.	A	147.	9.7
SULFUR			
1.00	26.5	ST**	28.6	7.8
1.25	49.2	ST**	52.2	6.0
1.39	65.5	A	71.7	9.4
	69.2	ST**		3.6
1.54	90.	A	95.3	5.8
	93.	ST**		2.4
	88.	CO**		8.2
1.94	173.	A	180.	4.1
	179.	ST**		0.6
2.29	289.	W**	284.	-1.8
2.50	355.	A	361.	1.6
3.38	787.	W**	816.	3.7
3.60	900.	A	967.	7.5
	949.	W**		1.9
3.93	1139.	W**	1220.	7.4
4.15	1350.	W**	1420.	4.9
4.37	1562.	W**	1620.	3.4
5.02	2100.	A	2340.	11.4
	210.	A	223.	6.0
5.18	221.	W**	241.	9.2

λ	X	A	%D
SULFUR (cont.)			
5.41	250.	W**	272.	8.7
6.98	500.	W**	533.	6.6
8.34	793.	W**	867.	9.3
	868.	H		-0.2
9.89	1270.	W**	1380.	8.4
	1390.	H		-0.9
11.91	2100.	A	2270.	8.1
13.35	3090.	H	3090.	0
16.00	4940.	H	4970.	0.6
18.32	6990.	H	7020.	0.5
23.57	13000.	H	13000.	-0.1
31.68	25400.	H	25000.	-1.7
44.60	47500.	H	47600.	0.2
67.90	69000.	H	74100.	7.4
CHLORINE			
1.00	30.5	A	32.6	6.8
1.39	76.9	Cl	81.2	5.6
1.54	102.7	Cl	108.	4.9
2.29	315.	Cl	319.	1.3
2.50	400.	A	405.	1.3
3.60	1020.	A	1060.	4.0
3.93	1256.	W	1370.	9.3
4.15	1476.	W	1590.	7.7
4.37	1798.	W	1820.	0.9
4.40	1830.	A	1840.	0.3
	178.	A	186.	4.2
5.18	277.	W	284.	2.6
5.41	311.	W	320.	2.9
6.98	610.	W	629.	3.1
8.34	960.	W	1020.	6.4
	1010.	H		1.1
9.89	1610.	H	1620.	0.7
	1570.	W		3.2
11.91	2500.	A	2660.	6.5
13.35	3300.	H	3570.	8.3
ARGON			
1.00	35.0	A	35.9	2.5
1.39	85.7	Cl	88.9	3.7
1.54	116.0	A	118.	1.5
	112.0	S		5.1
	114.0	Cl		3.3
	112.5	CO		4.6
	119.6	GA		-1.6
	118.0	B		-0.2
1.94	235.0	A	221.	-5.9
	274.0	B		-19.3
2.08	282.0	W	268.	-5.2
2.29	344.0	Sp	347.	1.0
	339.4	Cl		2.4
	354.0	W		-1.9
	422.0	B		-17.7
	362.0	WU		-4.0
2.50	475.0	A	441.	-7.2
	478.0	WU		-7.8
2.75	667.0	B	571.	-14.5
	631.0	WU		-9.6
3.60	1210.0	W	1180.	-2.4
	1192.0	WU		-0.9
	1470.0	B		-19.7
3.74	1320.0	Sp	1290.	-2.2

(continued)

λ	X		A	% D	λ	X		A	% D
ARGON (cont.)					ARGON (cont.)				
3.87	1465.0	Sp	1430.	−2.7	8.34	1157.	W	1160.	0.4
	1490.0	Wu		−4.3		1150.	B		1.0
	147.0	Sp	154.	4.6		1180.	H		−1.6
	146.0	Wu		5.3	9.00	1318.	L	1430.	8.2
3.93	152.7	W	160.	4.7	9.89	1865.	W	1840.	−1.4
3.96	153.0	Wu	162.	5.8		1770.	B		3.9
	151.0	Sp		7.2		1850.	H		−0.6
4.00	151.4	L	167.	10.3	12.00	2724.	L	3080.	13.0
	156.0	Wu		7.1	12.25	3200.	B	3250.	1.5
4.15	171.0	Sp	183.	7.1	13.35	4040.	B	4070.	0.7
	174.3	W		5.1		4070.	H		−0.1
	174.0	B		5.3	14.56	4600.	B	5080.	10.4
	179.0	Wu		2.4	15.00	4732.	L	5490.	16.0
4.37	202.0	W	208.	2.8	16.00	6390.	H	6460.	1.1
5.00	285.2	L	296.	3.7	18.32	8940.	H	9020.	0.9
	286.0	Wu		3.4	20.00	9886.	L	11100.	11.9
5.18	324.0	W	323.	−0.4	23.57	15900.	H	16100.	1.2
5.41	360.0	W	364.	1.0	25.00	16220.	L	18300.	12.8
	320.0	Wu		13.6	30.00	24800.	L	26700.	7.5
	315.0	B		15.4	31.68	30200.	H	29600.	−1.9
6.98	762.0	Sp	715.	−6.1	35.00	36300.	L	35600.	−2.0
	748.0	W		−4.4	44.60	45700.	D	45400.	−0.7
7.00	726.0	Wu	724.	−0.3		40740.	L		11.3
	683.9	L		5.8		45600.	H		−0.5
7.98	1030.	B	1020.	−0.6	46.50	46800.	L	48400.	3.4
					48.50	53090.	L	51500.	−2.9

TABLE III.　MASS ATTENUATION COEFFICIENTS VS WAVELENGTH
$\mu(cm^2/gram)$ vs. $\lambda(A)$

MASS ATTENUATION COEFFICIENTS

WAVELENGTH	SODIUM	MAGNESIUM	ALUMINUM	SILICON	PHOSPHORUS	SULFUR	CHLORINE	ARGON	ENERGY (EV)
1.0	9.	12.	15.	19.	23.	29.	33.	36.	12398.0
2.0	66.	87.	107.	137.	160.	196.	221.	240.	6199.0
3.0	208.	274.	332.	418.	486.	590.	660.	720.	4132.7
4.0	466.	610.	730.	910.	1060.	1280.	1440.	167.	3099.5
5.0	860.	1110.	1330.	1660.	1920.	2330.	261.	296.	2479.6
6.0	1410.	1810.	2160.	2690.	278.	357.	421.	479.	2066.3
7.0	2120.	2720.	3230.	339.	419.	540.	640.	720.	1771.1
8.0	3020.	3850.	361.	486.	600.	770.	910.	1040.	1549.8
9.0	4100.	5200.	498.	670.	830.	1070.	1260.	1430.	1377.6
10.0	5400.	510.	660.	890.	1100.	1420.	1670.	1890.	1239.8
11.0	6800.	660.	860.	1160.	1430.	1840.	2160.	2440.	1127.1
12.0	580.	830.	1090.	1470.	1810.	2320.	2720.	3080.	1033.2
13.0	720.	1040.	1360.	1830.	2250.	2880.	3370.	3800.	953.7
14.0	880.	1270.	1660.	2230.	2750.	3500.	4090.	4600.	885.6
15.0	1070.	1530.	2000.	2690.	3300.	4200.	4890.	5500.	826.5
16.0	1270.	1830.	2390.	3200.	3920.	4970.	5800.	6500.	774.9
17.0	1500.	2150.	2810.	3760.	4590.	5800.	6700.	7500.	729.3
18.0	1750.	2520.	3280.	4380.	5300.	6700.	7800.	8600.	688.8
19.0	2030.	2910.	3790.	5000.	6100.	7700.	8900.	9800.	652.5
20.0	2330.	3340.	4340.	5800.	7000.	8700.	10000.	11100.	619.9
21.0	2660.	3810.	4930.	6500.	7900.	9900.	11300.	12400.	590.4
22.0	3020.	4310.	5600.	7400.	8800.	11000.	12600.	13800.	563.5
23.0	3400.	4850.	6200.	8200.	9900.	12300.	13900.	15200.	539.0
24.0	3810.	5400.	7000.	9100.	10900.	13600.	15300.	16700.	516.6
25.0	4250.	6000.	7700.	10100.	12100.	14900.	16800.	18300.	495.9
26.0	4710.	6700.	8500.	11100.	13200.	16300.	18300.	19900.	476.8
27.0	5200.	7400.	9400.	12200.	14400.	17700.	19900.	21500.	459.2
28.0	5700.	8100.	10300.	13300.	15700.	19200.	21500.	23200.	442.8
29.0	6300.	8800.	11200.	14500.	17000.	20700.	23100.	24900.	427.5
30.0	6800.	9600.	12200.	15700.	18400.	22300.	24800.	26700.	413.3
31.0	7400.	10400.	13200.	16900.	19700.	23900.	26500.	28400.	399.9
32.0	8100.	11300.	14200.	18200.	21100.	25500.	28200.	30200.	387.4
33.0	8700.	12200.	15300.	19500.	22600.	27200.	30000.	32000.	375.7
34.0	9400.	13100.	16400.	20800.	24100.	28900.	31800.	33800.	364.6
35.0	10100.	14100.	17500.	22200.	25600.	30600.	33500.	35600.	354.2
36.0	10900.	15000.	18700.	23600.	27100.	32400.	35300.	37400.	344.4
37.0	11600.	16100.	19900.	25100.	28600.	34100.	37200.	39200.	335.1
38.0	12400.	17100.	21100.	26500.	30200.	35900.	39000.	41000.	326.3
39.0	13300.	18200.	22400.	28000.	31800.	37700.	40800.	36100.	317.9
40.0	14100.	19300.	23700.	29500.	33400.	39400.	42600.	37800.	309.9
42.0	15900.	21600.	26300.	32600.	36700.	43000.	46200.	41100.	295.2
44.0	17700.	24000.	29100.	35800.	40000.	46600.	49800.	44400.	281.8
46.0	19700.	26500.	31900.	39000.	43300.	50000.	45100.	47600.	269.5
48.0	21800.	29100.	34800.	42300.	46600.	54000.	48400.	51000.	258.3
50.0	23900.	31700.	37700.	45600.	50000.	57000.	52000.	54000.	248.0
52.0	26100.	34500.	40800.	48900.	53000.	61000.	55000.	•	238.4
54.0	28400.	37300.	43800.	52000.	57000.	54000.	58000.	•	229.6
56.0	30800.	40200.	46900.	56000.	60000.	57000.	61000.	•	221.4
58.0	33300.	43100.	50000.	59000.	63000.	60000.	64000.	•	213.8
60.0	35800.	46100.	53000.	62000.	66000.	63000.	66000.	•	206.6
62.0	38300.	49100.	56000.	65000.	69000.	66000.	•	•	200.0
64.0	41000.	52000.	59000.	69000.	72000.	69000.	•	•	193.7
66.0	43700.	55000.	63000.	72000.	64000.	72000.	•	•	187.8
68.0	46400.	58000.	66000.	75000.	66000.	74000.	•	•	182.3
70.0	49200.	61000.	69000.	78000.	69000.	77000.	•	•	177.1
72.0	52000.	65000.	72000.	81000.	72000.	79000.	•	•	172.2
74.0	55000.	68000.	75000.	84000.	74000.	82000.	•	•	167.5
76.0	58000.	71000.	78000.	87000.	77000.	•	•	•	163.1
78.0	61000.	74000.	81000.	90000.	79000.	•	•	•	158.9
80.0	64000.	77000.	84000.	92000.	81000.	•	•	•	155.0
82.0	66000.	80000.	87000.	81000.	83000.	•	•	•	151.2
84.0	69000.	83000.	90000.	83000.	85000.	•	•	•	147.6
86.0	72000.	86000.	92000.	87000.	87000.	•	•	•	144.2
88.0	75000.	90000.	95000.	88000.	89000.	•	•	•	140.9
90.0	78000.	93000.	98000.	90000.	91000.	•	•	•	137.8
92.0	81000.	96000.	100000.	93000.	•	•	•	•	134.8
94.0	84000.	98000.	103000.	95000.	•	•	•	•	131.9
96.0	87000.	101000.	105000.	97000.	•	•	•	•	129.1
98.0	90000.	104000.	108000.	99000.	•	•	•	•	126.5
100.0	93000.	107000.	110000.	101000.	•	•	•	•	124.0
105.0	101000.	114000.	116000.	105000.	•	•	•	•	118.1
110.0	108000.	121000.	104000.	109000.	•	•	•	•	112.7
115.0	115000.	127000.	109000.	112000.	•	•	•	•	107.8
120.0	122000.	133000.	113000.	115000.	•	•	•	•	103.3
125.0	129000.	138000.	117000.	•	•	•	•	•	99.2
130.0	135000.	144000.	120000.	•	•	•	•	•	95.4
135.0	142000.	149000.	123000.	•	•	•	•	•	91.8
140.0	148000.	132000.	126000.	•	•	•	•	•	88.6
145.0	154000.	136000.	129000.	•	•	•	•	•	85.5
150.0	159000.	140000.	131000.	•	•	•	•	•	82.7
155.0	165000.	143000.	133000.	•	•	•	•	•	80.0
160.0	170000.	147000.	134000.	•	•	•	•	•	77.5
165.0	175000.	150000.	135000.	•	•	•	•	•	75.1
170.0	179000.	152000.	•	•	•	•	•	•	72.9
175.0	183000.	155000.	•	•	•	•	•	•	70.8
180.0	187000.	157000.	•	•	•	•	•	•	68.9
185.0	191000.	159000.	•	•	•	•	•	•	67.0
190.0	195000.	160000.	•	•	•	•	•	•	65.3
195.0	171000.	162000.	•	•	•	•	•	•	63.6
200.0	174000.	163000.	•	•	•	•	•	•	62.0

TABLE III (continued)

MASS ATTENUATION COEFFICIENTS

WAVELENGTH	HELIUM	LITHIUM	BERYLLIUM	BORON	CARBON	NITROGEN	OXYGEN	FLUORINE	NEON	ENERGY (EV)
2.0		0.8	2.2	4.9	9.0	16.	26.	35.	51.	6199.0
3.0	0.7	2.8	7.6	17.0	33.9	56.	85.	116.	164.	4132.7
4.0	1.8	6.7	18.5	41.2	81.	132.	199.	265.	371.	3099.5
5.0	3.5	13.3	36.9	82.	159.	255.	379.	498.	690.	2479.6
6.0	6.0	23.3	65.	142.	274.	433.	640.	830.	1140.	2066.3
7.0	9.7	37.7	104.	227.	432.	680.	980.	1270.	1730.	1771.1
8.0	14.7	57.	157.	338.	640.	990.	1420.	1820.	2470.	1549.8
9.0	21.3	82.	226.	480.	900.	1370.	1960.	2500.	3360.	1377.6
10.0	29.6	114.	311.	660.	1210.	1840.	2610.	3300.	4420.	1239.8
11.0	40.1	154.	416.	870.	1590.	2390.	3370.	4230.	5700.	1127.1
12.0	53.	202.	540.	1120.	2030.	3030.	4240.	5300.	7100.	1033.2
13.0	68.	258.	690.	1410.	2530.	3760.	5200.	6500.	8600.	953.7
14.0	86.	325.	860.	1740.	3110.	4580.	6300.	7800.	10400.	885.6
15.0	107.	402.	1050.	2120.	3750.	5500.	7600.	9300.	780.	826.5
16.0	132.	491.	1270.	2540.	4470.	6500.	8900.	11000.	920.	774.9
17.0	160.	590.	1520.	3020.	5300.	7600.	10400.	12700.	1080.	729.3
18.0	192.	700.	1800.	3540.	6100.	8800.	11900.	14600.	1250.	688.8
19.0	228.	830.	2100.	4110.	7100.	10100.	13700.	960.	1450.	652.5
20.0	268.	970.	2440.	4730.	8100.	11500.	15500.	1100.	1660.	619.9
21.0	313.	1130.	2810.	5400.	9200.	13000.	17400.	1240.	1890.	590.4
22.0	363.	1300.	3210.	6100.	10300.	14600.	19500.	1400.	2140.	563.5
23.0	418.	1480.	3640.	6900.	11600.	16300.	21700.	1570.	2410.	539.0
24.0	479.	1680.	4110.	7700.	12900.	18000.	1250.	1760.	2700.	516.6
25.0	540.	1900.	4610.	8600.	14300.	19900.	1390.	1960.	3010.	495.9
26.0	620.	2140.	5100.	9600.	15800.	21900.	1530.	2170.	3340.	476.8
27.0	690.	2390.	5700.	10500.	17300.	23900.	1690.	2390.	3690.	459.2
28.0	780.	2670.	6300.	11600.	18900.	26100.	1850.	2630.	4060.	442.8
29.0	870.	2960.	7000.	12700.	20600.	28300.	2020.	2880.	4460.	427.5
30.0	970.	3270.	7600.	13800.	22400.	30700.	2210.	3140.	4880.	413.3
31.0	1070.	3600.	8400.	15100.	24300.	1550.	2400.	3420.	5300.	399.9
32.0	1180.	3950.	9100.	16300.	26200.	1690.	2600.	3710.	5800.	387.4
33.0	1300.	4320.	9900.	17600.	28200.	1820.	2810.	4020.	6300.	375.7
34.0	1430.	4710.	10700.	19000.	30300.	1970.	3030.	4340.	6800.	364.6
35.0	1560.	5100.	11600.	20400.	32400.	2120.	3260.	4680.	7300.	354.2
36.0	1710.	5600.	12500.	21900.	34600.	2270.	3500.	5000.	7800.	344.4
37.0	1860.	6000.	13400.	23400.	36900.	2440.	3750.	5400.	8400.	335.1
38.0	2020.	6500.	14400.	25000.	39200.	2600.	4010.	5800.	9000.	326.3
39.0	2190.	7000.	15400.	26600.	41700.	2780.	4280.	6200.	9600.	317.9
40.0	2360.	7500.	16500.	28300.	44100.	2960.	4560.	6600.	10300.	309.9
42.0	2750.	8600.	18700.	31800.	49300.	3340.	5100.	7400.	11600.	295.2
44.0	3170.	9800.	21100.	35600.	2270.	3740.	5800.	8400.	13000.	281.8
46.0	3620.	11100.	23600.	39500.	2550.	4180.	6500.	9400.	14500.	269.5
48.0	4120.	12500.	26300.	43600.	2840.	4630.	7200.	10400.	16100.	258.3
50.0	4670.	14000.	29100.	47900.	3140.	5100.	7900.	11500.	17800.	248.0
52.0	5300.	15600.	32100.	52000.	3460.	5600.	8700.	12700.	19600.	238.4
54.0	5900.	17300.	35200.	57000.	3790.	6200.	9600.	13900.	21400.	229.6
56.0	6600.	19100.	38500.	62000.	4140.	6700.	10500.	15200.	23300.	221.4
58.0	7300.	21000.	41900.	67000.	4510.	7300.	11400.	16500.	25300.	213.8
60.0	8000.	23000.	45500.	72000.	4890.	7900.	12400.	17900.	27300.	206.6
62.0	8900.	25000.	49200.	77000.	5300.	8600.	13400.	19400.	29400.	200.0
64.0	9700.	27200.	53000.	83000.	5700.	9200.	14400.	20900.	31500.	193.7
66.0	10600.	29500.	57000.	3130.	6100.	9900.	15500.	22500.	33800.	187.8
68.0	11600.	31900.	61000.	3390.	6600.	10700.	16700.	24100.	36000.	182.3
70.0	12600.	34400.	65000.	3650.	7000.	11400.	17900.	25800.	38400.	177.1
72.0	13700.	37000.	70000.	3920.	7500.	12200.	19100.	27500.	40800.	172.2
74.0	14800.	39700.	74000.	4190.	8000.	13000.	20400.	29300.	43200.	167.5
76.0	16000.	42500.	79000.	4480.	8500.	13800.	21700.	31100.	45700.	163.1
78.0	17200.	45400.	83000.	4770.	9000.	14700.	23000.	33000.	48200.	158.9
80.0	18500.	48400.	88000.	5100.	9500.	15600.	24400.	34900.	51000.	155.0
82.0	19900.	51000.	93000.	5400.	10100.	16500.	25900.	36900.	53000.	151.2
84.0	21300.	55000.	98000.	5700.	10600.	17400.	27400.	38900.	56000.	147.6
86.0	22700.	58000.	103000.	6000.	11200.	18400.	28900.	40900.	59000.	144.2
88.0	24200.	61000.	109000.	6300.	11800.	19400.	30400.	43000.	61000.	140.9
90.0	25800.	65000.	114000.	6700.	12400.	20500.	32000.	45100.	66000.	137.8
92.0	27400.	68000.	119000.	7000.	13000.	21500.	33700.	47300.	67000.	134.8
94.0	29100.	72000.	125000.	7400.	13700.	22600.	35300.	49500.	69000.	131.9
96.0	30800.	76000.	131000.	7700.	14300.	23700.	37000.	72000.	72000.	129.1
98.0	32600.	80000.	136000.	8100.	15000.	24900.	38800.	54000.	75000.	126.5
100.0	34500.	83000.	142000.	8500.	15700.	26000.	40600.	56000.	78000.	124.0
105.0	39400.	94000.	157000.	9400.	17500.	29100.	45200.	62000.	85000.	118.1
110.0	44700.	104000.	173000.	10400.	19400.	32300.	50000.	68000.	92000.	112.7
115.0	50000.	116000.	5400.	11500.	21300.	35700.	55000.	74000.	99000.	107.8
120.0	56000.	127000.	6100.	12600.	23400.	39300.	60000.	81000.	106000.	103.3
125.0	63000.	139000.	6700.	13700.	25600.	43000.	66000.	87000.	114000.	99.2
130.0	69000.	152000.	7400.	14900.	27900.	46900.	71000.	94000.	121000.	95.4
135.0	76000.	165000.	8100.	16100.	30300.	51000.	77000.	100000.	128000.	91.8
140.0	84000.	179000.	8800.	17400.	32800.	55000.	83000.	107000.	135000.	88.6
145.0	92000.	192000.	9600.	18800.	35400.	59000.	89000.	114000.	141000.	85.5
150.0	100000.	207000.	10300.	20100.	38100.	64000.	95000.	121000.	148000.	82.7
155.0	108000.	221000.	11100.	21500.	40900.	69000.	101000.	127000.	155000.	80.0
160.0	117000.	236000.	11900.	23000.	43900.	73000.	108000.	134000.	161000.	77.5
165.0	126000.	252000.	12700.	24500.	46900.	78000.	114000.	141000.	168000.	75.1
170.0	136000.	267000.	13600.	26100.	50000.	83000.	121000.	147000.	174000.	72.9
175.0	146000.	283000.	14400.	27700.	53000.	88000.	127000.	154000.	180000.	70.8
180.0	156000.	299000.	15300.	29400.	56000.	94000.	134000.	161000.	186000.	68.9
185.0	166000.	316000.	16200.	31100.	60000.	99000.	141000.	167000.	192000.	67.0
190.0	177000.	333000.	17100.	32800.	63000.	105000.	147000.	174000.	197000.	65.3
195.0	188000.	350000.	18000.	34700.	67000.	110000.	154000.	180000.	203000.	63.6
200.0	200000.	367000.	19000.	36500.	71000.	116000.	161000.	187000.	208000.	62.0

TABLE IV. MASS ATTENUATION COEFFICIENTS VS ENERGY
$\mu(cm^2/gram)$ vs. $E(eV)$

MASS ATTENUATION COEFFICIENTS

ENERGY (EV)	HELIUM	LITHIUM	BERYLLIUM	BORON	CARBON	NITROGEN	OXYGEN	FLUORINE	NEON	ENERGY (EV)
30.	863000.	.	79000.	161000.	290000.	380000.	395000.	299000.	283000.	30.
35.	657000.	.	54000.	118000.	221000.	312000.	348000.	324000.	270000.	35.
40.	509000.	.	45000.	90000.	172000.	255000.	303000.	297000.	253000.	40.
45.	402000.	.	35800.	71000.	137000.	210000.	262000.	269000.	270000.	45.
50.	322000.	.	29100.	57000.	111000.	175000.	226000.	242000.	251000.	50.
55.	262000.	457000.	24100.	46800.	91000.	146000.	196000.	217000.	233000.	55.
60.	215000.	390000.	20300.	39100.	76000.	124000.	170000.	195000.	215000.	60.
65.	179000.	335000.	17200.	33100.	64000.	105000.	148000.	175000.	198000.	65.
70.	150000.	290000.	14800.	28400.	55000.	91000.	130000.	157000.	183000.	70.
75.	127000.	253000.	12800.	24600.	47100.	79000.	114000.	141000.	168000.	75.
80.	108000.	221000.	11100.	21500.	40900.	69000.	101000.	127000.	155000.	80.
85.	93000.	195000.	9700.	19000.	35900.	60000.	90000.	115000.	143000.	85.
90.	81000.	172000.	8500.	16800.	31700.	53000.	80000.	104000.	131000.	90.
95.	70000.	153000.	7500.	15000.	28200.	47300.	72000.	94000.	121000.	95.
100.	61000.	137000.	6600.	13500.	25200.	42200.	65000.	86000.	112000.	100.
105.	54000.	123000.	5800.	12200.	22600.	37900.	58000.	78000.	104000.	105.
110.	47700.	110000.	5100.	11000.	20400.	34100.	53000.	72000.	96000.	110.
115.	42300.	100000.	166000.	10000.	18500.	30900.	47900.	66000.	89000.	115.
120.	37700.	90000.	152000.	9100.	16800.	28000.	43600.	60000.	83000.	120.
125.	33700.	82000.	140000.	8300.	15400.	25600.	39800.	55000.	77000.	125.
130.	30300.	75000.	129000.	7600.	14100.	23400.	36500.	51000.	71000.	130.
135.	27300.	68000.	119000.	7000.	13000.	21400.	33500.	47100.	66000.	135.
140.	24700.	62000.	110000.	6400.	12000.	19700.	30900.	43600.	62000.	140.
145.	22300.	57000.	102000.	5900.	11000.	18200.	28500.	40400.	58000.	145.
150.	20300.	52000.	95000.	5500.	10200.	16800.	26400.	37500.	54000.	150.
155.	18500.	48300.	86000.	5100.	9500.	15600.	24400.	34900.	51000.	155.
160.	16900.	44600.	82000.	4690.	8800.	14500.	22700.	32500.	47500.	160.
165.	15500.	41300.	77000.	4350.	8200.	13500.	21100.	30300.	44600.	165.
170.	14200.	38200.	72000.	4040.	7700.	12600.	19700.	28300.	41900.	170.
175.	13100.	35500.	67000.	3760.	7200.	11700.	18400.	26500.	39400.	175.
180.	12000.	33000.	63000.	3500.	6700.	11000.	17200.	24800.	37100.	180.
185.	11100.	30700.	59000.	3260.	6300.	10300.	16100.	23300.	34900.	185.
190.	10300.	28700.	55000.	86000.	6000.	9700.	15100.	21900.	32900.	190.
195.	9500.	26800.	52000.	82000.	5600.	9100.	14200.	20600.	31100.	195.
200.	8900.	25000.	49100.	77000.	5300.	8600.	13400.	19400.	29400.	200.
210.	7700.	22000.	43700.	70000.	4700.	7600.	11900.	17300.	26300.	210.
220.	6700.	19400.	39100.	63000.	4210.	6800.	10600.	15400.	23600.	220.
230.	5800.	17200.	35100.	57000.	3780.	6100.	9500.	13900.	21300.	230.
240.	5100.	15300.	31600.	52000.	3400.	5500.	8600.	12500.	19300.	240.
250.	4550.	13700.	28600.	47000.	3080.	5000.	7800.	11300.	17500.	250.
260.	4040.	12300.	25900.	42900.	2790.	4560.	7100.	10200.	15900.	260.
270.	3610.	11100.	23500.	39300.	2530.	4160.	6400.	9300.	14500.	270.
280.	3230.	10000.	21400.	36100.	2310.	3800.	5900.	8500.	13200.	280.
290.	2900.	9100.	19600.	33200.	51000.	3490.	5400.	7800.	12100.	290.
300.	2610.	8200.	18000.	30600.	47500.	3210.	4950.	7200.	11100.	300.
310.	2360.	7500.	16500.	28300.	44100.	2960.	4560.	6600.	10300.	310.
320.	2140.	6900.	15200.	26200.	41000.	2730.	4210.	6100.	9500.	320.
330.	1950.	6300.	14000.	24300.	38200.	2530.	3900.	5600.	8800.	330.
340.	1780.	5800.	12900.	22600.	35700.	2350.	3610.	5200.	8100.	340.
350.	1620.	5300.	12000.	21000.	33300.	2180.	3360.	4820.	7500.	350.
360.	1490.	4890.	11100.	19600.	31200.	2030.	3130.	4490.	7000.	360.
370.	1370.	4510.	10300.	18300.	29200.	1900.	2920.	4180.	6500.	370.
380.	1260.	4180.	9600.	17100.	27400.	1770.	2730.	3900.	6100.	380.
390.	1160.	3870.	9000.	16000.	25800.	1660.	2560.	3650.	5700.	390.
400.	1070.	3590.	8400.	15100.	24300.	1550.	2400.	3420.	5300.	400.
420.	920.	3110.	7300.	13300.	21600.	29600.	2120.	3010.	4670.	420.
440.	790.	2720.	6400.	11800.	19200.	26500.	1880.	2670.	4130.	440.
460.	690.	2380.	5700.	10500.	17200.	23800.	1680.	2380.	3670.	460.
480.	600.	2100.	5100.	9400.	15500.	21500.	1510.	2130.	3280.	480.
500.	530.	1860.	4500.	8400.	14000.	19500.	1360.	1910.	2940.	500.
520.	469.	1650.	4030.	7600.	12700.	17700.	1230.	1730.	2650.	520.
540.	416.	1470.	3620.	6900.	11500.	16200.	21600.	1570.	2390.	540.
560.	371.	1320.	3270.	6200.	10500.	14800.	19800.	1430.	2170.	560.
580.	332.	1190.	2960.	5700.	9600.	13600.	18200.	1300.	1980.	580.
600.	298.	1070.	2680.	5200.	8800.	12500.	16800.	1190.	1810.	600.
650.	231.	840.	2130.	4150.	7100.	10200.	13800.	970.	1460.	650.
700.	182.	670.	1720.	3380.	5900.	8500.	11500.	14100.	1200.	700.
750.	146.	540.	1400.	2790.	4880.	7100.	9700.	11900.	1000.	750.
800.	119.	445.	1160.	2320.	4100.	6000.	8200.	10100.	850.	800.
850.	98.	369.	970.	1960.	3480.	5100.	7000.	8700.	720.	850.
900.	82.	309.	820.	1660.	2970.	4390.	6100.	7500.	10000.	900.
950.	69.	262.	700.	1420.	2560.	3800.	5300.	6600.	8700.	950.
1000.	59.	223.	600.	1230.	2220.	3310.	4620.	5800.	7700.	1000.
1100.	43.3	166.	447.	930.	1700.	2560.	3590.	4510.	6000.	1100.
1200.	32.9	127.	344.	720.	1330.	2020.	2850.	3590.	4810.	1200.
1300.	25.5	99.	269.	570.	1060.	1620.	2300.	2910.	3910.	1300.
1400.	20.2	78.	215.	458.	860.	1310.	1880.	2390.	3220.	1400.
1500.	16.3	63.	174.	373.	700.	1080.	1560.	1990.	2690.	1500.
1600.	13.3	52.	143.	307.	580.	900.	1300.	1670.	2270.	1600.
1700.	11.0	42.8	118.	256.	487.	760.	1100.	1420.	1930.	1700.
1800.	9.2	35.6	99.	216.	412.	640.	940.	1210.	1660.	1800.
1900.	7.8	30.3	84.	183.	351.	550.	810.	1050.	1430.	1900.
2000.	6.7	25.8	72.	157.	302.	476.	700.	910.	1250.	2000.
2500.	3.4	13.0	36.0	80.	155.	249.	370.	487.	680.	2500.
3000.	1.9	7.4	20.5	45.5	90.	145.	219.	291.	406.	3000.
3500.	1.2	4.6	12.7	28.4	56.	92.	139.	187.	263.	3500.
4000.	0.8	3.1	8.4	18.8	37.5	62.	94.	127.	180.	4000.
4500.	0.6	2.2	5.9	13.1	26.2	43.	66.	90.	128.	4500.
5000.	.	1.6	4.2	9.5	19.0	32.	49.	66.	95.	5000.
5500.	.	1.2	3.2	7.0	14.2	24.	37.	50.	72.	5500.
6000.	.	0.9	2.4	5.4	10.8	18.	28.	39.	56.	6000.

B. L. Henke and R. L. Elgin

TABLE IV (continued)

MASS ATTENUATION COEFFICIENTS

ENERGY (EV)	SODIUM	MAGNESIUM	ALUMINUM	SILICON	PHOSPHORUS	SULFUR	CHLORINE	ARGON	ENERGY (EV)
35.	190000.	.	.	.	.	.	.	.	35.
40.	197000.	.	.	.	.	.	.	.	40.
45.	198000.	.	.	.	.	.	.	.	45.
50.	194000.	.	.	.	.	.	.	.	50.
55.	187000.	166000.	.	.	.	.	.	.	55.
60.	178000.	164000.	.	.	.	.	.	.	60.
65.	195000.	161000.	.	.	.	.	.	.	65.
70.	185000.	156000.	.	.	.	.	.	.	70.
75.	175000.	150000.	135000.	.	.	.	.	.	75.
80.	165000.	143000.	133000.	.	.	.	.	.	80.
85.	155000.	137000.	129000.	.	.	.	.	.	85.
90.	145000.	151000.	125000.	.	.	.	.	.	90.
95.	136000.	144000.	121000.	.	.	.	.	.	95.
100.	128000.	137000.	116000.	118000.	.	.	.	.	100.
105.	119000.	131000.	111000.	114000.	.	.	.	.	105.
110.	112000.	124000.	106000.	111000.	.	.	.	.	110.
115.	105000.	118000.	102000.	107000.	.	.	.	.	115.
120.	98000.	112000.	114000.	104000.	.	.	.	.	120.
125.	92000.	106000.	109000.	100000.	.	.	.	.	125.
130.	86000.	100000.	105000.	96000.	.	.	.	.	130.
135.	81000.	95000.	100000.	93000.	.	.	.	.	135.
140.	76000.	90000.	96000.	89000.	90000.	.	.	.	140.
145.	72000.	86000.	92000.	85000.	87000.	.	.	.	145.
150.	67000.	81000.	88000.	82000.	84000.	.	.	.	150.
155.	64000.	77000.	84000.	92000.	81000.	.	.	.	155.
160.	60000.	73000.	80000.	89000.	78000.	.	.	.	160.
165.	56000.	70000.	77000.	86000.	76000.	83000.	.	.	165.
170.	53000.	66000.	73000.	82000.	73000.	81000.	.	.	170.
175.	50000.	63000.	70000.	79000.	70000.	78000.	.	.	175.
180.	47600.	60000.	67000.	76000.	68000.	76000.	.	.	180.
185.	45000.	57000.	64000.	73000.	65000.	73000.	.	.	185.
190.	42600.	54000.	61000.	71000.	63000.	71000.	.	.	190.
195.	40400.	52000.	59000.	68000.	72000.	69000.	.	.	195.
200.	38300.	49100.	56000.	65000.	69000.	66000.	.	.	200.
210.	34600.	44700.	52000.	61000.	65000.	62000.	65000.	.	210.
220.	31200.	40700.	47400.	56000.	60000.	56000.	61000.	.	220.
230.	28300.	37200.	43700.	52000.	56000.	54000.	58000.	.	230.
240.	25700.	34400.	40200.	48300.	53000.	60000.	54000.	.	240.
250.	23500.	31200.	37100.	44900.	49300.	57000.	51000.	53000.	250.
260.	21400.	28700.	34300.	41700.	46100.	53000.	47900.	50000.	260.
270.	19600.	26400.	31800.	38900.	43200.	50000.	45000.	47500.	270.
280.	18000.	24300.	29500.	36200.	40500.	47100.	50000.	44800.	280.
290.	16600.	22500.	27300.	33800.	37900.	44400.	47600.	42400.	290.
300.	15300.	20800.	25400.	31600.	35600.	41800.	45000.	40000.	300.
310.	14100.	19300.	23700.	29500.	33400.	39400.	42600.	37800.	310.
320.	13100.	17900.	22100.	27600.	31400.	37200.	40300.	35700.	320.
330.	12100.	16700.	20600.	25900.	29500.	35100.	38200.	40200.	330.
340.	11200.	15500.	19200.	24300.	27800.	33200.	36200.	38200.	340.
350.	10400.	14500.	18000.	22800.	26200.	31300.	34300.	36300.	350.
360.	9700.	13500.	16900.	21400.	24700.	29700.	32500.	34400.	360.
370.	9100.	12600.	15800.	20200.	23300.	28100.	30900.	32900.	370.
380.	8500.	11800.	14900.	19000.	22000.	26600.	29300.	31300.	380.
390.	7900.	11100.	14000.	17900.	20800.	25200.	27900.	29800.	390.
400.	7400.	10400.	13200.	16900.	19700.	23900.	26500.	28400.	400.
420.	6600.	9200.	11700.	15100.	17700.	21500.	24000.	25800.	420.
440.	5800.	8200.	10400.	13500.	15900.	19500.	21800.	23500.	440.
460.	5200.	7300.	9300.	12200.	14400.	17600.	19800.	21500.	460.
480.	4630.	6600.	8400.	11000.	13000.	16100.	18000.	19600.	480.
500.	4150.	5900.	7600.	9900.	11800.	14600.	16500.	18000.	500.
520.	3740.	5300.	6900.	9000.	10800.	13300.	15100.	16500.	520.
540.	3380.	4820.	6200.	8200.	9800.	12200.	13900.	15200.	540.
560.	3070.	4380.	5700.	7500.	9000.	11200.	12700.	14000.	560.
580.	2790.	3990.	5200.	6800.	8200.	10300.	11800.	12900.	580.
600.	2550.	3650.	4730.	6300.	7600.	9500.	10900.	11900.	600.
650.	2050.	2940.	3830.	5100.	6200.	7800.	8900.	9900.	650.
700.	1680.	2410.	3140.	4190.	5100.	6500.	7500.	8300.	700.
750.	1390.	2000.	2610.	3490.	4270.	5400.	6300.	7000.	750.
800.	1170.	1670.	2190.	2940.	3600.	4580.	5300.	6000.	800.
850.	990.	1420.	1860.	2500.	3060.	3900.	4550.	5100.	850.
900.	850.	1210.	1590.	2140.	2630.	3360.	3920.	4410.	900.
950.	730.	1050.	1370.	1850.	2270.	2910.	3400.	3830.	950.
1000.	630.	910.	1190.	1610.	1980.	2540.	2970.	3350.	1000.
1100.	7300.	700.	920.	1240.	1530.	1960.	2300.	2610.	1100.
1200.	5800.	550.	720.	980.	1210.	1550.	1820.	2070.	1200.
1300.	4760.	444.	580.	780.	970.	1250.	1470.	1670.	1300.
1400.	3930.	5000.	475.	640.	790.	1020.	1200.	1370.	1400.
1500.	3280.	4190.	394.	530.	660.	850.	1000.	1130.	1500.
1600.	2770.	3540.	4210.	446.	550.	710.	840.	950.	1600.
1700.	2370.	3020.	3600.	379.	468.	600.	710.	810.	1700.
1800.	2030.	2600.	3100.	325.	401.	520.	610.	690.	1800.
1900.	1760.	2260.	2690.	3350.	347.	447.	530.	600.	1900.
2000.	1540.	1970.	2350.	2930.	303.	390.	459.	520.	2000.
2500.	840.	1090.	1300.	1620.	1880.	2280.	255.	289.	2500.
3000.	510.	660.	800.	1000.	1150.	1400.	1570.	181.	3000.
3500.	332.	434.	520.	660.	760.	930.	1040.	1130.	3500.
4000.	228.	300.	363.	457.	530.	650.	720.	790.	4000.
4500.	164.	216.	263.	331.	386.	470.	530.	570.	4500.
5000.	121.	161.	196.	248.	289.	353.	396.	431.	5000.
5500.	92.	123.	150.	190.	223.	272.	306.	333.	5500.
6000.	72.	96.	118.	150.	175.	214.	241.	263.	6000.

TABLE V.   ATOMIC CROSS SECTIONS VS ENERGY

ATOMIC CROSS SECTIONS

ENERGY (EV)	HELIUM	LITHIUM	BERYLLIUM	BORON	CARBON	NITROGEN	OXYGEN	FLUORINE	NEON	ENERGY (EV)
30.	5740000.		1190000.	2880000.	5780000.	8850000.	10490000.	9440000.	9490000.	30.
35.	4360000.		880000.	2120000.	4420000.	7250000.	9250000.	10220000.	9060000.	35.
40.	3390000.		670000.	1620000.	3440000.	5930000.	8040000.	9370000.	8480000.	40.
45.	2670000.		540000.	1270000.	2730000.	4890000.	6950000.	8490000.	9040000.	45.
50.	2140000.		435000.	1020000.	2210000.	4060000.	6000000.	7640000.	8620000.	50.
55.	1740000.	5270000.	361000.	840000.	1810000.	3400000.	5200000.	6860000.	7800000.	55.
60.	1430000.	4490000.	303000.	700000.	1510000.	2880000.	4520000.	6150000.	7210000.	60.
65.	1190000.	3860000.	258000.	590000.	1280000.	2450000.	3940000.	5510000.	6650000.	65.
70.	1000000.	3340000.	221000.	510000.	1090000.	2110000.	3450000.	4950000.	6120000.	70.
75.	840000.	2910000.	191000.	442000.	940000.	1830000.	3040000.	4450000.	5630000.	75.
80.	720000.	2550000.	166000.	387000.	820000.	1590000.	2690000.	4010000.	5190000.	80.
85.	620000.	2250000.	145000.	341000.	720000.	1400000.	2390000.	3630000.	4780000.	85.
90.	540000.	1990000.	127000.	302000.	630000.	1240000.	2130000.	3280000.	4410000.	90.
95.	466000.	1770000.	112000.	270000.	560000.	1100000.	1910000.	2980000.	4070000.	95.
100.	408000.	1560000.	99000.	242000.	500000.	980000.	1710000.	2710000.	3760000.	100.
105.	358000.	1410000.	87000.	218000.	451000.	860000.	1550000.	2470000.	3470000.	105.
110.	317000.	1270000.	77000.	197000.	407000.	790000.	1400000.	2260000.	3220000.	110.
115.	281000.	1150000.	2480000.	179000.	369000.	720000.	1270000.	2070000.	2980000.	115.
120.	251000.	1040000.	2270000.	163000.	336000.	650000.	1160000.	1900000.	2770000.	120.
125.	224000.	940000.	2090000.	149000.	307000.	590000.	1060000.	1750000.	2570000.	125.
130.	201000.	860000.	1930000.	137000.	281000.	540000.	970000.	1610000.	2390000.	130.
135.	181000.	780000.	1780000.	126000.	259000.	498000.	890000.	1490000.	2230000.	135.
140.	164000.	720000.	1650000.	115000.	238000.	458000.	820000.	1370000.	2080000.	140.
145.	148000.	660000.	1530000.	106000.	220000.	423000.	760000.	1270000.	1940000.	145.
150.	135000.	600000.	1420000.	98000.	204000.	391000.	700000.	1180000.	1820000.	150.
155.	123000.	560000.	1320000.	91000.	190000.	362000.	650000.	1100000.	1700000.	155.
160.	112000.	510000.	1230000.	84000.	176000.	336000.	600000.	1020000.	1590000.	160.
165.	103000.	476000.	1150000.	78000.	164000.	313000.	560000.	960000.	1490000.	165.
170.	94000.	441000.	1070000.	73000.	154000.	292000.	520000.	890000.	1400000.	170.
175.	87000.	409000.	1000000.	68000.	144000.	273000.	488000.	840000.	1320000.	175.
180.	80000.	380000.	940000.	63000.	135000.	255000.	457000.	780000.	1240000.	180.
185.	74000.	354000.	880000.	59000.	126000.	239000.	428000.	740000.	1170000.	185.
190.	68000.	330000.	830000.	1550000.	119000.	225000.	402000.	690000.	1100000.	190.
195.	63000.	308000.	780000.	1470000.	112000.	211000.	377000.	650000.	1040000.	195.
200.	59000.	288000.	740000.	1390000.	105000.	199000.	355000.	610000.	980000.	200.
210.	51000.	253000.	650000.	1250000.	94000.	177000.	316000.	540000.	880000.	210.
220.	44400.	224000.	590000.	1130000.	84000.	159000.	282000.	487000.	790000.	220.
230.	38900.	198000.	530000.	1020000.	75000.	143000.	253000.	437000.	710000.	230.
240.	34200.	177000.	473000.	930000.	68000.	129000.	228000.	394000.	650000.	240.
250.	30300.	158000.	427000.	840000.	61000.	117000.	206000.	356000.	590000.	250.
260.	26900.	142000.	387000.	770000.	56000.	106000.	187000.	323000.	530000.	260.
270.	24000.	128000.	352000.	710000.	51000.	97000.	171000.	294000.	485000.	270.
280.	21400.	115000.	321000.	650000.	46000.	88000.	156000.	268000.	444000.	280.
290.	19300.	104000.	293000.	600000.	1020000.	81000.	143000.	246000.	407000.	290.
300.	17400.	95000.	269000.	550000.	950000.	75000.	131000.	226000.	374000.	300.
310.	15700.	86000.	247000.	510000.	880000.	69000.	121000.	208000.	344000.	310.
320.	14200.	79000.	227000.	471000.	820000.	64000.	112000.	191000.	317000.	320.
330.	12900.	72000.	209000.	437000.	760000.	59000.	104000.	177000.	293000.	330.
340.	11800.	66000.	194000.	406000.	710000.	55000.	96000.	164000.	272000.	340.
350.	10800.	61000.	179000.	378000.	660000.	51000.	89000.	152000.	252000.	350.
360.	9900.	56000.	166000.	352000.	620000.	47300.	83000.	142000.	234000.	360.
370.	9100.	52000.	154000.	329000.	580000.	44100.	78000.	132000.	218000.	370.
380.	8400.	48100.	144000.	308000.	550000.	41200.	73000.	123000.	204000.	380.
390.	7700.	44600.	134000.	288000.	510000.	38500.	68000.	115000.	190000.	390.
400.	7100.	41400.	125000.	270000.	484000.	36100.	64000.	108000.	178000.	400.
420.	6100.	35900.	109000.	238000.	430000.	690000.	56000.	95000.	157000.	420.
440.	5300.	31300.	96000.	212000.	384000.	620000.	50000.	84000.	138000.	440.
460.	4590.	27400.	85000.	188000.	344000.	550000.	44600.	75000.	123000.	460.
480.	4010.	24200.	76000.	169000.	309000.	500000.	40000.	67000.	110000.	480.
500.	3530.	21400.	67000.	151000.	279000.	454000.	36000.	60000.	99000.	500.
520.	3120.	19000.	60000.	136000.	253000.	413000.	32600.	55000.	89000.	520.
540.	2760.	17000.	54000.	123000.	230000.	376000.	570000.	49400.	80000.	540.
560.	2460.	15200.	48900.	112000.	210000.	344000.	530000.	45000.	73000.	560.
580.	2200.	13700.	44200.	102000.	191000.	316000.	483000.	41100.	66000.	580.
600.	1980.	12400.	40100.	93000.	175000.	290000.	445000.	37600.	61000.	600.
650.	1530.	9700.	31900.	74000.	142000.	237000.	366000.	30600.	48900.	650.
700.	1210.	7700.	25700.	61000.	117000.	197000.	305000.	444000.	40200.	700.
750.	970.	6300.	21000.	50000.	97000.	165000.	257000.	375000.	33500.	750.
800.	790.	5100.	17300.	41700.	82000.	139000.	218000.	319000.	28300.	800.
850.	650.	4250.	14500.	35100.	69000.	119000.	187000.	274000.	24200.	850.
900.	540.	3560.	12200.	29900.	59000.	102000.	161000.	238000.	334000.	900.
950.	458.	3010.	10400.	25600.	51000.	88000.	140000.	207000.	292000.	950.
1000.	390.	2570.	8900.	22100.	44300.	77000.	123000.	182000.	257000.	1000.
1100.	288.	1910.	6700.	16700.	33900.	59000.	95000.	142000.	202000.	1100.
1200.	219.	1460.	5100.	13000.	26500.	46900.	76000.	113000.	161000.	1200.
1300.	170.	1140.	4030.	10200.	21100.	37600.	61000.	92000.	131000.	1300.
1400.	135.	900.	3210.	8200.	17100.	30600.	49900.	75000.	108000.	1400.
1500.	109.	730.	2600.	6700.	14000.	25200.	41300.	63000.	90000.	1500.
1600.	89.	600.	2140.	5500.	11600.	21000.	34600.	53000.	76000.	1600.
1700.	74.	494.	1770.	4600.	9700.	17700.	29200.	44700.	65000.	1700.
1800.	62.	413.	1490.	3870.	8200.	15000.	24900.	38300.	55000.	1800.
1900.	52.	350.	1260.	3290.	7000.	12800.	21400.	33000.	48000.	1900.
2000.	44.7	298.	1070.	2820.	6000.	11100.	18500.	28600.	41700.	2000.
2500.	22.8	150.	540.	1430.	3100.	5800.	9800.	15400.	22600.	2500.
3000.	13.3	86.	307.	820.	1790.	3380.	5800.	9200.	13600.	3000.
3500.	8.5	54.	191.	510.	1120.	2140.	3700.	5900.	8800.	3500.
4000.	5.8	36.3	127.	338.	750.	1430.	2500.	4010.	6000.	4000.
4500.	4.7	25.6	89.	236.	520.	1010.	1770.	2840.	4300.	4500.
5000.		18.9	64.	171.	380.	730.	1290.	2090.	3180.	5000.
5500.		14.3	48.	128.	284.	550.	970.	1580.	2410.	5500.
6000.		11.2	37.	98.	218.	423.	750.	1220.	1870.	6000.

TABLE V (continued)

ATOMIC CROSS SECTIONS

ENERGY (EV)	SODIUM	MAGNESIUM	ALUMINUM	SILICON	PHOSPHORUS	SULFUR	CHLORINE	ARGON	ENERGY (EV)
35.	7240000.								35.
40.	7530000.								40.
45.	7560000.								45.
50.	7410000.								50.
55.	7140000.	6690000.							55.
60.	6810000.	6630000.							60.
65.	7450000.	6490000.							65.
70.	7070000.	6290000.							70.
75.	6680000.	6050000.	6070000.						75.
80.	6290000.	5790000.	5940000.						80.
85.	5910000.	5510000.	5790000.						85.
90.	5540000.	6100000.	5600000.						90.
95.	5200000.	5820000.	5400000.						95.
100.	4870000.	5540000.	5200000.	5480000.					100.
105.	4560000.	5270000.	4980000.	5330000.					105.
110.	4270000.	5010000.	4770000.	5150000.					110.
115.	4000000.	4750000.	4560000.	5010000.					115.
120.	3750000.	4510000.	5100000.	4440000.					120.
125.	3520000.	4280000.	4890000.	4660000.					125.
130.	3300000.	4060000.	4680000.	4490000.					130.
135.	3100000.	3850000.	4480000.	4310000.					135.
140.	2910000.	3650000.	4290000.	4140000.	4610000.				140.
145.	2740000.	3460000.	4100000.	3980000.	4470000.				145.
150.	2570000.	3280000.	3920000.	3810000.	4320000.				150.
155.	2420000.	3110000.	3750000.	4310000.	4170000.				155.
160.	2280000.	2960000.	3590000.	4150000.	4030000.				160.
165.	2160000.	2810000.	3430000.	4000000.	3880000.	4420000.			165.
170.	2030000.	2670000.	3280000.	3850000.	3750000.	4290000.			170.
175.	1920000.	2540000.	3140000.	3700000.	3610000.	4160000.			175.
180.	1820000.	2410000.	3000000.	3560000.	3480000.	4020000.			180.
185.	1720000.	2290000.	2870000.	3430000.	3350000.	3900000.			185.
190.	1630000.	2180000.	2750000.	3300000.	3230000.	3770000.			190.
195.	1540000.	2080000.	2630000.	3170000.	3680000.	3650000.			195.
200.	1460000.	1980000.	2520000.	3050000.	3560000.	3530000.			200.
210.	1320000.	1800000.	2310000.	2830000.	3330000.	3300000.	3830000.		210.
220.	1190000.	1640000.	2130000.	2620000.	3110000.	3080000.	3610000.		220.
230.	1080000.	1500000.	1960000.	2430000.	2900000.	2880000.	3390000.		230.
240.	980000.	1370000.	1800000.	2250000.	2710000.	3200000.	3190000.		240.
250.	900000.	1260000.	1660000.	2090000.	2540000.	3010000.	3000000.	3530000.	250.
260.	820000.	1160000.	1540000.	1950000.	2370000.	2830000.	2820000.	3330000.	260.
270.	750000.	1070000.	1420000.	1810000.	2220000.	2670000.	2650000.	3150000.	270.
280.	690000.	980000.	1320000.	1690000.	2080000.	2510000.	2960000.	2970000.	280.
290.	630000.	910000.	1230000.	1580000.	1950000.	2360000.	2800000.	2810000.	290.
300.	580000.	840000.	1140000.	1470000.	1830000.	2230000.	2650000.	2650000.	300.
310.	540000.	780000.	1060000.	1360000.	1720000.	2100000.	2510000.	2510000.	310.
320.	498000.	720000.	990000.	1290000.	1620000.	1980000.	2370000.	2370000.	320.
330.	462000.	670000.	920000.	1210000.	1520000.	1870000.	2250000.	2670000.	330.
340.	429000.	630000.	860000.	1130000.	1430000.	1770000.	2130000.	2540000.	340.
350.	399000.	580000.	810000.	1060000.	1350000.	1670000.	2020000.	2410000.	350.
360.	372000.	550000.	760000.	1000000.	1270000.	1580000.	1920000.	2290000.	360.
370.	347000.	510000.	710000.	940000.	1200000.	1490000.	1820000.	2180000.	370.
380.	324000.	478000.	670000.	890000.	1130000.	1420000.	1730000.	2080000.	380.
390.	303000.	448000.	630000.	830000.	1070000.	1340000.	1640000.	1980000.	390.
400.	284000.	421000.	590000.	790000.	1010000.	1270000.	1560000.	1880000.	400.
420.	251000.	373000.	520000.	700000.	910000.	1150000.	1410000.	1710000.	420.
440.	222000.	331000.	467000.	630000.	820000.	1040000.	1280000.	1560000.	440.
460.	198000.	296000.	419000.	570000.	740000.	940000.	1170000.	1420000.	460.
480.	177000.	265000.	376000.	510000.	670000.	850000.	1060000.	1300000.	480.
500.	159000.	238000.	339000.	463000.	610000.	780000.	970000.	1190000.	500.
520.	143000.	215000.	307000.	420000.	550000.	710000.	890000.	1090000.	520.
540.	129000.	195000.	279000.	382000.	500000.	650000.	820000.	1010000.	540.
560.	117000.	177000.	254000.	348000.	462000.	600000.	750000.	930000.	560.
580.	107000.	161000.	231000.	319000.	423000.	550000.	690000.	860000.	580.
600.	97000.	147000.	212000.	292000.	389000.	500000.	640000.	790000.	600.
650.	78000.	119000.	171000.	237000.	318000.	414000.	530000.	660000.	650.
700.	64000.	97000.	141000.	196000.	263000.	343000.	439000.	550000.	700.
750.	53000.	81000.	117000.	163000.	219000.	288000.	369000.	465000.	750.
800.	44500.	68000.	98000.	137000.	185000.	244000.	313000.	395000.	800.
850.	37700.	57000.	83000.	116000.	158000.	208000.	268000.	339000.	850.
900.	32300.	49000.	71000.	100000.	135000.	179000.	231000.	293000.	900.
950.	27900.	42200.	61000.	86000.	117000.	155000.	200000.	254000.	950.
1000.	24200.	36700.	53000.	75000.	102000.	135000.	175000.	222000.	1000.
1100.	278000.	28300.	41100.	58000.	79000.	104000.	136000.	173000.	1100.
1200.	223000.	22300.	32400.	45500.	62000.	83000.	107000.	137000.	1200.
1300.	182000.	17900.	26000.	36600.	49900.	66000.	90000.	111000.	1300.
1400.	150000.	202000.	21300.	29900.	40800.	54000.	71000.	91000.	1400.
1500.	125000.	169000.	17700.	24800.	33800.	45000.	59000.	75000.	1500.
1600.	106000.	143000.	169000.	20800.	28400.	37800.	49300.	63000.	1600.
1700.	90000.	122000.	161000.	17700.	24100.	32100.	41800.	54000.	1700.
1800.	78000.	105000.	139000.	15100.	20600.	27500.	35900.	45900.	1800.
1900.	67000.	91000.	120000.	156000.	17900.	23800.	31000.	39700.	1900.
2000.	59000.	80000.	105000.	136000.	15600.	20700.	27000.	34600.	2000.
2500.	32100.	43900.	58000.	76000.	96000.	121000.	15000.	19200.	2500.
3000.	19500.	26800.	35700.	46500.	59000.	75000.	93000.	12000.	3000.
3500.	12700.	17500.	23500.	30700.	39200.	44300.	61000.	75000.	3500.
4000.	8700.	12100.	16300.	21300.	27300.	34400.	42700.	52000.	4000.
4500.	6300.	8700.	11800.	15400.	19800.	25000.	31000.	38000.	4500.
5000.	4630.	6500.	8800.	11600.	14900.	18800.	23300.	28600.	5000.
5500.	3530.	4950.	6700.	8900.	11400.	14500.	18000.	22100.	5500.
6000.	2750.	3870.	5300.	7000.	9000.	11400.	14200.	17400.	6000.

TABLE VI

EXTRAPOLATED ABSORPTION AND ABSORPTION-JUMP RATIOS
(r) AT K-EDGES

Element	$\lambda(A)$	$\mu-$	$\mu+$	r
4 Be	112.	179000.	(5000)	(35)
5 B	66.0	88400.	3130.	28.3
6 C	43.7	53900.	2230.	24.2
7 N	30.9	32800.	1540.	21.4
8 O	23.3	22400.	1160.	19.3
9 F	18.1	14800.	846.	17.5
10 Ne	14.3	10950.	687.	15.94
11 Na	11.56	7760.	525.	14.78
12 Mg	9.50	6000.	440.	13.63
13 Al	7.95	4500.	355.	12.68
14 Si	6.74	3640.	307.	11.89
15 P	5.77	2800.	251.	11.18
16 S	5.01	2340.	222.	10.52
17 Cl	4.38	1840.	185.	9.92
18 Ar	3.87	1440.	154.	9.34
19 K	3.44	1300.	148.	8.79
20 Ca	3.07	1120.	135.	8.28

TABLE VII.   CALCULATED COMPOUND MASS ATTENUATION COEFFICIENTS
(cm^2/gm)

WAVELENGTH	FORMVAR $(C5H7O2)X$	COLLODION $(C12H11O22N6)X$	POLYPROPYLENE $(CH2)X$	CELLULOSE ACETATE $(C10H21O15)X$	MYLAR $(C10H8O4)X$	TEFLON $(CF2)X$	ENERGY(EV)
2.0	14.	20.	8.	19.	14.	28.	6199.0
4.0	113.	156.	69.	150.	116.	220.	3099.5
6.0	372.	510.	234.	489.	384.	700.	2066.3
8.0	850.	1140.	550.	1100.	870.	1540.	1549.8
10.0	1580.	2110.	1040.	2020.	1630.	2800.	1239.8
12.0	2600.	3450.	1740.	3310.	2680.	4520.	1033.2
14.0	3920.	5200.	2660.	4950.	4040.	6700.	885.6
16.0	5600.	7300.	3830.	7000.	5800.	9400.	774.9
18.0	7500.	9800.	5200.	9400.	7800.	12600.	688.8
20.0	9900.	12800.	6900.	12300.	10200.	2740.	619.9
22.0	12500.	16200.	8800.	15500.	12900.	3540.	563.5
24.0	8200.	6400.	11000.	4850.	8500.	4430.	516.6
26.0	10100.	7900.	13500.	5900.	10400.	5400.	476.8
28.0	12000.	9400.	16200.	7100.	12400.	6500.	442.8
30.0	14300.	11100.	19200.	8400.	14700.	7800.	413.3
32.0	16700.	8200.	22400.	9900.	17200.	9100.	387.4
34.0	19300.	9500.	25900.	11400.	19900.	10600.	364.6
36.0	22100.	10800.	29600.	13100.	22800.	12100.	344.4
38.0	25000.	12300.	33600.	14900.	25800.	13800.	326.3
40.0	28200.	13900.	37800.	16800.	29100.	15600.	309.9
42.0	31500.	15500.	42200.	18700.	32500.	17500.	295.2
44.0	3250.	4540.	1940.	4370.	3350.	6900.	281.8
46.0	3640.	5100.	2180.	4900.	3760.	7800.	269.5
48.0	4050.	5600.	2430.	5400.	4170.	8600.	258.3
50.0	4450.	6200.	2690.	6000.	4590.	9500.	248.0
52.0	4910.	6800.	2960.	6600.	5100.	10500.	238.4
54.0	5400.	7500.	3240.	7200.	5600.	11500.	229.6
56.0	5900.	8200.	3540.	7900.	6100.	12500.	221.4
58.0	6400.	8900.	3860.	8600.	6600.	13600.	213.8
60.0	7000.	9700.	4190.	9300.	7200.	14800.	206.6
62.0	7500.	10500.	4540.	10100.	7800.	16000.	200.0
64.0	8100.	11300.	4880.	10900.	8400.	17300.	193.7
66.0	8700.	12100.	5200.	11700.	9000.	18600.	187.8
68.0	9400.	13100.	5700.	12600.	9700.	19900.	182.3
70.0	10000.	14000.	6000.	13500.	10300.	21300.	177.1
72.0	10700.	14900.	6400.	14400.	11100.	22700.	172.2
74.0	11400.	15900.	6800.	15400.	11800.	24200.	167.5
76.0	12200.	17000.	7300.	16300.	12500.	25700.	163.1
78.0	12900.	18000.	7700.	17300.	13300.	27200.	158.9
80.0	13600.	19100.	8100.	18400.	14100.	28800.	155.0
82.0	14500.	20200.	8600.	19500.	14900.	30500.	151.2
84.0	15300.	21400.	9100.	20600.	15800.	32100.	147.6
86.0	16100.	22600.	9600.	21700.	16600.	33800.	144.2
88.0	17000.	23700.	10100.	22900.	17500.	35500.	140.9
90.0	17800.	25000.	10600.	24100.	18400.	37300.	137.8
92.0	18800.	26300.	11100.	25300.	19400.	39100.	134.8
94.0	19700.	27600.	11700.	26500.	20300.	40900.	131.9
96.0	20600.	28900.	12200.	27800.	21300.	43000.	129.1
98.0	21600.	30300.	12800.	29200.	22300.	44600.	126.5
100.0	22600.	31700.	13400.	30500.	23300.	46300.	124.0
105.0	25200.	35300.	15000.	34000.	26000.	51000.	118.1
110.0	27900.	39100.	16600.	37600.	28800.	56000.	112.7
115.0	30700.	43000.	18200.	41300.	31600.	61000.	107.8
120.0	33600.	47000.	20000.	45100.	34600.	67000.	103.3
125.0	36800.	52000.	21900.	49600.	38000.	72000.	99.2
130.0	39800.	56000.	23900.	53000.	41100.	78000.	95.4
135.0	43200.	60000.	25900.	58000.	44600.	83000.	91.8
140.0	46700.	65000.	28100.	63000.	48200.	89000.	88.6
145.0	50000.	70000.	30300.	67000.	52000.	95000.	85.5
150.0	54000.	75000.	32600.	72000.	55000.	101000.	82.7
155.0	57000.	80000.	35000.	76000.	59000.	106000.	80.0
160.0	61000.	85000.	37600.	82000.	63000.	112000.	77.5
165.0	65000.	90000.	40200.	87000.	67000.	118000.	75.1
170.0	69000.	96000.	42800.	92000.	72000.	124000.	72.9
175.0	73000.	101000.	45400.	97000.	75000.	130000.	70.8
180.0	77000.	107000.	47900.	102000.	80000.	136000.	68.9
185.0	82000.	113000.	51000.	108000.	84000.	141000.	67.0
190.0	86000.	118000.	54000.	112000.	88000.	147000.	65.3
195.0	90000.	124000.	57000.	118000.	93000.	153000.	63.6
200.0	95000.	130000.	61000.	124000.	98000.	159000.	62.0

TABLE VII (continued).

MASS ATTENUATION COEFFICIENTS

WAVELENGTH	POLYSTYRENE (CH)X	NYLON (C12H22O3N2)X	VYNS (C22H33O2CL9)X	SARAN (C2H2CL2)X	ALUMINUM OXIDE AL2O3	QUARTZ (SIO2)X	ENERGY(EV)
2.0	9.	12.	114.	164.	68.	77.	6199.0
4.0	74.	96.	750.	1070.	480.	530.	3099.5
6.0	252.	318.	350.	375.	1440.	1600.	2066.3
8.0	590.	730.	780.	820.	860.	980.	1549.8
10.0	1120.	1370.	1440.	1520.	1580.	1810.	1239.8
12.0	1870.	2270.	2370.	2490.	2570.	2950.	1033.2
14.0	2870.	3440.	3590.	3760.	3840.	4400.	885.6
16.0	4120.	4910.	5100.	5400.	5500.	6200.	774.9
18.0	5600.	6700.	6900.	7200.	7300.	8400.	688.8
20.0	7500.	8800.	9000.	9300.	9600.	11000.	619.9
22.0	9500.	11100.	11400.	11800.	12100.	13800.	563.5
24.0	11900.	10600.	12800.	14400.	4290.	4920.	516.6
26.0	14600.	13000.	15500.	17300.	5200.	6000.	476.8
28.0	17400.	15500.	18400.	20400.	6300.	7200.	442.8
30.0	20700.	18400.	21400.	23700.	7500.	8500.	413.3
32.0	24200.	17300.	24700.	27100.	8700.	9900.	387.4
34.0	28000.	20000.	28100.	30800.	10100.	11300.	364.6
36.0	31900.	22800.	31600.	34400.	11500.	12900.	344.4
38.0	36200.	25800.	35400.	38200.	13100.	14500.	326.3
40.0	40700.	29100.	39100.	42100.	14700.	16200.	309.9
42.0	45500.	32500.	43100.	46000.	16300.	18000.	295.2
44.0	2090.	2730.	25700.	37000.	18100.	19800.	281.8
46.0	2350.	3060.	23600.	33600.	19900.	21700.	269.5
48.0	2620.	3400.	25300.	36100.	21800.	23600.	258.3
50.0	2900.	3750.	27300.	38800.	23700.	25500.	248.0
52.0	3190.	4130.	28900.	41100.	25700.	27500.	238.4
54.0	3500.	4540.	30600.	43400.	27700.	29400.	229.6
56.0	3820.	4950.	32200.	45700.	29800.	31800.	221.4
58.0	4160.	5400.	33900.	47900.	31800.	33700.	213.8
60.0	4510.	5800.	35100.	49500.	33900.	35600.	206.6
62.0	4890.	6300.	•	•	35900.	37500.	200.0
64.0	5300.	6800.	•	•	38000.	39900.	193.7
66.0	5600.	7300.	•	•	40600.	41900.	187.8
68.0	6100.	7900.	•	•	42800.	44000.	182.3
70.0	6500.	8400.	•	•	44900.	46000.	177.1
72.0	6900.	9000.	•	•	47100.	48000.	172.2
74.0	7400.	9600.	•	•	49300.	50000.	167.5
76.0	7800.	10200.	•	•	51000.	52000.	163.1
78.0	8300.	10800.	•	•	54000.	54000.	158.9
80.0	8800.	11400.	•	•	56000.	56000.	155.0
82.0	9300.	12100.	•	•	58000.	52000.	151.2
84.0	9800.	12800.	•	•	61000.	53000.	147.6
86.0	10300.	13500.	•	•	62000.	56000.	144.2
88.0	10900.	14200.	•	•	65000.	57000.	140.9
90.0	11400.	15000.	•	•	67000.	59000.	137.8
92.0	12000.	15700.	•	•	69000.	61000.	134.8
94.0	12600.	16500.	•	•	71000.	63000.	131.9
96.0	13200.	17300.	•	•	73000.	65000.	129.1
98.0	13800.	18100.	•	•	75000.	67000.	126.5
100.0	14500.	19000.	•	•	77000.	69000.	124.0
105.0	16100.	21100.	•	•	83000.	73000.	118.1
110.0	17900.	23400.	•	•	79000.	78000.	112.7
115.0	19600.	25800.	•	•	84000.	82000.	107.8
120.0	21600.	28300.	•	•	88000.	86000.	103.3
125.0	23600.	31000.	•	•	93000.	•	99.2
130.0	25700.	33600.	•	•	97000.	•	95.4
135.0	28000.	36500.	•	•	101000.	•	91.8
140.0	30300.	39400.	•	•	106000.	•	88.6
145.0	32700.	42400.	•	•	110000.	•	85.5
150.0	35100.	45600.	•	•	114000.	•	82.7
155.0	37700.	48900.	•	•	118000.	•	80.0
160.0	40500.	52000.	•	•	122000.	•	77.5
165.0	43300.	56000.	•	•	125000.	•	75.1
170.0	46100.	59000.	•	•	•	•	72.9
175.0	48900.	63000.	•	•	•	•	70.8
180.0	52000.	66000.	•	•	•	•	68.9
185.0	55000.	70000.	•	•	•	•	67.0
190.0	58000.	74000.	•	•	•	•	65.3
195.0	62000.	78000.	•	•	•	•	63.6
200.0	65000.	82000.	•	•	•	•	62.0

TABLE VII (continued).

MASS ATTENUATION COEFFICIENTS

WAVELENGTH	STEARATE CH3(CH2)16COO	ANIMAL PROTEINS C52.5%,H7%,S1.5% O22.5%,N16.5%	AIR (021%,N78% AR1%)	P 10 (CH4)10% AR90%	METHANE CH4	Q GAS (C4H10)1.3% HE98.7%	ENERGY(EV)
2.0	10.	16.	21.	230.	7.	1.	6199.0
4.0	84.	126.	148.	162.	60.	12.	3099.5
6.0	281.	361.	481.	467.	205.	41.	2066.3
8.0	650.	820.	1090.	1020.	479.	97.	1549.8
10.0	1220.	1530.	2020.	1850.	910.	185.	1239.8
12.0	2030.	2530.	3310.	3010.	1520.	313.	1033.2
14.0	3080.	3830.	4980.	4500.	2330.	484.	885.6
16.0	4410.	5500.	7100.	6400.	3350.	700.	774.9
18.0	6000.	7400.	9500.	8400.	4570.	970.	688.8
20.0	7900.	9700.	12400.	10900.	6100.	1300.	619.9
22.0	10100.	12300.	15700.	13500.	7700.	1670.	563.5
24.0	10000.	10300.	14100.	16400.	9700.	2110.	516.6
26.0	12200.	12600.	17100.	19600.	11800.	2620.	476.8
28.0	14600.	15100.	20400.	22800.	14100.	3160.	442.8
30.0	17300.	17800.	24000.	26300.	16800.	3780.	413.3
32.0	20300.	15500.	2290.	29700.	19600.	4460.	387.4
34.0	23400.	17900.	2650.	33300.	22700.	5200.	364.6
36.0	26800.	20400.	3040.	36900.	25900.	6000.	344.4
38.0	30300.	23100.	3460.	40500.	29300.	6900.	326.3
40.0	34100.	26000.	3810.	37600.	33000.	7800.	309.9
42.0	38200.	29100.	4270.	40900.	36900.	.8800.	295.2
44.0	2390.	3750.	4780.	42600.	1700.	2960.	281.8
46.0	2680.	4180.	5300.	45600.	1910.	3380.	269.5
48.0	2980.	4610.	5900.	48900.	2130.	3840.	258.3
50.0	3290.	5000.	6400.	52000.	2350.	4340.	248.0
52.0	3620.	5500.	6300.	.	2590.	4910.	238.4
54.0	3970.	5900.	7000.	.	2840.	5500.	229.6
56.0	4340.	.6400.	7600.	.	3100.	6100.	221.4
58.0	4730.	6900.	8200.	.	3380.	6700.	213.8
60.0	5100.	7500.	8900.	.	3660.	7400.	206.6
62.0	5600.	8100.	9700.	.	3970.	8200.	200.0
64.0	6000.	8700.	10400.	.	4270.	8900.	193.7
66.0	6400.	9300.	11200.	.	4570.	9700.	187.8
68.0	6900.	10000.	12100.	.	4940.	10600.	182.3
70.0	7400.	10600.	12900.	.	5200.	11500.	177.1
72.0	7900.	11300.	13800.	.	5600.	12500.	172.2
74.0	8400.	12000.	14700.	.	6000.	13500.	167.5
76.0	8900.	11600.	15600.	.	6400.	14600.	163.1
78.0	9500.	12300.	16600.	.	6700.	15600.	158.9
80.0	10000.	13000.	17600.	.	7100.	16800.	155.0
82.0	10600.	13800.	18600.	.	7600.	18000.	151.2
84.0	11200.	14600.	19700.	.	7900.	19300.	147.6
86.0	11800.	15400.	20800.	.	8400.	20500.	144.2
88.0	12400.	16200.	21900.	.	8800.	21900.	140.9
90.0	13100.	17000.	23100.	.	9300.	23300.	137.8
92.0	13700.	17900.	24200.	.	9700.	24700.	134.8
94.0	14400.	18800.	25400.	.	10300.	26200.	131.9
96.0	15100.	19700.	26700.	.	10700.	27800.	129.1
98.0	15800.	20700.	28000.	.	11200.	29400.	126.5
100.0	16600.	21600.	29200.	.	11800.	31000.	124.0
105.0	18400.	24100.	32700.	.	13100.	35400.	118.1
110.0	20400.	26700.	36200.	.	14500.	40100.	112.7
115.0	22500.	29300.	39900.	.	15900.	44800.	107.8
120.0	24600.	32200.	43800.	.	17500.	50000.	103.3
125.0	27000.	35300.	48000.	.	19200.	56000.	99.2
130.0	29300.	38200.	52000.	.	20900.	62000.	95.4
135.0	31800.	41500.	57000.	.	22700.	68000.	91.8
140.0	34400.	44800.	61000.	.	24600.	75000.	88.6
145.0	37000.	48200.	65000.	.	26500.	82000.	85.5
150.0	39800.	52000.	71000.	.	28500.	89000.	82.7
155.0	42600.	55000.	76000.	.	30600.	96000.	80.0
160.0	45700.	59000.	80000.	.	32900.	104000.	77.5
165.0	48600.	63000.	86000.	.	35100.	112000.	75.1
170.0	52000.	67000.	91000.	.	37400.	121000.	72.9
175.0	55000.	71000.	96000.	.	39700.	130000.	70.8
180.0	58000.	75000.	102000.	.	41900.	138000.	68.9
185.0	62000.	79000.	108000.	.	44900.	147000.	67.0
190.0	65000.	83000.	114000.	.	47200.	157000.	65.3
195.0	68000.	88000.	119000.	.	50000.	167000.	63.6
200.0	72000.	92000.	125000.	.	53000.	177000.	62.0

TABLE VIII.   TRANSMISSION THROUGH ATMOSPHERE
VERSUS
WAVELENGTH λ–A ALTITUDE H–METERS
MASS THICKNESS–M(h)–grams/cm^2

ALTITUDE WAVELENGTH	50,000	60,000	80,000	100,000	120,000	140,000	160,000	180,000	200,000	220,000
2.0	•	•	7.925E-01	9.433E-01	9.994E-01	9.998E-01	9.999E-01	1.000E 00	1.000E 00	1.000E 00
4.0	•	•	1.942E-01	9.534E-01	9.960E-01	9.988E-01	9.994E-01	9.997E-01	9.998E-01	9.999E-01
6.0	•	•	4.859E-03	8.564E-01	9.870E-01	9.962E-01	9.981E-01	9.989E-01	9.993E-01	9.995E-01
8.0	•	•	•	7.038E-01	9.709E-01	9.913E-01	9.957E-01	9.975E-01	9.984E-01	9.990E-01
10.0	•	•	•	5.215E-01	9.467E-01	9.840E-01	9.920E-01	9.953E-01	9.971E-01	9.981E-01
12.0	•	•	•	3.441E-01	9.142E-01	9.739E-01	9.869E-01	9.923E-01	9.952E-01	9.969E-01
14.0	•	•	•	2.009E-01	8.737E-01	9.610E-01	9.803E-01	9.884E-01	9.928E-01	9.953E-01
16.0	•	•	•	1.014E-01	8.249E-01	9.448E-01	9.720E-01	9.835E-01	9.897E-01	9.934E-01
18.0	•	•	•	4.680E-02	7.730E-01	9.269E-01	9.628E-01	9.780E-01	9.863E-01	9.911E-01
20.0	•	•	•	1.838E-02	7.145E-01	9.057E-01	9.517E-01	9.714E-01	9.822E-01	9.885E-01
22.0	•	•	•	6.344E-03	6.534E-01	8.821E-01	9.392E-01	9.640E-01	9.775E-01	9.854E-01
24.0	•	•	•	1.062E-02	6.823E-01	8.934E-01	9.452E-01	9.676E-01	9.797E-01	9.869E-01
26.0	•	•	•	4.040E-03	6.290E-01	8.723E-01	9.340E-01	9.608E-01	9.755E-01	9.841E-01
28.0	•	•	•	1.395E-03	5.752E-01	8.496E-01	9.217E-01	9.535E-01	9.708E-01	9.811E-01
30.0	•	•	•	4.370E-04	5.217E-01	8.255E-01	9.086E-01	9.455E-01	9.658E-01	9.778E-01
32.0	•	•	•	4.780E-01	9.398E-01	9.819E-01	9.909E-01	9.947E-01	9.967E-01	9.979E-01
34.0	•	•	•	4.257E-01	9.307E-01	9.790E-01	9.895E-01	9.938E-01	9.962E-01	9.975E-01
36.0	•	•	•	3.754E-01	9.209E-01	9.760E-01	9.879E-01	9.929E-01	9.956E-01	9.972E-01
38.0	•	•	•	3.278E-01	9.105E-01	9.727E-01	9.863E-01	9.919E-01	9.950E-01	9.968E-01
40.0	•	•	•	2.929E-01	9.019E-01	9.700E-01	9.849E-01	9.911E-01	9.945E-01	9.964E-01
42.0	•	•	•	2.525E-01	8.907E-01	9.665E-01	9.831E-01	9.901E-01	9.938E-01	9.960E-01
44.0	•	•	•	2.142E-01	8.785E-01	9.625E-01	9.811E-01	9.889E-01	9.931E-01	9.955E-01
46.0	•	•	•	1.812E-01	8.662E-01	9.585E-01	9.790E-01	9.877E-01	9.923E-01	9.950E-01
48.0	•	•	•	1.493E-01	8.522E-01	9.539E-01	9.767E-01	9.863E-01	9.915E-01	9.945E-01
50.0	•	•	•	1.271E-01	8.407E-01	9.501E-01	9.748E-01	9.852E-01	9.908E-01	9.940E-01
52.0	•	•	•	1.313E-01	8.430E-01	9.509E-01	9.751E-01	9.854E-01	9.909E-01	9.941E-01
54.0	•	•	•	1.047E-01	8.272E-01	9.456E-01	9.724E-01	9.838E-01	9.899E-01	9.935E-01
56.0	•	•	•	8.633E-02	8.138E-01	9.411E-01	9.701E-01	9.824E-01	9.890E-01	9.929E-01
58.0	•	•	•	7.115E-02	8.007E-01	9.366E-01	9.678E-01	9.810E-01	9.882E-01	9.923E-01
60.0	•	•	•	5.678E-02	7.856E-01	9.314E-01	9.651E-01	9.794E-01	9.872E-01	9.917E-01
62.0	•	•	•	4.387E-02	7.688E-01	9.254E-01	9.620E-01	9.776E-01	9.860E-01	9.910E-01
64.0	•	•	•	3.501E-02	7.543E-01	9.203E-01	9.593E-01	9.760E-01	9.850E-01	9.903E-01
66.0	•	•	•	2.705E-02	7.381E-01	9.144E-01	9.562E-01	9.742E-01	9.839E-01	9.896E-01
68.0	•	•	•	2.024E-02	7.203E-01	9.078E-01	9.528E-01	9.721E-01	9.826E-01	9.887E-01
70.0	•	•	•	1.564E-02	7.049E-01	9.021E-01	9.498E-01	9.703E-01	9.814E-01	9.880E-01
72.0	•	•	•	1.170E-02	6.879E-01	8.956E-01	9.464E-01	9.683E-01	9.802E-01	9.872E-01
74.0	•	•	•	8.756E-03	6.713E-01	8.692E-01	9.430E-01	9.662E-01	9.789E-01	9.863E-01
76.0	•	•	•	6.551E-03	6.551E-01	8.828E-01	9.396E-01	9.642E-01	9.776E-01	9.855E-01
78.0	•	•	•	4.746E-03	6.376E-01	8.758E-01	9.358E-01	9.620E-01	9.762E-01	9.846E-01
80.0	•	•	•	3.439E-03	6.206E-01	8.688E-01	9.321E-01	9.597E-01	9.748E-01	9.837E-01
82.0	•	•	•	2.491E-03	6.040E-01	8.619E-01	9.284E-01	9.575E-01	9.734E-01	9.827E-01
84.0	•	•	•	1.748E-03	5.862E-01	8.544E-01	9.243E-01	9.550E-01	9.718E-01	9.817E-01
86.0	•	•	•	1.226E-03	5.690E-01	8.469E-01	9.203E-01	9.526E-01	9.703E-01	9.807E-01
88.0	•	•	•	8.599E-04	5.523E-01	8.395E-01	9.162E-01	9.501E-01	9.687E-01	9.797E-01
90.0	•	•	•	5.841E-04	5.346E-01	8.315E-01	9.118E-01	9.475E-01	9.670E-01	9.786E-01
92.0	•	•	•	4.097E-04	5.189E-01	8.242E-01	9.078E-01	9.450E-01	9.655E-01	9.776E-01
94.0	•	•	•	2.783E-04	5.023E-01	8.163E-01	9.035E-01	9.424E-01	9.638E-01	9.765E-01
96.0	•	•	•	1.831E-04	4.849E-01	8.079E-01	8.988E-01	9.395E-01	9.620E-01	9.753E-01
98.0	•	•	•	1.204E-04	4.681E-01	7.995E-01	8.942E-01	9.367E-01	9.602E-01	9.741E-01
100.0	•	•	•	•	4.531E-01	7.919E-01	8.899E-01	9.340E-01	9.585E-01	9.730E-01
105.0	•	•	•	•	4.121E-01	7.701E-01	8.775E-01	9.264E-01	9.536E-01	9.698E-01
110.0	•	•	•	•	3.748E-01	7.488E-01	8.653E-01	9.189E-01	9.488E-01	9.667E-01
115.0	•	•	•	•	3.390E-01	7.270E-01	8.526E-01	9.110E-01	9.437E-01	9.633E-01
120.0	•	•	•	•	3.050E-01	7.047E-01	8.395E-01	9.027E-01	9.384E-01	9.598E-01
125.0	•	•	•	•	2.722E-01	6.814E-01	8.255E-01	8.939E-01	9.327E-01	9.561E-01
130.0	•	•	•	•	2.442E-01	6.600E-01	8.124E-01	8.856E-01	9.273E-01	9.525E-01
135.0	•	•	•	•	2.133E-01	6.342E-01	7.963E-01	8.753E-01	9.206E-01	9.480E-01
140.0	•	•	•	•	1.913E-01	6.142E-01	7.837E-01	8.672E-01	9.153E-01	9.445E-01
145.0	•	•	•	•	1.717E-01	5.949E-01	7.713E-01	8.591E-01	9.100E-01	9.410E-01
150.0	•	•	•	•	1.459E-01	5.670E-01	7.530E-01	8.471E-01	9.021E-01	9.357E-01
155.0	•	•	•	•	1.274E-01	5.448E-01	7.381E-01	8.373E-01	8.955E-01	9.313E-01
160.0	•	•	•	•	1.143E-01	5.277E-01	7.264E-01	8.295E-01	8.904E-01	9.278E-01
165.0	•	•	•	•	9.716E-02	5.030E-01	7.092E-01	8.180E-01	8.826E-01	9.226E-01
170.0	•	•	•	•	8.484E-02	4.833E-01	6.952E-01	8.085E-01	8.763E-01	9.183E-01
175.0	•	•	•	•	7.409E-02	4.644E-01	6.814E-01	7.991E-01	8.699E-01	9.140E-01
180.0	•	•	•	•	6.297E-02	4.426E-01	6.653E-01	7.879E-01	8.624E-01	9.089E-01
185.0	•	•	•	•	5.351E-02	4.219E-01	6.495E-01	7.770E-01	8.549E-01	9.038E-01
190.0	•	•	•	•	4.548E-02	4.022E-01	6.341E-01	7.662E-01	8.475E-01	8.988E-01
195.0	•	•	•	•	3.972E-02	3.864E-01	6.216E-01	7.573E-01	8.414E-01	8.946E-01
200.0	•	•	•	•	3.375E-02	3.683E-01	6.069E-01	7.467E-01	8.341E-01	8.896E-01
MASS	8.472E-01	2.393E-01	1.107E-02	3.223E-04	2.711E-05	7.991E-06	3.996E-06	2.337E-06	1.452E-06	9.363E-07

AUTHOR INDEX

A

Abeles, F., 199, 232
Aberg, T., 220, 235
Ackley, S., 526, 538
Adachi, T., 436, 451, 453
Adler, D., 164, 180
Adler, I., 40, 47, 330-341
Adlbridge, R.G., 390, 393, 397, 399, 404f
Alexander, L.E., 544, 549f, 566f, 575
Allison, S.K., 57, 67, 84, 91ff, 273, 286, 319, 329, 334, 341, 375, 381, 418-425
Anderegg, J.W., 613, 617, 619, 629
Andermann, G., 80-93, 118, 124
Anderson, C.A., 138, 157
Argo, H.V., 18, 25, 353, 371
Arnal, T., 85, 93
Arrhenius, G., 206, 233
Asaad, W.N., 15, 25
Aschaffenburg, R., 628, 631
Axelson, G., 401, 405
Azaroff, L.V., 248, 263, 270f, 273, 287, 436, 453

B

Baer, Y., 394, 396, 405
Baez, A.V., 320, 329
Bagus, P.S., 190, 231
Bailly, J.R., 610, 616
Baird, A.K., 26-48
Baker, D.W., 435-454
Baker, T.W., 456, 466, 619, 630
Barnes, S.W., 433f
Baro, R., 619, 629
Barrett, C.S., 436, 453, 488f, 505, 507, 523

Bates, D.R., 419, 425
Bates, S.R., 33, 46
Batt, A.P., 221, 235
Batyrev, V.A., 186, 231
Cauer, C.L., 526, 534, 538
Baun, W.L., 30, 34, 44, 46f, 49-67, 140ff, 144, 146, 151, 157, 160ff, 170f, 177, 179, 181, 186, 188, 205, 207, 211, 213, 215, 222f, 231, 233f, 237-247, 551, 571, 575
Baxter, A.J., 306ff, 311
Bearden, J.A., 217, 235, 273, 278, 286, 288, 358, 360ff, 321, 373, 381, 429, 434, 459, 461, 464ff
Beardon, K., 556, 576
Beeman, W.W., 613, 617, 619f, 629f
Bell, S. J., 294, 296, 309
Bellamy, B.A., 456, 466, 619, 630
Bello, J., 629, 631
Bennett, A.L., 508, 511, 524
Bennett, L.H., 225, 235
Bennett, N., 386, 389
Bergey, J.A., 18, 25, 353, 371
Bernstein, F., 30, 40, 45, 47
Bergmark, T., 390, 394, 396f, 399, 404f
Best, P.E., 199, 232
Bethe, H.A., 189, 231, 250, 271
Bhalla, A.A., 272-288
Birks, L.S., 118, 124, 211, 214, 221f, 234f, 279, 288
Blake, R.L., 18, 25, 352-372
Blanquet, P., 85, 93
Blokhin, M.A., 150, 158, 185, 189, 203f, 220, 224, 231, 233, 235

667

Boksenberg, A., 63f, 67
Boldt, E. A., 302, 306f, 311f
Bolsaitis, P., 214, 228f, 234
Bond, W.B., 503, 506
Bond, W.L., 355, 358, 365, 371,
    456, 462, 466
Bonnelle, C., 160, 163f, 179
Bonse, U., 619, 629
Bottazzi, H., 619, 629
Borovskii, I.B., 209, 234
Bose, S.M., 221, 235
Bowles, J.A., 382, 389
Bowyer, C.S., 306f, 312
Boyd, B.R., 85, 93
Bozorth, R.M., 459, 466
Bradt, H., 294, 296, 298, 310f,
    318, 328
Brantley, W.A., 526, 535, 538
Brecht, H., 393, 405
Brogen, G., 273, 287
Brooks, H., 164, 180
Brown, J.D., 27, 45, 81f, 92
Brown, R.L., 503, 506
Bruce, W.A., 507, 523
Brumberger, H., 619, 629
Brytov, I.A., 177, 181
Bucholtz, H., 469, 485
Bucklow, I.A., 633, 635ff
Buerger, M.J., 442, 454
Bühler, H., 469, 485
Bullen, T.G., 163, 180
Bunge, H.J., 436f, 442, 451,
    453f
Burcar, P.J., 610f, 616f
Burch, P., 220, 235
Burr, A.F., 217, 235, 426-434
Byram, E.T., 290, 297ff, 306ff,
    311f

C
Cady, W.G., 507, 523
Callaway, J., 225, 235
Campbell, W.J., 27, 30, 45, 81f,
    92
Canon, M., 33, 46
Carr-Brion, K.G., 96, 104
Carr, P.H., 507, 523
Carter, H.V., 118, 124
Cartmell, E., 227, 236

Caruso, A.J., 63, 67
Cauchois, Y., 173, 181
Causer, R., 456, 466, 619, 630
Cermak, J., 556, 576
Champion, K.P., 81, 83, 92
Chan, F.L., 105-124, 539-549
Charbnau, H.P., 456, 466
Chen, S.S., 509, 524
Cheremukhin, G.S., 331, 341
Chipman, D.R., 624, 630
Chivate, P., 249, 270
Chodil, G., 306f, 312
Chodos, A.A., 30, 45
Chopra, D.R., 160, 164, 179,
    373, 381
Christenson, A.L., 488f, 505
Christie, J.M., 436f, 444, 453
Chubb, T.A., 18, 25, 294, 298,
    309ff, 332, 341, 353, 365, 371
Claisse, F., 40, 47
Clift, J., 145, 157
Cocke, W. J., 294, 310
Cohen, M., 501, 503, 506
Cole, H., 458, 466
Collins, R.A., 294, 296, 309
Compton, A.H., 57, 67, 84, 91ff,
    273, 286, 319, 329, 334, 341,
    375, 381
Cooke, B.A., 633, 638
Cooper, A.S., 456, 466
Cooper, J.W., 251, 271
Corey, C.L., 607f
Coyle, R.A., 465, 467
Crisp, R.S., 59, 67
Criss, J.W., 211, 234
Cromer, D.T., 190, 217, 235
Cukor, P., 632-638
Cullen, T.J., 85, 93, 118, 174
Cullity, B.D., 488f, 505
Curry, B.P., 213, 234, 424f
Curry, C., 142f, 145, 157, 211,
    234
Cuthill, J.R., 138, 157, 204,
    210, 213, 230, 233f, 236
Cuttitta, F., 95, 104

D
Dadley, C.S., 392, 405
Damle, P.S., 249, 270

Das, B.N., 273, 287
Das Gupta, K., 207, 233, 551, 575
Davidge, R.W., 526, 535, 537
Davidson, F.D., 30, 34, 45f, 54, 66
Dean, J.A., 107, 124
DeBatist, R., 526, 535, 538
DeDominicus, C.T., 221, 235, 249, 251, 270
Delf, B.W., 574, 576
DelGrande, N.K., 19, 25, 646
Demaschun, G., 620, 630
Demekhin, V.F., 188, 218, 220, 231, 234f
Der, R.C., 424f
Desai, U.D., 306f, 312
Deslattes, R.D., 52, 56f, 63, 66f, 273, 287, 619, 629
Deubner, A., 535, 538
de Vries, J.L., 69, 79, 87, 93, 99, 104
d'Heurle, F. M., 456, 466
Dimond, R.K., 54, 57f, 66, 210, 213, 234
Disney, M.J., 294, 310
Dmitriev, A.B., 331, 341
Dodd, C.G., 173, 176f, 181, 199, 202, 232
Donachie, M.J., Jr., 488, 500, 505
Dowdey, J.E., 273, 287
Drewry, J., 628, 631
Dryer, H.T., 85, 93
Dubach, P., 610, 616
Duel, H., 610, 616
Dugdale, D.S., 470, 485
Dye, W.B., 85, 93

E
Ebel, H., 68-79
Ehlert, R.C., 34, 46, 53, 66, 365f, 372
Ehrenreich, H., 229, 236
Elbaum, C., 526, 535, 537
Elgin, R.L., 19, 25, 44, 48, 334, 341, 357, 361, 369, 371, 633, 638, 639-665
Engel, C.G., 30, 45

Ericson, U., 397, 399, 405
Ern, V., 150f, 158, 167f, 180
Ershov, O.A., 193, 201, 231f
Eshelby, J.D., 533, 538
Esquivel, A.L., 488, 505
Evans, J.C., 560f, 576
Evans, W.D., 18, 25, 353, 371
Evans, W.P., 488f, 505

F
Fabian, D.J., 54, 57f, 66, 189, 213, 218, 231, 234f
Faessler, A., 144, 157, 204, 233
Fahlman, A., 15, 25, 154f, 158, 199, 201, 203, 216, 222, 232, 234, 390, 393, 396f, 399ff
Falicov, L.M., 227, 236
Fay, M.J., 44, 48
Feder, R., 456, 466
Field, G.B., 306f, 312
Figgis, B.N., 229, 236
Fink, R.W., 334, 341
Fischer, D.W., 30, 34, 44, 46f, 140ff, 144, 146, 151, 157, 159-181, 188, 201, 205, 207, 211, 222f, 231, 233
Fishman, G.J., 294, 296, 310
Flanagan, F. J., 40, 47
Flemberg, H., 362f, 371
Fletcher, S.G., 503, 506
Fomichev, V.A., 177, 181, 190, 231
Forrest, P.G., 469, 484
Fortner, R.J., 424f
Fournet, G., 613, 617, 619, 624, 626, 629
Fowles, G.W.A., 227, 236
Fox, G.W., 507, 523
Franks, A., 386, 389, 574, 577
Freeman, A.J., 199, 232
Freeman, F.F., 353, 371
Frevel, L.K., 541, 549
Friedel, J., 193, 231
Friedman, H., 18, 25, 289-312, 332, 341, 353, 365, 371
Fritz, G., 294, 297ff, 201f, 306ff, 311f, 353, 365, 371
Fujino, N., 69, 77, 79
Furuta, T., 220, 235

G
Garcia, J.D., 360, 363, 371
Garmire, G., 302, 311, 318, 328
Garrod, R.I., 465, 467
Gates, W., 332, 341
Gavoret, J., 249, 251, 270
Gelius, U., 392, 394, 396, 401, 404f
Genkin, Ya.E., 203, 209, 233f
George, J.D., 456, 466, 619, 630
Giacconi, R., 313, 318, 324, 327ff, 342, 350
Gibbons, D.F., 464, 467
Gibbs, G.V., 177, 181, 186, 205f, 230, 233
Gigl, P.D., 186, 211, 230, 234
Gillam, E., 77, 79
Gillham, C.J., 550-597
Gingrich, N.S., 84, 92
Gjessing, E.T., 610, 616
Glaser, H., 278, 288
Glen, G.L., 173, 176f, 181, 199, 202, 232
Glencross, W.M., 382, 389
Glick, A.J., 221, 235
Gaganov, D.A., 193, 231
Gohshi, Y., 273, 287
Gold, T., 294, 309
Goldmann, J.B., 619, 629
Goodenough, J.B., 164, 167, 180
Gordon, R.B., 534, 538
Gorenstein, P., 302, 311, 313, 316ff, 328, 330-341
Gorski, L., 85, 93
Gould, R.W., 33, 46
Grader, R.J., 298, 302, 310
Graves, P.W., 62, 67
Gray, W., 373, 381
Green, D.W., 306ff, 311
Greenough, G.B., 488, 505
Griffing, G., 419, 425
Grodski, J.J., 60, 67
Grosskreutz, J.C., 469, 484
Grosso, J.S., 222, 235
Groven, L., 368, 372
Grube, W.L., 469, 485
Guard, R.W., 598, 601, 608
Gunier, A., 193, 231, 613, 617, 619, 624, 626f, 629, 631

Gunn, B.M., 44, 48
Gunn, E.L., 40, 47
Gursky, H., 302, 311, 313, 318, 327f, 330-341, 342, 350
Gwinner, E., 163, 180

H
Hach, J.T., 286, 288
Haessner, F., 436f, 442, 453
Hagström, S.B.M., 390, 392, 396, 404f
Hagstrum, H.D., 198, 232
Hamrin, K., 154f, 158, 390, 394, 396f, 399ff, 404f
Hanawalt, J.D., 540, 548
Hanzely, S., 160, 164ff, 179, 220, 223, 235, 366, 372, 373-301
Hardy, H.K., 607f
Harker, D., 629, 631
Harnden, F.R., Jr., 294, 296, 310
Harrison, W.A., 160, 180, 222, 235
Hart, M., 619, 629
Hartmann, R.J., 470, 478, 485
Haruta, K., 511, 514, 524
Harvey, C.E., 81, 92
Hasler, M.F., 81, 92
Hayakawa, S., 306, 311
Hayasi, T., 54, 66
Hayasi, Y., 54, 66
Hayes, C.E., 535, 538
Haymes, R.C., 294, 296, 310
Heady, H.H., 85, 93
Heal, H.T., 77, 79
Hearn, N.K., Jr., 508, 511, 524
Heckler, A.J., 436f, 453
Heden, P.F., 390, 392, 394, 396, 401f, 404f
Hedman, J., 390, 392, 394, 396f, 399, 401f, 404f
Heinrich, K.F.J., 128, 134, 633ff
Heise, R., 535, 538
Heller, L., 81f, 92
Heller, S.J., 609-617
Helmer, J.C., 406-417
Hempstead, C.F., 360, 371
Hendrickson, D.N., 400, 405
Henins, A., 273, 286, 464ff

Henke, B.L., 1-25, 31, 33ff, 39,
    43f, 46ff, 51, 63, 66f, 334,
    341, 357, 361, 365, 369, 371f,
    633, 638, 639-665
Henry, N.F.M., 447, 454
Henry, R.C., 294, 297ff, 301f,
    306ff, 310, 312
Herman, A., 273, 287
Hermann, F., 190, 228, 231, 235,
    258, 271
Hewish, A., 294, 296, 309
Hill, R.W., 298, 302, 310
Ho, C.Y., 472, 486
Ho, W.C., 75, 79
Hobby, M.G., 384, 389
Hollander, J.M., 400, 405
Holliday, J.E., 6, 24, 34, 46,
    54, 57f, 66, 136-158, 163,
    180, 203f, 233
Holly, W.L., 400, 405
Holt, S.S., 302, 306f, 311f
Hon, P.K., 634ff
Hopper, F.N., 134
Horger, O.J., 469, 484
Howe, C.E., 507, 523
Hsieh, K.A., 230, 236
Hubbell, J.H., 19, 25, 633,
    635ff, 646
Huffman, F.N., 358, 360ff, 371,
    373, 381
Huggins, F.G., 550-597
Hughes, G.D., 633, 635ff
Hunter, W.R., 357, 371
Hurley, J.W., 279, 288

I
Imelik, B., 619, 629
Ishimura, T., 249, 270
Itagaki, K., 526-538

J
Jaegle, P., 273, 287
James, R.W., 513, 524
Jauncey, G.E.M., 507, 523
Jaworowski, R.J., 632-638
Jenkins, R., 39, 46, 69, 79, 87,
    93, 99, 104
Johann, H.H., 624, 630
Johansson, G., 154f, 158, 390,
    394, 396, 399f, 404f

Johns, W.D., 98, 104
Johnson, G.G., Jr., 539-549
Johnson, J.L., 85, 93
Johnston, J.E., 163, 180
Jones, B.B., 353, 371
Jönsson, C., 347, 350
Jopson, R.C., 334, 341
Joshi, N.V., 249, 270

K
Kaelble, E.F., 98, 104
Kaesberg, P., 613, 617, 619, 629
Kalman, Z.H., 81f, 92
Kalnajs, J., 464, 467
Karlsson, S.E., 390, 399, 404
Kartha, G., 629, 631
Kato, T., 306, 311
Kato, W., 513, 524
Kavanagh, T.M., 424f
Kavish, S., 619, 630
Kemp, J.W., 81f, 92, 118, 124
Kerby, T.B., 223, 235
Kerr Del Grande, N., 633, 635ff
Kessler, J., 407, 417
Khan, J. M., 424f
Kim, H.H., 63, 67
King, H.W., 550-597
Kirby, T.B., 366, 372, 373-381
Kirkendall, T.D., 51, 65
Kirkpatrick, P., 320, 329
Kirkpatrick, S., 229, 236
Kittel, C., 227f, 236
Klein, M.P., 392, 405
Klug, H.P., 544, 549f, 566f, 575
Knausenberger, W.H., 211, 234
Koffman, D. M., 619, 630
Kohn, W., 230, 236
Koistinen, D.P., 488f, 493f,
    502f, 505f
Korff, S.A., 313, 328
Korsunskii, M.I., 203, 209, 233f
Koves, G., 468-486
Kozlenkov, A.I., 258, 271
Kratky, O., 613, 616, 620, 626,
    630f
Kreplin, R.W., 332, 341, 353,
    365, 371
Kristian, J., 294, 296, 310
Kudraytsev, I.Ya., 188, 220, 231,
    235, 469, 485

Kunosinski, T.F., 618-631
Kunze, G., 556, 558ff, 563, 575
Kuriyama, M., 508, 513, 524

L
Lachance, G.R., 44, 48
Ladell, J., 461, 467
Lamothe, R.E., 418-425
Lang, A.R., 508, 524
Larson, R.R., 95, 104
LaVilla, R.E., 63, 67
Lawrence, D.F., 199, 232
Lax, M., 228, 236, 252, 271
Ledingham, R.B., 19, 25, 44, 48,
    334, 341, 357, 361, 369, 371,
    633, 638f, 646
Lee, P.C.Y., 509, 515, 524
Lent, R.E., 7, 15, 19, 24f, 44,
    48, 63, 67, 334, 341, 357,
    361, 369, 371, 633, 638f,
    639, 646
Lewis, H.W., 419, 425
Lidiard, A.B., 533, 538
Liebhafsky, H.A., 97f, 104
Liefeld, R.J., 23, 54, 57, 66,
    137f, 157, 160, 164ff, 179,
    188, 220, 222ff, 230, 235,
    273, 287, 366, 372, 373-381
Lindberg, B.J., 390, 392, 397,
    399ff
Lindgren, I., 390, 399, 404
Lindqvist, I., 611, 616
Lindsay, G.M., 213, 234
Lindsey, K., 386, 389
Logothetis, E.M., 150, 158
Long, J.W., 384, 389
Longe, P., 221, 235
Lonsdale, K., 447, 454, 556, 576
Lubecki, A., 85, 93
Lublin, P., 632-638
Lucas, G.J., 633, 638
Lukirskii, A.P., 177, 181, 193,
    201, 231f
Luzatti, V., 619f, 624, 626,
    628ff
Lye, R.G., 150, 158
Lytle, F.W., 85, 93, 248-271

M
MacDonald, B.A., 487-506
MacFarlane, M., 294, 296, 310
Macherauch, E., 470, 478, 485,
    488f, 500, 505
Mack, J.E., 306f, 312, 360, 363,
    371
Mack, M., 461, 467, 551, 556, 560,
    563f, 569, 571, 574f, 577
Madden, A.J., 610, 616
Madlem, K.W., 40f, 44, 47
Magnusson, T., 365, 372
Mahan, G.D., 220, 235
Mallett, G.R., 44, 48, 630
Mallett, J.H., 19, 25, 633, 635ff,
    646
Mande, C., 249, 270
Mandelshtam, S.L., 306, 311, 331,
    341
Manley, O., 298, 310
Mannerandd, R., 394, 396, 405
Marburger, R.E., 488f, 494, 502,
    505f
March, N.H., 160, 180
Mariano, A.N., 286, 288
Mark, H., 306f, 312, 334, 341
Marks, C.L., 30, 45
Marshall, C.A.W., 213, 234
Martinelli, P., 85, 93
Martyshev, Yu. N., 526, 535, 538
Marzolf, J.G., 273, 286
Mason, W.P., 508, 514, 524
Massalski, T.B., 436, 453, 488f,
    505
Matsuoka, M., 306, 311
Mattson, R.A., 30, 45, 53, 66,
    365f, 372
Mayer, W., 294, 296, 310
McAlister, A.J., 138, 157, 204,
    210, 213, 233f
McEwan, G.J., 98, 104
McGee, J.F., 320, 329
McGuire, E.J., 19, 25, 645
McKinney, C.N., 125-135
McMaster, N.K., 19, 25
McMaster, W.H., 633, 635ff, 646
Meekins, J.F., 294,297ff, 301f,
    306ff, 310ff, 332, 341, 353,
    365, 371

Mehta, N.C., 610, 616
Meizer, W., 71, 79
Mendel, H., 173, 181
Men'shikov, A.Z., 203, 233
Merritt, L.L., Jr., 107, 124
Merzbacher, E., 419, 425
Metzer, A.E., 551, 575
Mickiewicz, S., 313, 316f, 328, 338, 341
Mihalison, J.R., 598-608
Miller, J.S., 294, 310
Mindlin, R.D., 508f, 516, 524f
Mitchell, B.J., 134f
Miyakawa, T., 508, 513, 524
Möllenstedt, G., 347, 350
Morgan, I.L., 220, 235
Morgan, P.D., 384, 389
Morin, F.J., 164, 170, 180f
Morris, P.R., 436f, 453
Moshin, R.W., 544, 549
Mott, D.L., 223, 235, 258, 271, 372, 373-381
Mott, N.F., 164, 180
Müller, J., 620, 630
Mueller, W.M., 44, 48
Muney, W.S., 318, 328
Murakami, Y., 469, 485

N
Nagel, D.J., 148, 158, 182-236
Naill, R.F., 51, 65
Naranan, S., 298, 311
Nather, R.E., 294, 296, 310
Neifert, H.R., 469, 484
Nemnonov, S.A., 203, 211, 233f
Neupert, W.M., 332, 341
Newey, C.W.A., 533, 538
Newkirk, J.B., 630
Nichols, M., 541, 549
Nicholson, J.B., 34, 46
Nicolaieff, A., 629, 631
Nikiforov, I.Ya., 185, 190, 211, 231, 234
Nishikawa, S., 507, 523
Nordberg, R., 15, 25, 390-405
Nordling, C., 15, 25, 154f, 158, 199, 201, 203, 216, 222, 232, 234, 390, 392, 394, 396f, 399ff, 404f

Norton, J.T., 488, 500, 505
Novick, R., 301, 311
Nowick, A.S., 456, 466
Nozieres, P., 221, 235, 249, 251, 270

O
O'Bryn, H.M., 159, 163, 172, 176, 179
Oda, M., 306, 311, 313, 317f, 328
Ogawara, Y., 306, 311
Ogier, W.T., 633, 638
Oke, J.B., 296, 310
Olbert, S., 298, 310
Osborne, P.J.H., 386, 389
Ovsyannikova, I.A., 209, 234

P
Paolini, F.R., 313, 318, 324, 327ff
Park, R.J., 633, 638
Parratt, L.G., 216f, 219, 221, 223f, 235, 273, 286, 360, 371, 461, 466
Parrish, W., 461, 465, 467, 551, 556, 559, 575, 560, 562ff, 566, 569, 571, 573f
Pasquine, D.L., 598f, 608
Paterson, M.S., 452, 454
Pattinson, E.J., 470, 485
Peacock, N.J., 384, 386, 389
Pearman, G.T., 507-525
Perez, P.R., 73, 79
Pessen, H., 618-631
Pfeiffer, H.G., 97f, 104, 630
Pike, E.R., 551, 556, 558, 564, 571, 575f
Pilkington, J.D.H., 294, 296, 309
Pilz, I., 620, 630
Pinckney, D.J., 609-617
Piret, E.L., 610, 616
Porod, G., 626, 631
Porteus, J.O., 273, 287
Posner, A.M., 611, 616
Potter, D.I., 607f
Pratt, P.L., 533, 538
Purcell, E.M., 407, 417

R

Ramberg, E., 433f
Ramquist, L., 151, 154f, 158
Rappaport, S., 294, 296, 298, 310f
Rassweiler, J.M., 469, 485
Redfern, J.H., 469, 484
Rees, M.J.,
Reidy, W.P., 313-329, 342, 350
Reilly, M.H., 195, 198, 200ff, 205, 232
Reimann, W.H., 483, 486
Remaut, G., 526, 535, 537
Renouprez, A., 619, 629
Renton, J.J., 551, 571, 575
Reuss, A., 502, 506
Richard, P., 220, 235
Richtmyer, F.K., 433f
Riley, D.P., 85, 92
Rinn, H.W.,
Rodberg, L.S., 256, 271
Roe, R.J., 436f, 454
Rooke, G.A., 185, 195, 199f, 207, 211, 213, 231ff
Rose, H.J., 40, 47
Rose, J.J., 95, 104
Rosenberg, A.S., 125-135
Rosenthanl, D., 469, 485
Rossi, B.B., 313, 327f
Roulet, B., 249, 251, 270
Rowland, E.S., 469, 484
Roy, R., 211, 234
Ruffa, A.R., 202, 233
Russell, C.M., 574, 577

S

Sachenko, V.P., 224, 235
Sagawa, T., 53, 66
Sakisaka, Y., 507, 523
Salmon, M.E., 94-104
Salpeter, E., 250, 271
Samson, C., 40, 47
Sander, W.C., 273, 286
Sargent, K.R., 483, 486
Savanick, G.A., 186, 211, 230, 234
Sawada, I., 58, 67
Sawada, M., 249, 270, 273, 287
Sayce, L.A., 386, 389

Sayers, D.E., 248-271
Saylor, W.P., 30, 45
Schmitz, P.J., 620, 630
Schnopper, H.W., 273, 287, 551, 575
Schulte, W.C., 469, 484
Schultz, J.M., 619, 630
Schwensfeir, R.J., Jr., 526, 535, 537
Scott, P.F., 294, 296, 309
Seebold, R.E., 222, 235
Segall, B., 195, 199, 231
Segmüller, A., 455-467, 556, 558ff, 563, 575
Seka, W., 199, 232
Seim, H.J., 85, 93
Semchyshen, M., 488f, 505
Serlemisos, P.J., 302, 311
Seward, F., 298, 302, 306f, 310, 312
Shatunova, A.V., 185f, 204, 231
Shaw, C.H., 273, 287
Sherman, J., 77, 79
Shields, A.R., 551, 575
Shiraiwa, T., 69, 77, 79, 249, 270
Shirley, D.A.S., 392, 405
Shmidt, V.V., 258, 271
Shuvaev, A.T., 150, 158, 203, 205, 209, 211, 233
Siegbahn, K., 15, 25, 199, 201, 203, 216, 222, 232, 234, 390, 392, 394, 396f, 399ff, 404ff, 408, 417, 428, 434
Sigmart, H., 469, 485
Sil'vestrova, I.M., 526, 535, 538
Simson, B., 52, 56, 63, 66f
Sines, G., 469, 485
Singer, S., 18, 25, 353, 371
Singh, J.N., 273, 287
Skala, Z., 626, 631
Skillman, S., 190, 228, 231, 236, 258, 271
Skinner, H.W.B., 140, 157, 159, 163, 172, 176, 179
Skolnick, L.P., 214, 228f, 234
Slater, J.C., 225, 227, 235f
Smakula, A., 464, 467

Smallbone, A.H., 85, 92
Smith, W.L., 520, 525
Smithells, 600, 608
Sokolowski, E., 390, 396, 404f
Solomon, J.S., 54, 63, 66f
Somer, T.A., 62, 67
Sorokin, L.S., 331, 341
Soules, J.A., 273, 287
Soven, P., 229, 236
Spada, G., 298, 311
Sparks, C.J., 33, 46
Speer, R.J., 382-389
Spencer, W.J., 507-525
Spicer, W.E., 199, 232
Spielberg, H., 34, 46
Sprialter, L., 544, 549
Sproull, R.L., 526, 535, 537
Sreekantan, B.V., 318, 328
Stanley, R.C., 42, 47
Staub, H.H., 313, 328
Sterk, A.A., 30, 35, 45f, 222,
  235
Stern, E.A., 214, 229f, 234,
  236, 248-271
Stever, K.R., 85, 93
Stewardson, E.A., 633, 638
Storms, E.K., 158
Stulen, F.B., 469, 484
Sugimoto, D., 306, 311
Sumoto, I., 507, 523
Sutula, C.L., 610, 616
Swartz, M., 332, 341
Sweeney, W.C., 279, 288
Swift, C., 306f, 312, 334, 341
Switendick, A.C., 150f, 158,
  167f, 180
Sykes, R.A., 520, 525

T
Taira, S., 469, 485
Taylerson, C.O., 560f, 576
Taylor, D.J., 294, 310
Taylor, D.L., 80-93
Taylor, J., 461, 467, 574, 577
Taylor, J.C., 81f, 92
Thaler, R.M., 256, 271
Thatcher, J.W., 30, 45
Thompson, B.J., 145, 157
Thomsen, J.S., 273, 286, 429, 434

Timasheff, S.N., 618-631
Timothy, A.F., 382, 389
Timothy, J.G., 382, 389
Tindo, I.P., 306, 311, 331, 341
Tippe, A., 535, 538
Tomboulian, D.H., 208, 234
Toor, A., 298, 306f, 310, 312
Traill, R.J., 44, 48
Trombka, J., 330-341
Tsutsirmi, K., 273, 287
Tucker, W.H., 301, 311, 344
Turner, R.M., 526, 535, 538
Tyren, F., 362f, 365, 368, 371

U
Ultriainen, J., 210, 235
Underwood, J.H., 318, 328
Unzicker, A.E., 18, 25, 353, 365,
  371
Urch, D.S., 199, 201f, 204, 232
Urusovskaya, A.S., 525, 535, 538

V
Vaiana, G.S., 313, 318, 324,
  327ff, 342, 350
Vand, V., 541f, 549
Vanden Bout, P., 301, 311
Van Gelder, S., 39, 46
Van Speybroeck, L.P., 313, 318,
  324, 327f, 342, 350
Van Wazer, J.R., 393, 405
Varadi, P.F., 51, 65
Varley, J.H.O., 211, 229, 234,
  236
Vassamillet, L.F., 550, 552,
  526f, 560, 562f, 565ff, 575
Vedam, K., 211, 234
Velicky, B., 229, 236
Vennik, J., 526, 535, 537
Voigt, W., 502, 506
Volborth, A., 40, 47
Von Dingen, E., 526, 535, 538
Von Grote, K.H., 347, 350

W
Waber, J.T., 190, 217, 225, 231,
  235
Walker, J.G., 459, 466
Waller, W.A., 386, 389

Walther, H.C., 610, 616
Wampler, E.J., 294, 310
Warner, B., 294, 296, 310
Wasilewski, M., 85, 93
Wasserman, G., 551, 575
Watson, L.M., 54, 57f, 66, 213, 234
Watson, R.E., 138, 157
Waters, J.R., 318, 328
Webb, M.B., 614, 617, 619, 629
Webb, W.W., 535, 538
Weich, G., 58, 67, 168, 176f, 180, 200, 232f
Weichert, N.H., 409-417
Weinberg, D.I., 620, 630
Weiss, L.E., 452, 454
Welday, E.E., 40ff, 44, 47f
Wenk, H.R., 436f, 442, 444, 453
Wenzel, K., 535, 538
Werme, L.O., 394, 396, 405
Wershaw, R.L., 609-617
Westbrook, J.H., 598, 601, 608
Weyerer, H., 465, 467
Whatley, T.A., 56, 67
White, E.Q., 56, 63, 67, 177, 181, 186, 205f, 211, 230, 233f, 237-247, 272-288
White, J.E., 507, 523
Whittem, R.N., 81f, 92
Whitworth, R.W., 526, 535, 537f
Wiegel, D., 619, 629
Wiewiorosky, J., 551, 575
Wiginton, B.J., 610, 616
Wilcock, P.D., 384, 389

Wilkinson, D.H., 313, 328
Willard, H.H., 107, 124
Williams, H.J., 459, 466
Williams, J.H., 273, 286
Williams, M.L., 138, 157, 204, 210, 213, 233
Willmore, A.P., 382, 389
Wilson, A.J.C., 556, 562, 566, 575f
Wilson, B.G., 306ff, 311f
Wilson, R., 353, 371
Winslow, E.H., 97f, 104
Wittry, D.B., 34, 46
Witz, J., 624, 628f, 631
Wolff, G.A., 286, 288
Wolff, R., 301, 311
Wolter, H., 320f, 329
Wood, W.A., 483, 486
Woodhouse, J.B., 633, 635ff
Wuilleumier, F., 644, 646
Wyckoff, R.W.G., 30, 34, 45f, 54, 66

Y
Yakowitz, H., 230, 236
Yin, L.I., 418-425
Young, R., 332, 341
Young, R.A., 508, 511, 524

Z
Zagoruiko, N.V., 526, 535ff
Zehnpfennig, T.F., 313, 318, 324, 327ff, 342-351
Zemany, P.D., 97f, 104
Ziegler, W., 211, 234

SUBJECT INDEX

A

Absorption, X-Ray,
  Atmospheric, 645
  Coefficients, 84, 87, 98-9,
    137, 214, 248ff, 639ff
  Edges, 88
  Extrapolation Function, 578ff
  Fine Structure, 255
    Temperature Correction, 258
  Heavy Elements, 645
  Light Elements, 646
  Self Absorption, 137, 164
Accuracy in Quantitative
  Analysis, 42
Acoustic Vibrations, 507ff
Aerobee, 150
  Rocket, 301
AlFe$_3$, 206
Al$_2$O$_3$, 195, 206
Albedo X-Ray Spectra, Lunar, 335
Alloys, Heat Resisting, 598ff
Alloy Theory, 210, 228
Alpha Excitation of Auger
  Spectra, 419
Aluminum Analysis, 125, 144,
  171, 199
  Foil Analysis, 634
Analyzer Crystals, 7, 33, 48,
  56, 162, 352, 357, 373ff
  Thermal Expansion of KAP, 358
Apollo Spacecraft, 330ff
Aqueous Solutions, Analysis of,
  87, 108
ASTM Powder Diffraction Data
  File, 539ff
Astronomy, X-Ray, 17, 289ff,
  313ff
  Instrumentation for, 313ff

Atmospheric Absorption of X-Ray,
  645
Atomic Arrangement, Local, 186,
  203ff
Atomic Energy Levels, 426
Atomic Number, Dependence of X-Ray
  Emission on, 82
Attenuation, Coefficients, 645
Auger Spectrum, 4, 218, 418
  Measurement Methods, 5
Automated Analysis Methods, 105ff

B

Big Bang Theory, 305
Black Body X-Ray Emission, 293, 298
Bond Theories, 192, 223
Bonding, Atomic
  Conductors, 153
  Effects on Emission Spectra, 202,
    209, 217, 390ff
  Insulators, 159
  Theories of, 191, 222, 226
Bond's Method, 455ff
Bounded Ejected Electron States,
  217
Bremsstrahlung Spectrum, 81, 298

C

Calcium Analysis, 85, 116, 125
Calibration Methods, Emission
  Spectra, 101, 129
Carbon Analysis, 39, 147
Centaurus XR-2, 300
Charge Distribution Detector, 63
Chemical Analysis, 208 (see also
  specific elements)
Chemical Bonding Effects (See
  Bonding)

677

Chlorine, Analysis, 116
Cobalt Analysis, 85
Compton Scattering, 80ff
Computer Control, Correction
    and Analysis Methods, 125,
    246, 267, 458
Coordination, Atomic (See
    Atomic Arrangement)
Copper Analysis, 85, 101, 145,
    255
Correction Procedures in Quan-
    titative Analysis, 44, 220
Cosmic X-Rays, 289ff
    Abundance, 301, 309
    Diffuse Background, 306
Crab Nebula, 290, 305
Cross-over Transitions, 170,
    173
Cross-Section, Photoelectric
    Absorption, 250
Cygnus XR-2, 300

D
Data Acquisition and Handling
    System, 243
Decay of X-Ray States, 214
De-excitation, 2, 4
Density of Electronic States,
    185, 189
Density of Thin Films, 74
Detection Methods, X-Ray,
    34ff, 49ff, 59ff, 242
    Limits, 96
    Windows in, 63, 117
    Cosmic X-rays, 334
Diffractometer
    Bragg-Brentano, 550ff
    Seeman-Bohlin, 550ff
    Error Sources, 555, 572
    Alignment Procedures, 564
Dislocations, 526ff
    Charge on, 526
    Surface Effects on, 533
    Impurity Effects on, 533
    Irradiation Pinning, 534
    Etch Pits, 535
Disordered Alloys, 210, 222
Dispersion of Soft X-Rays, 49
    Normal and Anomolous, 357ff

Double Crystal Spectrometer, 57

E
Electrical Conductivity, Relation
    to Emission, 205
Electron Cell Model of Alloys, 211,
    229
Electron Probe Analysis, 632
Electron Spectroscopy, 201
    Bonding Studies by, 390
    Photoelectron Spectrometer, 406ff
Electronic Structure, Effects on
    Emission Spectra, 136ff, 159ff,
    182ff, 190, 198
Electron Multiplier,
    Windowless, 59
    Magnetic, 60
    Channel, 61
Energy Levels, Atomic, Electronic,
    426ff
    Tables, 428
Error Estimation in Fluorescence
    Analysis, 130
Excitation Integral, 72
Excitation of Soft X-Rays, 49ff,
    53, 214, 219, 426
    Cobalt, 430
    Nickel, 432
    Proton, 241
Ex-Novae X-Ray Sources, 300
Extinction, X-Ray, 507
Extreme Ultraviolet Radiation, 356,
    371

F
Fatigue, 468ff
Ferromagnetism, 207
Films, For Absorption Coefficient
    Measurement, 632ff
Filters, 367
Flow Proportional Counter, 63, 242
Fluorescence Analysis, X-Rays, 68ff,
    94ff
Fluorine Analysis, 38
Forbidden Band Gaps in Oxides, 170

G
Galaxies, X-Ray, 302
    M-87 (Virgo A), 303, 304

Cygnus A, 303, 304
Centaurus A, 303, 304
Sco XR-1, 303
Seyfert, 304
  (See also specific galaxies)
Gallium Analysis, 85
Gamma Prime Phase, 598ff
Germanium, Absorption Fine
  Structure, 255
Gold, Photoelectronic Spec-
  trum, 416
  Film Analysis, 638
Goniometer, 1
  Pole Figure, 435ff
  Small Angle Scattering,
    618ff
Graphite Analyzer Crystal, 33
Grating Spectrometer, X-Ray,
  57, 342, 382ff
  Efficiency, 347
  Resolution, 348
  Grazing Incidence, 382
Gravitational Effects on X-Rays,
  300
Guinier Equation, 613, 628

H
Heat of Formation, 206
Heavy Element Analysis, By
  Gamma Rays, 85
Humic Acid, 609ff

I
Ice, Dislocations In, 526ff
Impact, Fatigue, 470
Intensity, X-Ray Emission, 97
  Scattered, Absolute-Scale,
    618ff
Interaction Coefficients, 19,
  352
Intergalactic Gas, 308
Interpulse, 295
Ionized State Effects on
  Emission, 190
Ion Neutralization Spec-
  troscopy, 198
Iron Analysis, 85, 125, 255
Isochromatic X-Rays, 426ff

K
K-Band Emission, 172ff

L
L-Band Emission, 163
Lamé Modes of Vibration, 516
Lattice Parameter Determination,
  455ff
Lifetimes, X-Ray State
Light Element Analysis, 26ff, 28
Line Broadening, X-Ray Line, 470
Low Energy Electrons, Applications,
  12
Low Energy X-Rays,
  Emission Bands, 148
  Applications, 12
  Absorption Tables, 639ff
Lunar Occultation, 293
Lunar X-Rays, 330
Lysozyme, Molecule Dimensions, 629

M
Magnesium Analysis, 38, 125, 142,
  171
Magnetic Susceptibility, 207
Magnetosphere, 294
Manganese Analysis, 85
Mass Absorption Coefficients, 639ff,
  632ff, 644
Mass Per Unit Area of Thin Films,
  73, 75
Matrix Effects, 40, 88, 125, 136
Mean Wavelength, 71
Metal Organic Compounds, 633
MgO, 206
Molybdenum Analysis, 116
Monochromator, X-Ray, 623
Molalithic Quartz Filters, 520
Museum Objects, Analysis of, 94ff

N
Neutron Star, 293
Nickel Analysis, 85, 136
  Film, 635
Nickel-Base Alloys, 598ff
Nova, 301

O
Open or Closed Universe?, 308
Optical Spectroscopy Compared to X-Ray and Electron, 198
Order, Long Range, 598ff
Ordered Materials, 222
Ore Analysis (Fe, Pb, Co, Cu, Zn), 81
Oxides, Metal, Emission From, 163, 166
Oxygen Analysis, 39, 369

P
Parratt Diagram, 216
Particle Size Effects, 81
Peak-Background Ratio Method, 81, 83, 87
Peak Position Location, 461
Peierls Potential, In Ice, 531
Phase Shift, Atomic
Phosphate Rock Analysis, 125
Phosphorous Analysis, 125
Photo-Auger Electrons
Photoelectric Cross Section, 19, 21
Photoelectron
    Absorption Cross Sections, 639, 644
    Spectrometer, 406ff
    Wave Functions, 251ff
Photon Excitation, 50
Photographic Detection, 59
Piezoelectricity, 507ff
Plasmas, Low Energy X-Rays from, 15, 206, 299, 303, 352, 384
Plutonium Analysis, 85
Polarization of Cosmic X-Rays, 301
Pole Figures, 435ff
Polonium Source of Alpha Particles, 420
Potassium Acid Phthalate, 373ff
    Rocking Curves, 376
    Percent Reflection, 379
Potassium Chloride, Absorption by, 87

Precision in Chemical Analysis, 42
Preferred Orientation, 435ff
Properties of Materials, Relation to Emission, 205
Proton Excitation, 241
Proton-Synchrotron Radiation, 296
Pseudopotentials, 251
Pulsar, 293

Q
Quartzite, 441
Quartz Vibrating Plates, 507ff

R
Radio Astronomy, 294
Rayleigh Scattering, 80ff
Recoil Factor, Breit-Dirac, 83
Residual Stress, 468ff
Ribonuclease, Molecular Dimensions, 629
Rocket Flights, 18
Rocking Curves, KAP, 376

S
Sample Preparation, 40
Scintillation Counter, 60
Scorpius XR-1, 297
Scylla I, 356
Self Absorption, 137, 164
SiC, 206
Silver, Analysis of
SiO_2, 195ff
Silicon Analysis, 125, 171
    Lattice Parameter, 464, 574
Skylark Rocket, 301
Slip Bands, Iron, 478
Small Angle Scattering, 609ff, 618ff
Soap Film Analyzer Crystals, 56
Sodium Analysis, 38
Sodium Humate, 609ff
Soft X-Rays, 26
    Absorption of, 639ff
    Generation of, 29, 30
    (See Low Energy X-Rays)
Solar Spectra, 349, 353, 369
Solid Solutions, Structure of, 539ff

Space Groups, 439, 444
Spatial Selectivity, X-Ray, 190, 213
Spectral Features, X-Ray, 218
Spectrometer, X-Ray
  Dispersive, 33, 57, 272
  Double Crystal, 57
  For X-Ray Astronomy, 313, 330, 342ff, 355
  Non-Dispersive, 35, 54
  Vacuum, 237ff, 277
Standards, Analytical, 44
Stearate Analyzer Crystal, 34
Steel, Stress Determination In, 487
Stress Analysis, 468ff, 487ff
Stress Factor, 500
Stretched String Model of Charged Dislocation Motion, 526ff
Strontium Analysis, 87
Sulfur Analysis, 85
Sunspots, 332
Superconductivity, 208
Supernova, 301, 305
Symmetry, 435ff
Synchrotron Radiation, 292

T
Take-off Angle Effects, 68ff
Tau XR-1, 302
Telescope, X-Ray, 318ff, 342
Tempering of Carburized Steel, 502
Texture Analysis, 435ff
Thin Films,
  Analysis of, 75

Density of, 74
Mass Per Unit Area, 73, 75
Titanium Analysis, 85, 136
Topography, X-Ray, 507ff
  Effects of Strain Gradient, 512
  Of Ice, 526
Transition Metal Emission, 163
Transition Probability, 189, 213
Transmission Gratings, 345
Tungsten, Lattice Parameter, 574
  Film Analysis, 634

U
Ultraviolet, Extreme, 354, 369
Universe, Open or Closed?, 308

V
Valence State Effects on Emission
  Spectra, 182ff, 187
  Conductors, 136ff, 211
  Insulators, 159, 202
  Intensity, 188
Vibration, Seen by X-Ray Topography, 507ff
  Modes in Quartz Plates, 518
  Lamé Modes, 516
  Dislocations in Ice, 526ff
Virgo A, 290

W
Wavelengths, 364
Windows in Detectors, 63, 117
Work Function, 431

Z
Zinc Analysis, 85, 116